Lecture Notes in Computer Science   8835

Commenced Publication in 1973
Founding and Former Series Editors:
Gerhard Goos, Juris Hartmanis, and Jan van Leeuwen

Chu Kiong Loo Keem Siah Yap
Kok Wai Wong Andrew Teoh
Kaizhu Huang (Eds.)

Neural
Information Processing

21st International Conference, ICONIP 2014
Kuching, Malaysia, November 3-6, 2014
Proceedings, Part II

 Springer

Volume Editors

Chu Kiong Loo
University of Malaya, Kuala Lumpur, Malaysia
E-mail: ckloo.um@um.edu.my

Keem Siah Yap
Universiti Tenaga Nasional, Selangor, Malaysia
E-mail: yapkeem@uniten.edu.my

Kok Wai Wong
Murdoch University, Murdoch, WA, Australia
E-mail: k.wong@murdoch.edu.au

Andrew Teoh
Yonsei University, Seoul, South Korea
E-mail: bjteoh@yonsei.ac.kr

Kaizhu Huang
Xi'an Jiaotong-Liverpool University, Suzhou, China
E-mail: kaizhu.huang@xjtlu.edu.cn

ISSN 0302-9743   e-ISSN 1611-3349
ISBN 978-3-319-12639-5   e-ISBN 978-3-319-12640-1
DOI 10.1007/978-3-319-12640-1
Springer Cham Heidelberg New York Dordrecht London

Library of Congress Control Number: 2014951688

LNCS Sublibrary: SL 1 – Theoretical Computer Science and General Issues

Typesetting: Camera-ready by author, data conversion by Scientific Publishing Services, Chennai, India

Printed on acid-free paper

Springer is part of Springer Science+Business Media (www.springer.com)

Preface

This volume is part of the three-volume proceedings of the 21st International Conference on Neural Information Processing (ICONIP 2014), which was held in Kuching, Malaysia, during November 3–6, 2014. The ICONIP is an annual conference of the Asia Pacific Neural Network Assembly (APNNA). This series of ICONIP conferences has been held annually since 1994 in Seoul and has become one of the leading international conferences in the area of neural networks.

ICONIP 2014 received a total of 375 submissions by scholars from 47 countries/regions across six continents. Based on a rigorous peer-review process where each submission was evaluated by at least two qualified reviewers, a total of 231 high-quality papers were selected for publication in the reputable series of *Lecture Notes in Computer Science* (LNCS). The selected papers cover major topics of theoretical research, empirical study, and applications of neural information processing research. ICONIP 2014 also featured a pre-conference event, namely, the Cybersecurity Data Mining Competition and Workshop (CDMC 2014) which was held in Kuala Lumpur. Nine papers from CDMC 2014 were selected for a Special Session of the conference proceedings.

In addition to the contributed papers, the ICONIP 2014 technical program included a keynote speech by Shun-Ichi Amari (RIKEN Brain Science Institute, Japan), two plenary speeches by Jacek Zurada (University of Louisville, USA) and Jürgen Schmidhuber (Istituto Dalle Molle di Studi sull'Intelligenza Artificiale, Switzerland). This conference also featured seven invited speakers, i.e., Akira Hirose (The University of Tokyo, Japan), Nikola Kasabov (Auckland University of Technology, New Zealand), Soo-Young Lee (KAIST, Korea), Derong Liu (Chinese Academy of Sciences, China; University of Illinois, USA), Kay Chen Tan (National University of Singapore), Jun Wang (The Chinese University of Hong Kong), and Zhi-Hua Zhou (Nanjing University, China).

We would like to sincerely thank Honorary Chair Shun-ichi Amari, Mohd Amin Jalaludin, the members of the Advisory Committee, the APNNA Governing Board for their guidance, the members of the Organizing Committee for all their great efforts and time in organizing such an event. We would also like to take this opportunity to express our deepest gratitude to all the technical committee members for their professional review that guaranteed high quality papers.

We would also like to thank Springer for publishing the proceedings in the prestigious LNCS series. Finally, we would like to thank all the speakers, authors, and participants for their contribution and support in making ICONIP 2014 a successful event.

November 2014

Chu Kiong Loo
Keem Siah Yap
Kok Wai Wong
Andrew Teoh
Kaizhu Huang

Organization

Honorary Chairs

Shun-Ichi Amari   RIKEN, Japan
Mohd Amin Jalaludin   University of Malaya, Malaysia

General Chair

Chu Kiong Loo   University of Malaya, Malaysia

General Co-chairs

Yin Chai Wang   University Malaysia Sarawak, Malaysia
Weng Kin Lai   Tunku Abdul Rahman University College,
     Malaysia

Program Chairs

Kevin Kok Wai Wong   Murdoch University, Australia
Andrew Teoh   Yonsei University, Korea
Kaizhu Huang   Xi'an Jiaotong-Liverpool University, China

Publication Chairs

Lakhmi Jain   University of South Australia, Australia
Chee Peng Lim   Deakin University, Australia
Keem Siah Yap   Universiti Tenaga Nasional, Malaysia

Registration Chair and Webmaster

Yun Li Lee   Sunway University, Malaysia

Local Organizing Chairs

Chong Eng Tan   University Malaysia Sarawak, Malaysia
Kai Meng Tay   University Malaysia Sarawak, Malaysia

Workshop and Tutorial Chairs

Chen Change Loy	Chinese University of Hong Kong, SAR China
Ying Wah Teh	University of Malaya, Malaysia
Saeed Reza	University of Malaya, Malaysia
Tutut Harewan	University of Malaya, Malaysia

Special Session Chairs

Thian Song Ong	Multimedia University, Malaysia
Siti Nurul Huda Sheikh Abdullah	Universiti Kebangsaan Malaysia, Malaysia

Financial Chair

Ching Seong Tan	Multimedia University, Malaysia

Sponsorship Chairs

Manjeevan Seera	University of Malaya, Malaysia
John See	Multimedia University, Malaysia
Aamir Saeed Malik	Universiti Teknologi Petronas, Malaysia

Publicity Chairs

Siong Hoe Lau	Multimedia University, Malaysia
Khairul Salleh Mohamed Sahari	Universiti Tenaga Nasional, Malaysia

Asia Liaison Chairs

ShenShen Gu	Shanghai University, China

Europe Liaison Chair

Wlodzislaw Duch	Nicolaus Copernicus University, Poland

America Liaison Chair

James T. Lo	University of Maryland, USA

Advisory Committee

Lakhmi Jain, Australia
David Gao, Australia
BaoLiang Lu, China
Ying Tan, China
Jin Xu, China.
Irwin King, Hong Kong, SAR China
Jun Wang, Hong Kong, SAR China
P. Balasubramaniam, India
Kunihiko Fukushima, Japan
Shiro Usui, Japan
Minho Lee, Korea
Muhammad Leo Michael Toyad
    Abdullah, Malaysia
Mustafa Abdul Rahman, Malaysia
Narayanan Kulathuramaiyer,
    Malaysia
David Ngo, Malaysia

Siti Salwah Salim, Malaysia
Wan Ahmad Tajuddin Wan Abdullah,
    Malaysia
Wan Hashim Wan Ibrahim, Malaysia
Dennis Wong, Malaysia
Nik Kasabov, New Zealand
Arnulfo P. Azcarraga, Phillipines
Wlodzislaw Duch, Poland
Tingwen Huang, Qatar
Meng Joo Err, Singapore
Xie Ming, Singapore
Lipo Wang, Singapore
Jonathan H. Chan, Thailand
Ron Sun, USA
De-Liang Wang, USA
De-Shuang Huang, China

Technical Committee

Ahmad Termimi Ab Ghani
Mark Abernethy
Adel Al-Jumaily
Leila Aliouane
Cesare Alippi
Ognjen Arandjelovic
Sabri Arik
Mian M. Awais
Emili Balaguer-Ballester
Valentina Emilia Balas
Tao Ban
Sang-Woo Ban
Younès Bennani
Asim Bhatti
Janos Botzheim
Salim Bouzerdoum
Ivo Bukovsky
Jinde Cao
Jiang-Tao Cao
Chee Seng Chan
Long Cheng
Girija Chetty

Andrew Chiou
Pei-Ling Chiu
Sung-Bae Cho
Todsanai Chumwatana
Pau-Choo Chung
Jose Alfredo Ferreira Costa
Justin Dauwels
Mingcong Deng
M.L. Dennis Wong
Hongli Dong
Hiroshi Dozono
El-Sayed M. El-Alfy
Zhouyu Fu
David Gao
Tom Gedeon
Vik Tor Goh
Nistor Grozavu
Ping Guo
Masafumi Hagiwara
Osman Hassab Elgawi
Shan He
Haibo He

Kingkarn Sookhanaphibarn
Indra Adji Sulistijono
Changyin Sun
Jun Sun
Masahiro Takatsuka
Takahiro Takeda
Shing Chiang Tan
Syh Yuan Tan
Ching Seong Tan
Ken-Ichi Tanaka
Katsumi Tateno
Kai Meng Tay
Connie Tee
Guo Teng
Andrew Teoh
Heizo Tokutaka
Dat Tran
Boris Tudjarov
Eiji Uchino
Kalyana C. Veluvolu
Michel Verleysen
Lipo Wang
Hongyuan Wang

Frank Wang
Yin Chai Wang
Yoshikazu Washizawa
Kazuho Watanabe
Bunthit Watanapa
Kevin Wong
Kok Seng Wong
Zenglin Xu
Yoko Yamaguch
Koichiro Yamauchi
Hong Yan
Wei Qi Yan
Kun Yang
Pcipei Yang
Bo Yang
Xu-Cheng Yin
Zhigang Zeng
Min-Ling Zhang
Chao Zhang
Rui Zhang
Yanming Zhang
Rui Zhang
Ding-Xuan Zhou

Table of Contents – Part II

Kernel and Statistical Methods

A New Ensemble Clustering Method Based on Dempster-Shafer
Evidence Theory and Gaussian Mixture Modeling 1
  Yi Wu, Xiabi Liu, and Lunhao Guo

Extraction of Dimension Reduced Features from Empirical Kernel
Vector 9
  Takio Kurita and Yayoi Harashima

Method of Evolving Non-stationary Multiple Kernel Learning 17
  Peng Wu, Qian Yin, and Ping Guo

A Kernel Method to Extract Common Features Based on Mutual
Information 26
  Takamitsu Araki, Hideitsu Hino, and Shotaro Akaho

Properties of Text-Prompted Multistep Speaker Verification Using
Gibbs-Distribution-Based Extended Bayesian Inference for Rejecting
Unregistered Speakers 35
  Shuichi Kurogi, Takuya Ueki, Satoshi Takeguchi, and Yuta Mizobe

Non-monotonic Feature Selection for Regression 44
  Haiqin Yang, Zenglin Xu, Irwin King, and Michael R. Lyu

Non-negative Matrix Factorization with Schatten p-norms
Reguralization 52
  Ievgen Redko and Younès Bennani

A New Energy Model for the Hidden Markov Random Fields 60
  Jérémie Sublime, Antoine Cornuéjols, and Younès Bennani

Online Nonlinear Granger Causality Detection by Quantized Kernel
Least Mean Square 68
  *Hong Ji, Badong Chen, Zejian Yuan, Nanning Zheng, Andreas Keil,
  and Jose C. Príncipe*

A Computational Model of Anti-Bayesian Sensory Integration in the
Size-Weight Illusion 76
  Yuki Ueyama

Unsupervised Dimensionality Reduction for Gaussian Mixture Model ... 84
  Xi Yang, Kaizhu Huang, and Rui Zhang

Graph Kernels Exploiting Weisfeiler-Lehman Graph Isomorphism Test
Extensions   93
    Giovanni Da San Martino, Nicolò Navarin, and Alessandro Sperduti

Texture Analysis Based Automated Decision Support System for
Classification of Skin Cancer Using SA-SVM   101
    Ammara Masood, Adel Al-Jumaily, and Khairul Anam

In-attention State Monitoring for a Driver Based on Head Pose and
Eye Blinking Detection Using One Class Support Vector Machine   110
    Hyunrae Jo and Minho Lee

An Improved Separating Hyperplane Method with Application to
Embedded Intelligent Devices   118
    Yanjun Li, Ping Guo, and Xin Xin

Fine-Grained Air Quality Monitoring Based on Gaussian Process
Regression   126
    *Yun Cheng, Xiucheng Li, Zhijun Li, Shouxu Jiang,
    and Xiaofan Jiang*

Retrieval of Experiments by Efficient Comparison of Marginal
Likelihoods   135
    Sohan Seth, John Shawe-Taylor, and Samuel Kaski

Evolutionary Computation and Hybrid Intelligent Systems

A New Approach of Diversity Enhanced Particle Swarm Optimization
with Neighborhood Search and Adaptive Mutation   143
    Dang Cong Tran, Zhijian Wu, and Hui Wang

Data Clustering Based on Particle Swarm Optimization with
Neighborhood Search and Cauchy Mutation   151
    Dang Cong Tran and Zhijian Wu

Accuracy Improvement of Localization and Mapping of ICP-SLAM via
Competitive Associative Nets and Leave-One-Out Cross-Validation   160
    *Shuichi Kurogi, Yoichiro Yamashita, Hikaru Yoshikawa,
    and Kotaro Hirayama*

Saliency Level Set Evolution   170
    Jincheng Mei and Bao-Liang Lu

Application of Cuckoo Search for Design Optimization of Heat
Exchangers   178
    Rihanna Khosravi, Abbas Khosravi, and Saeid Nahavandi

A Hybrid Method to Improve the Reduction of Ballistocardiogram
Artifact from EEG Data 186
    Ehtasham Javed, Ibrahima Faye, Aamir Saeed Malik,
    and Jafri Malin Abdullah

VLGAAC: Variable Length Genetic Algorithm Based Alternative
Clustering 194
    Moumita Saha and Pabitra Mitra

Social Book Search with Pseudo-Relevance Feedback 203
    Bin Geng, Fang Zhou, Jiao Qu, Bo-Wen Zhang, Xiao-Ping Cui,
    and Xu-Cheng Yin

A Random Key Genetic Algorithm for Live Migration of Multiple
Virtual Machines in Data Centers 212
    Tusher Kumer Sarker and Maolin Tang

Collaboration of the Radial Basis ART and PSO in Multi-Solution
Problems of the Hnon Map 221
    Fumiaki Tokunaga, Takumi Sato, and Toshimichi Saito

Reconstructing Gene Regulatory Network with Enhanced Particle
Swarm Optimization 229
    Rezwana Sultana, Dilruba Showkat, Mohammad Samiullah,
    and Ahsan Raja Chowdhury

Neural Network Training by Hybrid Accelerated Cuckoo Particle
Swarm Optimization Algorithm 237
    Nazri Mohd Nawi, Abdullah Khan, M.Z. Rehman,
    Maslina Abdul Aziz, Tutut Herawan, and Jemal H. Abawajy

An Accelerated Particle Swarm Optimization Based Levenberg
Marquardt Back Propagation Algorithm 245
    Nazri Mohd Nawi, Abdullah Khan, M.Z. Rehman,
    Maslina Abdul Aziz, Tutut Herawan, and Jemal H. Abawajy

Fission-and-Recombination Particle Swarm Optimizers for Search of
Multiple Solutions 254
    Takumi Sato and Toshimichi Saito

Fast Generalized Fuzzy C-means Using Particle Swarm Optimization
for Image Segmentation 263
    Dang Cong Tran, Zhijian Wu, and Van Hung Tran

Evolutionary Learning and Stability of Mixed-Rule Cellular
Automata 271
    Ryo Sawayama and Toshimichi Saito

Pattern Recognition Techniques

Radical-Enhanced Chinese Character Embedding   279
    Yaming Sun, Lei Lin, Nan Yang, Zhenzhou Ji, and Xiaolong Wang

Conditional Multidimensional Parameter Identification with
Asymmetric Correlated Losses of Estimation Errors   287
    Piotr Kulczycki and Malgorzata Charytanowicz

Short Text Hashing Improved by Integrating Topic Features
and Tags.   295
    *Jiaming Xu, Bo Xu, Jun Zhao, Guanhua Tian, Heng Zhang,
    and Hongwei Hao*

Synthetic Test Data Generation for Hierarchical Graph Clustering
Methods   303
    László Szilágyi, Levente Kovács, and Sándor Miklós Szilágyi

Optimal Landmark Selection for Nyström Approximation   311
    Zhouyu Fu

Privacy Preserving Clustering: A k-Means Type Extension   319
    Wenye Li

Stream Quantiles via Maximal Entropy Histograms   327
    Ognjen Arandjelović, Ducson Pham, and Svetha Venkatesh

A Unified Framework for Thermal Face Recognition   335
    *Reza Shoja Ghiass, Ognjen Arandjelović, Hakim Bendada,
    and Xavier Maldague*

Geometric Feature-Based Facial Emotion Recognition Using Two-Stage
Fuzzy Reasoning Model.   344
    Md. Nazrul Islam and Chu Kiong Loo

Human Activity Recognition by Matching Curve Shapes   352
    Poorna Talkad Sukumar and K. Gopinath

Sentiment Analysis of Chinese Microblogs Based on Layered Features. . .   361
    Dongfang Wang and Fang Li

Feature Group Weighting and Topological Biclustering.   369
    *Tugdual Sarazin, Mustapha Lebbah, Hanane Azzag,
    and Amine Chaibi*

A Label Completion Approach to Crowd Approximation   377
    Toshihiro Watanabe and Hisashi Kashima

Multi-label Linear Discriminant Analysis with Locality Consistency   386
    Yuzhang Yuan, Kang Zhao, and Hongtao Lu

Hashing for Financial Credit Risk Analysis 395
  Bernardete Ribeiro and Ning Chen

MAP Inference with MRF by Graduated Non-Convexity and Concavity
Procedure.. 404
  Zhi-Yong Liu, Hong Qiao, and Jian-Hua Su

Two-Phase Approach to Link Prediction 413
  Srinivas Virinchi and Pabitra Mitra

Properties of Direct Multi-Step Ahead Prediction of Chaotic Time
Series and Out-of-Bag Estimate for Model Selection 421
  Shuichi Kurogi, Ryosuke Shigematsu, and Kohei Ono

Multi-document Summarization Based on Sentence Clustering 429
  Hai-Tao Zheng, Shu-Qin Gong, Hao Chen, Yong Jiang,
  and Shu-Tao Xia

An Ontology-Based Approach to Query Suggestion Diversification...... 437
  Hai-Tao Zheng, Jie Zhao, Yi-Chi Zhang, Yong Jiang,
  and Shu-Tao Xia

Sensor Drift Compensation Using Fuzzy Interference System and
Sparse-Grid Quadrature Filter in Blood Glucose Control 445
  Péter Szalay, László Szilágyi, Zoltán Benyó, and Levente Kovács

Webpage Segmentation Using Ontology and Word Matching........... 454
  Huey Jing Toh and Jer Lang Hong

Continuity of Discrete-Time Fuzzy Systems 462
  Takashi Mitsuishi, Takanori Terashima, Koji Saigusa,
  Nami Shimada, Toshimichi Homma, Kiyoshi Sawada,
  and Yasunari Shidama

Sib-Based Survival Selection Technique for Protein Structure Prediction
in 3D-FCC Lattice Model 470
  Rumana Nazmul and Madhu Chetty

Tensor Completion Based on Structural Information.................. 479
  Zi-Fa Han, Ruibin Feng, Long-Ting Huang, Yi Xiao,
  Chi-Sing Leung, and Hing Cheung So

Document Versioning Using Feature Space Distances 487
  Wei Lee Woon, Kuok-Shoong Daniel Wong, Zeyar Aung,
  and Davor Svetinovic

Separation and Classification of Crackles and Bronchial Breath Sounds
from Normal Breath Sounds Using Gaussian Mixture Model   495
  Ali Haider, M. Daniyal Ashraf, M. Usama Azhar,
  Syed Osama Maruf, Mehdi Naqvi, Sajid Gul Khawaja, and
  M. Usman Akram

Combined Features for Face Recognition in Surveillance Conditions   503
  Khaled Assaleh, Tamer Shanableh, and Kamal Abuqaaud

Sparse Coding on Multiple Manifold Data   515
  Hanchao Zhang and Jinhua Xu

Mutual Information Estimation with Random Forests   524
  Mike Koeman and Tom Heskes

Out-Of-Vocabulary Words Recognition Based on Conditional Random
Field in Electronic Commerce   532
  Yanfeng Yang, Yanqin Yang, Hu Guan, and Wenchao Xu

Least Angle Regression in Orthogonal Case   540
  Katsuyuki Hagiwara

Evaluation Protocol of Early Classifiers over Multiple Data Sets   548
  Asma Dachraoui, Alexis Bondu, and Antoine Cornuéjols

Exploiting Level-Wise Category Links for Semantic Relatedness
Computing   556
  Hai-Tao Zheng, Wenzhen Wu, Yong Jiang, and Shu-Tao Xia

Characteristic Prediction of a Varistor in Over-Voltage Protection
Application   565
  Kohei Nagatomo, Muhammad Aziz Muslim, Hiroki Tamura,
  Koichi Tanno, and Wijono

Optimizing Complex Building Renovation Process with Fuzzy Signature
State Machines   573
  Gergely I. Molnárka and László T. Kóczy

News Title Classification with Support from Auxiliary Long Texts   581
  Yuanxin Ouyang, Yao Huangfu, Hao Sheng, and Zhang Xiong

Modelling Mediator Intervention in Joint Decision Making Processes
Involving Mutual Empathic Understanding   589
  Rob Duell

Author Index   597

A New Ensemble Clustering Method Based on Dempster-Shafer Evidence Theory and Gaussian Mixture Modeling

Yi Wu, Xiabi Liu, and Lunhao Guo

Beijing Lab of Intelligent Information Technology, School of Computer Science,
Beijing Institute of Technology, Beijing 100081, China
{wuyi,liuxiabi,guolunhao}@bit.edu.cn

Abstract. This paper proposes a new method based on Dempster-Shafer (DS) evidence theory and Gaussian Mixture Modeling (GMM) technique to combine the cluster results from single clustering methods. We introduce the GMM technique to determine the confidence values for candidate results from each clustering method. Then we employ the DS theory to combine the evidences supplied by different clustering methods, based on which the final result is obtained. We tested the proposed ensemble clustering method on several commonly used datasets. The experimental results confirm that our method is effective and promising.

Keywords: Data Clustering, Ensemble Clustering, Dempster-Shafer (DS) Evidence Theory, Gaussian Mixture Modeling (GMM).

1   Introduction

At present, there is no single clustering method can achieve robust results in all situations. It is promising to integrate various clustering methods for obtaining the better performance. This solution is usually called ensemble clustering, which has attracted more and more attentions in recent years. The existing ensemble clustering methods can be classified into two main categories: voting based and hyper-graph based [1].

- Voting based methods firstly solve the label correspondence problem and then find out the consensus partition. Tumer and Agogino [2] introduced the criterion of Average Normalized Mutual Information (ANMI) to measure the ensemble results. They assigned the data points to different clusters dynamically by a voting procedure for achieving the best ANMI. Dimitriadou et al. [3] presented a voting scheme for integrating fuzzy clustering algorithms. The main steps include creating a mapping between candidate clusterings, calculating the highest percentage of common points, and assigning the points to the common clusters. Wang et al. [4] proposed a soft-voting method to integrate the candidate soft clustering results and achieved acceptable results.
- In hyper-graph based methods, the ensemble clustering problem is transformed into a hyper-graph partitioning problem. The data points are represented as edges in a hyper-graph and the clusters as undirected hyper-edges. Under this idea there

C.K. Loo et al. (Eds.): ICONIP 2014, Part II, LNCS 8835, pp. 1–8, 2014.
© Springer International Publishing Switzerland 2014

are three representative methods which are Cluster-based Similarity Partitioning Algorithm (CSPA) [5], Hyper-Graph Partitioning Algorithm (HGPA) [6] and Meta-Clustering Algorithm (MCLA) [7]. The CSPA creates a binary similarity matrix for each single clustering. The entry-wise average of single similarity matrices yields an overall similarity matrix. They re-cluster the overall matrix and get the ensemble results. In the HPGA, the ensemble problem is formulated as partitioning the hyper-graph by cutting a minimal number of hyper-edges. In the MCLA, the idea is to group and collapse related hyper-edges and assign each object to the collapsed hyper-edge. It provides better performance than HPGA and retains low computational complexity.

In this paper, we propose a new ensemble clustering approach based on Dempster-Shafer (DS) evidence theory and Gaussian Mixture Modeling (GMM) technique. Each group of candidate clustering results can be regarded as an evidence to determine the final clustering results. Thus the ensemble clustering problem can be solved by using DS theory to combine the evidences from involved clustering methods. Based on this core idea, we introduce the GMM technique to calculate the confidence of assigning a data point to a candidate cluster for each clustering method. Then the orthogonal sum of confidences from different clustering methods is computed and used to decide the final result under the DS theory. We evaluate the effectiveness of our proposed approach by conducting the experiments on commonly used data sets.

2   Single Clustering Methods

In this paper we use the proposed ensemble method to combine single clustering methods based on dense Gaussian distributions. In this type of single clustering methods, we use the Expectation-Maximization (EM) algorithm [8] to fit a GMM with a large number of Gaussians to the data set. The data subset corresponding to each generated Gaussian component is taken as a minimum unit of data. Then, the classical clustering methods can be performed on these units to complete the clustering. This means that all the operations will be processed on Gaussian distributions, instead of on data points. Furthermore, each cluster can be seen as a GMM composed by dense Gaussians.

For completing such clustering, we need a measure of similarity between Gaussians. In this paper, the Gaussian Quadratic Form Distance (GQFD) [9] is used, which is defined as

$$GQFD_{f_s}(g_a, g_b) = \sqrt{(\omega^a | -\omega^b) A (\omega^a | -\omega^b)^T} , \qquad (1)$$

where g_a and g_b be two Gaussian distributions, $(\omega^a | -\omega^b)$ denotes the concatenation of weights from g_b and g_b, A is a matrix, each entry in which is the measure of similarity between two Gaussians. The GQFD has proved its effectiveness for modeling content-based similarity, for more details of which the reader is referred to Beecks et al. [9].

We summarize the algorithm framework of dense Gaussian distributions based single clustering method in Algorithm 1.

Algorithm 1. Dense Gaussian distributions based single clustering algorithm
---
Input: Data set
Output: Clustering Results
Steps:
    Step1. Fit a GMM with a large number of components to the whole data set by using the EM algorithm.
    Step2. Use a classical clustering method (such as k-means) configured with GQFD to cluster the generated Gaussians.
    Step3. The data points are grouped according to the clusters of Gaussians.

---

3 Ensemble Method

The core task of our ensemble clustering approach is to combine the confidences for candidate results from single clustering methods. We introduce the DS evidence theory to complete this task. Suppose we have N single clustering algorithms, each of them organizes data points into K clusters. Let $R_1, R_2, \cdots, R_N$ be the corresponding clustering results from each single algorithm. Then our DS theory based ensemble clustering approach is explained as follows. For the details of DS evidence theory itself the reader is referred to Shafer [10].

3.1 Evidence from Single Clustering Method

Under the DS evidence theory, we need to compute the evidence corresponding with candidate clustering results from each method for further combining them. For each $R_i\big|_{i=1}^{N}$, this evidence can be represented by the probabilities corresponding to each element in its power set. We calculate these probabilities based on the GMM.

Let $\mathbf{\Omega}_i = \left\{\mathbf{\Omega}_i^1, \mathbf{\Omega}_i^2, \cdots, \mathbf{\Omega}_i^T\right\}$ be the power set of R_i, where T be the number of elements in $\mathbf{\Omega}_i$. Obviously, $T = 2^K - 1$. The first K elements in $\mathbf{\Omega}_i$ are K clusters in R_i, which are generated by the single clustering method. The rest elements in R_i are the possible clusters combined by these K elements. For example, if R_i is $\{r_1, r_2, r_3\}$, where $r_j\big|_{j=1}^{3}$ represent the three clusters from the i-th single clustering method, then $\mathbf{\Omega}_i = \{\{r_1\}, \{r_2\}, \{r_3\}, \{r_1, r_2\}, \{r_1, r_3\}, \{r_2, r_3\}, \{r_1, r_2, r_3\}\}$.

For a large K, T will become a huge value and the following computation will become unfeasible if all the combinations in the power set $\mathbf{\Omega}_i$ are considered. In such cases, we just consider the clusters combined by at most 2 elements of R_i. In other words, the combinations with 3 or more elements are neglected for a large K. Denoeux and Masson [11] have proved that this strategy is suitable for big power sets.

We introduce the GMM to model each element in Ω_i. As explained before, the first K elements in Ω_i correspond to the GMMs generated by our single clustering method. The rest elements are composed by these generated GMMs. Let x be an arbitrary data point, $p(x|\Omega_i^j)$ be the resultant GMM for Ω_i^j, then we have

$$p(x|\Omega_i^j) = \sum_{k=1}^{m} \omega_k (2\pi)^{-\frac{d}{2}} |\Sigma_k|^{-\frac{1}{2}} \exp\left(-\frac{1}{2}(x-\mu_k)^T \Sigma_k^{-1}(x-\mu_k)\right) \qquad (2)$$

where m be the number of Gaussian components, ω_k, μ_k, Σ_k be the weight, the mean vector, and the covariance matrix of the k-th Gaussian component, respectively.

Based on the GMM, we use Bayesian rule to calculate the posterior probability of assigning x to the cluster Ω_i^j:

$$P(\Omega_i^j|x) = p(x|\Omega_i^j)P(\Omega_i^j) / \sum_{k=1}^{T} p(x|\Omega_i^k)P(\Omega_i^k), \qquad (3)$$

where the prior probability $P(\Omega_i^j)$ is assumed to be the same for each element.

3.2   Combining Evidences

The posterior probability $P(\Omega_i^j|x)$ can be seen as the confidence of the i-th clustering algorithm for associating x with the j-th cluster. We combine this kind of confidences from all the involved algorithms by using the orthogonal sum method. Let Ω^j be the combined result of the j-th cluster, then we have

$$P(\Omega^j|x) = \left(\sum_{\cap \Omega_i^j = \Omega^j} \prod_{i=1}^{N} P(\Omega_i^j|x)\right) \Big/ \left(\sum_{\cap \Omega_i^j \neq \phi} \prod_{i=1}^{N} P(\Omega_i^j|x)\right). \qquad (4)$$

As described above, $\Omega^j\big|_{j=1}^{K}$ is corresponding with K single clusters. To determine the final ensemble clustering results, we calculate the belief and plausibility values for these first K elements by

$$Bel(\Omega^j|x) = \sum_{A \subseteq \Omega^j} P(A|x), \quad j = 1,2,\cdots,K , \qquad (5)$$

and

$$Pl(\Omega^j|x) = 1 - Bel(\neg\Omega^j|x), \quad j = 1,2,\cdots,K , \qquad (6)$$

respectively. Based on the belief and plausibility functions, the final confidence of assigning x to the j-th cluster can be computed as the class probability. Let $|\Omega^j|$

and $|\Omega_i|$ denote the number of elements in Ω^j and Ω_i, respectively. Then the class probability is

$$f(\Omega^j|x) = Bel(\Omega^j|x) + \frac{|\Omega^j|}{|\Omega_i|}[Pl(\Omega^j|x) - Bel(\Omega^j|x)], \quad j = 1,2,\cdots,K, \quad (7)$$

Finally, the cluster with the maximum class probability is selected for x.

Algorithm 2 summarizes our ensemble clustering approach described above, the meaning of symbols used there are same with the counterparts above.

Algorithm 2. Ensemble clustering based on DS evidence theory

Input: $R_1, R_2, \cdots, R_N$
Output: Ensemble clustering results
Steps:

Step 1. For each $R_i|_{i=1}^{N}$,

    Step 1.1 Generate the power set of R_i, i.e., Ω_i

Step 2. For each data point x,

    Step 2.1 Calculate the posterior probabilities $P(\Omega_i^j|x)$ for each pair of i and j by using Eq. 3.

    Step 2.2 Calculate the orthogonal sum $P(\Omega^j|x)$ for each j by using Eq. 4.

    Step 2.3 Calculate the belief and plausibility values by using Eq. 5 and 6.

    Step 2.4 Compute the class probabilities for each Ω^j using Eq. 7 and assign x to the cluster with the maximum class probability.

---

4 Experiments

4.1 Experimental Setup

In the experiments, we select 4 commonly used data sets for testing our clustering method. The first two are 2-D data vector (http://cs.joensuu.fi/sipu/datasets/), including Flame (240 vectors with 2 clusters) and Jain (373 vectors with 2 clusters). The clustering results over them can be visualized intuitively. The third one is the Iris in UCI Machine Learning repository (http://arch-ive.ics.uci.edu/ml/). The Iris contains 3 classes of 50 instances each. Each instance has 4 attributes and the class label. So the clustering accuracy can be measured exactly. The last data set is KDD Cup 04Bio (http://www.sigkdd.org). It provides 145751 data points with 74 attributes. The number of clusters is given as 2000, but there is no class labeling. It can be used to test the clustering performance over large size and high dimensional data.

We implemented two single clustering algorithms by embedding k-means or spectral clustering method into dense Gaussian distributions based clustering method described in Section 2, i.e., using them in Step 2 of Algorithm 1, respectively. The two resultant methods are called k-means_G and spectal_G for short, respectively. The clustering results from them are combined by our ensemble approach and the voting based counterpart of Weingessel et al. [12], respectively. We call the method of Weingessel et al. as WDH voting for the convenience of descriptions.

We use the following criteria to evaluate the accuracy and efficiency of clustering approaches: Accuracy Rate (AR), Internal Quality (IQ) and External Quality (EQ) of clusters, Execution Time (ET) and Cost of Memory (CoM).

- The AR can reflect the accuracy of clustering on the data sets with class labeling. Let a_i be the number of correctly classified instances of the i-th cluster, n be the number of all instances in the data set, then $AR = \sum_{i=1}^{K} a_i / n$.

- The IQ and EQ of clusters [12] are widely used for data sets without labels. Let C_i be the i-th cluster, μ_i be its mean vector, $d(x, \mu_i)$ be the Euclidean distance between a data point x and μ_i, then $IQ = \sum_{i=1}^{K} \sum_{x \in C_i} d(x, \mu_i)$, $EQ = \sum_{1 \le j \le i \le K} d(\mu_i, \mu_j)$. The EQ is proportional to the degree of closeness between different clusters, so the bigger EQ is, the better the clustering quality is. While the IQ is inversely proportional to the degree of data closeness within a cluster, so the less IQ is preferred.

- The ET and CoM are useful for measuring the computational complexity. Since the topic of this paper is ensemble clustering, we only consider the ensemble procedure in the calculation of ET and CoM and neglect the cost consumed for performing each single clustering method. Notice that all the following experiments are performed on a computer with 3.4GHz CPU and 10GB inner memory

4.2   Experimental Results

The results of 4 algorithms on Flame and Jain distributions are shown in Fig. 1-2, respectively. In both of two figures, sub-figures (a) to (d) show the clustering results from 4 algorithms and (e) the true distribution. The ARs for each algorithm are given in the title of each figure. It can be discovered intuitively that our DS based ensemble results are close to the true distributions. On the Flame dataset, it performed better than both single clustering methods. On the Jain dataset, it behaved better than k-means_G but a little worse than spectral_G. These results demonstrate that our ensemble method cannot guarantee to achieve better results than the best single result, but it does improve the worse single results obviously. Thus the combination of results is effective. Furthermore, our method behaved better than the WDH voting. Compared with it, our DS based method brought 12.7% and 11.7% increase in the AR over two data sets, respectively.

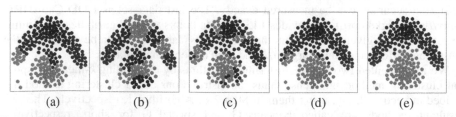

       (a)   (b)   (c)   (d)   (e)

Fig. 1. The results on the Flame distribution: (a) k-means_G (0.85); (b) spectral_G (0.71); (c) WDH voting (0.79); (d) our method (0.89); (e) true distribution

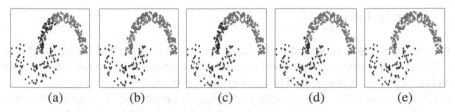

(a)   (b)   (c)   (d)   (e)

Fig. 2. The results on the Jain distribution: (a) k-means_G (0.83); (b) spectral_G (0.99); (c) WDH voting (0.85); (d) our method (0.94); (e) true distribution

For Iris data set, the AR from each algorithm are listed in Table 1. As shown in Table 1, both of two ensemble clustering algorithms can improve the worse single results, but our method still behaved better. Compared with the WDH voting, our method brought 10.8% increase in the AR. As for the efficiency, the ET of our method and the WDH voting on Iris data set are similar, which are 2.71s and 3.02s, respectively. The CoM is ignored since the data size is too small.

Table 1. Comparisons of ARs on Iris Data Set

k-means_G	Spectral_G	WDH Voting	Our method
84.3%	73.6%	75.1%	83.0%

For the clustering on KDD CUP 04Bio data set, the IQ and EQ are used to evaluate the effectiveness of the algorithms. The results are shown in Fig. 3. We can see that our method achieved the best IQ and the second best EQ on this data set. And the EQ from our method is very close to the best one from spectral_G. Compared with k-means_G, spectral_G and the WDH voting, the increase rates of IQ bought by our method are 10.3%, 23.8% and 19.1%, respectively. As for the EQ, the increase rates are 23.2%, 15.3% and -3.28% for k-means_G, WDH voting and spectral_G, respectively. Furthermore, the ET of our method on KDD CUP 04Bio is 13.22 minutes, which is a little better than 15.35 minutes of WDH voting. But the CoMs of ours are worse than that of WDH voting. The two values are 4.02MB and 2.88MB, respectively.

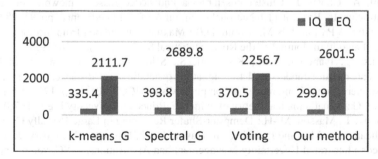

Fig. 3. IQ and EQ on KDD CUP 04Bio Data Set

5   Conclusions

In this paper, we have proposed a new ensemble clustering method based on Dempster-Shafer (DS) evidence theory and tested it for combining dense Gaussian distributions based clustering methods. The main contributions of this paper are:

(1) The Gaussian Mixture Modeling technique is introduced to compute the confidences of assigning each data point to candidate clusters, which reflects the evidences supplied by single clustering methods.

(2) The DS evidence theory is employed to combine evidences from various single clustering methods, based on which the final clustering result are obtained.

We tested the proposed approach on 4 commonly used data sets, including 2-D distributions (Flame and Jain) with intuitive visualization, Iris with exact class labeling and KDD Cup 04Bio with large size and high dimensional data. The results confirm that the proposed ensemble clustering method is effective and promising.

References

1. Vega-Pons, S., Ruiz-Shulcloper, J.: A survey of clustering ensemble algorithms. International Journal of Pattern Recognition and Artificial Intelligence 25(3), 337–372 (2011)
2. Tumer, K., Agogino, A.K.: Ensemble clustering with voting active clusters. Pattern Recognition Letters 29(14), 1947–1953 (2008)
3. Dimitriadou, E., Weingessel, A., Hornik, K.: A combination scheme for fuzzy clustering. International Journal of Pattern Recognition and Artificial Intelligence 16(7), 901–912 (2002)
4. Wang, H., Yang, Y., Wang, H., Chen, D.: Soft-Voting Clustering Ensemble. In: Zhou, Z.-H., Roli, F., Kittler, J. (eds.) MCS 2013. LNCS, vol. 7872, pp. 307–318. Springer, Heidelberg (2013)
5. Strehl, A.: Relationship-based clustering and cluster ensembles for high-dimensional data mining. Ph.D dissertation, The University of Texas at Austin (2002)
6. Karypis, G., Kumar, V.: A fast and high quality multilevel scheme for partitioning irregular graphs. SIAM Journal on Scientific Computing 20(1), 359–392 (1998)
7. Strehl, A., Ghosh, J.: Cluster ensembles-a knowledge reuse framework for combining partitioning. In: Proc. of 11th National Conf. on Artificial Intelligence, pp. 93–98 (2002)
8. Dempster, A.P., Laird, N.M., Rubin, D.B.: Maximum likelihood from incomplete data via the EM algorithm. Journal of the Royal Statistical Society 39(1), 1–38 (1977)
9. Beecks, C., Ivanescu, A.M., Kirchhoff, S., Seidl, T.: Modeling image similarity by Gaussian mixture models and the signature quadratic form distance. In: Proc. of 2011 IEEE International Conference on Computer Vision (ICCV 2011), pp. 1754–1761 (2011)
10. Shafer, G.: A mathematical theory of evidence. Princeton University Press (1976)
11. Denœux, T., Masson, M.-H.: Dempster-Shafer Reasoning in Large Partially Ordered Sets: Applications in Machine Learning. In: Huynh, V.-N., Nakamori, Y., Lawry, J., Inuiguchi, M. (eds.) Integrated Uncertainty Management and Applications. AISC, vol. 68, pp. 39–54. Springer, Heidelberg (2010)
12. Weingessel, A., Dimitriadou, E., Hornik, K.: An ensemble method for clustering. In: Proc. of the 3rd International Workshop on Distributed Statistical Computing (2003)
13. Rokach, L.: A survey of clustering algorithms. In: Data Mining and Knowledge Discovery Handbook, pp. 269–298. Springer US (2010)

Extraction of Dimension Reduced Features from Empirical Kernel Vector

Takio Kurita and Yayoi Harashima

Department of Information Engineering
Hiroshima University
1-4-1 Kagamiyama, Higashi-Hiroshima, 739-8527, Japan

Abstract. This paper proposes a feature extraction method from the given empirical kernel vector. We show the necessary condition for the feature extraction mapping to make the trained classifier by using the linear SVM with the extracted feature vectors equivalent to the one obtained by the standard kernel SVM. The proposed feature extraction mapping is defined by using the eigen values and eigen vectors of the Gram matrix. Since the eigen vector problem of the Gram matrix is closely related with the kernel Principal Component Analysis, we can extract a dimension reduced feature vector. This feature extraction method becomes equivalent to the kernel SVM if the full dimension is used. The proposed feature extraction method was evaluated by the experiments using the standard data sets. The cross-validation values of the proposed method were improved and the recognition rates were comparable with the original kernel SVM. The number of extracted features was very low compared to the number of features of the kernel SVM.

1 Introduction

Support vector machine (SVM) [15,12,3,7] has been successfully applied to many pattern recognition problems such as object detection[5] and image classification[4] etc. The nonlinear classifier with good generalization can be constructed by using kernel-trick and margin maximization.

However, the dimension of the empirical kernel vector in the kernel SVM increases as the number of training samples increases. Especially for big data, this makes the computation of the learning algorithm intractable. Also the generalization ability of the trained classifier probably decreases because the number of parameters increases as the number of training samples increases. It is well known that the complexity of the model used in the learning and the intrinsic dimension to describe the target classification problem should be the same to get the classifier with good generalization.

To reduce the difference between the complexities of the learning model and the the target classification problem, Nishida et al. proposed a method to select the important kernel features by using Boosting [10]. Also Nishida et al. proposed an algorithm called RANSAC-SVM in which the subsets of the training samples were randomly generated and the best subset was selected [11].

C.K. Loo et al. (Eds.): ICONIP 2014, Part II, LNCS 8835, pp. 9–16, 2014.

In this paper, we propose a method to extract new feature vector from the given empirical kernel vector by using kernel Principal Component Analysis (the kernel PCA).

At first, we show the necessary condition which has to be satisfied to make the classifier obtained by using the linear SVM with the extracted feature vector from the empirical kernel vector in the case of full dimension equivalent to the one trained by using the standard kernel SVM. Then a feature extraction mapping from the given empirical kernel vector is derived. The feature extraction mapping can be defined by using the eigen values and eigen vectors of the Gram matrix.

The eigen vector problem of the Gram matrix is closely related with the kernel PCA [1,13]. The eigen vectors corresponding to the first largest eigen values are the principal components and extract the dominant information from the Gram matrix. By combining the feature extraction mapping by Gram matrix and the dimension reduction by the kernel PCA, we can design feature extraction method in which the dominant information of the given empirical kernel vector is extracted but the unnecessary details are neglected. This feature extraction becomes equivalent to the kernel SVM if the full dimension is used.

By the experiments using the standard data sets, we found that the very few features were enough to achieve good recognition rates for test samples by using the proposed dimension reduced feature vectors. These results shows that the empirical kernel vector includes redundant information and there is margin to reduce the number of dimension.

As the related works, general theory of the kernel PCA whitening in kernel-based methods is shown in [13]. A relation between the kernel PCA and the least squares SVM (LS-SVM) is shown in [14]. Q. Chen et al. proposes a combination of the kernel PCA and LS-SVM and applied to time series prediction [2]. In this paper we experimentally evaluate the effect of the dimension reduction by using the kernel PCA whitening for the case of the kernel SVM. Also it is reported that the whitening using the covariance matrix of the HOG features can improve the recognition performance when it is used as the input of the linear SVM in [6]. The tendency of our experimental results agrees with the results of the whitening of the HOG features.

2   Feature Extraction from Empirical Kernels

In this paper we extract dimension reduced feature vector from the given empirical kernel vector by using a linear mapping. Then the extracted new feature vectors are used as the input of the linear SVM. We want to make the obtained classifier without dimension reduction identical to the one obtained by the original kernel SVM. To consider the constraints which should be satisfied by this linear mapping, we will briefly review the linear and kernel SVM.

2.1   Linear and Kernel SVM

The linear SVM determines the separating hyperplane with a maximal margin by using the given training samples $\{< \boldsymbol{x}_i, t_i > | i = 1, \ldots, n\}$, where $\boldsymbol{x}_i$ is the input feature vector and $t_i = \in \{+1, -1\}$ is the class label of the i-th sample. Then the classification function of the linear SVM is given as

$$y = \text{sign}(\boldsymbol{w}^T \boldsymbol{x} - h), \tag{1}$$

where $\boldsymbol{w}$ and h are the weight vector and the threshold, respectively. The function sign(u) is a sign function which outputs 1 when $u > 0$ and outputs -1 when $u \leq 0$. The soft-margin SVM is defined as an optimization problem for the following evaluation function

$$L(\boldsymbol{w}, \boldsymbol{\xi}) = \frac{1}{2}||\boldsymbol{w}||^2 + \gamma \sum_{i=1}^{n} \xi_i, \tag{2}$$

under the constraints $\xi_i \geq 0$, $t_i(\boldsymbol{w}^t \boldsymbol{x}_i - h) \geq 1 - \xi_i$, $i = 1, \ldots, n$, where ξ_i is the measure of the error for the training sample $\boldsymbol{x}_i$. The dual problem is obtained as the optimization problem that maximizes the object function

$$L_D(\boldsymbol{\alpha}) = \sum_{i=1}^{n} \alpha_i - \frac{1}{2} \sum_{i,j=1}^{n} \alpha_i \alpha_j t_i t_j \boldsymbol{x}_i^T \boldsymbol{x}_j \tag{3}$$

under the constraints $\sum_{i=1}^{n} \alpha_i t_i = 0$, $0 \leq \alpha_i \leq \gamma$, $i = 1, \ldots, n$.

By solving this optimization problem, the optimal classification function can be expressed as

$$y = \text{sign}(\sum_{i \in S} \alpha_i^* t_i \boldsymbol{x}_i^T \boldsymbol{x} - h^*), \tag{4}$$

where S is a set of support vectors and α_i^* and h^* are the optimal solutions.

By using the kernel-trick, this linear SVM can be extended to nonlinear (kernel SVM). In the kernel SVM, input vectors are mapped to higher dimensional feature space by non-linear function $\phi(\boldsymbol{x})$ and the linear SVM is applied to the mapped features. Since the linear SVM depends only on the inner products of the input vectors, we can define the object function as

$$L_D(\boldsymbol{\alpha}) = \sum_{i=1}^{n} \alpha_i - \frac{1}{2} \sum_{i,j=1}^{n} \alpha_i \alpha_j t_i t_j K(\boldsymbol{x}_i, \boldsymbol{x}_j), \tag{5}$$

where $K(\boldsymbol{x}_i, \boldsymbol{x}_j) = \phi(\boldsymbol{x}_i)^T \phi(\boldsymbol{x}_j)$ is the kernel function. Usually the kernel function $K(\boldsymbol{x}, \boldsymbol{y})$ is defined a priori. The polynomial function and the Radial Basis function are often used as the kernel function.

Then the optimal classification function of the kernel SVM can be derived as

$$y = \text{sign}(\sum_{i \in S} \alpha_i^* t_i \phi(\boldsymbol{x}_i)^T \phi(\boldsymbol{x}) - h^*) = \text{sign}(\sum_{i \in S} \alpha_i^* t_i K(\boldsymbol{x}_i, \boldsymbol{x}) - h^*). \tag{6}$$

2.2   Feature Extraction from Empirical Kernel Vector

The classification function of the kernel SVM given in equation (6) determines the separating hyperplane on the n dimensional feature vector

$$\boldsymbol{k}(\boldsymbol{x}) = (K(\boldsymbol{x}_1, \boldsymbol{x}), \ldots, K(\boldsymbol{x}_n, \boldsymbol{x}))^T. \tag{7}$$

This means that this n dimensional feature vector include enough information to construct the classification function of the kernel SVM. We call this n dimensional feature vector the empirical kernel vector.

To extract effective features from this empirical kernel vector, we consider the following linear feature extraction

$$g(\boldsymbol{x}) = U^T \boldsymbol{k}(\boldsymbol{x}). \tag{8}$$

Then we use this new feature vector $g(\boldsymbol{x})$ as the input of the linear SVM.

By substituting the $\boldsymbol{x}_i$ with the new feature $g(\boldsymbol{x}_i)$ in the equation (3), we have

$$L_D(\boldsymbol{\alpha}) = \sum_{i=1}^{n} \alpha_i - \frac{1}{2} \sum_{j=1}^{n} \alpha_j t_j \Gamma K U U^T \boldsymbol{k}(\boldsymbol{x}_j), \tag{9}$$

where $K = (K(\boldsymbol{x}_i, \boldsymbol{x}_j))_{i,j=1}^{n}$ and $\Gamma = \text{diag}(\alpha_1 t_1, \ldots, \alpha_n t_n)$. The matrix K is known as the kernel Gram matrix.

Similarly, the optimal classification function becomes

$$y = \text{sign}(\Gamma^* K U U^T \boldsymbol{k}(\boldsymbol{x}) - h^*), \tag{10}$$

where $\Gamma^* = \text{diag}(\alpha_1^* t_1, \ldots, \alpha_n^* t_n)$.

To get the same object function and the classification function with the kernel SVM, the coefficient matrix U must be satisfy the condition $U U^T = K^{-1}$. Since the kernel Gram matrix K is real symmetric, we can compute the U by using the eigen values λ_i and the corresponding eigen vectors $\boldsymbol{a}_i$ of the kernel Gram matrix K. The eigen values and the corresponding eigen vectors of the kernel Gram matrix K are given by

$$K A = A \Lambda \tag{11}$$

where $\Lambda = \text{diag}(\lambda_1, \ldots, \lambda_n)$ is the diagonal matrix with the eigen values and $A = (\boldsymbol{a}_1 \cdots \boldsymbol{a}_n)$ is the matrix of eigen vectors. Thus the coefficient matrix U in the feature extraction mapping to make the obtained classifier coincident with the kernel SVM can be given by

$$U = A \Lambda^{-\frac{1}{2}}, \tag{12}$$

where $\Lambda^{-1/2} = \text{diag}(\frac{1}{\sqrt{\lambda_1}}, \frac{1}{\sqrt{\lambda_2}}, \ldots, \frac{1}{\sqrt{\lambda_n}})$. Since the matrix A is orthogonal, we can confirm the condition of the inverse matrix as

$$K(U U^T) = K A \Lambda^{-\frac{1}{2}} \Lambda^{-\frac{1}{2}} A^T = A \Lambda \Lambda^{-1} A^T = A A^T = I. \tag{13}$$

Then the feature extraction which gives the same results with the kernel SVM is given by

$$g_{SVM}(\boldsymbol{x}) = \Lambda^{-\frac{1}{2}} A^T \boldsymbol{k}(\boldsymbol{x}). \tag{14}$$

This feature extraction is closely related with the kernel PCA and we can extract the dimension reduced features in terms of the kernel principal components.

2.3   Relation with Kernel Principal Component Analysis

Nonlinear extension of PCA using the kernel-trick is known as the kernel PCA. In the kernel PCA, the input vectors are also mapped to the higher dimensional feature space by non-linear function $\phi(x)$ and the linear PCA is applied to the mapped features.

For the given data set $\{x_1, \ldots, x_n\}$, the kernel PCA computes the principal score as

$$y(x) = U^T \phi(x). \tag{15}$$

Since the coefficient matrix can be represented by the linear combinations of the mapped feature vectors as

$$U = \sum_{j=1}^{n} \phi(x_j) \alpha_j^T, \tag{16}$$

the principal score vector can be given by

$$y(x) = \sum_{j=1}^{n} \alpha_j \phi(x_j)^T \phi(x) = \sum_{j=1}^{n} \alpha_j K(x_j, x). \tag{17}$$

The optimal solution can be obtained by taking the L eigen vectors $\tilde{A} = (\alpha_1 \cdots \alpha_L)$ of the kernel Gram matrix K corresponding to the L largest eigen values $\lambda_1, \ldots, \lambda_L$. The eigen vector equation for kernel PCA is given by

$$K\tilde{A} = \tilde{A}\tilde{\Lambda}, \tag{18}$$

where $\tilde{\Lambda} = \mathrm{diag}(\lambda_1, \ldots, \lambda_L)$.

By comparing the eigen vector equation (18) for the kernel PCA and the eigen vector equation (11) for the feature extraction from the empirical kernels, it is noticed that they are the same. Since the kernel PCA can extract dominant information from the data set and neglect the unnecessary details by taking the principal components, we can construct new dimension reduced features by taking the L eigen vectors of the kernel Gram matrix K corresponding to the L largest eigen values as

$$g_{PCA}(x) = \tilde{\Lambda}^{-\frac{1}{2}} \tilde{A}^T k(x). \tag{19}$$

This feature vector $g_{PCA}(x)$ can extract the dominant information of the training data set and neglect the unnecessary details. Also this feature vector can produce almost same result with the kernel SVM when this feature vector is used as the input of the linear SVM. Especially the result becomes the same as the kernel SVM if the full dimension, namely $L = n$, is used.

3   Experiments

The effectiveness of the proposed feature extraction defined in the equation (19) was evaluated by using seven standard data sets (**heart, iris, vowel, breast-cancer, glass,** and **vehicle**) from LIBSVM data sets [8]. The number of classes,

Table 1. The cross validation values (%), recognition rates (%) and the dimension of extracted feature vector for the standard data sets

data set		K-SVM	FE-PCA
heart	CV	85.33 (2.06)	**86.16** (1.32)
	training	**87.41** (2.06)	86.94 (1.32)
	test	78.52 (5.67)	**78.70** (5.03)
	dim.	216	18.6 (20.17)
iris	CV	97.50 (0.88)	**97.58** (1.00)
	training	97.66 (1.23)	**97.83** (1.05)
	test	**95.33** (2.81)	93.67 (4.83)
	dim.	120	4.8 (1.93)
vowel	CV	98.58 (0.45)	**98.70** (0.49)
	training	**99.98** (0.07)	99.95 (0.01)
	test	**98.58** (0.80)	**98.58** (0.67)
	dim.	677	168.7 (67.15)
breast-cancer	CV	97.20 (0.35)	**97.29** (0.42)
	training	**97.33** (0.36)	97.31 (0.37)
	test	**97.23** (1.74)	**97.23** (1.81)
	dim.	546	7.8 (8.66)
glass	CV	72.88 (3.20)	**74.10** (2.76)
	training	**92.75** (5.20)	89.59 (6.39)
	test	67.21 (8.09)	**68.14** (7.44)
	dim.	171	52.6 (29.61)
vehicle	CV	85.30 (1.10)	**85.89** (0.98)
	training	**93.24** (0.80)	91.57 (0.90)
	test	**83.41** (2.33)	83.29 (2.86)
	dim.	170	89.8 (10.76)

the number of samples, and the number of features are (2, 270, 13), (3, 150, 4), (11, 528, 10), (2, 683, 10), (6, 214, 9), and (4, 846, 18) respectively. For classification experiments, each data set was randomly divided into a training set (80% of all samples) and a test set (remaining samples). We performed 10 times with different partitions of training and test samples and the average and the standard deviation were measured. The Radial Basis functions $K(\boldsymbol{x}, \boldsymbol{y}) = \exp\left(-\frac{\|\boldsymbol{x}-\boldsymbol{y}\|^2}{2\sigma^2}\right)$ was used as the kernel function.

Table 1 shows the cross-validation values and the recognition rates for training samples and test samples of the proposed feature extraction methods, namely the feature extraction from empirical kernel vectors by using PCA (denoted as FE-PCA). In the proposed feature extraction method, the extracted feature vector was classified by using the linear SVM. The recognition rates of the standard kernel SVM (denoted as K-SVM) are also shown in Table 1 for comparison. In these experiments, the Radial Basis function is used as the kernel function. The parameters of the linear and the kernel SVM, i.e. the soft margin parameter γ and the kernel parameter σ^2 were determined by 10-fold cross validation (For the data

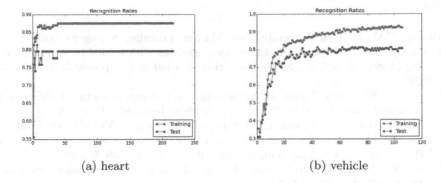

(a) heart   (b) vehicle

Fig. 1. Relation between the recognition rates and the extracted dimension

set **glass**, 5-fold cross validation was used because of the shortage of the samples of each class). Since the recognition rate depends on the number of extracted features in the proposed feature extraction methods, the best dimension was selected by the cross validation. The average and the standard deviation of the selected dimension are also shown in Table 1 (denoted as dim.).

From Table 1, the cross-validation values of the proposed feature extraction method gives a little bit better results for all data sets than the standard SVM while the recognition rates for training and test samples are comparable. This means that the proposed dimension reduction method can keep the generalization performance of the kernel SVM.

The number of extracted features is very low compared to the number of original features of the kernel SVM. For example, 546 was reduced to 7.8 for **breast-cancer**. This means that the empirical kernel vector includes very redundant information and there is margin to reduce the number of dimension.

Figure 1 shows the relation between the recognition rates and the number of dimension of the extracted feature vector for **heart** and **vehicle** data sets. It is noticed that there is almost flat regions from low to high dimension. This also shows that information included in the empirical kernel vector is very redundant and the proposed feature extraction method can extract intrinsic information from the empirical kernel vector.

Since the number of features of the kernel SVM increases as the number of training samples increases, the difference between the number of features of the kernel SVM and the intrinsic dimension of the target classification problem becomes large for large data. It is expected that the proposed feature extraction method can be used to reduce this gap.

Acknowledgment. The preliminary experiments on this work was mainly done by Mr. Kohei Takahashi. This work was partly supported by KAKENHI (23500211).

References

1. Bishop, C.M.: Pattern Recognition and Machine Learning. Springer (2006)
2. Chen, Q., Chen, X., Wu, Y.: Optimization Algorithm with Kernel PCA to Support Vector Machines for Time Series Prediction. Journal of Computers 5(3), 380–387 (2010)
3. Cristianini, N., Shawe-Taylor, J.: An Introduction to Support Vector Machines and other Kernel-based Learning Methods. Cambridge University Press (2000)
4. Csurka, G., Dance, C.-R., Fan, L., Willamowski, J., Bray, C.: Visual Categorization with Bag of Keypoints. In: Proc. of European Conference on Computer Vision 2004 Workshop on Statistical Learning in Computer Vision, pp. 59–74 (2004)
5. Dalal, N., Triggs, B.: Histogram of oriented gradients for human detection. In: Proc. of CVPR 2005 (2005)
6. Hariharan, B., Malik, J., Ramanan, D.: Discriminative decorrelation for clustering and classification. In: Fitzgibbon, A., Lazebnik, S., Perona, P., Sato, Y., Schmid, C. (eds.) ECCV 2012, Part IV. LNCS, vol. 7575, pp. 459–472. Springer, Heidelberg (2012)
7. Hastie, T., Tibshirani, R., Friedman, J.: The Elements of Statistical Learning -Data Mining, Inference, and Prediction, 2nd edn. Springer (2006)
8. Chang, C.-C., Lin, C.-J.: LIBSVM: a library for support vector machines (2001), http://www.csie.ntu.edu.tw/=sjlin/libsvm
9. Mika, S., Rätsch, G., Weston, J., Schölkopf, B., Smola, A., Müller, K.: Fisher discriminant analysis with kernels. In: Proc. IEEE Neural Networks for Signal Processing Workshop, pp. 41–48 (1999)
10. Nishida, K., Kurita, T.: Kernel Feature Selection to Improve Generalization Performance of Boosting Classifiers. In: The 2006 International Conference on Image Processing, Computer Vision, & Pattern Recognition, Monte Carlo Resort, Las Vegas, Nevada, June 26-29 (2006)
11. Nishida, K., Kurita, T.: RANSAC-SVM for Large-Scale Datasets. In: Proc. of International Conference on Pattern Recognition, December 8-11. Tampa Convention Center, Tampa (2008)
12. Scholköpf, B., Burges, C.-J.-C., Smola, A.-J.: Advances in Kernel Methods - Support Vector Learning. The MIT Press (1999)
13. Schölkopf, B., Mika, S., Burges, C.-J.-C., Knirsch, P., Müller, K.-R., Rätsch, G., Smola, A.-J.: Input Spcae Versus Feature Space in Kernel-Based Methods. IEEE Trans. on Neural Networks 10(5), 1000–1017 (1999)
14. Suykens, J.-A.-K., Gestel, T.V., Vandewalle, J., De Moor, B.: A Support Vector Machine Formulation to PCA Analysis and Its Kernel Version. IEEE Trans. on Neural Networks 14(2), 447–450 (2003)
15. Vapnik, V.-N.: Statistical Learning Theory. John Wiley & Sons (1998)
16. Yan, S., Xu, D., Zhang, B., Zhang, H.-J.: Graph embedding: a general framework for dimensionality reduction. In: Proc. of CVPR 2005 (2005)

Method of Evolving Non-stationary Multiple Kernel Learning

Peng Wu[1,2], Qian Yin[1], and Ping Guo[1,*]

[1] Image Processing and Pattern Recognition Laboratory
Beijing Normal University, Beijing 100875, China
[2] Shandong Provincial Key Laboratory of Network based Intelligent Computing
University of Jinan, Jinan 250022, China
pengwiseujn@gmail.com, yinqian@bnu.edu.cn, pguo@ieee.org

Abstract. Recently, evolving multiple kernel learning methods have attracted researchers' attention due to the ability to find the composite kernel with the optimal mapping model in a large high-dimensional feature space. However, it is not suitable to compute the composite kernel in a stationary way for all samples. In this paper, we propose a method of evolving non-stationary multiple kernel learning, in which base kernels are encoded as tree kernels and a gating function is used to determine the weights of the tree kernels simultaneously. Obtained classifiers have the composite kernel with the optimal mapping model and select the most appropriate combined weights according to the input samples. Experimental results on several UCI datasets illustrate the validity of proposed method.

Keywords: Genetic programming, non-stationary multiple kernel learning, composite kernel, gating function.

1 Introduction

Multiple kernel learning (MKL) has been a research hotspot, the essence of which is using multiple base kernels instead of a single one in SVM [1]. Recent theories and applications have proved that MKL is superior to SVM in lots of scenarios because using multiple kernels can enhance the interpret ability of decision functions [2] [3] [4] [5] [6].

The key problem of MKL is how to construct these base kernels and select a proper method to combine them. Earlier, some researchers constructed a specific number of base kernels, then learned a composite kernel by combining base kernels in a linear or non-linear manner [2] [3] [4]. This kind of MKL method is called array based MKL, the shortage of which is the base kernels may not be rich enough to represent the whole sample space because usually they are using some simple kernels such as linear, polynomial and Gaussian kernels. To address such limitations, a number of researchers have made some attempts to find better kernels by using genetic programming (GP) [7] [8] [9]. Starting from some simple base kernels,

---

* Corresponding author.

C.K. Loo et al. (Eds.): ICONIP 2014, Part II, LNCS 8835, pp. 17–25, 2014.

GP can evolve complex composite kernels, which can represent more rich mapping models in a high-dimensional feature space. By doing so, GP based MKL can find a suitable composite kernel from a richer feature space to represent the original sample space. Though GP based MKL exhibited superiority on some problems, it is not suitable to compute composite kernels in a stationary manner, i.e., computing just one composite kernel on the whole input space, in some cases such as for a dataset with locality-sensitive property.

In this paper, we present a hybrid algorithm to evolve non-stationary multiple kernel learning (ENMKL), in which base kernels are encoded as tree kernels and a gating function is used to determine the weights of the tree kernels simultaneously. The set of tree kernels joint with the gating function to produce different composite kernels, varying with regions of the input space. In this way constructed composite kernels can represent the original sample space better for a dataset with locality-sensitive property.

The rest of paper is organized as follow. Related works are given in Section 2. Section 3 describes proposed ENMKL method in details. Experimental results and discussion are given in Section 4. Section 5 gives the conclusion and further work.

2   Related Works

2.1   Tree Kernel of GP Based MKL

Considering a set of samples $D = (\mathbf{x}_i, y)^N$, where vector $\mathbf{x}_i$ denotes the feature of sample, $y \in [-1, 1]$ is the class index, and N is the number of samples. The most commonly used kernels are linear kernel, polynomial kernel and Gaussian kernel, which can be computed as follows:

$$k_{linear}(\mathbf{x}_i, \mathbf{x}_j) = <\mathbf{x}_i, \mathbf{x}_j>,$$
$$k_{poly}(\mathbf{x}_i, \mathbf{x}_j) = (<\mathbf{x}_i, \mathbf{x}_j>+1)^d,$$
$$k_{gaussian}(\mathbf{x}_i, \mathbf{x}_j) = \exp(-\frac{\|\mathbf{x}_i - \mathbf{x}_j\|^2}{2\delta^2}), \tag{1}$$

where d is a positive value degree of polynomial, δ is the width of Gaussian kernel. The theory foundation of GP based MKL is Mercer theorem, i.e., if $K_1 = k_1(\cdot, \cdot)$ and $K_2 = k_2(\cdot, \cdot)$ are kernels under the condition of Mercer, results computed by using (2)-(4) are also kernels.

$$k(\mathbf{x}_i, \mathbf{x}_j) = c_1 k_1(\mathbf{x}_i, \mathbf{x}_j) + c_2 k_2(\mathbf{x}_i, \mathbf{x}_j), \quad \text{for} \quad c_1, c_2 \geq 0, \tag{2}$$

$$k(\mathbf{x}_i, \mathbf{x}_j) = k_1(\mathbf{x}_i, \mathbf{x}_j) k_2(\mathbf{x}_i, \mathbf{x}_j), \tag{3}$$

$$k(\mathbf{x}_i, \mathbf{x}_j) = \exp(k_1(\mathbf{x}_i, \mathbf{x}_j)). \tag{4}$$

Where c_1, c_2 and a are scale coefficients. A more complex new kernel is encoded as a tree structure as shown in Fig. 1.

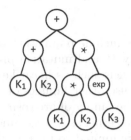

Fig. 1. A tree kernel with terminal set $T = \{K_1, K_2, K_3\}$, non-terminal set $N = \{+, *, exp\}$, where the expression of the tree kernel is $K_1 + K_2 + K_1 * K_2 * \exp(K_3)$

Such a tree is called as tree kernel, and each tree in the population of GP is a tree kernel. For the sake of simplicity, the composite kernel can be written as follow:

$$K_{com} = \{k_{com}(\cdot, \cdot) = h([k_1(\cdot, \cdot), k_2(\cdot, \cdot), ..., k_p(\cdot, \cdot)])\}, \qquad (5)$$

where h denotes a function applying such as addition and multiplication operations on the array of base kernels.

2.2   Non-stationary MKL

The earliest non-stationary MKL was reported in [10] by Lewis *et al.* A large-margin latent variable generative model within the maximum entropy discrimination framework is proposed in [10], which allows the base kernel weight to depend upon the properties of the example being classified. Later, Mehmet Gönen *et al.* proposed localized MKL (LMKL) [11]. LMKL uses a gating function for selecting the appropriate kernel function locally and achieves similar accuracy results as compared with other MKLs by storing less number of support vectors. However, these non-stationary MKLs use the array based base kernels, which might result in sub-optimal or decrease the performance of MKL. Therefore, we propose EN-MKL which combines the advantage of GP based MKL and non-stationary MKL, i.e., it achieves good kernels in a rich feature space and construct the optimal composite kernel using those good kernels according to the input sample. It is worth noting that if each composite kernel evolved by GP is exactly one of the base kernels, ENMKL degenerates into LMKL. So it can be said that LMKL is a special case of ENMKL.

3   Proposed Method

The main tasks of ENMKL are to find: 1) a group of optimal tree kernels, 2) a gating function which generates weights for the optimal tree kernels and 3) the hyper-parameters of classifier. To this end, we utilize a hybrid algorithm of gradient descent and SVM solver within the GP framework.

3.1 Initialization of GP

The parameters involved in GP include population size P_{size}, the iteration number N_{iter}, crossover probability P_c and mutation probability P_m. The terminal set consists of some simple kernels such as linear kernel, polynomial kernel and Gaussian kernel. We can also use other families of kernel containing less number of parameters, especially the family of arc-cosine kernel as discussed in [12]. To ensure the expression of tree kernel as a kernel under the condition of Mercer, the elements of terminal set are limited to $[+, *, exp]$.

3.2 Representation and Fitness of Individual in ENMKL

In order to find a set of composite kernels in searching space, each individual in ENMKL is a multi-gene consisting of more genes, and each of which is a tree kernel. Note that using a multi-gene rather than a single gene is a significant difference as compared with GP based MKLs [7] [8] [9]. We need to divide the dataset into three independent parts, i.e., training, validation and test datasets. To calculate the fitness of an individual in ENMKL, first, we use the gating function to compute the weights of each gene in an individual and get the composite kernel K_{final}, which is the weighted combination of multi-gene. Then, a classifier of canonical SVM is constructed by using K_{final}. The related parameters of the gating function and the classifier involved in the construction are learned by the hybrid algorithm of gradient descent and SVM solver on the training dataset, respectively. Finally, the accuracy of the classifier is verified on the test dataset, which is used as the fitness of an individual in ENMKL.

3.3 Optimizing the Gating Function and SVM

For the simplest case, consider the linear combination of multi-gene, we can write the gating function as follow [11]:

$$\beta_t(\mathbf{x}) = \frac{\exp(\mathbf{v}_t^T \mathbf{x} + v_{t0})}{\sum_{k=1}^{M} \exp(\mathbf{v}_k^T \mathbf{x} + v_{k0})}, \tag{6}$$

where $\mathbf{v}_t$ is the weight of input $\mathbf{x}$, v_{t0} is the bias term of input, and we refer to vector $\mathbf{v} = (\mathbf{v}_t, v_{t0})$ as the parameter of the gating function. M is the number of genes, vector $\mathbf{x}$ is the input sample. The classifier can be trained by solving the following formulation:

$$max \sum_{i=1}^{N} \alpha_i - \frac{1}{2} \sum_{i=1}^{N} \sum_{j=1}^{N} \alpha_i \alpha_j y_i y_j \sum_{m=1}^{M} \beta_m(\mathbf{x}_i) k_{com}^m(\mathbf{x}_i, \mathbf{x}_j) \beta_m(\mathbf{x}_j),$$

$$s.t. \sum_{i=1}^{N} \alpha_i y_i = 0, \quad C \geq \alpha_i \geq 0 \quad \forall i. \tag{7}$$

Where α is the vector of Lagrangian coefficients, b is the bias term of the separating hyperplane, N is the total number of training data, and C is predefined positive trade-off parameter between model complexity and classification error. For the following equation,

$$K_\beta = \sum_{m=1}^{M} \beta_m(\mathbf{x}_i)k_{com}^m(\mathbf{x}_i, \mathbf{x}_j)\beta_m(\mathbf{x}_j). \tag{8}$$

Given the value of $\mathbf{v}$, K_β is a kernel under the condition of Mercer. Then, the object function in (7) is equivalent to the problem solved of canonical SVM [1]. We have considered an alternative optimization scheme by using a hybrid algorithm of gradient descent and SVM solver. That is, 1) fix $\mathbf{v}$, solve α; 2) fix α, solve $\mathbf{v}$. We denote (7) by $J(\alpha, \mathbf{v})$ and take the derivatives of $J(\alpha, \mathbf{v})$ with respect to $\mathbf{v}_t$ and v_{t0},

$$\frac{\partial J(\alpha, \mathbf{v})}{\partial \mathbf{v}_t} = -\frac{1}{2}\sum_{i=1}^{N}\sum_{j=1}^{N}\sum_{m=1}^{M} \alpha_i\alpha_j y_i y_j \beta_m(\mathbf{x}_i)k_{com}^m(\mathbf{x}_i, \mathbf{x}_j)\beta_m(\mathbf{x}_j)$$
$$((\mathbf{x}_i + \mathbf{x}_j)\gamma_t^m - \beta_m(\mathbf{x}_i) - \beta_m(\mathbf{x}_j)), \tag{9}$$

$$\frac{\partial J(\alpha, \mathbf{v})}{\partial v_{t0}} = -\frac{1}{2}\sum_{i=1}^{N}\sum_{j=1}^{N}\sum_{m=1}^{M} \alpha_i\alpha_j y_i y_j \beta_m(\mathbf{x}_i)k_{com}^m(\mathbf{x}_i, \mathbf{x}_j)\beta_m(\mathbf{x}_j)$$
$$(2\gamma_t^m - \beta_m(\mathbf{x}_i) - \beta_m(\mathbf{x}_j)), \tag{10}$$

where if the value of m is equal to t, γ_t^m is 1 and 0 otherwise. Update $\mathbf{v}$ according to the following equations,

$$\mathbf{v}_t^{(i+1)} = \mathbf{v}_t^{(i)} - \mu\frac{\partial J(\alpha, \mathbf{v})}{\partial \mathbf{v}_t}, \tag{11}$$

$$v_{t0}^{(i+1)} = v_{t0}^{(i)} - \mu\frac{\partial J(\alpha, \mathbf{v})}{\partial v_{t0}}, \tag{12}$$

where μ is the step of updating. After updating the value of $\mathbf{v}$, we recompute K_β and required to update α by solving a single kernel SVM with K_β. The two steps run alternatively until convergence and the details of algorithm are shown in Algorithm 1, After the optimal $\mathbf{v}$ and α are obtained, the resulting discriminant function can be written as:

$$f(x) = \sum_{i=1}^{N}\sum_{m=1}^{M} \alpha_i y_i \beta_m(\mathbf{x}_i)k_{com}^m(\mathbf{x}_i, \mathbf{x})\beta_m(\mathbf{x}) + b, \tag{13}$$

where b is the bias term of the separating hyperplane.

Algorithm 1. Optimize **v** and α

Input: Training dataset X; A set of composite kernels $k_{com}^m, m = 1, ..., M$;
Output: Optimal **v** and α;
 1: Initialize $\mathbf{v}_t$ and $v_{t0}, t = 1, ..., M$;
 2: **repeat**
 3:     compute K_β by (8);
 4:     compute α of the SVM with K_β by SVM solver;
 5:     update $\mathbf{v}_t^{i+1}, t = 1, ..., M$ by (11);
 6:     update $v_{t0}^{i+1}, t = 1, ..., M$ by (12);
 7: **until** convergence
 8: return **v** and α;

3.4 Crossover and Mutation Operators

First, select two individuals as parents from the previous generation using some kind of selection strategy, e.g., tournament selection and roulette selection. The crossover operator is manipulated between pairs of genes from the parents based on the crossover probability P_c. The mutation operator is manipulated on one or more genes from one of the parents based on the mutation probability P_m.

Table 1. Datasets information

Dataset	#Classes	#Attrib.	#Instan.	Dataset	#Classes	#Attrib.	#Instan.
Breast	2	10	683	Control	6	61	600
Ionosphere	2	35	351	Ecoli	5	8	336
Heart	2	14	270	Glass	6	10	214
A1a	2	124	4217	Iris	3	5	150
A2a	2	124	2591	Parkinsons	2	23	195
Blood	2	5	748	Sonar	2	61	208
Soybean	15	36	266	Wine	3	14	178

4   Experimental Studies

To test the performance of ENMKL, we conduct some experiments on fourteen classification datasets available on UCI [13]. Table 1 lists the statistic information of the used datasets. All of the datasets are random divided to training, validate and test datasets in the ratio of 70%, 10% and 20%, respectively. All the training instances are normalized to be of zero mean and unit variance. The validate and test instances are also normalized using the same mean and variance as of the training data. For comparison, we have tested three respective MKL algorithms, i.e., eKoKs [9], simpleMKL [5] and LMKL [11], and the parameter settings of them are the same as values described in [5] [11]. To get stable results, for each dataset, we repeat each algorithm 10 times and compute the average results of the 10 runs. Table 2 lists the details of the parameters involved in ENMKL.

Table 2. Description and setting of related parameters of ENMKL

Parameter	Description and values
Population of GP	One population of 50 individuals
Number of iteration	Evolution ends after 30 generations
Number of genes	An individual consists of three genes
Nonterminal set	+, *, exp
Terminal set	Linear kernel, Polynomial kernel with d=1,
	arc-cosine kernel with n=[0,1,2]
	Gaussian kernel with $\delta = 2 * sqrt(\#dimension)$
Maximum depth of tree	6
Selection strategy	Tournaments selection with 5 participants
Crossover	Cross (prob. 0.8)
Mutation	Mutate (prob. 0.1)

Table 3. Classification results of different methods

Dataset	Classification accuracy			
	mean	mean (stand deviation)		
	eKoKs (MTS2-[0,1])	simpleMKL	LMKL	ENMKL
Breast	0.8889 [9]	0.9757 (0.0045)	0.9724 (0.0086)	**0.9840 (0.0079)**
Ionosphere	0.9803 [9]	0.9107 (0.0358)	0.9117 (0.0306)	**0.9872 (0.0128)**
Heart	0.8757 [9]	**0.9028 (0.0092)**	0.8959 (0.0384)	0.8963 (0.0248)
A1a	0.8438 [9]	0.8233 (0.0033)	0.8344 (0.0062)	**0.8543 (0.0121)**
A2a	0.8908 [9]	0.8301 (0.0020)	0.8295 (0.0159)	**0.8962 (0.0042)**
Blood	—	0.8032 (0.0044)	0.8061 (0.0208)	**0.8142 (0.0312)**
Control	—	0.9875 (0.0084)	0.9867 (0.0046)	**0.9950 (0.0075)**
Ecoli	—	0.8904 (0.0103)	0.8715 (0.0230)	**0.8985 (0.0258)**
Glass	—	0.6861 (0.0233)	0.6806 (0.0336)	**0.7044 (0.0203)**
Iris	—	0.9333 (0)	**1.0000 (0)**	0.9933 (0.0149)
Parkinsons	—	0.9103 (0.0257)	0.8905 (0.0225)	**0.9696 (0.0335)**
Sonar	—	0.8415 (0.0244)	0.8757 (0.0696)	**0.8863 (0.0772)**
Soybean	—	0.8268 (0.0135)	0.8337 (0.0168)	**0.8682 (0.0405)**
Wine	—	0.9703 (0.0022)	0.9882 (0.0161)	**1.0000 (0)**

Table 3 lists the mean accuracies and standard deviations of the compared methods on the test datasets, and boldface font in this table is used to express the best results. For eKoKs, we report the mean accuracies on five datasets, i.e., Breast, Heart, Ionosphere, A1a and A2a. And we make careful comparison of the performances of simpleMKL, LMKL and ENMKL on all of the fourteen datasets. As shown in Table 3, ENMKL outperforms over other methods on four out of five datasets (Breast, Heart, Ionosphere, A1a and A2a) and get a lower value, 0.8963, on Heart dataset as compared to the best result, 0.9026, owned by simpleMKL. The slight margin is acceptable. On the five datasets, ENMKL outperforms eKoKs and LMKL due to its combining both advantages of GP based MKL and non-stationary MKL. On the remaining datasets, ENMKL gets the best result on eight out of them and has a little lower performance on Iris dataset compared to LMKL. So as a whole, ENMKL is better than the others. It should be noted that the computation cost of training is a little higher than that of simpleMKL and LMKL, this is the cost that ENMKL gets the better classification results.

5   Conclusion and Future Work

We have proposed an evolutionary method of non-stationary MKL, which can find a set of good composite kernels in a high-dimension feature space and a gating function. Experimental results on UCI datasets validate the effectiveness of ENMKL. ENMKL is a superior choice for the cases pursuing high classification accuracy such as disease diagnosing. When it comes to large scale datasets that demand lower in real-time, we can make a choice to train ENMKL off-line. The further work may include applying other evolutionary algorithms to search good kernels in the feature space and trying with other gating functions, also try to develop an efficient algorithm to reduce the computational cost in dealing with large training set.

Acknowledgment. The research work in this paper was supported by the grants from National Natural Science Foundation of China (Project No. 61375045) and Beijing Natural Science Foundation (4142030).

References

1. Vapnik, V.: The nature of statistical learning theory. Springer (1999)
2. Lanckriet, G.R., Cristianini, N., Bartlett, P., et al.: Learning the kernel matrix with semidefinite programming. The Journal of Machine Learning Research 5, 27–72 (2004)
3. Bach, F.R., Lanckriet, G.R., Jordan, M.I.: Multiple kernel learning, conic duality, and the smo algorithm. In: Proceedings of the Twenty-first International Conference on Machine Learning, pp. 41–48. ACM (2004)
4. Sonnenburg, S., Rätsch, G., Schäfer, C., Schölkopf, B.: Large scale multiple kernel learning. The Journal of Machine Learning Research 7, 1531–1565 (2006)
5. Rakotomamonjy, A., Bach, F., Canu, S., Grandvalet, Y.: SimpleMKL. Journal of Machine Learning Research 9, 2491–2521 (2008)
6. Wu, P., Duan, F., Guo, P.: A pre-selecting base kernel method in multiple kernel learning. Neurocomputing (accepted, 2014)
7. Sullivan, K.M., Luke, S.: Evolving kernels for support vector machine classification. In: Proceedings of the 9th Annual Conference on Genetic and Evolutionary Computation, pp. 1702–1707. ACM (2007)
8. Methasate, I., Theeramunkong, T.: Kernel trees for support vector machines. IEICE Transactions on Information and Systems 90(10), 1550–1556 (2007)
9. Dioşan, L., Rogozan, A., Pecuchet, J.P.: Improving classification performance of support vector machine by genetically optimising kernel shape and hyper-parameters. Applied Intelligence 36(2), 280–294 (2012)

10. Lewis, D.P., Jebara, T., Noble, W.S.: Nonstationary kernel combination. In: Proceedings of the 23rd International Conference on Machine Learning, pp. 553–560. ACM (2006)
11. Gönen, M., Alpaydin, E.: Localized algorithms for multiple kernel learning. Pattern Recognition 46, 795–807 (2013)
12. Cho, Y., Saul, L.K.: Kernel methods for deep learning. In: Advances in Neural Information Processing Systems, pp. 342–350 (2009)
13. Bache, K., Lichman, M.: UCI machine learning repository. University of California, School of Information and Computer Science, Irvine (2013), http://archive.ics.uci.edu/ml

A Kernel Method to Extract Common Features Based on Mutual Information

Takamitsu Araki[1], Hideitsu Hino[2], and Shotaro Akaho[1]

[1] Human Technology Research Institute, National Institute of Advanced Industrial
Science and Technology,
Central 2, 1-1-1 Umezono, Tsukuba, Ibaraki 3058568, Japan
{tk-araki,s.akaho}@aist.go.jp
[2] Department of Computer Science, University of Tsukuba,
1-1-1 Tennodai, Tsukuba, Ibaraki 3058573, Japan
hinohide@cs.tsukuba.ac.jp

Abstract. Kernel canonical correlation analysis (CCA) aims to extract
common features from a pair of multivariate data sets by maximizing a
linear correlation between nonlinear mappings of the data. However, the
kernel CCA tends to obtain the features that have only small information of original multivariates in spite of their high correlation, because
it considers only statistics of the extracted features and the nonlinear
mappings have high degree of freedom. We propose a kernel method for
common feature extraction based on mutual information that maximizes
a new objective function. The objective function is a linear combination
of two kinds of mutual information, one between the extracted features
and the other between the multivariate and its feature. A large value of
the former mutual information provides strong dependency to the features, and the latter prevents loss of the feature's information related
to the multivariate. We maximize the objective function by using the
Parallel Tempering MCMC in order to overcome a local maximum problem. We show the effectiveness of the proposed method via numerical
experiments.

Keywords: Kernel canonical correlation analysis, mutual information,
Parallel Tempering.

1 Introduction

Recently, we can obtain data from multiple sources simultaneously such as electroencephalography (EEG) and near infra-red spectroscopy (NIRS) measurements of brain activity. Canonical correlation analysis is known as a linear
method to extract common features in which source-specific noise is reduced
from the original observations.

Kernel canonical correlation analysis (Kernel CCA; [1], [8], [3]) is a nonlinear
extension of canonical correlation analysis with positive definite kernels. Given
a pair of multivariates $\mathbf{x}$ and $\mathbf{y}$, the kernel CCA aims to extract the common
features from them by finding nonlinear mappings $f(\mathbf{x})$ and $g(\mathbf{y})$ such that

C.K. Loo et al. (Eds.): ICONIP 2014, Part II, LNCS 8835, pp. 26–34, 2014.

the correlation coefficient is maximized. The kernel CCA has been applied for extracting nonlinear relations between the multivariates in various data, e.g., genomic data and functional magnetic resonance imaging (fMRI) brain images. The kernel CCA is one of kernel methods ([11]) that use nonlinear mappings of the multivariates instead of linear transformations used in traditional multivariate analysis.

The kernel CCA uses flexible nonlinear mappings to extract nonlinear relation between the multivariates. However, since large degrees of freedom of the mappings are devoted to their correlation maximization, the *features*, the mappings of the multivariates, lose a large amount of information related to the multivariates in many cases. The kernel CCA evaluates only the relation between the features and fails to detect the relationship between the multivariates.

The kernel CCA also extracts the features that have no interdependency, even though the correlation coefficient of them is high. The correlation coefficient cannot evaluate the dependency correctly when the data follow a non-Gaussian distribution. Thus, the correlation coefficient is not an appropriate criterion that evaluates the dependency of the features of the data that have nonlinear structure.

In this study, we propose a kernel method for common feature extraction that maximizes a new objective function based on mutual information. The objective function is a linear combination of mutual information between the features and that between the multivariate and its feature for each data set. The mutual information can evaluate essential dependency between the variables distributed with any distribution, so that the mutual information is a suitable criterion for the relation between the features and between the feature and the multivariate. A large value of the former mutual information provides the highly interdependent features and the latter prevents the features from losing a large amount of information about the original multivariate.

We apply the proposed method to an analysis of synthetic data, and show that our method can extract the true nonlinear structure of the data, which cannot be extracted by the conventional kernel CCA.

2   Kernel Canonical Correlation Analysis

Suppose there is a pair of multivariates $\mathbf{x} \in \mathbb{R}^{n_x}$ and $\mathbf{y} \in \mathbb{R}^{n_y}$, the kernel CCA aims to find a pair of nonlinear mappings $f(\mathbf{x})$ and $g(\mathbf{y})$ such that their correlation coefficient is maximized, where f and g belong to the respective reproducing kernel Hilbert spaces (RKHS) $\mathcal{H}_x$ and $\mathcal{H}_y$.

Since the maximization is ill-posed when the dimensionalities of the RKHS $\mathcal{H}_x$ and $\mathcal{H}_y$ are large, we introduce a quadratic regularization term $\eta(\|f\|^2_{\mathcal{H}_x} + \|g\|^2_{\mathcal{H}_y})$, where $\eta > 0$ is a regularization parameter. The kernel CCA maximizes the objective function that consists of the correlation coefficient between $f(\mathbf{x})$ and $g(\mathbf{y})$ and the regularization term.

The kernel CCA does not always extract essential common structure of a pair of multivariate data. For example, a common factor underlying the synthetic data shown in Fig. 1 is not extracted by the kernel CCA. The synthetic data sets, $\{\mathbf{x}_i, \mathbf{y}_i\}_{i=1}^{50}$, were generated from the two dimensional circle-shaped distributions derived from common random angle (details of this data are in Section 5). However, the values of the common angle are mixed in the features extracted by the kernel CCA with Gaussian kernel (Fig. 2(a)). This is because the kernel CCA finds redundant nonlinear mappings that often produce features having small information about the multivariate (Fig. 2(b.1)-(b.2)).

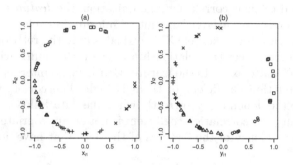

Fig. 1. Synthetic data sets, (a) $\mathbf{x}_i$, (b) $\mathbf{y}_i$. The data in each data set are grouped into five sections by intervals of the common angle value. The five groups are denoted by the five different marks.

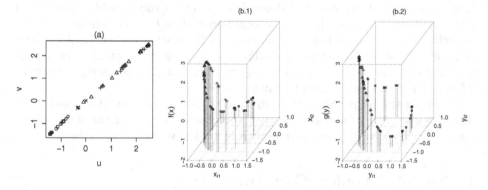

Fig. 2. (a) The features extracted by the kernel CCA, $u_i = \hat{f}(\mathbf{x}_i)$, $v_i = \hat{g}(\mathbf{y}_i)$, where $\hat{f}$ and $\hat{g}$ are the nonlinear mappings estimated by the kernel CCA. The marks correspond to those in Fig. 1. (b.1) $\hat{f}(\mathbf{x}_i)$, (b.2) $\hat{g}(\mathbf{y}_i)$.

This result shows that the kernel CCA dedicates the degree of freedom of nonlinear mappings to maximize the correlation of the features, and then the features' information on the multivariates is sacrificed. Therefore, the kernel CCA extracts only little information shared by the data set even if it obtains highly correlated features, because the features have little information about the multivariates.

The kernel CCA with another set of the regularization parameter and the Gaussian kernel's parameter extracts the non-informative features whose correlation coefficient is high (Fig. 3). This indicates that the correlation coefficient is not appropriate to evaluate the interdependence of the features.

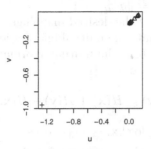

Fig. 3. The features extracted by the kernel CCA with other parameters

3 Mutual Information Based Objective Function

We propose a kernel method for common feature extraction based on mutual information that maximizes a new objective function. The objective function is constructed as follows.

We substitute the correlation coefficient between the features, $u = f(\mathbf{x})$ and $v = f(\mathbf{y})$, with the mutual information between them. The mutual information between two random variables, $x \in \mathcal{X}$ and $y \in \mathcal{Y}$, is $I(x, y) = \int_{\mathcal{Y}} \int_{\mathcal{X}} p(x, y) \log \frac{p(x,y)}{p(x)p(y)} dx dy$, where p denotes a density function. The mutual information quantifies interdependence between random variables, and is interpreted as a measure of information shared by the random variables from an information theoretic point of view.

Since the maximization of the mutual information between the features also causes information loss of the features related to the multivariate, we add a constraint that the mutual information between the feature and the multivariate is enough large for each data set. That is, we consider an optimization problem

$$\max_{f \in \mathcal{H}_x, g \in \mathcal{H}_y} I(u, v), \quad \text{subject to} \quad I(u, \mathbf{x}) \geq s_x, I(v, \mathbf{y}) \geq s_y, \tag{1}$$

where $u = f(\mathbf{x})$, $v = g(\mathbf{y})$ and $s_x, s_y > 0$.

The problem in (1) is solved by maximizing the objective function

$$L_\lambda(f, g) = I(u, v) + \lambda_x I(u, \mathbf{x}) + \lambda_y I(v, \mathbf{y}),$$

where $\lambda = (\lambda_x, \lambda_y)$ are regularization parameters and $\lambda_x, \lambda_y > 0$. The regularization parameters λ control the amount of the mutual information between the feature and the multivariate. The regularization term, $\lambda_x I(u, \mathbf{x}) + \lambda_y I(v, \mathbf{y})$, prevents loss of the information shared by the feature and the multivariate. The large

value of the mutual information between the features that have enough information on the multivariate provides the features capturing an essential nonlinear
relation between the multivariates.

The mutual information does not depend on scale of random variables, that is,
$I(cX, cY) = I(X, Y)$, for $c \neq 0$. Therefore, we maximize the objective function
under the constraint $\|f\|^2_{\mathcal{H}_x} = \|g\|^2_{\mathcal{H}_y} = 1$.

In practice, we have to find the desired mappings from finite amount of data,
$\{\mathbf{x}_i, \mathbf{y}_i\}^N_{i=1}$, so that we estimate the mutual information by Mean Nearest Neighbor (MeanNN) method ([4],[6]). The mutual information between $\mathbf{x}$ and $\mathbf{y}$ estimated by the MeanNN method is

$$\hat{I}(\mathbf{x}, \mathbf{y}) = \hat{H}(\mathbf{x}) + \hat{H}(\mathbf{y}) - \hat{H}(\mathbf{x}, \mathbf{y}),$$

where $\hat{H}(\mathbf{x}) = \frac{n_x}{N(N-1)} \sum_{i \neq j} \log \|\mathbf{x}_i - \mathbf{x}_j\| + \text{const}$.

Therefore, we maximize the objective function $\hat{L}_\lambda(f, g)$ under $\|f\|^2_{\mathcal{H}_x} =$
$\|g\|^2_{\mathcal{H}_y} = 1$, where $\hat{L}_\lambda(f, g) = \hat{I}(u, v) + \lambda_x \hat{I}(u, \mathbf{x}) + \lambda_y \hat{I}(v, \mathbf{y})$. The solution of this
problem is represented as $f(\cdot) = \sum^N_{i=1} \alpha_i k_x(\mathbf{x}_i, \cdot)$ and $g(\cdot) = \sum^N_{i=1} \beta_i k_y(\mathbf{y}_i, \cdot)$
by the Representer theorem ([10]), where k_x and k_y are kernel functions, and
$\alpha_i, \beta_i \in \mathbb{R}$. We assume that f's orthogonal part to the span of $k_x(\mathbf{x}_i, \cdot)$ is zero as
well as g. The objective function of $\alpha = (\alpha_1, \dots, \alpha_N), \beta = (\beta_1, \dots, \beta_N), \hat{L}_\lambda(\alpha, \beta)$,
is obtained by applying the representation of the solution to f, g. The constraint
is also represented by $\alpha^T K_x \alpha = \beta^T K_y \beta = 1$, where $(K_x)_{ij} = k_x(\mathbf{x}_i, \mathbf{x}_j)$ and
$(K_y)_{ij} = k_y(\mathbf{y}_i, \mathbf{y}_j)$. However, since the constraint space is too complex to find
the solution, we impose the simplified constraint $\|\alpha\|^2 = \|\beta\|^2 = 1$ in place of
the constraint above for the sake of computational efficiency.

4   Algorithm

Since the objective function $\hat{L}_\lambda(\alpha, \beta)$ has many local maximum points, simple
optimization methods such as a gradient method do not find a reasonable solution. To cope with this localization problem, we employ the Parallel Tempering
([5],[7]), which is one of the Markov chain Monte Carlo (MCMC) methods ([9]).

The MCMC methods efficiently generate samples from a target probability
distribution by simulating a Markov chain that converges to the distribution. The
Parallel Tempering introduces auxiliary distributions with a parameter called the
temperature, generates multiple MCMC samples from target and the auxiliary
distributions in parallel, and exchanges the positions of two samples.

The target distribution and the auxiliary distributions with inverse temperatures t_l are

$$\pi_{t_l}(\alpha_l, \beta_l) \propto \exp\left(t_l \hat{L}_\lambda(\alpha_l, \beta_l)\right), \quad l = 1, \dots, L,$$

where $t_1 > \cdots > t_L > 0$, $\pi_{t_1}(\alpha_1, \beta_1)$ is a target distribution, and the others are
the auxiliary distributions.

The Parallel Tempering executes either of the parallel step and the exchange step at each iteration. The parallel step generates the L samples according to $\pi_{t_l}(\alpha_l, \beta_l)$ for each by using a Metropolis algorithm. The Metropolis algorithm uses a proposal distribution that generates a sample candidate, which becomes the MCMC sample if accepted, and is rejected otherwise. We employ a von Mises-Fisher distribution, which is a probability distribution on the $(N-1)$-dimensional sphere in $\mathbb{R}^N$, as the proposal distribution. The proposal distribution enables us to directly generate samples of α on the constraint space, $\|\alpha\| = 1$, as well as β. The exchange step randomly chooses adjacent two samples and exchanges them with a Metropolis acceptance probability.

Since the target distribution is maximized if and only if the objective function is maximized, we find the solution from the samples generated by the Parallel Tempering.

5   Numerical Validation

Our method and the kernel CCA were applied to the circle data (analysed in Section 2) in order to show that our method can extract the essential nonlinear structure of a pair of data which the kernel CCA cannot extract.

The circle data, $\mathbf{x}_i, \mathbf{y}_i \in \mathbb{R}^2$, $i - 1, \ldots, 50$, were generated as follows. First θ_i is generated from the uniform distribution on $(1, 2\pi)$, and then $\mathbf{x}_i$ and $\mathbf{y}_i$ were generated by,

$$\mathbf{x}_i = \begin{pmatrix} \cos(\theta_i) \\ \sin(\theta_i) \end{pmatrix} + \epsilon_i^x \quad \text{and} \quad \mathbf{y}_i = \begin{pmatrix} \sin(\theta_i) \\ \cos(\theta_i) \end{pmatrix} + \epsilon_i^y,$$

where $\epsilon_i^x, \epsilon_i^y$ are independent two dimensional Gaussian noises with a mean 0 and a standard deviation 0.01.

In this experiment, we used the Gaussian kernel $k(\mathbf{x}_i, \mathbf{x}_j) = \exp\left(-\frac{\|\mathbf{x}_i - \mathbf{x}_j\|^2}{2\sigma^2}\right)$ both for $\mathbf{x}$ and $\mathbf{y}$. The Parallel Tempering was run for 5×10^5 iterations, and the first inverse temperature was $t_1 - 1$, the other parameters were determined by one simulation of the adaptive Parallel Tempering ([2]).

Our method extracted the features as reconstructions of the common factor θ_i between $\mathbf{x}_i$ and $\mathbf{y}_i$ (Fig. 4(a)), while the kernel CCA could not extract (Fig. 5-6(a)). The nonlinear mappings estimated by our method provide enough information about the multivariates to the features (Fig. 4 (b.1)-(b.2)). On the other hand, those obtained by the kernel CCA provide only a part of information about the multivariates to the features (Fig. 5-6 (b.1)-(b.2)).

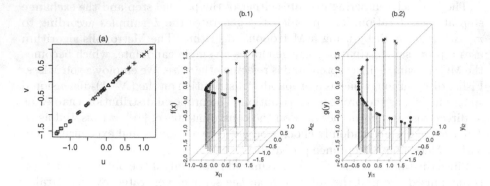

Fig. 4. (a) The features extracted by our method with parameters $\sigma = 0.45$ (the same value used in Section 2) and $\lambda_x = \lambda_y = 1$. (b.1)-(b.2) The nonlinear mappings estimated by our method, $\hat{f}(\mathbf{x}_i)$ (b.1) and $\hat{g}(\mathbf{y}_i)$ (b.2). (The marks are defined in Fig. 1 in Section 2.)

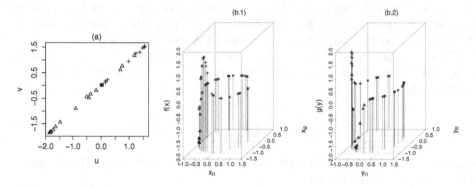

Fig. 5. (a) The features extracted by the kernel CCA with parameters $\sigma = 0.15$, $\eta = 3$. (b.1)-(b.2) The nonlinear mappings estimated by the kernel CCA.

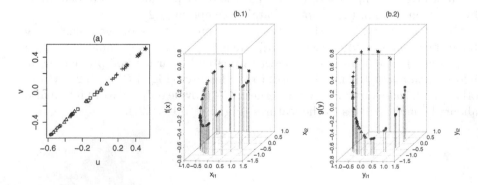

Fig. 6. (a) The features extracted by the kernel CCA with parameters $\sigma = 1.3$, $\eta = 7$. (b.1)-(b.2) The nonlinear mappings estimated by the kernel CCA.

6 Conclusions

We proposed the kernel method based on mutual information to extract common features from a pair of data sets. The proposed method maximizes a new objective function that consists of the mutual information between the features and those between the feature and the multivariate. The maximization of the objective function provides the features that represent nonlinear structure of a pair of the multivariate data set, because a large value of the mutual information between the feature and the multivariate provides the enough multivariate's information to the feature and the mutual information between the features is enlarged.

We also showed that our method can extract the common feature of the circle data which the kernel CCA cannot extract. This is because our method solves the essential problem of the kernel CCA that it tends to extract the features that have small information on the multivariates.

Our information-based method is difficult to apply to the extremely high dimensional data because estimation of entropy becomes unstable in such situation. However, our method is useful in adequate dimensional cases, in which it is difficult to extract nonlinear relations by the conventional kernel CCA.

The proposed method extracts only one component, so that we will extend the proposed method to the method that can extract multiple components. We will also develop faster algorithm that maximizes the proposed objective function, since the Parallel Tempering spends relatively large computational time.

Acknowledgments. Part of this work was supported by MEXT KAKENHI No.25120011 and JSPS KAKENHI No.25870811.

References

1. Akaho, S.: A kernel method for canonical correlation analysis. In: Proceedings of the International Meeting of the Psychometric Society (IMPS 2001) (2001)
2. Araki, T., Ikeda, K.: Adaptive Markov chain Monte Carlo for auxiliary variable method and its application to Parallel Tempering. Neural Networks 43, 33–40 (2013)
3. Bach, F.R., Jordan, M.I.: Kernel independent component analysis. The Journal of Machine Learning Research 3, 1–48 (2003)
4. Faivishevsky, L., Goldberger, J.: ICA based on a smooth estimation of the differential entropy. In: Advances in Neural Information Processing Systems, pp. 433–440 (2008)
5. Geyer, C.: Markov chain Monte Carlo maximum likelihood. In: Proc. 23rd Symp. Interface Comput. Sci. Statist., pp. 156–216 (1991)
6. Hino, H., Murata, N.: A conditional entropy minimization criterion for dimensionality reduction and multiple kernel learning. Neural Computation 22(11), 2887–2923 (2010)
7. Hukushima, K., Nemoto, K.: Exchange Monte Carlo method and application to spin glass simulations. Journal of the Physical Society of Japan 65(6), 1604–1608 (1996)

8. Melzer, T., Reiter, M.K., Bischof, H.: Nonlinear feature extraction using generalized canonical correlation analysis. In: Dorffner, G., Bischof, H., Hornik, K. (eds.) ICANN 2001. LNCS, vol. 2130, pp. 353–360. Springer, Heidelberg (2001)
9. Robert, C., Casella, G.: Monte Carlo Statistical Methods. Springer (2004)
10. Schölkopf, B., Herbrich, R., Smola, A.J.: A generalized representer theorem. In: Helmbold, D.P., Williamson, B. (eds.) COLT/EuroCOLT 2001. LNCS (LNAI), vol. 2111, pp. 416–426. Springer, Heidelberg (2001)
11. Schölkopf, B., Smola, A.: Learning with kernels. MIT Press, Cambridge (2002)

Properties of Text-Prompted Multistep Speaker Verification Using Gibbs-Distribution-Based Extended Bayesian Inference for Rejecting Unregistered Speakers

Shuichi Kurogi, Takuya Ueki, Satoshi Takeguchi, and Yuta Mizobe

Kyushu Institute of technology, Tobata, Kitakyushu, Fukuoka 804-8550, Japan
kuro@cntl.kyutech.ac.jp,
{ueki,takeguchi}@kurolab.cntl.kyutech.ac.jp
http://kurolab.cntl.kyutech.ac.jp/

Abstract. This paper presents a method of text-prompted multistep speaker verification for reducing verification errors and rejecting unregistered speakers. The method has been developed for our speech processing system which utilizes competitive associative nets (CAN2s) for learning piecewise linear approximation of nonlinear speech signal to extract feature vectors of pole distribution from piecewise linear coefficients reflecting nonlinear and time-varying vocal tract of the speaker. This paper focuses on rejecting unregistered speakers by means of multistep verification using Gibbs-distribution-based extended Bayesian inference (GEBI) in text-prompted speaker verification. The properties of GEBI and the comparison to BI (Bayesian inference) for rejecting unregistered speakers are shown and analyzed by means of experiments using real speech signals.

Keywords: Text-prompted multistep speaker verification, Gibbs-distribution-based extended Bayesian inference, Competitive associative net, Rejection of unregistered speakers.

1 Introduction

This paper examines a method of text-prompted multistep speaker verification presented in [1]. Here, from [2], text-prompted speaker verification has been developed to combat spoofing from impostors and digit strings are often used to lower the complexity of processing. However, since it would be simple for today's devices to record a person saying 10 digits and to produce digits by simply typing on a keypad, text-prompted modality itself is not enough for anti-spoofing. So, we would combine other information, such as a knowledge database, other biometrics, and so on, and a specific method would be designed depending on the application. From another point of view, the present method focuses on rejecting verification errors by means of multistep verification using Gibbs-distribution-based Bayesian inference (GEBI). Here, GEBI has been introduced for overcoming a problem of multistep Bayesian inference (BI) for rejecting unregistered speaker in speaker identification [3]. Although the effectiveness of GEBI in speaker identification in [3] and the reduction of verification error for text-prompted speaker verification in [1] has been shown so far, the properties for rejecting unregistered speakers in text-prompted speaker verification have not been clarified yet.

C.K. Loo et al. (Eds.): ICONIP 2014, Part II, LNCS 8835, pp. 35–43, 2014.

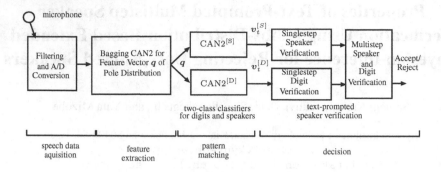

Fig. 1. Diagram of text-prompted speaker verification system using CAN2s

Here, note that the present method employs competitive associative nets (CAN2s). The CAN2 is an artificial neural net for learning efficient piecewise linear approximation of nonlinear function [4]. Recently, we have shown that feature vectors of pole distribution extracted from piecewise linear predictive coefficients obtained by the bagging (bootstrap aggregating) version of the CAN2 reflect nonlinear and time-varying vocal tract of the speaker [5]. Although the most common way to characterize the speech signal in the literature is short-time spectral analysis, such as Linear Prediction Coding (LPC) and Mel-Frequency Cepstrum Coefficients (MFCC), they extract spectral features of the speech from each of consecutive interval frames spanning 10-30ms [6]. Thus, a single feature vector of LPC and MFCC corresponds to the average of multiple piecewise linear predictive coefficients of the bagging CAN2. Namely, the bagging CAN2 learns more precise information on the speech signal.

In the next section, we show an overview and formulation of singlestep speaker and digit verification system using CAN2s. And then we introduce multistep speaker and digit verification to execute text-prompted speaker verification. In order to reject unregistered speaker, we introduce GEBI. In **3**, we show experimental results and examine the effectiveness of the present method.

2 Text-Prompted Speaker Verification System

2.1 Singlestep Digit and Speaker Verification

Fig. 1 shows an overview of our text-prompted speaker verification system using CAN2s. In the same way as general speaker recognition systems [6], it consists of four steps: speech data acquisition, feature extraction, pattern matching, and making a decision. In this research study, we use a feature vector of pole distribution obtained from a speech signal (see [5] for details). In order to achieve text-prompted speaker verification using digits, let $S = \{s_i | i \in I_S\}$ and $D = \{d_i | i \in I_D\}$ denote a set of speakers $s \in S$ and digits $d \in D$, respectively, where $I_S = \{1, 2, \cdots, |S|\}$ and $I_D = \{1, 2, \cdots, |D|\}$. Furthermore, let $\mathrm{RLM}^{[M]}$ for $M = S$ and M be a set of regression learning machines $\text{RLM}^{[m]}$ ($m \in I_M$), and each $\text{RLM}^{[m]}$ learns to approximate the following target function:

$$y^{[m]} = f^{[m]}(q) = \begin{cases} 1 \text{ if } q \in Q^{[m]} \\ -1 \text{ otherwise} \end{cases} \tag{1}$$

In the following, we apply the same procedures for both speaker and digit processing, and we use the variables m and M for representing s and S for speakers and d and D for digits. After the learning, singlestep verification or two class classification by binarizing the output $\hat{y}^{[m]} = \hat{f}^{[m]}(q^{[m]})$ of $\text{RLM}^{[m]}$ as

$$v^{[m]} = \begin{cases} 1 \text{ if } \hat{y}^{[m]} \geq y_\theta^{[m]} \\ -1 \text{ otherwise} \end{cases}. \tag{2}$$

Namely, we accept the speaker m if $v^{[m]} = 1$, and reject otherwise. Here, note that the threshold y_θ is tuned for stable verification as shown below.

2.2 Bayesian Inference (BI) and Gibbs-Distribution-Based Extended BI (GEBI)

BI and the Problem for Unregistered Speech. First, the probability $p(v^{[m_i]}|m)$ of the two class classifier $\text{RLM}^{[m_i]}$ for a given training dataset is obtained as

$$p(v^{[m_i]} = 1|m) = \frac{n(v^{[m_i]} = 1|m)}{n(v^{[m_i]} = 1|m) + n(v^{[m_i]} = -1|m)} \tag{3}$$

and $p(v^{[m_i]} = -1|m) = 1 - p(v^{[m_i]} = 1|m)$, where $n(v^{[m_i]} = 1|m)$ indicates the number of instances with the output of $\text{RLM}^{[m_i]}$ being $v^{[m_i]} = 1$ for the speech of a speaker or a digit $m \in M$. Now, for multistep inference, let us suppose that the joint probability of the output vector $v^{[M]} = (v^{[m_1]}, \cdots, v^{[m_{|M|}]})$ of all classifiers $m_i \in M$ for an input speech of $m \in M$ is given by $p(v^{[M]}|m) = \prod_{m_i \in M} p(v^{[m_i]}|m)$ because any two probabilities $p(v^{[m_i]}|m)$ and $p(v^{[m_j]}|m)$ for $i \neq j$ are supposed to be independent. Let $v_{1:t}^{[M]} = v_1^{[M]} v_2^{[M]} \cdots v_t^{[M]}$ be a sequence of $v^{[M]}$ obtained from a speech of a test speaker or a digit m, then we can execute naive Bayesian inference (BI) by

$$p_B(m|v_{1:t}^{[M]}) = \frac{p_B(m|v_{1:t-1}^{[M]}) \, p(v_t^{[M]}|m)}{\sum_{m_i \in M} p_B(m_i|v_{1:t-1}^{[M]}) \, p(v_t^{[M]}|m_i)}. \tag{4}$$

Here, we use the conditional independence assumption $p(v_t^{[M]}|s, v_{1:t-1}^{[M]}) = p(v_t^{[M]}|s)$, which is shown effective in many real world applications of naive Bayes classifier [8].

When the Bayesian probability $p_B(m|v_{1:t}^{[M]})$ for $m = m_i$ becomes larger than a threshold, say p_θ, for the increase of t, we would like accept the speech as of m_i. However, from the above Bayesian inference for $t = 1, 2, \cdots$, we have

$$p_B\left(m|v_{1:t}^{[M]}\right) = \frac{1}{Z_t} \exp\left(-t\left(\tilde{L}_{1:t}^{[m]} - \frac{1}{t}\log p_0(m)\right)\right), \tag{5}$$

where Z_t indicates the normalization constant for holding $\sum_{m \in M} p_B(m|\boldsymbol{v}_{1:t}^{[M]}) = 1$, $p_0(m) = p_B(m|\boldsymbol{v}_{1:0}^{[M]})$ denotes the prior, and $\tilde{L}_{1:t}^{[m]} \triangleq -\frac{1}{t}\left(\sum_{k=1}^{t} \log p\left(\boldsymbol{v}_k^{[M]}|m\right)\right)$ is the normalized negative log-likelihood. Since $\tilde{L}_{1:t}^{[m]}$ is the negative log of the geometric mean $(\prod_{k=1}^{t} p(\boldsymbol{v}_k^{[M]}|m))^{1/t}$, let us suppose that $\tilde{L}_{1:t}^{[m]}$ converges for the increase of t. Then, the ratio of the Bayesian probability for $m_i \in M$ to $m_\nu = \underset{m_i \in M}{\mathrm{argmax}}\, p_B(m_i|\boldsymbol{v}_{1:t}^{[M]})$ becomes

$$r_{B,i,\nu} \triangleq \frac{p_B(m_i|\boldsymbol{v}_{1:t}^{[M]})}{p_B(m_\nu|\boldsymbol{v}_{1:t}^{[M]})} = \frac{p_B(m_i|\boldsymbol{v}_{1:0}^{[M]})}{p_B(m_\nu|\boldsymbol{v}_{1:0}^{[M]})} \exp\left(-t(\tilde{L}_{1:t}^{[m_i]} - \tilde{L}_{1:t}^{[m_\nu]})\right) \rightarrow \begin{cases} 1, m_i = m_\nu \\ 0, m_i \neq m_\nu \end{cases}$$

$$(6)$$

for the increase of t, because $\sum_{m \in M} p_B(m|\boldsymbol{v}_{1:t}^{[M]}) = 1$. This indicates that we will have a large $p_B(m_\nu|\boldsymbol{v}_{1:t}^{[M]})$ for a registered speaker or digit m_ν even when the current speech is of an unregistered speaker or digit.

GEBI. Instead of (5), let us use the following probability given by Gibbs distribution:

$$p_G\left(m|\boldsymbol{v}_{1:t}^{[M]}\right) \triangleq \frac{1}{Z_t} \exp\left(-\beta\left(\tilde{L}_{1:t}^{[m]} + \frac{1}{t}\log p_B(m|\boldsymbol{v}_{1:0}^{[M]})\right)\right), \qquad (7)$$

where β is the parameter called inverse temperature. Then, for the increase of t, the ratio of $p_G(m|\boldsymbol{v}_{1:t}^{[M]})$ for $m = m_i$ to $m_\nu = \underset{m_i \in M}{\mathrm{argmax}}\, p_G(m_i|\boldsymbol{v}_{1:t}^{[M]})$ converges to a constant value less than 1 as follows;

$$r_{G,i} \triangleq \frac{p_G(m_i|\boldsymbol{v}_{1:t}^{[M]})}{p_G(m_\nu|\boldsymbol{v}_{1:t}^{[M]})} \rightarrow \exp\left(-\beta(\tilde{L}_{1:t}^{[m_i]} - \tilde{L}_{1:t}^{[m_\nu]})\right) \rightarrow c_i^\beta < 1. \qquad (8)$$

Thus, we can avoid the problem of BI described above. Here, from (7), the following recursive inference is obtained,

$$p_G\left(m|\boldsymbol{v}_{1:t}^{[M]}\right) = \frac{1}{Z_t} p_G\left(m|\boldsymbol{v}_{1:t-1}^{[M]}\right)^{\beta_t/\beta_{t-1}} p\left(\boldsymbol{v}_t^{[M]}|m\right)^{\beta_t}, \qquad (9)$$

where $\beta_t = \beta/t$ $(t \geq 1)$ and $\beta_0 = 1$. Note that the conventional BI is given by $\beta_t = 1$ $(t \geq 0)$, and the above inference is named GEBI.

2.3   Multistep Text-Prompted Speaker Verification Using GEBI

In order to execute text-prompted speaker verification, we employ multistep verification of digit and speaker parallelly and then combine the result. In order to verify whether the output sequence of all classifiers, $\boldsymbol{v}_{1:T}^{[M]} = \boldsymbol{v}_1^{[M]}\boldsymbol{v}_2^{[M]} \cdots \boldsymbol{v}_t^{[M]}$, is a response of a reference sequence $m_{1:T}^{[r]} = m_1^{[r]}m_2^{[r]} \cdots m_T^{[r]}$, we calculate two recursive posterior probabilities for $t = 1, 2, \cdots, T$ as follows,

$$p_{\mathrm{G}}\left(m_{1:t}^{[r]} \mid \boldsymbol{v}_{1:t}^{[M]}\right) = \frac{1}{Z_t} p_{\mathrm{G}}\left(m_{1:t-1}^{[r]} \mid \boldsymbol{v}_{1:t-1}^{[M]}\right)^{\beta_t/\beta_{t-1}} p\left(\boldsymbol{v}_t^{[M]} \mid m_t^{[r]}\right)^{\beta_t}, \quad (10)$$

$$p_{\mathrm{G}}\left(\overline{m_{1:t}^{[r]}} \mid \boldsymbol{v}_{1:t}^{[M]}\right) = \frac{1}{Z_t} p_{\mathrm{G}}\left(\overline{m_{1:t-1}^{[r]}} \mid \boldsymbol{v}_{1:t-1}^{[M]}\right)^{\beta_t/\beta_{t-1}} p\left(\boldsymbol{v}_t^{[M]} \mid \overline{m_t^{[r]}}\right)^{\beta_t}. \quad (11)$$

Here, Z_t is the normalization constant. and we employ

$$p\left(\boldsymbol{v}_t^{[M]} \mid \overline{m^{[r]}}\right) = \frac{1}{|M| - 1} \sum_{m \in M \setminus \{m^{[r]}\}} p\left(\boldsymbol{v}_t^{[M]} \mid m\right). \quad (12)$$

At $t = T$, we provide the decision of T-step digit verification by

$$V_{1:T}^{[D]} = \begin{cases} 1 \text{ if } p_{\mathrm{G}}\left(d_{1:T}^{[r]} \mid \boldsymbol{v}_{1:T}^{[D]}\right) \geq p_\theta^{[D]} \\ -1 \text{ otherwise} \end{cases} \quad (13)$$

and T-step speaker and digit verification, or text-prompted speaker verification, by

$$V_{1:T}^{[\mathrm{SD}]} = \begin{cases} 1 \text{ if } \left(V_{1:T}^{[D]} = 1\right) \wedge \left(p_{\mathrm{G}}\left(s_{1:T}^{[r]} \mid \boldsymbol{v}_{1:T}^{[S]}\right) \geq p_\theta^{[S]}\right) \\ -1 \text{ otherwise} \end{cases} \quad (14)$$

where $p_\theta^{[D]}$ and $p_\theta^{[S]}$ are thresholds. Here, $V_{1:T}^{[D]} = 1$ and $V_{1:T}^{[\mathrm{SD}]} = 1$ indicate the acceptance, and -1 the rejection. In order to use the same thresholds $p_\theta^{[D]}$ and $p_\theta^{[S]}$ for all speakers and digits in the experiments shown below, we have tuned the threshold $y_\theta = y_\theta^{[m]}$ in (2) for speakers $m = s \in S$ and digits $m = d \in D$, respectively, by the procedure shown in [1].

3 Experiments

3.1 Experimental Setting

We have used the speech data sampled with 8kHz of sampling rate and 16 bits of resolution in a silent room of our laboratory. They are from seven speakers (2 female and 5 male speakers): $S = \{\mathrm{fHS}, \mathrm{fMS}, \mathrm{mKK}, \mathrm{mKO}, \mathrm{mMT}, \mathrm{mNH}, \mathrm{mYM}\}$ for ten Japanese digits $D = \{/\mathrm{zero}/, /\mathrm{ichi}/, /\mathrm{ni}/, /\mathrm{san}/, /\mathrm{yon}/, /\mathrm{go}/, /\mathrm{roku}/, /\mathrm{nana}/, /\mathrm{hachi}/, /\mathrm{kyu}/\}$. For each speaker and each digit, ten samples are recorded on different times and dates among two months. We denote each spoken digit by $x = x_{s,d,l}$ for $s \in S, w \in W$ and $l \in L = \{1, 2, \cdots, 10\}$, and the given dataset by $X = (x_{s,d,l} | s \in S, d \in D, l \in L)$.

In order to evaluate the performance of the present method for untrained data and the data of unregistered speaker, we employ a combination of LOOCV (leave-one-out cross-validation) and OOB (out-of-bag) estimate [7]. Namely, for testing a speaker $s_u \in S$ as an unregistered speaker via LOOCV scheme, we prepare the dataset $X_{\overline{s_u}} = (x_{s,d,l} | s \in S \setminus s_i, d \in D, l \in L)$ for each s_u and execute the following training and testing for a reference speaker $s_i \in S$ via OOB estimate. First, we make a training dataset

$Z^{[s_i]}_{\overline{s_u}} = ((q(x), y^{[s_i]}(x))|x \in X_{\overline{s_u}})$, where $q(x)$ is the feature vector obtained from $x \in X_{\overline{s_u}}$, and $y^{[s_i]}(x) = 1$ if $x \in X_{\overline{s_u}}$ is of the speaker s_i, and -1 otherwise. Next, we execute resampling with replacement to make the bags $Z^{[s_i, \alpha n^{\#}, j]}_{\overline{s_u}}$ for $j \in J^{[\text{bg}]}$, where αn indicates the number of elements in the bag for a constant α called bagsize ratio, $n = |Z^{[s_i]}_{\overline{s_u}}|$ is the number of elements in $Z^{[s_i]}_{\overline{s_u}}$, and $J^{[\text{bg}]} = \{1, 2, \cdots, |J^{[\text{bg}]}|\}$ is an index set. Here, it is expected that $ne^{-\alpha}$ elements in $Z^{[s_i]}_{\overline{s_u}}$ are not in $Z^{[s_i, \alpha n^{\#}, j]}_{\overline{s_u}}$. Thus, we execute the OOB estimate of $y^{[s_i]}(x)$ by $\hat{y}^{[s_i, \text{ob}]}(x) = \langle \hat{y}^{[s_i, j]}(x) \rangle_{j \in J^{[s_i, \text{ob}]}_x}$, where $\hat{y}^{[s_i, j]}(x)$ is the output of $\text{RLM}^{[s_i, j]}$ which has learned $Z^{[s_i, \alpha n^{\#}, j]}_{\overline{s_u}}$, and $J^{[s_i, \text{ob}]}_x \triangleq \{j | (q(x), y^{[s_i]}(x)) \notin Z^{[s_i, \alpha n^{\#}, j]}_{\overline{s_u}}, j \in J^{[\text{bg}]}\}$. Here $\langle \cdot \rangle$ indicates the mean and the subscript indicates the range of the mean. Note that the experiments shown below are done for $|J^{[\text{bg}]}| = 300$, $\alpha = 1.6$, $n = |S - 1||D||L| = 600$. For regression learning machines we use CAN2s with the number of units being $N = 40$ and 100 for learning a speaker and a digit, respectively, with 36-dimensional feature vector q (see [5] for details of q). The OOB estimate for digit verification is done by the same procedure as above, where we do not have examined unregistered digits although we may have to examine it in our future research.

3.2 Experimental Results and Analysis

To examine the present method, we first show a statistical result of the verification of unregistered speakers for a number of datasets, where each dataset consists of 1000 pairs of T-length digit sequences of an unregistered test speaker and registered reference speakers. Precisely, for a test digit sequence $d_{1:T} = d_1 d_2 \cdots d_T$ of a test speaker s and the corresponding reference digit sequence $d^{[r]}_{1:T} = d^{[r]}_1 d^{[r]}_2 \cdots d^{[r]}_T$ of a reference speaker $s^{[r]}$, we select d_t, $d^{[r]}_t$, and $s^{[r]}$ randomly under the condition that $d_{1:T}$ involves correct digits satisfying $d_t = d^{[r]}_t$ with a ratio of $r^{[D]}_{\text{cor}} = n^{[D]}_{\text{cor}}/T$, where $n^{[D]}_{\text{cor}}$ represents the number of correct digits. Here note that for a digit $d \in D$ of a speaker $s \in S$, we use $x_{s,d,l}$ with l selected randomly.

The result is shown in Table 1. Here, in order to avoid spoofing from impostors, we suppose that the users have previously registered their secret sequences consisting of 5-digits and the corresponding questions to answer the sequences in the enrollment phase, and then a test speaker is prompted to answer three questions by uttering the sequences in the verification phase, where the questions are selected randomly. Thus, we use $T = 15 = 5 \times 3$. The thresholds $(p^{[D]}_\theta, p^{[S]}_\theta) = (0.80, 0.96)$ for GEBI and $(0.99, 0.80)$ for BI are chosen with the precision of 0.01 to (approximately) achieve EER (equal error rate) for $r^{[D]}_{\text{cor}} = 5/5$. Namely, in the table, FAR (False Acceptance Rate) for UR (UnRegistered incorrect) test speakers being 1.1 [%] for GEBI and 5.0 [%] for BI, respectively, is almost the same as FRR for RC (Registered Correct) test speakers being $100 - 98.8 = 1.2$ [%] for GEBI and $100 - 94.0 = 6.0$ [%] for BI. Thus, ERR is approximately $(1.1 + 1.2)/2 = 1.15$ [%] for GEBI and $(5.0 + 6.0)/2 = 5.5$ [%] for BI. Since smaller ERR is desirable, GEBI is better than BI for UR and RC speakers with $r^{[D]}_{\text{cor}} = 5/5$. From the point of view of UR and RI (Registered Incorrect) test speakers for $r^{[D]}_{\text{cor}} = 5/5$, FAR of UR speakers, $r^{[SD]}_{\text{acc}} = 1.0$ [%] for GEBI and 5.0

Table 1. Experimental result of percentage acceptance rates, $r^{[D]}_{acc}$ for digit verification and $r^{[SD]}_{acc}$ for text-prompted (speaker and digit) verification by the methods using GEBI and BI. The bold-face figures indicate TAR (True Acceptance Rate) and the others are FAR (False Acceptance Rate). It is desirable that FAR and FRR (= 100−TAR [%]; False Rejection Rate) are small.

| | UR (UnRegistered incorrect) test speakers | | | | RI (Registered Incorrect) test speakers | | | | RC (Registered Correct) test speakers | | | |
| | GEBI | | BI | | GEBI | | BI | | GEBI | | BI | |
$r^{[D]}_{cor}$	$r^{[D]}_{acc}$	$r^{[SD]}_{acc}$	$r^{[D]}_{acc}$	$r^{[SD]}_{acc}$	$r^{[D]}_{acc}$	$r^{[SD]}_{acc}$	$r^{[D]}_{acc}$	$r^{[SD]}_{acc}$	$r^{[D]}_{acc}$	$r^{[SD]}_{acc}$	$r^{[D]}_{acc}$	$r^{[SD]}_{acc}$
5/5	**99.0**	1.1	**98.2**	5.0	**99.1**	0.0	**97.2**	0.0	**99.0**	**98.8**	**97.7**	**94.0**
4/5	91.5	1.1	78.3	3.7	91.0	0.0	76.6	0.0	90.7	90.6	77.9	75.0
3/5	0.1	0.0	55.3	2.7	0.2	0.0	58.2	0.0	0.1	0.1	55.2	53.2
2/5	0.0	0.0	38.5	2.3	0.0	0.0	36.1	0.0	0.0	0.0	38.6	37.5

[%] for BI, is bigger than that of RI speakers, $r^{[SD]}_{acc} = 0.0$ [%] for GEBI and BI. Here, we can see that GEBI has achieved smaller FAR of UR speakers than BI for $r^{[D]}_{cor} = 5/5$.

Furthermore, we can see that FAR of UR speakers by GEBI has decreased to 0.0 [%] for incorrect digit sequences with $r^{[D]}_{cor} \leq 3/5$, while that by BI has not decreased so much. This is related to the result that FAR of RC speakers for GEBI has also decreased to 0.1 and 0.0 [%] for $r^{[D]}_{cor} = 3/5$ and 2/5, respectively, while that for BI has not decreased so much as 53.2 and 37.5 [%], respectively. As shown in [1], although $r^{[D]}_{acc}$ and $r^{[SD]}_{acc}$ of RC speakers for $r^{[D]}_{cor} = 4/5$ are undesirably very big for GEBI, the present method is supposed to work for avoiding spoofing, because it is not so easy for an impostor speaker to provide 4 or 5 correct digits in each 5-digit sequence.

In order to examine in detail, we show the multistep probabilities in Fig. 2. From (a), we can see that the probability curves of GEBI increase and the error range decreases with the increase of t. Furthermore, from the left hand side of (a), we can see that the uppermost curve for RC speakers can be separated from the other lower curves for UR and RI curves by the threshold $p^{[S]}_{\theta} = 0.8$. From the right hand side of (a), we can see that the curve for $r^{[D]}_{cor} = 3/5$ and its error range is almost below the threshold $p^{[D]}_{\theta} = 0.96$ at $t = 15$, which contribute to FAR=$r^{[D]}_{acc} = 0.1$ [%] for $r^{[D]}_{cor} = 3/5$. Here, note that we observe slightly bigger fluctuations of the curves and the error ranges than those of the corresponding ones shown in [1]. This may come from the verification performance for the digits of unregistered speakers, but it is not clear. We would like to clarify the reason in our future research study.

On the other hand, Fig. 2(b) for BI shows that the curves increase with the increase of t, and the error ranges are bigger than those for GEBI. Thus, we can understand the worse result of BI in Table 1. As shown in [1], the superiority of GEBI to BI is considered to be obtained from that the probability of GEBI, $p_G\left(m|v^{[M]}_{1:t}\right)$ given by (7), converges monotonically for the increase of t. On the other hand, the probability of BI, $p_B\left(s|v^{[M]}_{1:t}\right)$ given by (5), also converges monotonically but slowly. Namely, $p_G\left(m|v^{[M]}_{1:t}\right)$ reaches convergence at t satisfying $\tilde{L}^{[m]}_{1:t} \gg |\log p_0(m)|/t$, but $p_B\left(s|v^{[M]}_{1:t}\right)$ does not reach convergence at t satisfying $\tilde{L}^{[m]}_{1:t} \gg |\log p_0(m)|/t$.

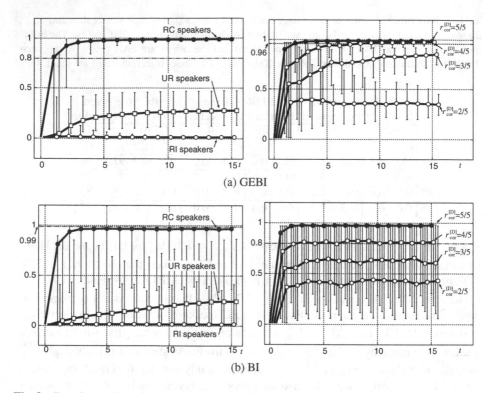

Fig. 2. Experimental result of multistep probability of (a) GEBI and (b) BI for speakers (left) and digits (right). The plus and minus error bars indicate RMS (root mean square) of positive and negative errors from the mean, respectively. The curves for different datasets are shifted slightly and horizontally to avoid crossovers.

4 Conclusion

We have presented a method of text-prompted speaker verification using GEBI, and examined the properties for rejecting unregistered speakers. The effectiveness of GEBI and the comparison to BI (Bayesian inference) are shown and analyzed by means of experiments using real speech signals.

This work was supported by JSPS KAKENHI Grant Number 24500276.

References

1. Kurogi, S., Ueki, T., Mizobe, Y., Nishida, T.: Text-prompted multistep speaker verification using Gibbs-distribution-based extended Bayesian inference for reducing verification errors. In: Lee, M., Hirose, A., Hou, Z.-G., Kil, R.M. (eds.) ICONIP 2013, Part III. LNCS, vol. 8228, pp. 184–192. Springer, Heidelberg (2013)
2. Beigi, H.: Fundamentals of speaker recognition. Springer-Verlag New York Inc. (2011)

3. Mizobe, Y., Kurogi, S., Tsukazaki, T., Nishida, T.: Multistep speaker identification using Gibbs-distribution-based extended Bayesian inference for rejecting unregistered speaker. In: Huang, T., Zeng, Z., Li, C., Leung, C.S. (eds.) ICONIP 2012, Part V. LNCS, vol. 7667, pp. 247–255. Springer, Heidelberg (2012)
4. Kurogi, S., Ueno, T., Sawa, M.: A batch learning method for competitive associative net and its application to function approximation. In: Proc. SCI 2004, vol. V, pp. 24–28 (2004)
5. Kurogi, S., Mineishi, S., Sato, S.: An analysis of speaker recognition using bagging CAN2 and pole distribution of speech signals. In: Wong, K.W., Mendis, B.S.U., Bouzerdoum, A. (eds.) ICONIP 2010, Part I. LNCS, vol. 6443, pp. 363–370. Springer, Heidelberg (2010)
6. Campbell, J.P.: Speaker Recognition: A Tutorial. Proc. the IEEE 85(9), 1437–1462 (1997)
7. Kurogi, S.: Improving generalization performance via out-of-bag estimate using variable size of bags. J. Japanese Neural Network Society 16(2), 81–92 (2009)
8. Zhang, H.: The optimality of naive Bayes. In: Proc. FLAIRS 2004 Conference (2004)

Non-monotonic Feature Selection for Regression

Haiqin Yang[1,2], Zenglin Xu[3,4], Irwin King[1,2], and Michael R. Lyu[1,2]

[1] Shenzhen Key Laboratory of Rich Media Big Data Analytics and Applications
Shenzhen Research Institute, The Chinese University of Hong Kong
[2] Computer Science & Engineering, The Chinese University of Hong Kong, Hong Kong
{hqyang,king,lyu}@cse.cuhk.edu.hk
[3] University of Electronic Science & Technology of China, Chengdu, Sichuan, China
[4] Purdue University, West Lafayette, IN, USA
zenglin@gmail.com

Abstract. Feature selection is an important research problem in machine learning and data mining. It is usually constrained by the budget of the feature subset size in practical applications. When the budget changes, the ranks of features in the selected feature subsets may also change due to nonlinear cost functions for acquisition of features. This property is called non-monotonic feature selection. In this paper, we focus on non-monotonic selection of features for regression tasks and approximate the original combinatorial optimization problem by a Multiple Kernel Learning (MKL) problem and show the performance guarantee for the derived solution when compared to the global optimal solution for the combinatorial optimization problem. We conduct detailed experiments to·demonstrate the effectiveness of the proposed method. The empirical results indicate the promising performance of the proposed framework compared with several state-of-the-art approaches for feature selection.

1 Introduction

Feature selection is an important task in machine learning and data mining. The goal of feature selection is to choose a subset of informative features from the input data so as to reduce the computational cost or save storage space for problems with high dimensional data. Feature selection has found applications in a number of real-world problems, such as data visualization, natural language processing, computer vision, speech processing, bioinformatics, sensor networks and so on [10]. More information can be found from the comprehensive survey paper [4] and references therein.

A general definition of selecting features from a learning task is to choose a subset of m features, denoted by $\mathcal{S}$, that maximizes a generalized performance criterion $\mathcal{Q}$. It is cast into the following combinatorial optimization problem:

$$\mathcal{S}^* = \arg\max \mathcal{Q}(\mathcal{S}) \quad \text{s. t.} \quad |\mathcal{S}| = m. \tag{1}$$

Here m is also called the budget of selected features. $\mathcal{Q}(\mathcal{S})$ is restricted to a performance measure for regression problems. More specifically, we adopt the dual objective function of Support Vector Regression (SVR), a popular regression model [7] in the literature, as is defined later in Eq. (3).

C.K. Loo et al. (Eds.): ICONIP 2014, Part II, LNCS 8835, pp. 44–51, 2014.

When the budget changes, the new feature subset may not be a subset or superset of the previous feature subset due to nonlinear cost functions for acquisition of features. This property is called non-monotonic feature selection [12]:

Definition 1 (Non-monotonic Feature Selection). *A feature selection algorithm $\mathcal{A}$ is monotonic if and only if it satisfies the following property: for any two different numbers of selected features, i.e., k and m, we always have $\mathcal{S}_k \subseteq \mathcal{S}_m$ if $k \leq m$, where $\mathcal{S}_m$ stands for the subset of m features selected by $\mathcal{A}$. Otherwise, it is called non-monotonic feature selection.*

Due to the dependance of feature selection on the budget of feature subsets, traditional feature selection methods may yield sub-optimal solutions. In order to tackle this problem, in this paper, we propose a **non-monotonic** feature selection for regression. Following the framework for classification derived in [8,12], we approximate the original combinatorial optimization problem of feature selection and formulate it closely related to multiple kernel learning (MKL) [1,6,11,13,14,17] framework, which yields the final optimization problem to be solved efficiently by a Quadratically Constrained Quadratic Programming (QCQP) problem. Differently, support vector regression [7] is selected as the regression model due to its power in solving real-world applications. We then present a strategy that selects a subset of features based on the solution of the relaxed problem and show the **performance guarantee**, which bounds the difference in the value of objective function between using the features selected by the proposed strategy and using the global optimal subset of features found by exhaustive search. Our empirical study shows that the proposed approach performs better than the state-of-the-arts for feature selection in tackling the regression task.

2 Model and Analysis

Suppose the training set includes N samples: $\mathcal{S} = \{(\mathbf{x}_i, y_i)\}_{i=1}^{N}$, where $\mathbf{x}_i \in \mathbb{R}^d$ represents the features of the i-th sample, and $y_i \in \mathbb{R}$ corresponds to the response. Let $\mathbf{e}_d \in \mathbb{R}^d$ be a d-dimensional vector with all elements being one and $\mathbf{I}_d$ be the $d \times d$ identity matrix. For a linear kernel, the kernel matrix $\mathbf{K}$ is written as: $\mathbf{K} = \mathbf{X}^\top \mathbf{X} = \sum_{i=1}^{d} \mathbf{x}_i \mathbf{x}_i^\top = \sum_{i=1}^{d} \mathbf{K}_i$, where a kernel $\mathbf{K}_i = \mathbf{x}_i \mathbf{x}_i^\top$ is defined for each feature. The goal of feature selection is to select a subset of $m < d$ features, i.e., to determine the value of $\mathbf{p}$ in the following form:

$$\mathbf{K}(\mathbf{p}) = \sum_{i=1}^{d} p_i \mathbf{x}_i \mathbf{x}_i^\top = \sum_{i=1}^{d} p_i \mathbf{K}_i, \tag{2}$$

where $p_i \in \{0, 1\}$ is a binary variable that indicates if the ith feature is selected, and $\mathbf{p} = (p_1, \ldots, p_d)$. As revealed in (2), to select m features, we need to find optimal binary weights p_i to combine the kernels derived from individual features. This observation motivates us to cast the feature selection problem into a multiple kernel learning problem.

Following the maximum margin framework with ε-insensitive loss function for support vector regression [7,15,16] and the derivation in [12], given a kernel matrix $\mathbf{K}(\mathbf{p}) = \sum_{i=1}^{d} p_i \mathbf{K}_i$, the regression model, $f(\mathbf{x}) = \sum_{i=1}^{d} w_i \mathbf{x}_i + b = \sum_{j=1}^{N}(\alpha_j - \alpha_j^*)\mathbf{K}(\mathbf{p})$, is found by solving the following optimization problem:

$$\tilde{\omega}(\mathbf{p}) := \begin{cases} \max_{\beta} \ 2\mathbf{v}^\top \beta - \beta^\top \mathbf{Q}(\mathbf{p})\beta \\ \text{s.t.} \ \ 0 \leq \beta \leq C, \ \mathbf{u}^\top \beta = 0 \end{cases} \tag{3}$$

where the variable $\beta = [\alpha; \alpha^*] \in \mathbb{R}^{2N}$, and $\alpha, \alpha^* \in \mathbb{R}^N$ are corresponding Lagrange multipliers used to push and pull $f(\mathbf{x})$ towards the outcome of y, respectively. b corresponds to the dual variable of $\mathbf{u}^\top \beta = 0$. The linear coefficient $\mathbf{v}$ is defined as $[\mathbf{v}_1; \mathbf{v}_2]$, where $\mathbf{v}_1 = [-\varepsilon \mathbf{e}_N + \mathbf{y}]$ and $\mathbf{v}_2 = [-\varepsilon \mathbf{e}_N - \mathbf{y}]$. $\mathbf{u}$ in the equality constraint is defined as $[\mathbf{e}_N^\top, -\mathbf{e}_N^\top]$. The matrix $\mathbf{Q}(\mathbf{p}) = [\mathbf{K}(\mathbf{p}), -\mathbf{K}(\mathbf{p}); -\mathbf{K}(\mathbf{p}), \mathbf{K}(\mathbf{p})] \in \mathbb{R}^{2N \times 2N}$.

By approximating the indicator vector $\mathbf{p}$ to a continuous indicator, we have to solve the following optimization problem:

$$\min_{0 \leq \mathbf{p} \leq 1} \ \tilde{\omega}(\mathbf{p}) \quad \text{s.t.} \quad \mathbf{p}^\top \mathbf{e}_d = m. \tag{4}$$

Following the derivation in [6,12], we can transform (3) to the following optimization problem:

$$\min_{\mathbf{p},t,\nu,\delta,\theta} \ t + 2C\delta^\top \mathbf{e}_{2N} \tag{5}$$

$$\text{s.t.} \quad \begin{pmatrix} \mathbf{Q}(\mathbf{p}) & \mathbf{v} + \nu - \delta + \theta\mathbf{u} \\ (\mathbf{v} + \nu - \delta + \theta\mathbf{u})^\top & t \end{pmatrix} \succeq 0,$$

$$\nu \geq 0, \ \delta \geq 0, \ \mathbf{p}^\top \mathbf{e}_d = m, \ 0 \leq \mathbf{p} \leq 1.$$

To further speedup the semi-definite programming (SDP) problem in (5), we show in the following theorem that (5) can be reformulated into a Quadratically Constrained Quadratic Programming (QCQP) problem similar to [12]:

Theorem 1. *The optimization problem in (4) can be reduced to the following optimization problem:*

$$\max_{\alpha,\alpha^*,\lambda,\gamma} \ 2(\mathbf{v}_1^\top \alpha + \mathbf{v}_2^\top \alpha^*) - m\lambda - \gamma^\top \mathbf{e}_N \tag{6}$$

$$\text{s.t.} \ \ \mathbf{e}_N^\top(\alpha - \alpha^*) = 0, \ \ 0 \leq \alpha, \alpha^* \leq C,$$

$$(\alpha - \alpha^*)^\top \mathbf{K}_i(\alpha - \alpha^*) \leq \lambda + \gamma_i, \ \ \gamma_i \geq 0.$$

The KKT conditions are

$$\mathbf{K}(\mathbf{p})(\alpha - \alpha^*) = \mathbf{v} + \nu - \delta + \theta\mathbf{u},$$

$$t = [\alpha; \alpha^*]^\top(\mathbf{v} + \nu - \delta + \theta\mathbf{u}),$$

$$[\alpha; \alpha^*] \circ \nu = 0, \ [\alpha; \alpha^*] \circ \delta = C\delta, \ \gamma \circ (\mathbf{e}_d - \mathbf{p}) = 0,$$

$$p_i(\lambda + \gamma_i - (\alpha - \alpha^*)^\top \mathbf{K}_i(\alpha - \alpha^*)) = 0, \ \ i = 1, \ldots, d. \tag{7}$$

Now, by observing (7), we rank the features in the descending order of $\tau_i = (\alpha - \alpha^*)^\top \mathbf{K}_i(\alpha - \alpha^*) = (\sum_{j=1}^{N} X_{i,j}(\alpha_j - \alpha_j^*))^2 = w_i^2$, where w_i is the weight computed for the i-th feature. We denote by $i_1, \ldots, i_d$ the ranked features, and by $k_{\min}$ and $k_{\max}$ the smallest and the largest indices such that $\tau_{i_k} = \tau_{i_m}$ for $1 \leq k \leq d$. We divide features into three sets and derive the properties of λ and $\mathbf{p}$ as follows:

$$\mathcal{A} = \{i_k | 1 \leq k < k_{\min}\}, \; \mathcal{B} = \{i_k | k_{\min} \leq k \leq k_{\max}\}, \; \mathcal{C} = \{i_k | k_{\max} < k \leq d\}. \quad (8)$$

Corollary 1. *We have the following properties for λ and $\mathbf{p}$:*

$$\lambda \in [\tau_{1+k_{\max}}, \tau_m], \quad p_i = \begin{cases} 1, i \in \mathcal{A}, \\ 0, i \in \mathcal{C}. \end{cases} \quad (9)$$

Remark. The above corollary can be derived by analyzing the KKT conditions in (7). We then can conduct our non-monotonic feature selection for regression, namely **NMMKLR**, by the following three steps: 1) Solve α, α^* in (6) by a linear objective function with quadratic constraints, a special case of the QCQP; 2) Compute $\tau_i = (\sum_{j=1}^{N} X_{i,j}(\alpha_j - \alpha_j^*))^2$; 3) Select the first m features with the largest τ_i.

Moreover, we provide the following theorem to show that the performance guarantee of the discrete solution constructed by our proposed algorithm and the combinatorial optimization problem in the form of (4).

Theorem 2. *The discrete solution constructed by our NMMKLR, denoted by $\mathbf{p}$, has the following performance guarantee for the combinatorial optimization problem defined in (1): $\frac{\omega(\mathbf{p})}{\omega(\widetilde{\mathbf{p}}^*)} \leq \frac{1}{1 - \sigma_{\max}(\mathbf{R}^{-1/2}\mathbf{B}\mathbf{R}^{-1/2})}$, where $\mathbf{R} = \mathbf{Q}(\mathbf{p}^*)$, $\mathbf{B} = \sum_{j \in \mathcal{B}} p_j^* \mathbf{K}_j$. The operator $\sigma_{\max}(\cdot)$ calculates the largest eigenvalue. $\mathbf{p}^*$ and $\widetilde{\mathbf{p}}^*$ denote the optimal solution of the relaxed optimization problem in (4) and the global optimal solution of the original combinatorial optimization problem defined in (1), respectively.*

The proof can be found in the long version of this paper. Theorem 2 indicates that by incorporating the required number of selected features, the resulting approximate solution could be more accurate than without it, which implies that the proposed NMMKLR algorithm produces a better approximation to the underlying combinatorial optimization problem (1).

3 Experiments

We conduct detailed experiments on both synthetic and real-world datasets and compare our proposed **NMMKLR** with the following state-of-the-art methods: 1) **Stepwise**: the forward stepwise feature selection method [2,3,4][1]; 2) **SVR-LW**: features are selected with the largest absolute weights $|w_i|$ computed by SVR [7]; 3) **LASSO-LW**: features are selected with the largest absolute weights $|w_i|$ computed by LASSO [9].

We adopt the following two metrics to measure the model performance: 1) **Mean Square Error (MSE)**: $MSE = \sum_{i=1}^{N}(y_i - \hat{y}_i)^2/N$, which measures the discrepancy of the predictive response and real response; 2) Q^2 **statistics**: $Q^2 = \frac{\sum_{i=1}^{N}(y_i - \hat{y}_i)^2}{(y_i - \bar{y}_i)^2}$, which is scaled by the variance of the response. where $\hat{y}_i$ is the prediction of y_i for the i-th test sample, and $\bar{y}$ is the mean of the actual response. Obviously, in both metrics, the smaller the corresponding value is, the better the performance is.

---

[1] http://www.robots.ox.ac.uk/~parg/software/fsbox_1_0.tar

Experiments on synthetic dataset. We first generate a toy dataset consisting of $d(=12)$ dimensions by an additive model similar to that in [2]: $y_i = \sum_{j=1}^{4} j x_{ji} + e^{x_{5i}}$, where y_i denotes the response for the i-th sample and $\mathbf{x}_j$ denotes the j-th feature, for $j = 1, \ldots, 12$. x_{ji} denotes the element of the j-th feature on the i-th sample. Here, only the first five features contribute to the response and each of them is generated from an independently and identically distributed normal distribution. The rest 7 features are generated as follows: the 6-th feature is $\mathbf{x}_6 = \mathbf{x}_1 + 1$, which is correlated to $\mathbf{x}_1$; the 7-th feature is $\mathbf{x}_7 = \mathbf{x}_2 \circ \mathbf{x}_3$, which is the element-wise product of $\mathbf{x}_2$ and $\mathbf{x}_3$; the rest five features, i.e., $\mathbf{x}_8, \ldots, \mathbf{x}_{12}$, generated by standard normal distribution, are totally irrelevant to the response y_i. For convenience, we also denote the irrelevant features by $NV_1, \ldots, NV_5$, respectively.

Table 1. Top 5 and 6 selected features (ordered) from four compared algorithms within 20 trials on the synthetic dataset. The stepwise and the LASSO-LW method include some irrelevant features when the number of selected features is greater than six.

Method	Times	Top 5 selected features	Times	Top 6 selected features
NMMKLR	19	4, 3, 2, 5, 1	20	4, 3, 2, 5, 1, 6
	1	4, 3, 2, 5, 6		
Stepwise			8	4, 3, 2, 5, 1
	10	4, 3, 2, 5, 1	3	4, 3, 5, 2, 6
			2	4, 3, 2, 5, 6, 8 (NV_1)
	6	4, 3, 2, 5, 6	1	3, 4, 2, 5, 6
			1	4, 3, 2, 5, 6
	2	4, 3, 5, 2, 6	1	4, 3, 2, 5, 1, 9 (NV_2)
			1	4, 3, 2, 5, 1, 11 (NV_4)
	2	3, 4, 2, 6, 5	1	4, 3, 2, 5, 6, 9 (NV_2)
			1	4, 3, 2, 5, 6, 10 (NV_3)
			1	4, 3, 2, 5, 6, 12 (NV_5)
SVR-LW	10	4, 3, 2, 5, 1	10	4, 3, 2, 5, 1, 6
	10	4, 3, 2, 5, 6	10	4, 3, 2, 5, 6, 1
LASSO-LW			4	4, 3, 2, 5, 1, 8 (NV_1)
			3	4, 3, 2, 5, 1, 10 (NV_3)
	15	4, 3, 2, 5, 1	3	4, 3, 2, 5, 1, 12 (NV_5)
			2	4, 3, 2, 5, 1, 9 (NV_2)
			2	4, 3, 2, 5, 1, 11 (NV_4)
	4	4, 3, 2, 5, 6	1	4, 3, 2, 5, 1, 7
			1	4, 3, 2, 5, 6, 8 (NV_1)
			1	4, 3, 2, 5, 6, 9 (NV_2)
			1	4, 3, 2, 5, 6, 10 (NV_3)
	1	4, 3, 2, 6, 5	1	4, 3, 2, 5, 6, 11 (NV_4)
			1	4, 3, 2, 6, 5, 12 (NV_5)

We conduct two batches of experiments, where the budget is set to 5 and 6, respectively. Then in each batch of experiments, we randomly generate 200 samples and hold out 50% of the samples for training while keeping the rest for test. Each exepriment is then repeated 20 times.

In order to examine the property of the selected features, we list the top 5 and 6 selected features returned for all algorithms in Table 1. Obviously, our method can stably select those important features while SVR-LW also selects features relatively stable. However, the selected features by the forward stepwise feature selection method and the LASSO-LW method are unstable, and some irrelevant features are included when the number of selected features is greater than 5.

To further evaluate the regression performance on the selected features, we employ Support Vector Regression (SVR) as the regression model. We tune the hyperparameters C and ε, of SVR through five-fold cross validation on the training data with the top 5, the top 6, and all the features. The hyperparameter of SVR, C, is chosen uniformly from the interval $[10^0, 10^3]$ on a logarithmic scale and ε is chosen in $[0.01, 0.1, 0.25, 0.5, 0.75, 1, 1.5, 2]$.

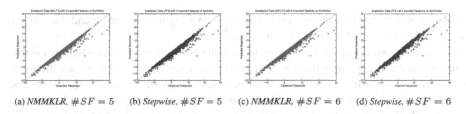

(a) NMMKLR, $\#SF = 5$ (b) Stepwise, $\#SF = 5$ (c) NMMKLR, $\#SF = 6$ (d) Stepwise, $\#SF = 6$

Fig. 1. Scatter plots of the pairs $(\mathbf{y}, \hat{\mathbf{y}})$. (a) and (c) show the plot of NMMKLR when the number of selected features equal to 5 and 6, respectively. (b) and (d) shows the plot of Stepwise regression when the number of selected features equal to 5 and 6, respectively. It can be observed that (a) is thinner than (b) and (c) is thinner than (d).

Finally, we show the evaluation results on four compared algorithms in Table 2. It can be observed that the proposed NMMKLR outperforms other three methods in both of the MSE and Q^2 measures in all cases. Moreover, the paired t-test with the confidence level of 95% indicates that the advantage of NMMKLR is significant. To better visualize the difference between the response values predicted by the feature selection algorithms, we plot the pairs of observed response and predicted response, i.e., $(\mathbf{y}, \hat{\mathbf{y}})$, for the NMMKLR and the Stepwise selection method. The results are shown in Figure 1. Ideally, if the MSE is zero, all the points should drop on the line $\mathbf{y} = \hat{\mathbf{y}}$. Thus a scatter plot with smaller areas will be better. We observe from Figure 1 that the proposed NMMKLR has a better performance in both cases.

Experiments on a real-world benchmark dataset. We employ a real-world benchmark dataset, the Boston Housing problem [5] to evaluate the above four feature selection algorithms. The Boston Housing problem [5] is a popular benchmark dataset for evaulating regression models. It consists of 506 instances with 13 continuous features, such as crime rate, lower status of the population, etc. The response variable is the median value of owner-occupied homes in $1000's,

In the experiment, we normalize the continuous features in the range of $[-1, 1]$ and hold out half of samples for training while keeping the rest for test. The parameters of SVRs are tuned on the training data with the top 5, the top 6, and all features, where C

Table 2. The test accuracy for the synthetic dataset evaluated by the performance measures (MSE and Q^2) on four algorithms. The best results are highlighted (achieved by the paired t-test with 95% confidence level).

#SF	NMMKLR		Stepwise	
	MSE	Q^2	MSE	Q^2
5	**1.1599** ± 0.6977	**0.0339** ± 0.0186	1.2156 ± 0.6893	0.0356 ± 0.0183
6	**1.1600** ± 0.6977	**0.0339** ± 0.0186	1.2352 ± 0.6787	0.0362 ± 0.0180

#SF	SVR-LW		LASSO-LW	
	MSE	Q^2	MSE	Q^2
5	1.2128 ± 0.7421	0.0353 ± 0.0198	1.2156 ± 0.6893	0.0356 ± 0.0183
6	1.2127 ± 0.7422	0.0353 ± 0.0198	1.2553 ± 0.6716	0.0368 ± 0.0178

is chosen uniformly from the interval $[10^0, 10^3]$ on a logarithmic scale and ε is chosen from $[0.01, 0.1, 0.5, 1:0.5:10]$, a Matlab notation.

Since the forward stepwise feature selection method can only select 5 features sometimes when the significance level is set to 0.05, for a fair comparison, we set the numbers of selected features to be 5 and 6 for two batch of experiments. We then calculate the MSE and Q^2 values of the SVRs trained in these selected features and report the results in Table 3. It can be observed that the regression results by NMMKLR are significantly better than those selected by SVR-LW, LASSO-LW, and the forward stepwise feature selection method in both cases.

Table 3. The results of two performance measures (MSE and Q^2) on the house dataset when varying the number of selected features by NMMKLR and stepwise feature selection, SVR-LW, and LASSO-LW method. The best results on feature selection are highlighted (achieved by the paired t-test with 95% confidence level).

#SF	NMMKLR		Stepwise	
	MSE	Q^2	MSE	Q^2
5	**25.65** ± 2.36	**0.3208** ± 0.0329	26.24 ± 2.41	0.3281 ± 0.0326
6	**25.07** ± 2.50	**0.3131** ± 0.0290	25.39 ± 2.69	0.3174 ± 0.0344

#SF	SVR-LW		LASSO-LW	
	MSE	Q^2	MSE	Q^2
5	26.95 ± 3.12	0.3365 ± 0.0368	26.25 ± 2.57	0.3283 ± 0.0345
6	26.75 ± 2.94	0.3342 ± 0.0360	25.83 ± 2.41	0.3232 ± 0.0344

4   Conclusion

This paper presents a new framework of non-monotonic feature selection for regression models. By fixing the number of selected features and adopting the SVR, we develop an efficient non-monotonic feature selection algorithm for SVR via the multiple kernel learning framework to approximately solve the original combinatorial optimization problem. We further propose a strategy to derive a discrete solution for the relaxed problem with performance guarantee. The empirical study on both synthetic and real-world datasets shows the effectiveness of the proposed algorithm.

Acknowledgement. The work described in this paper was fully supported by the National Grand Fundamental Research 973 Program of China (No. 2014CB340401 and No. 2014CB340405), the Basic Research Program of Shenzhen (No. JCYJ20120619152419087), a Project 985 grant from University of Electronic Science and Technology (No. A1098531023601041), the Research Grants Council of the Hong Kong Special Administrative Region, China (Project No. CUHK 413212 and CUHK 415113), and Microsoft Research Asia Regional Seed Fund in Big Data Research (Grant No. FY13-RES-SPONSOR-036).

References

1. Bach, F.R., Lanckriet, G.R.G., Jordan, M.I.: Multiple kernel learning, conic duality, and the SMO algorithm. In: ICML, pp. 41–48. ACM, New York (2004)
2. Bi, J., Bennett, K.P., Embrechts, M.J., Breneman, C.M., Song, M.: Dimensionality reduction via sparse support vector machines. J. Mach. Learn. Res. 3, 1229–1243 (2003)
3. Draper, N., Smith, H.: Applied Regression Analysis, 3rd edn. Wiley Interscience (1998)
4. Guyon, I., Elisseeff, A.: An introduction to variable and feature selection. Journal of Machine Learning Research 3, 1157–1182 (2003)
5. Harrison, D.J., Rubinfeld, D.L.: Hedonic housing prices and the demand for clean air. Journal of Environmental Economics and Management 5(1), 81–102 (1978)
6. Lanckriet, G.R.G., Cristianini, N., Bartlett, P., Ghaoui, L.E., Jordan, M.I.: Learning the kernel matrix with semidefinite programming. J. Mach. Learn. Res. 5, 27–72 (2004)
7. Smola, A.J., Schölkopf, B.: A tutorial on support vector regression. Statistics and Computing 14(3), 199–222 (2004)
8. Tan, M., Wang, L., Tsang, I.W.: Learning sparse svm for feature selection on very high dimensional datasets. In: ICML, pp. 1047–1054 (2010)
9. Tibshirani, R.: Regression shrinkage and selection via the lasso. J. Roy. Statist. Soc. Ser. B 58(1), 267–288 (1996)
10. Wolf, L., Shashua, A.: Feature selection for unsupervised and supervised inference: The emergence of sparsity in a weight-based approach. J. Mach. Learn. Res. 6, 1855–1887 (2005)
11. Xu, Z., Jin, R., King, I., Lyu, M.: An extended level method for efficient multiple kernel learning. In: NIPS, pp. 1825–1832 (2009)
12. Xu, Z., Jin, R., Ye, J., Lyu, M.R., King, I.: Non-monotonic feature selection. In: ICML, pp. 1145–1152 (2009)
13. Xu, Z., Jin, R., Zhu, S., Lyu, M.R., King, I.: Smooth optimization for effective multiple kernel learning. In: AAAI (2010)
14. Xu, Z., King, I., Lyu, M.R., Jin, R.: Discriminative semi-supervised feature selection via manifold regularization. IEEE Transactions on Neural Networks 21(7), 1033–1047 (2010)
15. Yang, H., Chan, L., King, I.: Support vector machine regression for volatile stock market prediction. In: Yin, H., Allinson, N.M., Freeman, R., Keane, J.A., Hubbard, S. (eds.) IDEAL 2002. LNCS, vol. 2412, pp. 391–396. Springer, Heidelberg (2002)
16. Yang, H., Huang, K., King, I., Lyu, M.R.: Localized support vector regression for time series prediction. Neurocomputing 72, 2659–2669 (2009)
17. Yang, H., Xu, Z., Ye, J., King, I., Lyu, M.R.: Efficient sparse generalized multiple kernel learning. IEEE Transactions on Neural Networks 22(3), 433–446 (2011)

Non-negative Matrix Factorization
with Schatten p-norms Reguralization

Ievgen Redko and Younès Bennani

Université Paris 13, Sorbonne Paris Cité - Laboratoire d'Informatique de Paris-Nord,
CNRS (UMR 7030) F-93430, Villetaneuse, France
name.surname@lipn.univ-paris13.fr

Abstract. In this paper we study the effect of regularization on clustering results provided by Non-negative Matrix Factorization (NMF). Different kinds of regularization terms were previously added to the NMF
objective function in order to produce sparser results and thus to obtain a more qualitative partition of data. We would like to propose the
general framework for regularized NMF based on Schatten p-norms. Experimental results show the effectiveness of our approach on different
data sets.

1 Introduction

There exists numerous unsupervised learning methods that were applied in many
contexts and by researchers in many disciplines. Typical applications of clustering are: statistics [1], pattern recognition [2], image segmentation and computer
vision [3], multivariate statistical estimation [4]. Clustering is also widely used
for data compression in image processing, which is also known as vector quantization [5]. Most general survey about clustering methods can be found in [6].

There exist lots of well-known clustering algorithms, notably: k-means, mixture
models, hierarchical clustering, non-negative matrix factorization etc. Among all
the methods used for clustering we will discuss the one called Non negative matrix
factorization (NMF).

NMF is a group of algorithms in machine learning where a data matrix is
factorized into (usually) two matrices with the property that all three matrices
have no negative elements. This non-negativity makes the resulting matrices
easier to interpret. We consider one of the matrices as a matrix containing the
prototypes of a data set and the other one as a data partition matrix. Since this
optimization problem is not convex in general, it is commonly approximated
numerically.

1.1 Background

Regularization methods are often used to prevent model's overfitting or in order to obtain sparse representations of features of a given data set. The most
common variants of regularization in machine learning are $\ell 1$ and $\ell 2$ regularizations, that can be added to a loss function $J(X_1, ..X_n)$ by instead minimizing

C.K. Loo et al. (Eds.): ICONIP 2014, Part II, LNCS 8835, pp. 52–59, 2014.

$J(X_1, ..X_n) + \sum_{i=1}^n \alpha_i \Psi(X_i)$, where $\Psi(X_i)$ is a regularization term of variable X_i and α_i is a parameter that is used to control sparsity of the corresponding term. Regularization is largely employed in logistic regression, neural nets, support vector machines, conditional random fields and some matrix decomposition methods. $\ell2$ regularization may also be called "weight decay", in particular in the setting of neural nets. $\ell1$ regularization is often preferred because it produces sparse models and thus performs feature selection within the learning algorithm, but since the $\ell1$ norm is not differentiable, it may require changes to learning algorithms, in particular gradient-based learners.[7][8]

1.2   Related Works

In [9] generalized model of regularized NMF was considered and multiplicative and additive update rules have been presented both for $\ell1$ and $\ell2$ norms. In [10] Schatten p-norms were used for low rank matrix recovery. Mixed ℓ_p and Schatten p-norms regularization for matrix completion can be found in [11]. Hessian Schatten p-norm regularization for image processing is studied in [12]. Works presented above clearly show that Schatten p-norms regularization has an ability to improve results on problems they were applied to.

1.3   Our Contributions

In our work we can highlight two main contributions:

- We studied the effectiveness of different regularizations for NMF
- We proposed a novel general approach based on Schatten p-norms for regularized NMF

The rest of this paper is organized as follows: in section 2 we will briefly introduce basic notations of Non-negative matrix factorizations and Schatten p-norms, in section 3 we are introducing multiplicative and additive update rules for Schatten p-norms regularized NMF. We will summarize the results in section 4. Finally, we will point out some ideas about the future extension of our method in section 5.

2   Preliminary Knowledge

2.1   Standard NMF

A standard NMF ([13]) seeks the following decomposition:

$$X \simeq FG^T, X \in \mathbb{R}^{n \times m}, F \in \mathbb{R}^{n \times k}, G \in \mathbb{R}^{m \times k}$$

$$X, F, G \geq 0,$$

where

- X is an input data matrix
- columns of F can be considered as basis vectors
- columns of G are considered as cluster assignments for each data object
- k is the desired number of clusters

2.2 $\ell 1$, $\ell 2$ and Schatten p-norms

Given a matrix $X \in \mathbb{R}^{n \times m}$ the Schatten p-norm $(0 < p < \infty)$ is defined as follows:

$$\|X\|_{S_p} = (\sum_{i=1}^{min(n,m)} \sigma_i^p)^{1/p} = tr((X^T X)^{p/2})^{1/p},$$

where σ_i is a ith singular value of X. We can observe that for $p = 1$ we obtain the following:

$$\|X\|_{S_1} = (\sum_{i=1}^{min(n,m)} \sigma_i) = \|X\|_1$$

that is known in the literature as a nuclear norm or Ky-Fan norm. Following the same idea, we can see that for $p = 2$ we obtain the Frobenius norm:

$$\|X\|_{S_2} = (\sum_{i=1}^{min(n,m)} \sigma_i^2)^{1/2} = \|X\|_F$$

and for $p = \infty$ Schatten norm is equal to the spectral norm.

3 Regularized NMF

In this section we would like to explore if the regularization with Schatten p-norms for $p \in [1; 2]$ can perform better than regularization with $\ell 1$ and $\ell 2$ norms that were used with NMF before.

3.1 General Update Rules for NMF with Regularization Terms

As it was said previously, update rules for NMF with regularization terms were presented in [9]. For a given optimization problem:

$$min\|X - FG\| + \alpha_F \Psi(F) + \alpha_G \Psi(G), X \in \mathbb{R}^{n \times m}, F \in \mathbb{R}^{n \times k}, G \in \mathbb{R}^{m \times k}$$

$$X, F, G \geq 0$$

we have the following update rules that decrease monotonically the objective function presented above:

$$F = F \circledast \frac{XG^T - \alpha_F \Phi(F)}{FGG^T}, G = G \circledast \frac{F^T X - \alpha_G \Phi(G)}{F^T FG},$$

where

$$\Phi(F) = \frac{\partial \Psi(F)}{\partial F}, \Phi(G) = \frac{\partial \Psi(G)}{\partial G}.$$

Subsequently, for $\ell 1$ norm we have:

$$F = F \circledast \frac{[XG^T - \alpha_F]_+}{FGG^T}, G = G \circledast \frac{[F^T X - \alpha_G]_+}{F^T FG}.$$

For $\ell 2$ norm the update rules take the following form:

$$F = F \circledast \frac{[XG^T - \alpha_F F]_+}{FGG^T}, G = G \circledast \frac{[F^T X - \alpha_G G]_+}{F^T FG}.$$

3.2 Update Rules for NMF with Schatten p-norm Regularization

Our minimization problem with Schatten p-norms based regularization terms takes the following form:

$$min\|X - FG\| + \alpha_F\|F\|_{S_p} + \alpha_G\|G\|_{S_p}, X \in \mathbb{R}^{n \times m}, F \in \mathbb{R}^{n \times k}, G \in \mathbb{R}^{k \times m}$$

$$X, F, G \geq 0.$$

Based on the lemma proved in [14] the gradient of Schatten p-norm for $p > 1$ can be calculated as:

$$\nabla\|X\|_{S_p} = U_X diag(\lambda)V_X^T \frac{1}{\|\sigma(X)\|_p^{p-1}},$$

where $\lambda_i = \sigma_i(X)^{p-1}$ and $X = U_X \Sigma(X)V_X^T$ is its singular value decomposition.

Finally for our problem we obtain the following multiplicative update rules:

$$F = F \circledast \frac{XG^T - \alpha_F U_F diag(\lambda_F)V_F^T \frac{1}{\|\sigma(F)\|_p^{p-1}}}{FGG^T},$$

$$G = G \circledast \frac{F^T X - \alpha_G U_G diag(\lambda_G)V_G^T \frac{1}{\|\sigma(G)\|_p^{p-1}}}{F^T FG}.$$

We can see that SVD is used to derive update rules for the proposed objective function. In order to avoid using time-consuming SVD at each iteration and following the example from [11] we would like to change slightly the regularization terms:

$$min\|X - FG\| + \alpha_F\|F\|_{S_p}^p + \alpha_G\|G\|_{S_p}^p, X \in \mathbb{R}^{n \times m}, F \in \mathbb{R}^{n \times k}, G \in \mathbb{R}^{k \times m}$$

$$X, F, G \geq 0.$$

In this case update rules take the following form:

$$F = F \circledast \frac{XG^T - \alpha_F \frac{p}{2}F(F^T F)^{\frac{p-2}{2}}}{FGG^T}, G = G \circledast \frac{F^T X - \alpha_G \frac{p}{2}G(G^T G)^{\frac{p-2}{2}}}{F^T FG}.$$

In our experimental tests we will use the update rules that do not involve computing SVD. Nevertheless we will provide a complexity study for both types of update rules.

3.3   Complexity

At each iteration of our algorithm we perform SVD which has a complexity $O(m^2n + mn^2 + n^3)$ for a given matrix $X \in \mathbb{R}^{m \times n}$. The complexity of NMF is of order $O(knm)$, where k - is a desired number of clusters. If t denotes the number of iterations used for NMF to converge (usually, $t \approx 100$), then we obtain the following order of complexity: $O(t(m^2k + k^2n + mk^2 + kn^2 + k^3 + n^3 + knm))$. For the second case the complexity is only $O(t(knm + k^2m + m^2k + n^2k + k^2n))$. However, in real-life tasks matrix operations can usually be easily paralleled so that the complexity will decrease. Another way to decrease the complexity of our algorithm is to use any arbitrary consensus technique after splitting the initial data set into parts and obtaining a solution for each of them.

4   Experimental Results

In this section we will study the effectiveness of our approach for some well-known images data sets such as: Yale (165 images of 15 individuals), ORL (400 images of 40 different individuals of size 32x32), original Olivetti (same as ORL but each image is of size 64x64), USPS (9298 images of 10 different digits of size 16x16) and PIE (41,368 images of 68 different individuals). We performed 10 fold cross-validation for each $p \in [1; 2]$ and evaluated the results using two different measures: entropy and purity [15]. These are the standard measures of clustering quality in supervised setting. Given a particular cluster S_r of size n_r, the entropy of this cluster is defined to be:

$$E(S_r) = -\frac{1}{q}\sum_{i=1}^{q} \frac{n_r^i}{n_r}log\frac{n_r^i}{n_r},$$

where q is the number of classes in the data set, and n_r^i is the number of elements of the ith class that were assigned to the rth cluster. Smaller entropy values indicate better clustering solutions.

Using the same mathematical definitions, the purity of a cluster is defined as:

$$Pu(S_r) = \frac{1}{n_r}\max_i n_r^i.$$

Larger purity values indicate better clustering solutions.

We did not perform the tuning of hyper-parameters α_F and α_G. As advised in [9], we set these parameters to a small constant that is 0.01 in our case. Results can be further improved if one will use an appropriate technique for hyper-parameters optimization. We recall that the goal of our work is rather to compare the effectiveness of different types of regularizations for NMF. In Figures 1-4 we present the results obtained using our approach and we summarize them in Table 1.

Table 1. NMF and Schatten-p regularized NMF purity performance on various data sets

Data set	NMF	Schatten NMF	p
Yale	0.352	0.373	1.8
ORL	0.368	0.383	1.5
Olivetti	0.374	0.388	1.7
Pie	0.15	0.169	1.9
USPS	0.42	0.46	2

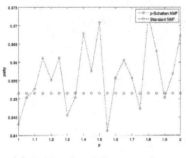

(a) Purity for different values p

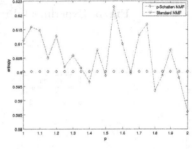

(b) Entropy values for different p

Fig. 1. Experimental results on Yale data set

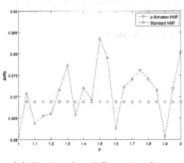

(a) Purity for different values p

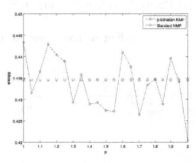

(b) Entropy values for different p

Fig. 2. Experimental results on ORL data set

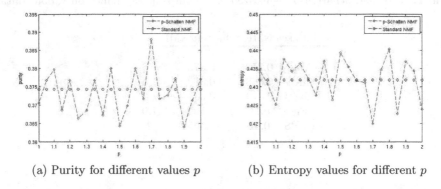

(a) Purity for different values p (b) Entropy values for different p

Fig. 3. Experimental results on Olivetti data set

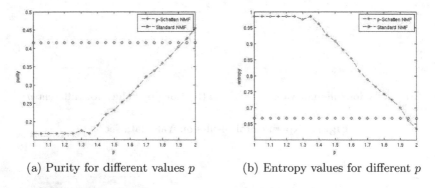

(a) Purity for different values p (b) Entropy values for different p

Fig. 4. Experimental results on USPS data set

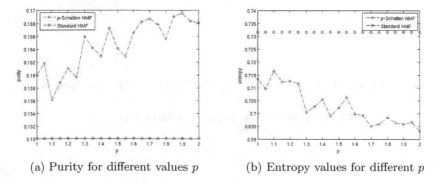

(a) Purity for different values p (b) Entropy values for different p

Fig. 5. Experimental results on PIE data set

The results presented above show that regularization with Schatten p-norms can be more efficient compared to standard regularization based on $\ell 1$ and $\ell 2$ norms. On the other hand, we see that effectiveness of regularization may depend on the size of a data set (for PIE data set regularized NMF performs better for all p) but does not depend a lot on the quality of the images (as it can be seen comparing Olivetti and ORL data sets).

5 Conclusion and Future Work

In this paper we proposed a new approach for regularized NMF based on Schatten p-norms. In future, we will extend our work in the multiple directions. First of all, we will start by creating an approach with Hessian Schatten p-norms regularization. Secondly, it would be useful to study the behavior of α_F and α_G in order to understand when part-based representations of prototypes obtained via NMF lose their interpretability due to the over-sparseness.

References

1. Arabie, P., Hubert, L.J.: An overview of combinatorial data analysis. World Scientific Publishing Co. (1996)
2. Duda, R., Hart, P.: Pattern Classification and Scene Analysis (1973)
3. Jain, A.K., Flynn, P.J.: Image segmentation using clustering. In: Advances in Image Understanding: A Festschrift for Azriel Rosenfeld, pp. 65–83. IEEE Press (1966)
4. Scott, D.W.: Multivariate Density Estimation. Wiley, New York (1992)
5. Gersho, A., Gray.: Vector Quantization and Signal Compression. Communications and Information Theory. Kluwer Academic Publishers (1992)
6. Han, J., Kamber, M.: Data Mining. Morgan Kaufmann Publishers (2001)
7. Galen, A., Jianfeng, G.: Scalable training of l1-regularized log-linear models. In: Proceedings of the 24th International Conference on Machine Learning, pp. 788–791 (2007)
8. Tsuruoka, Y., Tsujii, J., Ananiadou, S.: Stochastic gradient descent training for l1-regularized log-linear models with cumulative penalty. In: Proceedings of the AFNLP/ACL, pp. 788–791 (2009)
9. Cichocki, A., Phan, A.H., Zdunek, R.: Nonnegative Matrix and Tensor Factorizations: Applications to Exploratory Multi-way Data Analysis and Blind Source Separation. Wiley (2009)
10. Nie, F., Huang, H., Ding, C.: Low-Rank Matrix Recovery via Efficient Schatten p-Norm Minimization. In: AAAI Conference on Artificial Intelligence (2012)
11. Nie, F., Wang, H., Cai, X., Huang, H., Ding, C.: Low-Rank Matrix Recovery via Efficient Schatten p-Norm Minimization. In: ICDM, pp. 566–574 (2012)
12. Lefkimmiatis, S., Ward, J.P., Unser, M.: Hessian Schatten-Norm Regularization for Linear Inverse Problems. IEEE Transactions on Image Processing 22(5), 1873–1888 (2013)
13. Lee, D.D., Seung, H.S.: Learning the parts of objects by non-negative matrix factorization. Nature 401, 788–791 (1999)
14. Argyriou, A., Micchelli, C.A., Pontil, M.: On Spectral Learning. Journal of Machine Learning Research 11, 935–953 (2010)
15. Rendon, E., Abundez, I., Arizmendi, A., Quiroz, E.M.: Internal versus external cluster validation indexes. International Journal of Computers and Communications 5(1) (2011)

A New Energy Model for the Hidden Markov Random Fields

Jérémie Sublime[1,2], Antoine Cornuéjols[1], and Younès Bennani[2]

[1] AgroParisTech, INRA - UMR 518 MIA, F-75005 Paris, France
{jeremie.sublime,antoine.cornuejols}@agroparistech.fr
[2] Université Paris 13 - Sorbonne Paris Cité,
Laboratoire d'Informatique de Paris-Nord - CNRS UMR 7030,
99 Avenue Jean Baptiste Clément, 93430 Villetaneuse, France
Younes.bennani@lipn.univ-paris13.fr

Abstract. In this article we propose a modification to the HMRF-EM framework applied to image segmentation. To do so, we introduce a new model for the neighborhood energy function of the Hidden Markov Random Fields model based on the Hidden Markov Model formalism. With this new energy model, we aim at (1) avoiding the use of a key parameter chosen empirically on which the results of the current models are heavily relying, (2) proposing an information rich modelisation of neighborhood relationships.

Keywords: Markov Random Fields, Hidden Markov Models, Image Segmentation.

1 Introduction

A *Markov Random Fields network* is a graphical probabilistic model aiming at taking into consideration the neighborhood interactions between the data in addition to the observed a priori knowledge. Such model allows considering the explicit dependencies between the data and to weight their influence. In the case of image segmentation, these dependencies are the links between two neighbor pixels or segments (patches of pixels). Markov Random Fields networks rely on this notion of neighborhood, and are represented as non-oriented graphs the vertices of which are the data and the edges are the links between them. This additional information on the data has been shown to significantly improve the global results of the image segmentation process [1].

The *Hidden Markov Model* is also a probabilistic model in which a set of random variables $S = \{s_1, ..., s_N\}$, $s_i \in 1..K$ are linked to each other by neighborhood dependencies and are emitting observable data $X = \{x_1, ..., x_N\}$ where the x_i are the vectors containing the attributes of each observation. The goal is then to determine the optimal configuration for S, i.e. finding the values of the s_i in order to get the segmentation.

The *hidden Markov random field model* is the application of the hidden Markov model to the specific dependency structure of the Markov random fields.

C.K. Loo et al. (Eds.): ICONIP 2014, Part II, LNCS 8835, pp. 60–67, 2014.

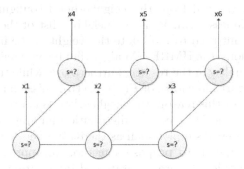

Fig. 1. An example of graph modelling the HMRF model

This model is quite common in image segmentation [2,3]. An example of a typical structure for the HMRF model is shown in Figure (1).

The main goal of the hidden Markov random field model applied to image segmentation is to find the right label for each pixel or segment of a picture in order to get a segmentation in homogeneous and meaningful areas. Doing so requires to infer the $s_i \in 1..K$, which is often achieved by using the maximum a posteriori criterion (MAP) to determine S such as:

$$S = \arg\max_S (P(X|S,\Theta)P(S)) \tag{1}$$

$$P(X|S,\Theta) = \prod_t P(x_t|s_t, \theta_{s_t}) \tag{2}$$

As it is often the case in image segmentation, we will consider that $P(x_t|s_t, \theta_{s_t})$ follows a Gaussian distribution of parameters $\theta_{s_t} = (\mu_{s_t}, \Sigma_{s_t})$.

The most common way to maximize $P(X|S,\Theta)P(S)$ is to use the two-step HMRF-EM couple [4]:

- Research of the prototype parameters Θ and maximization of $P(X|S,\Theta)$ using the Expectation Maximization algorithm (EM) [5].
- Local optimization of $P(X|S,\Theta)P(S)$ using the Iterated Conditional Modes Algorithm (ICM) [6].

The principle of the ICM algorithm is to iteratively perform a local optimization of $P(x|s,\theta)P(s)$. Such optimization is often done by minimizing an energy function with a form deriving from the logarithm of $P(x|s,\theta)P(s)$ is shown in equation (3).

$$U(s,x) = U_{obs}(s,x,\theta_s) + \beta \sum_{v \in V_x} U_{neighbor}(s_v, s) \tag{3}$$

U_{obs} and $U_{neighbor}$ are potential energy functions: U_{obs} is the potential energy derived from the observed data, also known as the data term, and $U_{neighbor}$

the potential energy derived from the neighborhood configuration, sometimes refered as the smoothness term. V_x is the neighbor list of the observed data. β is an empirical constant used to modulate the weight of the neighborhood.

In its current versions, the HMRF-EM algorithm heavily relies on the choice of β to determine the quality of the results. Furthermore, while the energy function U_{obs} usually is the logarithmic form of $P(x|s, \theta_s)$ the emission law, the exact form of $P(s)$ is unknown. For this reason, the neighborhood energy $U_{neighbor}$ is chosen as simple as possible: another gaussian distribution for complex models, and in most cases something even simpler such as the Kronecker delta function [4,7,8].

Our goal in this paper is to propose a method to approach the local value of $P(s)$ in order to provide a more accurate and information rich neighborhood energy function, and avoid using the empirical constant β.

2   Proposed Modifications to the HMRF-EM Algorithm

In this section we consider a set of linked random variables $S = \{s_1..., s_N\}$, $s_i \in 1..K$ that are emitting observable data $X = \{x_1, ..., x_N\}$, $x_i \in \mathbb{R}^d$. We assume the emission law of the x_i by the s_i to be a Gaussian distribution and we chose to consider a first order neighborhood for our Markov Random Fields. By convention, μ_s shall refer to the mean vector of a considered state s and Σ_s to its covariance matrix. Then, for each observation x associated to a state s we have the following data term as the observable energy:

$$U_{obs}(s, x) = \frac{1}{2}(x - \mu_s)^T \Sigma_s^{-1}(x - \mu_s) + log(\sqrt{|\Sigma_s|(2\pi)^d}) \qquad (4)$$

2.1   Proposed Energy Model

In the hidden Markov model formalism, the neighborhood relations are often described by a transition matrix $A = \{a_{i,j}\}_{K \times K}$ where each $a_{i,j}$ is the transition probability from one state to another between two neighboring data. We propose to keep this formalism and to adapt it to the HMRF model. Then, for $P(x|s, \theta_s)P(s)$, we get the following expression:

$$P(x|s, \theta_s)P(s) = P(x|s, \theta_s) \times \prod_{v \in V_x} P(s|s_v) = \frac{1}{Z} \times \mathcal{N}(\mu_s, \Sigma_s) \times \prod_{v \in V_x} a_{s_v,s}^{\frac{1}{|V_x|}} \qquad (5)$$

By considering the logarithm of equation (5) and the observable energy from equation (4), we get the following complete energy function to be minimized:

$$U_{HMRF}(s, x) = \frac{1}{2}(x - \mu_s)^T \Sigma_s^{-1}(x - \mu_s) + log(\sqrt{|\Sigma_s|(2\pi)^d}) - \sum_{v \in V_x} \frac{log(a_{s_v,s})}{|V_x|} \qquad (6)$$

Since the $a_{i,j}$ are values between 0 and 1, our neighborhood energy is a positive penalty function. This penalty function can take as many as K^2 distinct values instead of only two values (0 or β) for the regular neighborhood energy functions.

We note that this fitness function is as difficult to optimize as those from other models and that there is no warranty that the global optimum can be found.

It is important to highlight that in addition of providing a more flexible energy model for the HMRF-EM framework, the transition probability matrix also provides semantic information on the clusters: The diagonal of the matrix gives some insight on the clusters compactness, non-diagonal values provide information on cluster neighborhood affinities, but also incompatibilities for values near 0, or the inclusion of a cluster inside another one for values above 0, 5.

2.2 Proposed Algorithm

The main difficulty in our approach is that we have to evaluate the transition matrix A in order to run our HMRF-EM algorithm. Without some external knowledge on the neighborhoods of the different clusters, there is no way to get the exact values for the transition probabilities. It is however possible to approach them by evaluating the a posteriori transition probabilities after the EM algorithm and after each step of the ICM algorithm. This method is described in Algorithm (1).

We note $\delta_{n,m}$ the Kronecker delta function the value of which is 1 if n and m are identical and 0 otherwise. We note C the set of all the oriented binary cliques of the considered dataset (since the Markov Random Field Model is usually non-oriented, all links are represented twice in C to cover both directions of each dependency). The a posteriori transition probabilities $a_{i,j}$ are evaluated by browsing all the data and counting the number of transitions from variables whose state is i to variables in the state j, divided by the number of all links starting from variables whose state is i:

$$a_{i,j} = \frac{Card(c \in C | c = (i,j))}{Card(c \in C | c = (i, \#))} \tag{7}$$

Algorithm 1. EM+ICM with transition matrix update

Initialize S and Θ with the EM algorithm
Initialize A using Equation (7)
while *the algorithm has not converged* **do**
  for *each* $x \in X$ **do**
  | Minimize $U_{HMRF}(s, x)$ as defined in Equation (6)
  end
  Update A from the new distribution S using Equation (7)
end

The idea behind this computation is that the ICM algorithm is already making an approximation by considering that the states evaluated in its previous iterations can be used to evaluate the potential neighborhood energy in the current iteration. We are using this same approximation to approach the probability of a given transition between two states.

In our model each transition probability is computed with the goal of detecting statistically unlikely labels among the data depending on their neighborhood. While the classical energy formula penalty may give erratic results depending on both the picture and the penalty value β, our energy model proposes a more flexible penalty which is the result of a less coarse, but still highly improvable vision of how to model neighborhood relations in a Markov random field model.

The complexity of the regular HMRF-EM version is $O(i \times k \times d \times n \times |\bar{V}|)$. Since we have to browse the data twice instead of only once in order to update A with our proposed algorithm, we can expect that the optimization process will be up to twice slower. The complexity for our algorithm is $O(i \times (k \times d \times n \times |\bar{V}| + n \times |\bar{V}|))$, which is equivalent to the complexity of the original ICM algorithm.

3   Experiments

In our first experiment, we wanted to compare the performance of our energy model with the other models in term of raw segmentation results and cluster characteristics. To do so, we have been testing the HMRF-EM frameworks with three different energy models: An ICM algorithm with a basic energy formula based on the Potts model [7], see Equation (8), an HMRF-EM algorithm with one of the most commonly used energy formula, see Equation (9) [4,8], and the HMRF-EM framework with our Energy model, see Equation (6).

$$U_{POTTS}(s, x) = (1 - \delta_{s_x, s_x^{t-1}}) + \beta \sum_{v \in V_x} (1 - \delta_{s_v, s}) \qquad (8)$$

$$U_{MRF}(s, x) = (x - \mu_s)^T \Sigma_s^{-1} (x - \mu_s) + log(\sqrt{|\Sigma_s|(2\pi)^d}) + \beta \sum_{v \in V_x} (1 - \delta_{s_v, s}) \qquad (9)$$

Fig. 2. From left to right : Original picture 481×321 pixels, 3-clusters segmentation using energy formula (9) and $\beta = 0.66$, 3-clusters segmentation using our energy formula, 3-clusters segmentation using energy formula (8) and $\beta = 1.0$.

The results in Figure (2) and Table (1) show that our model achieves decent segmentation results that are similar to what can be done with other energy models using a nearly optimal value for β.

Furthermore, our method seem to be effective to agglomerate pixels in homogeneous areas while conserving a low internal variance for the clusters which tends to prove that the resulting areas are meaningful.

Table 1. Purity of the expected clusters and average internal variance (3 clusters)

	Cluster Purity	Internal Variance
U_{POTTS}, $\beta = 2/3$	73.42%	6.10
U_{POTTS}, $\beta = 1$	73.38%	6.13
U_{MRF}, $\beta = 2/3$	69.37%	5.83
U_{MRF}, $\beta = 2/3$	69.36%	5.84
Our U_{HMRF}	73.46%	5.86

Similar results have been found on another experiment on a satellite image, see Figure (3) (2014 Cnes/Spot Image, DigitalGlobe, Google), thus confirming that our approach is competitive when compared to what has already been done in this area.

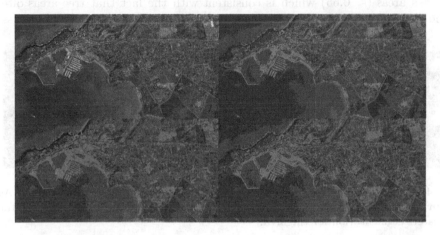

Fig. 3. From left to right, and from top to bottom : The original image 1131 × 575 pixels, the result of our algorithm, the result using energy equation (8) and $\beta = 1$, the result using energy equation (9) and $\beta = 1$

As can be seen on a color and full scale version of Figure (3), our energy model achieves a decent segmentation with a fair amount of easily visible elements such as roads, rocky areas and some buildings. On the other hand the classical HMRF-EM algorithm using Equation (9) fails to aggregate neighbor pixels from common elements despite a relatively high value for β (the same phenomenon can be observed in Figure (2)), while the algorithm based on the Potts model using Equation (8) tends to give too coarse results. This second experiment emphasizes again that finding a right value for β is a difficult task.

Our third experiment was on a satellite image from a more difficult dataset. In this dataset from [9], a very high resolution image from the city of Strasbourg (France) is described in 187.058 unlabeled segments having 27 numerical attributes either geometrical or radiometrical. Another particularity of this dataset is that the segments have irregular neighborhoods: each segment can have one to fifteen neighbors. The neighborhood irregular configuration of this dataset makes it impractical to find an optimal value for the constant β and the segmentation results using conventional HMRF-EM energy functions (8) and (9) were quite poor.

An extract of a 9-clusters segmentation on this dataset is available in Figure (4). The result shows that various elements of the city have been correctly detected such as sport areas, the river, and some buildings. On a full scale version of this image, we can see that some streets and houses have also been correctly labeled.

Furthermore, this experiment has also confirmed that our matrix representation for the neighborhood made sense, and that at a segment level it was possible to interpret some of the matrix elements (whereas it was difficult at a pixel level). For instance, on this dataset the matrix featured a low transition probability between building areas and water areas which is consistent with the satellite image. We also had an important transition probability from tree areas to grass areas (≈ 0.65) which is consistent with the fact that tree areas often come by patches instead of compact blocks and are often surrounded by grass.

Fig. 4. On the left: One of the original source image, ©CNES2012, Distribution Astrium Services / Spot Image S.A., France, All rights reserved. On the right: the resulting 9-clusters segmentation for the area.

In a last experiment, we have compared the computation times of the 3 energy models introduced in this article for 3 pictures of different sizes, see Table (2).

The results have confirmed what we had already hinted while describing the complexity of our modified HMRF-EM algorithm. Our version of the algorithm is slower, and it was to be expected because of the time needed to update the transition matrix after each iteration. Nevertheless, it is important to emphasize that it is still faster to run our version of HMRF-EM rather than trying several β values, or running an optimization algorithm for β, with another energy formula.

Table 2. Computation times in ms, Intel Core I5-3210M, 2.5GHz

	$404 \times 476px$ 4 clusters	$380 \times 270px$ 4 clusters	$1131 \times 575px$ 4 clusters	$1131 \times 575px$ 8 clusters
U_{Potts} (8)	4451ms	2453ms	13370ms	26920ms
U_{MRF} (9)	4933ms	2717ms	14771ms	29433ms
Our U_{HMRF} (6)	6632ms	3418ms	20810ms	42736ms

4 Conclusion and Future Work

We have proposed an improvement to the energy model of the Hidden Markov Random Field Model applied to image segmentation. In our model, the neighborhood energy formula is based on an approached transition matrix rather than an empirical penalty parameter. Our preliminary experiments have shown our model to give competitive results compared with other models based on more classical energy functions who rely on finding an optimal penalty value for the neighborhood energy.

Furthermore, while our contribution does not bring any significant improvement on the quality of the results for image segmentation, our neighborhood energy model using a transition matrix gives the opportunity to have a semantic rich representation of the interactions between the clusters.

In our future works, we will focus on using the information collected in these transition matrices with the goal of proposing collaborative frameworks such as the collaborative segmentation of similar pictures using several HMRF-based algorithms sharing their prototypes and transition matrices, or the segmentation of a sequence of pictures using an HMRF-HMM hybrid framework.

Acknowledgements. This work has been supported by the ANR Project COCLICO, ANR-12-MONU-0001.

References

1. Hernández-Gracidas, C., Sucar, L.E.: Markov Random Fields and Spatial Information to Improve Automatic Image Annotation. In: Mery, D., Rueda, L. (eds.) PSIVT 2007. LNCS, vol. 4872, pp. 879–892. Springer, Heidelberg (2007)
2. Zhang, L., Ji, Q.: Image segmentation with a unified graphical model. IEEE Transactions on Pattern Analysis and Machine Intelligence 32(8), 1406–1425 (2010)
3. Roth, S., Black, M.J.: Fields of experts, Markov Random Fields for Vision and Image Processing, pp. 297–310. MIT Press (2011)
4. Zhang, Y., Brady, M., Smith, S.: Segmentation of brain MR images through a hidden markov random field model and the expectation-maximization algorithm. IEEE Transactions on Medical Imaging 20(1), 45–57 (2001)
5. Dempster, A.P., Laird, N.M., Rubin, D.: Maximum Likelihood from Incomplete Data via the EM Algorithm. Journal of the Royal Statistical Society. Series B (Methodological) 39(1), 1–38 (1977)
6. Besag, J.: On the Statistical Analysis of Dirty Pictures. Journal of the Royal Statistical Society. Series B (Methodological) 48(3), 259–302 (1986)
7. Weinman, J.: A Brief Introduction to ICM (2008), http://www.cs.grinnell.edu/ weinman/courses/CSC213/2008F/labs/ 11-threads-icm.pdf
8. Kato, Z., Berthod, M., Zerubia, J.: A Hierarchical Markov Random Field Model and Multi-temperature Annealing for parallel Image Classification. Graphical Models and Image Processing 58(1), 18–37 (1996)
9. Rougier, S., Puissant, A.: Improvements of urban vegetation segmentation and classification using multi-temporal Pleiades images. In: 5th International Conference on Geographic Object-Based Image Analysis, Thessaloniki, Greece, p. 6 (2014)

Online Nonlinear Granger Causality Detection by Quantized Kernel Least Mean Square*

Hong Ji[1], Badong Chen[1], Zejian Yuan[1], Nanning Zheng[1],
Andreas Keil[2], and Jose C. Príncipe[2]

[1] The Institute of Artificial Intelligence and Robotics,
Xian Jiaotong University, 28 Xianning West Road, Xian 710049, China
[2] Center for the Study of Emotion and Attention,
University of Florida, Gainesville, Florida 32611, USA
{itsjihong,yzejian}@gmail.com
{chenbd,nnzheng}@mail.xjtu.edu.cn
{akeil,principe}@cnel.ufl.edu

Abstract. Identifying causal relations among simultaneously acquired signals is an important challenging task in time series analysis. The original definition of Granger causality was based on linear models, its application to nonlinear systems may not be appropriate. We consider an extension of Granger causality to nonlinear bivariate time series with the universal approximation capacity in reproducing kernel Hilbert space (RKHS) while preserving the conceptual simplicity of the linear model. In particular, we propose a computationally simple online measure by means of quantized kernel least mean square (QKLMS) to capture instantaneous causal relationships.

Keywords: Granger causality, kernel methods, quantized kernel least mean square(QKLMS), nonlinear time series.

1 Introduction

The problem of quantifying causal connectivity among simultaneously acquired time series has received considerable attention in the recent years due to its growing applicability in economy [1], neuroscience [2,3,4], medical and clinical science [5], and many others. One approach to evaluate causal relations between two time series is to examine if one series is better predicted by adding knowledge from the other. This was originally proposed by Wiener [6] and later formalized by Granger in the context of linear regression models of stochastic processes [1]. In particular, if the prediction error of the first time series is reduced by incorporating measurements from the second time series, then the second time series is said to have a causal inference on the first time series. By exchanging the roles of the two time series, the causal influence in the opposite direction can be addressed.

---

* This work was supported by NSFC grant No. 61372152.

C.K. Loo et al. (Eds.): ICONIP 2014, Part II, LNCS 8835, pp. 68–75, 2014.

As a technique to understand the directed connectivity of the underlying mechanisms, Granger causality has been well explored and construed in many different ways. Also, there is a freely available software toolbox incorporating these methods to facilitate its broadly application in neuroscience data analysis [7]. However, since Granger causality was formulated as linear regression, its application to nonlinear systems, such as brain signal that is highly nonlinear at many levels of description, may not be appropriate. There are several competing approaches to this problem. A simple solution [8] is to fit autoregressive coefficients to Taylor expansions of the data, but this method requires estimating a large number of parameters. Alternative approaches include the radial basis functions (RBFs) [9] and kernel methods. The kernel methods transform the data into a high dimensional reproducing kernel Hilbert space (RKHS) such that appropriate linear methods can be applied on the transformed data[10]. Most of these methods, however, assume the stationarity of the signals.

In this work, we propose a computationally simple online kernel method for causality detection, called the twin quantized kernel least mean square (twin-QKLMS) , which is able to capture the causal relations between nonlinear and non-stationary time series.

2   Granger Causality

2.1   Linear Modeling

Linear Granger causality is defined based on vector autoregressive model [1]. Let $X \equiv \{\overline{x_k}\}_{k=1,\dots,N}$ and $Y \equiv \{\overline{y_k}\}_{k=1,\dots,N}$ be two time series of N simultaneously measured quantities. Usually the stationarity of the time series is required. For $i = 1$ to M (where $M = N - m$, m being the order of the model), we denote $x_i = \overline{x_{i+m}}$, $y_i = \overline{y_{i+m}}$, $\mathbf{x}(i) = (\overline{x_{i+m-1}}, \overline{x_{i+m-2}}, \dots, \overline{x_i})$ and $\mathbf{y}(i) = (\overline{y_{i+m-1}}, \overline{y_{i+m-2}}, \dots, \overline{y_i})$ and treat these quantities as M realizations of the stochastic variables $(x, y, \mathbf{x}, \mathbf{y})$. The following model is then considered:

$$x = \mathbf{w}_1 \cdot \mathbf{x} + \xi_x$$
$$y = \mathbf{w}_2 \cdot \mathbf{y} + \xi_y \tag{1}$$

Here $\{\mathbf{w}\}$ being m-dimensional real vectors to be estimated from data, ξ_x and ξ_y being the residuals (prediction errors) for each time series when predicted solely based on the knowledge of its own past values. We denote their variance as $\epsilon_x = var(\xi_x)$ and $\epsilon_y = var(\xi_y)$ which are equal to the mean square prediction errors since zero mean has been guaranteed by pre-processing. The temporal dynamics of the two time series can be described by a bivariate autoregressive model:

$$x = \mathbf{w}_{11} \cdot \mathbf{x} + \mathbf{w}_{12} \cdot \mathbf{y} + \xi_{x|Y}$$
$$y = \mathbf{w}_{21} \cdot \mathbf{x} + \mathbf{w}_{21} \cdot \mathbf{y} + \xi_{y|X} \tag{2}$$

Similarly, we define $\epsilon_{x|y} = var(\xi_{x|Y})$ and $\epsilon_{y|x} = var(\xi_{y|X})$. If the prediction of x improves by incorporating the past values of series Y, that $\epsilon_{x|y}$ is smaller than

ϵ_x, then Y has a causal influence on X. Analogously, if $\epsilon_{y|x}$ is smaller than ϵ_y, then X has a causal influence on Y. The magnitude of this interaction can be measured by the log ratio of the prediction error variances:

$$F_{Y \to X} = ln \frac{\epsilon_x}{\epsilon_{x|y}}$$
$$F_{X \to Y} = ln \frac{\epsilon_y}{\epsilon_{y|x}} \tag{3}$$

The maximum of both terms

$$F_{XY} = max\{F_{Y \to X}, F_{X \to Y}\} \tag{4}$$

represents a simple measure for the strength of directional and/or bi-directional interaction.

2.2 Nonlinear Modeling in RKHS

The mapping from input to feature space is induced by a Mercer kernel which is a continuous, symmetric, and positive definite function $\kappa : \mathbb{X} \times \mathbb{X} \to \mathbb{R}, \mathbb{X} \subseteq \mathbb{R}^m$ [12,13]. The Gaussian kernel is widely used for its proved universal approximation property for any continuous function [14].

$$\kappa(\mathbf{x}, \mathbf{x}') = \exp(\frac{-\|\mathbf{x} - \mathbf{x}'\|^2}{2\sigma^2}) \tag{5}$$

where $\sigma > 0$ is the kernel size (or kernel bandwidth). According to the Mercer theorem [11,12], any Mercer kernel $\kappa(\mathbf{x}, \mathbf{x}')$ induces a mapping φ such that the inner product between the transformed input data (feature vectors) satisfies $\langle \varphi(\mathbf{x}), \varphi(\mathbf{x}') \rangle = \kappa(\mathbf{x}, \mathbf{x}')$.

We now construct the autoregressive model and bivariate model in transformed feature space.

$$x = \boldsymbol{\Omega_1} \cdot \varphi(\mathbf{x}) + \xi_x$$
$$y = \boldsymbol{\Omega_2} \cdot \varphi(\mathbf{y}) + \xi_y \tag{6}$$

with the corresponding prediction error variance ϵ_x and ϵ_y.

$$x = \boldsymbol{\Omega_{11}} \cdot \varphi(\mathbf{x}) + \boldsymbol{\Omega_{12}} \cdot \psi(\mathbf{y}) + \xi_{x|Y}$$
$$y = \boldsymbol{\Omega_{21}} \cdot \varphi(\mathbf{y}) + \boldsymbol{\Omega_{22}} \cdot \psi(\mathbf{x}) + \xi_{y|X} \tag{7}$$

where $\{\boldsymbol{\Omega}\}$ are the weight vectors in feature space (infinite dimensional for the Gaussian kernel case). The prediction errors to minimize are defined as:

$$\epsilon_{x|y} = \frac{1}{M} \sum_{i=1}^{M} [x_i - \boldsymbol{\Omega_{11}} \cdot \varphi(\mathbf{x}(i)) - \boldsymbol{\Omega_{12}} \cdot \psi(\mathbf{y}(i))]^2$$

$$\epsilon_{y|x} = \frac{1}{M} \sum_{i=1}^{M} [y_i - \boldsymbol{\Omega_{21}} \cdot \varphi(\mathbf{y}(i)) - \boldsymbol{\Omega_{22}} \cdot \psi(\mathbf{x}(i))]^2 \tag{8}$$

By using the kernel trick, we can efficiently compute the inner product output by kernel evaluation without knowing the exact form of mapping. In the following we take the bivariate model to predict x as an example, the prediction of y can be derived analogously:

$$f(\mathbf{x}) = \mathbf{\Omega_{11}} \cdot \boldsymbol{\varphi}(\mathbf{x}) = \sum_{i=1}^{l} \alpha_i \boldsymbol{\varphi}(\mathbf{c}_i)^T \boldsymbol{\varphi}(\mathbf{x}) = \sum_{p=1}^{l} \alpha_i \kappa(\mathbf{c}_i, \mathbf{x})$$

$$g(\mathbf{y}) = \mathbf{\Omega_{12}} \cdot \boldsymbol{\varphi}(\mathbf{y}) = \sum_{i=1}^{s} \beta_i \boldsymbol{\varphi}(\mathbf{c}_i')^T \boldsymbol{\varphi}(\mathbf{y}) = \sum_{i=1}^{l} \beta_i \kappa(\mathbf{c}_i', \mathbf{y})$$

$$(9)$$

where $\{\mathbf{c}_i, \alpha_i\}_{i=1}^{l}$ and $\{\mathbf{c}_i', \beta_i\}_{i=1}^{s}$ are the parameters to learn. Note we assume that x is the sum of a term depending solely on $\mathbf{x}$ and a term depending solely on $\mathbf{y}$ instead of the general bivariate model which are depending on the appending vector $(\mathbf{x} \ \mathbf{y})$:

$$x = f(\mathbf{x}) + g(\mathbf{y}) \tag{10}$$

It has been proposed that any prediction scheme providing a nonlinear extension of Granger causality should satisfy the following (P1) property [9]: *if* $\mathbf{y}$ *is statistically independent of* $\mathbf{x}$ *and* x, *then* $\epsilon_x = \epsilon_{x|y}$; *if* $\mathbf{x}$ *is statistically independent of* $\mathbf{y}$ *and* y, *then* $\epsilon_y = \epsilon_{y|x}$; Let us suppose $\mathbf{y}$ is statistically independent of $\mathbf{x}$ and x. Then $\boldsymbol{\varphi}(\mathbf{x})$ is uncorrelated with x and with $\boldsymbol{\psi}(\mathbf{y})$. It follows that

$$\begin{aligned}\epsilon_{x|y} &= var[x - \mathbf{\Omega_{11}} \cdot \boldsymbol{\varphi}(\mathbf{x}) - \mathbf{\Omega_{12}} \cdot \boldsymbol{\psi}(\mathbf{y})] \\ &= var[x - \mathbf{\Omega_{11}} \cdot \boldsymbol{\varphi}(\mathbf{x})] + var[\mathbf{\Omega_{12}} \cdot \boldsymbol{\psi}(\mathbf{y})]\end{aligned} \tag{11}$$

To minimize $\epsilon_{x|y}$ it follows that $\mathbf{\Omega_{12}} = 0$ which satisfy P1 property.

2.3   Twin-QKLMS Causality Detector

The QKLMS is one of the most simple and efficient online kernel adaptive filter algorithm. It natrally creates a growing radial-basis function network, learning network topology adaptively. It has been verified that a sufficient condition for mean square convergence and a bounded theoretical value of the steady-state excess mean square error [11]. Inheriting from KLMS, it does not need explicit regularization to obtain solutions that generalize appropriately [15]. In this section we propose a novel online detector for causality analysis and employ a twin QKLMS to the bivariate nonlinear model described in previous section. We name the model as "twin-QKLMS" to emphasize that two QKLMS filters work in parallel in an online mode. The reason to choose QKLMS comes from the key properties highlight below:

- QKLMS uses quantization to compress the input space so as to constrain the network size growth. Different from conventional sparsification methods normally discarding samples simply, the "redundant" data are used to locally update the coefficient of the closest center, which help to achieve better accuracy and a more compact network.

– The codebook is trained directly from online samples and is adaptively growing, unlike RBF constrains the network size to a fixed size (by cluster to n prototypes).

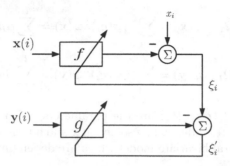

Fig. 1. Twin-QKLMS model as an estimator of mapping $x = f(\mathbf{x}) + g(\mathbf{y})$, filter $g(\cdot)$ is employed to predict the residual of $x - f(\mathbf{x})$ incorporating the knowledge from input y

The twin-QKLMS models a dual filtering structure in Fig.1. Denote $\varphi(i) = \varphi(\boldsymbol{u}(i))$, quantizing the input vector in the filter update equation. Using the basics of QKLMS, we propose the twin-QKLMS algorithm below.

$$
\begin{cases}
f_0 = 0 \\
g_0 = 0 \\
\xi_i = x_i - f_{i-1}(\mathbf{x}(i)) \\
f_i = f_{i-1} + \lambda \xi_i \kappa(\boldsymbol{Q}[\mathbf{x}(i)], \cdot) \\
\xi_i' = \xi_i - g_{i-1}(\mathbf{y}(i)) \\
g_i = g_{i-1} + \eta \xi_i' \kappa(\boldsymbol{Q}[\mathbf{y}(i)], \cdot)
\end{cases}
\tag{12}
$$

where ξ_i is the prediction error at iteration i by predicting x_i with filter f_{i-1} enclosed input $\boldsymbol{x}(i)$, ξ_i' is the prediction error that predict residual ξ_i with filter g_{i-1} that incorporate the knowledge of the other series $\boldsymbol{y}(i)$, λ and η are the step size, $\boldsymbol{Q}[.]$ denotes the quantization operator. f_i is the composition of $\boldsymbol{\Omega}_f$ and φ, that is $f_i = \boldsymbol{\Omega}_{11}(i)^T \varphi(\cdot)$, g_i is the composition of $\boldsymbol{\Omega}_{12}$ and ψ, that is $g_i = \boldsymbol{\Omega}_{12}(i)^T \psi(\cdot)$. They're calculated with kernel evaluation in original input space.

Notice the knowledge from the possible causal series is used to predict the residual as a measure to count its improvement to prediction power. The proposed twin-QKLMS framework is described in Algorithm 1.

3 Experiments

As a real example, we consider the physiological bivariate data (instantaneously acquired breath rate and heart rate) of a sleep human suffering from sleep apnea. The data can be downloaded: http://physionet.incor.usp.br/physiobank/

Algorithm 1. Twin-QKLMS Algorithm

Input: $\{\mathbf{x}(i) \in \mathbb{X} \subseteq \mathbb{R}^m, \mathbf{y}(i) \in \mathbb{Y} \subseteq \mathbb{R}^m, x_i \in \mathbb{R}\}$.

Initialization: Choose step size $\lambda, \eta > 0$, kernel width $\sigma_f, \sigma_g > 0$, the quantization size $\varepsilon_{\mathbb{X}}$, $\varepsilon_{\mathbb{Y}} \geq 0$
   and initialize the codebook (center set) $C_f(1) = \{\mathbf{x}(1)\}$, $C_g(1) = \{\mathbf{y}(1)\}$ and coefficient vector:
   $\alpha(1) = [\lambda x_1], \beta(1) = [0]$.

1: **while** $\{\mathbf{x}(i), \mathbf{y}(i), x_i\}$ $(i > 1)$ available **do**
2:   Compute the output of the filter f and g:

$$f_{i-1} = \sum_{j=1}^{size(C_f(i-1))} \alpha_j(i-1)\kappa(C_f(i-1), \mathbf{x}(i))$$

$$g_{i-1} = \sum_{j=1}^{size(C_g(i-1))} \beta_j(i-1)\kappa(C_g(i-1), \mathbf{y}(i))$$

3:   Compute the error: $\xi_i = x_i - f_{i-1}, \xi_i' = \xi_i - g_{i-1}$
4:   Compute the distance between $\mathbf{x}(i)$ and $C_f(i-1)$ and distance between $\mathbf{y}(i)$ and $C_g(i-1)$:
   $$dis(\mathbf{x}(i), C_f(i-1)) = \min_{1 \leq j \leq size(C_f(i-1))} \|\mathbf{x}(i) - C_f(i-1)\|$$
   $$dis(\mathbf{y}(i), C_g(i-1)) = \min_{1 \leq j \leq size(C_g(i-1))} \|\mathbf{y}(i) - C_g(i-1)\|$$

5:   **if** $dis(\mathbf{x}(i), C_f(i-1)) \leq \varepsilon_{\mathbb{X}}$ **then**
6:     Keep the codebook unchanged: $C_f(i) = C_f(i-1)$, and quantize $\mathbf{x}(i)$ to the closest center
     through updating the coefficient of that center $\alpha_{j*}(i) = \alpha_{j*}(i-1) + \lambda e_i$,
     where $j^* = \arg\min_{1 \leq j \leq size(C_f(i-1))} \|\mathbf{x}(i) - C_f(i-1)\|$
7:   **else**
8:     Assign a new center and corresponding new coefficient: $C_f(i) = \{C_f(i-1), \mathbf{x}(i)\}$ and
     $\alpha(i) = [\alpha(i-1), \lambda e_i]$:
9:   **end if**
10:   Similarly, repeat step 5-9 to update the codebook C_g.
11: **end while**

`database/santa-fe/` (data set B). Figure 2 clearly shows that bursts of the patient breath and cyclical fluctuations of heart rate are interdependent. Both time series have been normalized to be zero mean and unit variance in preprocessing.

We set the quantization size $\varepsilon_{\mathbb{X}}=1$ and $\varepsilon_{\mathbb{Y}}=0.5$ to constrain the network size in a reasonable range. The other parameters are adjusted empirically to minimize the training error:

$$\epsilon_{n+1} = \frac{n}{n+1}\epsilon_n + \frac{1}{(n+1)}\xi_{n+1}^T \xi_{n+1} \tag{13}$$

To detect the causal relation from breath to heart rate we set $\lambda=0.2$, $\eta=0.01$, $\sigma_f=1.6$, $\sigma_g=0.8$; for the reverse case we take $\lambda=0.02$, $\eta=0.02$, $\sigma_f=1.8$, $\sigma_g=1.1$. We also evaluate the influence of different m on causality detection and the results are shown in Figure 3. We may observe: 1) with proper m values, the causal influence of heart rate on breath is, obviously, stronger than the reverse and this coincide with the results in [9]; 2) if m is too small(e.g. $m = 2$), the detected causal relationship is controversial to the expectation since long memory dynamic structured can not be modelled; 3) the causality measures change over time, and track the non-stationary dynamical behavior of the time series well.

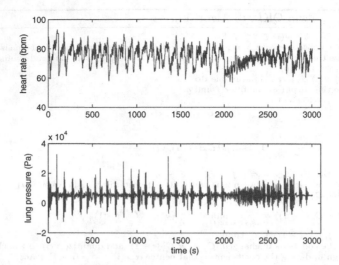

Fig. 2. Bivariate time series of the heart (upper) and breath signal (lower) of a patient suffering sleep apnea (Samples 2350-5350 of the data set are highly non-stationary and include abrupt changes in the last 1000 points). Data sampled at 2Hz.

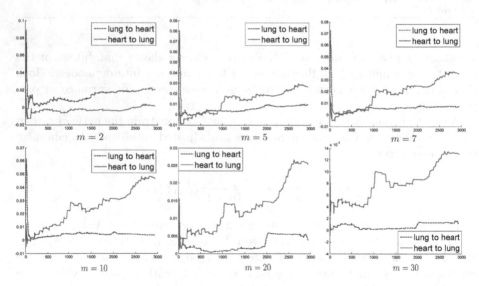

Fig. 3. Causality detection results by twin-QKLMS with different m values

4   Conclusions

We develop a computationally simple online algorithm with kernel method, called twin-QKLMS, to capture instantaneous causal relationships, which can be applied especially to nonlinear and non-stationary signals. There are several key problems that need to be addressed in the future: a) the current QKLMS algorithm does not discard old centers and hence the network size is growing especially in non-stationary situation, and this leads to increase computational burden; b) how to optimize the free parameters in twin-QKLMS, since they have significant influence on the causality detection result; c) more experiments need to be done to explore its applicability such as the causality analysis in neuroscience.

References

1. Granger, C.W.: Investigating causal relations by econometric models and cross-spectral methods. Econometrica: Journal of the Econometric Society 147, 424–438 (1969)
2. Rodriguez, E., George, N., Lachaux, J.P.: Perception's shadow: long-distance synchronization of human brain activity. Nature 397, 430–433 (1999)
3. Cadotte, A.J., DeMarse, T.B., He, P.: Causal measures of structure and plasticity in simulated and living neural networks. PloS one 3, e3355 (2008)
4. Keil, A., Sabatinelli, D., Ding, M.Z.: Re-entrant projections modulate visual cortex in affective perception: Evidence from Granger causality analysis. Human Brain Mapping 30, 532–540 (2009)
5. Akselrod, S., Gordon, D., Madwed, J.B.: Hemodynamic regulation: investigation by spectral analysis. American Journal of Physiology-Heart and Circulatory Physiology 249, H867–H875 (1985)
6. Wiener, N.: Modern mathematics for engineers. McGraw-Hill, New York (1956)
7. Seth, A.K.: A MATLAB toolbox for Granger causal connectivity analysis. Journal of Neuroscience Methods 186, 262–273 (2010)
8. Seth, A.K.: Measuring autonomy and emergence via Granger causality. Artificial Life 16, 179–196 (2010)
9. Ancona, N., Marinazzo, D., Stramaglia, S.: Radial basis function approach to nonlinear Granger causality of time series. Physical Review E 70, 056221 (2004)
10. Marinazzo, D., Liao, W., Chen, H.F.: Nonlinear connectivity by Granger causality. Neuroimage 58, 330–338 (2011)
11. Chen, B.D., Zhao, S.L., Zhu, P.P.: Quantized kernel least mean square algorithm. IEEE Transactions on Neural Networks and Learning Systems 23, 22–31 (2012)
12. Aronszajn, N.: Theory of reproducing kernels. Transactions of the American Mathematical Society, 337–404 (1950)
13. Burges, C.J.: A tutorial on support vector machines for pattern recognition. Data Mining and Knowledge Discovery 2, 121–167 (1998)
14. Steinwart, I.: On the influence of the kernel on the consistency of support vector machines. The Journal of Machine Learning Research 2, 67–93 (2001)
15. Liu, W.F., Pokharel, P., Príncipe, J.C.: The kernel least mean square algorithm. IEEE Transactions on Signal Processing 56, 543–554 (2008)

A Computational Model of Anti-Bayesian Sensory Integration in the Size-Weight Illusion

Yuki Ueyama

Research Institute of National Rehabilitation Center for Persons with Disabilities
4-1 Namiki, Tokorozawa, Saitama 359-8555, Japan
ueyama-yuki@rehab.go.jp

Abstract. We propose a computational model for anti-Bayesian sensory integration of human behavioral actions and perception in the size–weight illusion (SWI). The SWI refers to the fact that people judge the smaller of two equally weighted objects to heavier when lifted. Many aspects of human perceptual and motor behavior can be modeled with Bayesian statistics. However, the SWI cannot be explained on the basis of Bayesian integration, and the nervous system is thought to use two entirely different mechanisms to integrate prior expectations with current sensory information about object weight. Our proposed model is defined as a state estimator, combining a Kalman filter and a H^∞ filter. As a result, the model not only predicted the anti-Bayesian estimation of the weight but also the Bayesian estimation of the motor behavior. Therefore, we hypothesize that the SWI is realized by a H^∞ filter and a Kalman filter.

Keywords: Weight perception, H^∞ filter, Kalman filter, Bayesian integration, Motor control.

1 Introduction

Recent studies have suggested that Bayes' law explains several key features of perception, motor actions, and motor learning [1]. However, it has been suggested that the Bayes' law does not explain the perceptual biases characterizing the size–weight illusion (SWI) [2]. The SWI occurs when a person underestimates the weight of a large object when compared to a smaller object of the same mass. It is also known that similar illusions occur with differences in material and color: metal containers feel lighter than wooden containers of the same size and mass, and darker objects feel lighter than brighter objects of the same size and mass. These illusions can all be described as contrasts with the expected weight. In recent studies, it was shown that the lifting force adapts quickly to the actual mass of objects, but the SWI remains [3]. Thus, the illusion cannot be explained by the manner of lifting, and must be due to some perceptual recall, based on prior expectations. The rescaling has been described as "anti-Bayesian" or sub-optimal, in that the central nervous system (CNS) integrates prior expectations with current proprioceptive information in a manner that emphasizes the unexpected information, rather than taking an average of all information, as in the Bayesian estimation [3]. Moreover, the CNS is thought to use two

C.K. Loo et al. (Eds.): ICONIP 2014, Part II, LNCS 8835, pp. 76–83, 2014.
© Springer International Publishing Switzerland 2014

entirely different mechanisms to integrate prior expectations with current sensory information about object weights.

In this paper, we propose a computational model of anti-Bayesian sensory integration for human behavioral action and perception in the SWI. The model is defined as an expansion of optimal feedback control theory (OFC) [4]. OFC is combined with an optimal feedback controller and the Kalman filter, integrating sensory information according to Bayes' law, and can predict variability in movement phenomena [5, 6]. However, it was assumed that the SWI was not predicted by OFC, because it is not dependent on Bayes' law, but is an anti-Bayesian process. Thus, we supposed that the anti-Bayesian process may be predicted by the H^∞ filter [7]. The H^∞ filter is based on an assumption of worst-case uncertainty characterizing motor and sensory noises, and has higher robustness and sensitivity than the Kalman filter. In our model, the perceived weight, estimated by the H^∞ filter, was integrated with the OFC. Moreover, we simulated linear dynamical systems of hand postural control with a weight estimation, such as a lifting task for a mass-object, and evaluated the performance of our model. Consequently, we hypothesize that weight perception in the SWI is realized by an anti-Bayesian estimation, such as the H^∞ filter, and the process is integrated with OFC, based on a Bayesian estimation for sensorimotor control.

2   Model

We modeled SWI experiments using a feedback control with state estimators, and compared their performance as a model for the SWI. To examine the models, we assumed three kinds of simulation for a motor analog of the SWI according to an experimental study [3].

In simulation 1, a subject was presented with a 0.3-kg-mass cube (Fig. 2a). The subject placed his/her elbow on the table with their palm horizontal and facing up. First, the subject maintained a fixed posture with the right hand while supporting a single cube on the palm. Then, the cube was lifted rapidly. The supporting hand was raised after the lift. Even with a normal compensatory and appropriate anticipatory postural adjustment, the palm of the supporting hand rises some distance after the cube has been lifted. An accurate estimate of the weight being unloaded helps to reduce hand displacement. The experiment was used to compare our proposed model with other models, and to evaluate the effects of prior expectations.

In simulation 2, the subject maintained a posture with unknown-weight cube pairs totaling 0.6 kg on the palm. Then only one cube was lifted under the prior expectation that the mass of the lifted cube was equal to 0.3 kg (Fig. 2b). We carried out this simulation to evaluate the effects of the cube-mass variation under the same *a priori* expectation. Then, we assessed the relationship between the weight miss-estimation and positional displacement (motor SWI). We used the relationship to remap the motor SWI to the mass.

In simulation 3, the subject maintained a fixed posture with two 0.3-kg cubes on the palm, and one cube was lifted up under varying prior expectations of the mass of the cube (0-0.6 kg; Fig. 2c). Each simulation was carried out 100 times, and used simple Euler integration with a 10-ms sampling time. When the cubes were lifted up, the perceived masses were forced to change to the prior expectations.

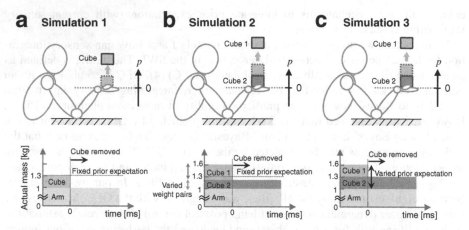

Fig. 1. Experimental setup for the size–weight illusion (SWI) assumed in this simulation study. Top panels illustrate the experiments. Bottom panels indicate the time profiles of the actual mass. (a) In simulation 1, a cube is lifted up while the subject maintains a fixed posture. (b) In simulation 2, the subject maintains the hand with two cubes of total weight 0.6 kg. Then, one of the unknown-mass cubes was removed, and the prior expectation of the cube weight was fixed at 0.3 kg. (c) In experiment 3, the subject maintains the hand position with two 0.3-kg cubes. Then, a 0.3-kg cube is removed from the palm under varying prior expectations.

2.1   Dynamics Model

We modeled the SWI tasks as a movement with one degree of freedom, such as multi-joint flexion and extension of the elbow and wrist joints to maintain the posture, as mass-point movements in the vertical axis, described by shifting the hand position, $p(t)$:

$$m\ddot{p}(t) = f(t) - b\dot{p}(t) - mg, \quad m = m^h + m^c, \tag{1}$$

where m^h and m^c are the hand and cube masses, respectively, and set equal to $m^h = 1.0$ [kg]. Additionally, g is the gravitational acceleration, set as 9.8 m/s². The combined action of all muscles is represented by the force $f(t)$ acting on the hand. The motor command $u(t)$ is transformed into a force $f(t)$ by adding control-dependent multiplicative noise and applying a second-order muscle-like low-pass filter of the form $\tau_1\tau_2\ddot{f}(t) + (\tau_1 + \tau_2)\dot{f}(t) + f(t) = (1 + \sigma_u\varepsilon(t))u(t)$, with time constants $\tau_1 = \tau_2 = 0.04$ [s]. The second-order muscle-like filter can be written as a pair of coupled first-order filters as $\dot{f}(t) = (a(t) - f(t))/\tau_2$, with $\dot{a}(t) = \{(1 + \varepsilon(t))u(t) - a(t)\}/\tau_1$, where $a(t)$ denotes a muscle state. The motor command $u(t)$ is disturbed by signal-dependent multiplicative noise that exists in the neural system, and plays an important role in motor planning [8]. The signal-dependent noise (SDN) is given by $\varepsilon(t) \sim N(0, \sigma_u^2)$, where σ_u is set to 0.4.

The dynamics model can be rewritten as a discrete-time system, using a state-space formulation by matrices $A(m_k) \in R^{5 \times 5}$, $B \in R^5$, and $G \in R^5$:

$$\mathbf{x}_{k+1} = A(m_k) \cdot \mathbf{x}_k + B(1 + \varepsilon_k)u_k + G \cdot n_k. \tag{2}$$

Here, we defined the physical state vector $\mathbf{p}_k = [p_k, \dot{p}_k, f_k, a_k]$, a state-space vector at time step k, $\mathbf{x}_k$ is represented as $x_k = [\mathbf{p}_k, m_k]^T$, where m_k is the perceptual weight representing the sum of hand and cube masses, and the perception is disturbed by the search noise $n_k \sim N(0, \sigma_n^2)$, as $m_{k+1} = m_k + n_k$, where σ_n is set equal to 10^{-2}.

In our model, the state variables cannot be observed fully. The sensory output $\mathbf{y}_k \in R^3$ is the position, velocity, and force disturbed by a sensory noise $\mathbf{v}_t \sim N(0, \boldsymbol{\sigma}_y \boldsymbol{\sigma}_y^T)$ with a feedback-delayed time step d, i.e., $\Delta \cdot d$ [ms], and given by

$$\mathbf{y}_k = [p_{k-d} \quad \dot{p}_{k-d} \quad f_{k-d}]^T + \mathbf{v}_k \quad (k \geq d) \quad \text{or} \quad \mathbf{y}_k = [p_0 \quad \dot{p}_0 \quad f_0]^T + \mathbf{v}_k \quad (k < d).$$

The diagonal matrix $\boldsymbol{\sigma}_y \in R^{3 \times 3}$ is defined by $\boldsymbol{\sigma}_y = diag(\sigma_p, \sigma_v, \sigma_f)$, where $\sigma_p = 0.02c$, $\sigma_v = 0.2c$, and $\sigma_f = c$ are, respectively, the standard deviations of the observed position, velocity and force, and c is a scaling parameter set equal to 0.4. Moreover, the sensory output is rewritten in a state space form:

$$\mathbf{y}_k = \begin{cases} C\mathbf{x}_{k-d} + \mathbf{v}_k & k \geq d \\ C\mathbf{x}_0 + \mathbf{v}_k & k < d \end{cases}, \text{ where } C = [\mathbf{I} \quad 0 \quad 0]. \tag{3}$$

Here, the state space equations could be extended to include past states to represent the system and output equations in the same state space. Thus, we defined the extended state vector $\mathbf{X}_k = [x_k{}^T, x_{k-1}{}^T, \dots, x_{k-d}{}^T]^T$, and matrices $\mathbf{A} \in R^{5(d+1) \times 5(d+1)}$, $\mathbf{B} \in R^{5(d+1)}$, $\mathbf{G} \in R^{5(d+1)}$, $\mathbf{C} \in R^{3 \times 5(d+1)}$, so the dynamics of the system can be rewritten as:

$$\begin{cases} \mathbf{X}_{k+1} = \mathbf{A}\mathbf{X}_k + \mathbf{B}(1 + \varepsilon_k)u_k + \mathbf{G}n_k \\ \mathbf{y}_k = \mathbf{C}\mathbf{X}_k + \mathbf{v}_k \end{cases}. \tag{4}$$

In all simulations, the initial state was $x_0 = [0, 0, m_0 g, m_0 g, m_0]^T$, and the feedback-delayed time step was set equal to $d = 3$, according to proprioceptive feedback delay (i.e., ~30 ms). Notably, m_0 was set equal to 1.3 kg in simulation 1, and 1.6 kg in simulations 2 and 3.

2.2   Controller Design

As in the linear quadratic Gaussian (LQG) problem (i.e., optimal feedback control with a state estimation problem solved by Kalman filtering), a solution of the control problem can be written in state feedback form: $u_k = -\mathbf{L}_k \hat{\mathbf{X}}_k$, where $\hat{\mathbf{X}}_k$ and $\mathbf{L}_k$ are the estimated state and feedback gains, respectively.

An optimal feedback controller also generates motor commands, thus forming state feedback. The feedback gain is computed to minimize the following cost function:

$$J(u) = \sum_{k=0}^{N-1} (\mathbf{X}_k{}^T \mathbf{Q} \mathbf{X}_k + u_k{}^2), \tag{5}$$

where $\mathbf{Q} = diag(w_p, w_v, w_f, 0, 0)$. Here, w_p, w_v, and w_f are the respective cost weights of the vertical hand position, velocity, and force, with the assigned values, $w_p = 10^6$, $w_v = 10^4$, and $w_f = 10$. The feedback gain is determined by solving the Ricatti Equation according to the dynamics programming approach. Then, we can set the feedback gain as a constant $\mathbf{L}_k \approx \mathbf{L}_0$, because the feedback gain was actually invariant, except before the terminal time of the simulation when the terminal time was set sufficiently long.

2.3   State Estimator Design

Similar to an actual biological system, our model could not observe the current state in real-time because of feedback delay. Then, we adapted state estimators to estimate the current state from sequences of past states and delayed feedback. Thus, the state vector was estimated from noisy observations using a state estimator expressed as a so-called Luenberger observer:

$$\hat{\mathbf{X}}_{k+1} = \mathbf{A}\hat{\mathbf{X}}_k + \mathbf{B}u_k + \mathbf{K}_k(\mathbf{y}_k - \mathbf{C}\hat{\mathbf{X}}_k), \tag{6}$$

where $\mathbf{K}_k$ is a time-varying filter gain matrix (i.e., a function of the uncertainty of the estimated state and the measurement noise). We determined the filter gain matrix using two approaches, the Kalman filter and the H^∞ filter.

Kalman Filter. We adapted a standard technique to calculate the Kalman gain, as follows:

$$\mathbf{K}_k = \mathbf{P}'_k\mathbf{C}^T(\mathbf{C}\mathbf{P}'_k\mathbf{C}^T + \mathbf{R})^{-1}, \tag{7}$$

$$\mathbf{P}'_k = \mathbf{A}\mathbf{P}_k\mathbf{A}^T + \mathbf{U}_k, \quad \mathbf{U}_k = (\mathbf{B}\sigma_u u_k)(\mathbf{B}\sigma_u u_k)^T + (\mathbf{G}\sigma_n)(\mathbf{G}\sigma_n)^T, \quad \mathbf{R} = \sigma_y\sigma_y^T,$$

where $\mathbf{P}_k$ is the predicted accuracy of the state estimation, given by $\mathbf{P}_{k+1} = (\mathbf{I} - \mathbf{K}_k\mathbf{C})\mathbf{P}'_k$. The Kalman gain is computed concurrently at each time step in the simulation, starting with the initial condition, $\mathbf{P}_0 = 10^{-3} \times \mathbf{I}$.

H∞ Filter. The H^∞ filter can be used to estimate system states that cannot be observed directly. In this, it operates similarly to a Kalman filter. However, only H^∞ filters are robust in the presence of unpredictable noise sources. The H^∞ filter is related to the mini-max problem of minimizing the estimation error for a maximized disturbance.

The H^∞ filter is a mini-max process for optimizing estimation accuracy with a worst-case scenario, assuming disturbance and the noise as deterministic signals. The filter gain is derived from the following:

$$\mathbf{K}_k = \mathbf{A}\mathbf{Z}_k\left(\mathbf{I} - \gamma^{-2}\mathbf{Z}_k + \mathbf{C}^T\mathbf{R}^{-1}\mathbf{C}\mathbf{Z}_k\right)^{-1}\mathbf{C}^T\mathbf{R}^{-1}. \tag{8}$$

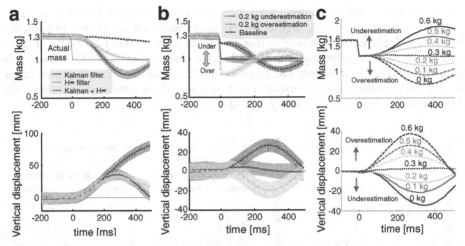

Fig. 2. Perceived weight (top row) and hand position (bottom row) profiles (mean ± standard deviation [SD]). The mass is changed at the onset time (i.e., 0 ms). (a) Comparison of the performances of the Kalman filter, H^∞ filter, and our proposed model in simulation 1. (b) Effects of different prior expectations on the proposed model in simulation 1. (c) Different masses under the same prior expectation in simulation 2. Solid and dotted lines indicate different cases in which the masses of the removed cube are 0 to 0.6 kg.

where γ is a scalar parameter representing the level of disturbance attenuation, set equal to $\gamma = 10^2$ in this study. Then, $\mathbf{Z}_k$ is given by $\mathbf{Z}_{k+1} = \mathbf{A}\mathbf{Z}_k \left(\mathbf{I} - \gamma^{-2}\mathbf{Z}_k + \mathbf{C}^T\mathbf{R}^{-1}\mathbf{C}\mathbf{Z}_k\right)^{-1} \mathbf{A}^T + \mathbf{U}_k$. However, the following condition must hold at each time step k for the estimator above to be a solution to the problem: $\mathbf{Z}_k^{-1} - \gamma^{-2}\mathbf{I} + \mathbf{C}^T\mathbf{R}^{-1}\mathbf{C} > 0$. The filter gain is computed concurrently at each time step in the simulation starting with the initial condition $\mathbf{Z}_0 = 10^{-3} \times \mathbf{I}$, similar to the Kalman filter.

Proposed Model. Our proposed model combines current states, estimated by the Kalman and H^∞ filters. In the model, the estimation process is divided into two components, Bayesian and anti-Bayesian estimations, realized with the Kalman and H^∞ filters, respectively. Here, the estimated states by the Kalman and H^∞ filters are represented, respectively, by the superscript 'K' and 'H' (i.e., $\hat{\mathbf{p}}_k^K$ and $\hat{\mathbf{p}}_k^H$). In our model, the Kalman and H^∞ filters estimate the states independently, and are combined to a state vector. The physical state variables and perceived weight are adapted from the Kalman and H^∞ filters, respectively. Thus, the current state is given by the combination, as $\hat{\mathbf{x}}_k \cong [\hat{\mathbf{p}}_k^K, \hat{m}_k^H]^T$. The combined state vector at each time step is used as the current estimated state.

3   Results

We compared the performances of the Kalman filter, the H^∞ filter, and the proposed model according to the results of experiment 1 (Fig. 2a). The Kalman filter adjusted the estimated mass slightly, and converged the actual mass at over 5 s. Furthermore, the vertical displacement was the largest. The H^{∞} filter could track the change in mass, and the vertical displacement was smaller than that of the Kalman filter because the H^{∞} filter is more sensitive to perturbations than the Kalman filter and ensures robustness for estimation of the miss-estimation. The proposed model tracked the changes in mass rapidly, and the estimation was undershot. The magnitude of the vertical displacement was almost equal to that of the H^{∞} filter (i.e., ~40 mm).

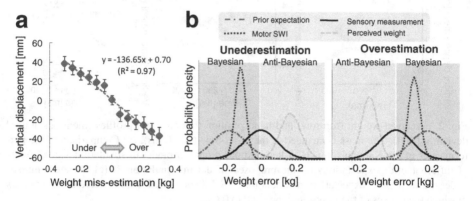

Fig. 3. Statistical specification of the model. (a) Relationship between the weight miss-estimations and vertical displacements (motor SWI) (mean ± SD). The data are from simulation 2. (b) Probability density of the SWI. In the left panel, prior expectations are lighter than the actual values (underestimation). In the right panel, prior expectations are heavier than the actual values (overestimation). The density functions were modeled with Gaussian functions.

We then evaluated the proposed model using different prior expectations (i.e., 0.2-kg underestimation and overestimation) for the same mass (Fig. 2b). The prior expectations were used to set forcibly the estimated mass at the onset time. The baseline, for which the prior expectation was the same as the actual mass, was flat in the estimated mass and vertical displacement after the onset time. The underestimation, for which the prior expectation was lighter than the actual mass, showed undershooting of the estimated mass, and the vertical height was increased. In contrast, overestimation, for which the prior expectation was heavier than the actual, showed opposite profiles in the mass estimation and vertical profiles. Moreover, in simulation 2, the proposed model also predicted opposite features between the underestimation and overestimation of the mass, the magnitudes of which were dependent on the estimation error (Fig. 2c).

From the results of simulation 2, we obtained the relationship between the weight miss-estimation and vertical displacement as the motor SWI (Fig. 3a). Values of the displacement were adapted to the maximal or minimal value of the profile. The trend could be approximated by a first-order linear regression equation ($R^2 = 0.97$). Then, we modeled the results of simulation 3 as the density function using a Gaussian function (Fig. 3b). The weight corresponding to the motor SWI was computed from the approximate equation given by simulation 2. Moreover, the variance of the sensory measurement was assumed to be due to the generated force. Thus, it was computed as $(\sigma_f^2 + b^2\sigma_u^2)/g^2$. As a result, the densities of the perceived weight were distributed opposite to the prior expectations across sensory measurements. In contrast, the mass density of the motor SWI was plotted between the prior expectations and sensory measurements. Thus, the distribution of motor behavior was thought to follow a Bayesian integration. However, the perceived weight distribution differed from the process, and was similar to an anti-Bayesian integration.

4   Conclusions

The SWI cannot be explained on the basis of Bayesian integration, and the brain maintains independent representations of object weight for sensorimotor control and perception [2]. In this study, we proposed a computational model that could explain the features of SWI by combining the Kalman and H^∞ filters. The Kalman filter estimated the physical state in the Bayesian integration, and the H^∞ filter estimated the perceived weight. The combined model could predict the perceived weight, and acted similarly to an anti-Bayesian integration in the SWI. Thus, we suggest that weight perception in the SWI is realized by the H^∞ filter, as an anti-Bayesian estimation, and the process is integrated with a Bayesian estimation for sensorimotor control.

References

1. Kording, K.P., Wolpert, D.M.: Bayesian integration in sensorimotor learning. Nature 427, 244–247 (2004)
2. Flanagan, J.R., Bowman, M.C., Johansson, R.S.: Control strategies in object manipulation tasks. Current Opinion in Neurobiology 16, 650–659 (2006)
3. Brayanov, J.B., Smith, M.A.: Bayesian and "Anti-Bayesian" biases in sensory integration for action and perception in the size-weight illusion. J. Neurophysiol. 103, 1518–1531 (2010)
4. Todorov, E., Jordan, M.I.: Optimal feedback control as a theory of motor coordination. Nat. Neurosci. 5, 1226–1235 (2002)
5. Izawa, J., Shadmehr, R.: On-Line Processing of Uncertain Information in Visuomotor Control. J. Neurosci. 28, 11360–11368 (2008)
6. Ueyama, Y., Miyashita, E.: Optimal Feedback Control for Predicting Dynamic Stiffness during Arm Movement. IEEE Trans. Ind. Electron. 61, 1044–1052 (2014)
7. Simon, D.: Optimal state estimation: Kalman, H infinity, and nonlinear approaches. John Wiley & Sons (2006)
8. Harris, C.M., Wolpert, D.M.: Signal-dependent noise determines motor planning. Nature 394, 780–784 (1998)

Unsupervised Dimensionality Reduction for Gaussian Mixture Model

Xi Yang, Kaizhu Huang, and Rui Zhang

Xi'an Jiaotong-Liverpool University,
SIP, Suzhou,215123, China
Xi.Yang07@studnet.xjtlu.edu.cn, {Kaizhu.Huang,Rui.Zhang02}@xjtlu.edu.cn

Abstract. Dimensionality reduction is a fundamental yet active research topic in pattern recognition and machine learning. On the other hand, Gaussian Mixture Model (GMM), a famous model, has been widely used in various applications, e.g., clustering and classification. For high-dimensional data, previous research usually performs dimensionality reduction first, and then inputs the reduced features to other available models, e.g., GMM. In particular, there are very few investigations or discussions on how dimensionality reduction could be interactively and systematically conducted together with the important GMM. In this paper, we study the problem how unsupervised dimensionality reduction could be performed together with GMM and if such joint learning could lead to improvement in comparison with the traditional unsupervised method. Specifically, we engage the Mixture of Factor Analyzers with the assumption that a common factor loading exist for all the components. Such setting exactly optimizes a dimensionality reduction together with the parameters of GMM. We compare the joint learning approach and the separate dimensionality reduction plus GMM method on both synthetic data and real data sets. Experimental results show that the joint learning significantly outperforms the comparison method in terms of three criteria for supervised learning.

1 Introduction

Dimensionality Reduction (DR) has been an important and fundamental research topic in pattern recognition and machine learning. Over the last fifty years, there have been many famous proposals in this area. Among them are Principal Component Analysis (PCA), Independent Component Analysis (ICA), Fisher Discriminant Analysis (FDA), Latent Diriclet Analysis (LDA), Maxi-Min Discriminant Analysis (MMDA) [6], and 1-norm based feature selection approach. In the context of classification or regression, DR could be conducted in the supervised style by utilizing certain supervised information (e.g., class labels) so as to find a subspace where different classes of data could be separated as far as possibly. These methods include the above mentioned FDA and MMDA. On the other hand, when the class information is not available, DR is performed in an unsupervised way. This family of approaches includes the famous PCA

C.K. Loo et al. (Eds.): ICONIP 2014, Part II, LNCS 8835, pp. 84–92, 2014.

and ICA. On the other hand, Gaussian Mixture Model (GMM) has achieved big success in both supervised learning, e.g., classification and regression, and unsupervised learning, e.g., clustering.

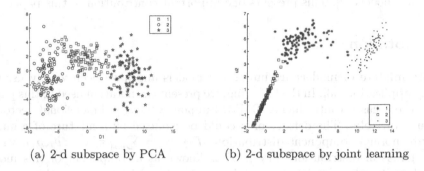

(a) 2-d subspace by PCA   (b) 2-d subspace by joint learning

Fig. 1. Comparison of DR by PCA and the joint learning on simulated data (see Sect. 4.1). Data points with the same shape are supposed to be clustered together.

When GMM is used for practical data, it is usually to perform dimensionality reduction beforehand. The purpose is both to reduce the computational time for high dimensional data and to find a suitable subspace where better clustering or classification performance could be achieved due to the removal of possible noisy features. In this setting, the optimal subspace and the following optimal parameters of GMM are searched separately or independently. Apparently, the optimal subspace obtained by the independent DR may not be appropriate for the following GMM. This is particularly the case in the context of unsupervised learning, e.g., clustering. In supervised learning, class labels could be used for deriving a good subspace, whilst in unsupervised learning, the principles used for DR (e.g., maximization of variance in PCA) may not be appropriate for GMM [4]. Figure 1 (a) illustrates the best 2-dimensional subspace obtained by PCA in one synthetic data. Clearly, the original clustering information among data was less obvious after PCA.

To handle unsupervised dimensionality reduction for GMM, we argue that both the optimal subspace and the parameters for GMM should be jointly learned. This is significantly different from the traditional setting that the two steps are usually conducted separately. Specifically, we engage the Mixture of Factor Analyzers (MFA) [5] where a common factor loading is assumed to exist for all latent factors. Importantly, when this special MFA called MCFA is optimized via the modified EM algorithm, the common factor loading could be regarded as the dimensionality reduction matrix, while the mixture of latent factors can be regarded as the GMM. When GMM is used for unsupervised clustering, its joint learning with the DR subspace would make the clustering properties clearly reserved and even clear. To see the advantages, we also show in Figure 1 (b) the subspace obtained by the joint learning method. Obviously, it could lead to much better clustering performance, especially compared with PCA.

It should be noted that although mixture of factor analysis has been earlier discussed literatures such as [1], it was presented from the viewpoint of data analysis rather than dimensionality reduction. More importantly, the idea of using common loadings, or the joint learning, could also be applied in other mixture models [2]. This presents one important contribution of this paper.

2 Notation

Finite mixture of models are important models and have been widely used in many applications [3]. In the following, we present the notation used in this paper with the focus on introducing GMM. Suppose $\mathbf{y}$ be a p-dimensional vector of feature variables. The density of $\mathbf{y}$ could be modeled by a mixture of g multivariate normal component distributions $P(\mathbf{y}; \theta) = \sum_{i=1}^{g} \pi_i \mathcal{N}(\mathbf{y}; \mu_i, \sigma_i)$, where each gaussian distribution $\mathcal{N}(\mathbf{y}; \mu, \sigma)$ is known as a component of this model and describes the p-variate normal density function with mean μ and covariance matrix σ. The unknown parameter vector θ consists of the mixture weight π_i, the means of component μ_i, and the covariance of component matrices $\sigma_i (i = 1, \ldots, g)$. This vector can be estimated by maximizing the log-likelihood function: $\log L(\theta) = \sum_{j=1}^{n} \log P(\mathbf{y}_j; \theta)$, where $\{\mathbf{y}_j\}$ $(j = 1, \ldots, n)$ is an observed random sample set. By using the Expectation-Maximization (EM) algorithm [1], the local maximizers of log-likelihood function can be obtained as follows:

$$\pi_i^{(k+1)} = \frac{1}{n} \sum_{(i=1)}^{n} P^{(k)}(\omega_i \mid \mathbf{y}_j; \theta), \quad \mu^{(k+1)} = \frac{\sum_{(i=1)}^{n} P^{(k)}(\omega_i \mid \mathbf{y}_j; \theta)\mathbf{y}_j}{\sum_{(i=1)}^{n} P^{(k)}(\omega_i \mid \mathbf{y}_j; \theta)}$$

$$\sigma^{(k+1)} = \frac{\sum_{(i=1)}^{n} P^{(k)}(\omega_i \mid \mathbf{y}_j; \theta)(\mathbf{y}_j - \mu^{(k)})(\mathbf{y}_j - \mu^{(k)})^T}{\sum_{(i=1)}^{n} P^{(k)}(\omega_i \mid \mathbf{y}_j; \theta)}.$$

In the above, ω_i represent the i-th latent component category that each sample $\mathbf{y}_j$ belongs to. With the Bayes theorem, the posterior distribution $P(\omega_i \mid \mathbf{y}_j; \theta)$ can be expressed as $P(\omega_i \mid \mathbf{y}_j; \theta) = \frac{\pi_i \mathcal{N}(\mathbf{y}_j; \mu_i, \sigma_i)}{\sum_{h=1}^{g} \pi_h \mathcal{N}(\mathbf{y}_j; \mu_h, \sigma_h)}, i = 1, \ldots, g; j = 1, \ldots, n$. Here, as the categories ω_i of each sample $\mathbf{y}_j$ are unknown, the latent variable is the indicator variable ω, $\omega = \{0, 1\}, \pi_i = P(\omega_i = 1)$. A data point could be assigned to the component that has the highest estimated posterior probability.

3 Unsupervised Dimensionality Reduction with MCFA

In this section, we will present the mixtures of factor analyzers with common factor loadings (MCFA). This model learns jointly the dimensionality reduction and the parameters of GMM. We will first describe the model definition, and then introduce the involved optimization.

3.1   Model Description

Suppose $Y = (Y_1, \ldots, Y_p)^T$ be generated by linear combination with q-dimensional vector of (unobservable) factors $\mathbf{Z}_{i1}, \ldots, \mathbf{Z}_{in}$. In MFA, the mixture weight $\pi_i, (i = 1, \ldots, g)$ is modeled as

$$\mathbf{Y}_j - \mu_i = \mathbf{\Lambda}\mathbf{Z}_{ij} + e_{ij}, \tag{1}$$

where $\mathbf{\Lambda}$ is called the factor loading vector, the factors $\mathbf{Z}_{ij}$ are distributed independently as $\mathcal{N}(0, I_q)$, e_{ij} is random noise distributed independently under $\mathcal{N}(0, \mathbf{D}_i)$. Here $\mathbf{D}_i$ is a $q \times q$ positive definite symmetric matrix $(i = 1, \ldots, g)$. MCFA further assumes the additional restrictions:

$$\mu_i = \mathbf{A}\xi_i; \quad \sigma_i = \mathbf{A}\mathbf{\Omega}_i\mathbf{A}^T + \mathbf{D}; \quad \mathbf{D}_i = \mathbf{D}; \quad \mathbf{\Lambda}_i = \mathbf{A}\mathbf{K}_i, \tag{2}$$

where $\mathbf{A}$ is a $p \times q$ matrix, ξ_i is a q-dimensional vector $(i = 1, \ldots, g)$, and $\mathbf{D}$ is a diagonal $p \times p$ matrix. Hence the distribution of $\mathbf{Y}_j$ is modeled as

$$\mathbf{Y}_j = \mathbf{A}\mathbf{Z}_{ij} + \mathbf{e}_{ij}, \tag{3}$$

where the (unobservable) factors $\mathbf{Z}_{ij}$ are distributed independently under $\mathcal{N}(\xi_i, \mathbf{\Omega}_i)$, $\mathbf{e}_{ij}$ is random noise distributed independently under $\mathcal{N}(0, \mathbf{D})$, and $\mathbf{D}$ is a diagonal matrix. Here the common loading $\mathbf{A}$ can easily be seen as the transformation matrix, reducing p-dimensional to a latent q-dimensional space.

With the above definitions, the MCFA model can be written as

$$P(\mathbf{y}; \theta) = \sum_{i=1}^{g} \sum_{j=1}^{n} \pi_i \mathcal{N}(\mathbf{y}_j; \mathbf{A}\xi_i, \mathbf{A}\mathbf{\Omega}_i\mathbf{A}^T + \mathbf{D}).$$

Assume we have a mixture of g components by the component-indicator labels ω_i, where ω_i is one or zero depending on whether or not $\mathbf{y}_j$ belongs to the i-th component of the model. The likelihood function can then be written as

$$\mathcal{L}(\mathbf{y}) = \prod_{i=1}^{g} \prod_{j=1}^{n} P(\mathbf{y}_j \mid \mathbf{Z}_{ij}, \omega_i) P(\mathbf{Z}_{ij} \mid \omega_i) P(\omega_i).$$

Since the factors are distributed independently $N(\xi_i, \mathbf{\Omega}_i)$, we have $P(\mathbf{Z}_{ij} \mid \omega_i) = \mathcal{N}(\mathbf{Z}_{ij} \mid \xi_i, \mathbf{\Omega}_i)$. Then, the log-likelihood function is given by

$$\log L_c(\theta) = \sum_{i=1}^{g} \sum_{j=1}^{n} \omega_{ij}\{\log \pi_i + \log \mathcal{N}(\mathbf{y}_j; \mathbf{A}\mathbf{u}_{ij}, \mathbf{D}) + \log \mathcal{N}(\mathbf{Z}_{ij}; \xi_i, \mathbf{\Omega}_i)\}. \tag{4}$$

In the next subsection, we will introduce how to use EM to find the dimensionality reduction matrix $\mathbf{A}$ as well as the parameters of GMM.

3.2   Optimization

Maximization of (4) can be conducted by the famous EM algorithm, or in particularly, the alternating expectation-conditional maximization algorithm (AECM) [5].

E-step. At this step, we need to compute expectations of the hidden variables τ_{ij}, $E(\mathbf{Z} \mid \mathbf{y}_j, \omega_i)$ and $E(\mathbf{ZZ}' \mid \mathbf{y}_j, \omega_i)$ that appear in the log-likelihood for all data point $j = 1, \ldots, n$ and mixture components $i = 1, \ldots, g$. It is easily verified that

$$E(\mathbf{Z} \mid \mathbf{y}_j, \omega_i) = \xi_i + \gamma_i^T(\mathbf{y}_j - \mathbf{A}\xi_i), \tag{5}$$
$$E(\mathbf{ZZ}' \mid \mathbf{y}_j, \omega_i) = (I_q - \gamma_i^T A)\mathbf{\Omega}_i + E(\mathbf{Z} \mid \mathbf{y}_j, \omega_i)E(Z \mid \mathbf{y}_j, \omega_i)', \tag{6}$$

where $\gamma_i = (\mathbf{A}\mathbf{\Omega}_i\mathbf{A}^T)^{-1}\mathbf{A}\mathbf{\Omega}_i$.

At each iteration, it is also necessary to compute the conditional expectation of (4) denoted by $\mathbf{Q}(\theta; \theta^{(k)})$. Given the observed data $\mathbf{y}$ and $\theta^{(k)}$, we have

$$\mathbf{Q}(\theta; \theta^{(k)}) := P(Z^k \mid \mathbf{y}^k; \theta). \tag{7}$$

The conditional expectation of the component labels $\omega_{ij}(i=1,\ldots,g; j=1,\ldots,n)$ can be written as $\mathbf{E}_\theta\{\omega_{ij} \mid \mathbf{y}_j\} = Pr_\theta\{\omega_{ij} = 1 \mid \mathbf{y}_j\} = \tau_i(\mathbf{y}_j; \theta)$, where $\tau_i(\mathbf{y}_j)$ is the posterior probability that $\mathbf{y}_j$ belongs to the ith component. From (2), it can then be obtained

$$\tau_i(\mathbf{y}_j; \theta) = \frac{\pi_i\phi(\mathbf{y}_j; \mathbf{A}\xi_i, \mathbf{A}\mathbf{\Omega}_i\mathbf{A}^T + D)}{\sum_{h=1}^g \pi_h\phi(\mathbf{y}_j; \mathbf{A}\xi_h, \mathbf{A}\mathbf{\Omega}_h\mathbf{A}^T + D)}, \tag{8}$$

Denoting $\tau_{ij}^{(k)} = \tau_i(\mathbf{y}_j; \theta^{(k)})$, we can transform (7) as

$$Q(\theta; \theta^{(k)}) = \sum_{i=1}^g \sum_{j=1}^n \tau_{ij}^{(k)}\{\log \pi_i + E_{\theta^{(k)}}\{\log \mathcal{N}(y_j; Az_{ij}, D)|y_j, \omega_{ij} = 1\}$$
$$+ E_{\theta^{(k)}}\{\log N(z_{ij}; \xi_i, \mathbf{\Omega}_i)|y_j, \omega_{ij} = 1\}\}.$$

M-step. At the $(k+1)$-th iteration of the EM algorithm, the M-step consists of calculating the updated estimates $\pi_i^{(k+1)}$, $\xi_i^{(k+1)}$, $\mathbf{\Omega}_i^{(k+1)}$, $\mathbf{A}^{(k+1)}$ and $D^{(k+1)}$ by maximization of $Q(\theta; \theta^{(k)})$. The updated estimates of the mixing proportions π_i are derived in the case of the normal mixture model by $\pi_i^{(k+1)} = \frac{1}{n}\sum_{j=1}^n \tau_{ij}^{(k)}$, $(i = 1, \ldots, g)$. Concerning the other parameters, we have the following

$$\xi_i^{(k+1)} = \xi_i^{(k)} + \frac{\sum_{j=1}^n \tau_{ij}^{(k)}\varphi^{(k)}}{\sum_{j=1}^n \tau_{ij}^{(k)}}, \mathbf{\Omega}_i^{(k+1)} = \frac{\sum_{j=1}^n \tau_{ij}^{(k)}\varphi^{(k)}\varphi^{(k)T}}{\sum_{j=1}^n \tau_{ij}^{(k)}} + (I_q - \varphi^{(k)})\mathbf{\Omega}_i^{(k)},$$
$$\varphi^{(k)} = (\mathbf{A}^{(k)}\mathbf{\Omega}_i^{(k)}\mathbf{A}^{(k)T} + D^{(k)})^{-1}\mathbf{A}^{(k)}\mathbf{\Omega}_i^{(k)}(y_j - \mathbf{A}^{(k)}\xi_i^{(k)}).$$

The updated estimates $D^{(k+1)} = diag(D_1^{(k)} + D_2^{(k)})$, where

$$D_1^{(k)} = \frac{\sum_{i=1}^g \sum_{j=1}^n \tau_{ij}^{(k)}D^{(k)}(I_p - \beta_i^{(k)})}{\sum_{i=1}^g \sum_{j=1}^n \tau_{ij}^{(k)}}, \beta_i^{(k)} = (\mathbf{A}^{(k)}\mathbf{\Omega}^{(k)}\mathbf{A}^{(k)T} + D^{(k)})^{-1}D^{(k)},$$
$$D_2^{(k)} = \frac{\sum_{i=1}^g \sum_{j=1}^n \tau_{ij}^{(k)}\beta_i^{(k)T}(y_j - \mathbf{A}^{(k)}\xi_i^{(k)})(y_j - \mathbf{A}^{(k)}\xi_i^{(k)})^T\beta_i^{(k)}}{\sum_{i=1}^g \sum_{j=1}^n \tau_{ij}^{(k)}}.$$

We also have $\mathbf{A}^{(k+1)} = (\sum_{i=1}^{g} \mathbf{A}_{1i}^{(k)})(\sum_{i=1}^{g} \mathbf{A}_{2i}^{(k)})^{-1}$, where $\mathbf{A}_{1i}^{(k)} = \sum_{i=1}^{g} \tau_{ij}^{(k)} \{y_j E^{(k)}(\mathbf{Z} \mid \mathbf{y}_j, \omega_i^{(k)})\}; \mathbf{A}_{2i}^{(k)} = \sum_{i=1}^{g} \tau_{ij}^{(k)} \{E^{(k)}(\mathbf{ZZ'} \mid \mathbf{y}_j, \omega_i^{(k)})\}$.

4 Experiments

In this section, we evaluate the performance of the joint learning approach MCFA on one simulation and three real data sets (obtained from UCI machine learning repository) in comparison with the PCA followed by GMM. Following previous research, we report the error rate (ERR), the adjust rand index (ARI), and the Bayesian information criterion (BIC) to compare different algorithms. Note that, although we did not use any labeled information in clustering, the clustering result for each sample is known beforehand in the data sets used. Hence we could exploit ERR as the evaluation metric for clustering.

Table 1. Comparison among the MCFA and PCA-GMM on Simulated Data

	MCFA				PCA			
Cluster	DIM	ERR	BIC	ARI	Cluster	DIM	ERR	BIC	ARI
2	2	0.3333	4173	0.5600	2	2	0.3333	3153	0.5553
3	2	0.0100	4105	0.9702	3	2	0.0300	3080	0.9126

Simulation Data. To validate the effectiveness of the joint learning approach MCFA, we first performed a simulation experiment. We generated 300 random vectors from each of $g = 3$ different three-dimensional multivariate normal distributions. The three distributions have respectively means $\mu_1 = (0,0,0)^T, \mu_2 = (2,2,6)^T, \mu_3 = (8,8,8)^T$, and covariance matrices

$$\Sigma_1 = \begin{pmatrix} 4 & -1.8 & -1 \\ -1.8 & 2 & 0.9 \\ -1 & 0.9 & 2 \end{pmatrix}, \Sigma_2 = \begin{pmatrix} 4 & 1.8 & 0.8 \\ 1.8 & 2 & 0.5 \\ 0.8 & 0.5 & 2 \end{pmatrix}, \Sigma_3 = \begin{pmatrix} 4 & 0 & -1 \\ -1.8 & 2 & 0.9 \\ -1 & 0.9 & 2 \end{pmatrix}.$$

We compared the performance of MCFA with PCA, and plot the unsupervised feature reduction results on Figure 1. It is obvious the joint learning approach leaded to better data separation. To quantitatively evaluate the clustering performance, we compute the ERR, ARI and BIC with the PCA followed by GMM and the joint learning MCFA. These results are shown in Table 1. From the table, the lowest BIC of both approaches are pointed to 3 clusters, indicating that 3 is the best cluster number. Moreover, in case of 3 cluster number, the joint learning MCFA outperformed PCA followed by GMM significantly in terms of the other two criteria.

Table 2. Comparison among MCFA and PCA+GMM on User Knowledge Data

User Knowledge Modeling								
MCFA					PCA			
Cluster	DIM	ERR	BIC	ARI	Cluster	DIM	ERR	BIC	ARI
	2	0.3891	-117	0.4474		2	0.4358	187	0.2469
2	3	0.3891	-87	0.4474	2	3	0.4514	212	0.2896
	4	0.3891	-48	0.4474		4	0.4553	150	0.2442
	2	0.3074	-126	0.4190		2	0.3735	210	0.3001
3	3	0.3035	-121	0.4242	3	3	0.3969	232	0.2924
	4	0.3074	-22	0.4477		4	0.4786	159	0.1781
	2	0.1634	-142	0.6456		2	0.4008	225	0.2771
4	3	0.1868	-92	0.6240	4	3	0.4591	230	0.2791
	4	0.2451	-86	0.5901		4	0.4669	216	0.2593

User Knowledge Modeling Data. This data set consists of $n = 403$ samples and 5 attribute information. The classes are four knowledge levels of the students. As we usually do not know the cluster number, we have compared the joint learning MCFA and PCA+GMM in case of various cluster number and different dimensionality ranged from 2 to the feature number. We report the comparison results in Table 2. Again, it is observed that almost in all the cases, the joint learning demonstrated the better performance than PCA+GMM. Furthermore, the best estimated cluster number of MCFA is 4 and 2 factors according to the lowest BIC. This setting also achieved the lowest ERR, and the highest ARI. It is significantly better than the best case of PCA+GMM.

Table 3. Comparison among the MCFA and PCA+GMM on Physical Data

Physical Data								
MCFA					PCA			
Cluster	DIM	ERR	BIC	ARI	Cluster	DIM	ERR	BIC	ARI
	2	0.3146	7398	0.4717		2	0.3258	4142	0.3963
	3	0.2921	7134	0.5397		3	0.3202	5106	0.4219
	4	0.2921	7010	0.5298		4	0.3202	5945	0.4820
2	5	0.2753	6962	0.5711	2	5	0.3258	6452	0.4088
	6	0.2697	6973	0.5820		6	0.3315	6922	0.3916
	7	0.2697	6986	0.5820		7	0.2865	7255	0.5499
	8	0.2697	7045	0.5820		8	0.2921	7487	0.5397
	2	0.0562	7384	0.8298		2	0.2978	4130	0.3827
	3	0.0225	7096	0.9295		3	0.2697	5109	0.4302
	4	0.0225	6922	0.9309		4	0.1401	5872	0.6170
3	5	0.0169	6935	0.9485	3	5	0.0730	6413	0.7822
	6	0.0056	6881	0.9817		6	0.0562	6905	0.8319
	7	0.0056	6948	0.9817		7	0.0618	7253	0.8185
	8	0.0056	6944	0.9832		8	0.0449	7462	0.8708
	2	0.0618	7411	0.8145		2	0.2978	4165	0.3600
	3	0.0393	7106	0.8792		3	0.2865	5143	0.3977
	4	0.0169	6999	0.9470		4	0.1404	5863	0.6479
4	5	0.0169	6988	0.9470	4	5	0.1124	6445	0.7531
	6	0.0169	7018	0.9551		6	0.1180	6992	0.7436
	7	0.0056	7099	0.9833		7	0.0899	7327	0.8355
	8	0.0056	7121	0.9900		8	0.1011	7505	0.8264

Physical Data. This data set contains results of a chemical analysis of wines grown in the same region in Italy but derived from three different cultivars. It consists of 178 samples with 13 constituents and 3 classes. Again, we have compared the joint learning MCFA and PCA+GMM in Table 3 in cases of various cluster number and different dimensionality ranged from 2 to the feature number. Obviously, in almost all the cases, the joint learning leaded to better performance than PCA+GMM in terms of ERR and ARI. Furthermore, the best estimated cluster number of MCFA is 3 according to the lowest BIC. This also matches the class number in this data set. Such setting again achieved the lowest ERR, and the highest ARI, which outperformed that of PCA+GMM.

Iris Data. The data set contains 3 classes of 150 3-dimensional instances. Using the similar setting in the previous data, we present the performance of the joint learning model MCFA and PCA+GMM approaches in Table 4. Once again, MCFA demonstrated better performance than PCA+GMM. In particular, when cluster number is set to 2, MCFA performed the same as PCA+GMM, while it outperformed significantly in cases of 3 and 4 cluster numbers.

Table 4. Comparison among the MCFA and PCA+GMM on Iris Data

Iris									
MCFA					PCA				
Cluster	DIM	ERR	BIC	ARI	Cluster	DIM	ERR	BIC	ARI
2	2	0.3333	624	0.5681	2	2	0.3333	672	0.5681
	3	0.3333	571	0.5681		3	0.3333	717	0.5681
3	2	0.0200	654	0.9410	3	2	0.0267	672	0.9222
	3	0.0200	571	0.9410		3	0.0267	733	0.9222
4	2	0.0200	692	0.9410	4	2	0.0533	706	0.8700
	3	0.0200	628	0.9410		3	0.0800	755	0.8570

5   Conclusion

This paper mainly introduced a method learning jointly both the optimal subspace and the parameters for GMM. This is significantly different from traditional unsupervised dimensionality reduction for GMM, where the dimensionality reduction and parameter learning are usually conducted independently. A series of experiments on 1 synthetic and 3 real data sets showed that the engaged joint learning approach consistently outperformed the competitive model.

Acknowledgement. The research was partly supported by the National Basic Research Program of China (2012CB316301) and Jiangsu University Natural Science Research Programme (14KJB520037).

References

1. Baek, J., McLachlan, G.J., Flack, L.K.: Mixtures of factor analyzers with common factor loadings: Applications to the clustering and visualization of high-dimensional data. IEEE Transactions on Pattern Analysis and Machine Intelligence 32(7), 1298–1309 (2010)
2. Figueiredo, M., Jain, A.K.: Unsupervised learning of finite mixture models. IEEE Transaction on Pattern Analysis and Machine Intelligence 24(3), 381–396 (2002)
3. Huang, K., King, I., Lyu, M.R.: Finite mixture model of bound semi-naive bayesian network classifier. In: Kaynak, O., Alpaydın, E., Oja, E., Xu, L. (eds.) ICANN 2003 and ICONIP 2003. LNCS, vol. 2714, pp. 115–122. Springer, Heidelberg (2003)
4. Huang, K., Yang, H., King, I., Lyu, M.R.: Machine Learning: Modeling Data Locally and Gloablly. Springer (2008) ISBN 3-5407-9451-4
5. Mclanchlan, G.J., Peel, D., Bean, R.W.: Modelling high-dimensional data by mixtures of factor analyzers. Computational Statistics & Data Analysis 41, 379–388 (2003)
6. Xu, B., Huang, K., Liu, C.-L.: Maxi-min discriminant analysis via online learning. Neural Networks 34, 56–64 (2012)

Graph Kernels Exploiting Weisfeiler-Lehman Graph Isomorphism Test Extensions

Giovanni Da San Martino, Nicolò Navarin, and Alessandro Sperduti

Department of Mathematics, University of Padova, Padova, Italy
{dasan,nnavarin,sperduti}@math.unipd.it

Abstract. In this paper we present a novel graph kernel framework inspired the by the Weisfeiler-Lehman (WL) isomorphism tests. Any WL test comprises a relabelling phase of the nodes based on test-specific information extracted from the graph, for example the set of neighbours of a node. We defined a novel relabelling and derived two kernels of the framework from it. The novel kernels are very fast to compute and achieve state-of-the-art results on five real-world datasets.

1 Introduction

In many real world learning problems, input data are naturally represented as graphs [18][20]. A typical approach for solving machine learning tasks on structured data is to project the input data onto a vectorial feature space and then perform learning on such space. Ideally, a good projection should ensure non-isomorphic data to be represented by different vectors in feature space, i.e. to be injective. When high dimensional data, such as graphs, is involved, specific challenges arise, especially from the computational point of view.

Kernel methods are considered to be among the most successful machine learning techniques for structured data. They replace the explicit projection in feature space with the evaluation of a symmetric semidefinite positive similarity function, called the kernel function. A major advantage of kernel methods is that very large, possibly infinite, feature spaces can be utilized by the learning algorithm with a computational burden dependent on the complexity of the kernel function and not on the size of the feature space. Unfortunately, any kernel function for graphs, whose correspondent feature space projection is injective, is as hard to compute as deciding whether two graphs are isomorphic [8], which is believed to be a NP-Hard problem.

As a consequence, in order to have computationally tractable kernel functions for graph data, a certain amount of information loss is inevitable. Most kernel functions for graphs associate specific types of substructures to features. The evaluation of the kernel function is then related to the number of common substructures between two input graphs. Such substructures include walks [11] [12] [16], paths [1] [9], specific types of subgraphs [3] [15] and tree structures [5]. Such kernels, with the exception of the ones in [5] and [9], are computationally too demanding to be used with large datasets and are effective when the correspondent features are relevant for the current task. Recently, the Fast Subtree Kernel

C.K. Loo et al. (Eds.): ICONIP 2014, Part II, LNCS 8835, pp. 93–100, 2014.
© Springer International Publishing Switzerland 2014

has been proposed [14]. It has linear complexity (in the number of edges) and its features are subtree patterns of the input graphs. The kernel computes a rough approximation of the one-dimensional Weisfeiler-Lehman isomorphism test [19], with the explicit goal of being fast to compute.

In this paper we present two kernel functions for graphs inspired by extensions of the Weisfeiler-Lehman isomorphism test. We define kernels whose feature space is much larger than the Fast Subtree Kernel with a modest increase in computational complexity.

2   Weisfeiler-Lehman Isomorphism Test and Extensions

Some notation is first introduced. A graph is a triplet $G = (V, E, L)$, where V is the set of nodes and $|V|$ its cardinality, E the set of edges and $L()$ a function returning the label of a node. A graph is undirected if $(v_i, v_j) \in E \Leftrightarrow (v_j, v_i) \in E$, otherwise it is directed. A path of length $n-1$ in a graph is a sequence of distinct nodes $v_1, \ldots, v_n$ such that $(v_i, v_i + 1) \in E$ for $1 \leq i < n$; if $v_1 = v_n$ the path is a cycle. The distance $d(v_i, v_j)$ between the nodes v_i, v_j is the length of any shortest path connecting them.

We can now describe the Weisfeiler-Lehman isomorphism test and a few extensions [2] [6], which are all based on a relabelling process of the nodes of a graph $G = (V, E, L)$. We introduce two functions which, instantiated, determine the isomorphism test: $\pi(G, v)$, where $v \in V$, and $h()$ with the constraint that the codomain of $\pi(G, v)$ must coincide with the domain of $h()$. The role of $\pi(G, v)$ is to extract specific information from G: for example in the one-dimensional WL (1-dim WL) test $\pi(G, v)$ extracts the set of neighboring nodes of v: $\pi(G, v) = \{u | u \in V, d(u, v) = 1\}$. The function $h()$ associates a unique numerical value (colour in the mathematical jargon) to each $\pi(G, v)$ and $h(\pi(G, v))$ will be used as novel label for v. In order for $h()$ to be well defined, a canonical representation for elements in its domain has to be defined, which practically boils down to defining a partial ordering between $\pi(G, v)$ elements. For example, in the 1-dim WL test the elements of $\pi(G, v)$ are sorted alphabetically according to their labels.

The algorithm for computing the isomorphism test proceeds by iteratively relabelling G nodes by means of a family of functions $L_\pi^i()$:

$$L_\pi^i(v) = h(\pi(G^{i-1}, v)), \tag{1}$$

where $G^0 = (V, E, L)$ and $G^i = (V, E, L_\pi^i)$ for $i > 0$. The functions $L_\pi^i(v)$ are constructed for all $i \leq i^*$, where i^* is the lowest index for which, $\forall v \in V$, $L_\pi^{i^*}(v) = L_\pi^{i^*+1}(v)$. Note that $i^* \leq |V|$ for the 1-dim WL test [13]. By applying the relabelling in eq. (1) to graphs G and G', we obtain two multisets of node labels: $\{L_\pi^{i^*}(v) | v \in V\}$ and $\{L_\pi^{j^*}(v') | v' \in V'\}$. If such multisets are different, then the two graphs are not isomorphic. On the contrary, if the two multisets are identical, there is not enough information to tell whether the two graphs are isomorphic.

Extensions to the 1-dim WL test have been proposed to increase the discriminative power of the test. Their idea is to enrich the type of information used in the relabelling phase [6], [13]. The extension proposed by Miyazaki [13] considers the colour of the nodes up to distance K: $\pi(G, v) = \{(l, u)|u \in V, d(v, u) = l \leq K\}$; $\pi(G, v)$ elements, i.e. the tuples (l, u), are ordered according to the relation $(l, u) < (l', u') \Leftrightarrow l < l' \vee (l = l' \wedge L_\pi(u) < L_\pi(u'))$, where $L_\pi()$ is a generic labelling function. In the extension of Oliveira et al. [6], $h()$ is defined on paths, which are ordered according to the sequence of labels of the nodes in the path. Specifically, $\pi(G, v)$ extracts, for each $u \in V$, the shortest path between v, u having lower $h()$ value: let $s(v, u)$ be the set of shortest paths connecting u and v, $\pi(G, v) = \cup_{u \in V} \arg\min_{p \in s(v, u)} h(p)$.

3 Weisfeiler-Lehman Kernel Framework

Let us consider a function $\pi_r(G, v)$ depending on a parameter r, with $1 \leq r \leq K$. Given a graph $G = (V, E, L)$, the application of eq. (1), for a fixed r value at the i-th iteration, yields the graph $G_r^i = G(V, E, L_{\pi_r}^i)$, which differs from the original graph only in the labelling function.

Definition 1. *Let $k()$ be any kernel for graphs that we will refer to as the base kernel. Then the Extended Weisfeiler-Lehman kernel with h iterations, depth K and base kernel $k()$ is defined as:*

$$WL_h^K(G, G') = \sum_{r=1}^{K} \sum_{i=0}^{h} k(G_r^i, G_r'^i).$$ (2)

Since the functions in eq. (1) are well defined and the Extended Weisfeiler-Lehman kernel of eq. (2) is a finite sum of positive semidefinite functions, it is also positive semidefinite.

Let us now present the main contribution of the paper, i.e. two novel kernels which are instances of eq. (2). For both kernels the function $\pi(G, v)$ returns the following Directed Acyclic Graph (DAG) rooted at v: $D_r(v) = (V_r, E_r, L)$ where $V_r = \{u \in V|d(v, u) \leq r\}$ and F_r consists in all edges of G that appear in any of the shortest path connecting v and any $u \in V_r$ (see Fig. 1-b for an example). In order to have a canonical representation for the DAG $D_r(v)$, the ordering for DAG nodes described in [5] is used. The function $h()$ assigns a unique numerical value to each DAG, and it can be implemented efficiently as presented in [4]. Let the maximum number of nodes of each DAG $D_r(v)$ be $|D_r|$. Then it can be shown that $|D_r|$ is $O(\rho^r)$ [5], where ρ is the maximum node outdegree. Computing all the indices $L_{\pi_r}^i()$ for a graph G has worst-case time complexity $O(|D_r||V| \log |D_r||V|)$ (see [5] for details). Assuming ρ constant (a condition that usually holds in real-world datasets) the worst-case time complexity reduces to $O(|V| \log |V|)$.

In the first proposed kernel, that we will refer to as WL_{NS-DDK}, the base kernel is defined as

$$k(G_r^i, G_r'^i) = \sum_{v \in V} \sum_{v' \in V'} \delta(L_{\pi_r}^i(v), L_{\pi_r}^i(v')),$$ (3)

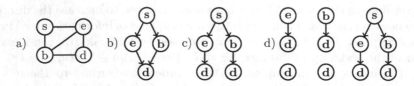

Fig. 1. Steps for obtaining some of the features of the WL_{DDK} kernel: a) an input graph G; b) the DAG resulting from the application of $\pi(G, v)$ where v is the node labelled as **s**; c) the tree visit $T(v)$; d) the features of the ST kernel related to $T(v)$

where δ is the Kronecker's delta function. Note that computing the kernel is equivalent to performing a hard match between the DAGs encoded by $L^i_{\pi_r}(v)$ and $L^i_{\pi_r}(v')$. If we order the list of indices $\{L^i_{\pi_r}(v)|v \in V\}$ and $\{L^i_{\pi_r}(v')|v' \in V'\}$, then eq. (3) can be computed in $O(|V|\log|V|)$ time.

The second kernel we propose, referred to as WL_{DDK}, differs from the first one only in the base kernel $k()$. Let $T(v)$ be the function that, first computes the DAG $\pi_r(G, v)$ and then returns the tree resulting from the breadth-first visit of the DAG starting from v (see Fig. 1-c for an example). Finally, $k()$ can be defined as any kernel for trees applied to $T(v)$ and $T(v')$, for example the subtree kernel (ST) [17]:

$$k(v, v') = \sum_{v \in V} \sum_{v' \in V'} k_{ST}(T(v), T(v')). \tag{4}$$

The ST kernel counts the number of matching proper subtrees of $T(v)$ and $T(v')$, where a proper subtree of a tree T rooted at u is the subtree composed by u and all of its descendants (in Fig. 1-d are listed the set of proper subtrees of the tree in Fig. 1-c). The complexity of $k_{ST}(T, T')$ is $O(n\log n)$ where $n = \min(|T|, |T'|)$. Assuming ρ constant, $O(|T(v)|) = O(|D_r(v)|)$. By using the algorithm described in [5], the complexity of computing eq. (4) is $O(|V|\log|V|)$.

There are a number of kernels in literature that are instances of eq. (2). The Fast Subtree Kernel (FS) counts the number of identical subtree patterns of depth h [14]. It can be obtained from eq. (2) by setting: *i)* $K = 1$; *ii)* $\pi(G, v) = \{u|u \in V, d(v, u) = 1\}$ and then ordering $\pi(G, v)$ elements alphabetically according to their labels; *iii)* the base kernel $k()$ is the one in eq. (3). The ODD-ST$_h$, described in [5], is an instance of the WL_{DDK} of eq.(4) and it is obtained setting $h = 0$ in eq. (2).

4   Experimental Results

In this section, we compare the two kernels presented in Section 3 against other state-of-the-art kernels on five real-world datasets.

We considered the Fast Subtree kernel [14], the ODD-ST$_h$ kernel [5] (described in section 3) and the NSPDK kernel [3] , that computes the exact matches between pairs of subgraphs with controlled size and distance. For the assessment of the performance of the proposed kernels, we considered five real-world datasets:

Table 1. Average accuracy results ± standard deviation in nested 10-fold cross validation for the Fast Subtree, the Neighborhood Subgraph Pairwise Distance, the K_{ODD-ST_h}, WL_{NS-DDK} and WL_{DDK} kernels obtained on CAS, CPDB, AIDS, NCI1 and GDD datasets. The rank of the kernel is reported between brackets.

$Kernel$	CAS	CPDB	AIDS	NCI1	GDD	AVG Rank
FS	81.05 (5) ($\pm$0.50)	73.22 (5) ($\pm$0.78)	75.61 (5) ($\pm$1.00)	84.77 (3) ($\pm$0.31)	76.21 (2) ($\pm$1.15)	4
NSPDK	83.60 (2) ($\pm$0.34)	76.99 (2) ($\pm$1.15)	82.71 (3) ($\pm$0.66)	83.46 (4) ($\pm$0.46)	74.09 (5) ($\pm$0.91)	3.2
$ODD-ST_h$	83.34(3) ($\pm$0.31)	76.44 (4) ($\pm$0.62)	81.51(4) ($\pm$0.74)	82.10 (5) ($\pm$0.42)	75.23(4) ($\pm$0.70)	4
WL_{NS-DDK}	82.96 (4) ($\pm$0.49)	<u>77.03</u> (1) ($\pm$1.18)	82.80 (2) ($\pm$0.66)	84.79 (2) ($\pm$0.36)	<u>77.20</u> (1) ($\pm$0.65)	2
WL_{DDK}	<u>83.91</u> (1) ($\pm$0.29)	76.52 (3) ($\pm$1.16)	<u>82.93</u>(1) ($\pm$0.71)	<u>84.90</u> (1) ($\pm$0.33)	75.45 (3) ($\pm$0.86)	1.8

CAS[1], CPDB [10], AIDS [20], NCI1 [18] and GDD [7]. All the datasets represent binary classification problems. The first four datasets involve chemical compounds, represented as graphs where the nodes represent the atoms (labelled according to the atom type) and the edges the bonds between them. In chemical compounds, there are no self-loops. GDD is a dataset of proteins, where each protein is represented by a graph, in which the nodes are amino acids and two nodes are connected by an edge if they are less than 6° Angstroms apart. CAS and NCI1 are the largest datasets, with 4337 and 4110 examples, respectively. For more information about the datasets, please refer to [5].

All the kernels have been employed together with a Support Vector Machine. The C parameter of the SVM has been selected in the set $\{0.01, 0.1, 1, 10, 100\}$. For all the experiments, the values of the parameters of the ODD-ST$_h$ kernel have been restricted to: $h = \{1, 2, \ldots, 8\}$ $\lambda = \{0.1, 0.2, \ldots, 2.0\}$ (λ is a parameter of K_{ST}); for the Fast Subtree Kernel we optimized the only parameter of the kernel $h = \{1, 2, \ldots, 10\}$; for the NSPDK kernel we optimized the parameters $r = \{1, 2, \ldots, 8\}$ and $d = \{1, 2, \ldots, 8\}$. Concerning the two kernels presented in this article, their parameters are $K = \{1, 2, 3, 4\}$, $h = \{0, 1, 2, \ldots, 8\}$ and $\lambda = \{0.1, 0.2, \ldots, 2.0\}$. The parameters range has been selected in such a way that the computational time needed for the calculation of the kernel matrices is roughly comparable, i.e. at most one hour on a modern PC. For parameter selection we adopt a technique commonly referred to as *nested* K-fold cross validation following [14]. All the experiments have been repeated 10 times and the average results (with standard deviation) are reported.

Table 1 summarizes the average accuracy results of the proposed kernels and the state-of-the-art ones on the considered datasets. The mean accuracy is reported with the standard deviation. Between brackets, the ranking of the specific

---

[1] http://www.cheminformatics.org/datasets/bursi

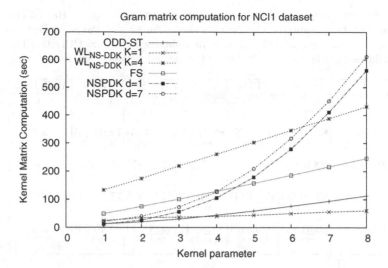

Fig. 2. Comparison between the time needed for computing the Gram matrix on the NCI1 dataset for the different kernels, as a function of the parameter: h for FS and WL_{NS-DDK}, K for $ODD - ST$, r for NSPDK

kernel on the dataset is reported. In the rightmost column, the average ranking value on all the datasets for each kernel is reported. When considering single datasets, there is no dataset where NSPDK or FS kernels rank first. On all the considered datasets, either WL_{NS-DDK} or WL_{DDK} outperforms the other kernels. If we look at the average ranking, the situation is clearer. The best average ranking of the competing kernels is the one of $NSPDK$, with a value of 3.2. The WL_{NS-DDK} has an average ranking of 2. WL_{DDK} performs slightly better, with an average ranking value of 1.8. These results clearly show that, on the considered datasets, the WL kernel family performs better than the other kernels present in literature.

Figure 2 reports the computational time, in seconds, needed from the WL_{NS-DDK} kernel and the competing ones to compute the Gram matrix for the NCI1 dataset. The computation time required by WL_{DDK} is very similar and thus omitted.

5 Conclusions and Future Work

This paper proposed a new framework for the definition of graph kernels based on a generalization of the 1-dimensional WL test. The framework can be instantiated with any kernel for graphs as a base kernel. In particular, we analyzed two instances inspired by the Decompositional DAGs graph kernels [5]. The two kernels show state-of-the-art predictive performance on five real-world datasets, with a computational burden that, on such datasets, grows only linearly with respect to the kernel parameters. As a future work, we will explore other members of the framework.

Acknowledgments. This work was supported by the University of Padova under the strategic project BIOINFOGEN.

References

1. Borgwardt, K.M., Kriegel, H.-P.: Shortest-Path Kernels on Graphs. In: ICDM, pp. 74–81 (2005)
2. Cai, J.-Y., Furer, M., Immerman, N.: An optimal lower bound on the number of variables for graph identification. Combinatorica 12(4), 389–410 (1992)
3. Costa, F., De Grave, K.: Fast neighborhood subgraph pairwise distance kernel. In: ICML (2010)
4. Da San Martino, G., Navarin, N., Sperduti, A.: A memory efficient graph kernel. In: The 2012 International Joint Conference on Neural Networks (IJCNN). IEEE (June 2012)
5. Da San Martino, G., Navarin, N., Sperduti, A.: A Tree-Based Kernel for Graphs. In: SDM, pp. 975–986 (2012)
6. de Oliveira Oliveira, M., Greve, F.: A new refinement procedure for graph isomorphism algorithms. Electronic Notes in Discrete Mathematics 19, 373–379 (2005)
7. Dobson, P.D., Doig, A.J.: Distinguishing Enzyme Structures from Non-enzymes Without Alignments. Journal of Molecular Biology 330(4), 771–783 (2003)
8. Gartner, T., Flach, P., Wrobel, S.: On Graph Kernels: Hardness Results and Efficient Alternatives. In: Schölkopf, B., Warmuth, M.K. (eds.) COLT/Kernel 2003. LNCS (LNAI), vol. 2777, pp. 129–143. Springer, Heidelberg (2003)
9. Heinonen, M., Välimäki, N., Mäkinen, V., Rousu, J.: Efficient Path Kernels for Reaction Function Prediction. Bioinformatics Models, Methods and Algorithms (2012)
10. Helma, C., Cramer, T., Kramer, S., De Raedt, L.: Data mining and machine learning techniques for the identification of mutagenicity inducing substructures and structure activity relationships of noncongeneric compounds. Journal of Chemical Information and Computer Sciences 44(4), 1402–1411 (2004)
11. Kashima, H., Tsuda, K., Inokuchi, A.: Marginalized kernels between labeled graphs. In: ICML, pp. 321–328. AAAI Press (2003)
12. Mahé, P., Vert, J.-P.: Graph kernels based on tree patterns for molecules. Machine Learning 75(1), 3–35 (2008)
13. Miyazaki, T.: The complexity of McKay's canonical labeling algorithm. In: Groups and Computation II (1997)
14. Shervashidze, N., Borgwardt, K.: Fast subtree kernels on graphs. In: NIPS, pp. 1660–1668 (2009)
15. Shervashidze, N., Mehlhorn, K., Petri, T.H., Vishwanathan, S.V.N., Borgwardt, K.: Efficient graphlet kernels for large graph comparison. AISTATS 5, 488–495 (2009)
16. Vishwanathan, S.V.N., Borgwardt, K.M., Schraudolph, N.N.: Fast Computation of Graph Kernels. In: NIPS, pp. 1449–1456 (2006)

17. Vishwanathan, S.V.N., Smola, A.J.: Fast Kernels for String and Tree Matching. In: Becker, S., Thrun, S., Obermayer, K. (eds.) NIPS, pp. 569–576. MIT Press (2002)
18. Wale, N., Watson, I., Karypis, G.: Comparison of descriptor spaces for chemical compound retrieval and classification. Knowledge and Information Systems 14(3), 347–375 (2008)
19. Weisfeiler, B.: On construction and identification of graphs. Lecture Notes in Mathematics (1976)
20. Weislow, O.S., Kiser, R., Fine, D.L., Bader, J., Shoemaker, R.H., Boyd, M.R.: New soluble-formazan assay for HIV-1 cytopathic effects: application to high-flux screening of synthetic and natural products for AIDS-antiviral activity. Journal of the National Cancer Institute 81(8), 577–586 (1989)

Texture Analysis Based Automated Decision Support System for Classification of Skin Cancer Using SA-SVM

Ammara Masood, Adel Al-Jumaily, and Khairul Anam

School of Electrical, Mechanical and Mechatronic Engineering,
University of Technology, Sydney, Australia
{ammara.masood,khairul.Anam}@student.uts.edu.au,
Adel.Ali-Jumaily@uts.edu.au,

Abstract. Early diagnosis of skin cancer is one of the greatest challenges due to lack of experience of general practitioners (GPs). This paper presents a clinical decision support system aimed to save lives, time and resources in the early diagnostic process. Segmentation, feature extraction, and lesion classification are the important steps in the proposed system. The system analyses the images to extract the affected area using a novel proposed segmentation method H-FCM-LS. A set of 45 texture based features is used. These underlying features which indicate the difference between melanoma and benign images are obtained through specialized texture analysis methods. For classification purpose, self-advising SVM is adapted which showed improved classification rate as compared to standard SVM. The diagnostic accuracy obtained through the proposed system is around 90% with sensitivity 91% and specificity 89%.

Keywords: Skin Cancer, diagnosis, feature extraction, classification, self-advising support vector machine.

1 Introduction

Malignant melanoma is one of the deadliest forms of skin cancer. A rapid increase in melanoma cases is observed in Europe, North America, and Australia over the last decade. Over 76,250 new cases of invasive melanoma were diagnosed in the US in 2012 [1]. An estimated 1,890 Australians die from skin cancer each year [2]. From treatment point of view, skin cancer is one of the most expensive forms of cancer, but early diagnosis can make the situation better as melanoma has near 95% cure rate if diagnosed and treated in early stages [1].

A Computer Aided Diagnostic (CAD) system for diagnosis of skin cancer is aimed to find the exact boundaries of a lesion automatically and also to provide an estimate of the probability of a disease. There are various diagnostic systems proposed in literature [3-5] but as we discussed in [6] more research is required to make the best choice and for setting the benchmarks for diagnostic system development and validation. This paper presents a part of our research being carried out to come up with the best combination of segmentation, feature extraction and classification algorithms which can consequently form the basis of a more generalized and efficient skin cancer diagnostic system. The diagnostic model proposed here is shown in Fig. 1.

C.K. Loo et al. (Eds.): ICONIP 2014, Part II, LNCS 8835, pp. 101–109, 2014.

The proposed decision support system uses adaptive median filter for pre-processing of image to reduce the ill effects and various artifacts like hair that may be present in the images. It is followed by the detection of the lesion by our Histogram based fuzzy C means thresholding algorithm presented in [7]. This algorithm provided efficient segmentation results as compared to other segmentation methods used in literature; the comparative analysis is presented in [8]. Once the lesion is localized, texture based features are quantified. Finally, Self-advisable Support Vector machine (SA-SVM) is used for classification of cancerous and non-cancerous skin lesions.

This paper is organized as follows: Section 2 describes the computer-aided diagnosing (CAD) system which consists of pre-processing, segmentation, features extraction and classification stages. Section3 presents the experimental results, comparative analysis and discussion, and final section for conclusions and future work.

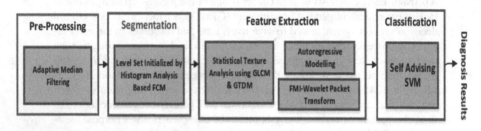

Fig. 1. Computer Aided Diagnostic Support System

2   Proposed CAD System

2.1   Pre-processing

Skin images have certain extraneous artifacts such as skin texture, dermoscopic gel and hair that make border detection a bit difficult. It is necessary to pre-process the images with a smoothing filter like adaptive median filter. The median filter also performs well as long as the spatial density of the impulse noise is not too large. However the adaptive median filtering has a better capability to handle impulse noise with even larger probabilities. An additional benefit of the adaptive median filter is that it seeks to preserve details while smoothing the non-impulse noise [9]. Considering the high level of noise that may be present in skin lesion images and the need of preserving structural details, the adaptive median algorithm performed quite well.

2.2   Segmentation

In one of our previous work [10], We proposed a segmentation algorithm, histogram analysis based fuzzy C mean algorithm for Level Set initialization (H-FCM-LS) as presented in figure 2. In the proposed method, histogram analysis of image was done to see average intensity distribution in the images and then the hard threshold was selected between classes with dominant intensity values. This method was further used as an initializing step for complex segmentation method like Level set having spatial information. Segmentation results for some of the skin lesion images are shown in figure 2. For details of algorithm refer to [10].

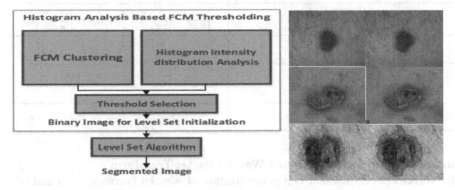

Fig. 2. Block Diagram & Results of Segmentation Algorithm (H-FCM-LS)

2.3   Feature Extraction

Texture analysis can potentially expand the visual skills of the expert eye by extracting features that are relevant to diagnostic problem and not necessarily visually extractable. Three types of methods for texture based feature extraction are used here 1) model-based 2) statistical and 3) transform-based.

Gray Level Co-occurrence Matrix (GLCM)
GLCM was introduced by Haralick [11] provides one of the most popular statistical methods in analysis of grey tones in an image. The GLCM functions characterize the texture of image by calculating how often pairs of pixel with specific values and in a specified spatial relationship occur in an image, creating a GLCM, and then extracting statistical measures from this matrix. Details for GLCM based features that are used for classification stage can be found in [12].

Grey-Tone Difference Matrix (GTDM)
GTDM was suggested by Amadasun [13] in an attempt to define texture measures correlated with human perception of textures. A GTDM matrix is a column vector containing G elements. Its entries are calculated based on the difference between intensity level of a pixel and average intensity computed over a square, while sliding window centered at the pixel. Suppose the image intensity level $I_L(x,y)$ at location (x,y) is I, i=0,1,... G-1. The average intensity over a window centered at (x,y) will be $I_L(i) = I(x,y) = \frac{1}{W-1}\sum_{m=-K}^{K}\sum_{n=-K}^{K} f(x+m, y+n)$, where K specifies the window size and W= $(2K+1)2$. The ith entry of GTDM x is s(i)=$\sum_{x=0}^{M-1}\sum_{y=0}^{N-1} |i - f_i|$ for all pixel having intensity level I, otherwise, s(i)=0. Mathematical formulae for GTDM based features used here are given in Table 1.

Table 1. GTDM based Features

Feature	Mathematical Equation
Coarseness	$\left(\varepsilon + \sum_{i=0}^{G-1} p_i\, s(i)\right)^{-1}$
Contrast	$\left[\dfrac{1}{N_t(N_t-1)}\sum_{i=0}^{G-1}\sum_{j=0}^{G-1} p_i p_j(i-j)^2\right]\left[\dfrac{1}{n}\sum_{i=0}^{G} s(i)\right]$
Business	$\dfrac{\sum_{i=0}^{G-1} p_{i\,s(i)}}{\sum_{i=0}^{G-1}\sum_{j=0}^{G-1}\lvert ip_i - jp_j\rvert}\quad p_i \neq 0, p_j \neq 0$
Complexity	$\sum_{i=0}^{G-1}\sum_{j=0}^{G-1}\dfrac{\lvert i-j\rvert}{n(p_i+p_j)}\left[p_i s(i) + p_j\, s(j)\right]\quad p_i \neq 0, p_j \neq 0$
Texture Strength	$\dfrac{\sum_{i=0}^{G-1}\sum_{j=0}^{G-1}(p_i+p_j)(i-j)^2}{\varepsilon+\sum_{i=0}^{G-1} s(i)}\quad p_i \neq 0, p_j \neq 0$

Fuzzy-Mutual Information Based Wavelet Packet Transform

The wavelet packet method is a generalization of wavelet decomposition and offers a richer signal analysis. Different extensions of wavelet packet transform are present in literature for different applications [14,15]. It is observed that features extracted using wavelet transforms provide significant increase in the classification accuracy. After converting skin images to corresponding vectors, following Fuzzy mutual-information based wavelet packet transform (FMI-WPT) is used:

1) For each original image vector, perform a full WPT decomposition to the maximum level J (taken as 3 here). For all j = 0, 1,.., J and k = 0, 1, . . . , 2j − 1, construct features according to relation $E_{\Omega j.k} = \log\left(\frac{\sum_n (w^T_{j,k,n}\, x)^2}{N/2^j}\right)$. where Ωj,k is the decomposition subspace with j denoting scale and k denoting sub-band index within the scale [15].

2) Construct associated fuzzy sets and compute fuzzy entropies and mutual information. Then evaluate classification ability of n number of features using fuzzy-set based criterion F_i where $F_i = I(C; f_i)/H(f_i)$ for i = 1, 2, . . . , n.

Note: I (f;C) = H(f) − H(f |C) where H(f) is marginal entropy of f and H(f |C) is conditional entropy of f and C [16].

3) Determine the optimal WPT decomposition X, being the one that corresponds to the maximum value of F.

4) The set X is the final FMIWPT-based decomposition.

Autoregressive Modeling Based Features

Autoregressive modeling is an all pole modeling which is widely used to get a robust spectral estimation of one dimensional signal. Yule-walker method is the most widely used method to estimate autoregressive coefficients. In order to determine the coefficients, it uses Levinson-Durbin algorithm to minimize error. This estimation method minimizes square of forward prediction error and finds out autoregressive parameters by solving autocorrelation function (1) as expressed in [16].

$$\sigma^2 = \frac{1}{N} \sum_{n=-\infty}^{\infty}\left|x(n) - \sum_{k=1}^{p} a(k)x(n-k)\right|^2 \tag{1}$$

Where x(n) is input and a(k) demonstrate autoregressive parameters. As images are 2-dimensional signals, so for doing autoregressive modeling, skin lesion images were converted into corresponding vectors. Estimated autoregressive parameters which are poles of 1-dimensional signals are used as features vectors extracted from images.

2.4   Classification

Classification of the lesion as cancer or non-cancer is the final step. For classification of skin lesions, an improved version of support vector machine (SVM), named Self Advising SVM is adapted here. SVM is a well-known machine learning method proposed by Vapnik [17] and a lot of literature is available for applications of SVM. The idea of SVM is to construct a maximized separating hyper plane that can separate data in the feature space. The classic SVM ignores the training data that has not been separated linearly by the kernels during the training phase. Thus, if data that is similar or identical to this misclassified data appears in the test set, it will be classified wrongly. This misclassification is not reasonable and it can be handled if the available data and information in the training phase has not been ignored by SVM algorithm.

In this study a non-iterative self-advising approach [18] for SVM is adapted that extracts subsequent knowledge from training phase. The misclassified data can come from two potential sources 1) outliers 2) data that have not been linearly separated by using any of the types of kernels. SA-SVM deals with the ignorance of SVM from the knowledge that can be acquired from misclassified data by generating advice weights based on use of misclassified training data, and through use of these weights together with decision values of SVM in the test phase. These weights also help the algorithm to eliminate the outlier data. The details of SA-SVM algorithm are as follows:

Training Phase

1. Finding hyperplane by using decision function $f(x) = sign(\sum_{\alpha_i > 0} y_i \alpha_i \, k(x, x_i) + b)$ i.e. the normal SVM training. Note here that the kernel function we used is radial Basis Function so $K(x_i x_j) = e^{-\gamma |x_i - x_j|^2}$.

2. To benefit from the misclassified data of the training phase, the misclassified data sets (MD) in the training phase is determined as

$$MD = U_{i=1}^{N} x_i \, | y_i \neq sign(\sum_{\alpha_j > 0} y_i \, \alpha_j \, k(x_i, x_j) + b) \qquad (2)$$

The MD set can be null, but experimental results have revealed that the occurrence of misclassified data in training phase is a common occurrence. Note that xi is the input vector corresponding to the ith sample and labeled by yi depending on its class and αi is the nonnegative Lagrange multiplier as used in standard SVM [19].

3. If the MD is null, go to the testing phase else compute neighborhood length (NL) for each member of MD. NL is given as

$$NL(x_i) = minimum_{x_j} \, (\|x_i - x_j\| \, | \, y_i \neq y_j) \qquad (3)$$

Where xj , j=1, …., N are the training data that do not belong to the MD set. Here the training data is mapped to a higher dimension, the distance between xi and xj is be computed according to the following equation with reference to the related kernel k (RBF).

$$\|\theta(x_i) - \theta(x_j)\| = (k(x_i, x_i) + k(x_j, x_j) - 2k((x_i, x_j))^{0.5} \tag{4}$$

Testing Phase

1. Compute the advised weight $AW(x_k)$ for each x_k, from the test set. Where AW is computed as (5), These AWs represent how close the test data to the misclassified data is.

$$\begin{cases} 0 & \forall x_i \in MD, \|x_k - x_i\| > NL(x_i) \text{ or } MD = NUL, \\ \sum 1 - \dfrac{\sum_{x_i} \|x_k - x_i\|}{\sum_{x_i} NL(x_i)} & x_i \in MD, \|x_k - x_i\| \le NL(x_i) \end{cases} \tag{5}$$

2. Compute the absolute value of the SVM decision values for each xk from the test set and scale it to [0, 1].

3. For each xk from the test set, If $(AW(x_k) <$ decision value (x_k) then $y_k = \text{sign} (\sum_{\alpha_j>0} y_j \alpha_j k(x_k, x_j) + b)$ which is normal SVM labeling, otherwise $y_k = y_i \mid (\|x_k - x_i\| \le NL(x_i) \text{ and } x_i \in MD)$

3   Experimental Results

A clinical database of dermoscopic and clinical view lesion images were obtained from different sources but most of the images came from Sydney Melanoma Diagnostic Centre, Royal Prince Alfred Hospital. A total of 168 images (56 benign and 112 mela-noma) were included in experimental data set. All the images were rescaled to a resolu-tion of 720x 472 with bit depth 24 and size around 526KB. After pre-processing and segmentation, a total of 45 features (15 GLCM, 5 GTDM, 15 FMI_WPT, and 10 Auto-regressive) were extracted for each image. For training the SA_SVM, 84 images are used and 84 images were used for testing. The whole process is implemented using MATLAB software R2013 and simulated by a system with corei5 3.10 GHz processor and 4 GB memory under Windows7 operating system.

The constructed feature sets are used separately as well as in different combina-tions for feeding the classifier. The contribution of each feature extraction method used as well as of the proposed set of 45 texture based features for classification can be seen from figure 3. It can be seen clearly that use of GLCM and GTDM based features resulted in better sensitivity ($\frac{True\ Positive}{True\ Positive+False\ Negative}$) but lower specificity ($\frac{True\ Negative}{True\ Negative+False\ Positive}$). On the other hand use of autoregressive and FMI-WPT based features lead to better specificity but relatively lower sensitivity. However, the expe-rimental analysis clearly indicated that a proposed combination of (45 features) using these four feature extraction methods resulted in a feature set that formed a good basis for classification using SA_SVM. The proposed diagnostic system achieved an over-all accuracy of around 90%, with sensitivity 91% and specificity 89%.

For cross validating the results, hold-out validation a specific type of k-fold cross validation is used. For each fold, the skin images are randomly divided into two equal sized sets S1 and S2. Then SA-SVM is trained using S1 and tested on S2. This is followed by training using S2 and testing using S1. This has the advantage that our training and test sets are both large, and each data point is used for both training and validation on each fold. In order to ensure better validation of classifier performance the hold-out validation was repeated 5 times and each time the experimental data set was shuffled and split it into two parts.

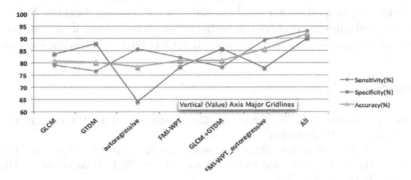

Fig. 3. Comparative Diagnosis results of SA-SVM using proposed feature set

The performance of diagnostic system is analyzed using statistical parameters such as sensitivity, specificity and accuracy. The classification results of SA-SVM are compared with standard SVM (both linear and kernel based). Higher value of both sensitivity and specificity shows better performance of the system. The experimental analysis also shows that the results obtained by the self-advising SVM are significantly better than the results of traditional SVM.

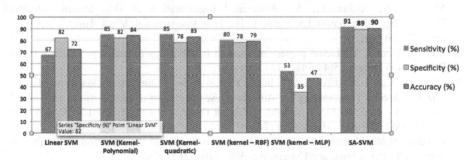

Fig. 4. Relative Performance Measure of SA-SVM

4   Conclusion and Future Work

In this paper, an automated skin cancer diagnostic system is proposed based on SA-SVM. SA-SVM uses information generated from misclassified data in the training phase and thus, improves performance by transferring more information from training phase to the test phase. Features used for differentiating melanoma and benign images are extracted using four texture analysis methods. A set of features is proposed that worked best for SA-SVM. The diagnostic results obtained are quite satisfactory with sensitivity of 91% and specificity of 89%.

Despite the high accuracy that can be achieved by the proposed system, for developing a more reliable diagnostic system we intend to test multiple classifier based systems as well to undo the chances of any misclassification due to classifier limitations. We intend to do experiments combining different classification algorithms like neural networks, support vector machine and extreme learning machine. Such tools may serve as diagnostic adjuncts for medical professionals. And it will provide the opportunity of implementing more accurate, faster and reliable diagnostic systems.

References

1. Siegel, R., Naishadham, D., Jemal, A.: Cancer statistics. CA: A Cancer Journal for Clinicians 62(1), 10–29 (2012)
2. Causes of death, C.W.o. Australia, Editor, Australian Bureau of Statistics: Canberra (2010)
3. Ganster, H., et al.: Automated melanoma recognition. IEEE Transactions on Medical Imaging 20(3), 233–239 (2001)
4. Rubegni, P., et al.: Automated diagnosis of pigmented skin lesions. International Journal of Cancer 101(6), 576–580 (2002)
5. Ruiz, D., et al.: A decision support system for the diagnosis of melanoma: A comparative approach. Expert Systems with Applications 38(12), 15217–15223 (2011)
6. Masood, A., Jumaily, A.: Computer Aided Diagnostic Support System for Skin Cancer: A Review of Techniques & Algorithms. International Journal of Biomedical Imaging, 22 (2013)
7. Masood, A., Jumaily, A.: FCM Thresholding based Level Set for Automated Segmentation of Skin Lesions. Journal of Signal & Information Processing 4(3B), 66–71 (2013)
8. Masood, A., Al-Jumaily, A.: Automated segmentation of skin lesions: Modified FCM thresholding based LS method. In: 16th International Multi Topic Conference, pp. 201–206 (2013)
9. Sun, T., Neuvo, Y.: Detail-preserving median based filters in image processing. Pattern Recognition Letters 15(4), 341–347 (1994)
10. Masood, A., Al-Jumaily, A.A., Maali, Y.: Level Set Initialization Based on Modified Fuzzy C Means Thresholding for Automated Segmentation of Skin Lesions. In: Lee, M., Hirose, A., Hou, Z.-G., Kil, R.M. (eds.) ICONIP 2013, Part III. LNCS, vol. 8228, pp. 341–351. Springer, Heidelberg (2013)
11. Haralick, R.M., Shanmugan, K.: Textural features for image classification. IEEE Transactions on Systems, Man, and Cybernetics 3(6), 610–621 (1973)
12. Nitish Zulpe, V.P.: GLCM Textural Features for Brain Tumor Classification. International Journal of Computer Science Issues 9(3), 354–359 (2012)
13. Amadasun, M., King, R.: Textural Features Corresponding to Textural Properties. IEEE Transactions on System, Man Cybernetics 19(5), 1264–1274 (1989)
14. Khushaba, R.N., Al-Jumaily, A., Al-Ani, A.: Novel feature extraction method based on fuzzy entropy and wavelet packet transform for myoelectric Control. In: International Symposium on Communications and Information Technologies, pp. 352–357 (2007)

15. Khushaba, R.N., Kodagoa, S., Lal, S., Dissanayake, G.: Driver Drowsiness Classification Using Fuzzy Wavelet-Packet-Based Feature-Extraction Algorithm. IEEE Transactions on Biomedical Engineering 58(1), 121–131 (2011)
16. Proakis, J.G., Manolakis, D.G.: Digital Signal Processing Principles, Algorithms, and Applications. Prentice-Hall, New Jersey (1996)
17. Vapnik, V.N.: The nature of statistical learning theory, 2nd edn. Springer, New York (2000)
18. Maali, Y., Al-Jumaily, A.: Self-advising support vector machine. Knowledge-Based Systems 52, 214–222 (2013)

In-attention State Monitoring for a Driver Based on Head Pose and Eye Blinking Detection Using One Class Support Vector Machine

Hyunrae Jo and Minho Lee[*]

School of Electronics Engineering, Kyungpook National University
1370 Sankyuk-Dong, Puk-Gu, Taegu 702-701, South Korea
hrjo@ee.knu.ac.kr, mholee@knu.ac.kr

Abstract. This paper proposes a model to detect inattention cognitive state of a driver during various driving situations. The proposed system predicts driver's inattention state based on the analysis of eye blinking patterns and head pose direction. The study uses an infrared camera and several feature extraction stages such as modified census transform (MCT) to reduce the effect of light source change in real traffic environment. Also, we propose a new eye blinking detection using the difference between center and surround of Hough circle transform image. The local linear embedding (LLE) is used to extract real-time features of head movement. Finally, the driver's cognitive states can be estimated by the one-class support vector machines (OCSVMs) using both eyes blinking patterns and head pose direction information. We implement a prototype of the proposed driver state monitoring (DSM) system. Experimental results show that the proposed system using OCSVM works well in real environment compared to the system that employs SVM.

Keywords: One class SVM, Inattention detection, Eye blink, Driver state monitoring.

1 Introduction

According to "European accident research and safety report 2013", human error is the major contributing factor in 90% of the accidents involving Volvo trucks. Human error in particular constitutes driver's inattention and distraction. The cognitive vacuum due to inattention even for a second, at the speed of 100 km/h, can create defenseless situation and may cause a terrible accident. Several researchers and automobile manufacturers have studied these problems over the last decade. Recently, the trend of integrating intelligence systems with vehicles is seen as a solution to this problem and development of such systems is rapidly gathering pace.

Especially, in computer vision domain, existing research issues such as face detection, landmark localization, etc. are useful to grasp features of driver's cognitive state. Orazio et al. [1] proposed a probabilistic model to recognize driver's inattention and

---

[*] Corresponding author.

C.K. Loo et al. (Eds.): ICONIP 2014, Part II, LNCS 8835, pp. 110–117, 2014.

sleepiness by analyzing the eye occurrence. They used mixture of Gaussian models that are learned by EM algorithm and analyzed driver's behavior and inattention. Pohl et al. [2] used head pose and eye gaze factor for modeling driver's distraction level. It was represented as time dependent and nonlinear signal. Mbouna et al. [3] presented a visual analysis method of eye state and head pose for monitoring of alertness of a driver. They used SVM to classify alert or non-alert driving events.

In this paper, we use eye blinking and head-pose features from images and propose inattention mental state detection using one class support vector machine (OCSVM) [4]. It assumes that all training instances have only one-class label and discriminative boundary is learned around the normal instances. If any test instance does not fall within the learned boundary, it can be declared to be an inattention. We show that proposed one-class setting based model is more efficient in variety of real situations. We also discuss several existing methods that are adapted to implement our model. Modified census transform (MCT) feature based AdaBoost [5] is used to detect face landmark and to reduce effect of light source. Also, we propose a new eye blinking detection method based on the difference between center and surround of Hough circle transform image [6]. We use the local linear embedding (LLE) [7] to estimate head-pose features in image.

The rest of this paper is organized as follows: Section 2 discusses the proposed eye blink and head pose feature detection methods. Section 3 also discusses how detect driver's in-attention state using the OCSVM. Experimental results are presented in Section 4 and Section 5 concludes the paper.

2   Eye Blink and Head Pose Pattern Detection

2.1   MCT-Based AdaBoost to Detect Face Landmarks

In a driving situation, the driver's seat in the car is exposed to the outside light. Therefore, it is essential that the DSM system should be able to counter this problem so as to reduce the external influences. So, we choose the MCT-based AdaBoost that proposed by Froba et al. [5] for area detection of face and another landmarks.

Boosting algorithm is a machine learning method that makes a strong classifier by combining various weak classifiers. Generally, Boosting-training procedure is implemented according to the type of weak classifier. Viola and Jones [8] proposed Adapted Boosting method that used Haar-like features.

MCT is a non-parametric local structured feature and is a modified version of the census transform. It is a representation of contrast of pixel intensities in a local area and shows which pixels have greater intensity than the mean pixel intensity. It has advantages of being robust to the variations in light and is memory efficient too. In general, MCT features are calculated by 3×3 kernel and are represented with 511 patterns. MCT feature doesn't use the decision boundary since it has discrete characteristic. So MCT-based classifier is designed using lookup table.

2.2 Eye Blinking Pattern Extraction

Eye blinking state is closely related to driver's fatigue [9]. We use following process to classify open and closed eye. We use detected face images as an input for this stage. The sampling time window size is T at each instance. First, we detect eye's location and size using the MCT-based AdaBoost. After detecting the region of eyes using the MCT-based Adaboost, we need to distinguish between open and closed states of eyes. At first, we use self-quotient image (SQI) [10] as a preprocessing step in the eye detected regions. It is a kind of high pass filter that has robust characteristic from change of light source. The main difference between open eye and closed eye is whether the iris exists in the eye detected images. Usually, iris has a circle shape, and thus the Hough circle transform can be a good candidate to detect the iris existence in the eye detected images. Since the center points with maximum value in the Hough circle transform can be detected in the closed eyes, the Hough circle transform itself may not enough to detect the eye states whether eye is open or close.

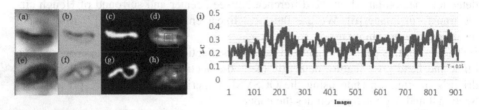

Fig. 1. Eye blink detection: (a) closed eye image, (e) open eye images, (b), (f) self-quotient images, (c), (g) binary images, (d), (h) Hough transform images, red box indicate center and surround area, (i) $CS(i_x, i_y)$ values at BioID database.

On the other hand, the center and surround difference is usually used for construction of visual saliency maps, which can reinforce a center area with different characteristics in a surround region. Since eye open images usually brighter at the center area in the Hough circle transformed images than the surround areas, we apply the center and surround difference operator to the Hough circle transformed images, which is useful to detect the eye states whether they are open or close. Fig. 1 shows the results of an example image to get the eye states. Eq. (1) and (2) are the operator for center and surround difference and the final binary output of eye blink, respectively, where L is intensity level in gray scale image. The center area pixels, $c(i, j)$ where i and j are indexes of x and y position in the image, is defined as a preset region around a maximum value of Hough circle transformed image in Fig. 1 (d), (h),   the surround area pixels, $s(i, j)$, is defined as twice background area around the center area. As shown in Eq. (1), the center and surround difference value, $CS(i_x, i_y)$, at a center of iris candidate at i_x and i_y in an image is obtained by the difference of normalized center area pixels and the normalized surround area pixels. If the $CS(i_x, i_y)$ is over a threshold, we define it as eye open state. Fig. 1 (i) and Table 1 show the detection results of eye blink according to time in 1000 non-glasses images of BioID database.

$$CS(i_x, i_y) = \frac{1}{L} \left\| \frac{\sum_{i,j \in C} C(i,j)}{C_{width} \times C_{Height}} - \frac{\sum_{i,j \in S} S(i,j)}{S_{width} \times S_{Height}} \right\| \tag{1}$$

$$EB = \begin{cases} 1, & CS(i_x, i_y) > T \\ 0, & elsewher \end{cases} \tag{2}$$

Table 2 shows the features related with eye blinking and head pose states, which are used for the OCSVM as input features.

Table 1. Comparison accuracy

Approach	Acc (%) (BioID)
Z. Liu (2008) [11]	97.14%
E. Cheng (2009) [12]	94.00%
Inho Choi (2011) [13]	96.00%
Proposed method	96.30 %

Table 2. Feature Variables

Features	Variable	Description
Head-pose	HP_MEAN	Mean head-pose vectors
	HP_STDEV	Standard deviation head-pose
Eye blinking	EB_PEROP	Time portion that the eyes are open
	EB_STDEV	Standard deviation of eye blinking
	EB_DUCLOS	Duration of longest eye closure

2.3 Head-Pose State Extraction

Dimensionality reduction techniques are needed in those applications where dimensionality is an issue. Locally linear embedding (LLE) [7] was presented at the same time as Isomap [14], which is a type of dimension reduction method on manifold. LLE begins by finding a set of nearest neighbors of each instances and a set of weights that best reconstructs each instances as a linear combination of its neighbors. Finally, it finds the low-dimensional embedding of instances using eigenvector based optimization technique. Each data are still described with the same linear combination of its neighbors. The following are the steps involved:

1. Find neighbors in X space
2. Solve reconstruction weights W
3. Compute embedding coordinates Y using weights W

We used LLE to estimate head-pose information from a face image. Then we found two principal axes on the manifold space. Further, we normalized these two axes to [0-1]. HP_MEAN and HP_STDEV can be calculated by projected values on manifold axes. (Table 1)

3   In-attention State Detection of a Driver

This paper proposes driver's inattention mental state detection model and the overall block diagram of the proposed system is shown in Fig. 2. It is hard to recognize inattention state of a diver using only one image. Therefore, driver state should be continuously analyzed by observing a series of situation flow. We assume that the input of our model is consecutive images set $I = \{i_1, i_2, ..., i_T\}$ during T time window. We find the face and eye location using MCT-based Adaboost. The binary value for the eye blinking $EB(i)$ and head pose direction $HP(i)$ are obtained by center-surround method and LLE, respectively. The eye blinking and head-pose features mentioned in Table I are calculated using the $HP(i)$ and $EB(i)$ at each sampling instance during T time window. Then, the feature vectors with five dimensions are constructed and used as training and test samples. For training, we randomly select video data for normal driving states. Remaining data, which are not used for training, are used for test procedure. The test cases may include the drowsiness state and distraction situations in real-time video. The decision boundary for detecting the driver's inattention states is determined by training the OCSVM.

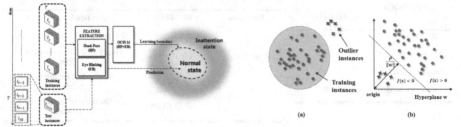

Fig. 2. Driver's inattention state detection model

Fig. 3. One-class SVM. (a) Input space, and (b) projected feature space.

3.1   Inattention Detection Using One-Class Support Vector Machine (OCSVM)

One-class SVM (OCSVM) was proposed by Scholkopf et al. [4] to estimate the support of individual classes. SVM can classify two classes with labels. However, OCSVM separates all the data points from the origin in feature space. It maximizes the distance from this hyperplane to the origin. This results in a binary function, which captures regions in the input space where the probability density of the data exists. Thus the function returns +1 in a region that represents the training data and -1 elsewhere.

Fig. 3 presents a simple example of how outliers can be separated from the training data. Fig. 3a indicates the distribution of the training data and outliers in the input space. Then, the kernel function makes the data to project onto the feature space (Fig. 3b). The hyperplane w separates the training data from the origin by a maximal margin $\rho/\|w\|$.

Training data $\{x_{1,...,}x_n\} \in \chi$ are given from feature extraction step. Then, we solve the quadratic programming minimization function to find the optimal hyperplane w.

$$\min_{w,\xi_i,\rho} \frac{1}{2}\|w\|^2 + \frac{1}{vn}\sum_{i=1}^{n}\xi_i - \rho \tag{3}$$

Subject to:

$$(w \cdot \phi(x_i)) \geq \rho - \zeta_i, \quad \text{for all } i=1,...,n$$

$$\zeta_i \geq 0, \quad \text{for all } i=1,...,n$$

This optimization problem can be solved by Lagrangian multipliers and the kernel is used for projection. The decision function is written as:

$$f(x) = \text{sgn}((w \cdot \phi(x_i)) - \rho) \tag{4}$$

$$= \text{sgn}(\sum_{i=1}^{n} \alpha_i K(x, x_i) - \rho) \tag{5}$$

The parameter v sets an upper bound on the outliers and a lower bound on the training instances used as support vector. We can control generalization performance by the parameter v. and α_i are the Lagrange multipliers that are weighted in the decision function.

4   Experiment and Result

In an experiment to test the performance of our proposed method, we used grayscale webcam with 640*480 resolution. IR was used because image based DSM systems cannot be used at night. IR webcam was installed in front of driver's seat and it could extract driver's features vector using the developed software in real-time.

To achieve the best inattention detection model, we employed the following steps: First, we make use of five values such as: HP_MEAN, HP_STDEV, EB_PEROP, EB_STDEV, EB_DUCLOS. The input vector x consists of these five values. Second, the training examples were trained using an OCSVM classifier. Over 2000 continuous images that are considered normal state by humans were used for training. Then we applied OCSVM algorithm for inattention detection applications, training examples whose decision outputs were greater than zero were considered normal, while the others were detected as inattention. RBF (Radial basis function) was applied for OCSVM. SVM needs a kernel function that transforms the input space to higher dimension. This kernel function is given by (6):

$$K(x, x') = \exp(-\frac{\| x - x' \|^2}{2\sigma^2}) \tag{6}$$

Since there is no public database available about driving situations, we made our own videos in real driving situation and simulated several situations. The videos had ten minutes run time and we made the ground truth data manually to test the performance. We made 10 video data sets for experiments. 5 data sets were used for learning process and rest were used for test. Table 3 and Fig. 4 are the comparison results between two class SVM and OCSVM. It shows that the performance of OCSVM is better than two class SVM if the model is trained with same data sets. This phenomenon can occur due to lack of training data or noise of data by inaccuracy of driver's inattention data.

Table 3. Comparison between SVM and OCSVM

Data sets	Accuracy	
	SVM	OCSVM
1	76.4	91.4
2	77.3	85.8
3	88.1	89.2
4	85	89.3
5	84.3	94.2

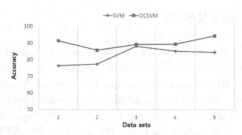

Fig. 4. Graph of comparison between SVM and OCSVM

5   Conclusion and Discussion

In this paper, we implemented driver's inattention state detection system using OCSVM with eye blinking and head pose information. As we discussed in the introduction, there are three major considerations to implement efficient DSM system. First, a method is needed to reduce the effect of light sources in external environment. For this purpose, we used MCT based AdaBoost algorithm. Second is to determine feature that is useful to understand driver's state. We propose a new eye blinking detection method based on the difference between center and surround of Hough circle transform image and it have strength of simple but useful performance. Third is to define driver's normal cognitive state and classification to distinguish between normal and in-attention state of the driver.

We tested the proposed DSM system in real environment. Despite the fact that it is too difficult to obtain abnormal sample data in real environment, overall the performance of our system was evidently better than two class SVM in real driving situations. However, there are few challenges that need to be addressed. For example, when the external light source is too bright or sun shines on the car at a steep angle (in case of sunrise or sunset), the camera gets saturated. In this case, the analysis from image becomes useless.

In our future work, we want to consider more robust features and method for changing of light source. And the study to implement system that conflates external risk factor awareness and internal driver state recognition will be done.

Acknowledgment. This research was supported by the MSIP(Ministry of Science, ICT & Future Planning), Korea, under the C-ITRC(Convergence Information Technology Research Center) support program (NIPA-2014-H0401-14-1004) (50%) and 'Software Convergence Technology Development Program' (S1002-13-1014) (50%).

References

1. D'Orazio, T., et al.: A visual approach for driver inattention detection. Pattern Recognition 40(8), 2341–2355 (2007)
2. Pohl, J., Birk, W., Westervall, L.: A driver-distraction-based lanekeeping assistance system. J. Syst. Control Eng. 221(14), 541–552 (2007)
3. Mbouna, R.O., Kong, S.G., Chun, M.-G.: Visual analysis of eye state and head pose for driver alertness monitoring. IEEE Transactions on Intelligent Transportation Systems 14(3), 1462–1469 (2013)
4. Schölkopf, B., Platt, J.C., Shawe-Taylor, J., Smola, A.J., Williamson, R.C.: Estimating the upport of a highdimensional distribution. Neural Computation 13(7), 1443–1471 (2001)
5. Froba, B., Ernst, A.: Face Detection with the Modified Census Transform. In: Proceedings of the Sixth IEEE International Conference on Automatic Face and Gesture Recognition, pp. 91–96 (2004)
6. Hough, P.V.: Machine analysis of bubble chamber pictures. In: International Conference on High Energy Accelerators and Instrumentation, vol. 73 (1959)
7. Roweis, S.T., Saul, L.K.: Nonlinear dimensionality reduction by locally linear embedding. Science 290(5500), 2323–2326 (2000)
8. Viola, P., Jones, M.J.: Robust real-time face detection. International Journal of Computer Vision 57(2), 137–154 (2004)
9. Caffier, P.P., Erdmann, U., Ullsperger, P.: Experimental evaluation of eye-blink parameters as a drowsiness measure. European Journal of Applied Physiology 89(3-4), 319–325 (2003)
10. Wang, H., Li, S.Z., Wang, Y.: Face recognition under varying lighting conditions using self quotient image. In: Proceedings of Sixth IEEE International Conference on Automatic Face and Gesture Recognition, pp. 819–824. IEEE (2004)
11. Liu, Z., Ai, H.: Automatic eye state recognition and closed-eye photo correction. In: 19th International Conference on Pattern Recognition, ICPR 2008, pp. 1–4. IEEE (2008)
12. Cheng, E., et al.: Eye state detection in facial image based on linear prediction error of wavelet coefficients. In: IEEE International Conference on Robotics and Biomimetics, ROBIO 2008, pp. 1388–1392. IEEE (2009)
13. Choi, I., Han, S., Kim, D.: Eye detection and eye blink detection using adaboost learning and grouping. In: 2011 Proceedings of 20th International Conference on Computer Communications and Networks (ICCCN), pp. 1–4. IEEE (2011)
14. Tenenbaum, J.B., De Silva, V., Langford, J.C.: A global geometric framework for nonlinear dimensionality reduction. Science 290(5500), 2319–2323 (2000)

An Improved Separating Hyperplane Method
with Application to Embedded Intelligent Devices

Yanjun Li[1], Ping Guo[1,2], and Xin Xin[1,*]

[1] School of Computer Science and Technology, Beijing Institute of Technology,
Beijing 100081, China
[2] Image Processing and Pattern Recognition Laboratory, Beijing Normal University,
Beijing 100875, China
liyanjunzgz@126.com, pguo@ieee.org, xxin@bit.edu.cn

Abstract. Classification is a common task in pattern recognition. Classifiers used in embedded intelligent devices need a good trade-off between prediction accuracy, resource consumption and prediction speed. Support vector machine(SVM) is accurate but its run-time complexity is higher due to the large number of support vectors. A new separating hyperplane method (NSHM) for the binary classification task was proposed. NSHM allows fast classification. However, NSHM is order-sensitive and this affects its classification accuracy. Inspired by NSHM, we propose CSHM, a combining separating hyperplane method. CSHM combines all optimal separating hyperplanes found by NSHM. Experimental results on UCI Machine Learning Repository show that, compared with NSHM and SVM, CSHM achieves a better trade-off between prediction accuracy, resource consumption and prediction speed.

Keywords: separating hyperplane, embedded intelligent devices, support vector machines.

1 Introduction

Classification is a common task in pattern recognition. Support vector machine (SVM) is a promising method for classification [1–3]. Compared with other classifiers one significant advantage of SVM is that SVM can use kernel trick.

As information technology rapidly evolves, embedded intelligent devices are widely used. For example, medical embedded device for individualized care, face detection in video sensor networks and intrusion detection at network routing hubs. In these applications embedded devices are expected to analyze the data and make on-site intelligent decisions (classifications) instead of transmitting the data.

Classifiers used in embedded intelligent devices need a good trade-off between prediction accuracy, resource consumption and prediction speed, because the prediction of new data points must be done at run time with limited resources (computational power and memory of embedded devices). SVM, though accurate, is not suitable for embedded intelligent devices. Because the run-time complexity, as explained in [4], can be

---

* Corresponding author.

C.K. Loo et al. (Eds.): ICONIP 2014, Part II, LNCS 8835, pp. 118–125, 2014.
© Springer International Publishing Switzerland 2014

higher due to the large number of support vectors (SVs). When the number of SVs increases, the time and memory needed to predict new data points increase.

[5] proposed a new separating hyperplane method (NSHM) which is suitable for embedded intelligent devices. NSHM generalizes well, allows fast classification and can use kernel trick.

However, NSHM is order-sensitive, *i.e.* different training data input orders may generate different optimal separating hyperplanes which have the same minimum misclassification rate on the training set. This affects its classification accuracy.

Inspired by NSHM we propose CSHM, a combining separating hyperplane method. CSHM combines all optimal separating hyperplanes found by NSHM into one separating hyperplane. Compared with SVM and NSHM, CSHM has a better trade-off between prediction accuracy, resource consumption and prediction speed, because CSHM solves the problem of order-sensitivity of NSHM and the number of SVs of CSHM is far less than SVM. The relation between the number of SVs and training data is shown in Table 1. Here, we use Pima Indians Diabetes data set which can be used for medical embedded device for individualized care. We note that in CSHM the relation between the number of SVs and training data is not linear. This may be caused by the following reasons. (1) CSHM searches the optimal separating hyperplane by exhaustive search. (2) Increasing the number of data points may lead to the optimal separating hyperplanes change.

Table 1. The relation between SVs and training data number

# of training data	200	220	240	260	280	300	320	340	360	380	400
# of SVs of SVM	154	167	178	180	192	205	217	225	231	244	254
# of SVs of CSHM	4	6	4	4	4	10	6	4	6	10	8
# of SVs of NSHM	2	2	2	2	2	2	2	2	2	2	2

The remainder of this paper is organized as follows. First, we provide an overview of NSHM in Section 2. Then we put forward CSHM in Section 3. After experimental results are shown in Section 4, we draw the conclusions in Section 5.

2   NSHM Overview

NSHM is a separating hyperplane method. The optimal separating hyperplane found by NSHM has the minimum misclassification rate on the training set. The NSHM searches the optimal separating hyperplane by exhaustive search.

Suppose that the training set is $\mathbf{Tr} = \{(\mathbf{x}_l, y_l)\}_{l=1}^{m}$, where $\mathbf{x}_l$ is a $d-$dimensional feature vector and $y_l \in \{-1, +1\}$ is the class label of $\mathbf{x}_l$. The decision function generated by NSHM can be defined as

$$f(\mathbf{x}) = sgn(y_i \mathbf{x}_i^T \mathbf{x} + y_j \mathbf{x}_j^T \mathbf{x} + b), \tag{1}$$

where $b = -(y_i \mathbf{x}_i^T \mathbf{x}_k + y_j \mathbf{x}_j^T \mathbf{x}_k)$ and $y_i y_j = -1$. The geometrical interpretation of decision function (1) is that the separating hyperplane is orthogonal to lines connecting a pair of data points from the different classes of training set and includes at least one point in the training set. The dot product $\mathbf{x}_i^T \mathbf{x}$ can be replaced with kernel function $K(\mathbf{x}_i, \mathbf{x})$ to achieve non-linear classification. The parametric representation of the separating hyperplane defined by (1) is $\mathbf{w}^T\mathbf{x} + b = 0$, where $\mathbf{w} = y_i\mathbf{x}_i + y_j\mathbf{x}_j$ is normal vector and $b = -\mathbf{w}^T\mathbf{x}_k$ is intercept. Classification rules are if $(\mathbf{w}^T\mathbf{x} + b) \geq 0$ then $f(\mathbf{x}) = 1$ else $f(\mathbf{x}) = -1$. For convenience, we call $\mathbf{x}_i$ and $\mathbf{x}_j$ the corresponding vectors of $\mathbf{w}$, and we call the corresponding vectors of $\mathbf{w}$ support vectors of NSHM. With the use of kernel function, (1) becomes

$$f(\mathbf{x}) = sgn(\sum_{l=1}^{n_{NSHM}} y_l K(\mathbf{x}_l, \mathbf{x}) + b), \tag{2}$$

where n_{NSHM} is the number of SVs of NSHM. NSHM has two SVs, so $n_{NSHM} = 2$.

3 Approach

In this section, we first analyze the order-sensitive problem of NSHM, then propose CSHM.

3.1 Theoretical Analysis of Order-Sensitivity

NSHM is sensitive to training data input orders. Different training data input orders may generate different separating hyperplanes, which have the same minimum misclassification rate on the training set. Given a training set with m data points, there are $m^+ m^- m$ separating hyperplanes. m^+ and m^- are the number of data points from class $y = 1$ and class $y = -1$, respectively. The number of classification accuracy is $m + 1$. Therefor, for the decision function defined by (1), there may be several separating hyperplanes, which meet the constraint of minimizing the misclassification rate on the training set. These separating hyperplanes may be different in intercept, norm vector or intercept and norm vector, as shown in Fig. 1. Fig. 1 (a) and (b) show two separating hyperplanes which are different in norm vector. Fig. 1 (c) and (d) show two separating hyperplanes which are different in intercept. Fig. 1 (e) and (f) show two separating hyperplanes which are different in intercept and norm vector. When we make predictions, these separating hyperplanes have different classification accuracies (detailed experimental analysis see 4.1).

Intuitively, we can solve the order-sensitive problem of NSHM by setting a validation set, *i.e.* we split a validation set from the training set and choose the classifier attaining best classification performance on the validation set as final classifier. However, when the training data set is relatively small this method has two problems.

One is, when the training set is split into a subtraining set and a validation set, the optimal separating hyperplane learned by NSHM on the subtraining set may be different from the optimal separating hyperplane learned on the original training, because some data points are not included in the subtraining set and NSHM searches the optimal separating hyperplane by exhaustive search. The other is that the separating hyperplane learned by NSHM will be partial to validation set. When the data distribution of validation set is inconsistent with test set, NSHM generalizes poor (detailed experimental analysis see 4.2).

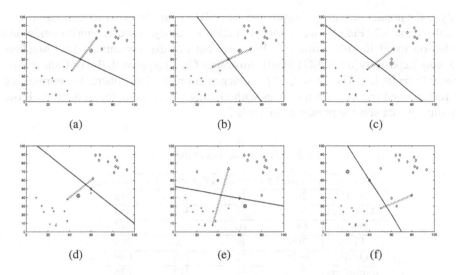

Fig. 1. Different separating hyperplanes. Data point marked with circle is misclassified. Solid line corresponds to separating hyperplane. Dotted line corresponds to norm vector.

3.2 CSHM

Our method (CSHM) is combining all optimal separating hyperplanes found by NSHM, instead of only using one of these separating hyperplanes. First we find all optimal separating hyperplanes, which have the same minimum misclassification rate on the training set. These separating hyperplanes are denoted by $\langle \mathbf{w}_l, b_l \rangle$, $l = 1, \ldots, p$, where p is the number of separating hyperplane. Then we combine these optimal separating hyperplanes into one separating hyperplane. Our decision function is defined as

$$f(\mathbf{x}) = sgn(\sum_{l=1}^{p}(\mathbf{w}_l^T \mathbf{x} + b_l)) \tag{3}$$

Classification rules are if $f(\mathbf{x}) = sgn(\sum_{l=1}^{p}(\mathbf{w}_l^T \mathbf{x} + b_l)) \geq 0$ then $f(\mathbf{x}) = 1$ else $f(\mathbf{x}) = -1$. For convenience, we call the corresponding vectors of $\mathbf{w}_l$ support vectors of CSHM, where $l = 1, 2, \ldots, p$. By computing and using kernel function, (3) becomes

$$f(\mathbf{x}) = sgn(\sum_{l=1}^{n_{CSHM}} y_l K(\mathbf{x}_l, \mathbf{x}) + b_{CSHM}), \tag{4}$$

where n_{CSHM} is the number of SVs of CSHM and $b_{CSHM} = \sum_{l=1}^{p} b_l$ is a constant. Each separating hyperplane has two SVs, so $n_{CSHM} = 2p$.

4 Experiments

In this section, we provide detailed empirical evidences. First, we analyze the order-sensitive problem of NSHM from experimental viewpoints in Subsection 4.1. Then we analyze the impact of setting a validation on NSHM from the experimental viewpoints in Subsection 4.2. Finally, we report prediction accuracy, resource consumption and prediction speed in Subsection 4.3. We carry out our experiments on UCI Machine Learning Repository[6]: SPECT Heart, Australian Credit Approval, Breast Cancer Wisconsin, Pima Indians Diabetes, QSAR biodegradation and Ionosphere. We normalized the features and removed 16 instances with missing values from Breast data set. Basic statistics of data sets are presented in Table 2.

Table 2. Basic statistics of data sets

Data set	# of total	# of features	# of class 1	# of class 2
SPECT	267	44	55	212
Australian	690	14	307	383
Breast	683	10	239	444
Pima	768	8	500	268
QSAR	1055	41	356	699
Ionosphere	351	34	126	225

4.1 Experimental Analysis of Order-Sensitivity

SPECT and Australian data set are used in this subsection. For convenience, we do not use kernel function, because this makes our experiments have no turning parameters related to classification accuracy. The SPECT data set has been divided into training set and test set. For Australian data set, 150 randomly selected samples of each class are used for training and the rest are used for testing.

We search all possible separating hyperplanes found by NSHM. Then we evaluate classification accuracy of these separating hyperplanes on the test set. We found 8 separating hyperplanes on SPECT data set and 6 separating hyperplanes on Australian data set. Experimental results are shown in Table 3 and Table 4, respectively. We also evaluate the classification accuracy of CSHM.

Table 3. Classification accuracy of possible separating hyperplanes on SPECT data set

Separating hyperplanes	1	2	3	4	5	6	7	8
Accuracy(%)	75.40	74.87	73.26	66.84	70.59	68.98	80.75	72.73

Table 4. Classification accuracy of possible separating hyperplanes on Australian data set

Separating hyperplanes	1	2	3	4	5	6
Accuracy(%)	84.87	83.59	80.51	80.00	80.26	81.54

Table 3 and Table 4 show that, given same training set and different training data input orders, the classification accuracy of NSHM fluctuates in a large range (66.84%

to 80.75% on SPECT data set, 80% to 84.87% on Australian data set). The average classification accuracies of these separating hyperplanes on SPECT and Australian data set are 72.93% and 81.80%. They are all lower than CSHM, whose classification accuracies on SPECT and Australia data set are 73.80% and 84.62%, respectively. CSHM efficiently solves the order-sensitive problem and generalizes well.

4.2   Experimental Analysis of Setting a Validation Set

We use same experimental settings as Subsection 4.1, except that 20% data points of training set are used for validation. We repeated the experiment 10 times with different randomly selected validation set. Experimental results on SPECT and Australian data set are shown in Table 5 and Table 6, respectively.

Table 5. Classification accuracy of setting a validation set on SPECT data set

Experimental sequence	1	2	3	4	5	6	7	8	9	10
Accuracy(%)	52.41	17.65	90.37	49.20	73.26	85.03	16.04	71.12	59.36	70.05

Table 6. Classification accuracy of setting a validation set on Australian data set

Experimental sequence	1	2	3	4	5	6	7	8	9	10
Accuracy(%)	66.92	13.85	46.92	61.28	51.28	85.13	84.87	14.87	38.97	27.18

Table 5 and Table 6 show that, when the training data set is relatively small and we split the training set into a subtraining set and a validation set, the classification accuracy of NSHM fluctuates in a large range (16.04% to 90.37% on SPECT data set, 13.85% to 85.13% on Australian data set).

When the data distribution of validation set is consistent with test set NSHM obtains a very high classification accuracy (90.37% on SPECT data set, 85.13% on Australian data set). When the data distribution of validation set is inconsistent with test set NSHM obtains a very low classification accuracy (16.04% on SPECT data set, 13.85% on Australian data set). The average classification accuracies of setting a validation set on SPECT and Australia data set are 58.45% and 49.13%, respectively. They are all lower than CSHM, whose classification accuracies on SPECT and Australia data set are 73.80% and 84.62%, respectively.

4.3   Prediction Accuracy, Resource Consumption and Prediction Speed

We compare prediction accuracy, resource consumption and prediction speed of CSHM, NSHM and SVM on SPECT, Australian, Breast, Pima , QSAR and Ionosphere data set. For each data set, we randomly choose same number of data points from each class as training and the rest are used as testing. The number of data points from each class are 40 (SPECT), 100 (Ionosphere), 150 (Australian), 150 (Pima), 150 (Breast) and 200

(QSAR), respectively. Experiments on each data set were repeated ten times with randomly selected training and test data. The mean and standard deviation of classification accuracy and the number of SVs were recorded.

Prediction speed and resource consumption are related to the calculated amount of decision function. The decision function of CSHM is $f(\mathbf{x}) = sgn(\sum_{l=1}^{n_{CSHM}} y_l K(\mathbf{x}_l, \mathbf{x})+b_{CSHM})$, where n_{CSHM} is the number of SVs of CSHM and b_{CSHM} is a constant. The decision function of NSHM is $f(\mathbf{x}) = sgn(\sum_{l=1}^{n_{NSHM}} y_l K(\mathbf{x}_l, \mathbf{x}) + b)$, where n_{NSHM} is the number of SVs of NSHM and b is a constant. The decision function of SVM is $f(\mathbf{x}) = sgn(\sum_{l=1}^{n_{SVM}} y_l \alpha_l K(\mathbf{x}_l, \mathbf{x}) + b_{SVM})$, where n_{SVM} is the number of SVs of SVM and b_{SVM} is a constant. The calculated amount of decision function is related to the number of SVs. Therefore, we use the number of SVs to measure the prediction speed and resource consumption.

We use RBF Kernel ($k(\mathbf{x}_i, \mathbf{x}_j) = \exp(-\gamma\|\mathbf{x}_i - \mathbf{x}_j\|^2)$). Experimental parameters are selected by cross-validation. Parameters are obtained by grid-search. Parameter C is searched on grid $\{0.01, 0.05, 0.1, 0.5, 1, 5, 10, 50, 100\}$ and parameter γ is searched on grid $\{\frac{1}{2^7}, \frac{1}{2^5}, \frac{1}{2^3}, \frac{1}{2}, 1, 2, 2^3\}$. Experimental results of classification accuracy and the number of SVs are shown in Table 7 and Table 8, respectively.

Table 7. Classification accuracy

Data Set	SPECT	Australian	Breast	Pima	QSAR	Ionosphere
CSHM(%)	74.43±4.58	84.97±0.81	96.66±0.83	71.50±2.79	82.70±1.62	86.69±2.87
NSHM(%)	73.21±4.32	84.10±1.47	96.58±0.82	70.94±3.08	80.99±1.8	85.56±2.62
SVM(%)	74.49±3.04	85.02±1.39	96.76±0.41	71.71±1.37	84.48±1.69	86.95±2.83

Table 8. The number of SVs

Data Set	SPECT	Australian	Breast	Pima	QSAR	Ionosphere
CSHM(#)	10.97±3.40	5.4±2.98	12.6±9	5.4±3.37	4.4±2.45	4.8±2.12
NSHM(#)	2±0.00	2±0.00	2±0.00	2±0.00	2±0.00	2±0.00
SVM(#)	69.6±7.00	184.6±56.67	84.9±34.3	214.1±47.64	196.4±33.2	160.8±3.42

Table 7 shows that except QSAR data set the classification accuracy of CSHM is close to SVM. Compared with NSHM, CSHM improves the classification accuracy by solving the order-sensitive problem.Table 8 shows that the number of SVs of SVM is far more than CSHM. As far as applying CSHM to embedded devices is concerned, the increased number of SVs is acceptable. From Table 7 and 8 we can conclude that, compared with NSHM and SVM, CSHM achieves a better trade-off between prediction accuracy, resource consumption and prediction speed.

5   Conclusions

Classifiers used in embedded intelligent devices need a good trade-off between prediction accuracy, resource consumption and prediction speed. SVM is accurate but it is not suitable for embedded intelligent devices due to that the large number of support vectors increase the run-time complexity. NSHM is suitable for embedded intelligent devices. However, NSHM is order-sensitive and this affects its classification accuracy. Based on NSHM this paper proposes a new method called CSHM that combines all the optimal separating hyperplanes found by NSHM into one separating hyperplane. Experimental results on UCI Machine Learning Repository show that, compared with NSHM and SVM, CSHM achieves a better trade-off between prediction accuracy, resource consumption and prediction speed.

Acknowledgements. The work described in this paper was mainly supported by National Natural Science Foundation of China (No. 61300076), National Natural Science Foundation of China (No. 61375045), Ph.D. Programs Foundation of Ministry of Education of China (No. 20131101120035) and Beijing Natural Science Foundation(4142030).

References

1. Pang, S., Kim, D., Bang, S.Y.: Face membership authentication using SVM classification tree generated by membership-based LLE data partition. IEEE Transactions on Neural Networks 16(2), 436–446 (2005)
2. Tuia, D., Volpi, M., Mura, M.D., Rakotomamonjy, A., Flamary, R.: Automatic feature learning for spatio-spectral image classification with sparse SVM. IEEE Transactions on Geoscience and Remote Sensing 52(10), 6062–6074 (2014)
3. Wu, J.: Efficient HIK SVM learning for image classification. IEEE Transactions on Image Processing 21(10), 4442–4453 (2012)
4. Renjifo, C., Barsic, D., Carmen, C., Norman, K., Peacock, G.S.: Improving radial basis function kernel classification through incremental learning and automatic parameter selection. Neurocomputing 72(1-3), 3–14 (2008)
5. Owczarczuk, M.: New separating hyperplane method with application to the optimisation of direct marketing campaigns. Pattern Recognition Letters 32(3), 540–545 (2011)
6. Bache, K., Lichman, M.: UCI Machine Learning Repository (2013),
   http://archive.ics.uci.edu/ml

Fine-Grained Air Quality Monitoring Based on Gaussian Process Regression

Yun Cheng*, Xiucheng Li*, Zhijun Li, Shouxu Jiang, and Xiaofan Jiang

Harbin Institute of Technology
{chengyun.hit,xiucheng90,fxjiang}@gmail.com,
{lizhijun_os,jsx}@hit.edu.cn

Abstract. Air quality is attracting more and more attentions in recent years due to the deteriorating environment, and $PM_{2.5}$ is the main contaminant in a lot of areas. Existing softwares that report the level of $PM_{2.5}$ can provide only the value in the city level, which may indeed varies greatly among different areas in the city. To help people know about the exact air quality around them, we deployed 51 carefully designed devices to measure the $PM_{2.5}$ at these places and present a Gaussian Process based inference model to estimate the value at any place. The proposed method is evaluated on the real data and compared to some related methods. The experimental results prove the effectiveness of our method.

Keywords: $PM_{2.5}$ concentration monitoring, non-linear regression, Gaussian Process.

1 Introduction

The air pollution is considered a major and serious problem globally, especially in some cities of developing countries, such as Beijing and New Delhi. Among the various dimensions of air quality, particulate matter (PM) with diameters less than 2.5 micron, or $PM_{2.5}$, has gained a lot of attention recently. Medical studies have shown that $PM_{2.5}$ can be easily absorbed by the lung, and high concentrations of $PM_{2.5}$ can lead to respiratory disease [1] or even blood diseases [2]. Due to its close relation to public health, it has gained a lot of attention.

Nowadays, people are looking for better ways to monitor the quality of air in their immediate environment. There are many web or smartphone applications that report publicly-available air quality data at the city or district level, however, they cannot tell the actual air quality people breath-in, which is more relevant and valuable. Actually, there is probably a significant difference between the values of $PM_{2.5}$ concentration at different locations at the district level, which has been attested by the real data as shown in Section 4. Therefore, it is necessary to develop a fine-grained air quality monitoring system.

---

* Corresponding authors.

C.K. Loo et al. (Eds.): ICONIP 2014, Part II, LNCS 8835, pp. 126–134, 2014.

In order to estimate the air quality of any location, two major classical ways are proposed in the past of years. One is classical dispersion models, such as Gaussian Plume models, Operational Street Canyon models, and Computational Fluid Dynamics. These models are normally a function of meteorology, street geometry, receptor locations, traffic volumes, and emission factors (e.g., g/km per single vehicle), based on a number of empirical assumptions that might not be applicable to all urban environments and parameters which are also difficult to obtain precisely [3]. The other is interpolation using reports from nearby air quality monitor stations. This method is usually employed by public websites releasing the air quality index (AQI). Recently, big data reflecting city dynamics have become widely available and a group of researchers seek to infer the air quality using machine learning and data mining techniques. In the "U-Air" paper by Yu [4], the authors infer air quality based on AQIs reported by public air quality stations and meteorological data, taxi trajectories, road networks, and POIs (Point of Interests). Since there are only a few public monitor stations in a city, their training dataset is insufficient to train a commonly used supervised learning model. Therefore they propose a co-training-based semi-supervised learning model to tackle the data sparsity problem.

To overcome the drawbacks of existing methods shown above, we present a $PM_{2.5}$ monitoring system composing of a sensor network and a inference model. The sensor network that is deployed among the area to be monitored provides the values of $PM_{2.5}$ at these places. So we are in a much different situation compared with "U-Air", since we have designed our $PM_{2.5}$ monitoring devices and deployed them at a much higher density (51 monitor stations over an 30 km × 30 km urban area), which provides much more sufficient data for air quality estimation. Although our $PM_{2.5}$ devices cannot achieve the same measurement precision as the expensive public monitor stations, they are precise enough for the air quality estimation after calibration. Therefore we simply treat their readings as ground truth, and we mainly focus on the development of an effective model to estimate the value of $PM_{2.5}$ at any place using the acquired data at such an relatively higher deployment density. The paper is organized as follows: in Section 2 a description of the system is given. The inference model based on Gaussian Process is detailed in Section 3. The experiment setup and evaluation results are given in Section 4, and the conclusions are drawn in Section 5.

2   System

The system architecture of AirCloud is shown in Fig. 1. The system mainly contains two parts: 1) $PM_{2.5}$ monitoring system, which contains the AQM monitoring front-end and the backend data collection module; 2) Inference platform, which will be used to infer the unknown $PM_{2.5}$ concentrations of locations where there is not monitoring equipment. We will describe these two parts in the following subsections.

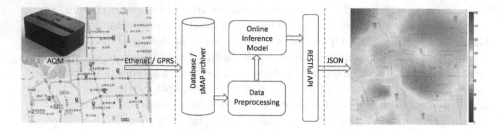

Fig. 1. The system architecture of Air-Cloud

2.1 $PM_{2.5}$ Monitoring System

The $PM_{2.5}$ monitroing system is used to collect, process and store the AQM sensor data.

$PM_{2.5}$ concentrations vary significantly over space, especially for metropolitan cities where pollution sources are multi-faceted. In addition, as we can observe from official $PM_{2.5}$ monitoring station data, $PM_{2.5}$ concentration changes at an hourly rate. As a result, direct monitoring is necessary. To solve our problem, we designed and built our own Internet-connected $PM_{2.5}$ monitors, AQM, as shown in Fig. 1. AQM station contains $PM_{2.5}$ concentration sensor, temperature and humidity sensor, the mechanical structure of the hardware is carefully designed and all the monitoring station will do the hardware calibration to remove initial hardware variations. We take an approach of using inexpensive sensors at the front-end and deployed at certain density, but rely on the reference model on the cloud to infer the $PM_{2.5}$ concentrations on the whole area.

To make the monitoring system more stable and scalable, we choose sMAP [5] as the data representation and storage system. We defined the standard specification for physical $PM_{2.5}$ sensor data, which contains the location information and the sensor readings, and use the database designed for time-series data, plus a powerful query language, provided by sMAP, as shown in Fig. 1. We use different communication approaches , Ethernet or GPRS, to connect AQM with the cloud server, and store all the data in sMAP archiver by the minute, which will be used by the inference model.

2.2 Inference Platform

The Inference platform is used to infer $PM_{2.5}$ concentrations at locations where there is not monitoring stations. We deploy the AQM monitor stations at certain density to get a general idea of the $PM_{2.5}$ concentrations around, however, to get the $PM_{2.5}$ concentration at locations where there is not monitoring stations, we have to rely on the inference platform, with the help of the Gaussian Process Inference model, we can get the accurate and fine-grained $PM_{2.5}$ concentration estimation of the whole area.

3 Air Quality Inference Model Based on Gaussian Process

In this section, we detail the Gaussian Process based inference module that estimates the value of $PM_{2.5}$ using the data from the monitoring network. First, we model the inference module as a regression after some necessary definitions. Then the problem is solved as a Gaussian Process regression and the details are presented.

3.1 Problem Definition and Regression Model

Using $\mathbf{x}_i$ to denote the coordinates of the i-th monitoring station and y_i the value of $PM_{2.5}$ at this place, the objective of the inference module is to inference the value of $PM_{2.5}$ y at any place given the data from all monitoring stations $\mathcal{D} = \{(\mathbf{x}_i, y_i), i = 1, \ldots, n\} \subseteq \mathcal{R}^2$ and the coordinate of the place to be estimate $\mathbf{x}$. This is a typical regression problem which is usually formulated as

$$y_i = f(\mathbf{x}_i) + \epsilon_i, \; \epsilon_i \sim \mathcal{N}(0, \delta^2), \tag{1}$$

where ϵ_i is the noise term and the objective is to learn a proper f from $\mathcal{D}$ which can predict a proper y for any given $\mathbf{x}$.

In our system we select Gaussian Process regression to model it, which is also known as Kirging in the spatial statistics field. Gaussian Process is a non-parametric Bayesian approach with sufficient flexibility to capture the complex and non-linear properties of the model. It has been proved to be a powerful tool in many areas and applied widely in practice. Since it is a fully probabilistic model, the objective is to learn a proper distribution of y instead of its value. We would like to detail how it is used to estimate the value of $PM_{2.5}$ at a specific place given its coordinate in following.

3.2 Gaussian Process Regression

Gaussian Process for Regression. A Gaussian process is a collection of random variables, any finite number of which have consistent joint Gaussian distributions. In Gaussian process regression problems, latent function f behaves following a Gaussian distribution (Normal distribution) when conditioning on $\mathbf{x}$

$$\mathbf{P}(f_1, f_2, \ldots, f_n | \mathbf{x}_1, \mathbf{x}_2, \ldots, \mathbf{x}_n) = \mathcal{N}(0, K)$$

where $f_i = f(\mathbf{x}_i)$ is latent function and K is a covariance matrix with entries given by the covariance function, $K_{ij} = k(\mathbf{x}_i, \mathbf{x}_j)$. $k(\mathbf{x}_1, \mathbf{x}_2)$ can be any valid kernel function satisfying Mercer's condition [6].

In inference, the training and test latent values is denoted as $\mathbf{f} = [f_1, f_2, \ldots, f_n]$, $\mathbf{f}_* = [f_{*1}, f_{*2}, \ldots, f_{*n}]$ separately, we combine the prior with the likelihood function via Bayes rule obtaining the posterior distribution:

$$\mathbf{P}(\mathbf{f}, \mathbf{f}_* | \mathbf{y}) = \frac{\mathbf{P}(\mathbf{f}, \mathbf{f}_*) \mathbf{P}(\mathbf{y}|\mathbf{f})}{\mathbf{P}(\mathbf{y})} \tag{2}$$

The desired posterior predictive distribution can be produced by marginalizing out the training set latent variables $\mathbf{f}$ in equation (2):

$$P(\mathbf{f}_*|\mathbf{y}) = \frac{1}{P(\mathbf{y})} \int P(\mathbf{f}, \mathbf{f}_*)P(\mathbf{y}|\mathbf{f}) \, d\mathbf{f} \tag{3}$$

since the prior and the likelihood function are mutually independent and both follow Gaussian distribution as

$$\begin{bmatrix} \mathbf{f} \\ \mathbf{f}_* \end{bmatrix} \sim \mathcal{N} \left(0, \begin{bmatrix} K_{\mathbf{f},\mathbf{f}} & K_{*,\mathbf{f}} \\ K_{\mathbf{f},*} & K_{*,*} \end{bmatrix} \right), \quad \mathbf{y}|\mathbf{f} \sim \mathcal{N}(\mathbf{f}, \delta^2 I) \tag{4}$$

where δ^2 is the noise variance and I is the identity matrix. Then the integral in equation (3) can be computed in close form and the result is also a Gaussian distribution [7] with $\mathbf{f}_*|\mathbf{y} \sim \mathcal{N}(\mu_*, \Sigma_*)$.

$$\mu_* = K_{*,\mathbf{f}}(K_{\mathbf{f},\mathbf{f}} + \delta^2 I)^{-1}\mathbf{y} \tag{5}$$

$$\Sigma_* = K_{*,*} - K_{*,\mathbf{f}}(K_{\mathbf{f},\mathbf{f}} + \delta^2 I)^{-1}K_{\mathbf{f},*} \tag{6}$$

where μ_* is the predictive mean and Σ_* is the corresponding covariance which indicate us the uncertain of the predictive value in the locations (we use $*$ as shorthand for f_*). In our scenario, μ_{*i} will be used as the predictive value y_i.

4   Experiment and Results

In experiment the real deployment dataset of more than one month was used to evaluate the performances of Gaussian Process Inference. There are totally 51 monitor stations deployed in an area with the size of 30 km × 30 km and each station reports its measurements every 30 minutes, the deployment map is shown in Figure 2-($\mathbf{A}$). We deliberately remove one station as ground truth and infer its

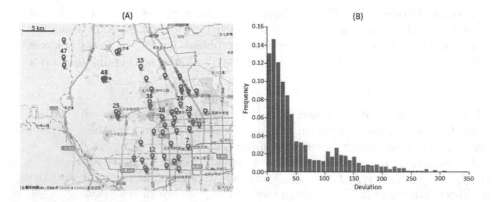

Fig. 2. ($\mathbf{A}$) The deployment map of the monitor stations; ($\mathbf{B}$) The distribution of the deviation between station S_{26} and S_{27} over one month

value using the remaining stations' reading at each timestamp. U-Air provides a public online visualization of its inference result in [8]. However, the AQI inferred by U-Air is just five standard levels specified by United States Environmental Protection Agency, at most cases the inferred results all stay in the same level in our deployment region, therefore it is unmeaningful to compare with it, and the linear and cubic spline interpolation are selected as the baseline methods. The following parts are organized as following: 1) we first discussed the setting of the covariance function as well as its parameter; 2) then the comparison between GP and the baseline methods was presented.

The covariance function plays a significantly import role in Gaussian Process, the training points that are close to a test point should be informative about the prediction at that point. From the Gaussian process view it is the covariance function that defines nearness or similarity [9]. In experiment we mainly investigated the following squared exponential covariance functions

$$k(\mathbf{x}_1, \mathbf{x}_2) = \exp\left(-\frac{1}{2\ell^2}\|\mathbf{x}_1 - \mathbf{x}_2\|_2\right)$$

Here ℓ is the horizontal scale over which the function changes. When the horizontal scale ℓ becomes large, the corresponding feature dimension is deemed irrelevant and the contrary is also true. If a relatively larger ℓ provides us a better inference result, it would imply that the distribution of the $PM_{2.5}$ concentrations in space is much smoother otherwise it would mean that the change of $PM_{2.5}$ concentrations among space is rapid. Therefore a suitable value of ℓ could reflect the variation degree of the $PM_{2.5}$ concentrations among the urban.

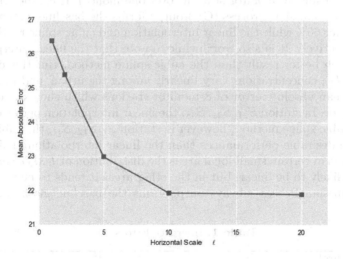

Fig. 3. The relationship between horizontal scale and mean absolute error

In Figure 3 we present the relationship between the covariance function parameter ℓ and the mean absolute error (using the data of all monitor stations).

With the increment of ℓ the error goes from 27.6 down to 21.9. This implies that the distribution of $PM_{2.5}$ concentrations is not smooth and the concentration in one location would highly differ from the one which departing in a long distance away from it. The Figure 2-(**B**) also shows the distribution of deviation between our two monitor stations, S_{26} and S_{28}, from May. 1, 2014 to Jun. 1, 2014. The geospatial distance of the two stations is about 6 km shown in Figure 2-(**A**), over 21% cases have a deviation greater than 100.

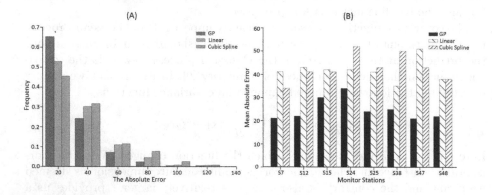

Fig. 4. (**A**)-The distribution of absolute error (using all dataset); (**B**)-The mean absolute error of eight monitor stations over one month

The Figure 4-(**A**) shows the absolute error distribution of the three methods (generated using all monitor stations over one month). It can be seen quite clearly that the Gaussian Process (GP) outperforms the baseline methods with small errors over 65% while the linear interpolation and cubic spline reaches 52% and 46% respectively. It is also worthwhile to note that the linear interpolation achieves a much better result than the cubic spline method. But this does not imply the $PM_{2.5}$ concentrations vary linearly among the urban, and Figure 4-B depicts the mean absolute error of 8 monitor stations which showing relatively large mean error. In station S_{24}, S_{25}, S_{38}, the linear interpolation performs much better than cubic spline method, however in station S_7, S_{12}, S_{47}, the cubic spline shows a more desirable performances than the linear interpolation. This might indicate us that in certain small local areas the distribution of $PM_{2.5}$ concentrations is more likely to be linear, but in the other areas it tends to be non-linear. Since the Gaussian Process always outperforms the baseline methods, it also

Table 1. Inference Errors

Measure Method	$\|x\|_1$	$\frac{1}{n}\|x\|_1$	$\|x\|_2$	RMSE	$\|x\|_\infty$
Linear	929135.20	33.54	5991.73	36.00	266.92
Cubic Spline	975562.63	35.21	6379.94	38.33	266.92
Gaussian Process	585229.04	21.12	4101.57	24.64	154.14

proves the flexibility of the Gaussian Process in spatial inference of $PM_{2.5}$ concentrations.

Table 1 lists the inference errors of the three methods measured via different rules(assume that x is the absolute error vector). Gaussian Process beats all baseline methods, especially the Chebyshev norm $\|x\|_\infty$ achieved by both the linear and cubic spline interpolation could be as large as 266.92 while the Gaussian Process obtains a much smaller value 154.14, which proves that the Gaussian Process is much more stable in the inference of $PM_{2.5}$ concentrations.

5 Conclusion

In this paper, we present an ambient $PM_{2.5}$ concentrations monitoring and estimation system using Gaussian Process Inference model. We deployed 51 our designed air quality measuring devices among the area to be monitored and the $PM_{2.5}$ at these places are continuously sent back. We use the Gaussian Process Inference model to estimate the $PM_{2.5}$ concentrations at locations where monitor stations are unavailable and the proposed method is compared with two baseline models: linear and cubicle spline interpolation. The result shows that GP Inference model performs much better in different situations than the other two baseline models, which proves the flexibility of Gaussian Process in spatial inference and that it is indeed suitable for estimation of $PM_{2.5}$ concentration among the urban area. Since an exact inference in Gaussian Process involves computing K^{-1}, the computation cost is $O(n^3)$ (n is the number of the training cases), so when the deployment scale growing out of 1000 stations, we will resort to the approximation schemes, such as sparse approximations [7][9]. Additionally, we also analyzed the effect on inference of altering the parameter in covariance function, the results indicated us that the distribution of $PM_{2.5}$ concentrations is not that smooth and it is necessary to resort to dense deployment in order to monitor the fine-granularity $PM_{2.5}$ pollution.

Acknowledgement. We thank the anonymous reviewers for their valuable comments. This work was jointly supported by the National Natural Science Foundation of China under Grant No. 61300210 and No. 61370214.

References

1. Boldo, E., Medina, S., Le Tertre, A., Hurley, F., Mücke, H.G., Ballester, F., Aguilera, I.: Apheis: Health impact assessment of long-term exposure to pm2. 5 in 23 european cities. European Journal of Epidemiology 21(6), 449–458 (2006)
2. Sørensen, M., Daneshvar, B., Hansen, M., Dragsted, L.O., Hertel, O., Knudsen, L., Loft, S.: Personal pm2. 5 exposure and markers of oxidative stress in blood. Environmental Health Perspectives 111(2), 161 (2003)
3. Vardoulakis, S., Fisher, B.E., Pericleous, K., Gonzalez-Flesca, N.: Modelling air quality in street canyons: a review. Atmospheric Environment 37(2), 155–182 (2003)

4. Zheng, Y., Liu, F., Hsieh, H.P.: U-air: when urban air quality inference meets big data. In: Proceedings of the 19th ACM SIGKDD International Conference on Knowledge Discovery and Data Mining, pp. 1436–1444. ACM (2013)
5. Dawson-Haggerty, S., Jiang, X., Tolle, G., Ortiz, J., Culler, D.: smap: a simple measurement and actuation profile for physical information. In: Proceedings of the 8th ACM Conference on Embedded Networked Sensor Systems, pp. 197–210. ACM (2010)
6. Murphy, K.P.: Machine learning: a probabilistic perspective. MIT Press (2012)
7. Quiñonero-Candela, J., Rasmussen, C.E.: A unifying view of sparse approximate gaussian process regression. The Journal of Machine Learning Research 6, 1939–1959 (2005)
8. Zheng, Y.: Urban air, http://urbanair.msra.cn/
9. Rasmussen, C.E.: Gaussian processes for machine learning (2006)

Retrieval of Experiments by Efficient Comparison of Marginal Likelihoods

Sohan Seth[1], John Shawe-Taylor[2], and Samuel Kaski[1,3]

[1] Helsinki Institute for Information Technology HIIT,
Department of Information and Computer Science, Aalto University, Finland
[2] Centre for Computational Statistics and Machine Learning,
University College London, UK
[3] Helsinki Institute for Information Technology HIIT,
Department of Computer Science, University of Helsinki, Finland
{sohan.seth,samuel.kaski}@hiit.fi, j.shawe-taylor@ucl.ac.uk

Abstract. We study the task of retrieving relevant experiments given a query experiment. By experiment, we mean a collection of measurements from a set of 'covariates' and the associated 'outcomes'. While similar experiments can be retrieved by comparing available 'annotations', this approach ignores the valuable information available in the measurements themselves. To incorporate this information in the retrieval task, we suggest employing a retrieval metric that utilizes probabilistic models learned from the measurements. We argue that such a metric is a sensible measure of similarity between two experiments since it permits inclusion of experiment-specific prior knowledge. However, accurate models are often not analytical, and one must resort to storing posterior samples which demands considerable resources. Therefore, we study strategies to select informative posterior samples to reduce the computational load while maintaining the retrieval performance. We demonstrate the efficacy of our approach on simulated data with simple linear regression as the models, and real world datasets.

Keywords: information retrieval, experiments, ranking, classification.

1 Introduction

An experiment is an organized procedure for validating a hypothesis, and usually comprises measurements over a set of variables that are either varied (covariates or independent variables) or studied (outcomes or dependent variables). For example, in the study of genome-wide association, one explores the association between 'traits' (controlled variable) and common genetic variations (response variables), or in the study of functional genomics covariates can be the species, disease state, and cell type, whereas outcome can be microarray measurements.

Traditionally, similar experiments have been retrieved from qualitative assessment of related scientific documents without explicitly handling the experimental data. Recent technological advances have allowed researchers to both acquire measurements in an unprecedented scale throughout the globe, and to release

C.K. Loo et al. (Eds.): ICONIP 2014, Part II, LNCS 8835, pp. 135–142, 2014.
© Springer International Publishing Switzerland 2014

these measurements for public use after curation, e.g., [1]. However, exploring similar experiments still relies on comparing the manual annotations which suffer extensively from variations in terminology, and incompleteness in annotations (see e.g., [2]). The global effort of availing researchers with wealth of data invites the need for sophisticated retrieval systems that look beyond annotations in comparing related experiments to improve accessibility.

The next step toward this goal is to compare the *knowledge* acquired from experimental measurements rather than just annotations. From a Bayesian perspective, one can quantify knowledge as the posterior distribution of parameters given the measurements. The posterior distribution captures both the information content of the measurements, in terms of the *likelihood function*, as well as the experience and expertise of the experimenter in terms of the *prior distribution* over parameters. We study the future scenario where researchers have submitted (Bayesian generative) models learned on their experiment along with measurements and annotations. We explicitly assume that we have access to such database and develop efficient approaches for retrieving relevant experiments. Developing a successful retrieval engine is a first step toward realizing the future scenario.

We suggest the *marginal likelihood* (1) as a similarity metric, where the underlying idea is to evaluate the likelihood of the query experiment on Bayesian models from (individual) existing experiments. Here the underlying idea is that an experiment is relevant to a query if models learned from it are good for describing the query data. Bayesian models usually need to be stored as a collection of samples from the posterior distribution since the posterior distribution itself might not be available in closed form. The suggested metric (1) then can be efficiently estimated as the average likelihoods over the posterior samples (2). However, this approach has two issues: storing the posterior samples requires considerable resource, and evaluating each marginal likelihood can be computationally demanding (in particular for latent variable models for which the latent components cannot be integrated out in closed form). This paper deals with selecting *informative* posterior samples to reduce both storage and computational requirements while maintaining the retrieval performance.

We achieve this by approximating the marginal likelihood as a *weighted average* of individual likelihoods over posterior samples (3). The weights are then learned to preserve the relative order of experiments in a training set (section 2.1). This is done while imposing a suitable sparsity constraint which allows us to only consider posterior samples with non-zero weights when computing the likelihood of a query sample, thus reducing the storage and computational burden considerably.

2   Method

Assume a set of experiments $\{\mathcal{E}_d\}_{d=1}^{D}$. Each experiment is defined as a collection of measurements over covariates and outcomes, i.e., $\mathcal{E}_d = \{(\mathbf{x}_{di}, \mathbf{y}_{di})\}_{i=1}^{n_d}$. We assume that each experiment $\mathcal{E}_d$ has been modeled by a model $\mathcal{M}_d$, producing a set of posterior MCMC samples $\{\theta_{dk}\}_{k=1}^{m_d}$ from each model. Our general objective is to rank the experiments $\mathcal{E}_d$—actually the models $\mathcal{M}_d$ in the database—according to their relevance to a new query experiment $\mathcal{E}_q$ which is not in the database.

We suggest retrieving similar experiments ranked according to the marginal likelihood they produce for the query, i.e. [1],

$$\text{ML}_{q|d} = p(\mathcal{E}_q|\mathcal{E}_d). \tag{1}$$

This metric has been previously discussed in the context of document retrieval where its use is motivated by capturing the user's intent in terms of the likelihood of a set of keywords $\mathcal{E}_q$ being generated by a document $\mathcal{E}_d$ [3]. In the context of document retrieval the marginal likelihood is usually computed by jointly modeling the whole document database. However, we cannot evaluate this metric by modeling multiple experiments jointly, since we explicitly allow experimenters to submit their models to the database (however, query does not need to be modeled). Therefore, we utilize individual models, represented by posterior distributions, $p(\cdot|\mathcal{E}_d) \propto p(\mathcal{E}_d|\cdot)\pi_d(\cdot)$ to evaluate the marginal likelihood as $\text{ML}_{q|d} = \mathbb{E}_{p(\cdot|\mathcal{E}_d)}p(\mathcal{E}_q|\cdot)$, where π_d is the prior information specific to experiment d. The likelihood can be approximated using posterior samples $\{\theta_{dk}\}_{k=1}^{m_d} \sim p(\cdot|\mathcal{E}_d)$ as

$$\widehat{\text{ML}}_{q|d} \approx \frac{1}{m_d}\sum_{k=1}^{m_d} p(\mathcal{E}_q|\theta_{dk}). \tag{2}$$

However, this approach is computationally demanding: even if one has access to a closed form likelihood function without latent components, this scales up as $O(\sum_d m_d n_q p)$ where p is the number of parameters for the model (assuming the models are in the same exponential family). Additionally, if the latent variables cannot be explicitly integrated out then the samples have to computationally approximate $\int p(\mathbf{x}, \mathbf{z}|\theta)d\mathbf{z}$ as well. The technical contribution of this paper is to address this issue by selecting *fewer* posterior samples that are essential in the retrieval task, i.e., discriminative between experiments. We achieve this by approximating the marginal likelihood as

$$\widetilde{\text{ML}}_{q|d} \approx \frac{1}{m_d}\sum_{k=1}^{m_d} w_{dk} \prod_{i=1}^{n_d} p((\mathbf{x}_{qi}, \mathbf{y}_{qi})|\theta_{dk}) \tag{3}$$

where $\mathbf{w}_d = [w_{d1}, \ldots, w_{dm_d}]$ is a vector of *sparse non-negative weights*. In this way, the posterior samples for which the corresponding weights are zero can be safely ignored. Since we are effectively estimating the *weighted mean* of a set of values, ideally speaking, $\mathbf{w}_d$ should be a *stochastic* vector: positive values that sum to one. However, we observe that even without explicitly imposing this constraint we can achieve favorable performance, and this simplifies the optimization problem considerably.

---

[1] Marginal likelihood is often used to refer to the model evidence. In our case the model is defined by the data set $\mathcal{E}_d$, and the data in computing the likelihood is the query data $\mathcal{E}_q$. We retrieve the data set for which model evidence is the largest.

2.1 Preserving Ranking of Experiments

To learn the weights for each experiment, we adapt the concept of *learning to rank* which is a well explored research problem in information retrieval [4]. However, while this approach is usually applied for learning a function over document-query pairs, we utilize the concept in learning weights over posterior samples for all experiments ("documents") together. Assume, without loss of generality, that given a query q and two experiments i_1 and i_2 in the database, i_1 ranks higher than i_2, i.e., $\widehat{\mathrm{ML}}_{q|i_1} > \widehat{\mathrm{ML}}_{q|i_2}$. Therefore, while learning the weights $\mathbf{w}_{i_1}$ and $\mathbf{w}_{i_2}$, we need to ensure that

$$\sum_k w_{i_1 k} p(\mathcal{E}_q|\theta_{i_1 k}) > \sum_k w_{i_2 k} p(\mathcal{E}_q|\theta_{i_2 k})$$

i.e., we learn the weights to preserve the relative ranks of the experiments with respect to the unweighted metric. When each experiment in the training set is used as a query q, preserving the relative ranks of each pair $\{i_1, i_2\} \subset \{1, \ldots, D\} \setminus \{q\}$ translates to needing to satisfy $D(D-1)(D-2)$ binary constraints for learning the weight vectors $\mathbf{w}_1, \ldots, \mathbf{w}_d$. Fortunately not all of the constraints are usually required since a user is often interested in retrieving only the top (say, top K) experiments rather than all experiments. Therefore, we reformulate our approach and, given a query q, focus on preserving the order of top K experiments. Given any experiment q we select the K closest experiments, $I_q^K = \{i_{j_1}, \ldots, i_{j_K}\}$, and compare them pairwise with the rest of the $(D-2)$ experiments in the database. Intuitively, this preserves the relative orders among the top K experiments I_q^K, and also ensures that these experiments are ranked higher compared to the rest of the $\{1, \ldots, D\} \setminus \{q \cup I_q^K\}$ experiments. This reduces the set of constraints to $KD(D-2)$ where $K \ll D$. Notice that it is certainly feasible to choose different K for different queries.

2.2 Optimization Problem

Satisfying the binary constraints can be formalized as a classification problem $\{(\mathbf{X}_l, y_l)\}_{l=1}^L$ with a highly sparse design matrix $\mathbf{X}$ of dimension $L \times m$, with $L = KD(D-2)$ realizations and $m = \sum_d m_d$ features for learning a combined weight vector $\mathbf{w} = [\mathbf{w}_1, \ldots, \mathbf{w}_d]$, i.e., to satisfy $(\mathbf{X}_l \mathbf{w} + b)y_l > 0$ for all l. Each row of $\mathbf{X}$ belongs to a triplet (q, i_1, i_2), and in that row only the columns associated with posterior samples from i_1 and i_2 are non-zero, and have values $\{p(\mathcal{E}_q|\theta_{i_1 k})\}_{k=1}^{m_{i_1}}$ and $\{-p(\mathcal{E}_q|\theta_{i_2 k})\}_{k=1}^{m_{i_2}}$ respectively. The label associated with this entry is 1 if $\widehat{\mathrm{ML}}_{q|i_1} > \widehat{\mathrm{ML}}_{q|i_2}$, and zero otherwise. An important aspect of this construction is that the label is not absolute, i.e., we can change the sign of a row in the design matrix, i.e., assign the values $\{-p(\mathcal{E}_q|\theta_{i_1 k})\}$ and $\{p(\mathcal{E}_q|\theta_{i_2 k})\}$ to the row instead, and switch the label accordingly. Actually, for each row we randomly pick one of these scenarios to maintain class balance, i.e., we have similar numbers of zeros and ones. Since we are solving a classification problem, each row of the design matrix can be normalized without effecting the class label. This helps solve scaling issues: Instead of likelihoods p_l, we can classify log likelihoods $\ln p_l$, and compute the normalized entries as $\pm \exp(\ln p_l - \max_l \ln p_l)$. These values are in $[-1, 1]$. We use the library liblinear [5] to solve this optimization problem. We use the logistic cost with l_1 regularization, and set the regularization value to 1.

3 Related Works

If one models the query experiment as well, then there are other possible approaches of evaluating similarity between two experiments. Posterior samples $\{\theta_{dk}\}$ have recently been modeled [6] sequentially with Dirichlet process mixtures of normal distributions using particle filtering. Once this model (over posterior samples) has been learned, the similarity between two experiments can be evaluated through similarity of the cluster assignments of the respective posterior samples. Given models of the query and the existing experiments, one can also evaluate their similarity in terms of probabilistic distances or kernels [7]. However, both these approaches have the limitation that the models have to belong to the same family for the similarity to be defined whereas the proposed approach does not require that. Moreover, the distances or kernels between models are not tailored to assist in the user's task, in our case retrieval.

Another possible approach for measuring similarity between experiments is to model the measurements together in a multi-task learning framework [8]. However, off-the-shelf methods for modeling multiple experiments together utilize the same prior and likelihood for all experiments which restricts the generality, and will not exploit the benefit of the knowledge available at the experimenter's disposal. That said, the true purpose of multi-task learning is to utilize knowledge from similar tasks to improve the learning of a new task, which is fundamentally different than retrieval. Also, treating each experiment or model separately rather than as part of a unified model provides well desired modularity to separate the modeling and retrieval task that can be handled by respective experts which is achieved by the proposed set-up.

A similar problem has been explored before by [9] where the authors aimed at retrieving a single data vector given a query vector. This was done by modeling all data together using latent Dirichlet allocation. Retrieving an experiment given a query experiment, however, is conceptually very different since a single data point cannot capture the experimental variability that one would be interested in which is achieved by the proposed approach. That said, retrieval of experiments as discussed in this article allows one to also query with a single observation to find the closest experiment which could have generated that particular sample. This approach has an intriguing characteristic that it enables assigning different parts of the query experiment to different models.

4 Experiments

We demonstrate the performance of the proposed approach on four real world datasets: landmine [8], computer [10], restaurant [11], and LINCS (described below). The first two are standard in the multi-task learning genre. For landmine,

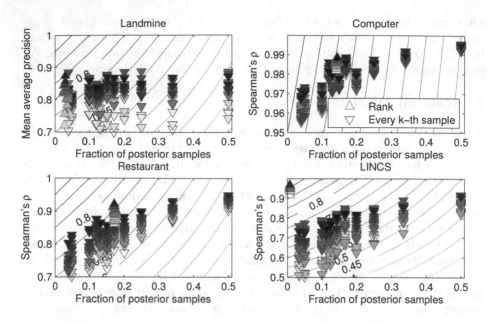

Fig. 1. Comparison of the proposed approach and a simpler metric (evaluating $\widehat{\mathrm{ML}}_{q|d}$ by choosing every k-th posterior samples without any optimization) on real datasets. For landmine we present mean average precision MAP as we have access to labels of each experiment, while for the other two datasets we present the performance compared to $\widehat{\mathrm{ML}}_{q|d}$ estimated with all posterior samples. Each gray shade corresponds to a random partition of the dataset in database and queries. The proposed approach shows improved performance compared to storing every k-th sample for LINCS with respect to (1-sparsity) x retrieval-performance (contours), and performs equally well otherwise.

we have access to class labels of each experiment, and we evaluate the performance of our approach in terms of mean average precision MAP, while on the other two datasets we use correlation with respect to the ranking given by $\widehat{\mathrm{ML}}_{q|d}$ with all posterior samples. We present the results collectively in Fig. 1. For landmine, we train binary probit regression models, while for the other datasets we use normal regression models with non-sparse gamma priors over the weight precisions. For each experiment we generate 100 (1000 for LINCS) posterior samples. For each dataset we randomly split it 3:1 into the database and queries. We observe whether we can preserve retrieval performance after selecting informative posterior samples.

Landmine. The data consist of 29 experiments: each experiment is a classification task for detecting the presence of either landmine (1) or clutter (0) from 9 input features. Each experiment has been collected from either a highly foliated region or a desert-like region. Thus they can be split in two classes (16-13). We observe that this is a relatively simple problem in the sense that the classes are well separated, and thus a few posterior samples are sufficient for good retrieval performance. Due to the same reason, the proposed approach is able to retain the retrieval performance using only very few posterior samples.

Computer. The data consist of 200 experiments: each experiment is a prediction task of how a student rates 20 computers in the scale 0-10. Each computer is described in 13 binary features. Thus, each experiment $\mathbb{R}^{13} \to \mathbb{R}$ has about 20 samples (some entries missing). Since there are no obvious ground truth labels, we measure how well the proposed approach can reduce the number of posterior samples while preserving rankings. We observe that the problem is relatively simple since even a few posterior samples have been able to preserve the ranking with respect to $\widehat{ML}_{q|d}$. However, the number of samples stored is larger than in the previous example since there is no clear clustering.

Restaurant. The data consist of 119 experiments: each experiment is a prediction task of how a customer rates 130 restaurants in the scale 1-3. All customers do not rate all available restaurants, and so the number of observations in each experiments varies, from 3-18. We select 7 categorical features for each experiment and binarize them, resulting in a $\mathbb{R}^{22} \to \mathbb{R}$ regression problem. We observe that this problem is more difficult in the sense that performance drops when the number of samples is decreased. However, the proposed approach has been able to collect essential samples to preserve the true rank better.

LINCS. The LINCS (Library of Integrated Network-based Cellular Signatures) data consist of 65 experiments, each measuring post-treatment gene expression values in response to a specific drug[2]. The model for each experiment is a prediction model from the post-treatment gene expression values to drug toxicity: 959 gene expression[3] values have been measured over 26-44 cell lines for each drug, thus the equivalent regression problem is $\mathbb{R}^{959} \to \mathbb{R}$. Drug toxicity values were acquired from CTD[2] (Cancer Target Discovery and Development). We observe that we achieve distinctively better performance over random sampling.

5   Discussion

This paper is intended to be a proof of concept towards a potentially highly useful community effort of extending experiment databanks to include also knowledge of the experimenters in a rigorously reusable form, as models. As of now, this is highly non-standard yet would be beneficial since the experimenter alone is best acquainted with his/her measurements and is able to train the most sensible model by incorporating his/her experience as prior knowledge. Storing models of experiments can, however, be cumbersome since most often they are not expressed in an analytic form. A widely applicable alternative is to store samples of the posterior; we suggested approaches to select the most informative posterior

---

[2] Personal communication with Dr. Subramanian, Broad Institute.
[3] Originally 978.

samples to store. Notice that posterior samples can be generated also when one has an analytic posterior. We have presented a set of convincing results on simulated data with regression as a task, as well as on standard real datasets.

Acknowledgments. This project is partly supported by the Academy of Finland (Finnish Centre of Excellence in Computational Inference Research COIN, 251170), and the Aalto University MIDE (Multidisciplinary Institute of Digitalisation and Energy) research programme. The calculations presented above were performed using computer resources within the Aalto University School of Science "Science-IT" project. Collection of gene expression values in LINCS data has been supported by the Broad Institute LINCS center 5U54HG006093 from the NIH Common Fund. The authors thank Suleiman Ali Khan for his help with the LINCS experiment preparation.

References

1. Rustici, G., Kolesnikov, N., Brandizi, M., Burdett, T., Dylag, M., Emam, I., Farne, A., Hastings, E., Ison, J., Keays, M., Kurbatova, N., Malone, J., Mani, R., Mupo, A., Pedro Pereira, R., Pilicheva, E., Rung, J., Sharma, A., Tang, Y.A., Ternent, T., Tikhonov, A., Welter, D., Williams, E., Brazma, A., Parkinson, H., Sarkans, U.: ArrayExpress update–trends in database growth and links to data analysis tools. Nucleic Acids Research 41, D987–D990 (2013)
2. Baumgartner Jr., W.A., Cohen, K.B., Fox, L.M., Acquaah-Mensah, G., Hunter, L.: Manual curation is not sufficient for annotation of genomic databases. Bioinformatics 23, i41–i48 (2007)
3. Buntine, W., Lofstrom, J., Perkio, J., Perttu, S., Poroshin, V., Silander, T., Tirri, H., Tuominen, A., Tuulos, V.: A scalable topic-based open source search engine. In: Proceedings of IEEE/WIC/ACM International Conference on Web Intelligence, pp. 228–234 (2004)
4. Burges, C.J.C., Shaked, T., Renshaw, E., Lazier, A., Deeds, M., Hamilton, N., Hullender, G.N.: Learning to rank using gradient descent. In: ICML, pp. 89–96 (2005)
5. Fan, R.E., Chang, K.W., Hsieh, C.J., Wang, X.R., Lin, C.J.: Liblinear: A library for large linear classification. J. Mach. Learn. Res. 9, 1871–1874 (2008)
6. Dutta, R., Seth, S., Kaski, S.: Retrieval of experiments with sequential Dirichlet process mixtures in model space. arXiv:1310.2125 [cs, stat] (2013)
7. Muandet, K., Fukumizu, K., Dinuzzo, F., Schlkopf, B.: Learning from distributions via support measure machines. arXiv e-print 1202.6504 (2012)
8. Xue, Y., Liao, X., Carin, L., Krishnapuram, B.: Multi-task learning for classification with Dirichlet process priors. Journal of Machine Learning Research 8, 35–63 (2007)
9. Caldas, J., Gehlenborg, N., Faisal, A., Brazma, A., Kaski, S.: Probabilistic retrieval and visualization of biologically relevant microarray experiments. Bioinformatics 12, i145–i153 (2009)
10. Argyriou, A., Evgeniou, T., Pontil, M.: Multi-task feature learning. In: Schölkopf, B., Platt, J., Hoffman, T. (eds.) Advances in Neural Information Processing Systems 19, pp. 41–48. MIT Press, Cambridge (2007)
11. Vargas-Govea, B., González-Serna, J.G., Ponce-Medellín, R.: Effects of relevant contextual features in the performance of a restaurant recommender system. In: Workshop on Context Aware Recommender Systems (CARS) (2011)

A New Approach of Diversity Enhanced Particle Swarm Optimization with Neighborhood Search and Adaptive Mutation

Dang Cong Tran[1,3], Zhijian Wu[1], and Hui Wang[2]

[1] State Key Laboratory of Software Engineering, School of Computer,
Wuhan University, Wuhan 430072, China
[2] School of Information Engineering, Nanchang Institute of Technology,
Nanchang 330099, China
[3] Vietnam Academy of Science and Technology, Hanoi, Vietnam
trandangcong@gmail.com, zhijianwu@whu.edu.cn

Abstract. Like other stochastic algorithms, particle swarm optimization algorithm (PSO) has shown a good performance over global numerical optimization. However, PSO also has a few drawbacks such as premature convergence and low convergence speed, especially on complex problem. In this paper, we present a new approach called AMPSONS in which neighborhood search, diversity mechanism and adaptive mutation were utilized. Experimental results obtained from a test on several benchmark functions showed that the performance of proposed AMP-SONS algorithm is superior to five other PSO variants, namely CLPSO, AMPSO, GOPSO, DNLPSO, and DNSPSO, in terms of convergence speed and accuracy.

Keywords: particle swarm optimization, neighborhood search, adaptive mutation, global optimization.

1 Introduction

Particle swarm optimization (PSO) proposed by J. Kennedy et al. [1] is a population based algorithm. PSO has superior performance over global numerical optimization. However, It has some drawbacks such as premature convergence and low convergence speed, especially on complex problem. In 2002, M. Clerc et al. [2] proposed new variant for finding optimal regions of complex search spaces through the interaction of individuals in a population of particle. For solving complex multimodal problem, J. J. Liang et al. [3] proposed CLPSO algorithm, where a novel learning strategy whereby all other particles' historical best information was used to update a particle's velocity. For this problem, Wang et al. [4] presented a method called AMPSO, in which three different mutation operators including Gaussian, Cauchy, and Levy were utilized. In order to improve the convergence speed of PSO, in [5] an approach denoted GOPSO was proposed, where the method generalized opposition-based learning (GOBL) and Cauchy mutation were employed. Attracted by neighborhood search strategy,

C.K. Loo et al. (Eds.): ICONIP 2014, Part II, LNCS 8835, pp. 143–150, 2014.
© Springer International Publishing Switzerland 2014

various methods were presented, such Md Nasir et al. [6] proposed a method called DNLPSO, which was improved from the idea of CLPSO. Furthermore, H. Wang et al. [7] presented a DNSPSO algorithm, in which the neighborhood search of local and global strategies were utilized. In our previous research [8], by using the improved neighborhood search, a new approach called EPSODNS was presented.

In order to improve the PSO algorithm over the drawback and motivated by the efficiency of those PSO variants, especially the neighborhood search strategy, the hybrid of PSO with mutation operation, in this paper we propose new approach AMPSONS by introducing neighborhood search strategy, diversity mechanism, and adaptive mutation into PSO. The performance of the proposed algorithm is evaluated by the test over benchmark functions and compared to several powerful algorithms.

Remain of this paper is structured as follows: the particle swarm optimization algorithm is briefly reviewed in Section 2. The methodology of the proposed algorithm will be described in Section 3. The benchmark functions, parameters setting and results will be demonstrated in Section 4. Finally, in Section 5 the conclusions are drawn.

2   Particle Swarm Optimization

In PSO [1, 2], a particle in PSO has a velocity vector (V) and a position vector (X). PSO remembers both the best position found by all particles and the best positions found by each particle in the search process. For a search problem in D-dimensional space, a particle represents a potential solution. The velocity v_{ij} and position x_{ij} of the jth dimension of the ith particle are updated according to Eqs. (1) and (2) as follows:

$$v_{ij} = w \cdot v_{ij} + c_1 \cdot rand1_{ij} \cdot (pbest_{ij} - x_{ij}) + c_2 \cdot rand2_{ij} \cdot (gbest_j - x_{ij}) \qquad (1)$$

$$x_{ij} = x_{ij} + v_{ij} \qquad (2)$$

where the particle index i =1,2, ...,NP, NP is the population size, x_i is the position of the ith particle, v_i represents the velocity of ith particle, $pbest_i$ is the best previous position yielding the best fitness value for the ith particle and $gbest$ is the global best particle found by all particles so far, $rand1_{ij}$ and $rand2_{ij}$ are two random numbers independently generated within the range of [0, 1], c_1 and c_2 are two learning factors which control the influence of the social and cognitive components, w is the inertia factor. The inertia weight w in Eq. (1) was introduced by Y. Shi et al. [9], a w linearly decreasing with the iterative generations was proposed as Eq. (3).

$$w_k = w_0 - \frac{(w_0 - w_1) \cdot k}{Max\_Gen} \qquad (3)$$

where k is the generation index, w_0 and w_1 are maximum and minimum inertia weight values, respectively. Max_Gen is the maximum number of generation.

3   The Methodology

As above mentioned, in this paper we propose a new approach, in which the hybridization of neighborhood search, diversity mechanism, and adaptive mutation operation are employed. In this Section the methodology of the proposed algorithm will be described.

3.1   Neighborhood Search

By employment of local neighborhood search and global neighborhood search strategies, H. Wang et al. [7] proposed DNSPSO approach to enhance PSO algorithm. To improve the ability of exploitation, a local neighborhood search (LNS) strategy is proposed. During searching the neighborhood of a particle P_i, a trial particle $L_i = (LX_i, LV_i)$ is generated by Eqs. (4, 5).

$$LX_i = r_1 \cdot X_i + r_2 \cdot pbest_i + r_3 \cdot (X_c - X_d) \tag{4}$$

$$LV_i = V_i \tag{5}$$

where X_c and X_d are the position vectors of two random particles in the k-neighborhood radius of P_i, $c, d \in [i - k, i + k] \wedge c \neq d \neq i$, r_1, r_2 and r_3 are three uniform random numbers within (0,1), and $r_1 + r_2 + r_3 = 1$.

Besides the LNS, a global neighborhood search (GNS) strategy is proposed to enhance the ability of exploration. When searching the neighborhood of a particle P_i, another trial particle $G_i = (GX_i, GV_i)$ is generated by Eqs. (6, 7).

$$GX_i = r_4 \cdot X_i + r_5 \cdot gbest + r_6 \cdot (X_e - X_f) \tag{6}$$

$$GV_i = V_i \tag{7}$$

where X_e and X_f are the position vectors of two random particles chosen for the entire swarm, $e, f \in [1, NP] \wedge e \neq f \neq i$, r_4, r_5 and r_6 are three uniform random numbers within (0, 1), and $r_4 + r_5 + r_6 = 1$.

3.2   Diversity Mechanism

Like DNSPSO [7], the diversity mechanism was employed, where for each particle $P_i(t)$ a new particle $P_i(t + 1)$ is generated by the PSO's velocity and position updating equations. By recombining $P_i(t)$ and $P_i(t + 1)$, a trial particle $TP_i(t + 1) = (TX_i(t + 1), TV_i(t + 1))$ is generated as follows:

$$TX_{ij}(t+1) = \begin{cases} X_{ij}(t+1) & \text{if } rand_j(0,1) < P_r \\ X_{ij}(t) & \text{otherwise} \end{cases} \tag{8}$$

$$TV_{ij}(t+1) = V_{ij}(t+1) \tag{9}$$

where P_r is a user-defined value of greedy selection probability. After recombination, a greedy selection is used as follows:

$$P_i(t+1) = \begin{cases} TP_i(t+1) & \text{if } f(TP_i(t+1)) < f(P_i(t+1)) \\ P_i(t+1) & \text{otherwise} \end{cases} \tag{10}$$

3.3   Adaptive Mutation

In the lecture, various approaches employed mutation operation such as Gaussian, Cauchy, etc [11–13]. In [4], H. Wang et al. suggested that Gaussian mutation is good for local search, while Cauchy, Levy are beneficial for global search. In addition, the Cauchy and Levy mutations are better than the Gaussian for the large mutation sizes. In AMPSO [4], H. Wang et al. proposed a method of switching automatically which mutation more suitable and apply on *pbest* and *gbest* as follows:

$$pbest'_{ij} = pbest_{ij} + mutation_j() \tag{11}$$

$$gbest'_j = gbest_j + mutation_j() \tag{12}$$

where *mutation* is the mutation operation selected among Gaussian, Cauchy, Levy according to the number of successful mutations. The selection mechanism is builded based on a roulette wheel, where a mutation with larger selection ratios wins more chances to be selected (details are described in [4]).

3.4   The Proposed Algorithm

Firstly, in order to improve the exploitation ability of the algorithm, the neighborhood search is improved by using the best particle of neighbour in local search strategy. Therefore, the trial LNS particle which was generated by Eq. (4), is now calculated as follows:

$$LX_i = r_1 \cdot X_i + r_2 \cdot (pbest_i - X_i) + r_3 \cdot nbest_i \tag{13}$$

where $nbest_i$ is the best particle of X_i neighborhood.

Secondly, to enhance the performance of the algorithm for complex functions (especially for multimodal functions), the selection mechanism is still employed to select adaptively mutation operation among Gaussian, Cauchy, Levy mutation operations and applied on *gbest*.

Finally, similar to the scout bee of artificial bee colony (ABC) [14], particle is re-initialized when the number of relative *pbest* fitness not changed is more than the pre-defined number (called *limit*). By using re-initialized particle, the exploration ability of the algorithm can be improved.

The main steps of the proposed algorithm are listed in Table 1, where *NP* is the population size, *lastpbest* and *lastgbest* record the last fitness values of *pbest* and *gbest*, respectively. $Monitor[i]$ and *gbestmonitor* record the successive number of iterations where the fitness values of $pbest_i$ and *gbest* do not change, respectively. *FEs* is the number of fitness evaluations, and *MaxFEs* is the maximum number of fitness evaluations. P_{ns} is the probability value to implement the neighborhood search strategy, *limit* is the pre-defined number.

4   Experimental Results

To evaluate the performance of the algorithm for global numerical optimization, seventeen well-known test functions are used, the results obtained from AMPSONS and five other PSO variants, namely CLPSO[3], AMPSO[4], GOPSO[5], DNLPSO[6], and DNSPSO[7] will be demonstrated and analyzed.

Table 1. The main steps of AMPSONS

1 Initialize the population and parameters;
2 **While** $FEs \leq MaxFEs$ **do**
3   Update inertia weight w according to Eq. (3);
4   **For** $i = 1$ to NP **do**
5   Update the velocity and position according to Eq. (1, 2);
6   Generate a new trial particle TP_i by Eqs. (8, 9);
7   Select a fitter one between P_i and TP_i as the new P_i by Eq. (10);
8   Update $pbest$ and $gbest$;
9   **If** $f(pbest) = f(lastpbest)$ **then** $monitor[i] + +$ **else** $monitor[i] = 0$;
10   **If** $f(gbest) = f(lastgbest)$ **then** $gbestmonitor + +$ **else** $gbestmonitor = 0$;
11   **End for**
12   **For** $i = 1$ to NP **do**
13   **If** $rand(0, 1) \leq P_{ns}$ **then**
14   Select the best particle $nbest$ from the local neighborhood of particle;
15   Generate a trial particle L_i according to Eqs. (13, 5);
16   Generate a trial particle G_i according to Eqs. (6, 7);
17   Select the best one among P_i, L_i, and G_i as the new P_i;
18   Implement similar to steps from 8 to 10;
19   **End if**
20   **End for**
21   **If** $gbestmonitor \geq m$ **then**
22   Determine which mutation to be conducted according to the selection ration;
23   Conduct a mutation on $gbest$ according to Eq. (12);
24   **End if**
25   **If** $monitor[k] \geq max(monitor[j], j = 1, .., NP)$ and $monitor[k] \geq limit$ **then**
26   Reinitialize the kth particle;
27 **End while**

4.1 Benchmark Functions

Seventeen well-known benchmark functions including thirteen classical functions from f_1 to f_{13} (due to the paper space limit, details can be found in Table II of [8] and the first thirteen functions in Table I of [15]) and four rotated functions from f_{14} to f_{17} are Rotaled Ackley, Rotated Griewank, Rotated Weierstrass, and Rotated Rastrigin, respectively (details can be found in [3–7]) are used to evaluate the performance of the algorithm.

4.2 Parametric Settings

For the sake of fair comparison, in this test, all of algorithms run over 25 times, the population size is set to 40, the maximum number of fitness evaluation ($MaxFEs$) is set to 2e+5 [4, 5, 7]. The parameters of five other competitive algorithms are set according to their experiments.

The parameters for AMPSONS algorithm were empirically set as follows: $w_0 = 0.9$, $w_1 = 0.4$, $c_1 = c_2 = 1.49$, $P_r = 0.9$, $P_{ns} = 0.6$. And $m, limit$ are set to 10, 20, respectively.

4.3   Comparison of AMPSONS with Other PSO Variants

This Section will demonstrate the results obtained from AMPSONS and five other competitive PSO-based variants CLPSO, AMPSO, GOPSO, DNLPSO, and DNSPSO over 25 times on 17 test functions with search space $D = 30$. The average error values $f(x) - f(x^0)$ ($f(x^0)$ is the global optimum of $f(x)$) are recorded and given in Table 2, where Mean indicates the average function error value, Dev indicates standard deviation value, in table the best results are written in bold, and '$w/t/l$' means that AMPSONS wins in w functions, ties in t functions and loses in l functions, compared with its competitors.

Table 2. The comparison of results of AMPSONS with five other PSO variants

Func	CLPSO mean/dev	AMPSO mean/dev	GOPSO mean/dev	DNLPSO mean/dev	DNSPSO mean/dev	AMPSONS mean/dev
f_1	6.5887e−11	1.6391e−85	9.3796e−97	1.9172e−98	8.1845e−175	**0**
	4.5005e−11	6.9060e−85	4.4016e−96	9.2787e−98	0.0000e+00	**0**
f_2	4.7252e−08	3.1840e−35	1.5348e−53	3.1553e−27	1.7096e−89	**9.2427e−179**
	1.3091e−08	1.3875e−34	4.4310e−53	1.5772e−26	4.2478e−89	**0**
f_3	1.4346e+03	4.8000e+03	1.2962e−22	3.0518e−08	3.4625e−109	**1.1111e−294**
	3.7903e+02	3.9476e+03	6.4517e−22	1.4359e−07	1.6416e−108	**0**
f_4	6.4390e+00	1.8823e−05	2.9108e−26	1.6092e−04	4.1989e−68	**1.7366e−165**
	9.0374e−01	4.6106e−05	7.5345e−26	1.9116e−04	9.6100e−68	**0**
f_5	4.2818e+01	9.7004e+00	1.0176e+01	1.1897e+01	1.8142e+01	**8.6793e−05**
	2.1893e+01	7.2206e+00	9.1401e−01	1.3016e+01	4.1286e−01	**1.3244e−04**
f_6	0	0	0	0	0	0
	0	0	0	0	0	0
f_7	7.1231e−03	7.5760e−01	1.5199e−03	2.6427e−03	2.3463e−04	**2.6189e−05**
	1.9787e−03	1.9788e+00	8.8604e−04	1.2101e−03	1.2642e−04	**3.2625e−05**
f_8	0	3.4591e+03	3.7709e+03	1.8579e+03	3.0932e+03	2.8281e+03
	0	7.7651e+02	5.7681e+02	4.3209e+02	7.1797e+02	1.2040e+03
f_9	2.2742e+00	1.0444e+02	9.9646e−01	2.9849e+01	0	0
	1.5878e+00	3.8880e+01	2.5874e+00	7.9285e+00	0	0
f_{10}	7.6373e−07	1.6634e+00	3.9968e−15	1.7639e−14	1.8652e−15	**4.4409e−16**
	2.0441e−07	2.8718e+00	8.0513e−31	1.0692e−14	1.7764e−15	**1.0064e−31**
f_{11}	2.4151e−07	2.2069e−02	0	9.7453e−03	0	0
	3.5659e−07	2.4950e−02	0	9.5839e−03	0	0
f_{12}	2.9112e−12	1.5093e−01	**3.0166e−17**	3.0166e−17	3.0166e−17	3.0166e−17
	1.9739e−12	2.1421e−01	**1.8870e−32**	1.8904e−32	1.8870e−32	1.8870e−32
f_{13}	**2.6152e−11**	1.5264e−01	3.4779e−03	1.3185e−03	1.0108e−02	6.6221e−03
	1.6799e−11	7.1780e−01	5.9998e−03	3.6441e−03	2.0779e−02	2.0311e−02
f_{14}	1.2521e−05	3.5284e−14	1.2536e−12	8.2563e−09	6.3568e−16	**5.3628e−16**
	2.1500e−05	1.6953e−15	3.2658e−19	6.3254e−08	0	**1.2658e−32**
f_{15}	8.2598e−02	2.5680e−16	0	1.2548e−15	0	0
	1.2568e−05	1.2587e−18	0	2.3587e−17	0	0
f_{16}	3.1248e+00	4.2147e+00	**3.1256e−13**	2.8975e+00	2.5669e+00	1.2547e+00
	3.9874e+00	3.9854e+00	**7.2354e−14**	3.2548e+00	2.8954e−01	5.3624e−01
f_{17}	5.5684e+01	4.8726e+01	0	0	0	0
	5.0124e+01	5.3648e+01	0	0	0	0
w/t/l	14/1/2	16/1/0	11/4/2	12/3/2	11/6/0	

The results in Table 2 show that the proposed AMPSONS approach has the best performance on thirteen functions, three other functions belong to CLPSO and GOPSO algorithms. In order to compare the performance of multiple algorithms on the test suite, we conduct Friedman test according to the suggestions of [16]. The results of Friedman test shows the average ranking of six competitive algorithms, the highest ranking which belongs to AMPSONS, namely the ranks of CLPSO, AMPSO, GOPSO, DNLPSO, DNSPSO, and AMPSONS are 4.74, 4.91, 3.03, 3.85, 2.65, 1.82, respectively. The convergence curves of four representative functions are illustrated in Fig. 1.

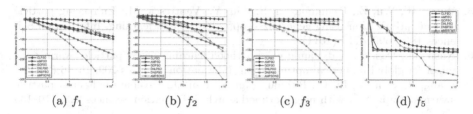

(a) f_1 (b) f_2 (c) f_3 (d) f_5

Fig. 1. The convergence curves on four test functions f_1, f_2, f_3, and f_5

For comparison of computational time, the average computation time for each test functions is counted, from our experimental results, CLPSO consumes the smallest time in total, then DNSPSO is second, the proposed AMPSONS algorithm is third, and the last one is GOPSO.

5   Conclusions

According to the experimental results, it appears that our proposition outperforms the existing methods in terms of convergence speed and accuracy. In the proposed algorithm, the improvement of neighborhood search strategy, diversity mechanism and adaptive mutation enhances the exploitation and exploration abilities of the algorithm including the multimodal and rotated problems. In addition, exploration ability is improved by re-initialization of particle when the *pbest* fitness value of particle does not change in pre-defined number of iterations. By using the hybridization of the techniques, our approach can be adapted for more kinds of problems.

In the future, the proposed approach will be applied in solving other domains of application such as data clustering, image segmentation, etc.

Acknowledgments. This work was supported by the National Natural Science Foundation of China (No.: 61070008 and 61364025).

References

1. Kenedy, J., Eberhart, R.: Particle swarm optimization. In: Proceedings of IEEE International Conference on Neuron Networks Conference Proceedings, Perth, Australia, pp. 1942–1948 (1995)
2. Clerc, M., Kennedy, J.: The Particle Swarm-Explosion, Stability, and Convergence in a Multidimensional Complex Space. IEEE Trans on Evol. 6(1), 58–73 (2002)
3. Liang, J., Qin, A., Suganthan, P., Baskarr, S.: Comprehensive learning particle swarm optimizer for global optimization of multimodal functions. Trans on Evol. 10, 281–295 (2006)
4. Wang, H., Wang, W., Wu, Z.: Particle swarm optimization with adaptive mutation for multimodal optimization. Applied Mathematics and Computation 221, 296–305 (2013)
5. Wang, H., Wu, Z., Rahnamayan, S., Liu, Y., Ventresca, M.: Enhancing particle swarm optimization using generalized opposition-based learning. Information Sciences 181, 4699–4714 (2011)
6. Nasir, M., Das, S.: A dynamic neighborhood learning based particle swarm optimizer for global numerical optimization. Information Sciences 209, 16–36 (2012)
7. Wang, H., Sun, S., Li, C., Rahnamayan, S., Pan, J.: Diversity enhanced particle swarm optimization with neighborhood search. Information Sciences 223, 119–135 (2013)
8. Tran, D.C., Wu, Z., Wang, H.: A novel enhanced particle swarm optimization method with diversity and neighborhood search. In: IEEE International Conference on Systems, Man, and Cybernetics (SMC 2013), pp. 180–187 (2013)
9. Shi, Y., Eberhart, R.: A Modified Particle Swarm Optimizer. In: Proceedings of the 1998 Congress on Evolutionary Computation (CEC 1998), pp. 69–73 (1998)
10. Kennedy, J.: Small Worlds and Mega-Minds: Effects of Neighborhood Topology on Particle Swarm Performance. In: Proceedings of the 1999 Congress on Evolutionary Computation, pp. 1931–1938 (1999)
11. Stacey, S., Jancic, M., Grundy, I.: Particle swarm optimization with mutation. In: Proceedings of IEEE Congress on Evolutionary Computation, pp. 1425–1430 (2003)
12. Higashi, N., Iba, H.: Particle swarm optimization with Gaussian mutation. In: Proceeding of IEEE Swarm Intelligence Symposium, Indianapolis, pp. 72–79 (2003)
13. Krohling, R.: Gaussian particle swarm with jumps. In: Proceedings of IEEE Congress on Evolutionary Computation, pp. 1226–1231 (2005)
14. Karaboga, D.: An idea based on honey bee swarm for numerical optimization. Technical report-TR06, Erciyes University, Engineering Faculty, Comput. Eng. Dep. (2005)
15. Brest, J., Greiner, S.: Self-adapting control parameters in differential evolution: A comparative study on numerical benchmark problems. IEEE Trans on Evol. Comput. 10, 646–657 (2006)
16. Derrac, J.: A practical tutorial on the use of nonparametric statistical tests as a methodology for comparing evolutionary and swarm intelligence algorithms. Swarm and Evolutionary Computation 1, 3–18 (2011)

Data Clustering Based on Particle Swarm Optimization with Neighborhood Search and Cauchy Mutation

Dang Cong Tran[1,2] and Zhijian Wu[1]

[1] State Key Laboratory of Software Engineering, Computer School,
Wuhan University, Wuhan 430072, China
[2] Vietnam Academy of Science and Technology, Hanoi, Vietnam
trandangcong@gmail.com, zhijianwu@whu.edu.cn

Abstract. K-means is one of the most popular clustering algorithm, it has been successfully applied in solving many practical clustering problems, however there exist some drawbacks such as local optimal convergence and sensitivity to initial points. In this paper, a new approach based on enhanced particle swarm optimization (PSO) is presented (denoted CMPNS), in which PSO is enhanced by new neighborhood search strategy and Cauchy mutation operation. Experimental results on fourteen used artificial and real-world datasets show that the proposed method outperforms than that of some other data clustering algorithms in terms of accuracy and convergence speed.

Keywords: data clustering, K-means, particle swarm optimization.

1 Introduction

Data clustering is the process of identifying natural groupings or clusters, within multidimensional data, based on some similarity measure. The K-means clustering algorithm was developed by J.A. Hartigan [1] which is one of the most popular and widely used clustering techniques because it is easy to implement and very efficient, with linear time complexity. However, its main drawbacks are that it converges to arbitrary local optima as well as at local maxima and saddle points and that it cannot deal well with non-spherical shaped clusters [2]. The performance of the K-means algorithm depends on the initial choice of the cluster centers. In order to tackle the drawback of initialization, in [3] a method called K-means++ was presented, where a new initial method was presented. An alternative approach is applying evolutionary algorithms (EAs) in clustering, yielding EA-based clustering algorithms. Unlike K-means clustering, they simultaneously optimize a population of candidate solutions, which give them the ability to escape from local optima. Various EA-based clustering algorithms have been developed, including genetic algorithms, differential evolution, ant colony optimization, artificial bee colony, and particle swarm optimization [4–6].

C.K. Loo et al. (Eds.): ICONIP 2014, Part II, LNCS 8835, pp. 151–159, 2014.

Remain of this paper is structured as follows: some preliminaries of K-means and PSO algorithms are briefly reviewed in Section 2. The proposed CMPNS algorithm will be described in Section 3. The benchmark datasets, parameters setting and results will be demonstrated in Section 4. Finally, in Section 5 the conclusions will be drawn.

2 Preliminaries

2.1 K-means Clustering Algorithm

In partitioning clustering problems, we need to divide a set of N objects into K clusters. Let $O(o_1, o_2, ..., o_N)$ be the set of N objects of data set. Each object has D features, and each feature is quantified with a real-value. Let $S_{N \times D}$ be the feature data matrix. It has N rows and D columns. Each row S_i presents a data vector and s_{ij} corresponds to the jth feature of ith data vector ($i=1,2,...,N$, $j=,1,2,...,D$). Let $C = (C_1, C_2, ..., C_K)$ be the K clusters. Then $C_i \neq \phi$, $C_j \cap C_i \neq \phi$, $\cup_{j=1}^{K} C_i = O$, $i, j = 1, 2, ..., K$, $i \neq j$. The goal of clustering algorithm is to find such a C that makes the objects in the same clusters are as similar as possible while other objects in the different clusters as dissimilar, which can be measured by some criterions.

K-means clustering [1] groups data vectors into a pre-specified number of clusters, based on Euclidean distance as similarity measure. The classical K-means algorithm is summarized as follows:

Step 1. Randomly choose K cluster centroids from N objects.

Step 2. For each data vector, assign the vector to the cluster with the closest centroid, where the distance to the centroid is determined by Eq. (1).

$$d(S_i, Z_j) = \sqrt{\sum_{p=1}^{D} (S_{ip} - Z_{jp})^2} \tag{1}$$

Step 3. Recalculate the cluster centroids, using Eq. (2) as follows:

$$Z_j = \frac{1}{N_{C_j}} \sum_{\forall S_p \in C_j} S_p \tag{2}$$

where N_{C_j} is the number of data vectors in cluster j and C_j is the subset of data vectors that form cluster j, return *Step 2* if stopping criterion is not satisfied.

2.2 Particle Swarm Optimization

Each particle in PSO [7, 8] has a velocity vector (V) and a position vector (X). PSO remembers both the best position found by all particles and the best positions found by each particle in the search process. For a search problem in D-dimensional space, a particle represents a potential solution. The velocity and position of particle are updated according to Eqs. (3) and (4).

$$v_{ij} = w \cdot v_{ij} + c_1 \cdot rand1_{ij} \cdot (pbest_{ij} - x_{ij}) + c_2 \cdot rand2_{ij} \cdot (gbest_j - x_{ij}) \tag{3}$$

$$x_{ij} = x_{ij} + v_{ij} \tag{4}$$

where the particles index $i = 1, 2, ..., NP$, NP is the population size, x_i is the position of the ith particle, v_i represents the velocity of ith particle, $pbest_i$ is the best previous position yielding the best fitness value for the ith particle, and $gbest$ is the global best particle found by all particles so far, $rand1_{ij}$ and $rand2_{ij}$ are two random numbers independently generated within the range of $[0, 1]$, c_1 and c_2 are two learning factors which control the influence of the social and cognitive components, w is the inertia factor. The inertia weight w in Eq. (3) was introduced by Y. Shi et al. [9], a w linearly decreasing with the iterative generations was proposed as Eq. (5).

$$w_k = w_0 - \frac{(w_0 - w_1) \cdot k}{Max\_Gen} \tag{5}$$

where k is the kth generation index, w_0 and w_1 are maximum and minimum inertia weight value, respectively.

3   Proposed Method

To improve the performance of K-means over the drawbacks and enhance the algorithm in terms of convergence speed and accuracy, in this paper we present a new approach based on improved PSO, where PSO is introduced into K-means.

3.1   Neighborhood Search

By employment of local neighborhood search and global neighborhood search strategies with ring topology and radius is equal to 2, H. Wang et al. [10] proposed DNSPSO approach to enhance PSO algorithm, in which a local neighborhood search (LNS) and global search (GNS) strategies were proposed. To improve the exploitation ability of the local search strategy, the best particle of local neighbour is employed to generate the trial particle LNS. The neighborhood of a particle P_i, a trial particle $L_i = (LX_i, LV_i)$ is generated by Eqs. (6, 7).

$$LX_i = r_1 \cdot X_i + r_2 \cdot (pbest_i - X_i) + r_3 \cdot nbest_i \tag{6}$$

$$LV_i = V_i \tag{7}$$

where X_c and X_d are the position vectors of two random particles in the k-neighborhood radius of P_i, $c, d \in [i - k, i + k] \wedge c \neq d \neq i$, r_1, r_2 and r_3 are three uniform random numbers within $(0,1)$, and $r_1 + r_2 + r_3 = 1$, and $nbest_i$ is the best particle of X_i neighborhood.

Besides the LNS, a global neighborhood search (GNS) strategy is proposed to enhance the ability of exploration. When searching the neighborhood of a particle P_i, another trial particle $G_i = (GX_i, GV_i)$ is generated by Eqs. (8, 9).

$$GX_i = r_4 \cdot X_i + r_5 \cdot gbest + r_6 \cdot (X_e - X_f) \tag{8}$$

$$GV_i = V_i \tag{9}$$

where X_e and X_f are the position vectors of two random particles chosen for the entire swarm, $e, f \in [1, NP] \wedge e \neq f \neq i$, r_4, r_5 and r_6 are three uniform random numbers within $(0, 1)$, and $r_4 + r_5 + r_6 = 1$.

3.2 Diversity Mechnism

Like DNSPSO [10], the diversity mechanism was employed, where for each particle $P_i(t)$ a new particle $P_i(t+1)$ is generated by the PSO's velocity and position updating equations. By recombining $P_i(t)$ and $P_i(t+1)$, a trial particle $TP_i(t+1) = (TX_i(t+1), TV_i(t+1))$ is generated as follows:

$$TX_{ij}(t+1) = \begin{cases} X_{ij}(t+1) & \text{if } rand_j(0,1) < P_r \\ X_{ij}(t) & \text{otherwise} \end{cases} \tag{10}$$

$$TV_{ij}(t+1) = V_{ij}(t+1) \tag{11}$$

where P_r is a user-defined value of greedy selection probability. After recombination, a greedy selection is used as follows:

$$P_i(t+1) = \begin{cases} TP_i(t+1) & \text{if } f\left(TP_i(t+1)\right) < f\left(P_i(t+1)\right) \\ P_i(t+1) & \text{otherwise} \end{cases} \tag{12}$$

3.3 Cauchy Mutation

Aim to improve the convergence speed, in each iteration the global best particle is mutated by Cauchy distribution function [11] as follows:

$$gbest_j = gbest_j + Cauchy() \tag{13}$$

3.4 Reinitialization

Similar to the scout bee of artificial bee colony (ABC) [12], particle is reinitialized randomly if the number of relative *pbest* fitness not changed is more than the pre-defined number (called *limit*, in this case the particle may be trapped into local optima). By this technique, the exploration ability of the algorithm can be enhanced.

3.5 Proposed Algorithm

Firstly, particle is encoded according to Eqs. (14), (15). Each particle is a potential candidate solution for the optimal center centroids. In this case, solving data clustering problem can be seen as solving the global optimization with fitness function is the validity index of SED calculated by Eq. (16).

$$X_i = (X_{i1}, ..., X_{ij}, ..., X_{iK}) \tag{14}$$

$$V_i = (v_{i,1}, v_{i,2}, ..., v_{i,K \times D}) \tag{15}$$

Table 1. The main steps of CMPNS

1 Initialize each particle by randomly selecting from dataset;
2 **While** $FEs \leq MaxFEs$ **do**
3   Update inertia weight w according to Eq. (5);
4   **For** $i = 1$ to NP **do**
5   Update the velocity and position according to Eq. (3, 4);
6   Generate a new trial particle TP_i by Eqs. (10, 11);
7   Select a fitter one between P_i and TP_i as the new P_i by Eq. (12);
8   Update *pbest* and *gbest*;
9   **If** $f(pbest) = f(lastpbest)$ **then** $monitor[i] + +$ **else** $monitor[i] = 0$;
10   **End for**
11   **For** $i = 1$ to NP **do**
12   **If** $rand(0,1) \leq P_{ns}$ **then**
13   Generate a trial particle L_i according to Eqs. (6, 7);
14   Generate a trial particle G_i according to Eqs. (8, 9);
15   Select the best one among P_i, L_i, and G_i as the new P_i;
16   Update *pbest* and *gbest*;
17   **If** $f(pbest) = f(lastpbest)$ **then** $monitor[i] + +$ **else** $monitor[i] = 0$;
18   **End if**
19   **End for**
20   Mutate *gbest* according to Eq. (13);
21   **If** $monitor[k] \geq max(monitor[j], j = 1, .., NP)$ and $monitor[k] \geq limit$ **then**
22   Reinitialize the kth particle from random K distinct data objects of dataset;
23 **End while**

where D-dimensional vector $X_{ij} = (x_{i,1j}, x_{i,2j}, ..., X_{i,Dj})$ represents the $j$cluster centroid of ith particle.

L. Kaufman et al. [13] suggested that Sum of Euclid Distance (SED) is better than Mean Squared Error (MSE) for measuring cluster analysis results. In this paper we also use SED, which is calculated by Eq. (16), is used as the fitness function.

$$SED = \sum_{j=1}^{K} \sum_{S_i \in C_j} \|S_i - X_j\| \tag{16}$$

The main steps of the proposed algorithm are listed in Table 1, where NP is the population size, K is the number of clusters, *lastpbest* records the last fitness values of *pbest*. *monitor[i]* records the successive number of iterations where the fitness values of *pbest$_i$* does not change. FEs is the number of fitness evaluations, and $MaxFEs$ is the maximum number of fitness evaluations. P_{ns} is the probability to implement the neighborhood search strategy, *limit* is the pre-defined number. The fitness function is SED function calculated by Eq. (16).

3.6   Measure Criterions

Two metrics were used in our experiments, the first measure is the fitness value, the sum of Euclid distance SED, as defined in Eq. (16). The second metric is the clustering accuracy, which is the percentage of the objects that are correctly recovered in a clustering result (called classification accuracy percentage CAP) defined in Eq. (17).

$$CAP = 100 \times \frac{\#\text{of correctly classified examples}}{\text{size of test data set}} \tag{17}$$

4 Experimental Results

To evaluate the performance of the proposed algorithm, fourteen benchmark datasets including four artificial datasets and ten real-world datasets were used. In addition, four data clustering algorithms K-means[1], K-means++[3], KPSO[4], and PSOK[5] were compared to the proposed algorithm in terms of fitness value SED and accuracy CAP.

Table 2. The main properties of artificial datasets

Data set	Size	Features	No of clusters	Data set	Size	Features	No of clusters
Dataset1	400	3	4	Dataset3	300	2	6
Dataset2	250	2	5	Dataset4	500	2	10

4.1 Benchmark Datasets

The details of properties of artificial datasets are described in Table 2 [14], the properties of ten real-world datasets Iris, Wine, Glass, Ecoli, Liver disorder, Vowel, Vowel 2, Pima, WDBC, and CMC can be found in [15].

4.2 Parametric Settings

In this test, the parameters of four other competitive algorithms K-means, K-means++, KPSO, PSOK are set according to their experiments. For the sake of fair comparison, the population size $NP=100$. The maximal number of fitness evaluations $MaxFEs$ was set to 10e+04 for all algorithms, all algorithms were run on each of the 14 datasets over 25 times and their mean value of SED, accuracy percentage. For CMPNS, other parameters were empirically set as follows: $w_0 = 0.9$, $w_1 = 0.4$, $c_1 = c_2 = 1.49$, $P_r = 0.9$, $P_{ns} = 0.6$, $limit = 50$.

4.3 Comparison of Results

The results of SED are shown in Tables 3, where the best values are written in bold. The results in Table 3 indicate that the proposed CMPNS algorithm has the best results of SED on 12 of 14 datasets, two other datasets of Vowel2 and Ecoli belong to K-means and K-means++, respectively. In order to compare the performance of multiple algorithms on the test suite, we conduct Friedman test [16], the highest ranking belongs to CMPNS, namely the ranks of K-means,

Table 3. Comparison of SED results

Data set	K-means	K-means++	KPSO	PSOK	CMPNS
Dataset1	8.5182e+02	7.4998e+02	8.1271e+02	7.4997e+02	**7.4961e+02**
Dataset2	3.2838e+02	3.2816e+02	4.1106e+02	3.2841e+02	**3.2644e+02**
Dataset3	4.4943e+02	4.2890e+02	4.4795e+02	3.7449e+02	**3.7361e+02**
Dataset4	9.4805e+02	8.7136e+02	1.1241e+03	8.8793e+02	**8.6534e+02**
Iris	1.0502e+02	9.8663e+01	1.0685e+02	9.7272e+01	**9.6691e+01**
Wine	1.6838e+04	1.7339e+04	1.7078e+04	1.6364e+04	**1.6299e+04**
Glass	2.2470e+02	2.3202e+02	2.4546e+02	2.1866e+02	**2.1773e+02**
Ecoli	6.4785e+01	**6.3604e+01**	6.7063e+01	6.4673e+01	6.4697e+01
Liver dis	1.0213e+04	1.0222e+04	1.0262e+04	9.8829e+03	**9.8519e+03**
Vowel	1.5306e+05	1.5304e+05	1.7413e+05	1.5119e+05	**1.5069e+05**
Vowel2	**7.0912e+02**	7.0980e+02	8.6285e+02	7.2348e+02	7.1489e+02
Pima	5.2072e+04	5.2072e+04	5.0867e+04	4.7832e+04	**4.7564e+04**
WDBC	1.5295e+05	1.5295e+05	1.5215e+05	1.4985e+05	**1.4953e+05**
CMC	5.5133e+03	5.5142e+03	6.1808e+03	5.5140e+03	**5.5103e+03**

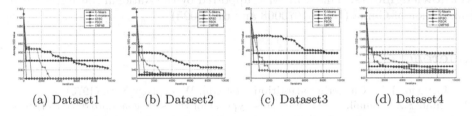

(a) Dataset1   (b) Dataset2   (c) Dataset3   (d) Dataset4

Fig. 1. The convergence curves on artificial datasets

Table 4. Comparison of results of CAP on artificial datasets (in percentage)

Data set	K-means	K-means++	KPSO	PSOK	CMPNS
Dataset1	96.25±9.16	100.00±0	100.00±0	100.00±0	100.00±0
Dataset2	94.00±0	94.68±1.09	85.44±5.07	95.80±1.62	**96.46±1.50**
Dataset3	89.17±9.79	92.50±8.51	97.95±4.04	100.00±0	100.00±
Dataset4	89.13±4.81	91.01±5.38	92.31±4.84	93.76±5.23	**94.85±4.28**
Iris	82.27±10.48	87.83±4.99	88.30±2.98	89.33±0.43	**89.97±0.15**
Wine	69.97±0.62	69.24±1.18	71.29±1.03	70.84±0.25	**71.52±0.32**
Glass	56.80±2.51	55.33±3.26	59.30±0.86	59.37±3.40	**60.09±2.81**
Ecoli	80.51±2.76	**81.49±1.99**	78.81±3.01	81.12±2.20	80.79±3.03
Liver	**57.97±0**	**57.97±0**	**57.97±0**	**57.97±0**	**57.97±0**
Vowel	58.29±2.78	58.86±2.21	57.65±2.68	58.27±1.66	**59.39±2.70**
Vowel2	37.43±2.36	36.66±1.83	34.99±2.72	36.94±1.96	**37.44±1.93**
Pima	65.10±0	65.10±0	65.10±0	**66.02±0**	**66.02±0**
WDBC	85.41±0	85.41±0	86.29±0.79	86.41±0.38	**86.820**
CMC	45.34±0.40	45.11±0.38	44.95±0.91	45.26±0.38	**45.58±0.13**

K-means++, KPSO, PSOK, and CMPNS are 3.29, 3.07, 3.18, 2.36, and 1.29, respectively. The representative convergence curves of artificial datasets are illustrated in the Fig. 1.

The CAP results of CAP average of 25 times on each of all datasets are listed in Tables 4, where the best results be written in bold. The results in Tables show that the proposed CMPNS algorithm has the best accuracy percentage in majority of benchmark datasets, only on Ecoli dataset the best result belongs to K-means++ algorithm.

5   Conclusions

In this study, we propose a new data clustering approach in order to improve K-means algorithm by enhanced PSO algorithm. Aiming to overcome the shortcoming of K-means, enhanced PSO approach by employing the proposed neighborhood search strategy and combining with diversity mechanism, Cauchy mutation operation, and reinitialization was introduced into K-means. The results obtained from testing on fourteen benchmark datasets including artificial and real-world datasets the proposed CMPNS algorithm is also good at data clustering in compared with some data clustering algorithms. So that, CMPNS can be an alternative for solving data clustering problems and other relevant problems.

Acknowledgments. This work was supported by the National Natural Science Foundation of China (No.: 61070008 and 61364025).

References

1. Hartigan, J.A.: Clustering algorithms, 1st edn. Wiley, New York (1975)
2. Selim, S.Z., Ismail, M.A.: K-means type algorithms: a generalized convergence theorem and characterization of local optimality. IEEE Trans. Patfern Anal. Mach. Intell. 6, 81–87 (1984)
3. Arthur, Vassilvitskii: K-means++: The advantages of careful seeding. In: Proceedings of the Eighteenth Annual ACM-SIAM Symposium on Discrete Algorithms (SODA 2007), pp. 1027–1035 (2007)
4. Merwe, D., Engelbrecht, A.: Data clustering using particle swarm optimization. In: Proceedings of the Congress on Evolutionary Computation 2003 (CEC 2003), pp. 215–220 (2003)
5. Neshat, M., Yazdi, S.F., et al.: A New Cooperative Algorithm Based on PSO and K-Means for Data Clustering. Journal of Computer Science 8(2), 188–194 (2012)
6. Kao, Y., Lee, S.: Combining K-means and Particle Swarm Optimization for Dynamic Data Clustering Problems. In: Proceedings of IEEE International Conference on Intelligent Computing and Intelligent Systems 2009, pp. 757–761 (2009)
7. Kennedy, J., Eberhart, R.: Particle swarm optimization. In: Proceedings of IEEE International Conference on Neuron Networks Conference Proceedings, Perth, Australia, pp. 1942–1948 (1995)
8. Clerc, M., Kennedy, J.: The Particle Swarm-Explosion, Stability, and Convergence in a Multidimensional Complex Space. IEEE Trans on Evol. 6(1), 58–73 (2002)
9. Shi, Y., Eberhart, R.: A Modified Particle Swarm Optimizer. In: Proceedings of the 1998 Congress on Evolutionary Computation (CEC 1998), pp. 69–73 (1998)
10. Wang, H., Sun, S., Li, C., Rahnamayan, S., Pan, J.: Diversity enhanced particle swarm optimization with neighborhood search. Information Sciences 223, 119–135 (2013)
11. Wang, H., Wu, Z., Rahnamayan, S., Liu, Y., Ventresca, M.: Enhancing particle swarm optimization using generalized opposition-based learning. Information Sciences 181, 4699–4714 (2011)

12. Karaboga, D.: An idea based on honey bee swarm for numerical optimization. Technical report-TR06, Erciyes University, Engineering Faculty (2005)
13. Kaufman, L., Rousseeuw, P.: Finding Groups in Data: An Introduction to Cluster Analysis. John Wiley and Sons, New York (1990)
14. Bandyopadhyay, S.: Artificial data sets for data mining, http://www.isical.ac.in/~sanghami/data.html
15. UCI Repository of Machine Learning Databases: retrieved from the World Wide Web, http://www.ics.uci.edu/~mlearn/MLRepository.html
16. Derrac, J.: A practical tutorial on the use of nonparametric statistical tests as a methodology for comparing evolutionary and swarm intelligence algorithms. Swarm and Evolutionary Computation 1, 3–18 (2011)

Accuracy Improvement of Localization and Mapping of ICP-SLAM via Competitive Associative Nets and Leave-One-Out Cross-Validation

Shuichi Kurogi, Yoichiro Yamashita, Hikaru Yoshikawa, and Kotaro Hirayama

Kyushu Institute of Technology, Tobata, Kitakyushu, Fukuoka 804-8550, Japan
kuro@cntl.kyutech.ac.jp,
{yamashita,yoshikawa}@kurolab.cntl.kyutech.ac.jp
http://kurolab.cntl.kyutech.ac.jp/

Abstract. This paper presents a method to improve the accuracy of localization and mapping obtained by ICP-SLAM (iterative closest point - simultaneous localization and mapping) algorithm. The method uses competitive associative net (CAN2) for learning piecewise linear approximation of the cloud of 2D points obtained by the LRF (laser range finder) mounted on a mobile robot. To reduce the propagation error caused by the consecutive pairwise registration by the ICP-SLAM algorithm, the present method utilizes leave-one-out cross-validation (LOOCV) and tries to minimize the LOOCV registration error. The effectiveness is shown by analyzing the real experimental data.

Keywords: ICP-SLAM, LRF range data, Propagation error owing to pairwise registration, Competitive associative nets, Leave-one-out cross-validation.

1 Introduction

This paper presents a method to improve the accuracy of localization and mapping obtained by ICP-SLAM algorithm. Here, we consider the ICP (iterative closest point) method [1] to find the transformation between two successive point clouds obtained by the LRF (laser range finder), and the SLAM (simultaneous localization and mapping) (e.g. [2]) utilizing the ICP method for a mobile robot to execute the localization and mapping. The result (localization and mapping) of the ICP-SLAM algorithm involves the propagation error owing to the successive processing. Here, note that the LRF range data are characterized as involving lack of data called black spots, quantization errors owing to the range (distance) resolution (e.g. 10mm), and a large number of data owing to high angular resolution (e.g. $0.25°$). To deal with such data, we have developed a plane extraction method using CAN2 (competitive associative net) [3], where the CAN2 is an artificial neural net for learning piecewise linear approximation of nonlinear functions and provides several advantages such as data compression, noise reduction, availability of approximated piecewise line segments of LRF range data in this application.

We have utilized these advantages in range image registration [4], where we use particle filter (PF) and loop-closing for reducing the propagation error. From the surveys of

C.K. Loo et al. (Eds.): ICONIP 2014, Part II, LNCS 8835, pp. 160–169, 2014.
© Springer International Publishing Switzerland 2014

range image registration [6,7], our method in [4] is classified into statistic techniques for cycle minimization studied for SLAM problems. As an another idea, we have presented a multiview registration method for 3D range images in [5], where we have introduced leave-one-image-out cross-validation (LOOCV) error to reduce the propagation error of successive pairwise registration. In this paper, we apply the above idea to improve the result of the ICP-SLAM algorithm. From the survey [7], the idea is closely related to the metaview strategies of analytic techniques for cycle minimization to reduce the propagation error. Although the conventional metaview methods incrementally register and merge the consecutive images into a metaview, the present method generate all LOOCV metaviews, and tries to reduce the registration error between every pair of LOOCV metaview and the remaining cloud of points. Here, the correspondence of the points between the LOOCV metaview and the remaining image is obtained by means of using the distance as well as the orientation of the tangent lines at the points, which is easy to be possible by the piecewise line segments approximated by the CAN2 in this research. Here, the method using the distance and the orientation of tangent line, which we call point-to-line registration, is a variant of the point-to-plane registration [8] which shows better performance than the point-to-point registration used in many ICP methods (see [6,7]).

In the next section, we show the present method to improve the accuracy of localization and mapping obtained by the ICP-SLAM application by means of using CAN2 and LOOCV, and then the effectiveness of the method is evaluated in 3.

2   Accuracy Improvement of ICP-SLAM Via CAN2 and LOOCV

2.1   Localization and Mapping Estimated by ICP-SLAM Application

We have used a pioneer 3-AT mobile robot from MobileRobots Inc. with a Hokuyo UTM-30LX LRF mounted on the top to execute ICP-SLAM as shown in Fig. 1(a). The specification of UTM-30LX is as follows; the distance scanning range is 0.1 to 30m and the angular scanning range is $\phi_{\max} = 270°$, while the accuracy is ± 30mm for 0.1 to 10m and ± 50mm for 10 to 30m, and the angular resolution is $\Delta_\phi = 0.25°$. We use the ICP-SLAM application with icp-classic option provided by MRPT (Mobile Robot Programming Toolkit) [9]. It is an offline application which provides an estimated trajectory of the robot poses (positions and orientations) and an estimated 2D map after a running of the robot, where the map is represented as a cloud of points obtained from the LRF as shown Fig. 1(c).

Let $\widehat{\boldsymbol{x}}_t = (\widehat{x}_t, \widehat{y}_t, \widehat{\theta}_t)^T$ be the pose of the robot at a discrete time $t \in I^{\text{time}} = \{1, 2, \cdots\}$ obtained by the ICP-SLAM application, and $D_t = \{r_t^{(i)}) \mid i \in I^{\text{scan}}\}$ be the range (distance) dataset obtained by the LRF at $t \in I^{\text{time}}$, where $r_t^{(i)}$ for $I^{\text{scan}} = \{0, 1, 2, \cdots, i_{\max} = \phi_{\max}/\Delta_\phi\}$ is the ith scanning range data. From D_t, we obtain $Z_t = \{\boldsymbol{p}_t^{(i)} = (x_t^{(i)}, y_t^{(i)})^T = (r_t^{(i)} \cos(i\Delta_\phi - \phi_r), r_t^{(i)} \sin(i\Delta_\phi - \phi_r))^T \mid i \in I^{\text{scan}}\}$ denoting the cloud of points on the tth LRF coordinate system (see Fig. 1(b)). Here, by means of using ϕ_r denoting the robot orientation on the LRF coordinate system, the y-axis of $\boldsymbol{p}_t^{(i)}$ directs to the orientation of the robot. We call Z_t the tth submap in the following.

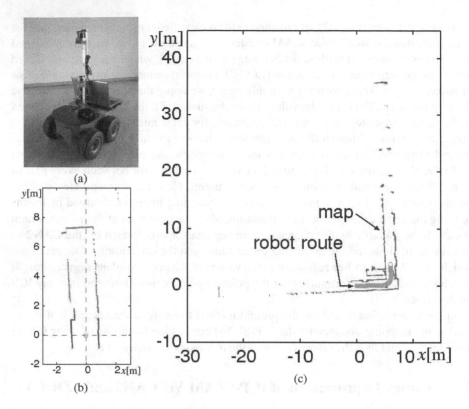

Fig. 1. (a) Mobile robot (P3-AT), LRF (Hokuyo UTM-30LX) mounted on the top to scan the environment horizontally, and a note PC on the robot used in the experiment, (b) a cloud of points obtained by a single 2D scan of the LRF, and (c) a map and a robot route estimated by the ICP-SLAM application.

We transform the point $\boldsymbol{p}_t^{(i)}$ in Z_t on the tth LRF coordinate system to

$$\widehat{\boldsymbol{p}}_{t-1,t}^{(i)} = \widehat{\boldsymbol{R}}_{t-1,t}\boldsymbol{p}_t^{(i)} + \widehat{\boldsymbol{t}}_{t-1,t} \tag{1}$$

on the $(t-1)$th LRF coordinate system. Here, the parameter $(\widehat{\boldsymbol{R}}_{t-1,t}, \widehat{\boldsymbol{t}}_{t-1,t})$ is obtained by the poses of the robot via the ICP-SLAM application, i.e. the rotation matrix $\boldsymbol{R}_{t-1,t} = \boldsymbol{R}\left(\widehat{\theta}_{t-1} - \widehat{\theta}_t\right)$ and the translation vector $\widehat{\boldsymbol{t}}_{t-1,t} = (\widehat{x}_{t-1} - \widehat{x}_t, \widehat{y}_{t-1} - \widehat{y}_t)^T$, where $\boldsymbol{R}(\theta) = \begin{pmatrix} \cos\theta & \sin\theta \\ -\sin\theta & \cos\theta \end{pmatrix}$. By means of applying this relationship recursively, we can transform $\boldsymbol{p}_t^{(i)}$ to the 1st LRF coordinate system as $\widehat{\boldsymbol{p}}_{1,t}^{(i)} = \widehat{\boldsymbol{R}}_{1,t}\boldsymbol{p}_t^{(i)} + \widehat{\boldsymbol{t}}_{1,t}$, where

$$(\widehat{\boldsymbol{R}}_{1,t}, \widehat{\boldsymbol{t}}_{1,t}) = (\widehat{\boldsymbol{R}}_{1,t-1}\widehat{\boldsymbol{R}}_{t-1,t}, \widehat{\boldsymbol{t}}_{1,t-1} + \widehat{\boldsymbol{R}}_{1,t-1}\widehat{\boldsymbol{t}}_{t-1,t}). \tag{2}$$

The map represented as a cloud of points as $\widehat{Z} = \{\widehat{\boldsymbol{p}}_{1,t}^{(i)} | i \in I^{\text{scan}}, t \in I^{\text{time}}\}$.

1. **Algorithm   search_function_region**(D_t, i_0, i_1):
2. $\phi_c := \pi/2 - (i_1 - i_0)\Delta_\phi/2$ $//\Delta_\phi = 0.25°$: angular resolution of LRF
3. **for** $i := i_0$ **to** i_1 **do**
4. $\qquad x_+ := (r_t^{(i)} + \Delta_r) * \cos(i\Delta_\phi + \phi_c)$ $//\Delta_r = 10mm$: allowable range error
5. $\qquad x_- := (r_t^{(i)} - \Delta_r) * \cos(i\Delta_\phi + \phi_c)$
6. $\qquad$ **if** $i \neq i_0$ **and** $\min\{x_+, x_-\} > x_0$ **then**   **return** $i - 1$
7. $\qquad x_0 := \max\{x_+, x_-\}$
8. **endfor**
9. **return** $i = i_1$

(a)

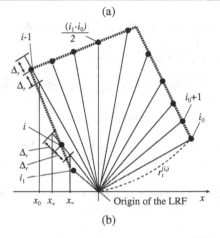

(b)

Fig. 2. (a) Algorithm to search the region where the data can be represented as a function, and (b) the relationship between the range data $r_t^{(i)}$ and the index i, where $x_i = r_t^{(i)} \cos(i\Delta_\phi + \phi_c)$ decreases with the increase of i (with allowable error Δ_r) in the region

2.2 Preprocessing for Function Approximation

In order to apply the CAN2 for function approximation, the data should be represented as a function. However, the relationship from $x_t^{(i)}$ to $y_t^{(i)}$ of $\boldsymbol{p}_t^{(i)} = (x_t^{(i)}, y_t^{(i)})^T$ in Z_t is not always possible to be represented as a function as $y_t^{(i)} = f(x_t^{(i)})$. However, the original range data $r_t^{(i)}$ can be represented as a function as $r_i = r(\phi_i)$. If $r = r(\phi)$ is continuous around ϕ_c, we can derive $r \simeq r_c + r'(\phi_c)\Delta\phi$ for $r = r(\phi)$, $r_c = r(\phi_c)$ and $\Delta\phi = \phi - \phi_c$. Then, by means of using $(x, y) = (r\cos(\Delta\phi + \pi/2), r\sin(\Delta\phi + \pi/2))$, we have $(x, y) \simeq (-r\Delta\phi, r)$ for $|\Delta\phi| \ll 1$. Thus, y is represented by a function of x as $y \simeq r_c - r'(\phi_c)x/(r_c + r'(\phi_c)\Delta\phi)$. This indicates that for a small angular region, the LRF data with the rectangular form can be represented as a function. So, we divide the range dataset D_t into a number of subsets $D_t^{(l)}$ for $l = 1, 2, \cdots$ so that the data in each $D_t^{(l)}$ can be represented as a function. To have this done, we use two algorithms, one is shown in Fig. 2 to search the region in which the data can be represented as a function, and the other is shown in Fig. 3 to divide the dataset by means of using the former algorithm to search the function region. From the subset

```
 1. Algorithm    divide_dataset($D_t$):
 2. $\Delta_i := i_{max}/L_0$        // $L_0$: default number of the regions to be divided
 3. $i_1 := 0$
 4. $L := 0$
 5. for   $l = 1$   to   $L_0$   do
 6.         $i_0 := i_1$
 7.         repeat
 8.                 $i_1 := \min\{i_1 + \Delta_i, i_{max}\}$
 9.                 $i := $ search_function_region $(D_t, i_0, i_1)$
10.         until $i = i_1$ and $i_1 < i_{max}$
11.         $i_1 := i$
12.         if   $i_1 - i_0 \geq i_a$    then   // $i_a$ (= 4) is a threshold
13.                 $L := L + 1$
14.                 $D_t^{(L)} := \{r_t^{(i)} \mid i \in I_t^{(L)} \triangleq [i_0, i_1]\}$
15.         else
16.                 $r_t^{(i)} := -1$    for   $i \in [i_0, i_1]$   // inactivate data
17.         endif
18.         if   $i_1 \geq i_{max}$    then    breakfor
19. endfor
20. return $D_t^{(l)}$ for $l \in I^{region} \triangleq \{1, 2, \cdots, L\}$   // return the regions
```

Fig. 3. Algorithm to divide D_t into subsets $D_t^{(l)}$ ($l \in I^{region}$) to be represented as as a function

$D_t^{(l)} = \{r_t^{(i)} | i \in I_t^{(l)} = [i_{min,t}^{(l)}, i_{max,t}^{(l)}]\}$ divided by the algorithms from $D_t^{(l)}$, we obtain the rectangular dataset $Z_t^{(l)} = \{p_t^{(i)} = (x_t^{(i)}, y_t^{(i)})^T = (r_t^{(i)} \cos\phi_t^{(i)}, r_t^{(i)} \sin\phi_t^{(i)})^T$ $| \phi_t^{(i)} = \left(i + \frac{i_{max,t}^{(l)} - i_{min,t}^{(l)}}{2}\right)\Delta_\phi + \frac{\pi}{2}, i \in I_t^{(l)}\}$. Then, we can apply the CAN2 to learn to extract piecewise line segments from the cloud of points in $Z_t^{(l)}$.

2.3 Piecewise Linear Approximation by the CAN2

We use a CAN2 with N units for learning to approximate $Z_t^{(l)}$. The jth unit has a weight vector (scalar in this application) $w_t^{(l,j)} = w_{1,t}^{(l,j)}$ and an associative matrix (row vector) $M_t^{(l,j)} = (M_{0,t}^{(l,j)}, M_{1,t}^{(l,j)})$ for $j \in I^{CAN2} = \{1, 2, \cdots, N\}$. After learning $Z_t^{(l)} = \{(x_t^{(i)}, y_t^{(i)})^T | i \in I_t^{(l)}\}$ as a function $y = f(x)$ for $x = x_t^{(i)}$ and $y = y_t^{(i)}$, the CAN2 divide the input space into Voronoi regions $V_t^{(l,j)} = \{x | j = \underset{i}{\operatorname{argmin}}\{\|x - w_t^{(l,i)}\|\}$ for $j \in I^{CAN2}$, and performs linear approximation $y = M_t^{(l,j)}\tilde{x}$ in each region, where $\tilde{x} = (1, x)^T$. As a result, the submap $Z_t^{(l)}$ is approximated by piecewise line segments. Namely, the equation of the line is given by $(n_t^{(l,j)})^T p_t = \alpha_t^{(l,j)}$ and the center of the line segment is given by $p_t = q_t^{(l,j)} = (w_{t,1}^{(l,j)}, M_t^{(l,j)} \tilde{w}_{t,1}^{(l,j)})^T$, where $\tilde{w}_{t,1}^{(l,j)} = (1, w_{t,1}^{(l,j)})$, and the normal vector $n_t^{(l,j)} = (n_{x,t}^{(l,j)}, n_{y,t}^{(l,j)})^T$ and the distance to the origin $\alpha_t^{(l,j)}$ are given by

$$\left(\left(n_t^{(l,j)} \right)^T, \alpha_t^{(l,j)} \right) = \frac{\left(-M_{1,t}^{(l,j)}, 1, M_{0,t}^{(l,j)} \right)}{\sqrt{(M_{1,t}^{(l,j)})^2 + 1}}. \tag{3}$$

Here, note that $n_{y,t}^{(l,j)} = 1/\sqrt{(M_{1,t}^{(l,j)})^2 + 1} > 0$ or the normal vector directs forward from the origin of the tth LRF coordinate system. As an approximation of $Z_t^{(l)}$, we use $Z_t^{(l),\text{CAN2}} = \{ q_t^{(l,j)}, n_t^{(l,j)}, \alpha_t^{(l,j)} \mid j \in I^{\text{CAN2}} \}$. For each approximated point $(x, y)^T \in Z_t^{(l),\text{CAN2}}$ for $l \in I^{\text{region}} \triangleq \{1, 2, \cdots, L\}$, we reconstruct the cloud of approximated points in the LRF coordinate system as $(x_t^{(i')}, y_t^{(i')})^T = \{ (r_t^{(i')} \cos(i'\Delta_\phi - \phi_r), r_t^{(i')} \sin(i'\Delta_\phi - \phi_r))^T$, where $i' = (\tan^{-1}(y/x) - \phi_c^{(l)})/\Delta_\phi$ and $r_t^{(i')} = \sqrt{x^2 + y^2}$, where $\phi_c^{(l)} = (i_{\max,t}^{(l)} - i_{\min,t}^{(l)})\Delta_\phi/2 + \pi/2$. Here, i' does not have to be an integer. From the above procedure, we reconstruct the tth approximated submap $Z_t^{\text{CAN2}} = \bigcup_{l \in I^{\text{region}}} Z_t^{(l),\text{CAN2}}$.

From Z_t^{CAN2}, we remove the following data to obtain the ROI (Region of Interest) dataset $Z_t^{(l),\text{CAN2}} = \{ q_t^{(l,j)}, n_t^{(l,j)}, \alpha_t^{(l,j)} \mid j \in I^{\text{CAN2ROI}} \}$.

(i) (Remove jump edge) The data on the jump edge hold $(n_t^{(j)})^T q_t^{(j)} = 0$. So, we remove the data with $|(n_t^{(j)})^T q_t^{(j)}|/\|q_t^{(j)}\| < \cos(\pi/2 - \psi_{\text{je}})$, where $\psi_{\text{je}}(= 5°)$ indicates allowable error.

(ii) (Remove unreliable piecewise line segments) We remove the data in the Voronoi region of the unit which involves less than 4 data because the data are unreliable.

2.4   Range Data Registration Using LOOCV

We try to improve the transformation $(\widehat{R}_{1,t}, \widehat{t}_{1,t})$ in (2) obtained by the ICP-SLAM application into the transformation $(R_{1,t}, t_{1,t})$ as follows.

[Algorithm: Range Data Registration Using LOOCV]

Step 1. (Initialize Transformation) Let $(R_{1,t}, t_{1,t}) := (\widehat{R}_{1,t}, \widehat{t}_{1,t})$ for all $t \in I^{\text{time}}$.

Step 2. (Obtain corresponding points) Obtain $q_{1,t}^{(j)} := R_{1,t} q_t^{(j)} + t_{1,t}$ for all $q_t^{(j)} \in Z_t^{\text{CAN2ROI}}$. Search $q_{1,\bar{t}}^{(l_j)}$ being the closest point to $q_{1,t}^{(j)}$ for all $\bar{t} \in I^{\text{time}} \setminus \{t\}$ and let $Z_{1,t}^{\text{CAN2ROI}} := \{ q_{1,t}^{(j)} \mid j \in I_{1,t}^{\text{CAN2ROI}} \}$ be the set of $q_{1,t}^{(j)}$ satisfying a distance condition given by $\|q_{1,t}^{(j)} - q_{1,\bar{t}}^{(l_j)}\| \le d_{\text{th}}$ and an orientation condition given by $\left(n_{1,t}^{(j)} \right)^T n_{1,\bar{t}}^{(l_j)} \ge \cos \psi_{\text{th}}$ for the thresholds d_{th} and ψ_{th}.

Step 3. (Refine transformation) In order to improve the accuracy, obtain the point $\xi_{1,\bar{t}}^{(l_j)} := q_{1,\bar{t}}^{(l_j)} + \left(\alpha_{1,\bar{t}}^{(l_j)} - (n_{1,\bar{t}}^{(l_j)})^T q_{1,t}^{(j)} \right) n_{1,\bar{t}}^{(l_j)}$ which is on the tangent line of $q_{1,\bar{t}}^{(l_j)}$ and closest to $q_{1,t}^{(j)}$, and introduce

$$\eta_{1,\bar{t}}^{(l_j)} := \begin{cases} \xi_{1,\bar{t}}^{(l_j)} & \text{if } \|q_{1,t}^{(j)} - q_{1,\bar{t}}^{(l_j)}\| \le d_{\text{p2l}}, \\ q_{1,\bar{t}}^{(l_j)} & \text{otherwise}, \end{cases} \tag{4}$$

where d_{p2l} indicates a threshold of point-to-line distance. Now, we introduce the registration error given by

$$J \triangleq \langle (\Delta Z_{t,\bar{t}})^2 \rangle_{t \in I^{\text{time}}} = \left\langle \| \Delta R_t \, q_{1,t}^{(j)} + \Delta t_t - \eta_{1,\bar{t}}^{(l_j)} \|^2 \right\rangle_{q_{1,t}^{(j)} \in Z_{1,t}^{\text{CAN2ROI}}, c \in I^{\text{time}}}. \tag{5}$$

This indicates the LOOCV MSE (mean-square-error) of the current registration with $(\Delta R_t, \Delta t_t) = (I, 0)$. From ICP techniques, the error J is expected to be reduced by $(\Delta R_t, \Delta t_t) := (U_t V_t^T, \overline{\eta_{1,\bar{t}}^{(l_j)}} - \Delta R_t \overline{q_{1,t}^{(j)}})$, where U_t and V_t are the left and the right singular matrices of the cross-covariance matrix of $\eta_{1,t}^{(l_j)}$ and $q_{1,t}^{(j)}$ with the mean vectors $\overline{\eta_{1,\bar{t}}^{(l_j)}}$ and $\overline{q_{1,t}^{(j)}}$, respectively. We update the transformation as $(R_{1,t}, t_{1,t}^{(j)}) := (\Delta R_t R_{1,t}, \Delta R_t t_{1,t}^{(j)} + \Delta t_t)$ for all t and repeat **Step 2** and **3** until convergence.

3   Experimental Results

We have run the robot around a corner of a corridor in our department building as shown in Fig. 1(c), and obtained odometry and LRF data with sampling rate 10Hz and 40Hz, respectively, for 45s duration of time. As a result of applying the ICP-SLAM

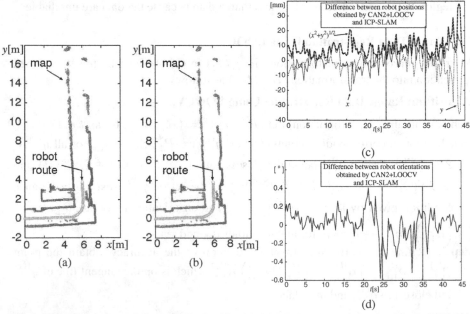

Fig. 4. The robot route and the map obtained by (a) the ICP-SLAM and (b) the CAN2+LOOCV. Difference between the robot poses ((c) position and (d) orientation) at each time t obtained by the ICP-SLAM and the CAN2+LOOCV

application, we have 450 pairs of robot poses and LRF scanning data for the sampling
rate 10Hz. For examining the present method, which we denote CAN2+LOOCV in the
following, we show the result using one out of every three pairs of data. Namely, we
use 150 pairs of the pose $\widehat{x}_t = (\widehat{x}_t, \widehat{y}_t, \widehat{\theta}_t)^T$ and the LRF dataset D_t for the discrete
time $t \in I^{\text{time}} = \{1, 2, \cdots, 150\}$ for the sampling rate $T_s = 0.3$s.

We show magnified pictures of the robot route and the map obtained by the two
methods in Fig. 4(a) and (b), where the CAN2+LOOCV employs p2l-only registration
using the thresholds as $(d_{\text{p2l}}, d_{\text{th}}, \psi_{\text{th}}) = (0.3\text{m}, 0.3\text{m}, 45°)$. From the pictures, we may
say that there is not a big difference, exept that the area of cloud points seem slightly
reduced by the CAN2+LOOCV. So, we examine numerical results as shown in (c) and
(d). From (c), we can see that the maximum distance of the positions obtained by the
two methods is about 40mm which occurs after the gradual and fluctuating increase of
the difference in the y-direction for $t > 35$s, which may indicate the propagation error
of the ICP-SLAM. Let us divide 45s duration of time into 3 (rough) periods: the first
period is 0–25s corresponding to the robot going straight in the x-direction, the second
one is 25–35s corresponding to the turning around the corner, and the last one is 35–45s

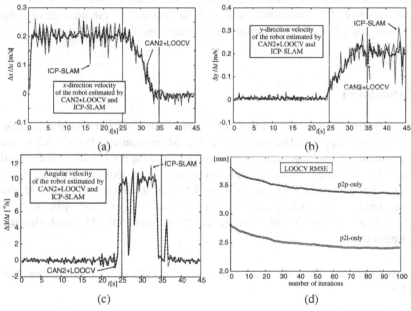

Fig. 5. The velocity in (a) the x-direction, (b) y-direction and (c) orientation obtained by the
CAN2+LOOCV and the ICP-SLAM. Here, the velocity is obtained by $\left(\dfrac{\Delta x}{\Delta t}, \dfrac{\Delta y}{\Delta t}, \dfrac{\Delta \theta}{\Delta t}\right) =$
$\left(\dfrac{x_t - x_{t-1}}{T_s}, \dfrac{y_t - y_{t-1}}{T_s}, \dfrac{\theta_t - \theta_{t-1}}{T_s}\right)$ for the estimated robot pose $x_t = (x_t, y_t, \theta_t)^T$ for $t \in$
I^{time} and the sampling period $T_s = 0.3$s. (d) Reduction of the LOOCV RMSE (root mean
square error $J^{1/2}$) by the CAN2+LOOCV method with p2l-only (point-to-line only) registration
using $(d_{\text{p2l}}, d_{\text{th}}, \psi_{\text{th}}) = (0.3\text{m}, 0.3\text{m}, 45°)$ and p2p-only (point-to-point only) registration using
$(d_{\text{p2l}}, d_{\text{th}}, \psi_{\text{th}}) = (0\text{m}, 0.3\text{m}, 45°)$.

corresponding to going straight in the y-direction, respectively. From (c) and (d), we can see that the difference in the x position, the y position and the angle seems big for the first, third and second periods, respectively.

To examine this result much more, we show the velocity of the robot obtained by the two methods in Fig. 5(a), (b) and (c). We can see that the fluctuation of the velocity in the x-direction, the y-direction, and the orientation is big in the first, the third and the second period, respectively. This indicates that the fluctuation is big along for the moving direction or orientation. From the point of view of the comparison between two methods, the CAN2+LOOCV has provided smaller magnitude of the velocity fluctuation along with the x- and y-direction for all periods than the ICP-SLAM. Since the actual movement of the robot had not involve fluctuations, the CA2+LOOCV is considered to have provided better result. From the point of view of angular velocity fluctuation, the CAN2+LOOCV has also provided smaller magnitude in the 2nd period but bigger one in the first and the third period. This is supposed to be because the LOOCV method tries to reduce the registration error for all periods and the error is bigger in the second period than in other periods. As a result, the CAN2+LOOCV method seems to improve the accuracy of the localization and the mapping obtained by the ICP-SLAM.

From Fig. 5(f), we can see that the CAN2+LOOCV reduces the LOOCV RMSE (root mean square error $J^{1/2}$) monotonically with the increase of the number of iteration. Furthermore, the p2l-only registration has provided smaller RMSE than the p2p-only registration.

4   Conclusion

We have presented a method using CAN2 and LOOCV to improve the accuracy of localization and mapping obtained by ICP-SLAM application. The CAN2 is used to obtain a number of line segments as a piecewise linear approximation of the cloud of 2D points. The center points and the normal vectors of the approximated line segments are utilized for multiview registration using LOOCV technique. The method reduces the propagation error caused by the consecutive pairwise registration by the ICP-SLAM algorithm. We have shown the effectiveness of the present method by analyzing the real experimental data.

Acknowledgement. This work was supported by JSPS KAKENHI Grant Number 24500276.

References

1. Besl, P.J., McKay, H.D.: A method for registration of 3-D shapes, IEEE Transactions on Pattern Analysis and Machine Intelligence (1992)
2. Thrun, S., Burgard, W., Fox, D.: Probabilistic robotics (Intelligent robotics and autonomous agents). The MIT Press (2005)
3. Kurogi, S., Koya, H., Nagashima, R., Wakeyama, D., Nishida, T.: Range image registration using plane extraction by the CAN2. In: Proc. of CIRA 2009, pp. 346–550 (2009)

4. Kurogi, S., Nagi, T., Nishida, T.: Registration using particle filter and competitive associative nets. In: Wong, K.W., Mendis, B.S.U., Bouzerdoum, A. (eds.) ICONIP 2010, Part II. LNCS, vol. 6444, pp. 352–359. Springer, Heidelberg (2010)
5. Kurogi, S., Nagi, T., Yoshinaga, S., Koya, H., Nishida, T.: Multiview range image registration using competitive associative net and leave-one-image-out cross-validation error. In: Lu, B.-L., Zhang, L., Kwok, J. (eds.) ICONIP 2011, Part III. LNCS, vol. 7064, pp. 621–628. Springer, Heidelberg (2011)
6. Salvi, J., Matabosh, C., Foli, D., Forest, J.: A review of recent range image registration methods. Image and Vision Computing 25, 578–596 (2007)
7. Batlle, E., Matabosch, C., Salvi, J.: Summarizing image/Surface registration for 6DOF robot/Camera pose estimation. In: Martí, J., Benedí, J.M., Mendontca, A.M., Serrat, J. (eds.) IbPRIA 2007. LNCS, vol. 4478, pp. 105–112. Springer, Heidelberg (2007)
8. Chen, Y., Medioni, G.: Object modeling by registration of multiple range images. In: IEEE International Conference on Robotics and Automation, pp. 2724–2729 (1991)
9. http://www.mrpt.org/

Saliency Level Set Evolution

Jincheng Mei[1] and Bao-Liang Lu[1,2,*]

[1] Center for Brain-Like Computing and Machine Intelligence
Department of Computer Science and Engineering
Key Laboratory of Shanghai Education Commission for Intelligent Interaction and
Cognitive Engineering
[2] Shanghai Jiao Tong University
800 Dong Chuan Road., Shanghai 200240, China
bllu@sjtu.edu.cn

Abstract. In this paper, we consider saliency detection problems from a unique perspective. We provide an implicit representation for the saliency map using level set evolution (LSE), and then combine LSE approach with energy functional minimization (EFM). Instead of introducing sophisticated segmentation procedures, we propose a flexible and lightweight LSE-EFM framework for saliency detection. The experimental results demonstrate our method outperforms several existing popular approaches. We then evaluate several computation strategies independently. The comparisons results indicate their effectiveness and strong abilities in combatting saliency confusions.

Keywords: Saliency Detection, Level Set Evolutionm, Computer Vision.

1 Introduction

Saliency detection has drawn much attentions in computer vision society. It aims to extract anomalous objects and informative structures from images. Recently, it has been widely employed in computer vision and multimedia applications [1].

In traditional saliency detection, objects are usually extracted through post-processing of saliency map. For example, Hou and Zhang [2] proposes a simple thresholding strategy to extract proto-objects from images. Achanta et al. [3] suggest a widely adopted adaptive thresholding methods one step further. Unlike the methods mentioned above, Gopalakrishnan et al. [4] formulate saliency detection as a foreground and background labeling problem.

In this paper, we notice the interpretability of level set methods, which enable us to readily compute the saliency posterior probability. From this perspective, we propose a novel method, named as Saliency Level Set Evolution (SLSE), for separating salient objects and contexts of an image. As a classical method for shape representation, level set uses the *zero level set* of a 3D level set function (LSF) to represent a closed 2D curve [5]. This implicit representation enables numerical computations without curve and surface parametrization.

---

* Corresponding author.

C.K. Loo et al. (Eds.): ICONIP 2014, Part II, LNCS 8835, pp. 170–177, 2014.
© Springer International Publishing Switzerland 2014

(a)   (b)   (c)   (d)

Fig. 1. Saliency level set evolution. (a) An input image and its ground truth; (b) The initial LSF and its binary map. The red contour marks the zero level set; (c) The LSF and binary map after 5 iteration rounds; (d) After 30 rounds.

As shown in Fig.1(b), we firstly initialize the level set with *real number* values. This enables us to evolve the LSF without explicitly determining the binary labels. After several rounds of evolution, we obtain the final LSF as shown in Fig.1(c)-(d). Saliency map is novelly defined as the final normalized LSF.

The main contributions of this paper include:

1) We first provide level set representation for saliency and propose saliency detection method via LSE. This framework is interpretably plausible and provides interfaces for further adaptive improvements.

2) Experimental results show that SLSE outperforms many recently proposed methods in MSAR-1000 [3] and SED1 [6] datasets.

The rest is organized as follows: Section 2 introduces the LSE-EFM formulation. Section 3 demonstrates the experiment results and provides some discussions. Finally, Section 4 gives the conclusion about our work.

2   LSE-EFM Framework

The interpretability of combining LSE with saliency comes from two points: *on the one hand*, because of the ill-posed nature of saliency, it is questionable to directly appoint each pixel a binary label. A more appropriate representation is to utilize a binary random variable to provide a probability description. *On the other hand*, saliency detection can also be viewed as a contour selection *i.e.*, to exactly segment out salient objects is equivalent to select contours which accurately envelope the object regions. Based on these two points, we introduce the level set framework [5].

2.1 Level Set Representation

In our formulation, every pixel $\mathbf{z}$ of an image I is represented as a 5D vector $I(\mathbf{z}) = [G\ P]^\top = [L, a, b, x, y]^\top$ where $G = [L, a, b]$ is the value in the CIE Lab color space [7], and $P = [x, y]$ denotes the position of $\mathbf{z}$. Background and foreground regions are denoted as R_0, R_1, respectively. Using the level set terminology, we get:

$$\begin{cases} \mathbf{z} \in R_0 & \text{if } \phi(\mathbf{z}) < 0, \\ \mathbf{z} \in R_1 & \text{if } \phi(\mathbf{z}) \geq 0, \end{cases} \tag{1}$$

where ϕ is the *level set function* (LSF) [5]. The set $\mathcal{B} = \{\mathbf{z}|\phi(\mathbf{z}) = 0\}$ is called the *zero level set*, which indicates the boundary partitioning of R_0 and R_1. With the mapping rule:

$$\mathcal{C}(\mathbf{z}) = i \quad \text{if } \mathbf{z} \in R_i,\ i \in \{0, 1\}, \tag{2}$$

we get a level set representation based on ϕ instead of explicitly assigning each pixel a binary label.

2.2 Energy Functional Minimization

We firstly review the Bayesian framework [8], in which the saliency $\mathcal{S}$ of a given image I is defined as a posterior probability:

$$\mathcal{S} = p(\mathcal{C}|I), \tag{3}$$

where $\mathcal{C}$ is the salient region matrix with each element a binary random variable. We take the negative logarithm of the post probability and get the energy functional E [9] as:

$$E(\mathcal{C}; I) \sim -\log p(\mathcal{C}|I) \tag{4}$$

$$\sim \underbrace{-\log p(I|\mathcal{C})}_{\text{likelihood term}} \underbrace{-\log p(\mathcal{C})}_{\text{prior term}} \tag{5}$$

where $\log p(I)$ is independent of $\mathcal{C}$ thus omitted. Following [10], the energy functional is defined as a mutual information with a curve length penalty term,

$$E(\mathcal{C}; I) = \underbrace{-|\Omega|\hat{I}(\mathcal{C}; I)}_{\text{data term}} + \underbrace{\alpha \oint_{\mathcal{B}} ds}_{\text{penalty term}} \tag{6}$$

where $|\Omega|$ denotes the number of pixels in I, and $\mathcal{B}$ is a closed boundary. $\hat{I}$ is the mutual information which can be expanded as:

$$\hat{I}(\mathcal{C}; I) = H(I) - \sum_{i=0,1} \frac{|R_i|}{|\Omega|} H(I|\mathcal{C} = i) \tag{7}$$

Note the entropy $H(I)$ is also independent of $\mathcal{C}$ thus dropped. $H(I|\mathcal{C} = i)$ can be approximated by applying the weak law of large numbers and a nonparametric kernel density estimation (KDE) [11],

$$H(I|\mathcal{C} = i) = -\frac{1}{|R_i|} \int_{R_i} \log \hat{p}_i(I(\mathbf{z})) d\mathbf{z} \tag{8}$$

$$= -\frac{1}{|R_i|} \int_{R_i} \log \left(\int_{R_i} K(I(\hat{\mathbf{z}}) - I(\mathbf{z})) d\hat{\mathbf{z}} \right) d\mathbf{z} \tag{9}$$

where $\hat{p}_i(I(\mathbf{z})) \triangleq p(I(\mathbf{z})|\mathbf{z} \in R_i)$ indicates the likelihood of observing $I(\mathbf{z})$ when the pixel $\mathbf{z}$ belongs to region R_i, $\hat{p}_i(I(\mathbf{z}))$ is estimated employing a fast improved Gaussian transform [12], and $K(x) \propto \exp\left\{-\frac{x^2}{2\sigma^2}\right\}$ is a Gaussian kernel function.

2.3 Gradient Flow

Since both the energy explanation and the contour explanation are reasonable in the level set framework, the energy functional can be denoted as $E(\mathcal{B})$ as well as $E(\mathcal{C}; I)$. Thus the gradient flow of $E(\mathcal{B})$ in (4) is

$$\frac{\partial E(\mathcal{B})}{\partial t} = \left[\log \frac{\hat{p}_0(I(\mathbf{z}))}{\hat{p}_1(I(\mathbf{z}))} - \alpha \kappa \right] \mathbf{N} = v\mathbf{N}, \tag{10}$$

where $\mathbf{N} = \nabla\phi/|\nabla\phi|$ is the outward unit normal vector of $\mathcal{B}$, and $\kappa = \nabla \cdot (\nabla\phi/|\nabla\phi|)$ is the curvature of $\mathcal{B}$, which is defined as the divergence of $\mathbf{N}$.

2.4 Saliency Computation

As shown in Fig.1(b), we can initialize the LSF as a central box function. An alternative is to moderately narrowing the scope by thresholding. This leads to faster evolution. For thresholding, we adopt OTSU method [13] here.

During the evolution procedure, there exists different strategies. Recall in Section 2.1, every pixel is represented as a 5D vector. Actually, luminance contrast does not contribute as much information as color contrast for saliency [14]. This leads to weight tuning for luminance, *i.e.*, suppressing the luminance weight during the evolution. Another cue is that salient objects generally favor more compact spatial distributions than contexts [15]. This leads to weight tuning for position. Totally, the $\hat{p}_i(I(\mathbf{z}))$ in (8) is modified as:

$$\hat{p}_i(I(\mathbf{z})) \triangleq p(I(\mathbf{z})|\mathbf{z} \in R_i) \tag{11}$$

$$= p([\omega_L L(\mathbf{z}), a(\mathbf{z}), b(\mathbf{z}), \omega_{Pi} P(\mathbf{z})]^\top |\mathbf{z} \in R_i) \tag{12}$$

In the experiments, we set $\omega_L = 0.15$, $\omega_{P0} = 0$, and $\omega_{P1} = 0.25$. In each iteration, gradient flow (10) is employed without step searching:

$$\phi(k + 1) = \phi(k) - \Delta \cdot v \tag{13}$$

where $\Delta = 10$ empirically is the step rate. We set the maximum iteration number as 30, since the convergence is generally very fast in the experiments. Finally, we normalize the LSF to obtain the saliency map.

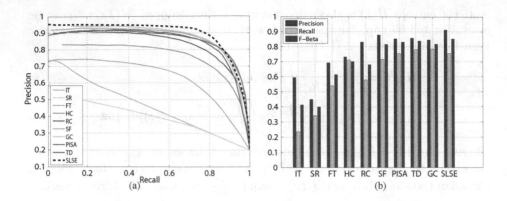

Fig. 2. (a) Average PR curves of different approaches on the MSRA-1000 dataset; (b) Average precision, recall, and F_β-measure using the adaptive thresholding. The proposed SLSE method achieves the best performance.

3   Experiments

3.1   Performance Evaluation

We evaluate our SLSE approach on two benchmarks. One is MSRA-1000, which contains 1000 natural images selected from the MSRA dataset [16] and accurate human-labeled ground truth [3]. The other is SED1 [6]. We compare SLSE with nine baselines following the selection criteria: citation number (IT [17], SR [2], FT [3]), recency (PISA [18], TD [19], GC[20]), and relation with our approach (HC, RC [1], SF [15], PISA [18]).

In particular, HC, RC [1] focus on color contrast, SF [15] proposed spatial distribution cue, and PISA [18] ensemble color and spatial feature. The proposed SLSE method combines color, spatial and luminance information.

Following [3], we evaluate the performance of each method using two metrics: the Precision-Recall (PR) curve and the F_β-measure where $\beta^2 = 0.3$ as suggested in [3]. In the first evaluation, each saliency map is segmented by thresholds varying within $[0, 255]$ and then compared to the ground truth to get PR values. An average PR curve is generated for each method. In the second evaluation, each saliency map is segmented by an adaptive threshold. The PR values is used to calculate their harmonic mean value F_β-score.

Fig.2 and Fig.3 demonstrate the comparisons on MSRA-1000 and SED1, respectively. As shown in Fig.2, the proposed approach outperforms most methods, including recently proposed HC, RC [1], TD [19] and GC [20]. Notably, SLSE performs better than SF [15] and PISA [18] within a wide (near 90%) recall range, and has comparable performance at high recall rates. While on the SED1 dataset, SLSE also achieves better performance than PISA in both metrics.

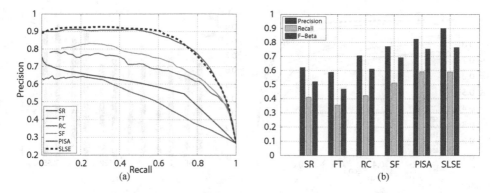

(a)

(b)

Fig. 3. Evaluation results on the SED1 dataset

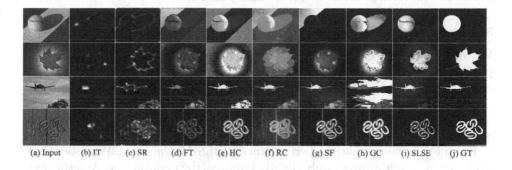

(a) Input   (b) IT   (c) SR   (d) FT   (e) HC   (f) RC   (g) SF   (h) GC   (i) SLSE   (j) GT

Fig. 4. Visual comparison of the existing approaches to our method (SLSE) and ground truth (GT). Here we compare with IT [17], SR [2], FT [3], HC, RC [1], SF [15] and GC [20]. PISA [18] and TD [19] are not included since there is no public implementation or result. SLSE consistently generates saliency maps closest to ground truth (GT). See the text for discussion.

3.2   Visual Comparison

Fig.4 presents a visual comparison. SLSE provides the best results overall. In particular, we notice that for images with compact bright or shadow regions, such as the first two examples, all of the existing methods used fail to detect the object or falsely highlight contexts while SLSE extracts the object accurately. This validates the consideration about weight tuning for luminance. The third example contains two seemingly similar objects. Only SLSE correctly highlight the ground truth object, attributing to the different constraints on object and background. The last image have relatively more salient context. Most methods falsely detect the vertical line as a salient object while SLSE not, owing to the benefits of level set method, which can hold the LSF contour close to the natural object boundaries.

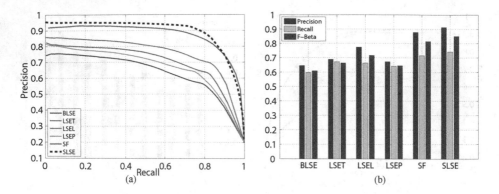

Fig. 5. Experimental comparison of different methods on the MSRA-1000 dataset. (a) PR curves; (b) Average precision, recall, and F_β-measure.

3.3 Strategy Comparison

We briefly introduce several computation strategies in Section 2.4. For completeness, we experimentally verify their effects. In Fig.5, we compare results of employing different strategies on the MSRA-1000 dataset. BLSE denotes basic LSE without using any strategy; LSET denotes using only thresholding; LSEL denotes using only luminance tuning; LSEP denotes using only position tuning; and SLSE represents using all the computation strategies.

From Fig.5, we see that the performance of BLSE is the worst. We notice that utilizing any single strategy only slightly improves the performance. And the best result is achieved by SLSE (outperforms the baseline SF [15] method). Following the comparison, we observe that the strategies assist LSE and are complementary to each other, providing effective detection cues to visual saliency.

3.4 Discussion

The LSE framework can easily ensemble different cues for saliency detection, and the computation is independent of the feature dimensionality [12]. We exploit color, luminance and spatial information here, all of which are low-level features. Some high-level features, such as shape, text and face should be prudently considered. How to ensemble both high-level and low-level information in our framework is an open question.

4   Conclusions

In this paper, we have proposed a flexible LSE-EFM framework for saliency detection. The proposed framework is different from the existing methods and provides a level set representation for saliency and transforms the evolution into an EFM problem. The proposed SLSE method is extensively evaluated on two public datasets and achieves good performance. We have further validated the computation strategies and the results demonstrate their effectiveness.

Acknowledgments. This work was partially supported by the National Natural Science Foundation of China (Grant No. 61272248), the National Basic Research Program of China (Grant No. 2013CB329401), and the Science and Technology Commission of Shanghai Municipality (Grant No. 13511500200).

References

1. Cheng, M.M., Zhang, G.X., Mitra, N.J., Huang, X., Hu, S.M.: Global contrast based salient region detection. In: CVPR (2011)
2. Hou, X., Zhang, L.: Saliency detection: A spectral residual approach. In: CVPR (2007)
3. Achanta, R., Hemami, S., Estrada, F., Süsstrunk, S.: Frequency-tuned salient region detection. In: CVPR (2009)
4. Gopalakrishnan, V., Hu, Y., Rajan, D.: Random walks on graphs to model saliency in images. In: CVPR (2009)
5. Osher, S., Fedkiw, R.: Level set methods and dynamic implicit surfaces, vol. 153. Springer (2003)
6. Alpert, S., Galun, M., Basri, R., Brandt, A.: Image segmentation by probabilistic bottom-up aggregation and cue integration. In: CVPR (2007)
7. Hunt, R.W.G., Pointer, M.R.: Measuring colour. John Wiley & Sons (2011)
8. Zhang, L., Tong, M.H., Marks, T.K., Shan, H., Cottrell, G.W.: Sun: A bayesian framework for saliency using natural statistics. Journal of Vision 8(7) (2008)
9. Chang, J., Fisher, J.: Efficient mcmc sampling with implicit shape representations. In: CVPR (2011)
10. Kim, J., Fisher III, J.W., Yezzi, A., Çetin, M., Willsky, A.S.: A nonparametric statistical method for image segmentation using information theory and curve evolution. IEEE Trans. Image Process. 14(10), 1486–1502 (2005)
11. Parzen, E.: On estimation of a probability density function and mode. The Annals of Mathematical Statistics 33(3), 1065–1076 (1962)
12. Morariu, V.I., Srinivasan, B.V., Raykar, V.C., Duraiswami, R., Davis, L.S.: Automatic online tuning for fast gaussian summation. In: NIPS (2008)
13. Otsu, N.: A threshold selection method from gray-level histograms. Automatica (1975)
14. Einhäuser, W., König, P.: Does luminance-contrast contribute to a saliency map for overt visual attention? European Journal of Neuroscience 17(5), 1089–1097 (2003)
15. Perazzi, F., Krahenbuhl, P., Pritch, Y., Hornung, A.: Saliency filters: Contrast based filtering for salient region detection. In: CVPR (2012)
16. Liu, T., Yuan, Z., Sun, J., Wang, J., Zheng, N., Tang, X., Shum, H.-Y.: Learning to detect a salient object. IEEE Trans. Patt. Anal. and Mach. Intell. 33(2), 353–367 (2011)
17. Itti, L., Koch, C., Niebur, E.: A model of saliency-based visual attention for rapid scene analysis. IEEE Trans. Patt. Anal. and Mach. Intell. 20(11), 1254–1259 (1998)
18. Shi, K., Wang, K., Lu, J., Lin, L.: Pisa: Pixelwise image saliency by aggregating complementary appearance contrast measures with spatial priors. In: CVPR (2013)
19. Scharfenberger, C., Wong, A., Fergani, K., Zelek, J.S., Clausi, D.A.: Statistical textural distinctiveness for salient region detection in natural images. In: CVPR (2013)
20. Cheng, M.-M., Warrell, J., Lin, W.-Y., Zheng, S., Vineet, V., Crook, N.: Efficient salient region detection with soft image abstraction. In: ICCV (2013)

Application of Cuckoo Search for Design Optimization of Heat Exchangers

Rihanna Khosravi, Abbas Khosravi, and Saeid Nahavandi

Centre for Intelligent Systems Research, Deakin University, Geelong, Australia
{rkhosrav,abbas.khosravi,saeid.nahavandi}@deakin.edu.au

Abstract. A wide variety of evolutionary optimization algorithms have been used by researcher for optimal design of shell and tube heat exchangers (STHX). The purpose of optimization is to minimize capital and operational costs subject to efficiency constraints. This paper comprehensively examines performance of genetic algorithm (GA) and cuckoo search (CS) for solving STHX design optimization. While GA has been widely adopted in the last decade for STHX optimal design, there is no report on application of CS method for this purpose. Simulation results in this paper demonstrate that CS greatly outperforms GA in terms of finding admissible and optimal configurations for STHX. It is also found that CS method not only has a lower computational requirement, but also generates the most consistent results.

Keywords: Heat exchanger, optimization, genetic algorithm, cuckoo search.

1 Introduction

Often a multi-objective function is considered for the design optimization of shell and tube heat exchangers (STHXs). The optimization process tries to find a trade-off between maximizing efficiency (related to thermodynamics) and minimizing capital and operational costs (related to economics). The objective function considered for STHX design optimization is nonlinear, discontinues, and nondifferentiable with respect to the design parameters. Besides, it has several local optimums (multimodal). According to these, gradient descent based methods cannot be applied for solving it (either maximization or minimization). Also, the manual optimization is extremely time-consuming, if not impossible. This is due to the massiveness of the search space and the multimodality of objective functions. The common practice in industry is to evaluate a large number of different STHX geometries and then selecting those which satisfy both thermodynamics and economics requirements/constraints. This is a complex trial-and-error process. It is obvious that this approach leads to sub-optimal solutions considering the limited time and budget for the design engineering [1] [2].

Due to inherent complexities of STHX design and critical drawbacks and shortcomings of traditional optimization methods, evolutionary algorithms have been widely used in the last few years for optimal design of different types of

C.K. Loo et al. (Eds.): ICONIP 2014, Part II, LNCS 8835, pp. 178–185, 2014.

heat exchangers [1] [3] [4] [5]. Evolutionary algorithms such as genetic algorithm (GA), particle swarm optimization, and artificial bee colony have the capability to quickly find the optimal geometries for STHXs which satisfy all required geometric and operational constraints. No doubt, the application of these evolutionary algorithms makes the optimal design of heat exchangers much more efficient and less labour-intensive. The whole process can be automated, so thousands and even millions of design geometries can be evaluated in a short time and modified in an intelligent way with the purpose of finding the optimal design parameters.

The point to highlight here is that different evolutionary algorithms demonstrate different performance for different applications. This is due to the fact that performance of methods such as GA, particle swarm optimization, and artificial bee colony closely depends on fine-tunning of several controlling parameters. If these specific parameters are not fine-tuned, the computational burden of the method significantly increases (requiring to continue iterations/generations for a prolonged period) or sub-optimal solutions (local optimums) are returned [1].

Considering these shortcomings and issues, this research aims at using cuckoo search (CS) method for design optimization of STHX. To the best of our knowledge, this is the first time that the CS optimization method is applied for design of STHX. The focus of existing literature is more on how traditional evolutionary optimization methods such as GA can be adopted and applied to find the optimal values for the design parameters. Compared to GA, CS has only one controlling parameters and has a much better local and global search ability. As part of experiments in this study, we use both GA and CS method to optimally determine seven design parameters of a STHX with the purpose of maximizing its efficiency considering several thermodynamic constraints. It is expected that the outcomes and findings of this research will introduce more advanced and efficient optimization techniques to chemical and process engineering society.

The rest of this paper is organized as follows. Section 2 briefly introduces the STHX model used in this study. GA and CS are briefly described in Section 3. Section 4 represents simulations results. Finally, conclusions are provided in Section 5.

2 Modelling Shell and Tube Heat Exchanger

The efficiency of the TEMA E-type STHX is calculated as,

$$\epsilon = 2 \left(1 + C^* + \sqrt{1 + C^{*2}} \frac{1 - e^{-NTU\sqrt{1+C^{*2}}}}{1 + e^{-NTU\sqrt{1+C^{*2}}}} \right)^{-1} \tag{1}$$

where the heat capacity ratio (C^*) is calculated as,

$$C^* = \frac{C_{min}}{C_{max}} = \frac{min(C_s, C_t)}{max(C_s, C_t)} = \frac{min\left((\dot{m}c_p)_s, (\dot{m}c_p)_t\right)}{max\left((\dot{m}c_p)_s, (\dot{m}c_p)_t\right)} \tag{2}$$

where subscripts s and t stand for shell and tube respectively. The number of transfer units is defined as,

$$NTU = \frac{U_o A_t}{C_{min}} \tag{3}$$

where $C_{min} = min(C_h, C_c)$ and $C_{max} = max(C_h, C_c)$ and C_h and C_c are the hot and cold fluid heat capacity rates, i.e., $C_h = (\dot{m}c_p)_h$ and $C_c = (\dot{m}c_p)_c$. $\dot{m}$ is the fluid mass flow rate. Specific heats c_p are assumed to be constant.

The overall heat transfer coefficient (U_o) in (3) is then computed as,

$$U_o = \left(\frac{1}{h_o} + R_{o,f} + \frac{d_o\, ln(d_o/d_i)}{2k_w} + R_{i,f}\frac{d_o}{d_i} + \frac{d_o}{h_i d_i} \right)^{-1} \tag{4}$$

where L, N_t, d_i, d_o, $R_{i,f}$, $R_{o,f}$, and k_w are the tube length, number, inside and outside diameter, tube and shell side fouling resistances and thermal conductivity of tube wall respectively. h_i and h_o are heat transfer coefficients for inside and outside flows, respectively.

The tube side heat transfer coefficient (h_i) is calculated as,

$$h_i = 0.024 \frac{k_t}{d_i}\, Re_t^{0.8}\, Pr_t^{0.4} \tag{5}$$

for $2500 < Re_t < 124000$. k_t and Pr_t are tube side fluid thermal conductivity and Prandtl number respectively.

The average shell side heat transfer coefficient is calculated using the Bell-Delaware method correlation,

$$h_s = h_k\, J_c\, J_l\, J_b\, J_s\, J_r \tag{6}$$

where h_k is the heat transfer coefficient for an ideal tube bank.

J_c, J_l, J_b, J_s, and J_r in (6) are the correction factors for baffle configuration (cut and spacing), baffle leakage, bundle and pass partition bypass streams, bigger baffle spacing at the shell inlet and outlet sections, and the adverse temperature gradient in laminar flows.

The STHX total cost is made up of capital investment (C_{inv}) and operating (C_{opr}) costs [3],

$$C_{total} = C_{inv} + C_{opr} \tag{7}$$

There are several methods for determining the price of STHX. Here we use the Halls method for estimation of the investment cost as detailed in [6] (alternative cost estimation methods can be found in [7]). C_{inv} as a function of the total tube outside heat transfer surface area (A_t) is defined as,

$$C_{inv} = 8500 + 409\, A_t^{0.85} \tag{8}$$

where the construction materials are carbon and stainless steel.

The total discounted operating cost associated to pumping power is computed as follows [3],

$$C_{opr} = \sum_{k=1}^{N_y} \frac{C_0}{(1+i)^k} \tag{9}$$

where i and N_y are the annual discount rate (%) and the STHX life time in year. C_0 is the annual operating cost.

More details about the STHX model used in this study can be found in [4].

3   Optimization Algorithms

3.1   Genetic Algorithm

GA is a stochastic search method initially proposed by Holland in US [8] and Rechengerg in Germany [9]. As a population-based algorithm, GA has been developed by simplification of natural evolutionary processes. It is simply a biological metaphor of Darwinian evolution. The fitter an individual is, the more chance it will have to survive and produce offspring. Elitism, crossover, and mutation mechanisms are used in the GA as means for evolving candidate solutions towards the optimal values. A population of individuals are iteratively updated in GA. On each iteration, the fitness (objective) function is calculated for all individuals. A new population of individuals is generated by probabilistically selecting better individuals from the current population. Best individuals are directly admitted to the next generation (elitism operator). Other individuals in the population are subjected to crossover and mutation operators to form new offspring. A wide variety of genetic operators have been proposed in literature with the purpose of improving the GA performance [10] [11].

The pseudo code for GA including three genetic operators is displayed in Fig. 1. Detailed discussion about GA and its operators can be found in basic reading sources such as [8] and [12].

3.2   Cuckoo Search

The CS method is a nature-inspired metaheuristic optimization method which was proposed by Yand and Deb in 2009 [13]. The reproduction strategy of cuckoos is the core idea behind the CS method. The CS method has been developed based on three idealized assumptions: (i) each cuckoo lays one egg at a time and deposits it at a random chosen nest, (ii) the best nests with the highest quality eggs are carried to the next generations, and (iii) the number of host nests for depositing eggs are fixed. Eggs laid by a cuckoo are discovered by the host bird with a pre-set fraction probability, $p_a \in [0, 1]$. In case of discovering alien eggs, the host bird may simply through away them or abandon the nest and build a completely new one.

In terms of optimization implementation, eggs in nests represent solutions. The idea is to replace not-so-good solutions in the nests with new and potentially better solutions. Based on the three idealized assumption, Fig. 2 shows the pseudo code for implementation of the CS method. The method applies two exploration methods. Some solutions are generated in the neighborhood of the current best solution (a Lévy walk). This speeds up the local search. At the same time, a major fraction of new solutions are generated by far field randomization

```
Begin
  Generate initial population of individuals
  Evaluate the fitness of the individuals
  while (termination criteria)
      Select the best individuals to be used by GA operators
      Generate new individuals using crossover and mutation operators
      Evaluate the fitness of new individuals
      Replace the worst individuals by the best new individuals
  end while
  Postprocess results and visualization
end
```

```
Begin
  Objective function f(x), x = (x₁, ..., x_d)ᵀ
  Generate initial population of n host nests x_i (i = 1, 2, ..., n)
  while (termination criteria)
      Get a cuckoo randomly by Levy flights
      Evaluate its quality/fitness F_i
      Choose a nest among n (say. j) randomly
      if (F_i > F_j ),
          replace solution j by the new solution;
      end
      A fraction (p_a) of worse nests are abandoned and new ones are built
      Keep the best solutions (or nests with quality solutions)
      Rank the solutions and find the current best
  end while
  Postprocess results and visualization
end
```

Fig. 1. Pseudo code for GA   **Fig. 2.** Pseudo code for CS method

and whose locations are far away from the current best solution location. This is done to make sure the method is not trapped in a local optimum.

A Levy flight is considered when generating new solutions x^{t+1} for the ith cuckoo,

$$x_i^{t+1} = x_i^t + \alpha \oplus L\acute{e}vy \tag{10}$$

where α is the step size which depends on the scales of the problem of interest. Often, $\alpha = O(L/100)$ satisfies the search requirements for most optimization problems. L represents the difference between the maximum and minimum valid value of the problem of interest. The product $\oplus$ means entry-wise multiplication.

Further discussion about CS method and its details can be found in [14] and [15].

4   Simulation Results

STHX model used in simulations is identical to one described and analyzed in [4]. There are seven design parameters to be optimized using evolutionary algorithms. Both the population size and the number of generations (optimization iterations) are set to 30 for GA and CS methods. This provides both methods with ample search and diversification capacity to find admissible solution, to avoid local optima, and to locate the globally optimal solution. The discovery rate of alien eggs (p_a) in the CS method is set to 0.25. Note that this is the only parameter that we need to set in CS optimization method. To make results statistically significant, we repeat the optimization process 50 times.

As per results shown in Fig. 3, GA finds an optimal configuration leading to maximum efficiency in only 16 out of 50 runs (32%). In other cases, either it cannot find an admissible solution (26 out of 50 runs, 52%) or the found configuration is suboptimal (8 out of 50 runs, 16%). In contrast, the CS method finds globally optimal solutions in all optimization runs. The method is quite robust against the random initialization and can quickly locate admissible solutions and then optimize them with the purpose of maximizing efficiency.

The convergence profile of the CS optimization method is displayed in Fig. 4 for an experiment. According to this, only a few iterations are required for the

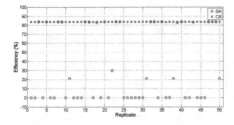

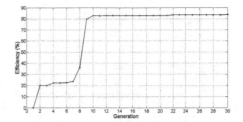

Fig. 3. STHX efficiency optimization using GA and CS for 50 runs

Fig. 4. The convergence behavior of CS method for maximizing efficiency

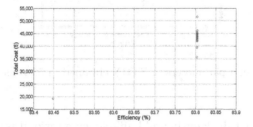

Fig. 5. The efficiency and total cost for 50 solutions found by CS method

method to find the optimal solution. Note that the initial efficiency is zero which means that all 30 initial solutions are inadmissible. This clearly demonstrate the effectiveness of the Levy flight search method in CS optimization for exploring the search space. For this specific experiment, CS finds the optimal solution in 10 generations. Similar results are obtained in other experiments as well.

As the performance of GA for optimal design of STHX is inferior, we hereafter just report the optimization results for the CS method. It is important to note that the week performance of the GA is not something to be rectified purely by increasing the number of optimization generations or the population size. Even if the performance is improved, the computational burden for finding globally optimal solutions will be massive.

The scatter plot of efficiency and total cost for 50 runs of the optimization process is shown in Fig. 5. According to this figure, while 49 out of 50 designs have identical efficiencies (83.80%), their corresponding total costs are significantly different. The total cost approximately varies between $35K to $52K. It is also important to note that the total cost is less than $20K for a design with an 83.45% efficiency. This simply means that achieving the last 0.5% efficiency increases the total cost as a minimum by more than two times.

The index for seven design parameters obtained using the CS optimization method in 50 runs are displayed in Fig. 6. Note the tube arrangement is shown in its actual values (30°, 45°, and 90°). For the case of tube arrangement, the 90° arrangement is the optimal value returned in the majority of runs (29 out of 50 cases, 58%). There is no special preference in the selection optimal values for tube

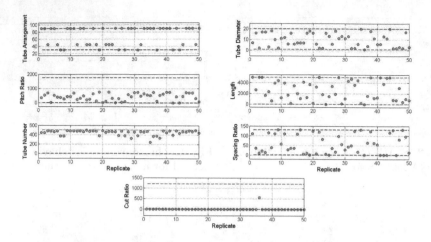

Fig. 6. Seven design parameters of STHX obtained in 50 runs of CS method

diameter, pitch ratio, tube length, and baffle spacing ratio. It is important to note that the CS method often selects smaller pitch ratios and greater tube numbers. This statistically imply that these two design parameters have moderate negative and positive correlation with the STHX efficiency. The most consistent behavior is observed for the baffle cut ratio. The CS method returns a unique index (the smallest one) in 49 out of 50 runs. As per this, the smaller the baffle cut ratio, the greater the performance.

The statistics for time required to finish one optimization run of GA and CS method are reported in Table 1. On average, each run terminates in less than 1.5 seconds which is quite small considering design time scales in the world of process engineering. The mean and median values of elapsed times indicate that the computational burden of both GA and CS methods are almost identical.

Table 1. Statistics of elapsed time (sec) for running the optimization process

Measure	GA	CS
Mean	1.53	1.49
Median	1.46	1.48
Standard deviation	0.21	0.02

5   Conclusion

As per obtained results in this paper, more advanced evolutionary algorithms such as cuckoo search method have a significantly better performance than genetic algorithm for design optimization of shell and tube heat exchangers. Starting from a randomly selected set of parameters, genetic algorithm often fails to find admissible solutions. In contrast, cuckoo search method can always locate admissible solutions and then find the globally optimal solution in a limited number of iterations. A fine balance of randomization and intensification and fewer number of optimization parameters are of the primary reasons why cuckoo search method outperforms genetic algorithm for design optimization of shell and tube heat exchangers.

References

1. Xie, G., Sunden, B., Wang, Q.: Optimization of compact heat exchangers by a genetic algorithm. Applied Thermal Engineering 28(8-9), 895–906 (2008)
2. Rao, R.V., Patel, V.: Multi-objective optimization of heat exchangers using a modified teaching-learning-based optimization algorithm. Applied Mathematical Modelling 37(3), 1147–1162 (2013)
3. Caputo, A.C., Pelagagge, P.M., Salini, P.: Heat exchanger design based on economic optimisation. Applied Thermal Engineering 28(10), 1151–1159 (2008)
4. Sanaye, S., Hajabdollahi, H.: Multi-objective optimization of shell and tube heat exchangers. Applied Thermal Engineering 30, 1937–1945 (2010)
5. Mariani, V.C., Duck, A.R.K., Guerra, F.A., Coelho, L.D.S., Rao, R.V.: A chaotic quantum-behaved particle swarm approach applied to optimization of heat exchangers. Applied Thermal Engineering 42, 119–128 (2012)
6. Hall, S., Ahmad, S., Smith, R.: Capital cost targets for heat exchanger networks comprising mixed materials of construction, pressure ratings and exchanger types. Computers & Chemical Engineering 14(3), 319–335 (1990)
7. Taal, M., Bulatov, I., Klemes, J., Stehlik, P.: Cost estimation and energy price forecasts for economic evaluation of retrofit projects. Applied Thermal Engineering 23(14), 1819–1835 (2003)
8. H., H.J.: Adaptation in Natural and Artificial Systems. Michigan Press (1975)
9. Rechenberg, I.: Evolutionsstrategie. Fromman-Hozboog Verlag (1973)
10. Hasancebi, O., Erbatur, F.: Evaluation of crossover techniques in genetic algorithm based optimum structural design. Computers and Structures 78(1-3), 435–448 (2000)
11. Kaya, M.: The effects of two new crossover operators on genetic algorithm performance. Applied Soft Computing 11(1), 881–890 (2011)
12. Goldberg, D.E.: Genetic Algorithm in Search, Optimization, and Machine Learning. Addision-Wesley, Reading (1989)
13. Yang, X.S., Deb, S.: Cuckoo search via levy flights. In: World Congress on Nature & Biologically Inspired Computing, pp. 210–214 (2009)
14. Yang, X.S.: Nature-inspired metaheuristic algorithms. Luniver Press (2008)
15. Yang, X.-S., Deb, S.: Cuckoo search via levy flights. In: World Congress on Nature and Biologically Inspired Computing, pp. 210–214 (2009)

A Hybrid Method to Improve the Reduction of Ballistocardiogram Artifact from EEG Data

Ehtasham Javed[1,2], Ibrahima Faye[1,3], Aamir Saeed Malik[1,2],
and Jafri Malin Abdullah[4]

[1] Centre for Intelligent Signal & Imaging Research
[2] Department of Electrical & Electronics Engineering
[3] Department of Fundamental and Applied Sciences
Universiti Teknologi PETRONAS, Perak, Malaysia
[4] Centre for Neuroscience Services and Research
Hospital Universiti Sains Malaysia, Kelantan, Malaysia
{rajaehti1,brainsciences}@gmail.com
{ibrahima_faye,aamir_saeed}@petronas.com.my

Abstract. Simultaneous recordings of functional magnetic resonance imaging (fMRI) and electroencephalography (EEG) allow acquisition of brain data with high spatial and temporal resolution. However, the EEG data get contaminated by additional artifacts such as Gradient artifact and Ballistocardiogram (BCG) artifact. The BCG artifact's dynamics appear to be more challenging and it hinders in the assessment of the neuronal activities. In this paper, a reference-free method is implemented in which Empirical Mode Decomposition (EMD) and Principal Component Analysis (PCA) has been combined to reduce the BCG artifact while preserving the neuronal activities. The qualitative analysis of the proposed method along with three existing methods demonstrates that the proposed method has improved the quality of the reconstructed data. Moreover, it does not require any reference signal to extract BCG artifact.

Keywords: Ballistocardiogram artifact, Simultaneous EEG & fMRI, Principal Component Analysis, Empirical Mode Decomposition.

1 Introduction

Simultaneous non-invasive Electroencephalography (EEG) and functional Magnetic Resonance Imaging (fMRI) has been used to acquire neuronal activities. Such acquisitions provide further insight to the functions of the brain due to high temporal and spatial resolution [1]. The neuronal activities measured directly with EEG have a good temporal resolution as it acquires neural function on the order of milliseconds [2]. On the other hand, fMRI acquires the neuronal activity indirectly by using the Blood Oxygenation Level Dependent (BOLD) response with spatial resolution of few millimeters. Combining both modalities, brain activities with high temporal and spatial resolution can be analyzed at the same time. In real-time environment, the combination means the concurrent acquisition of EEG and fMRI but such acquisitions also

C.K. Loo et al. (Eds.): ICONIP 2014, Part II, LNCS 8835, pp. 186–193, 2014.

get contaminated by additional artifacts compared to the EEG recording alone [3]. The additional artifacts are because of magnetic field inside the scanner. The magnetic field related artifacts which contaminate EEG data are indicated as Equation (1).

$$EEG_{raw} = EEG + Artifact_{GA} + Artifact_{BCG} \tag{1}$$

The $Artifact_{GA}$ represents the gradient artifact also known as imaging artifact. It arises due to the gradient magnetic fields used for spatial encryption of fMRI data [4]. It is periodic in nature and has amplitude of 10 to 100 times compared to the EEG amplitude and can be easily removed using subtraction methods [5]. The artifact that has dynamic impact on the EEG data that can hinder the correct assessment is the Ballistocardiogram (BCG) artifact indicated as $Artifact_{BCG}$. In the first publication on concurrent EEG-MRI acquisition by [6], the BCG artifact was already reported. Its influence on the amplitudes of EEG is up to 200μV at 3T and on the range of 0.5-13 Hz in frequency domain [7]. The periodic occurrence of the BCG artifact after every heart-beat is considered as its key characteristic.

Several methods have been proposed in the literature to remove the BCG artifacts: subtraction method [8], blind source separation (BSS) such as principal component analysis (PCA) [9] and independent component analysis (ICA) [10] and others like [11,12]. Despite several existing methods, the residuals of the BCG artifact are left behind [13]. The main limitations of above mentioned methods are: dependency over a reference (ECG) signal, stability assumption in temporal BCG waveforms, and selecting the number of components that represents the BCG artifact in BSS methods. Finding a suitable method that can reduce the BCG artifact from all types of dataset of EEG is still an ongoing issue [7]. The aim of this paper is to present a reference-free reduction method to extract the BCG artifact and to qualitatively assess the reconstructed artifact-free EEG data.

The structure of the paper is as it follows. Section II provides insight to the datasets used as well as the working phenomenon of proposed algorithm along with mathematical illustration. The results are presented in the section III and their discussion is in section IV. Section V concludes the work done in this study.

2   Data and Methods

2.1   EEG and fMRI Acquisition

The compatible EEG cap with the magnetic field (128-channel HydroCel Geodesic Sensor Net [14]) was used to acquire the EEG data. Two normal subjects participated in this study. They had no history of brain diseases and are normal or corrected to normal vision. The informed consent form was given prior to the experimentation and was signed. The study was approved by local ethics committee of Universiti Teknologi Petronas (UTP), Malaysia and was conducted according to the given guidelines. Two types of datasets were recorded. In the first, the subjects were instructed to stay relaxed and keep their eyes closed, while in the second, an oddball paradigm of two stimuli was presented on a magnetic-compatible projector [15]. Stimuli were divided into target stimuli (10% of the content) and standard stimuli (40% of the content). For the remaining 50% of the content, a fixation symbol appeared between every stimulus. The subjects were instructed to press a button when they saw the target stimulus and do nothing when they saw a standard stimulus. The fMRI scanner used for simultaneous EEG-fMRI acquisition has the magnetic field of 3.0 T. T2*-weighted echo planar imaging (EPI) sequence was used for the acquisitions. In addition, the same tasks have been repeated outside the fMRI scanner and were considered as EEG datasets at 0 T.

2.2   Pre-Processing and Data Analysis

The contaminations of gradient artifact and ocular artifacts (when eyes were not closed) were removed in the pre-processing. To remove the gradient artifact, EGI's Net Station 4.5 EEG software was used with default settings in which Average Artifact Subtraction (AAS) is implemented. Then, the data was filtered using bandpass filter of 0.3-40 Hz and exported to Matlab. The scalp regions used in this study for analysis are: Frontal (F): (19, 4, 24, 124, 27, 123, 33, 122, 32, 1 and 11), Central (C): (30, 105, 36, 104, 41 and 103), Temporal (T): (45, 108, 44, 114, 108, 34, 116, 38 and 121), Occipital (O): (70, 83 and 75), and Parieto-Occipital (PO): (67, 77, 65 and 90). The division of electrodes into the regions was based on 10/10 international system of electrode placement.

2.3   Methods

A reference-free hybrid method EMD-PCA is presented in this paper. In this method, Empirical Mode Decomposition (EMD) is used to decompose the BCG contaminated signal into set of components named Intrinsic Mode Functions (IMFs). The IMFs have different frequency with varying amplitudes [16]. The process to compute the IMFs is known as sifting process. The advantages of using the EMD are: it is adaptive to temporal changes in the original signal, and unlike other decomposition methods, it does not require prior information about the sources mixed in the signal [16]. The PCA is applied then to decompose the correlated set of signals into linearly uncorrelated components i.e. Principal components (PCs) [9]. The components are generally arranged in descending order with respect to the computed variances. The procedure of extracting BCG artifact using the proposed method is as follows:

1. Extract the first four frequency bands using band pass filter (the range 0.3-25 Hz is the typical BCG frequency range [10]).
2. Select first frequency band
3. Decompose the contaminated signal into different components of variable frequency and amplitude called as IMF.
4. Compute the PCs by implementing the PCA over set of IMFs.
5. Calculate the similarity index of each PC with the contaminated signal.
6. From all PCs, select the PC which has maximum similarity index (computed in step 5) as the BCG related component.
7. Repeat the steps 3 to 6 for all bands.

8. Add all the extracted BCG related components to get the BCG template.
9. Subtract the extracted template from the contaminated data to get the artifact free EEG.

The mathematical illustration of proposed method is as follows:

Let E_j, j $= 1, 2, 3, ..., $n be the band of the contaminated EEG data of sample n. $I = [I_{i,j}]$ of dimension N × n is the decomposed matrix of IMFs, where N is the total number of IMFs. The residual left behind after the EMD decomposition, symbolized by r is a row vector i.e.

$$r = [r_1, r_2, r_3, ..., r_n]$$

The decomposed contaminated data into IMFs and the residue can be recovered using following equation

$$E_i = \left(\sum_{i=1}^{N} I_{i,j}\right) + r_j, j = 1, 2, 3, ..., n \quad (2)$$

For further analysis, the set of IMFs and residual were arranged in a matrix form

$$C = \begin{bmatrix} I \\ r \end{bmatrix}$$

where C has $M = N+1$ rows and n columns. The PCA method is then implemented on the matrix C, in which Eigen decomposition is used to compute the eigenvalues and eigenvectors of the covariance matrix of C. Then C can be denoted as

$$C = [V^T][PC]$$

where V^T is the transpose of the matrix of eigenvalues; PC is the matrix containing the eigenvectors. The uncorrelated components or projected signals can be obtained by matrix multiplication of PC with C i.e.

$$S = [PC][C]$$

From the set of uncorrelated components S, one component is selected as a BCG related component i.e. the component that has the maximum similarity index with the contaminated data. Hence, from four bands, a total of four such components were extracted and summed to get the BCG artifact template. Then the extracted BCG template was subtracted from the contaminated data to get the artifact-free EEG data.

$$EEG_{artifact-free} = Contaminated\_data - Extracted\ BCG \quad (3)$$

In addition to the proposed method, three existing methods; Average Artifact Subtraction (AAS) [8], Optimal Basis Set (OBS) [9] and Statistical Feature Extraction for artifact removal (SFE) [17], have been implemented. The implementation of AAS and OBS is done using open source *EEGLAB 10.2.2.4b* toolbox plugin named *FMRIB1.21* and SFE using the *MATLAB* code, provided by the authors at *http://www.amri.ninds.nih.gov/software.html*. The comparison with these implemented methods will help in providing the practical significance of the proposed method.

3   Results

The results of the evaluated parameters after reducing the BCG artifact via proposed method, AAS, OBS and SFE have been presented in this section. In this study, the reconstructed EEG datasets (eyes closed and ERP) has been assessed using the qualitative parameters. The qualitative measures help in the evaluation of original brain activities preserved while reducing the BCG artifact. The parameters are: power spectrum, sample entropy and extraction of ERP. Power spectrum and sample entropy are measured using eye closed dataset, while ERP is extracted from dataset of oddball paradigm.

The power spectrum of eyes closed data has been calculated using Fast Fourier Transform (FFT) in which the welch window of 500 samples with an overlap of 50% and 1024 points are used. Since, the alpha power increases in eyes closed data [18], this parameter is used to analyse whether the power of alpha band is preserved while reducing BCG artifact or not. Two scalp regions have been analysed where the alpha waves are dominant than the other scalp regions which are Occipital and Parieto-occipital region. Figure 1 presents the power spectrum of two subjects.

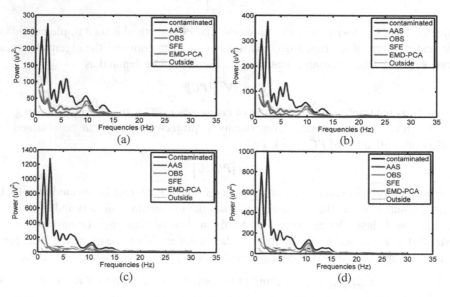

Fig. 1. Power spectrum of two subjects: (a) Occipital region, (b) Parieto-occipital of subject 1 and similarly (b), (d) of subject 2, using contaminated data, reconstructed data via four methods (including proposed method) and dataset recorded outside the scanner

Another way to evaluate the quality of the reconstructed eye closed data i.e. alpha rhythm is to measure the complexity or randomness in the signals [18]. The sample entropy (SE) appears to be a more influential tool to measure the variations in the data and is computed by

$$SE = -\ln\left(\frac{P^{n+1}(r)}{P^n(r)}\right) \qquad (4)$$

where, n is the number of samples and in this study it is equal to 500 (2 seconds), threshold is denoted by r i.e. 2*σ, where, σ is the standard deviation of the sequence. Pn(r) and Pn+1(r) are the probabilities that m and m+1 sample are same respectively. The SE is calculated using 1 minute data from reconstructed and outside the scanner datasets. The average SE of two subjects has been presented in Fig. 2.

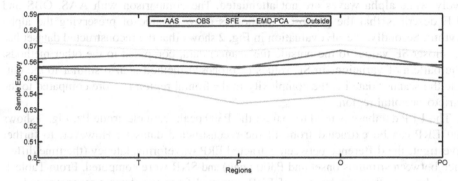

Fig. 2. Sample Entropy of reconstructed datasets via AAS (Blue), OBS (Magenta), SFE (Yellow), EMD-PCA (Red) and outside the scanner data (Green) at five scalp regions

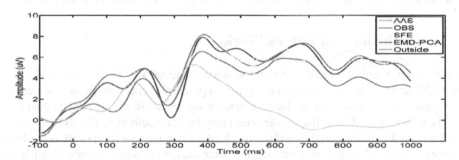

Fig. 3. The ERP waveform extracted from reconstructed datasets (AAS in blue, OBS in magenta, SFE in yellow and EMDPCA in red) and data outside the scanner in green color

The extraction of Event-Related Potential (ERP) is a well-known and widely used parameter in the existing studies to assess the quality of the reconstructed data. The ERP in which P300 has been analyzed at electrode Pz is presented in Fig. 3. The ERP is extracted after averaging 60 trials of target stimuli and then the average of both subjects using reconstructed and outside the scanner datasets. Table 1. describes the parameters (Latency and SNR) to compare the extraction of P300 among the methods.

Table 1. The latency and signal to noise ratio (SNR) of P300

Methods / Parameters	AAS	OBS	SFE	EMDPCA	OS
Latency (ms)	384	388	400	380	356
SNR (dB)	9.67	8.64	16.37	15.14	13.17

4   Discussion

The capability of the proposed method in preserving the brain signal after reduction of the BCG artifact has been evaluated using three qualitative parameters. First, the power spectrum in Fig. 1 shows that the method reduces the BCG influence effectively while alpha waves are not attenuated. The comparison with AAS, OBS and SFE describes that the proposed method is more capable of preserving the alpha rhythm. Secondly, the SE evaluation in Fig. 2 shows that the reconstructed dataset has the closer SE values to the outside the scanner data, compared to the other methods. The pattern of variations in SE values of proposed method is also similar to the outside the scanner data i.e. the complexity in the frontal region is more compared to the Parieto-occipital region.

The ERP database is used to extract the P300 peak from electrode Pz. Fig. 3 shows that ERP can be extracted from all the reconstructed datasets. However, to further investigate the difference between extracted ERP waveforms, latency (the time difference between stimulus onset and P300 peak) and SNR were computed. From Table 1, it can be seen that the latency of ERP extracted from the dataset reconstructed via EMDPCA has less delay compared to others and is closer to the latency of outside the scanner data. However, SFE method has a little high SNR than the proposed method.

5   Conclusion and Future Work

In this research article, the BCG artifact which contaminates EEG data recorded inside the fMRI scanner has been reduced using the proposed reference-free reduction method. The quality of the reconstructed data evaluated in the results shows that the method efficiently reduces the BCG artifact while preserving the brain activities. The method is also compared to the existing studies and the results show that the reduction of BCG has been improved. In addition, the prior information about the data or any reference signal is not required for the extraction of the BCG artifact. The method will be further tested on a large datasets to provide the significant improvement in the utilization of simultaneous EEG and fMRI.

Acknowledgement. The authors gratefully acknowledge Ministry of Education Malaysia, for Fundamental Research Grant Scheme (FRGS/1/2014/SG04/UTP/02/1).

References

1. Levan, P., Maclaren, J., Herbst, M., Sostheim, R., Zaitsev, M., Hennig, J.: Ballistocardiographic artifact removal from simultaneous EEG-fMRI using an optical motion-tracking system. Neuroimage, 1–11 (March 1, 2013)
2. Ahmed, R.F., Malik, A.S., Kamel, N., Tharakan, J.: Predicting the epileptic seizure with beta rhythm of brain: A real-time approach. European Journal of Neurology 246 (2009)

3. Sun, L., Rieger, J., Hinrichs, H.: Maximum noise fraction (MNF) transformation to remove ballistocardiographic artifacts in EEG signals recorded during fMRI scanning. Neuroimage, 144–153 (2009)
4. Nakamura, W., Anami, K., Mori, T., Saitoh, O., Cichocki, A., Amari, S.: Removal of ballistocardiogram artifacts from simultaneously recorded EEG and fMRI data using independent component analysis. IEEE Trans. Biomed. Eng., 1294–308 (2006)
5. Laufs, H., Daunizeau, J., Carmichael, D.W.: Kleinschmidt a. Recent advances in recording electrophysiological data simultaneously with magnetic resonance imaging. Neuroimage, 515–528 (April 1, 2008)
6. Ives, J.R., Warach, S., Schmitt, F., Edelman, R.R., Schomer, D.: Monitoring the patient's EEG during echo planar MRI. Electroencephalogr. Clin. Neurophysiol., 417–420 (1993)
7. Grouiller, F., Vercueil, L., Krainik, A., Segebarth, C., Kahane, P., David, O.: A comparative study of different artefact removal algorithms for EEG signals acquired during functional MRI. Neuroimage, 124–137 (2007)
8. Allen, P.J., Polizzi, G., Krakow, K., Fish, D.R., Lemieux, L.: Identification of EEG events in the MR scanner: the problem of pulse artifact and a method for its subtraction. Neuroimage, 229–239 (October 1998)
9. Niazy, R.K., Beckmann, C.F., Iannetti, G.D., Brady, J.M., Smith, S.: Removal of FMRI environment artifacts from EEG data using optimal basis sets. Neuroimage, 720–737 (November 15, 2005)
10. Debener, S., Mullinger, K.J., Niazy, R.K., Bowtell, R.: Properties of the ballistocardiogram artefact as revealed by EEG recordings at 1.5, 3 and 7 T static magnetic field strength. Int. J. Psychophysiol., 189–199 (March 2008)
11. Dyrholm, M., Goldman, R., Sajda, P., Brown, T.R.: Removal of BCG artifacts using a non-Kirchhoffian overcomplete representation. IEEE Trans. Biomed. Eng., 200–204 (February 2009)
12. Ertl, M., Kirsch, V., Leicht, G., Karch, S., Olbrich, S., Reiser, M., et al.: Avoiding the ballistocardiogram (BCG) artifact of EEG data acquired simultaneously with fMRI by pulse-triggered presentation of stimuli. J. Neurosci. Methods, 231–241 (February 15, 2010)
13. Ferdowsi, S., Sanei, S., Abolghasemi, V., Nottage, J., O'Daly, O.: Removing ballistocardiogram artifact from EEG using short- and long-term linear predictor. IEEE Trans. Biomed. Eng., 1900–1911 (July 2013)
14. HydroCel Geodesic Sensor Net 128 Channel Map,
    ftp://ftp.egi.com/pub/support/Documents/net_layouts/
    hcgsn_128.pdf
15. Stevens, A.A., Skudlarski, P., Gatenby, J.C., Gore, J.: Event-related fMRI of auditory and visual oddball tasks. Magn. Reson. Imaging, 495–502 (2000)
16. Wang, T., Zhang, M., Yu, Q., Zhang, H.: Comparing the applications of EMD and EEMD on time–frequency analysis of seismic signal. J. Appl. Geophys., 29–34 (August 2012)
17. Liu, Z., de Zwart, J.A., van Gelderen, P., Kuo, L.-W., Duyn, J.: Statistical feature extraction for artifact removal from concurrent fMRI-EEG recordings. Neuroimage, 2073–2087 (February 1, 2012)
18. Amin, H.U., Malik, A.S., Badruddin, N., Chooi, W.-T.: EEG Mean Power and Complexity Analysis during Complex Mental Task. In: Int. Conf. Complex Med. Eng., Beijing / China (2013)

VLGAAC: Variable Length Genetic Algorithm Based Alternative Clustering

Moumita Saha and Pabitra Mitra

Department of Computer Science and Engineering,
Indian Institute of Technology Kharagpur, West Bengal - 721302, India
{moumitasaha,pabitra}@cse.iitkgp.ernet.in

Abstract. Complex and heterogeneous data sets can often be interpreted as having multiple clustering, each of which are valid but distinct from the others. Several algorithms involving multiple objective functions have been reported for such alternative clusterings. We propose a genetic algorithm based approach for obtaining valid but diverse clustering. A variable length genetic algorithm approach is used to enable varying number of clusters in each interpretation. A suitable method for population initialization and appropriate crossover and mutation operators are also used. Experimental results on benchmark data sets show that the method is comparable with related alternative clustering techniques.

Keywords: Alternative clustering, non-redundant clustering, multi-view clustering, genetic algorithm.

1 Introduction

Data clustering is an important task of exploratory data mining, which deals with finding homogeneous patterns within the data objects. Conventional clustering methods are focused on creating single good clustering result. However, most real datasets are high dimensional and can be interpreted in different ways. There are more than a single pattern in dataset, each is reasonable and interesting in different perspectives. Alternative clustering aims to find all different groupings of data such that every clustering solutions are of high quality and distinct from each other.

Current approaches for alternative clustering can be broadly divided into two classes: (a) objective-function oriented approach, wherein alternative clusterings are evaluated based on the designed objective function which drives the search away from one or more existing clusterings, (b) data transformation-oriented approach in which the alternative clusterings are found by transforming the data into alternate projections such that transformed data will preserve most statistical characteristics of original data. Most of the approaches for alternative clustering works with fixed number of clusters, which is unworthy in most situations where number of clusters are not known as a prior. The paper proposes

C.K. Loo et al. (Eds.): ICONIP 2014, Part II, LNCS 8835, pp. 194–202, 2014.

a novel objective function-oriented approach for alternative clustering using genetic algorithm based optimization. It utilizes variable length genetic algorithm which gives alternative clusterings with optimal number of clusters. Two objectives must be accomplished to generate alternative clusterings– clustering quality should be good and distinctness between different alternative clusterings should be high. We utilize davies bouldin index to quantify clustering quality, and jaccard and rand index to judge the distinctness between different clusterings. We propose a clustering algorithm $VLGAAC$ (**V**ariable **L**ength **G**enetic **A**lgorithm based **A**lternative **C**lustering) that optimizes both objectives of quality and distinctness along with finding optimal number of clusters. Some of the salient features of $VLGAAC$ are as follows: (a) utilizes variable length GA which gives optimal number of clusters in a clustering, (b) initial population is generated using GA which gives much promising result for further execution of the proposed algorithm, (c) recombination operator are newly defined to accompany variable length chromosomes, (d) elitism is incorporated which guarantees passing of best solution over next generation, (e) objective function is designed to optimize dual objectives of quality and distinctness.

We present some of the related work on alternative clustering in Section 2 and discuss the proposed method in Section 3. We analyse and illustrate the experimental results in Section 4 and finally conclude in Section 5.

2   Related Work

In objective function based alternative clustering, Vinh and Epps [1] proposed an information theoretic approach, using havrda-charvat's α structural entropy to evaluate alternative clustering. Dang and Bailey [2] proposed a hierarchical information theoretic approach to evaluate alternative clustering by maximizing the mutual information between cluster labels and data, and minimizing the information sharing between different clusterings. Bae and Bailey proposed COALA [3], an agglomerative hierarchical clustering which attempts to satisfy the cannot link constraints generated from a pre-existing clustering.

In the transformation approach, Cui et al. [4] proposed extraction of non redundant clustering via orthogonalization. The first approach orthogonalized data based on existing clustering solutions, while the second seeks orthogonality in feature space. Davison and Qi proposed a method [5], that projects the original data to a new space such that the data points in same cluster in pre-defined clustering lie at a distance to obtain good alternative clusterings.

Genetic algorithm (GA) had been widely used in alternative clustering problems. Maulik and Bandyopadhyay [6] proposed a GA based clustering technique to search for appropriate clusters by optimizing the similarity metric of resulting clusters. In [7], chromosomes were taken of different length to frame different number of cluster centres. Variable length chromosomes were used for proper clustering with unknown number of clusters. Bandhopadhyay and Maulik [8] used variable length GA to optimize the number of hyperplanes to build the pattern classifier. Truong and Battiti [9] proposed multi-objective GA algorithm, that performs the recombination operation at cluster level instead of object level.

3   Methodology

3.1   Objective Measures for Alternative Clustering

Davies Bouldin Index (*DBI*): *DBI* is utilized to evaluate clustering quality. It is expressed as the ratio of intra-cluster distance to inter cluster distance. Lower value is preferred and is described by Equation 1.

$$DBI = \frac{1}{n} \sum_{i=1}^{n} max_{i \neq j} \left(\frac{\sigma_i + \sigma_j}{d\,(c_i, c_j)} \right) \tag{1}$$

where n is the number of clusters, c_x is the centroid of cluster x, σ_x is the average distance of all elements in cluster x from centroid c_x, $d\,(c_i, c_j)$ is the distance between centroids c_i and c_j.

Rand Index (*RI*) and Jaccard Index (*JI*): *RI* and *JI* are used to quantify the similarity between two clusterings. Lower value is preferable for two alternative clusterings. It is shown by Equation 2 and 3 respectively.

$$RI\,(C^+, C^-) = \frac{TP + TN}{TP + FP + FN + TN} \tag{2}$$

$$JI\,(C^+, C^-) = \frac{|\,C^+ \cap C^-\,|}{|\,C^+ \cup C^-\,|} = \frac{TP}{TP + FP + FN} \tag{3}$$

TP is number of pair of data points in the same cluster in clusterings C^+ and C^-, *TN* is number of pair of points in the different clusters in C^+ and C^-, *FP* is number of pair of points in the same cluster in C^+ and different clusters in C^-, *FN* is pair of points in the different clusters in C^+ and same cluster in C^-.

DBI is taken as fitness function and is minimized over run of *GA* to build initial population of clusterings with variable number of clusters. Proposed algorithm *VLGAAC* takes combination of *DBI*, *RI* and *JI* as the fitness function and minimize the measure to obtain a set of alternative clusterings. The fitness objective function is defined by Equation 4.

$$Fitness\ function = \alpha \cdot DBI + \frac{(1 - \alpha)}{2} \cdot RI + \frac{(1 - \alpha)}{2} \cdot JI \tag{4}$$

where α is the trade-off weightage between clustering quality and distinctness, which is ascertained empirically.

3.2   Initial Population by Variable Length Chromosome Encoding

Chromosomes are framed by incorporating the cluster centres of a clustering, where each chromosome represents a clustering. A random number L_i representing number of clusters in clustering ($2 \leqslant L_i \leqslant \sqrt{p}$) is generated, where p denotes number of objects. The chromosomes are built by selecting L_i number of data points as cluster centres from data. GA based optimization method produces set of variable length chromosomes having different number of clusters, used as initial population to *VLGAAC*.

3.3   Genetic Operators for Alternative Clustering

Cluster-Oriented Crossover Operator: Chromosomes are variable length so two things need to be focussed– (a) cross-over point is chosen between the cluster centres, as cluster centres are indivisible, (b) discard the child that exceeds $\sqrt{p}$ allowed number of clusters. Two parent chromosomes having L_1 and L_2 number of cluster centres are chosen by roulette wheel selection. Single-point crossover is performed with crossover probability β_c, where crossover point is selected as $cp_i = rand()modL_i$ to produce children chromosomes. It helps in selecting randomized cross-over point which results in good clustering solution.

Cluster-Oriented Mutation Operator: Cluster-oriented mutation operation is defined by three methods and one of them is chosen randomly with mutation probability β_m– (a) adding random number of new cluster centres to the chromosomes, (b) deletion of randomly chosen random number of cluster centres of the chromosomes, (c) modifying certain cluster centres of the chromosome.

Elitism: Top ranked fittest chromosomes are carried forward from the current to next generation, unaltered. It helps to hold back the best chromosomes over generation and attain the global optimum quickly.

3.4   Proposed *VLGAAC*

The main objective of proposed alternative clustering algorithm *VLGAAC* is to generate a set of alternative clusterings of high quality and distinctness.

The population size is taken as half the number of objects in the data set. The values assigned to β_c, β_m, α are *0.7, 0.2* and *0.35* empirically. Model runs the whole process until there is no further improvement in fitness function. The flow diagram of *VLGAAC* is presented in Figure 1.

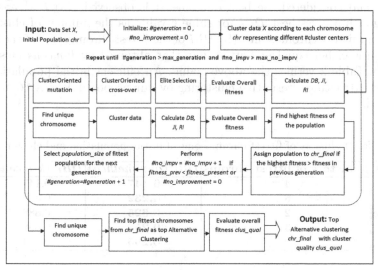

Fig. 1. Flow diagram of proposed method *VLGAAC*

4   Result and Analysis

4.1   Data sets

VLGAAC approach is applied on two synthetic data and three real image data: *CMU* face image [10], bird and flower escher images [9]. Two synthetic data are built by 4 and 6 gaussian clusters respectively, each with 20 2-*D* points. *CMU* face data consists of 624, 32 × 30 pixels images of 20 persons, taken in various pose– straight, left, right, up, expression– neutral, happy, sad, angry and eyes– wearing sun-glass or not. Escher images of bird and flower having different interpretations to human eye perception are 106 × 111 and 120 × 119 pixels respectively. Data sets are described in Table 1.

Table 1. Data sets description

Data sets	# of objects	# of features	# of classes
Syndata1	80	2	4
Syndata2	120	2	6
CMU face	624	960	4
Bird	11766	2	2
Flower	14280	2	2

4.2   Results for Population Initialization

Initial population is evaluated in terms of *DBI* of the fittest chromosome of the population. To validate our result, we execute *k-means* clustering algorithm over data with different values of *k*. We have executed k-means for a number of iterations and considered the best outcome. It is observed– (a) *K-means* clustering algorithm gives best *DBI* for same *k* value as the number of clusters *m* found in our fittest chromosome, (b) *DBI* of the fittest chromosome of the initial population is either equal or better than that obtained by *k-means* clustering algorithm. Table 2 verifies the above claims. An efficient initial population, effective for extracting alternative clustering by our proposed *VLGAAC* method is obtained.

Table 2. No. of clusters of fittest chromosome of initial population and cluster quality

Data set	#clusters of fittest chroms	DBI (*our method*)	DBI (*k-means*)
Syndata1	4	**0.147**	0.147
Syndata2	6	**0.232**	0.389
CMU face	2	**0.625**	0.888
Bird	3	**0.252**	0.258
Flower	2	**0.295**	0.314

4.3   Results for Final Alternative Clusterings by *VLGAAC*

Synthetic Data: The bounded area depicts a particular cluster of the clustering. The resultant clusterings for syndata1 and syndata2 are shown in Figure 2 and 3 respectively. The result clearly shows that *VLGAAC* correctly identify a set of alternative clusterings, even with different number of clusters in different clusterings. The clusters obtained in different alternative clusterings are rational and comprehensive.

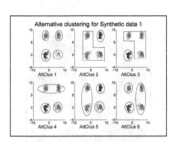

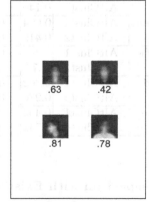

Fig. 2. Alternative clustering for syndata1   **Fig. 3.** Alternative clustering for syndata2

CMU Face: Two most important alternative clusterings of CMU face data with respect to ground truth are presented. There are twenty different persons' face in data, the first alternative clustering groups images into twenty clusters, each represents face of an individual. The second alternative clustering groups images according to pose of faces namely, straight, left, right, and up. The results are shown in Figure 4 and 5. Each cluster represents average image of the images grouped in the corresponding cluster and the value represents the fraction of belongingness of images to particular cluster. It is a confidence measure of

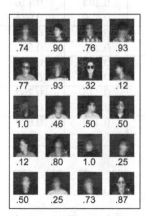

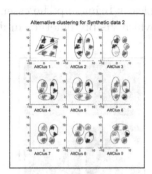

Fig. 4. Average clustered images with confidence measure, grouped by persons   **Fig. 5.** Average clustered images with confidence measure, grouped by head pose

identification of correctly clustered images. *VLGAAC* has identified alternative clusterings with correctly determined number of clusters in each clustering from different knowledge domain.

Escher Data Set: Four alternative clusterings for escher image of bird and flower are shown in Figure 6 and 7 respectively. Different gray shades represent different clusters of a particular clustering. Different alternative clusterings with varied number of clusters, help to view the escher image from different perspective. It clearly brings up the essence of escher image– different views of the same image, which visibly emerges in different alternative clusterings.

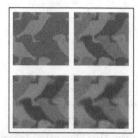   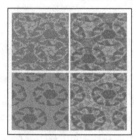

Fig. 6. Alternative clustering of bird   **Fig. 7.** Alternative clustering of flower

4.4   Clustering Quality and Clustering Distinctness

Alternative clusterings are evaluated in terms of cluster quality (*DBI*) and distinctness (*JI* and *RI*). The results are depicted for best two alternative clustering for each data set and shown in Table 3.

Table 3. Clustering quality: *DBI* and clustering distinctness: *JI* and *RI*

Data	Clusterings	DBI	JI (best)	JI (worst)	RI (best)	RI (worst)
Syndata1	AltClust1	0.388	0.281	0.591	0.463	0.746
	AltClust2	0.140	0.381	0.880	0.620	0.971
Syndata2	AltClust1	0.624	0.215	0.548	0.495	0.775
	AltClust2	0.419	0.202	0.659	0.202	0.926
Cmu face	AltClust1	1.07	0.102	0.529	0.497	0.796
	AltClust2	1.76	0.067	0.465	0.494	0.617
Bird	AltClust1	0.262	0.019	0.570	0.445	0.752
	AltClust2	0.270	0.032	0.719	0.672	0.900
Flower	AltClust1	0.453	0.020	0.444	0.469	0.679
	AltClust2	0.741	0.109	0.372	0.498	0.948

4.5   Comparison with Existing Algorithms

VLGAAC is compared with standard approaches, mentioned in literature, following two different principles of evaluating alternative clusterings. First, objective-function oriented approach– minCEntropy [1], and second, data-transformation oriented approach– clustering via orthogonalization(ortho-method1 and ortho-method2) [4]. The results for syndata1, syndata2 and CMU face image data are shown in Figure 8. The bar graph shows $VLGAAC$ gives comparable results with existing approaches in terms of clustering quality (DBI) and clustering distinctness (JI and RI).

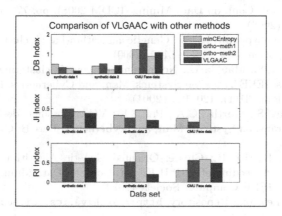

Fig. 8. Comparison of $VLGAAC$ with minCEntropy, ortho-method1, ortho-method2

5   Conclusion

We present a genetic algorithm approach $VLGAAC$ for alternative clustering. $VLGAAC$ frames the clustering problem with variable length chromosomes and design a fitness function for the purpose. The fitness function has two objectives, namely maximizing the clustering quality and minimizing the clustering distinctness. Genetic operators are also modified. $VLGAAC$ yields alternative clusterings with variable number of clusters. The results of $VLGAAC$ are better or comparable with other existing state of art methods in terms of clustering quality and distinctness. In future, other fitness functions may be evaluated and multi-objective genetic algorithm can be used to improve the performance.

References

1. Vinh, N.X., Epps, J.: minCEntropy: A Novel information theoretic Approach for the generation of alternative clusterings. In: Proceeding of 10th IEEE Int. Conf. on Data Mining (ICDM 2010), pp. 521–530. IEEE Computer Society (2010)
2. Dang, X.H., Bailey, J.: Hierarchical Information Theoretic Technique for the Discovery of Non Linear Alternative Clusterings. In: Proceeding of 16th ACM SIGKDD Int. Conf. on Knowledge Discovery and Data Mining, pp. 573–582. ACM (2010)

3. Bae, E., Bailey, J.: COALA: A novel approach for the extraction of an alternate clustering of high quality and high dissimilarity. In: Proceeding of 6th International Conference on Data Mining (ICDM 2006), pp. 53–62. IEEE Computer Society (2006)
4. Cui, Y., Fern, X.Z., Dy, J.G.: Non-redundant Multi-view Clustering via Orthogonalization. In: Proceeding of 7th Int. Conf. on Data Mining (ICDM 2007). pp. 133–142. IEEE Computer Society (2007)
5. Davidson, I., Qi, Z.: Finding Alternative Clusterings using Constraints. In: Proceeding of 8th Int. Conf. on Data Mining (ICDM 2008). pp. 773–778. IEEE Computer Society (2008)
6. Maulik, U., Banbyopadhyay, S.: Genetic algorithm-based clustering technique. Pattern Recognition 33(9), 1455–1465 (2000)
7. Bandyopadhyay, S., Maulik, U.: Nonparametric genetic clustering: comparison of validity indices. IEEE Trans. on Systems, Man, and Cybernetics, Part C: Applications and Reviews. 31(1), 120–125 (2001)
8. Bandyopadhyay, S., Murthy, C.A., Pal, S.K.: VGA-Classifier: design and applications. IEEE Trans. on Systems, Man, and Cybernetics, Part B: Cybernetics 30(6), 890–895 (2000)
9. Truong, D.T., Battiti, R.: Cluster-Oriented Genetic Algorithm for Alternative Clustering. In: Proceeding of 12th IEEE Int. Conf. on Data Mining (ICDM 2012), pp. 1122–1127. IEEE Computer Society (2012)
10. UCI Machine Learning Repository, http://archive.ics.uci.edu/ml/

Social Book Search with Pseudo-Relevance Feedback*

Bin Geng, Fang Zhou**, Jiao Qu, Bo-Wen Zhang, Xiao-Ping Cui,
and Xu-Cheng Yin**

Department of Computer Science and Technology,
School of Computer and Communication Engineering,
University of Science and Technology Beijing, Beijing 100083, China
zhoufang@ies.ustb.edu.cn, xuchengyin@ustb.edu.cn

Abstract. Massive books with social information, e.g. reviews, rates and tags, have emerged in large numbers on the web. However, there are several limitations in traditional search methods for social books, as social books include complicated and various social information. Relevance feedback is always an important and concerned technique in information retrieval. Therefore in this paper we propose a search system based on pseudo-relevance feedback (PRF) for expanding and enriching the social information of queries. In our system, First, Galago is used to get the initial rank list. Then relevance models are performed to select candidate high-frequent words that can be benefit to queries. Next, the original queries and these selected words are combined into new queries by linear smoothing. With evaluation on the INEX2012 / 2013 Social Book Search Track database, our proposed system has an encouraged performance (nDCG@10) compared to several state-of-the-art (contest) systems.

Keywords: Social Book Search, Social Information, Pseudo-Relevance Feedback, Query Expansion.

1 Introduction

With the fashion of social networks, people are more likely to share their reading reviews and obtain book information from the social networks. The key difference between social books and traditional books is that social books have not only traditional descriptions, but also contain complex social information [1].

For traditional books, there exist many retrieval methods. However in social book search, the social information is so important that investigating new methods is very meaningful. The INEX Social Book Search Track [1] which focuses on the social information of books is a typical topic about such issues.

Existing methods for the social book search can be roughly categorized into four groups: preprocessing the original documents and building index with different strategies, changing the query's structures, combining different query fields as query

---

* The research was partly supported by the National Natural Science Foundation of China (61105018, 61175020).
** Corresponding authors.

C.K. Loo et al. (Eds.): ICONIP 2014, Part II, LNCS 8835, pp. 203–211, 2014.

keywords, using social features to rerank. However, these methods do not show much enhance in Social Book Search. Therefore we propose a method, which combines documents preprocessing and uses social features to rerank. But we choose pseudo-relevance feedback to extend each query. Experiments show that our approach is more effective. In INEX 2012 Social Book Track dataset, we achieve a considerable increase in nDCG@10, (enhance 15.38% than the champion) which proves the pseudo-relevance feedback is worth of investigating.

In this paper, we propose a social book search system with pseudo-relevance feedback (PRF) to extend query topics more completely and accurately. The rest of this paper is organized as follows: In section 2, we introduce several used methods detailedly in social book search. Section 3 describes the specific pseudo-relevance feedback strategy for social book search. Section 4 presents experiments on the overall system performance on several public databases.

2   Related Work

As described above, the method of reranking with social features has demonstrated promising performance in the social book search. However, the books have many social features such as user rating, tags, authorship information, etc. The selection of those features could be important. Hence, we should conduct different combination strategies of the social features to verify the effectiveness of results, and then select the better ones.

The method utilizes book similarities as well as personalized re-ranking strategies to rerank the initial list. Book similarity calculates a new retrieval score for each book. The score is calculated by a linear combination of the initial score and the combined contributions of all other documents. The combined contributions are weighted by their similarity to the book. Then the relevant documents are pushed in the initial ranked result list through the new score. [2] In another method, the new personalized score is a linear combination of the original retrieval score. The book similarity reranking examine the usefulness of different information sources for calculation these books similarities [3]. Personalized re-ranking did not work as well as the non-personalized methods, which is likely due its inappropriate for the recommendation task [3].The results of the INEX 2012 Social Book Search Track shows that using social feature to rerank does not bring effective performances. Therefore, in our paper, we just use pseudo-relevance feedback algorithms [4, 5, 6, 7, 8] to rerank which have better performance than the others.

There are many ways to change query's structures. The Sequential Dependence Model (SDM) is one of them. This model concerns three groups: single word, successive two meaningful words, and successive three meaningful words. All in all, SDM separate the query into three classes according to the language database [9]. How to select the successive two words or three words is difficult, as a lot of noise will be brought in. In our method, we not only change the query's construction but also use pseudo-relevance feedback to enlarge the query's contents.

Certainly, there are also several pseudo-relevance feedback techniques in IR [10, 15]. For example, using latent concepts [11],query-regularized estimation[12], additional features other than term distributions[13], and a clustered-based re-sampling method for generating pseudo-relevant documents. [14] However, for social information, and social book search topics, there are very few related methods with pseudo-relevance feedback. Consequently we propose a social book search system with pseudo-relevance feedback. It has an effective enhancement in nDCG@10 value. We also extend our work and design a competitive system for participating the INEX 2014 Social Book Search evaulation [16, 17].

3   Social Book Search with PRF

By incorporating several key improvements over traditional PRF methods, we design a specific pseudo-relevance feedback strategy for social book search. The structure of proposed system is presented in figure 1.

3.1   System Overview

The index is built with Galago. When creating the index, stopwords are removed and Porter Stemmer is used for stemming. The words in title, group, query, narrative (TQGN) fields of each topic are used as query keywords.

Galago is used to get the initial ranked list, and the top k documents are chosen as feedback collection. In order to get relevant words, relevance model is used to select expanding terms. The extending words are combined with original queries using linear smoothing, by a parameter α, and then we will get the final results.

Fig. 1. The structure of our proposed system

3.2 Pseudo-Relevance Feedback (PRF) Strategy

By a variety of techniques, query expansion method reassembles extending words associated with the original query into a new query to describe the original query more completely and accurately. Therefore, the information retrieval system will get more relevant documents. Pseudo-relevance feedback has been proved to be an effective method for query expansion.

Pseudo-relevance feedback assumes that the top k sorted documents in the initial list are relevant [17]. There are two steps. First, get the initial rank list and choose the top k result books as relevant collection. Second, the documents in the collection are processed and the top w relevant words are taken as feedback words, those words are combined with the original query as new query keywords to retrieve again. Many studies have shown that pseudo-relevance feedback is extremely helpful in improving query precision and recall rates.

The initial rank list is obtained with the following Equation by Galago

$$\log P(Q \mid D) = \sum_{i=1}^{n} \log \frac{f_{q_i,D} + \mu \frac{c_{q_i}}{|C|}}{|D| + \mu}, \tag{1}$$

Where m means the q_i appearance number in document collection D, c_{q_i} indicates the number of query terms appear in the dataset, $|C|$ is the total number of all the words appear in the collection, μ is uniform over the set, always choose 1000-2000. The score are processed by Dirichlet smoothing.

For different retrieval model, pseudo-relevance feedback has different strategies. In our system, relevance model is used within the language model. A correlation model is estimated from query, then the language model and the document model are compared directly, documents will be sorted according to the similarity of both the document model and relevance model. There is a KL-dispersion to compare the two models, which is defined as follows

$$KL(P \parallel Q) = \sum_{x} P(x) \log \frac{P(x)}{Q(x)}, \tag{2}$$

Where P is given the real probability distribution, Q is used to estimate the probability distribution P.

If we assume that the true distribution is the query (R) correlation model, and the approximate distribution is the document language model (D), then the scores of document are calculated as follows

$$score(w, D) = \sum_{w \in C} P(w \mid R) \log P(w \mid D), \tag{3}$$

Where C is the top w words in feedback collection.

The relevant model is considered as a multinomial distribution which estimates the likelihood of word w when a query Q is given. The query words $q_1 \ldots q_n$ and word w in relevant documents are sampled identically and independently from a distribution R. Hence, the probability of a word in the distribution R is estimated as follows

$$P(w \mid R) = \sum_{D \in C} P(D)P(W \mid D)P(Q \mid D) \tag{4}$$

where $P(D)$ is a priori probability, $P(Q \mid D)$ is query likelihood score of document D.
The following equation is used to calculate the finally result,

$$P(w \mid Q') = \alpha P(w \mid Q) + (1 - \alpha)P(w \mid D) \tag{5}$$

The social book search system with PRF is summarized in Figure 2.

```
Input: Query (one topic)
Output: Search and suggestion books
Procedure:
Begin
    1: Rank documents with Equation (1) and get the initial list.
    2: Choose top k documents.
    3: Select top w words with Equation (4) from k documents.
    4: Update the query by combining w words with Equation (5).
    5: Re-rank with the new query and get the final results.
End
```

Fig. 2. The social book search flowchart with pseudo-relevance feedback

4   Evaluation

4.1   Evaluation Setup

Our experimental datasets are from the INEX social book search track organizers. We use INEX 2011 (contains 211 topics) and 2012 (contains 96 topics) query topics as training set for testing the effectiveness of parameters,2012 and 2013 query topics (contains 286 topics) as test set. Each topic consists of three short descriptions (title, query and group) and a detailed description (narrative) of user request.

In this experiment, nDCG@10 is our main evaluation measure, which focuses on only the first 10 books of the result list.

4.2   Search Results in INEX 2012 Dataset

In this experiment, our system trains the parameters in INEX 2011 set (the official training set for INEX 2012) and evaluates the results in INEX 2012 set. Table 1 shows the comparison of our results (test set) and the top3 results ("p4.xml_social. fb.10.50", "p54.run2.all-topic-fields.all-doc-fields.baseline" and "p54.run3.all-topic-fields.QIT.alpha0.99") in INEX 2012. For "p4.xml" team, only title (T) field in topics is used as query keywords and SDM retrieval model is used as their retrieval model; for "p54"team, they choose all fields (TQGN) as query keywords. The run2 result is obtained by using all social features to reranking and therun3 result is obtained by using QIT (quotation, similar products, tag) features to reranking. For our method, we just use PRF method to get the result it can be seen that our method has a great improvement in nDCG@10.

Table 1. Result comparison on INEX 2012 dataset

Run	nDCG@10	P@10
Our result	**0.1674**	**0.1872**
p4.xml_social.fb.10.50	0.1456	0.1376
p54.run2.all-topic-fields.all-doc-fields.baseline	0.1452	0.1248
p54.run3.all-topic-fields.QIT.alpha0.99	0.1447	0.1248

Then the INEX 2012 query topic set are chosen as training set for training parameters of INEX 2013, and we will introduce the training process in detail. The TQGN fields are used as our query keywords. First, the stopwords from topic fields' title, group, query and narrative are removed. Then the TQGN fields are stemmed by Porter Stemmer.

The initial rank list is obtained by language retrieval model with Galago. The top 10 and 20 books of each topic in initial rank list are chosen as feedback collections. Experiments results in Table 2 shows that it is important for choosing the number of feedback document (k) on the prediction quality of nDCG@10. It is obvious that the result performs better, when k equals 20.

Table 2. Comparison of different feedback collection

feedback document	Best nDCG@10
k=10	0.1652
k=20	0.1680

In the training (INEX 2012 topic set) set we obtain the optional parameters of our pseudo-relevance feedback system, including feedback words (w) in the collection, the α value. Especially, the word select of the query extension is important. For each topic, the most frequent words from 0 to 40 are separately chosen as feedback words. Calculate the correlation between each word and document collection. Then take the top w related words as query terms. From Table 3, it can be seen that the most effective w is 15, achieves a considerable increase in the nDCG@10 value. It also improves the P@10 and MAP@10 value.

In query expansion, the original query terms that reflects the user's query intent is always the most important part. Expansion words are just the semantics complement of original query. Therefore, our experiments train the balance factor α between original query and expansion words from 0.85 to 0.99 step of 0.02. The results of different α are showed in Table 4.

Table 3. Results of feedback word selection

w	nDCG10	P@10	MAP@10
0	0.1648	0.1858	0.0911
5	0.1677	0.1876	0.0921
10	0.1675	0.1932	0.0921
15	**0.1680**	**0.1892**	**0.0971**
20	0.1660	0.1871	0.0906
25	0.1663	0.1860	0.0909
30	0.1651	0.1850	0.0903
35	0.1646	0.1850	0.0894
40	0.1621	0.1814	0.0882

Table 4. Comparison the results of different α values

α	0.85	0.87	0.89	0.91	0.93	0.95	0.97	**0.99**
nDCG@10	0.1521	0.1515	0.1513	0.1528	0.1536	0.1523	0.1637	**0.1680**
MAP@10	0.0805	0.0809	0.0808	0.0644	0.081	0.0792	0.0886	**0.0971**
P@10	0.1799	0.1777	0.1802	0.1819	0.1795	0.1772	0.1865	**0.1892**

4.3   Search Results in INEX 2013 Dataset

From the training set we obtain the optional parameters of our feedback system. (1) Get initial rank book lists with Galago, and choose the top 20 books of each topic as our feedback collection. (2) Select the most frequent words of each topic and calculate the correlation between each word and document collection. Use the most related 15 words as query extension. (3) The final expanded queries are combined with the original query using linear interpolation, weighted by a parameter α. The α is set as 0.99. The result comparison of our method and other methods in INEX 2013 are shown in table 5, where "RSLIS - run3.all-plus-query.all-doc-fields", "UAms_ILLC - inex13SBS.ti_qu_gr_na.bayes_avg", and "UAms_ILLC - inex13SBS.ti_qu.bayes_avg" are top 3 results in the INEX 2013 contest.

Table 5. Result comparison on INEX 2013

Run	nDCG@10	MAP@10
RSLIS - run3.all-plus-query.all-doc-fields	0.1361	0.0653
Our result	**0.1332**	**0.0882**
UAms_ILLC - inex13SBS.ti_qu_gr_na.bayes_avg	0.1331	0.0771
UAms_ILLC - inex13SBS.ti_qu.bayes_avg	0.1331	0.0771

We see that, the feedback result compared with baseline (nDCG@10 0.1305, MAP@10 0.0841) get nDCG@10 0.1332, with improvement of 2.1%, and the MAP@10 value is 0.0882 with the improvement of 4.9%. Compared with other methods, our MAP@10 value gets an encourage improvement.

5   Conclusion

In this paper, we propose an effective social book search system with pseudo-relevance feedback. We use relevant model to obtain the optional words. The new queries are constructed by original queries and optional words using learned parameters. The experimental results show that our system is stable and effective.

References

1. Koolen, M., Kazai, G., Kamps, J., Preminger, M., Doucet, A.: Overview of the INEX 2012 Social Book Search Track. In: INEX 2012 Workshop Pre-proceedings, pp. 78–92 (2012)
2. Geva, S., Kamps, J.: INEX 12 Workshop Pre-proceedings (2012)
3. Bogers, T., Larsen, B.: RSLIS at INEX 2012: Social Book Search Track. In: INEX 2012 Workshop Pre-proceedings, pp. 97–108 (2012)
4. Buckley, C., Salton, G., Allan, J.: Automatic query expansion using SMART: TREC 3. pp. 69–69 (1995)
5. Robertson, S.E., Walker, S., Jones, S.: Okapi at TREC-3 (1995)
6. Lavrenko, V., Bruce Croft, W.: Relevance based language models. In: Proceedings of the 24th Annual International ACM SIGIR Conference on Research and Development in Information Retrieval (SIGIR 2001), pp. 120–127 (2001)
7. Zhai, C., Lafferty, J.: Model-based feedback in the language modeling approach to information retrieval. In: Proceedings of the Tenth ACM International Conference on Information and Knowledge Management (CIKM 2001), pp. 403–410 (2001)
8. Xu, J., Bruce Croft, W.: Improving the effectiveness of information retrieval with local context analysis. ACM Transactions on Information Systems (TOIS) 18(1), 79–112 (2000)
9. Benkoussas, C., Bellot, P.: Book Recommendation based on social information. In: INEX 12 Workshop Pre-proceedings (2012)
10. Xu, Y., Jones, G.J.F., Wang, B.: Query dependent pseudo-relevance feedback based on Wikipedia. In: Proceedings of the 32nd International ACM SIGIR Conference on Research and Development in Information Retrieval (SIGIR 2009) (2009)
11. Metzler, D., Bruce Croft, W.: Latent concept expansion using markov random fields. In: Proceedings of the 30th Annual International ACM SIGIR Conference on Research and Development in Information Retrieval (SIGIR 2007), pp. 311–318 (2007)
12. Tao, T., Zhai, C.: Regularized estimation of mixture models for robust pseudo-relevance feedback. In: Proceedings of the 29th Annual International ACM SIGIR Conference on Research and Development in Information Retrieval (SIGIR 2006), pp. 126–129 (2006)
13. Cao, G., Nie, J.-Y., Gao, J., Robertson, S.: Selecting good expansion terms for pseudo-relevance feedback. In: Proceedings of the 31st Annual International ACM SIGIR Conference on Research and Development in Information Retrieval (SIGIR 2008), pp. 243–250 (2008)
14. Lee, K., Bruce Croft, W., Allan, J.: A cluster-based resampling method for pseudo-relevance feedback. In: Proceedings of the 31st Annual International ACM SIGIR Conference on Research and Development in Information Retrieval (SIGIR 2008), pp. 235–242 (2008)

15. Butman, O., Shtok, A., Kurland, O., Carmel, D.: Query-Performance Prediction Using Minimal Relevance Feedback. In: Proceedings of the 2013 Conference on the Theory of Information Retrieval (ICTIR 2013), pp. 14–21 (2013)
16. Zhang, B.-W., Yin, X.-C., Cui, X.-P., Qu, J., Geng, B., Zhou, F., Hao, H.-W.: USTB at INEX 2014: Social Book Search. In: INEX 2014 Workshop Pre-proceedings (2014)
17. Zhang, B.-W., Yin, X.-C., Cui, X.-P., Qu, J., Geng, B., Zhou, F., Song, L., Hao, H.-W.: Social Book Search Reranking with Generalized Content-Based Filtering. In: Proceedings of ACM International Conference on Information and Knowledge Management (CIKM 2014) (2014)

A Random Key Genetic Algorithm for Live Migration of Multiple Virtual Machines in Data Centers

Tusher Kumer Sarker and Maolin Tang

School of Electrical Engineering and Computer Science
Queensland University of Technology
2 George Street, Brisbane, QLD 4000, Australia
{t.sarker,m.tang}@qut.edu.au

Abstract. Live migration of multiple Virtual Machines (VMs) has become an integral management activity in data centers for power saving, load balancing and system maintenance. While state-of-the-art live migration techniques focus on the improvement of migration performance of an independent single VM, only a little has been investigated to the case of live migration of multiple interacting VMs. Live migration is mostly influenced by the network bandwidth and arbitrarily migrating a VM which has data inter-dependencies with other VMs may increase the bandwidth consumption and adversely affect the performances of subsequent migrations. In this paper, we propose a Random Key Genetic Algorithm (RKGA) that efficiently schedules the migration of a given set of VMs accounting both inter-VM dependency and data center communication network. The experimental results show that the RKGA can schedule the migration of multiple VMs with significantly shorter total migration time and total downtime compared to a heuristic algorithm.

Keywords: Live migration, virtual machine, migration time, downtime, random key genetic algorithm.

1 Introduction

Live migration of Virtual Machine (VM) is a prominent feature of system virtualization where a VM moves across the Physical Machines (PMs) in a data center without stopping its services. Though live migration is almost a transparent phenomenon, a small downtime cannot be avoided. The performance of a live migration is characterised by two time related metrics - migration time and migration downtime. A bulk of studies have been devoted to improve the performance of live migration of VM. Liu et al. [1] developed trace and replay approach (CR/TR-Motion) to reduce the amount of memory transfer during migration. The effects of network bandwidth and VM memory modification rate on migration time and migration downtime were studied in [2]. Ye et al. [3] investigated the virtual cluster, a group of dependent VMs, migration with different migration granularities. Deshpande et al. [4] introduced live gang migration where

C.K. Loo et al. (Eds.): ICONIP 2014, Part II, LNCS 8835, pp. 212–220, 2014.

correlated VMs on the same PM were migrated simultaneously. However, these works do not consider the necessity of migration scheduling when multiple VMs are required to be migrated concurrently. In a data center a large number of VMs are required to be migrated periodically or frequently for power saving, load balancing, server maintenance and failure recovery. Furthermore, a number of VMs that host scientific application, MapReduce application and application components of a multi-tier application, exhibit inter-VM data dependency and a non-trivial amount of traffic flow takes place among these correlated VMs. Therefore, arbitrarily migrating such a correlated VM may increase the network bandwidth consumption for its increased inter-VM traffic flow through the data center network, and which may negatively affect the subsequent migrations due to less availability of network bandwidth. As a result, a proper sequencing of migrations becomes mandatory. In this paper, we propose a Random Key Genetic Algorithm (RKGA) that efficiently schedules the migrations of a set of VMs. The experimental results show that our proposed RKGA can schedule the migrations of a large number of VMs with significantly shorter total migration time and total downtime compared to a heuristic algorithm [5].

The remainder of this paper is structured as follows. Section 2 formulates the problem. Section 3 describes our proposed RKGA. Section 4 presents experimental results. Finally, section 5 concludes the paper.

2   The Multiple VMs Live Migration Problem

2.1   Live Migration Attributes

As the proposed RKGA considers parallel migrations of multiple VMs from different PMs, a migration sequence, S can be comprised the sets of parallel migrations, i.e. $S = \langle S_1, S_2, \ldots, S_j, \ldots, S_q \rangle$, where S_j is a set of VMs that can be migrated in parallel, $S_j \neq \emptyset$, $S_j \subseteq C$, $\bigcup_{j=1}^{q} S_j = C$, $q \geq 1$ and C is the set of candidate VMs that are selected for migration. As the Total Migration Time (TMT) and Total Downtime (TDT) depend on the sequence of migrations, both are defined as functions of a migration sequence, S.

TMT is the time required to migrate all the candidate VMs to their final target PMs and is calculated as the summation of migration times of all parallel migration sets, $MT(S_j)$. Equation (1) calculates the TMT for a sequence, S.

$$TMT(S) = \sum_{j=1}^{q} MT(S_j) \ . \tag{1}$$

$MT(S_j)$ is equal to the migration time of the VM in S_j that requires maximum time to complete its migration and is calculated using (2)

$$MT(S_j) = \max\{MT_{vm_i} | vm_i \in S_j\} \ . \tag{2}$$

and Migration time of vm_i, MT_{vm_i}, is the elapsed time between the start of its migration and the time when vm_i resumes its operation at target PM by terminating its all dependency with source PM. MT_{vm_i} is calculated by (3) following the mathematical model given by Mann et al. [6].

$$MT_{vm_i} = \begin{cases} \left[\frac{r_{vm_i}}{b_{vm_i}-f_{vm_i}}\right] \times \left[1 - \left(\frac{f_{vm_i}}{b_{vm_i}}\right)^h\right] + lat_{pm_l} & \text{if } b_{vm_i} > f_{vm_i} \\ h \times \frac{r_{vm_i}}{b_{vm_i}} + lat_{pm_l} & \text{otherwise} \end{cases} \quad (3)$$

where h is the number of iterations required to complete the migration (h is a user setting), b_{vm_i} is the allocated bandwidth to vm_i during migration, f_{vm_i} is the memory modification rate of vm_i, lat_{pm_l} is the latency at target PM, pm_l and r_{vm_i} is the remaining memory on source PM to be transferred to pm_l, initially $r_{vm_i} = m_{vm_i}$, where m_{vm_i} is the size of vm_i.

TDT of a migration sequence, S is the cumulative sum of migration downtimes of all VMs in the candidate set, C, for that S. TDT is calculated as follows.

$$TDT(S) = \sum_{i=1}^{|C|} DT_{vm_i}(S) \ . \quad (4)$$

and Migration downtime of vm_i, DT_{vm_i}, is the duration of time for which the services of vm_i are completely unavailable to the user and this is equivalent to the time required in the last iteration. The mathematical model for calculating DT_{vm_i} provided by Mann et al. [6] is given by (5).

$$DT_{vm_i} = \begin{cases} \frac{r_{vm_i} \times (f_{vm_i})^{(h-1)}}{(b_{vm_i})^h} + lat_{pm_l} & \text{if } b_{vm_i} > f_{vm_i} \\ \frac{r_{vm_i}}{b_{vm_i}} + lat_{pm_l} & \text{otherwise} \end{cases} \quad (5)$$

As DT_{vm_i} is very small compared to MT_{vm_i}, it is less likely that some VMs will go to down state simultaneously. Moreover, as downtime of vm_i affects its dependent VMs directly, it is required to calculate the cumulative sum of downtimes of all VMs to evaluate the effectiveness of migration scheduling algorithm.

2.2   Problem Formulation

Input:

1. A set of m PMs, $P = \{pm_1, pm_2, \ldots, pm_i, \ldots, pm_m\}$, where each PM, pm_i is attributed by its initial CPU capacity, $c^*_{pm_i}$ and memory capacity, $m^*_{pm_i}$ and its current utilization of CPU, c_{pm_i} and memory, m_{pm_i}.

2. A set of n VMs, $V = \{vm_1, vm_2, \ldots, vm_i, \ldots, vm_n\}$. Each VM, vm_i is characterized by its CPU requirement, c_{vm_i}, memory requirement, m_{vm_i} and memory modification rate, f_{vm_i}.

3. A set C, of candidate VMs selected for migration, $C \subseteq V$ with their current placement and target placement to where they will be migrated finally.

4. A VM dependency weighted undirected graph, $G = (V, E)$, where V is the set of VMs and E is the set of edges. An edge $(vm_i, vm_j) \in E$ indicates that there is inter-VM traffic flow between vm_i and vm_j, where $1 \leq i, j \leq |V|$, $i \neq j$.

5. A data center network represented by a weighted undirected graph, $D = (U, L, B)$, where $U = \{u_i\}$ is the set of nodes consisting of PMs and switches, L is the set of communication links $\{e_{i,j}\}$, where $e_{i,j}$ is an edge connecting nodes u_i and u_j, and $B_{u_i, u_j} : L \to B$ is a bandwidth function between u_i and u_j.

Constraints: The following three resource constraints must be satisfied whenever migrating a VM, vm_i, from a PM, pm_k, to another PM, pm_l:

1. Available CPU resource on pm_l must be adequate to meet the CPU requirement of vm_i, which is represented as $c_{vm_i} \leq |c^*_{pm_l} - c_{pm_l}|$.

2. Available memory resource on pm_l must be adequate to meet the memory requirement of vm_i, which is represented as $m_{vm_i} \leq |m^*_{pm_l} - m_{pm_l}|$.

3. The inter-VM traffic flow between vm_i and its connected VM, vm_j, on any PM, pm_t, other than pm_l, d_{vm_i, vm_j} must not exceed the available bandwidth capacity between pm_l and pm_t, B_{pm_l, pm_t}, i.e. $d_{vm_i, vm_j} \leq B_{pm_l, pm_t}$.

Output: A sequence, S, of subsets of VMs in C, such that both $TMT(S)$ and $TDT(S)$ are minimal.

3 The RKGA

We use RKGA to solve the VMs migration scheduling problem. RKGA efficiently handles the sequencing problems without generating any infeasible solution [7]. Random numbers in the range $[0, 1]$, used as sort keys, are generated to represent the alleles of genes. The genes of a chromosome are ranked based on the sort key values and two sort keys of same value get the consecutive ranks. Therefore, any genetic operation applied on the chromosomes always yields feasible solutions. In the following sub-sections, we describe our chromosome encoding scheme, genetic operators, fitness function and proposed RKGA in details.

3.1 Chromosome Representation

A chromosome is represented by x genes, where each gene corresponds to a candidate VM. The values of genes are real numbers generated randomly in the range $[0, 1]$, used as sort keys. The VMs are ranked based on these sort keys and with the lowest value of the sort key, the VM gets the highest priority to migrate first. These ranks represent the initial migration priorities of VMs. However, a VM with lower priority can be migrated in parallel with other VMs of higher ranks if all the resource constraints for that VM are satisfied. Similarly, it can be the case that a VM with the highest priority will not be migrated until the target PM does not meet the resource requirements of the VM.

Our RKGA scans all the genes starting from the highest priority to the lowest and selects the queued VMs to migrating in parallel for which the resource constraints are satisfied. The process iterates until all the VMs are not scheduled for migration. Therefore, our proposed RKGA always finds a feasible solution even though the initial sequence of migration cannot be followed.

3.2   Genetic Operators

Selection: Elitist strategy is used, where a percentage of individuals with the highest fitness values are replicated to the next generation. This approach ensures that the elite solutions get improvement from one generation to the next.

Crossover: Parameterized uniform crossover is employed, where a new offspring is generated from two parents of old generation. After selection of two parents a random number between 0 and 1 is generated against each gene. If this random number is less than a predefined value (for example, 0.7) then the allele from the second parent is chosen; otherwise allele of the first parent retains.

Mutation: Mutation operator is applied to create genetic diversity. Multiple genes of a selected parent are altered with the random numbers between 0 and 1.

3.3   Fitness Function

As the aim is to minimize TMT and TDT, these two attributes are used to calculate the fitness value of a solution. As discussed previously a population does not contain any infeasible solution, no penalty is given to the fitness value. Fitness value of a solution, X, is calculated using (6).

$$F(X) = 1 - \left[\frac{TMT(X)}{TMT(X_{\max}^{Gen})} \times w1 + \frac{TDT(X)}{TDT(X_{\max}^{Gen})} \times w2 \right] . \tag{6}$$

where $TMT(X_{\max}^{Gen})$ and $TDT(X_{\max}^{Gen})$ are respectively the maximum TMT and maximum TDT obtained upto current generation, Gen, $w1$ and $w2$ are weights, and $w1 + w2 = 1$. Equation (6) ensures that the solution that requires less TMT and TDT always has the highest fitness value.

3.4   Algorithm Description

Algorithm 1 is the description of our proposed RKGA. Step 1 generates the initial population. Steps 2 to 21 find the best solution in all generations. Steps 3 to 13 calculate the total migration time and total downtime for each solution in a population. Step 7 invokes a heuristic procedure that eliminates the migration loop by finding a suitable intermediate PM. When the VMs cannot be migrated to their target PMs until one or more VMs are not migrated to the PMs rather than their target PMs, a migration loop is created. Another procedure that controls the migrations and dynamically allocates the relinquished bandwidth by the migrated VM(s) to the continuing migrations is invoked in step 10. The detail descriptions of these algorithms are not given in this paper because of space limitation. Individuals' fitness values are calculated in step 14 using (6). Steps 16 to 18 apply genetic operators.

4   Experimental Results

As the proposed RKGA is designed to deal with the migrations of hundreds of VMs, it is essential to evaluate it on a large-scale virtualized data center infrastructure. However, it is difficult to conduct large-scale experiments on a real

Algorithm 1. Our RKGA for scheduling VMs migration

```
 1  generate initial population
 2  while termination criteria are false do
 3  │  for each individual in the population do
 4  │  │  while a complete migration sequence is not found for an individual do
 5  │  │  │  find the set of VMs, Sⱼ that can be migrated in parallel
 6  │  │  │  if Sⱼ = ∅ then
 7  │  │  │  │  invoke the heuristic procedure to eliminate migration loop
 8  │  │  │  end
 9  │  │  │  else
10  │  │  │  │  invoke the migration control procedure
11  │  │  │  end
12  │  │  end
13  │  end
14  │  calculate the fitness values of all individuals in a population
15  │  if termination criteria are false then
16  │  │  select predefined amount of fittest individuals for the next generation
17  │  │  apply the parameterized uniform crossover operation
18  │  │  apply mutation operation
19  │  │  replace current generation by the new generation
20  │  end
21  end
22  output the TMT and TDT of the fittest individual in all generations
```

data center, especially when it is necessary to reproduce the experiment with the same conditions to compare different algorithms. Therefore, simulation experiments were conducted to show the effectiveness and scalability of our RKGA. Since there are no benchmarks available for the research problem, we randomly generated the test problems. The CPU and memory requirements of a VM were arbitrarily chosen from the set $\{1000, 2000, \ldots, 5000\}$ MIPS and from the range $[500, 2000]$ MB respectively. In the experiments we created a fat tree [8] data center network consisting of 432 PMs with a capacity of housing at least 3 VMs on each PM, although our RKGA works for data centers of any network topology. The memory modification rate of a VM was between 1% and 6% of its size, a VM cluster was composed of 0 to 5 VMs and the inter-VM traffic rate between two dependent VMs was a value from the set $\{0, 1, 2\}$ Mbps. The randomly generated attributes of PMs and VMs, i.e. CPU capacities, memory capacities and VMs memory modification rates were same through out the experiments.

In the experiments, we fixed both $w1$ and $w2$, which are two parameters used in the fitness function (6), to 0.5. With regards to the parameters for our RKGA, through a series of trials we chose the following set of parameters for our RKGA: the population size and maximum generation count were 100 and 300 respectively with the crossover, mutation and selection probabilities of 0.85, 0.07 and 0.08 respectively. The RKGA terminates when the best solutions of 50 consecutive generations remain unchanged or the number of generations becomes

300. We implemented both the RKGA and the heuristic algorithm in Java. The experiments were conducted on a desktop computer with 2.80 GHz Intel Core i7-2640M CPU and 8.00 GB RAM.

4.1 Experiments on the Scalability of the RKGA

We show the scalability of our RKGA by varying the number of candidate VMs between 20 and 200 with an increment of 20 VMs in each test problem. The total number of VMs and the total number of PMs were kept constant for all experiments to 1000 and 432 respectively. As we followed the fat tree architecture the number of PMs became 432. The link speed was 1 Gbps. The candidate VMs for each test problem were arbitrarily chosen from the pool of 1000 VMs in the data center and migration plans were random as well. In this experiment we measured the computation time required by our RKGA to schedule the migrations of different numbers of VMs. Due to stochastic nature of RKGA, we ran each test problem for 20 times and the average of the 20 runs was taken to evaluate the result. Table 1 shows the average computation time for each test problem. Figure 1 shows that the computation time for scheduling the migrations increases almost linearly with increase of the number of candidate VMs. The near linear trend of computation time graph indicates the good scalability of our proposed RKGA in scheduling the migrations of different numbers of VMs.

Table 1. Experimental results

Test problems	RKGA			Heuristic Algorithm	
	TMT (s)	TDT (ms)	Computation Time (s)	TMT (s)	TDT (ms)
20	128.69	235.95	16.18	189.63	236.00
40	171.86	444.90	35.85	261.38	446.00
60	232.91	783.95	82.49	443.06	815.00
80	188.97	922.95	141.70	396.26	927.00
100	342.03	1548.20	180.16	640.26	1574.00
120	301.96	2141.60	309.67	687.44	2377.00
140	379.35	5355.05	369.33	897.48	5857.00
160	386.50	3834.10	426.05	716.83	4597.00
180	440.05	3340.05	480.11	953.13	3607.00
200	463.72	6902.95	545.32	957.06	7422.00

4.2 Experiments on the Effectiveness of the RKGA

In this set of experiments we illustrate the quality of solutions produced by our RKGA. To show this we compare the total migration time and total downtime calculated by our RKGA with those by the heuristic algorithm. As the heuristic algorithm is a deterministic one, the result of only one run was taken for comparison where the average of 20 runs was taken for the RKGA. The results of the experiments are presented in Table 1. For all test problems our RKGA schedules

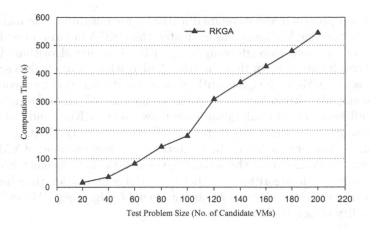

Fig. 1. Computation time of RKGA for scheduling the migrations of different numbers of VMs

the migrations of the candidate VMs with significantly shorter total migration time compared to the heuristic algorithm. Moreover, with the increase of the size of the test problems the difference of total migration time between the RKGA and the heuristic algorithm increases. This illustrates the effectiveness of our proposed RKGA for scheduling the migrations of a large number of VMs. However, the total migration time does not increase linearly with the increase of the size of test problems. Table 1 shows that the test problem of 80 candidate VMs gives shorter total migration time than the test problem of 60 candidate VMs. This is because for each test problem the candidate VMs were chosen arbitrarily from the pool of 1000 VMs in the data center giving the chance of selecting some or all distinct candidate VMs for each test problem. This may result in a test problem consists of a large number of small size VMs. Consequently, the test problem comprises small size VMs gives shorter total migration time.

The performance of the RKGA, in terms of total downtime, gives a little improvement over the heuristic algorithm. The reason behind this is that, the total downtime is calculated by adding the migration downtimes of all candidate VMs using (4) and the same dynamic bandwidth adaptation algorithm is used for both the RKGA and the heuristic algorithm. In dynamic bandwidth adaptation approach, the freed bandwidth by the migrated VM(s) in a batch is allocated to the ongoing migration(s). As a result, all VMs get a significant amount of bandwidth, nearly to the link speed, during the last iteration in both the strategies. Moreover, a larger choice of h in (5), for example, $h = 30$ in our experiments, causes a very small amount of remaining memory to be transferred in the last iteration. Therefore, a VM goes to down state almost for the same duration in both the migration scheduling strategies.

5   Conclusion

This paper has proposed and developed a RKGA for scheduling the migrations of multiple interacting VMs, and has evaluated the RKGA in terms of scalability and quality of solutions through comparing with a heuristic algorithm. The experimental results have shown that the RKGA algorithm always finds a schedule for a given set of VMs with significantly shorter total migration time compared to the heuristic algorithm. Moreover, with the increase of number of migrating VMs, the difference of total migration time between the RKGA and the heuristic algorithm increases which illustrates the effectiveness of the RKGA over the heuristic algorithm for scheduling the migrations of a large number of VMs. The total downtime calculated by the RKGA gives a little improvement over that found by the heuristic algorithm. Furthermore, the computation time increases almost linearly with the increase of the number of migrating VMs indicating good scalability of our RKGA.

Acknowledgments. This research work was funded by International Post-graduate Research Scholarship (IPRS) and Australian Post-graduate Award (APA).

References

1. Liu, H., Jin, H., Liao, X., Hu, L., Yu, C.: Live migration of virtual machine based on full system trace and replay. In: Proceedings of the 18th ACM International Symposium on High Performance Distributed Computing (HPDC), pp. 101–110 (2009)
2. Akoush, S., Sohan, R., Rice, A., Moore, A.W., Hopper, A.: Predicting the performance of virtual machine migration. In: The 18th Annual IEEE International Symposium on Modeling, Analysis and Simulation of Computer and Telecommunication Systems (MASCOTS), Florida, USA, pp. 37–46 (2010)
3. Ye, K., Jiang, X., Ma, R., Yan, F.: VC-Migration: Live Migration of Virtual Clusters in the Cloud. In: 13th ACM/IEEE International Conference on Grid Computing (GRID), Beijing, China, pp. 209–218 (2012)
4. Deshpande, U., Wang, X., Gopalan, K.: Live gang migration of virtual machines. In: Proceedings of the 20th International Symposium on High Performance Distributed Computing (HPDC), San Jose, California, pp. 135–146 (2011)
5. Sarker, T.K., Tang, M.: Performance-driven live migration of multiple virtual machines in datacenters. In: Proceedings of the 2013 IEEE International Conference on Granular Computing (GrC), Beijing, China, pp. 253–258 (2013)
6. Mann, V., Gupta, A., Dutta, P., Vishnoi, A., Bhattacharya, P., Poddar, R., Iyer, A.: Remedy: Network-aware steady state VM management for data centers. In: Bestak, R., Kencl, L., Li, L.E., Widmer, J., Yin, H. (eds.) NETWORKING 2012, Part I. LNCS, vol. 7289, pp. 190–204. Springer, Heidelberg (2012)
7. Bean, J.C.: Genetic algorithms and random keys for sequencing and optimization. ORSA Journal on Computing 6, 154–160 (1994)
8. Al-Fares, M., Loukissas, A., Vahdat, A.: A scalable, commodity data center network architecture. In: Proceedings of the ACM SIGCOMM 2008 conference on Data communication (SIGCOMM), Seattle, Washington, USA, pp. 63–74 (2008)

Collaboration of the Radial Basis ART and PSO in Multi-Solution Problems of the Hénon Map

Fumiaki Tokunaga, Takumi Sato, and Toshimichi Saito

Hosei University, Koganei, Tokyo, 184–8584 Japan
tsaito@hosei.ac.jp

Abstract. This paper studies collaboration of the ART and PSO in application to a multi-solution problem for analysis of the Hénon map. In our algorithm, the PSO gives candidates of solutions which have no labels. Applying the candidates as inputs, the ART classifies the candidates, labels the categories, and clarify the number of solutions. Performing fundamental numerical experiments, the algorithm efficiency is investigated.

Keywords: ART, PSO, Multi-solution problems, Hénon map.

1 Introduction

This paper studies collaboration of the adaptive resonance theory system (ART, [1]-[4]) and the particle swarm optimized (PSO, [5]) in multi-solution problems (MSPs, [5] [6]). The ART is a learning system that can make categories based on incremental learning of a kernel object. We have studied the radial basis ART (RB-ART) with a circle kernel for application to combinatorial optimization problems [4]. The PSO is a population-based paradigm for solving optimization problems inspired by flocking behavior of living beings [5]. Referring to their past history, the particles communicate to each other and try to find an optimal solution of an objective problem. Engineering applications are many, including design of signal processors, filters, and switching power converters [7] [9].

In order to consider the collaboration of the RB-ART and the PSO, we use an MSP for analysis of the Hénon map. The Hénon map is a simple nonlinear dynamical system that can generate various periodic/chaotic phenomena [10]. The Hénon map has been studied as an important example and has contributed to develop the study of chaos and bifurcation. We consider search of periodic points of the Hénon Map and the search problem is described as an MSP. In our algorithm, the PSO gives candidates of solutions which have no labels. Applying the candidates as inputs, the ART classifies the candidates, labels the categories, and clarifies the number of solutions. That is, our algorithm identifies all the solutions of the MSP automatically. Performing fundamental numerical experiments, the algorithm efficiency is investigated. Note that it is hard to clarify the number of solutions in the case of the number is unknown [11]. Note also that this is the first paper of collaboration of ART and PSO, and its application to analysis of dynamical systems.

C.K. Loo et al. (Eds.): ICONIP 2014, Part II, LNCS 8835, pp. 221–228, 2014.

2   Multi-Solution Problem

Here we define the objective MSP based on the Hénon map:

$$\begin{cases} x_1(n+1) = 1 - ax_1(n)^2 + x_2(n) \\ x_2(n+1) = bx_1(n) \end{cases} \text{ ab. } \boldsymbol{x}(n+1) = \boldsymbol{F}(\boldsymbol{x}(n)) \tag{1}$$

where $(x_1(n), x_2(n)) \equiv \boldsymbol{x}(n)$ is the 2-dimensional state variable vector at discrete time n. This is a typical example of discrete-time dynamical systems and can exhibit a variety of periodic/chaotic phenomena depending on the parameters (a, b) [10]. We define periodic points.

A point $\boldsymbol{p}$ is said to be a periodic point with period m if $\boldsymbol{p} = \boldsymbol{F}^m(\boldsymbol{p})$ and $\boldsymbol{p} \neq \boldsymbol{F}^l(\boldsymbol{p})$ for $0 \leq l < m$ where $\boldsymbol{F}^m$ is the m-fold composition of $\boldsymbol{F}$. A periodic point with period 1 is referred to as a fixed point.

Search of periodic points is a basic problem to analyze the dynamics. We define a basic function

$$G_m(\boldsymbol{x}) \equiv \| \boldsymbol{F}^m(\boldsymbol{x}) - \boldsymbol{x} \| \geq 0, \ \boldsymbol{x} \in S_A \equiv \{\boldsymbol{x} | X_{Lj} \leq x_j \leq X_{Rj}, \ j = 1, 2\} \tag{2}$$

where S_A is the search space. $\| \cdot \|$ denotes a distance and the Euclidean distance is used in this paper. $G_m(\boldsymbol{x}) = 0$ gives periodic points with period m and divisors of m. For simplicity, this paper considers a basic problem:

$$G_4(\boldsymbol{x}) = 0, \ \boldsymbol{x} \in S_A, \quad \text{for } a = 1, b = 0.3 \tag{3}$$

In this case, the Hénon map has 7 periodic points as shown by blue crosses in Fig. 1: four periodic points with period 4, two periodic points with period 2, and one fixed point. Eq. (3) has 7 solutions corresponding to the 7 periodic points. Our MSP is search and identification of all the solutions of Eq. (3) when the number of solutions is unknown.

3   Search Algorithm

The search algorithm consists of the PSO- and ART-subroutines. First, we define the PSO-subroutine. Let $\boldsymbol{p}_i^t \equiv (\boldsymbol{x}_i^t, \boldsymbol{v}_i^t)$ denote the i-th particle characterized by its position x_i^t and velocity v_i^t, where $i = 1 \sim N_p$ and N_p is the number of particles. In order to solve the two-dimensional MSP of Eq. (3), we consider the two dimensional search S_A in which the positions are potential solutions:

$$\boldsymbol{x}_i^t \equiv (x_{i1}^t, x_{i2}^t) \in S_A, \ \boldsymbol{v}_i^t \equiv (v_{i1}^t, v_{i2}^t) \tag{4}$$

The position is evaluated by the fitness function G_4. Here we introduce a threshold parameter C_t

$$0 \leq G_4(\boldsymbol{x}_i^t) < C_t, \ \boldsymbol{x}_i^t \in S_A \tag{5}$$

If this is satisfied, the particle position $\boldsymbol{x}_i^t$ is referred to as a solution candidate. Let $\boldsymbol{x}_i^P$ denote the personal best (Pbest) of the i-th particle that gives the smallest (the best) value of G_4 in its own past history. Let $\boldsymbol{x}_i^L$ denote the local best (Lbest) of the i-th particle that is the best of the Pbest in the neighbor of the i-th particle. Although the neighbor depends on the topology of the particle swarm and we adopt the ring topology in which the both sides particles construct the neighbor. It is known that, in the MSP, the ring topology can realize better performance than the complete graph. The PSO-subroutine is defined by the following 5 steps.

Step 1 (Initialization): Let search step $t = 0$. The particle positions $\boldsymbol{x}_i^0$ are assigned randomly S_A. Let $v_i^0 = 0$.

Step 2: Positions and velocities are updated.

$$x_{ij}^t \leftarrow x_{ij}^t + v_{ij}^t, \ v_{ij}^t \leftarrow wv_{ij}^t + c_1(x_{ij}^P - x_{ij}^t) + c_2(x_{ij}^L - x_{ij}^t) \tag{6}$$

Step 3: Pbest and Lbest are renewed.

$$\begin{aligned} \boldsymbol{x}_i^P &\leftarrow \boldsymbol{x}_i^t(t) \ \text{if } G_4(\boldsymbol{x}_i^t) < G_4(\boldsymbol{x}_i^P) \\ \boldsymbol{x}_i^L &\leftarrow \boldsymbol{x}_j^t(t) \ \text{if } G_4(\boldsymbol{x}_j^P) < G_4(\boldsymbol{x}_i^L) \ \text{for } \boldsymbol{x}_j^P \in B_i \end{aligned} \tag{7}$$

where B_i is the neighbor of the i-th particle.

Step 4: Let $t \leftarrow t + 1$, go to Step 2, and repeat until $t = t_{max}$.

Step 5: If $G_4(\boldsymbol{x}_j^P) < C_s$ then the Pbest x_i^P is declared as a solution candidate where C_s is a criterion for the solution candidates. Let N_s denote the number of solution candidates and let $\boldsymbol{X}_1 = (X_1, Y_1)$ to $\boldsymbol{X}_{N_s} = (X_{N_s}, Y_{N_s})$ be the solution candidates that will be inputs of the ART-subroutine.

Next, we define the ART-subroutine. Let $S_o \equiv S_A$ be the 2-dimensional input space. The ART map consists of categories the number of which is time variant. Let $\boldsymbol{X}_j = (X_i, Y_i)$ be the i-th input (i-th solution candidate) where $i = 1 \sim N_s$. Let N_c^τ denote the number of categories at discrete time τ. The j-th category $\boldsymbol{W}_j^\tau$ at discrete time τ is characterized by a circle at center $\boldsymbol{x}_j$ with radius r_j

$$\boldsymbol{W}_j^\tau = (\boldsymbol{x}_j^\tau, r_j^\tau), \ \boldsymbol{x}_j^\tau \equiv (x_j^\tau, y_j^\tau), \ j = 1 \sim N_c^\tau \tag{8}$$

The ART-subroutine is defined by the following 5 steps.

Step 1: Let $\tau = 0$, $N_c^\tau = 1(j = 1)$ and let $r_j^\tau = 0$. The category center $\boldsymbol{x}_j$ is initialized randomly in S_o.

Step 2 (winner selection): An input $\boldsymbol{X}_i$ is selected. If the input belongs to a category then go to Step 5. Otherwise, we find the winning category $\boldsymbol{W}_c$ that is the nearest to the input:

$$T_c = \min_j T_j, \ T_j = \| \boldsymbol{x}_j - \boldsymbol{X}_i \| - k r_j^\tau \tag{9}$$

where $k \in [-1, 1]$ is a selection parameter. If $T_c > \rho$ then go to Step 3 where ρ is the vigilance parameter. If $T_c \leq \rho$ then go to Step 4.

Step 3 (Birth of a new category): Let $N_c^\tau \leftarrow N_c^\tau + 1$ and $j \leftarrow j + 1$. A new category with radius 0, $\boldsymbol{W}_{N_c^\tau} = (X_{N_c^\tau}, Y_{N_c^\tau}, 0)$, is inserted and go to Step 5.

Step 4 (Category enlargement): The winning category $\boldsymbol{W}_c^\tau$ is enlarged such that the input $\boldsymbol{P}_i$ is on the border and go to Step 5.

Step 5: Let $\tau \leftarrow \tau + 1$, go to Step 2 and repeat until the maximum time limit τ_{max}.

After this subroutine, N_c at $\tau = \tau_{max}$ is the number of categories. If N_c is equal to the number of solutions N_s then the MSP is said to be solved successfully.

4   Numerical Experiments

We have applied the algorithm to the search problem defined by Eq. (3): search of 7 periodic points of the Hénon map. We select the vigilance parameter ρ and the number of particles N as control parameters. Other parameters are fixed after trial-and-errors:

$$N = 100, \ \omega = 0.7, \ c_1 = c_2 = 1.4, \ t_{max} = 100, \ C_s = 0.03$$
$$k = -0.5, \ \rho = 0.35, \ \tau_{max} = N_s.$$

In the ART-subroutine, all the solution candidates are input once and the order of application is randomized. Figures 1 and 2 show several snapshots in the successful and failure search processes, respectively. In the successful case, we can see that the PSO-subroutine has found solution candidates ($N_s = 69$) near all the seven regions; however, the algorithm has not recognized the seven regions yet. In the failure case, we can see that the PSO-subroutine has not found solution candidates near 7 solutions but near 6 solutions ($N_s = 67$). In both the cases, the ART-subroutine has constructed one category each in all the solution regions (7 categories in Fig. 1 and 6 categories in Fig. 2) . It suggests that the ART-subroutine can recognize the number and approximate value of solutions if the PSO-subroutine can find solution candidates near all the solutions.

Next, we have performed 100×100 trials for all the combinations of the parameter values

$$\text{PSO} : N \in \{60, 80, 100, 120, 140, 160\}$$
$$\text{ART} : \rho \in \{0.2, 0.24, 0.28, 0.32, 0.36, 0.4\}, \ k \in \{-1, -0.5, 0\}$$

The 100×100 means 100 different initial particle positions for the PSO and 100 different order of N_s inputs for the ART. The results of the success rate (SR) are summarized in Tables 1 to 3 where SR means rate of successful runs in all the trials. SR$> 50\%$ is achieved around $(k, N, \rho) = (-0.5, 140, 0.32)$ as suggested by bold figures. We have confirmed that almost all the failure runs are due to the failure of the PSO as suggested in Fig. 2. The ART-subroutine can identify almost all the solution regions given by the PSO if ρ is selected suitably as suggested in Figs. 1 and 2. Note that we have used a standard version of the local-best PSO without improvement. If we apply the ART-subroutine to some improved PSO, the algorithm efficiency must be improved.

Table 1. Success rate (SR) in 10000 trials for $k = -1$

$\rho \backslash N$	60	80	100	120	140	160
0.2	16.3	26.0	21.9	18.7	9.9	12.5
0.24	20.9	32.2	29.8	27.4	24.0	20.4
0.28	24.8	34.8	31.6	32.6	30.3	28.3
0.32	24.9	36.1	32.6	36.6	41.5	33.5
0.36	24.3	34.0	35.3	39.2	45.8	36.7
0.4	14.2	19.8	20.4	20.2	26.0	21.4

Table 2. Success rate (SR) in 10000 trials for $k = -0.5$

$\rho \backslash N$	60	80	100	120	140	160
0.2	18.9	31.9	26.9	26.3	23.4	21.8
0.24	23.8	36.0	31.5	30.0	36.7	28.5
0.28	25.0	36.2	33.4	36.3	48.8	35.4
0.32	25.5	35.2	34.2	39.5	**55.0**	38.7
0.36	22.3	32.6	34.3	39.2	**55.0**	38.0
0.4	8.3	13.8	17.3	18.6	26.0	16.7

Table 3. Success rate (SR) in 10000 trials for $k = 0$

$\rho \backslash N$	60	80	100	120	140	160
0.2	30.8	34.1	27.4	36.5	33.3	28.1
0.24	32.0	35.9	32.2	42.9	45.0	34.5
0.28	32.0	36.4	34.7	45.7	49.2	38.7
0.32	31.4	31.9	33.4	**50.2**	**56.4**	40.5
0.36	28.2	28.6	30.0	45.1	49.3	37.3
0.4	8.4	11.0	9.7	16.4	16.3	11.9

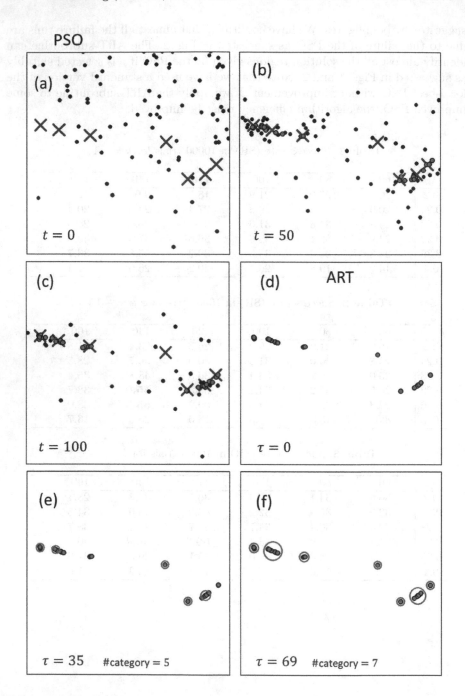

Fig. 1. Successful run. (a) to (c): PSO-subroutine. Black points denote particle positions and red points denote solution candidates ($N_s = 69$) that attain the criterion. (d) to (f): ART-subroutine. Green circles denote categories for solutions identification.

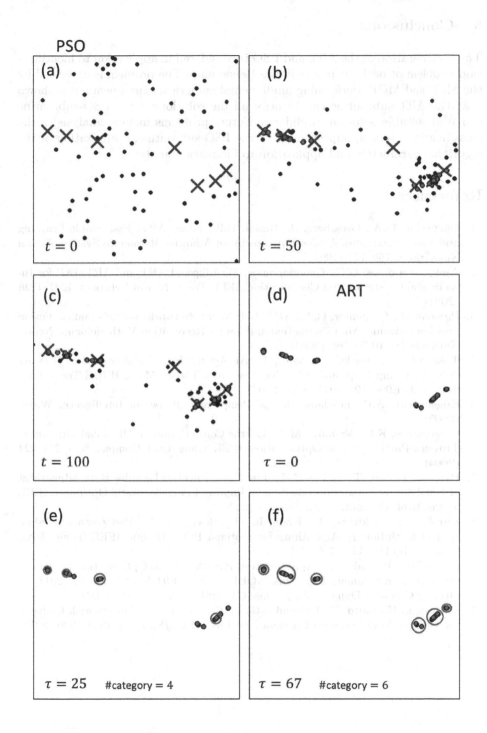

Fig. 2. Failure run. (a) to (c): PSO-subroutine ($N_s = 67$). (d) to (f): ART-subroutine.

5   Conclusions

The collaboration of the ART and PSO is considered in application to identification problem of periodic points in the Hénon map. The problem is described by the MSP and MOP. Performing fundamental numerical experiments, it is shown that the ART-subroutine can identify all the solutions if the PSO-subroutine can find suitable solution candidates. Future problems include analysis of the classification process, improvement of the PSO-subroutine, optimization of the algorithm parameters and application to bifurcation analysis.

References

1. Carpenter, G.A., Grossberg, S., Rosen, D.B.: Fuzzy ART: Fast Stable Learning and Categorization of Analog Patterns by an Adaptive Resonance System. Neural Networks 4, 759–771 (1991)
2. Anagnostopoulos, G.C., Georgiopoulos, M.: Ellipsoid ART and ARTMAP for Incremental Clustering and Classification. IEEE Trans. Neural Networks, 1221–1226 (2001)
3. Parsons, O., Carpenter, G.A.: ARTMAP Neural Networks for Information Fusion and Data Mining: Map Production and Target Recognition Methodologies. Neural Networks 16, 1075–1089 (2003)
4. Takanashi, M., Torikai, H., Saito, T.: An Approach to Collaboration of Growing Self-Organizing Maps and Adaptive Resonance Theory Maps. IEICE Trans. Fundamentals E90-A(9), 2047–2050 (2007)
5. Engelbrecht, A.P.: Fundamentals of Computational Swarm Intelligence. Willey (2005)
6. Parsopoulos, K.E., Vrahatis, M.N.: On the Computation of All Global Minimizers Through Particle Swarm Optimization. IEEE Trans. Evol. Comput. 8(3), 211–224 (2004)
7. Hsieh, S.-T., Sun, T.-Y., Lin, C.-L., Liu, C.-C.: Effective Learning Rate Adjustment of Blind Source Separation Based on an Improved Particle Swarm Optimizer. IEEE Trans. Evol. Comput. 12(2), 242–251 (2008)
8. Vural, R.A., Yildirim, T., Kadioglu, T., Basargan, A.: Performance Evaluation of Evolutionary Algorithms for Optimal Filter Design. IEEE Trans. Evol. Comput. 16(1), 135–147 (2012)
9. Matsushita, H., Saito, T.: Application of Particle Swarm Optimization to Parameter Search in Dynamical Systems. NOLTA, IEICE E94-N(10), 458–471 (2011)
10. Ott, E.: Chaos in Dynamical Systems. Cambridge Univ. Press, (1993)
11. Maruyama K., Saito, T.: Deterministic Particle Swarm Optimizers with Collision for Discrete Multi-Solution Problems. In: Proc. IEEE/SMC, pp.1335–1340 (2013)

Reconstructing Gene Regulatory Network with Enhanced Particle Swarm Optimization

Rezwana Sultana[1], Dilruba Showkat[2], Mohammad Samiullah[1],
and Ahsan Raja Chowdhury[1]

[1] Dept. of Computer Science and Engg., University of Dhaka, Dhaka, Bangladesh
[2] Dept. of Computer Science and Engg., BRAC University, Dhaka, Bangladesh
rezwana.sultana@gmail.com, dilruba@bracu.ac.bd,
samiullah@cse.univdhaka.edu, farhan717@univdhaka.edu

Abstract. Inferring regulations among the genes is a well-known and significantly important problem in systems biology for revealing the fundamental cellular processes. Although computational models can be used as tools to extract the probable structure and dynamics of such networks from gene expression data, capturing the complex nonlinear system dynamics is a challenging task. In this paper, we have proposed a method to reverse engineering Gene Regulatory Network (GRN) from microarray data. Inspired from the biologically relevant optimization algorithm 'Particle Swarm Optimization' (PSO), we have enhanced the PSO incorporating two genetic algorithm operators, namely crossover and mutation. Furthermore, Linear Time Variant (LTV) Model is employed to modeling the GRN appropriately. In the evaluation, the proposed method shows superiority over the state-of-the-art methods when tested with synthetic network, both for the noise free and noise in data. The strength of the proposed method has also been verified by analyzing the real expression data set of SOS DNA repair system in *Escherichia coli*.

Keywords: Genetic Network, Microarray, Linear Time Variant.

1 Introduction

After the introduction of genomic sequence, the gene expression data are provided on the basis of DNA microarrays (DNA chips). Primarily, these data are used for revealing the mechanism of gene regulations and protein interactions. Therefore, researchers are trying to uncover the underlying regulation circuits using model-based identification methods. The GRN reconstruction is the way of modeling genetic network from the genetic information. The study of GRN is made much easier because of the microarray technology, where expression levels for a very large number of genes can be measured simultaneously.

There are two important aspects of reverse engineering GRN: i) selecting an appropriate model ii) selecting an efficient optimizer. Among numerous linear formalism, Linear Time Variant (LTV) model is the only one that can discover non-linear interactions among genes similar to the non-linear models, even with

C.K. Loo et al. (Eds.): ICONIP 2014, Part II, LNCS 8835, pp. 229–236, 2014.

noisy gene expression profiles. Hence, in this research, LTV model [1] is used to represent the GRN during the reverse engineering process. On the other hand, Particle Swarm Optimization (PSO), an emerging evolutionary computation proposed by Kennedy and Eberhart [2], has been successfully applied on many real life applications e.g., job shop scheduling, vehicle routing, object recognition etc. Due to PSO's promising performance in the same kind of problems solved by GAs, it has received enormous attention to the researchers. Furthermore, PSO has overcome the difficulties of GA and possesses additional facilities over GA, as mentioned in [2]. In this paper, we have proposed an enhanced PSO algorithm, or PSO*, by including the GA operators: Crossover and Mutation, to further strengthen the efficiency of PSO for solving complex and multimodal problems.

Rest of the paper is organized as follows: Section 2 discusses about various modeling techniques along with standard PSO algorithm. The proposed PSO* is thoroughly discussed in Section 3. Reverse engineering GRN with a LTV model and the proposed PSO* is discussed in Section 4. Evaluation of the proposed method is shown in Section 5 along with discussions. Section 6 concludes the paper.

2   Literature Review

A variety of models have been proposed in literature for GRN reconstruction where each of them have their own strength and weakness. Among the available models, linear and non-linear models [1], Boolean networks [3], Bayesian and dynamic Bayesian networks [4], the system of differential equations are well-known and widely used for their own merits and application domain [5–7]. However, estimating the appropriate model parameters is the most challenging part of the reconstruction procedure and normally formulated as optimization problems. Based on gene expression microarray time series data, these optimization techniques enable genetic network architectures to be reconstructed. To optimize the network parameters and to capture the dynamics in gene expression data, Evolutionary Computation (EC) is becoming a popular approach that refers to a class of population-based stochastic optimization search algorithms [8]. Among the varieties of ECs, evolutionary optimization is becoming prevalent in solving critical and real-world problems in industry, medicine, and defense [8]. Amongst the EAs that aim to learn the parameters of a GRN, GA, Differential Evolution (DE), Particle Swarm Optimization (PSO) are suitable for handling complex, multidimensional and multi-modal problems.

Particle Swarm Optimization (PSO) [2], a relatively newer addition to a class of population based search technique for solving numerical optimization problems, is used in this research for training the LTV model parameters. PSO is a simple and elegant nature-inspired metaphor for heuristic search that has achieved good performance across a broad range of optimization problems. PSO has been used extensively in the last decade on various real applications, and successfully solved many difficult optimization problems. It is very efficient because of the simple conceptual framework.

In PSO, a solution is represented as a particle, and a swarm of particles is considered as a population where each particle has two main properties: position and velocity. Each particle moves to a new position using the velocity, and once a new position is reached, the best position of each particle and the best position of the swarm are updated as needed. The velocity of each particle is then adjusted based on the experiences of the particle. The process is repeated until a stopping criterion is met. The computational steps for PSO are given below:

$$vid = w * vid + c1 * rand * (lbest - xid) + c2 * rand * (gbest - xid) \quad (1)$$
$$xid = vid + xid \quad (2)$$

The first part of Eqn. (1) represents the inertia of the previous velocity, the second part is the cognition part which implies the personal thinking of the particle, the third part represents the cooperation among particles and is therefore named as the social component [2]. Acceleration constants $c1$, $c2$ [2] and inertia weight w [9] are predefined by the user and $r1$, $r2$ are the uniformly generated random numbers in the range [0, 1]. For basic PSO algorithm, the number of required constants is three: $c1$, $c2$ and w.

3   PSO*: The Improved PSO

We propose an enhanced PSO algorithm which is applied to reconstructing GRN from microarray data. The traditional PSO is extended incorporating the mechanism of simulated binary crossover (SBX) [10] and polynomial mutation [11]. SBX is an effective operator used in well-known Genetic Algorithm (GA), while mutation, another important operator of GA, is used to maintain the diversity in the population. Previously, SBX has been used in the optimization proposed in [12] to infer the target GRN from gene expression data sets. To the best of our knowledge, both SBX and polynomial mutation are introduced with PSO as the first endeavor in literature.

3.1   Simulated Binary Crossover (SBX)

The crossover operator is considered to be the key operator of GA. Since PSO has no genetic operator, the key idea of this work was to investigate the performance of PSO with the presence of genetic operator(s). A number of crossover operators exist in the GA literature (e.g., binary coded crossover (BCC), SBX etc), where the search power to reach to the solution differs from one crossover to another. BCC is successful at solving problems with discrete search space shown in [10], whereas, SBX operator is successful at continuous search space. We describe the SBX [10] with the following equations:

$$c_{1,k} = 0.5 * [(1-\beta_k)p_{1,k} + (1+\beta_k)p_{2,k}], c_{2,k} = 0.5 * [(1+\beta_k)p_{1,k} + (1-\beta_k)p_{2,k}] \quad (3)$$

where, $c_{i,k}$ is the ith child with the kth component, $p_{i,k}$ is the selected parent and $\beta_k (\geq 0)$ is a sample from a random number generated having the density:

$$C(\beta) = \begin{cases} 0.5 * (n+1)\beta^n & \text{if } 0 \leq \beta \leq 1 \\ 0.5 * (n+1) * (1/\beta^{n+2}) & \text{if } \beta > 1 \end{cases} \quad (4)$$

This distribution can be obtained from a uniformly sampled random number u between $(0, 1)$. n is the distribution index for crossover. β is calculated using the following equations:

$$\beta(u) = \begin{cases} (2u)^{1/(n+1)} & \text{if } u \leq 0.5 \\ (2(1-u))^{-1/(n+1)} & \text{otherwise } u > 0.5 \end{cases} \tag{5}$$

Thus, the basic steps to create two children solutions ($c_{1,k}$ and $c_{2,k}$) from two parent solutions ($p_{1,k}$ and $p_{2,k}$) are i) select a random number $u \ \epsilon \ [0, 1)$ ii) calculate β using Eqn. (5), and iii) compute children solutions using Eqn. (3).

3.2 Polynomial Mutation

We define the polynomial mutation [11] using the following equation

$$c_k = p_k + (p_k^u - p_k^l)\delta_k \tag{6}$$

Here, c_k is the child and p_k is the parent with p_k^u being the upper bound on the parent component, p_k^l is the lower bound and δ_k is small variation which is calculated from a polynomial distribution by using the following equation:

$$\delta_k = \begin{cases} (2r_k)^{1/(n_m+1)} - 1 & \text{if } r_k < 0.5 \\ 1 - [2(1-r_k)]^{1/(n_m+1)} & \text{otherwise } r_k \geq 0.5 \end{cases} \tag{7}$$

Here, r_k is an uniformly sampled random number between $(0, 1)$ and n_m is mutation distribution index.

4   Reverse Engineering GRNs with PSO* and LTV Model

The Linear Time Variant (LTV) model is used in this research work which can represent a much wider class of time-series data than linear time-invariant(LTI) models. This model is very simple and can be applied to very large scale networks. All the methods based on LTI-based models have fundamental limitations to infer certain non-linear network interactions. However, the LTV-based model can provide meaningful insight in capturing the nonlinear dynamics of genetic networks [1] and revealing genetic regulatory interactions. Let, the total regulatory input to a gene-i of an N gene GRN is:

$$Z_i(t) = \sum_{j=1}^{N} W_{i,j}(t)X_j(t) \tag{8}$$

Here, $Z_i(t)$ is the total regulatory input to gene-i, at time t, $X_j(t)$ is the expression level or mRNA level of gene-i, at time t and $W_{i,j}(t)$ indicates the strength (weight coefficient) of the influence of gene-j on the regulation of gene-i at time t. The weight matrix W provides all the essential interactions among genes in the underlying network. A positive value of $W_{i,j}$ represents gene-j is inducing gene-i whereas a negative value is the sign of repression. On the other hand, a zero value on W implies that gene-j does not influence the transcription of gene-i. According to [1] $W_{i,j}(t)$ can be written as a finite sum of Fourier series given as the following:

$$W_{i,j}(t) = \alpha_{i,j} \sin(\omega_i t + \psi_{i,j}) + \beta_{i,j}, \forall i \tag{9}$$

Here, $\alpha_{i,j}, \beta_{i,j}, \psi_{i,j}, \omega_i$ are the constants to be determined $\forall i, j = 1 \ldots N$. These are the model parameters for the LTV model. Thus, the LTV model is defined by the parameters set $\{\alpha, \beta, \psi, \omega\}$. $\beta_{i,j}$ represents the linear part of the interactions and the sinusoidal term approximates the nonlinear terms in the interactions. The response of the gene-i to the regulatory input is the expression level at time $t + 1$, i.e. $X_i(t + 1)$. Thus, the value of $X_i(t + 1)$ is obtained by normalizing Z_i using the following "squashing" function:

$$X_i(t + 1) = (1 + \exp^{-Z_i(t)})^{-1} \tag{10}$$

where $X_{i,j}$ lies between 0 and 1. We have used MSE based fitness function, also used in [1]. Now, we propose the reverse engineering GRN using PSO* and LTV model in the following algorithm (Algorithm 1).

Algorithm 1. *PSO*_with_LTV(N)*

Input: $N-$ Number of genes
Output: Optimal LTV parameters

1: Initialize the Swarm
2: Initialize the PSO parameters ($c1$, $c2$, w)
3: **While** *until stopping criteria is reached* **do**
4:      **For** *each particle* **do**
5:          Update Velocity and Location using Eqn. (1) & (2), respectively
6:          Apply SBX and mutation on the updated location using Eqn. (3)
                                              & (6), respectively
7:          Calculate fitness value
8:          Keep the location with better fitness
9:          Update personal best fitness
10:         Update global best fitness
11:     **End For**
12: **End While**

---

5   Experimental Results and Discussions

The proposed PSO* is tested with a synthetic network (with and without noise in microarray data) and a real-life network called *SOS DNA* repair network of the *Escherichia coli*. The proposed method is also evaluated with state-of-the-art optimization methods, e.g., Genetic Algorithm (GA), standard Particle

Swarm Optimization (PSO) and PSO incorporated with Differential Evolution (PSODE). For the synthetic network, we have calculated the well-known performance metrics sensitivity (S_n) and specificity (S_p) to verify the efficacy of the proposed method, while known inferred regulations are reported for the real-life network. In LTV, each individual is represented by the parameter set $\{\alpha, \beta, \psi, \omega\}$. The parameters for each of the individuals are set randomly within a given range [1] as follows: α = [-5.0, 5.0], β = [-4.0, 4.0], ψ and ω = [-90.0, 90.0]. We have used the common parameters for all training algorithms, such as number of generations G=5000, population size=50. For PSO, PSODE and PSO* the common parameters are $c1$=$c2$=2 and w. w changes dynamically with respect to each iteration. The crossover probability and mutation probability for PSODE and PSO* are set to 0.9 and 1/n respectively, where n is the number of decision variables, which is 80.

We have used the 5-gene synthetic network of Kikuchi et al. [13], and using their given parameter set, the 5 sets synthetic target data sets were generated, each having 11 samples. For the noisy data, 5% and 10% Gaussian noise has been added at the time data were generated. After inferring the target networks using the proposed method, the sensitivity and specificity values are shown in Table 1. We note that, the specificity values are also satisfactory for all three situations (i.e., 0 noise, 5% noise, and 10% noise). Next, we consider the well-studied SOS DNA repair network within *Escherichia coli* (*E. coli*). The entire DNA repair system of *E.coli* involves more than 100 genes [4], however, only its 30 genes contribute towards key regulations at the transcription level. We make use of the expression data set collected by Ronen *et al.* [14], which contains information about 8 genes namely *uvrD*, *lexA*, *umuD*, *recA*, *uvrA*, *uvrY*, *ruvA*, and *polB*. The data sets are obtained from four different experiments under various UV light conditions, with the gene expression levels being measured at 50 instants evenly spaced at a 6-minute interval. As the exact ground truth for this network is not precisely known, it is not possible to calculate the two performance metrics, i.e., sensitivity, specificity. However, from the functional description of each gene in the original paper [14], it is generally recognized that suppression of all genes from *lexA* and activation to *lexA* from *recA* are considered as true regulations. From the four data-set, Exp3 and Exp4 are used for our experiment.

Table 1. Summary of the investigations on the synthetic data

	Noiseless	5% Noisy	10% Noisy
S_n	1.000	1.000	1.000
S_p	0.923	0.846	0.769

The expression profiles for 5 genes SOS DNA repair system, shown in Figure 1, implies that the proposed PSO* infers the target expressions very closely for all the 5 genes. Table 2 shows the summary of the known regulations and predicted regulations identified by the proposed algorithm from SOS DNA repair network.

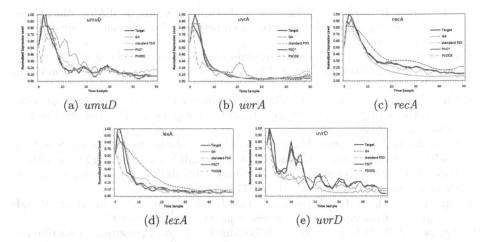

(a) *umuD*   (b) *uvrA*   (c) *recA*

(d) *lexA*   (e) *uvrD*

Fig. 1. Target and inferred expression profiles SOS DNA repair system

Table 2. Predicted gene regulations for SOS DNA repair system

Gene	Inferred Regulations	References
uvrD	*uvrD* ⊣ *uvrD*, *lexA* → *uvrD*	[15–17]
lexA	*uvrD* → *lexA*, *lexA* ⊣ *lexA*, *umuD* → *lexA*, *recA* ⊣ *lexA*	[16–18]
umuD	*lexA* ⊣ *umuD*, *recA* ⊣ *umuD*	[15, 18, 19]
recA	*lexA* ⊣ *recA*, *umuD* → *recA*, *recA* ⊣ *recA*	[17, 18]
uvrA	*lexA* ⊣ *uvrA*, *uvrA* ⊣ *uvrA*	[15]

6 Conclusion

This paper presents an effective use of Simulated Binary Crossover and Polynomial Mutation operators in Particle Swarm Optimization algorithm for inferring
the transcriptional regulations in a biochemical network represented based on
the Linear Time Variant model. The proposed is tested with widely used 5-gene
synthetic network, both in noise free and noisy environment, and to reconstruct
the real-life *E. coli* network. For both the evaluations, we observe superior performance of the proposed method over state-of-the-art methods.

References

1. Kabir, M., Noman, N., Iba, H.: Reverse engineering gene regulatory network from
   microarray data using linear time-variant model. BMC Bioinformatics 11(S-1), 56
   (2010)
2. Kennedy, J., Eberhart, R., et al.: Particle swarm optimization. In: IEEE International Conference on Neural Networks, vol. 4, pp. 1942–1948 (1995)
3. Wang, R.-S., Saadatpour, A., Albert, R.: Boolean modeling in systems biology: an
   overview of methodology and applications. Physical Biology 9(5), 055001 (2012)

4. B.E. Perrin, L. Ralaivola, A. Mazurie, S. Bottani, J. Mallet, F. d'Alche-Buc: Gene networks inference using dynamic Baycsian networks. Bioinformatics 19(suppl. 2), ii138–ii148 (2003)
5. Chowdhury, A.R., Chetty, M.: An improved method to infer gene regulatory network using S-system. In: IEEE Congress on Evolutionary Computation, pp. 1012–1019 (2011)
6. Chowdhury, A.R., Chetty, M., Vinh, N.X.: Adaptive regulatory genes cardinality for reconstructing genetic networks. In: IEEE Congress on Evolutionary Computation, pp. 1–8 (2012)
7. Chowdhury, A.R., Chetty, M., Vinh, N.X.: Incorporating time-delays in S-system model for reverse engineering genetic networks. BMC Bioinformatics 14, 196 (2013)
8. Fogel, D.B.: The advantages of evolutionary computation. In: Biocomputing and Emergent Computation, pp. 1–11 (1997)
9. Shi, Y., Eberhart, R.: A modified particle swarm optimizer. In: IEEE World Congress on Computational Intelligence, pp. 69–73. IEEE (1998)
10. Agrawal, R.B., Deb, K., Agrawal, R.B.: Simulated binary crossover for continuous search space (1994)
11. Raghuwanshi, M., Kakde, O.: Survey on multiobjective evolutionary and real coded genetic algorithms. In: Asia Pacific Symposium on Intelligent and Evolutionary Systems, pp. 150–161 (2004)
12. Showkat, D., Kabir, M.: Inference of genetic networks using multi-objective hybrid spea2+ from microarray data. In: ICCI*CC, pp. 195–202 (2013)
13. Kikuchi, S., Tominaga, D., Arita, M., Takahashi, K., Tomita, M.: Dynamic modeling of genetic networks using genetic algorithm and S-system. Bioinformatics 19(5), 643–650 (2003)
14. Ronen, M., Rosenberg, R., Shraiman, B.I., Alon, U.: Assigning numbers to the arrows: Parameterizing a gene regulation network by using accurate expression kinetics. National Academy of Sciences 99(16), 10555–10560 (2002)
15. Kimura, S., Shiraishi, Y., Hatakeyama, M.: Inference of genetic networks using linear programming machines: application of a priori knowledge. In: International Joint Conference on Neural Networks, pp. 694–701 (2009)
16. Cho, D.Y., Cho, K.H., Zhang, B.T.: Identification of biochemical networks by s-tree based genetic programming. Bioinformatics 22, 1631–1640 (2006)
17. Kimura, S., Sonoda, K., Yamane, S., Maeda, H., Matsumura, K., Hatakeyama, M.: Function approximation approach to the inference of reduced ngnet models of genetic networks. BMC Bioinformatics 9(1), 23 (2008)
18. Bansal, M., Gatta, G.D., di Bernardo, D.: Inference of gene regulatory networks and compound mode of action from time course gene expression profiles. Bioinformatics 22(7), 815–822 (2006)
19. Gardner, T.S., di Bernardo, D., Lorenz, D., Collins, J.J.: Inferring genetic networks and identifying compound mode of action via expression profiling. Science 301(5629), 102–105 (2003)

Neural Network Training by Hybrid Accelerated Cuckoo Particle Swarm Optimization Algorithm

Nazri Mohd Nawi[1], Abdullah Khan[1], M.Z. Rehman[1], Maslina Abdul Aziz[2],
Tutut Herawan[3,4], and Jemal H. Abawajy[2]

[1]Faculty of Computer Science and Information Technology
Universiti Tun Hussein Onn Malaysia
86400 Parit Raja, Johor, Malaysia
[2]School of Information Technology, Deakin University, Waurn Ponds,
Geelong, VIC, Australia
[3]University of Malaya, 50603 Kuala Lumpur, Malaysia
[4]AMCS Research Center, Yogyakarta, Indonesia
nazri@uthm.edu.my, hi100010@siswa.uthm.edu.my,
zrehman862060@gmail.com, tutut@um.edu.my,
{mabdula,jemal.abawajy}@deakin.edu.au

Abstract. Metaheuristic algorithm is one of the most popular methods in solving many optimization problems. This paper presents a new hybrid approach comprising of two natures inspired metaheuristic algorithms i.e. Cuckoo Search (CS) and Accelerated Particle Swarm Optimization (APSO) for training Artificial Neural Networks (ANN). In order to increase the probability of the egg's survival, the cuckoo bird migrates by traversing more search space. It can successfully search better solutions by performing levy flight with APSO. In the proposed Hybrid Accelerated Cuckoo Particle Swarm Optimization (HACPSO) algorithm, the communication ability for the cuckoo birds have been provided by APSO, thus making cuckoo bird capable of searching for the best nest with better solution. Experimental results are carried-out on benchmarked datasets, and the performance of the proposed hybrid algorithm is compared with Artificial Bee Colony (ABC) and similar hybrid variants. The results show that the proposed HACPSO algorithm performs better than other algorithms in terms of convergence and accuracy.

Keywords: Metaheuristic algorithm, Neural network, Cuckoo search, Particle Swarm Optimization, Optimization.

1 Introduction

Artificial Neural Network (ANN) is one of the widely used techniques for dataset classification in data mining. It is also known as a powerful prediction tool and has been applied in a wide variety of areas such as engineering, finance, military, telecommunication and so on. In general, the ANN presents and emulates the actual biological nervous systems with layers of interconnected individual artificial neurons. The ANN are structured by their number of layers (architecture), types of topology

C.K. Loo et al. (Eds.): ICONIP 2014, Part II, LNCS 8835, pp. 237–244, 2014.
© Springer International Publishing Switzerland 2014

and learning process. One of the most popular supervised network classifiers is the Multilayer Perceptron (MLP) or the feed-forward ANN [1]. The MLP consists of three layers architecture. The first layer receives input, the second layer is the hidden layer, and the third layer produces output. Each of these layers contains nodes. Every node in these layers is connected to every adjacent layer [2]. There are different types of datasets: the training (supervised learning) and testing data set. The trained data sets results provide the flexibility to the learning process and can be used to make projections by using the pre-set target value. The MLP utilizes the Error Back Propagation (EBP) learning algorithm that is mainly used for solving classification tasks by sorting patterns in the datasets. The Back Propagation Neural Network (BPNN) [3-5] learning algorithm uses the gradient-descent technique in adjusting weights and biases for training an ANN in many domains. Along various advantages of ANN there are some downsides. Unfortunately, the EBP learning algorithm is not efficient enough in handling such large learning problems. The classifiers tend to make wrong selection of the characteristics for a specific task. Therefore, the whole network will be affected. It is also not easy to find the appropriate ANN architectures. Also, as ANN generate multifaceted error-planes with multiple local minimum, the BPNN fell prey to local minima instead of converging to global minimum [6].

A number of research studies have attempted to overcome these problems by introducing different techniques to analyze the performance of the standard steepest descent algorithm. These methods include the gradient descent with adaptive learning rate, gradient descent with momentum, gradient descent with momentum and adaptive learning rate, the resilient algorithm and standard steepest descent [6-8]. However, one limitation of gradient-descent technique is that it requires a differentiable neuron transfer function. Also, as neural networks generate complex error surfaces with multiple local minimum, the BPNN fall into local minima instead of a global minimum [7, 9-10].

In this paper, we solved the above optimization problems by improving the accuracy and decrease number of training errors with a fast convergence rate by using hybrid metaheuristic algorithms. Among the various metaheuristic populations based search algorithms for training BPNN are the Artificial Bee Colony (ABC) [1], Genetic Algorithm (GA) [11], Particle Swarm Optimization-Back Propagation (PSO-BP) [12], Ant Colony Optimization (ACO) [13] Cuckoo Search (CS) Algorithm [14-21], Bat Algorithm [22] and so on.

This paper proposes an improved Accelerated Particle Cuckoo Swarm Optimization (HACPSO) algorithm. In the proposed HACPSO algorithm the communication ability for the cuckoo bird has been added. The Accelerated Particle Swarm Optimization (APSO) algorithm searches a better place with the best nest and share the information with Cuckoo bird. The convergence behavior and performance of the proposed HACPSO on classification datasets is analyzed. The results are compared with artificial bee colony using BPNN algorithm, and similar hybrid variants.

The remaining paper is organized as follows: Section 2 describes the proposed method. Section 3 describes result and discussion. Finally, the paper is concluded in the Section 4.

2   The Proposed HACPSO Algorithm

In this section, we examine the detail of the proposed hybrid algorithm. As mentioned above, the nature of cuckoo birds is that instead building their own nests, they lay their eggs in the nest of the other host bird. If the host bird detects an unknown egg, it will either throw the egg away or simply abandon its nest and build a new nest elsewhere. Thus, the cuckoo birds are always looking for a better place in order to reduce the chance of their eggs to be discoursed. In this proposed hybrid algorithm, the CS initializes the population for the nest, and randomly selects the best nest via levy flight. In addition, the communication ability for the cuckoo birds has been added, where the APSO search better place that has the best nest, and share the information with cuckoo search. Then cuckoo search selected the best nest among all via levy flight using the Equation (1) as follow

$$x_i^{t+1} = x_i^t + \alpha \oplus levy(\lambda) + v_i^{t+1}, \tag{1}$$

where v_i^{t+1} is the velocity vector generated from Equation (2) as follow

$$v_i^{t+1} = v_i^t + \alpha \varepsilon_n + \beta(g^* - x_i^t), \tag{2}$$

where v_i^t and x_i^t be the position vector for the particle i , α is the learning parameter or accelerating constant, ε_n is random vector which is draw from $N(0, 1)$. The pseudo code of the proposed algorithm is given as follow.

Generate initial population of N host nest   for *i= 1,...,n*
While (f$_{min}$<MaxGeneration) or (stop criterion)
Get a cuckoo bird randomly
Choose randomly a nest j among n.
Move cuckoo bird using Equation (1) and (2)
Calculate the fitness   F$_i$
IfF$_i$ > F$_j$
Then, Replace j by the new solution.
End if
A fraction (probability (pa)) of worse nest are abandoned and new ones are built.
Keep the best solutions (or nest with quality solutions).
Rank the solutions and find the current best.
End while

3   Results and Discussion

3.1   Preliminary Study

The performance of the proposed Hybrid Cuckoo Particle Swarm Optimization (HACPSO) algorithm was tested on benchmark classification datasets. The proposed HACPSO algorithm is compared Artificial Bee Colony (ABC) with Levenberg Marquardt algorithm (ABCLM), Artificial Bee Colony Back Propagation (ABCBP) algorithm and standard BPNN based on Mean Squared Error (MSE), no of epochs, accuracy and CPU time. The minimum error for benchmark classification datasets is set to 0.00001 and a total of 20 trials is run for each case. The network results are stored in a separate file for each trial.

3.2   7 Bit Parity Dataset

The first dataset used for this experimentation is the 7 Bit Parity datasets that consist a mixture of even and odd parity tuples. The network structural design has 7 inputs 5 hidden neurons in the hidden layer, and 1 output in the output layer. It has forty connection weights and six biases. Due to the structure of this dataset, the learning process for a neural network is quite complex and slow. Tables 1 shows the CPU time, number of epochs, the MSE, and accuracy for the 7 Bit Parity dataset.

Table 1. CPU Time, Epochs, MSE Accuracy, and Standard deviation for 7-bit Parity dataset

Algorithm	BPNN	ABCBP	ABCLM	HACPSO
CPUTIME	22.07	183.3	134.88	409.24
EPOCHS	1000	1000	1000	1000
MSE	0.263	0.217	0.083	0.001
SD	0.014	0.008	0.0124	0.0001
Accuracy (%)	85.12	82.12	69.137	95.09

From the Table 1, it is clear that the proposed HACPSO converged to a global minimum with an MSE of 0.001 with an average accuracy of 95.09. While, the standard deviation (SD) for all the MSE's was recorded to be a mere 0.0001for all 20 trials.

ABCLM, BPNN and ABCBP algorithms failed to provide a smaller MSE within 1000 epochs throughout the entire cycle of 20 trials. Figure 1 illustrates the convergence performance of the proposed HACPSO and other algorithms used in this study.

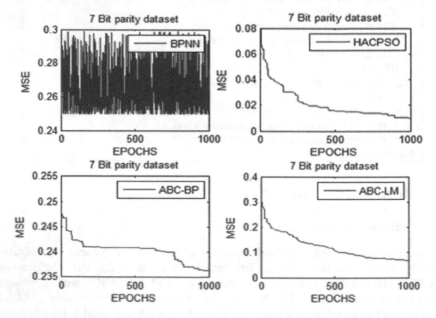

Fig. 1. Convergence Performance for7 Parity Bit Classification dataset on selected algorithms

3.3   IRIS Classification Dataset

The Iris classification datasets is one of the best pattern recognition multivariate data-set used since 1936. Iris dataset consists of 150 samples from three Iris species, i.e. Iris setosa, Iris virginica and Iris versicolor. The classification of IRIS involves the length and width of sepal and petals from each sample of three selected species. The selected network configuration for Iris classification dataset comprises of 4 inputs, 5 hidden, and 3 outputs nodes. A total of 95 instances are used for training the algorithms, while the rest is used for testing purpose. Table 2 illustrates the MSE, SD, Epochs, CPU time and accuracy of the proposed HACPSO algorithms when compared with simple BPNN and other hybrid variants.

Table 2. CPU Time, Epoch, MSE Accuracy, and Standard deviation for IRIS dataset

Algorithm	BPNN	ABCBP	ABCLM	HACPSO
CPUTIME	28.47	156.43	171.52	163.36
EPOCH	1000	1000	1000	1000
MSE	0.312	0.155	0.058	0.0009
SD	0.022	0.023	0.005	0.0001
Accuracy (%)	87.19	86.88	79.559	99.01

Based on the simulation results in Table 2, it is clear that the proposed HACPSO model has better performance in terms of MSE, SD, and accuracy. HACPSO converged with the MSE of 0.0001 and an accuracy of 99.01, compared to other algorithms. Figure 2 shows the MSE convergence performance of the proposed HACPSO and the other compared models used in this study. It can be clearly seen that the proposed HACPSO algorithm has far better performance than the BP, ABCBP, and ABCLM, in terms of MSE.

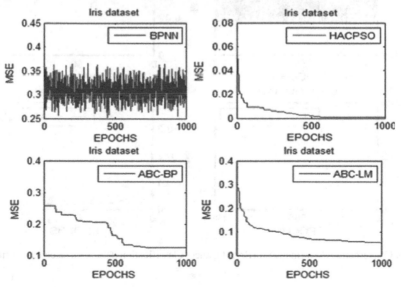

Fig. 2. Convergence Performance forIRIS Classification dataset on selected algorithms

3.4 Breast Cancer (Wisconsin) Classification Dataset

This dataset consists of a total 699 instances. The selected network architecture used for the breast cancer classification problem consists of 9 inputs nodes, 5 hidden nodes and 2 output nodes. Table 3, shows the simulation results of all the algorithms used in this study. From the Table 3, one can easily understand that the proposed HACPSO algorithm achieves better results than the ABCBP, ABCLM, and BPNN in terms of MSE, SD, epochs and accuracy. The proposed HACPSO converged to global minima with an MSE of 0.0004, a 9.9E-05 SD and 98.08 percentile of accuracy. While the comparison algorithms such as BPNN, ABCBP, and ABCLM, showed bigger MSE's; as they converged to global minima with 0.271, 0.184, and 0.0139, MSE's respectively. The SD values of 0.05, 0.0011, and 0.01 respectively, which show less performances than the proposed methods in term of MSE, SD, and accuracy.

Table 3. CPU Time, Epochs, MSE Accuracy, and Standard deviation for Breast Cancer dataset

Algorithm	BPNN	ABCBP	ABCLM	HACPSO
CPUTIME	95.46	1482.9	1880.64	1335.83
EPOCH	1000	1000	1000	1000
MSE	0.271	0.184	0.0139	0.0004
SD	0.01	0.05	0.0011	9.90E-05
Accuracy (%)	90.71	92.02	93.831	98.08

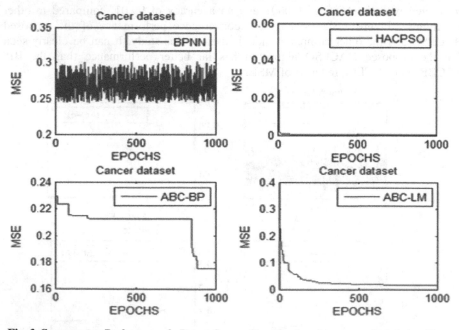

Fig. 3. Convergence Performance forBreast Cancer Classification dataset on selected algorithms

From the Table 3, it is clearly seen that proposed HACPSO outperforms the other algorithms in terms of MSE, and convergence accuracy. Figure 3 represents the MSE convergence of the proposed HACPSO algorithm compared with the previous methods.

4 Conclusion

This paper presents a combination of Cuckoo Search (CS) and Accelerated particle swarm optimization (APSO) for training neural network. The approach is proposed to achieve faster convergence rate, minimize the training error, and to increase the accuracy. In standard CS, cuckoo bird gets anew place by levy flight. In the proposed HACPSO algorithm, the communication ability for the cuckoo birds has been added by introducing APSO, thus giving cuckoo bird the ability to search the best nest that has higher survival chances. The experiments were carried out on benchmarked datasets. Finally, as a comparison between the performance of the proposed HACPSO algorithm with Artificial Bee Colony (ABC) and other similar hybrid variants, it can be concluded that in terms of convergence rate and accuracy, the HACPSO algorithm outperforms the other individual and hybrid variants algorithms.

Acknowledgement. This work is supported by University of Malaya High Impact Research Grant no vote UM.C/625/HIR/MOHE/SC/13/2 from Ministry of Higher Education Malaysia.

References

1. Kumbhar, S.K.P.Y.: Use of artificial bee colony (ABC) algorithm in artificial neural network synthesis. International Journal of Advanced Engineering Sciences and Technologies 11, 162–171 (2011)
2. Nawi, N.M., Rehman, M.Z.: Improving the Accuracy of Gradient Descent Back Propagation Algorithm (GDAM) on Classification Problems. International Journal on New Computer Architectures and Their Applications (IJNCAA) 1, 838–847 (2011)
3. Abid, S., Fnaiech, F., Najim, M.: Fast feedforward training algorithm using a modified form of the standard back propagation algorithm. IEEE Transactions on Neural Networks 12, 424–430 (2001)
4. Gori, A.T.M.: On the problem of local minima in back-propagation. IEEE Transactions on Pattern Analysis and Machine Intelligence 14, 76–86 (1992)
5. Yu, X., Onder Efe, M., Kaynak, O.: A general back propagation algorithm for feed forward neural networks learning. IEEE Transactions on Neural Networks 13, 251–259 (2002)
6. Nawi, M.N., Ransing, R.S., Hamid, A.: BPGD-AG: A New Improvement of Back-Propagation Neural Network Learning Algorithms with Adaptive Gain. Journal of Science and Technology 2 (2011)
7. Nawi, N.M., Ransing, R.S., Salleh, M.N.M., Ghazali, R., Hamid, N.A.: An improved back propagation neural network algorithm on classification problems. In: Zhang, Y., Cuzzocrea, A., Ma, J., Chung, K.-i., Arslan, T., Song, X. (eds.) DTA and BSBT 2010. CCIS, vol. 118, pp. 177–188. Springer, Heidelberg (2010)

8. Nawi, N.M., Ghazali, R., Salleh, M.N.M.: The development of improved back-propagation neural networks algorithm for predicting patients with heart disease. In: Zhu, R., Zhang, Y., Liu, B., Liu, C. (eds.) ICICA 2010. LNCS, vol. 6377, pp. 317–324. Springer, Heidelberg (2010)
9. Khan, A., Nawi, N.M., Rehman, M.Z.: A New Levenberg-Marquardt based Back-propagation Algorithm trained with Cuckoo Search. In: ICEEI 2013 Procedia Technology 8C, pp. 18–24 (2013)
10. Nawi, N.M., Khan, A., Rehman, M.Z.: A New Cuckoo Search Based Levenberg-Marquardt (CSLM) Algorithm. In: Murgante, B., Misra, S., Carlini, M., Torre, C.M., Nguyen, H.-Q., Taniar, D., Apduhan, B.O., Gervasi, O. (eds.) ICCSA 2013, Part I. LNCS, vol. 7971, pp. 438–451. Springer, Heidelberg (2013)
11. Sexton, J.J.R.D.R.: Optimization of neural networks:A comparative analysis of the genetic algorithm and simulated annealing. European Journal of Operational Research 114, 589–601 (1999)
12. Zhang, J.Z.T.L.J., Lyu, M.: A hybrid particle swarm optimization back propagation algorithm for neural network training. Applied Mathematics and Computation 185, 1026–1037 (2007)
13. Blum, C., Socha, K.: Training feed-forward neural networks with ant colony optimization. In: An Application to Pattern Classification, pp. 233–238 (2005)
14. Nawi, N.M., Khan, A., Rehman, M.Z.: A New Back-Propagation Neural Network Optimized with Cuckoo Search Algorithm. In: Murgante, B., Misra, S., Carlini, M., Torre, C.M., Nguyen, H.-Q., Taniar, D., Apduhan, B.O., Gervasi, O. (eds.) ICCSA 2013, Part I. LNCS, vol. 7971, pp. 413–426. Springer, Heidelberg (2013)
15. Deb, S., Yang, X.S.: Cuckoo search via Lévy flights. In: Proceeings of World Congress on Nature & Biologically Inspired Computing, pp. 210–214 (2009)
16. Milan Tuba, M.S., Stanarevic, N.: Modified cuckoo search algorithm for unconstrained optimization problems. In: Proceedings of the European Computing Conference, pp. 263–268 (2011)
17. Yang, X.S., Deb, S.: Engineering Optimisation by Cuckoo Search. Int. J. of Mathematical Modelling and Numerical Optimisation 1, 330–343 (2010)
18. Yang, X.-S.: Nature-Inspired Metaheuristic Algorithms. Luniver Press (2008)
19. Yang, X.-S., Deb, S., Fong, S.: Accelerated Particle Swarm Optimization and Support Vector Machine for Business Optimization and Applications. In: Fong, S. (ed.) NDT 2011. CCIS, vol. 136, pp. 53–66. Springer, Heidelberg (2011)
20. Yang, X.-S.: Engineering Optimization: An Introduction with Metaheuristic Applications. John Wiley & Sons (2010)
21. Walton, S., Hassan, O., Morgan, K., Brown, M.: Modified cuckoo search: A new gradient free optimisation algorithm. J. Chaos, Solitons & Fractals 44, 710–718 (2011)
22. Nawi, N.M., Rehman, M.Z., Khan, A.: A New Bat Based Back-Propagation (BAT-BP) Algorithm. In: Świątek, J., Grzech, A., Świątek, P., Tomczak, J.M. (eds.) Advances in Systems Science. AISC, vol. 240, pp. 395–404. Springer, Heidelberg (2014)

An Accelerated Particle Swarm Optimization Based Levenberg Marquardt Back Propagation Algorithm

Nazri Mohd Nawi[1], Abdullah Khan[1], M.Z. Rehman[1], Maslina Abdul Aziz[2],
Tutut Herawan[3], and Jemal H. Abawajy[2]

[1]Faculty of Computer Science and Information Technology
Universiti Tun Hussein Onn Malaysia
86400 Parit Raja, Johor, Malaysia
[2]School of Information Technology, Deakin University
Waurn Ponds, Geelong, VIC, Australia
[3]University of Malaya, 50603 Kuala Lumpur, Malaysia
[4]AMCS Research Center, Yogyakarta, Indonesia
nazri@uthm.edu.my, hi100010@siswa.uthm.edu.my,
zrehman862060@gmail.com, tutut@um.edu.my,
{mabdula,jemal.abawajy}@deakin.edu.au

Abstract. The Levenberg Marquardt (LM) algorithm is one of the most effective algorithms in speeding up the convergence rate of the Artificial Neural Networks (ANN) with Multilayer Perceptron (MLP) architectures. However, the LM algorithm suffers the problem of local minimum entrapment. Therefore, we introduce several improvements to the Levenberg Marquardt algorithm by training the ANNs with meta-heuristic nature inspired algorithm. This paper proposes a hybrid technique Accelerated Particle Swarm Optimization using Levenberg Marquardt (APSO_LM) to achieve faster convergence rate and to avoid local minima problem. These techniques are chosen since they provide faster training for solving pattern recognition problems using the numerical optimization technique.The performances of the proposed algorithm is evaluated using some bench mark of classification's datasets. The results are compared with Artificial Bee Colony (ABC) Algorithm using Back Propagation Neural Network (BPNN) algorithm and other hybrid variants. Based on the experimental result, the proposed algorithms APSO_LM successfully demonstrated better performance as compared to other existing algorithms in terms of convergence speed and Mean Squared Error (MSE) by introducing the error and accuracy in network convergence.

Keywords: Artificial Neural Networks, Particle Swarm Optimization, Levenberg Marquardt Back Propagation, Meta-heuristic optimization, Nature inspired algorithms.

1 Introduction

Artificial Neural Networks (ANN) is one of the best approaches in Machine Learning. An ANN is modeled and designed based on the actual human brain concept with

C.K. Loo et al. (Eds.): ICONIP 2014, Part II, LNCS 8835, pp. 245–253, 2014.

interconnected neurons. It simulates the exact way of a processing information. Unlike other conventional techniques, the key element of an ANN is known as supervised learning, in which a set of input/output complex patterns is analyzed and classified [1-7]. As a multilayer perceptron feed-forward network, an ANN is used for random nonlinear function approximation and information processing which other techniques do not have [8]. There are many different types of ANNs depending on their structure and training model, but we will only focus on the most basic one is the Back Propagation Neural Network (BPNN) [9]. A Back-Propagation (BP) algorithm is designed to reduce error between the actual output and the desired output and adjust the ANN weights (and biases) of the network in a gradient descent manner.

However, despite of its reputation, simple architecture and easy to understand learning process, the BPNN has few limitations. The limitations are the risk of getting trapped in a local minima [10-12], possibility of overshooting the minima of the error surface [13-16], slow rate of convergence and so on. Therefore to get faster and more efficient trainings process, second order learning algorithms have to be used.

The Levenberg Marquardt (LM) algorithm is one of the most successful algorithm in speeding up the convergence rate of the ANN with Multilayer Perceptron (MLP) architectures [17]. It is ranked as one of the most efficient training algorithms for small and medium sized patterns. The LM algorithm was developed only for layer-by-layer ANN topology, which is far from optimal [18]. It combines Gauss–Newton Algorithm (GNA) with gradient descent [19]. It inherits speed from Newton method but it also has the convergence capability of steepest descent method. It suits specially in training neural network in which the performance index is calculated in Mean Squared Error (MSE). Unfortunately, along with its benefits, LM algorithm can also suffer the problem of local minimum entrapment [20-23].

Alternatively, there is no one size fits all solution exist. Researchers have been trying to find the optimal solution by proposing different approaches and robust algorithm. Therefore, our focus is finding the best and efficient algorithm(s) of optimizing the neural network using genetic algorithms. In order to overcome the drawback of the Levenberg Marquardt Back Propagation (LMBP), the solution will be converged to the meta-heuristic algorithm. Among the examples of the meta-heuristic nature inspired algorithms are Evolutionary Algorithm (EA) [24], Ant Colony Optimization (ACO) Algorithm [25], Artificial Bee Colony (ABC) Algorithm [26], Genetic Algorithm (GA) [27], Cuckoo Search (CS) Algorithm [6], Bat Algorithm (BA) [28], Particle Swarm Optimization (PSO) Algorithm [29].

Specifically for this paper, The LMBP is combined with APSO, which was originally proposed by Yang in 2008 [30]. In this paper, the convergence behavior and performance of the proposed APSO_LM algorithm is analyzed on selected benchmark classification datasets obtained from UCI machine learning repository. This method is based on the imitation of the social behavior of bird flocking and fish schooling. The LM and scaled conjugate gradient based back-propagation training algorithms are used to train the network. These two training algorithms have been chosen since they provide faster training for solving pattern recognition problems using the numerical optimization technique [13]. Their classification performances with different network architecture are reported in the result section. The results are compared with Artificial Bee Colony (ABC) Algorithm using BPNN algorithm, and other similar hybrid variants. The objective of the optimization is to minimize the computational cost and to accelerate the learning process using a hybridization method.

The outline of this article is as follows. In section 2, the proposed APSO_LM algorithm is explained, and simulation results are discussed in Section 3. Finally, the conclusion of this work is presented in Section 4.

2   The Proposed APSO_LM Algorithm

The APSO is a population based optimization global search algorithm, which has strong ability to find global optimistic result, the LM algorithm has the strong ability to find local optimistic result, but its ability to find the global optimistic result is weak. By combining the APSO with LM, a new algorithm referred to as APSO_LM hybrid algorithm is formulated. Similar to many other meta-heuristic algorithms, APSO starts with a random initial population. The searching process is also started from initialization a group of random particle. First all particle are update according to the Equation (5), and (6), until a new generation set of particle are generated, and then those new particle are used to search the global best position in solution space. Finally the LM algorithm is used to search around the global optimum. In this way the hybrid algorithm may find as optimum more quickly.

In the proposed Accelerated Particle Swarm Optimized Levenberg-Marquardt (APS_LM) algorithm, each best particle or solution represents a possible solution (i.e., the weight space and the corresponding biases for NN optimization in this study) to the considered problem and the size of a population represents the quality of the solution. The initialization of weights is compared with output and the best weight cycle is selected by APSO. The APSO will continue searching until the last cycle to find the best weights for the network. The main idea of this combined algorithm is that APSO algorithm is used at the beginning stage of searching for the optimum to select the best weights. Then, the training process is continued with the LM algorithm using the best particle as weights of APSO algorithm. The LM algorithm interpolate between the Newton method and gradient descent method. The pseudo code for the ASPO-LM algorithm is given as follow.

1. **Initialized** APSO population size, dimensions, and NN structure.
2. Evaluate each initialized particle is fitness value, and x_i is set as the position of the current particle, while g^* is set as the best position of initialized particle.
3. Load training data
4. While (MSE<Stopping criteria)
5. Pass the current best particle as weights to the network.
6. Present all inputs to the network and compute the corresponding network outputs and errors using Equation (1) over all inputs. And compute sum of square of error over all input.

$$E(t) = \frac{1}{2}\sum_{i=1}^{N} e_i^{\,2}(t), \tag{1}$$

7. The sensitivity of one layer is calculated from its previous one and the calculation of the sensitivity start from the last layer of the network and move backward.

8. Compute the Jacobin matrix using Equation (2).

$$J(t) = \begin{bmatrix} \dfrac{\partial v_1(t)}{\partial t_1} & \dfrac{\partial v_1(t)}{\partial t_2} & \cdots\cdots & \dfrac{\partial v_1(t)}{\partial t_n} \\[2ex] \dfrac{\partial v_2(t)}{\partial t_1} & \dfrac{\partial v_2(t)}{\partial t_2} & \cdots\cdots & \dfrac{\partial v_2(t)}{\partial t_n} \\ & \cdot & & \\ & \cdot & & \\ & \cdot & & \\ \dfrac{\partial v_n(t)}{\partial t_1} & \dfrac{\partial v_n(t)}{\partial t_2} & \cdots\cdots & \dfrac{\partial v_n(t)}{\partial t_n} \end{bmatrix} \tag{2}$$

9. Solve Equation (3) to obtain ∇t .

$$\nabla t = -[J^T(t)J(t) + \mu I]^{-1} J(t)e(t) \tag{3}$$

10. Recomputed the sum of squares of errors using Equation (3) if this new sum of squares is smaller than that computed in Step 6, then reduce μ by $\lambda=10$, update weight using $w(k+1) = w(k) - \nabla w$ and go back to Step 6. If the sum of squares is not reduced, then increase μ by $\lambda=10$ and go back to Step 8.

11. The algorithm is assumed to have converged when the norm of the gradient Equation (4) is less than some prearranged value, or when the sum of squares has been compact to some error goal.

$$\nabla E(t) = J^T(t)e(t) \tag{4}$$

12. Chose the particle with the best fitness value of all the particle as gbest
13. For each particle
14. Calculate particle velocity according Equation (5)

$$v_i^{t+1} = v_i^t + \alpha\varepsilon_n + \beta(g^* - x_i^t), \tag{5}$$

15. Update particle position according Equation (6)

$$x_i^{t+1} = x_i^t + v_i^{t+1}. \tag{6}$$

End
16. APSO keep on calculating the best possible weight at each epoch until the network is converged.
End while

3   Results and Discussion

Basically, the main focus of this paper is to compare the performance of different algorithms introducing the error and accuracy in network convergence. Some simulation results, tools and technologies, network topologies, testing methodology and the classification problems used for the entire experimentation will be discussed further in the this section.

3.1   Wisconsin Breast Cancer Classification Problem

This problem tried to diagnosis of Wisconsin breast cancer by trying to classify a tumor as either benign or malignant based from continues clinical variable. This dataset consist of 9 inputs and 2 outputs with 699 instances. The input attribute are, for instance, the clump thickness, the uniformity of cell size, the uniformity of cell shape, the amount of marginal adhesion, the single epithelial cell size, frequency of bare nuclei, bland chromatin, normal nucleoli, and mitoses. The selected network architecture is used for the breast cancer classification problem is consists of 9 inputs nodes, 5 hidden nodes and 2 output nodes.

Table 1. Summary of algorithms performance for breast cancer classification problem

Breast Cancer Benchmark Classification Problem			
Algorithms	Accuracy	MSE	SD
ABC-BP	92.02	0.184	0.459
ABC-LM	93.83	0.0139	0.001
ABCNN	88.96	0.014	0.0002
BPNN	90.71	0.271	0.017
APSO_LM	99.9	2.40E-06	2.80E-06

Table 1, illustrate that the proposed algorithm (APSO_LM), shows superior performance than BPNN, ABC-BP, ABCNN, and ABC-LM. The proposed models such as APSO_LM, have achieve small MSE (2.4E-06) and SD (2.8E-06) with 99.95 percent of accuracy. While the other algorithms such as ABCNN, BPNN, ABC-BP, and ABC-LM fall behind of the proposed algorithms with large MSE (0.014, 0.271, 0.184, and 0.013), and SD (0.0002, 0.017, 0.459, and 0.001) and low accuracy. Similarly, Figure1 shows the performances of MSE convergence for the used algorithms. The proposed APSO_LM algorithm convergences only in 3 epochs. While the other algorithm take more epochs for their convergence. From the simulation results its can easily understand that the proposed algorithms such as APSO_LM shows better performance than the BPNN, ABC-BP, and ABC-LM, algorithms in term of MSE, SD and accuracy.

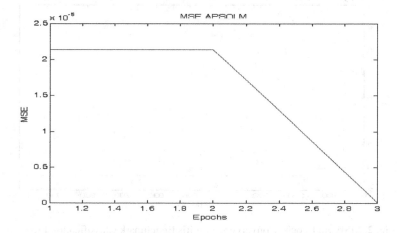

Fig. 1. MSE via Epochs Convergence for breast cancer classification problem

3.2   IRIS Classification Problem

The Iris classification dataset was created by Fisher. Who used it to demonstrate the values of differentiate analysis. This is maybe the best famous database to be found in the pattern recognition literature. There were 150 instances, 4 inputs, and 3 outputs in this dataset. The classification of Iris dataset involving the data of petal width, petal length, sepal length, and sepal width into three classes of species, which consist of Iris Santos, Iris Vermicular, and Iris Virginia. The selected network structure for Iris classification dataset is 4-5-3. Which consist of 4 inputs nodes, 5 hidden nodes and 3 outputs nodes. 75 instances are used for training dataset and the rest as for testing dataset.

Table 2. Summary of algorithms performance for Iris Benchmark Classification Problem

Iris Benchmark Classification Problem			
Algorithms	Accuracy	MSE	SD
ABC-BP	86.87	0.155	0.022
ABC-LM	79.55	0.058	0.0057
ABCNN	80.23	0.048	0.004
BPNN	87.19	0.311	0.022
APSO_LM	99.99	1.21E-05	1.84E-06

Table 2 shows the comparison performances of the proposed algorithm such as APSO_LM, with the BPNN, ABCNN, ABC-BP, ABC-LM algorithms in term of MSE, SD, and accuracy. From the table 2 it's clear that the proposed APSO_LM models have better performances achieved less MSE, SD, and high accuracy than the BPNN, ABCNN, ABC-BP, ABC-LM algorithms. Meanwhile, the Figure 2 illustrates the MSE's convergence performances of the algorithm. Form these figure it's clear that the proposed algorithms show high performances than the other algorithms in term of MSE, Standard deviation (SD), and accuracy.

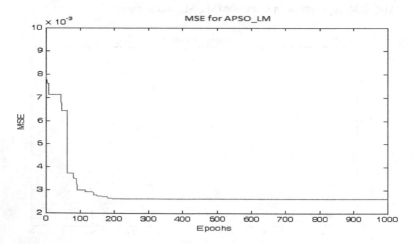

Fig. 2. MSE via Epochs Convergence on Iris Benchmark Classification Problem

4   Conclusion

APSO algorithm is one of the latest addition among the meta-heuristic nature inspired algorithms, which provide derivative-free solutions to solve complex problems. This paper studies the data classification problem using the dynamic behavior of LMBP, trained by nature inspired meta-heuristic APSO algorithm, in-order to achieve fast convergence rate and to avoid local minima problem. The performances of the proposed models APSO_LM is compared with the Artificial Bee Colony using BPNN algorithm, and other hybrid variants. Specifically, 7-Bit Parity, and some selected benchmark classification datasets are used for training and testing the network. The simulation results show that the proposed APSO_LM is far better than the previous methods in terms of convergence rate, and achieved higher accuracy and less MSE on all the designated datasets.

Acknowledgement. This work is supported by University of Malaya High Impact Research Grant no vote UM.C/625/HIR/MOHE/SC/13/2 from Ministry of Higher Education Malaysia.

References

1. Zheng, H., Meng, W., Gong, B.: Neural Network and its Application on Machine fault Diagnosis. In: ICSYSE 1992, pp. 576–579 (1992)
2. Kosko, B.: Neural Network and Fuzzy Systems, 1st edn. Prentice Hall of India (1992)
3. Basheer, I.A., Hajmeer: Artificial Neural Networks: fundamentals, computing, design and application. Journal of Microbiological Methods 43(1), 3–31 (2000)
4. Krasnopolsky, V.M., Chevallier: Some Neural Network applications in environmental sciences. Part II, advancing computational efficiency of environmental numerical models. Neural Networks 16(3), 335–348 (2003)
5. Coppin, B.: Artificial Intelligence Illuminated. Jones and Bartlet illuminated Series, USA, pp. 291–324 (2004)
6. Nawi, N.M., Khan, A., Rehman, M.Z.: A new cuckoo search based levenberg-marquardt (CSLM) algorithm. In: Murgante, B., Misra, S., Carlini, M., Torre, C.M., Nguyen, H.-Q., Taniar, D., Apduhan, B.O., Gervasi, O. (eds.) ICCSA 2013, Part I. LNCS, vol. 7971, pp. 438–451. Springer, Heidelberg (2013)
7. Nawi, N.M., Khan, A., Rehman, M.Z.: A new back-propagation neural network optimized with cuckoo search algorithm. In: Murgante, B., Misra, S., Carlini, M., Torre, C.M., Nguyen, H.-Q., Taniar, D., Apduhan, B.O., Gervasi, O. (eds.) ICCSA 2013, Part I. LNCS, vol. 7971, pp. 413–426. Springer, Heidelberg (2013)
8. Contreras, J., et al.: ARIMA models to predict next-day electricity prices. IEEE Transactions on Power Systems 18, 1014–1020 (2003)
9. Leung, C.T., Chow, T.: A hybrid global learning algorithm based on global search and least squares techniques for back propagation networks. In: International Conference on Neural Networks, pp. 1890–1895 (1997)
10. Zhang, J.-R., et al.: A hybrid particle swarm optimization–back-propagation algorithm for feedforward neural network training. Applied Mathematics and Computation 185, 1026–1037 (2007)

11. Ahmed, W.A.M., Saad, E.S.M., Aziz, E.S.: Modified Back Propagation Algorithm for Learning Artificial Neural Networks. In: Proceedings of the Eighteenth National Radio Science Conference, pp. 345–352 (2001)
12. Nawi, N.M., et al.: Countering the Problem of Oscillations in Bat-BP Gradient Trajectory by Using Momentum. In: Proceedings of the First International Conference on Advanced Data and Information Engineering (DaEng 2013), pp. 103–110. Springer Singapore (2014)
13. Jin, W., et al.: The improvements of BP neural network learning algorithm. In: Proceedings of 5th International Conference on Signal Processing WCCC-ICSP 2000, vol. 3, pp. 1647–1649 (2000)
14. Nawi, N.M., Ransing, R.S., Salleh, M.N.M., Ghazali, R., Hamid, N.A.: An Improved Back Propagation Neural Network Algorithm on Classification Problems. In: Zhang, Y., Cuzzocrea, A., Ma, J., Chung, K.-i., Arslan, T., Song, X. (eds.) DTA and BSBT 2010. CCIS, vol. 118, pp. 177–188. Springer, Heidelberg (2010)
15. Mohd Nawi, N., et al.: BPGD-AG: A New Improvement Of Back-Propagation Neural Network Learning Algorithms With Adaptive Gain. Journal of Science and Technology 2 (2011)
16. Nawi, N.M., Ghazali, R., Salleh, M.N.M.: The Development of Improved Back-Propagation Neural Networks Algorithm for Predicting Patients with Heart Disease. In: Zhu, R., Zhang, Y., Liu, B., Liu, C. (eds.) ICICA 2010. LNCS, vol. 6377, pp. 317–324. Springer, Heidelberg (2010)
17. Hagan, M.T., Menhaj, M.B.: Training feedforward networks with the Marquardt algorithm. IEEE Transactions on Neural Networks 5, 989–993 (1994)
18. Wilamowski, B., et al.: Neural network trainer with second order learning algorithms. In: 11th International Conference on Intelligent Engineering Systems, INES 2007, pp. 127–132 (2007)
19. Lourakis, M.I.: A brief description of the Levenberg-Marquardt algorithm implemented by levmar (2005)
20. Xue, Q., et al.: Improved LMBP algorithm in the analysis and application of simulation data. In: 2010 International Conference on Computer Application and System Modeling (ICCASM), vol. 6, pp. 545–547 (2010)
21. Yan, J., et al.: Levenberg-Marquardt algorithm applied to forecast the ice conditions in Ningmeng Reach of the Yellow River. In: Fifth International Conference on Natural Computation, ICNC 2009, pp. 184–188 (2009)
22. Ozturk, C., Karaboga, D.: Hybrid artificial bee colony algorithm for neural network training. In: 2011 IEEE Congress on Evolutionary Computation (CEC), pp. 84–88 (2011)
23. Nawi, A.K.N.M., Rehman, M.Z.: A New Levenberg-Marquardt based Back-propagation Algorithm trained with Cuckoo Search. In: ICEEI 2013 Procedia Technology 8C, pp. 18–24 (2013)
24. Bäck, T., Schwefel, H.-P.: An overview of evolutionary algorithms for parameter optimization. Evolutionary Computation 1, 1–23 (1993)
25. Blum, C., Socha, K.: Training feed-forward neural networks with ant colony optimization: An application to pattern classification. In: Fifth International Conference on Hybrid Intelligent Systems, HIS 2005, p. 6 (2005)
26. Karaboga, D., Akay, B., Ozturk, C.: Artificial bee colony (ABC) optimization algorithm for training feed-forward neural networks. In: Torra, V., Narukawa, Y., Yoshida, Y. (eds.) MDAI 2007. LNCS (LNAI), vol. 4617, pp. 318–329. Springer, Heidelberg (2007)
27. Montana, D.J., Davis, L.: Training Feedforward Neural Networks Using Genetic Algorithms. In: IJCAI, pp. 762–767 (1989)

28. Nawi, N.M., Rehman, M.Z., Khan, A.: A New Bat Based Back-Propagation (BAT-BP) Algorithm. In: Świątek, J., Grzech, A., Świątek, P., Tomczak, J.M. (eds.) Advances in Systems Science. AISC, vol. 240, pp. 395–404. Springer, Heidelberg (2014)
29. Zhang, J.Z.T.L.J., Lyu, M.: A hybrid particle swarm optimization back propagation algorithm for neural network training. Applied Mathematics and Computation 185, 1026–1037 (2007)
30. Yang, X.-S.: Nature-inspired metaheuristic algorithms. Luniver Press (2010)

Fission-and-Recombination Particle Swarm Optimizers for Search of Multiple Solutions

Takumi Sato and Toshimichi Saito

Hosei University, Koganei, Tokyo, 184–8584 Japan
tsaito@hosei.ac.jp

Abstract. This paper presents the fission-and-recombination particle swarm optimizer (FRPSO) and its application to search of periodic points of a nonlinear dynamical system. The search problem is translated into a multi-solution problem evaluated in a multi-objective problem. The FRPSO is based on the ring topology. The FRPSO is effective to escape from trap of partial/local solutions and to find all the solutions. Performing basic numerical experiments, the algorithm efficiency is investigated.

Keywords: particle swarm optimizers, multi-solution problems, multi-objective problems, Hénon map.

1 Introduction

The particle swarm optimizer is a population-based paradigm for solving optimization problems inspired by flocking behavior of living beings [1]-[3]. The particles correspond to potential solutions and construct a swarm. Referring to their past history, the particles communicate to each other and try to find an optimal solution of an objective problem. The PSO is simple in concept, is easy to implement and has been applied to optimization problems in various systems, e.g., signal processors, artificial neural networks, and power electronics [4]-[8].

This paper presents the fission-and-recombination particle swarm optimizer (FRPSO) and its application to search of periodic points of the Hénon map [9]. This map can exhibit various interesting bifurcation phenomena [10] and search of periodic points is the first step to analyze the phenomena. In our algorithm, the search problem is translated into a multi-solution problem (MSP [11]-[13]) evaluated in a multi-objective problem (MOP). The MOP consists of plural cost functions and logical operation. The FRPSO is based on the ring topology and has a lifetime parameter that controls timing of fission-and-recombination. If the parameter is selected suitably, the FRPSO is effective to escape from traps of partial/local solutions and to find all the solutions of the MSP. Performing basic numerical experiments, the algorithm efficiency is investigated.

Note that the FRPSO can realize a global search. Although there exist several tools for search of periodic points of dynamical systems, the tools employ the Newton-Raphson method whose initial value setting is trial-and-errors. The tools have been developed for the local search and are not available for the global

C.K. Loo et al. (Eds.): ICONIP 2014, Part II, LNCS 8835, pp. 254–262, 2014.

search [10]. This is the first paper of application of PSO with variable number of particles to combination problem of MSP and MOP in continuous search space.

2 The Multi-Solution Problem

The Hénon map is described by

$$\begin{cases} x_1(n+1) = 1 - ax_1^2(n) + x_2(n) \\ x_2(n+1) = bx_1(n) \end{cases} \text{ ab. } \boldsymbol{x}(n+1) = \boldsymbol{F}_H(\boldsymbol{x}(n)) \tag{1}$$

where n denotes discrete time and $(x_1(n), x_2(n)) \equiv \boldsymbol{x}(n)$. Depending on the parameters a and b, this map can exhibit various chaotic/periodic phenomena: The Hénon map is an important example of nonlinear dynamical systems [9]. We define the MSP for this map. Replacing $\boldsymbol{F}_H$ with an N-dimensional map, the definition can be generalized. A point $\boldsymbol{p}$ is said to be a periodic point with period m if $\boldsymbol{p} = \boldsymbol{F}_H^m(\boldsymbol{p})$ and $\boldsymbol{p} \neq \boldsymbol{F}_H^l(\boldsymbol{p})$ for $0 \leq l < m$ where $\boldsymbol{F}_H^m$ is the $m-$ fold composition of $\boldsymbol{F}_H$. Search of desired periodic points is basic to analyze bifurcation phenomena. First, we define the basic function

$$G_m(\boldsymbol{x}) \equiv \|\boldsymbol{F}_H^m(\boldsymbol{x}) - \boldsymbol{x}\| \geq 0, \ \boldsymbol{x} \in S_A \equiv \{\boldsymbol{x} \mid x_j \in [X_{Lj}, \ X_{Rj}]\} \tag{2}$$

where $j = 1, 2$ and S_A is the search space. $\|\cdot\|$ denotes distance and the Euclidean distance is used in this paper. If m is not a prime number, we have the prime factorization: $m = m_1^{n_1} \times \cdots \times m_K^{n_K}$ where $m_k \geq 2$ is the $k-th$ submultiple and $k = 1 \sim K$. The periodic points with period m are given by the solutions (minima) of

$$G_m(\boldsymbol{x}) = 0, \ G_{m/m_k}(\boldsymbol{x}) \neq 0, \ k = 1 \sim K, \ \boldsymbol{x} \in S_A \tag{3}$$

In general, Eq. (3) has multiple-solutions. Let N_s denote the number of solutions. For simplicity, we fix parameters $a = 1$ and $b = 0.3$. In this case, the map has

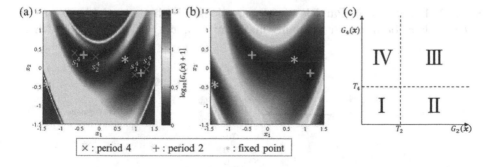

Fig. 1. (a) Contour map of G_4 of the Hénon map. (b) Contour map of G_2 of the Hénon map. (c) Regions I to IV on the evaluation plane.

four periodic points with period 4 that are solutions of

$$G_4(x) = 0, \ G_2(x) \neq 0, x \in S_A \tag{4}$$

where $X_{Li} = -X_{Ri} = 1.5$ and $i = 1, 2$. Figure 1 (a) and (b) show contour maps of G_4 and G_2, respectively. Search of he four solutions ($N_s = 4$) is the objective MSP. The solutions are denoted by s_1^4, s_2^4, s_3^4 and s_4^4 hereafter.

3   Fission and Recombination PSO

Here we present the FRPSO where the number and topology can vary depending on the situation of the search process. The FRPSO has two kinds of particle swarms: one main-swarm and its sub-swarms. Depending on the search situation, a sub-swarm can be generated from the main-swarm (fission) and the sub-swarm can be return into the main-swarm (recombination). The fission and recombination are controlled by the life-time parameters.

First, we define the algorithm for the main swarm (FRPSO-main) for the MSP (4) of two fitness functions G_4 and G_2. Replacing (G_4, G_2) with $(G_m, G_{m/m_k})$, the algorithm can be generalized. Let P_i^t be the i-th particle at a discrete time t and let N^t denote the number of particles that is time-variant. This particle is characterized by the position $x_i^t \equiv (x_{i1}^t, x_{i2}^t)$ and the velocity $v_i^t \equiv (v_{i1}^t, v_{i2}^t)$. The particles are updated based on the personal best position (Pbest, x_i^P) and local best position (Lbest, x_i^L).

The Pbest evaluation is based upon two fitness functions (G_4, G_2) as defined afterward. In usual PSO, the evaluation is based on a single fitness function. The Lbest is the best of Pbest within the neighbor. Let $V(i)$ be the neighbor of the i-th particle position. The neighbor depends on the topology of particles. For simplicity, we adopt the ring topology that is more suitable than the complete graph in the MSP. As shown in Fig. 2, a particle and both sides particles construct the neighbor: $V(i) = \{x_{i-1}^t, x_i^t, x_{i+1}^t\}$ ($i \bmod N^t$). The i-th particle has a life time parameter L_i^t (time-variant) that controls the fission. The FRPSO-main is defined by the following 6 steps.

Step M1: Let $t = 0$. Particle positions, velocities, Pbests, and Lbests are all initialized: x_i^t be assigned randomly in S_A, $v_i^t = 0$, $x_i^P = x_i, and x_i^L = x_i^t$ where

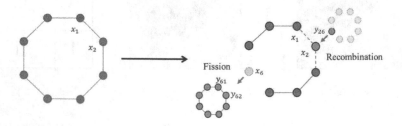

Fig. 2. Ring topology, fission and recombination

$i = 1 \sim N$. The life-time parameter L_i^t is also initialized. Let L_0 denote the initial value.

Step M2: The particle position x_i^t is declared as an approximate solution if

$$G_4(x_i^t) \leq T_4, \ G_2(x_i^t) \geq T_2 \tag{5}$$

where T_4 and T_2 are thresholds for the approximation. For convenience, we divide the G_4-vs-G_2 plane into four regions as shown in Fig. 1 (c): I: $G_4(x_i^t) \leq T_4$, $G_2(x_i^t) < T_2$; II: Eq. (3); III: $G_4(x_i^t) > T_4$, $G_2(x_i^t) < T_2$; IV: $G_4(x_i^t) > T_4$, $G_2(x_i^t) \geq T_2$.

Step M3: Pbests and Lbests are updated in the two cases.
Case 1: $G_4(x_i^t) > T_4$ OR $G_2(x_i^t) \geq T_2$ (Region II $\cup$ III $\cup$ IV in Fig. 1(c).)

$$x_i^P = \begin{cases} x_i^t & \text{if } G_4(x_i^t) < G_4(x_i^P) \\ x_i^P & \text{otherwise} \end{cases} \quad x_i^L = \begin{cases} x_c^t & \text{if } G_4(x_c^t) < G_4(x_i^L) \\ x_i^L & \text{otherwise} \end{cases} \tag{6}$$

where $G_4(x_c^t)$ is the minimum on $V(i)$ in Case 1.
Case 2: $G_4(x_i^t) \leq T_4$ AND $G_2(x_i^t) < T_2$ (Rregion I in Fig. 1(c).)

$$x_i^P = \begin{cases} x_i^t & \text{if } G_2(x_i^t) > G_2(x_i^P) \\ x_i^P & \text{otherwise} \end{cases} \quad x_i^L = \begin{cases} x_c^t & \text{if } G_2(x_c^t) > G_2(x_i^L) \\ x_i^L & \text{otherwise} \end{cases} \tag{7}$$

where $G_2(x_c^t)$ is the maximum on $V(i)$ in Case 2. In Case 2, the function(s) G_2 is the object to increase and the function G_4 can increase. Such flexible movement can be effective to escape form the local/partial solution(s).

Step M4 (Lifetime and fission): If Pbest is not updated then $L_i^t \leftarrow L_i^t - 1$.
If $L_i^t = 0$ then the i-th particle is deactivated ($N^t \leftarrow N^t - 1$) and the i-th sub-swarm of ring topology is generated. The sub-swarm is governed by an algorithm defined afterward, If recombination occurs in the sub-swarm, the i-th particle is activated ($N^t \leftarrow N^t + 1$).

Step M5: The positions and velocities are updated.

$$x_i^t \leftarrow x_i^t + v_i^t, \ v_i^t \leftarrow wv_i + c_1 r_1 \left(x_i^P - x_i^t \right) + c_2 r_2 \left(x_i^L - x_i^t \right) \tag{8}$$

where r_1 and r_2 are random numbers in $[0,1]$. w, c_1 and c_2 are parameters.

Step M6: Let $t \leftarrow t + 1$, return to Step M2 and repeat until $t = t_{max}$.

If the fission occurs in the i-th (min) particle in Step M4 then a sub-swarm is generated. We define the algorithm for the sub-swarm (FRPSO-sub). Let y_{ij}^t and u_{ij}^t be the j-th particle position and velocity, respectively. For simplicity, we omit the subscript i hereafter: $y_j^t \equiv y_{ij}^t$ and $u_j^t \equiv u_{ij}^t$. Let N_s^t be the number of particles. Let y_j^P and y_j^L be the Pbest and Lbest, respectively. The j-th particle has a life time parameter L_j^t that controls the recombination. The FRPSO-sub is defined by the following 4 steps.

Step S1: Let $N_s^t = M$ where M is an initial number of particles. Since the initial time t means the time of fission in the main swarm. Particle positions y_j^t, $j = 1 \sim M$ are assigned randomly in the square with width E^t around the deactivated i-th particle of the main swarm where the width is time-variant: $E^t = E_{max} \times (t_{max} - t)/t_{max}$. The velocities, Pbest, Lbest and lifetime are all initialized as is Step M1.

Step S2: The particle position x^t is declared as an approximate solution if the Condition (5) is satisfies. Pbest and Lbest are updates as is Step M3.

Step S3(Lifetime and recombination): If Pbest is not updated then $L_j^t \leftarrow L_j^t - 1$. If $L_j^t = 0$ then the j-th particle is deactivated ($N_s^t \leftarrow N_s^t - 1$). If $N_s^t = 1$ then the i-th sub-swarm is deactivated and the particle is returned into the main-swarm as the i-th particle. This is the recombination at the last of Step M4.

Step S4: Positions and velocities of particles are updated as is Step M5. Let $t \leftarrow t + 1$, return to Step S2 and repeat until $t = t_{max}$.

4   Numerical Experiments

We apply the FRPSO to the MSP of Eq. (4), We have selected the initial life-time L_0 as the control parameter and fix the other parameters after trial-and-errors:

$$N^0 = 10, \ t_{max} = 100, \ w = 0.7, \ c_1 = c_2 = 1.4$$
$$E_{max} = 1.5, \ M = 5, T_4 = T_2 = 0.03,$$

Fig. 3 shows snapshots of Pbests in a typical run. On the evaluation plane, we can see that the four approximate solutions have found at time $t = 29$, 33, 39, and 45, respectively. In the early stage of the run, almost all particles exist in region III. As time elapses, several particles reach region II of the approximate solutions. However, some particles fall into region I where G_2 is too small. G_2 increase in the region I and the particle can reach region II as shown in the figure.

In order to consider the search process, Fig. 4 shows trajectories and G_4 of Pbests that reaches the four approximate solutions. The initial particles exist in region III In (a) the Pbest reaches region II direcrly. In (b) the fission to the sub-swarm is effective to avid trapping into region II and the sub-particle can reach region II. In (c) the main particle is trapped into local minimum in the early stage. In (d) the main particle enters into region I, retuns to region III, and is trapped into local mimimum. In these cases, the fission is effective for escape from the trap and the sub-particle can reach region II.

We have performed 1000 trials of different initial particle positions and the results are summarized in Table 1 with the following measures.

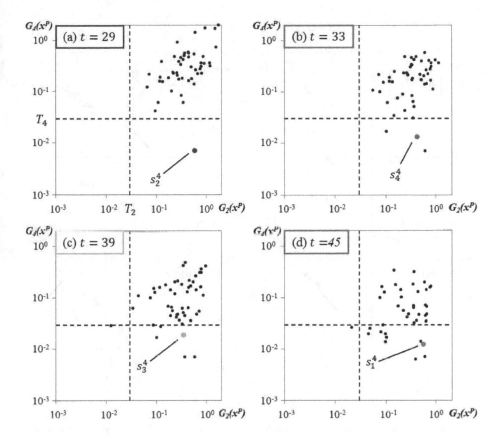

Fig. 3. Search process in evaluation plane. Black dots denote the personal bests.

SR: the average rate of successful runs where all the four solutions are found.
#SOL: the average number of solutions found until the time limit t_{max}.
#ITE: the average number of iterations until all solutions are found in successful runs.
P_{av}: the average number of particles per one discrete time until all the solutions are found in successful runs.

In the table, we can see that initial lifetime $L_0 = 20$ gives the highest SR and the largest #SOL are highest. Table 2 shows results of RPSO that is defined by removing fission and recombination of the FRPSO. The number of particles N is time-invariant and is selected as a control parameter. For $N \in \{10, 20, 30, 40\}$, SR is lower than the FRPSO of $L_0 \in \{8, 10, 20\}$ where P_{av} is at most 30. These results may suggest effectiveness of fission and recombination.

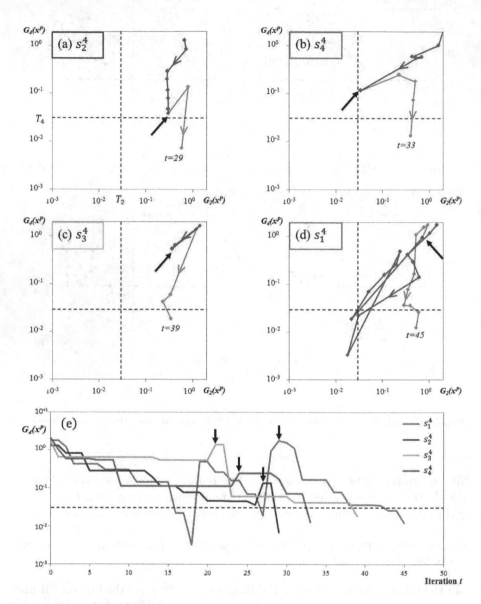

Fig. 4. Trajectories of four Pbests that reaches the four approximate solutions. Blue and red trajectories means main- and sub-swarms, respectively. Black arrows show timing of fission.

Table 1. Basic results of FRPSO for MSP

L_0	1	2	4	6	8	10	20	40	60	80	100
SR[%]	21.1	26.7	39.8	53.8	56.4	62.5	65.5	47.6	33.2	6.7	0.4
#SOL	2.9	3.0	3.2	3.4	3.5	3.5	3.6	3.3	3.0	2.3	1.8
#ITE	79.9	78.3	76.7	72.2	69.3	66.3	60.2	67.2	83.2	90.5	48.3
P_{av}	18.3	24.9	27.9	28.8	28.9	28.8	27.1	21.5	16.2	11.6	10.0

Table 2. Basic results of RPSO for MSP

N	10	20	30	40	50
SR	0.4	10.8	38.0	54.8	69.4
#SOL	1.8	2.6	3.2	3.5	3.7
#ITE	48.2	43.8	39.1	36.2	31.5

5 Conclusions

The FRPSO and its application to search of periodic points are considered in this paper. The number and topology of the particle swarm can vary by fission and recombination controlled by the lifetime parameter. Performing fundamental numerical experiments, the algorithm efficiency is confirmed.

Future problems include analysis of the search process, analysis of parameters' effects and application to bifurcation analysis.

References

1. Engelbrecht, A.P.: Fundamentals of computational swarm intelligence. Willey (2005)
2. Maruyama, K., Saito, T.: Deterministic Particle Swarm Optimizers with Collision for Discrete Multi-solution Problems. In: Proc. IEEE/SMC, pp. 1335–1340 (2013)
3. Snasel, V., Kroemer, P., Abraham, A.: Particle Swarm Optimization with Protozoic Behaviour. In: Proc. IEEE/SMC, pp. 2026–2030 (2013)
4. Hsieh, S.-T., Sun, T.-Y., Lin, C.-L., Liu, C.-C.: Effective learning Rate Adjustment of Blind Source Separation Based on an Improved Particle Swarm Optimizer. IEEE Trans. Evol. Comput. 12(2), 242–251 (2008)
5. Vural, R.A., Yildirim, T., Kadioglu, T., Basargan, A.: Performance Evaluation of Evolutionary Algorithms for Optimal Filter Design. IEEE Trans. Evol. Comput. 16(1), 135–147 (2012)
6. van Wyk, A.B., Engelbrecht, A.P.: Overfitting by PSO Trained Feedforward Neural Networks. In: Proc. IEEE/CEC, pp. 2672–2679 (2010)
7. Kawamura, K., Saito, T.: Design of Switching Circuits based on Particle Swarm Optimizer and Hybrid Fitness Function. In: Proc. IEEE/IECON, pp. 1099–1103 (2010)
8. Matsushita, H., Saito, T.: Application of Particle Swarm Optimization to Parameter Search in Dynamical Systems. NOLTA, IEICE E94-N(10), 458–471 (2011)
9. Ott, E.: Chaos in Dynamical Systems. Cambridge Univ. Press (1993)
10. Tsumoto, K., Ueta, T., Yoshinaga, T., Kawakami, H.: Bifurcation Analyses of Nonlinear Dynamical Systems: from Theory to Numerical Computations. NOLTA, IEICE 3(4), 458–476 (2012)

11. Parsopoulos, K.E., Vrahatis, M.N.: On the Computation of All Global Minimizers through Particle Swarm Optimization. IEEE Trans. Evol. Comput. 8(3), 211–224 (2004)
12. Li, X.: Niching without Niching Parameters: Particle Swarm Optimization Using a Ring Topology. IEEE Trans. Evol. Comput. 14(1), 150–169 (2010)
13. Sano, R., Shindo, T., Jin'no, K., Saito, T.: PSO-based Multiple Optima Search Systems with Switched Topology. In: Proc. IEEE/CEC, pp. 3301–3307 (2012)

Fast Generalized Fuzzy C-means Using Particle Swarm Optimization for Image Segmentation

Dang Cong Tran[1,2], Zhijian Wu[1], and Van Hung Tran[3]

[1] State Key Laboratory of Software Engineering, Computer School,
Wuhan University, Wuhan 430072, China
[2] Vietnam Academy of Science and Technology, Hanoi, Vietnam
[3] University of Transport and Communications, Hanoi, Vietnam
trandangcong@gmail.com, zhijianwu@whu.edu.cn

Abstract. Fuzzy C-means algorithms (FCMs) incorporating local information has been widely used for image segmentation, especially on image corrupted by noise. However, they cannot obtain the satisfying segmentation performance on the image heavily contaminated by noise, sensitivity to initial points, and can be trapped into local optima. Hence, optimization techniques are often used in conjunction with algorithms to improve the performance. In this paper, Particle Swarm Optimization (PSO) is introduced into fast generalized FCM (FGFCM) incorporating with local spatial and gray information called PFGFCM, where the membership degree values were modified by applying optimal-selection-based suppressed strategy. Experimental results on synthetic and real images heavily corrupted by noise show that the proposed method is superior to other fuzzy algorithms.

Keywords: particle swarm optimization, image segmentation, local information, Fuzzy c-means, fuzzy algorithm.

1 Introduction

Image segmentation is one of the most important tasks in image analysis and computer vision. In the literature, various methods have been proposed for object segmentation and feature extraction [1]. However, it remains a challenge due to overlapping intensities, low contrast of images, and noise perturbation. The image segmentation can be divided into four categories: clustering, thresholding, edge detection, and region extraction. In this paper, clustering method for image segmentation will be considered. In the last decades, fuzzy segmentation methodologies, especially the fuzzy C-means algorithms (FCM) [2], have been widely studied and applied in image segmentation. Their fuzzy makes the clustering procedure able to retain more original image information than the crisp or hard clustering methodologies [3]. Although the conventional FCM algorithm works well on most noise-free images, it is very sensitive to noise and other imaging artifacts, since it does not consider any information about spatial context.

C.K. Loo et al. (Eds.): ICONIP 2014, Part II, LNCS 8835, pp. 263–270, 2014.

J. L. Fan et al. [4] proposed a method called S-FCM to speed up the convergence of FCM. L. Szilagyi et al. [5] proposed an enhanced FCM algorithm (called EnFCM) to accelerate the image segmentation process by performing on the basis of gray level histogram instead of pixels of the summed image. In 2007, Cai et al. [6] proposed the fast generalized FCM algorithm (FGFCM) which incorporates the spatial information, the intensity of the local pixel neighborhood and the number of gray levels in an image. The computational time of FGFCM is very small, since clustering is performed on the basis of the gray level histogram. The quality of the segmented image is well enhanced.

Apart from the drawbacks of FCMs above, they also often get stuck at local minima and the result is largely dependent on the choice of the initial cluster centers. In addition, a fuzzy clustering problem is a combinatorial optimization problem [9] that is hard to solve and obtaining optimal solutions to large problems can be quite difficult. The methods above are no exception. In order to deal with these issues, there are several approaches such as combination of heuristic method, chaos, etc... Particle Swarm Optimization (PSO) [10,11] is a heuristic technique suited for search of optimal solutions and based on the concept of swarm. Using PSO with image segmentation is not new, however, the method and heuristics used are unique. In this paper, the method PFGFCM is to incorporate PSO with FGFCM algorithm [6]. In addition, in order to improve the convergence speed of FCM, a optimal-selection-based suppressed strategy is applied. To evaluate the performance of the proposed methods, the synthetic, and real images were used in experiments.

The rest of paper is organized as follows: Section 2 briefly describes the algorithms of FCM, PSO, and related fuzzy clustering algorithms. Two proposed PEFCM and PFGFCM methods are introduced in Section 3. Experimental results are presented in Section 4 and conclusions are drawn in Section 5.

2   Preliminaries

2.1   Fuzzy C-means Algorithm

Fuzzy C-means clustering algorithm (FCM) was proposed by Bezdek [2] to deal with the problem of clustering N multivariate data points of data set X into K clusters. In the original FCM algorithm, the fuzzy objective function that need to be minimized is given as follows:

$$J_m = \sum_{i=1}^{N} \sum_{k=1}^{K} u_{ki}^m \cdot d_{ik}^2 = \sum_{i=1}^{N} \sum_{k=1}^{K} u_{ki}^m \cdot \|y_i - \mu_k\|^2 \qquad (1)$$

where N is the number of image pixels, K is the number of clusters, u_{ki} is the membership of pixel y_i to the kth cluster identified by its center μ_k, and m is the weighting exponent controlling the fuzziness of the membership. The membership of pixel y_i to the kth cluster identified by its center μ_k is defined according to Eq. (2), and the cluster centers are iteratively updated as Eq. (3).

$$u_{ki} = 1/\sum_{j=1}^{K} \left(\frac{d_{ik}^2}{d_{jk}^2} \right)^{1/(m-1)} \tag{2}$$

$$\mu_k = \sum_{i=1}^{N} (u_{ki}^m \cdot y_i) / \sum_{i=1}^{N} u_{ki}^m \tag{3}$$

with the constraint $\sum_{k=1}^{K} u_{ki} = 1$.

2.2   Fast Generalized Fuzzy C-means Clustering

W. Cai et al. [6] proposed the fast generalized fuzzy C-means algorithm (FGFCM) to improve clustering results. FGFCM exploits a local similarity measure that combine both spatial and gray level image information, in terms of

$$S_{ij} = \begin{cases} e^{-\max(|p_i - p_j|, |q_i - q_j|)/\lambda_s \ -\|x_i - x_j\|/\lambda_g \sigma_i^2} & i \neq j \\ 0 & i = j \end{cases} \tag{4}$$

where the ith pixel is the center of the local window and the jth pixel represents the set of the neighbors falling into window centered at ith pixel. (p_i, q_i) are the coordinates of ith pixel and x_i is its gray level value. λ_s and λ_g are two scale factors playing a role similar to factor α in EnFCM, and σ_i is defined as follows:

$$\sigma_i = \sqrt{\sum_{j \in N_i} \|x_i - x_j\|^2 / N_R} \tag{5}$$

FGFCM incorporates local and gray level information into its objective function generating in advance a new image ξ as follows:

$$\xi_i = \sum_{j \in N_i} S_{ij} \cdot x_j / \sum_{j \in N_i} S_{ij} \tag{6}$$

where ξ_i denotes the gray level value of the ith pixel of the image ξ, x_j represents the gray level value of neighbors of x_i, N_i is the set of neighbors falling in the local window centered at ith pixel, and S_{ij} is the local similarity measure between the ith and jth pixel.

the objective function in this case is defined as follows:

$$J_m = \sum_{i=1}^{M} \sum_{j=1}^{K} \gamma_i \cdot u_{ki}^m \cdot (\xi_i - \mu_k)^2 \tag{7}$$

where μ_k represents the prototype of the jth cluster, u_{ki} represents the fuzzy membership of gray level value i with respect to cluster j, M denotes the number of gray levels of image ξ, which is generally much smaller than N, and γ_i is the number of pixels having gray level value equal to i. The membership degree values and the cluster centers are calculated as follows:

$$u_{ki} = 1/\sum_{j=1}^{K} \left(\frac{\xi_i - \mu_k}{\xi_i - \mu_j} \right)^{2/(m-1)} \tag{8}$$

$$\mu_k = \sum_{i=1}^{M} (\gamma_i \cdot u_{ki}^m \cdot \xi_i) / \sum_{i=1}^{M} (\gamma_i \cdot u_{ki}^m) \tag{9}$$

The iterative process of the FGFCM algorithm is similar to FCM, but it is applied to the new image ξ by using Eqs. (8) and (9). Due to in fact that the gray level value of the pixels is generally encoded with 8 bit resolution (256 gray levels), the number M of gray levels is generally much smaller than the size N of the image. Thus, the execution time is significantly reduced.

2.3   Particle Swarm Optimization

Particle swarm optimization algorithm (PSO) was proposed by J. Kennedy [10,11], in which each particle has a velocity vector (V) and a position vector (X). PSO remembers both the best position found by all particles and the best positions found by each particle in the search process. For a search problem in D-dimensional space, a particle represents a potential solution. The velocity and position of particle are updated according to Eqs. (10) and (11) as follows:

$$v_{ij} = w \cdot v_{ij} + c_1 \cdot rand1_{ij} \cdot (pbest_{ij} - x_{ij}) + c_2 \cdot rand2_{ij} \cdot (gbest_j - x_{ij}) \tag{10}$$

$$x_{ij} = x_{ij} + v_{ij} \tag{11}$$

where $i = 1, 2, ..., NP$ is the particles index, NP is the population size, x_i is the position of the ith particle, v_i represents the velocity of ith particle, $pbest_i$ is the best previous position yielding the best fitness value for the ith particle, and $gbest$ is the global best particle found by all particles so far, $rand1_{ij}$ and $rand2_{ij}$ are two random numbers independently generated within the range of $[0, 1]$, c_1 and c_2 are two learning factors which control the influence of the social and cognitive components, w is the inertia factor.

3   The Methodology

In this Section, two proposed methods PEFCM and PFGFCM will be described in details. Two proposed PEFCM and FGFCM approaches are respectively improved from EnFCM and FGFCM and perform on gray level image. Firstly, the new image ξ is formed. Then PSO algorithm introduced into FCM is applied in new image ξ to segment image, where the membership degree values u_{ki} is modified by using optimal-selection-based strategy.

In S-FCM algorithm, the membership degree values are modified for all data point in image X are modified during each iteration step. According to [13], F. Zhao et al. pointed out that if the biggest membership degree value of x_i is close to 0.5, this data point may be located in the boundary of some clusters. Once the membership degree values of x_i were compulsively altered in this iteration step, it may be wrongly grouped into a specific cluster. Therefore, it is unreasonable to compulsively modify the membership degree values for all the data points during each iteration step. To overcome this drawback of S-FCM F. Zhao et al. [13] proposed the method called optimal-selection-based suppressed strategy, where membership degree values are modified as follows:

$$u_{pi} = 1 - \beta \sum_{k \neq p} u_{ki} = 1 - \beta + \beta \cdot u_{pi}$$
$$u_{ki} = \beta, \ k \neq p \qquad\qquad x_i \in X_r \qquad\qquad (12)$$

where X_r is a data set formed as follows: firstly, all the data points in X are ranked based on their biggest membership degree values. Then the membership degree values of only top r ranked data points.

Due to S-FCM algorithm's sensitivity to the factor β that affects the relationship between HCM's fast convergence speed and FCM's good clustering performance, in our approach the factor β is calculated based on Gaussian distribution as follows:

$$\beta = Gaussian\,(x, \mu, \sigma) \qquad\qquad (13)$$

where x is uniform random number between 0 and 1, μ is the mean value equal to 0.5, the variance value $\sigma^2 = 0.2$. By this method, the factor β obtained by Gaussian distribution makes the algorithm more flexible and effective.

In the proposed approach, PSO is introduced into FCM, each particle is a potential candidate solution for the optimal cluster centers. A particle is encoded according to Eqs. (14), (15) for position and velocity vectors, respectively. In this case, clustering based on clustering can be seen as solving the global optimization with fitness function is the objective functions J_m.

$$x_i = (x_{i1}, x_{i2}, ..., x_{iK}) \qquad\qquad (14)$$

$$v_i = (v_{i1}, v_{i2}, ..., v_{iK}) \qquad\qquad (15)$$

where x_{ij} represents the jth cluster center centroid of ith particle, K is the number of clusters, $i = 1, 2, ..., NP$, NP is the size of the population. In this paper, the proposed approach of image segmentation is based on clustering algorithm and for gray level image with 8 bit resolution (256 gray levels).

3.1   The Proposed PFGFCM Approach

Firstly, the new image ξ is in advance formed from the original image and its local neighbor average image according to Eq. (6). Then, the PSO algorithm is applied to the image ξ, where the fitness function is the objective function calculated

Table 1. The main steps of PEFCM

1 Initialize parameter values used in the algorithm;
2 Compute the new image ξ according to Eq. (6) in advance;
3 Initialize the population by selecting randomly from 256 gray levels;
4 **While**(*iteration* $\leq$ *MaxIteration*)
5   **For** $i = 1$ to *NP*
6   Update the ith particle by using Eqs. (10), (11);
7   Calculate the fitness value of ith particle by using Eq. (7)
where u_{ki} is first calculated by Eq. (8), then modified by Eq. (12);
8   **End for**
9   Update *pbest* and *gbest* ;
10 *iteration* $++$;
11 End while

by Eq. (7), the membership degree values u_{ki} are modified by the optimal-selection-based suppressed strategy Eq. (12). The main steps of the algorithm are described in Table 1, where *MaxIteration* is the maximum number of iterations, *NP* is the population size.

4   Experimental Results

In this Section, we perform segmentation experiments on synthetic and real images and adopt FCM [2], S-FCM [4], EnFCM [5], and FGFCM [6] as comparative methods. In order to make a fair comparison, similar to the proposed methods PFGFCM the size of the neighbor window for S-FCM, EnFCM,FGFCM is set 3×3 ($NR = 8$). The parameter α for EnFCM is set to 6. According to the results and the parameter analysis of FGFCM preseneted by W. Cai et al. [8], the parameters λ_s and λ_g of FGFCM and PFGFCM are set to 3 and 6, respectively. The parameter r used in Eq. (12) for PEFCM and PFGFCM is set to 158 [13]. The population size for PEFCM and PFGFCM is experimentally set to 40. Furthermore, the fuzzyness index m for all methods is set to 2. The PSO inertia factor $w = 0.7298$, the maximum number of iterations *MaxIteration* for all methods is set to 500, each experiment has been performed over 20 times with different random initialization.

4.1   Experiments on Synthetic Image

In this section, a synthetic images with size of 256×256 is used for evaluating the performance of competitive algorithms. It includes four clusters with the corresponding gray values as 0, 85, 175 and 255, respectively. The noisy image corrupted Gaussian noise (0 means, 0.024 normalized variance)is presented in Fig. 1(a). The segmentation accuracy (SA) [12] is used to evaluate the algorithm performance in terms of accuracy. The segmented images with Gaussian noise is present in Figs. 1(b)–(f). Results of SA demonstrate that the proposed method

PFGFCM has results better than other comparative algorithms. Visually, from Fig. 1, PFGFCM, EnFCM and FGFCM remove most of the noise, while FCM and S-FCM are much poorer.

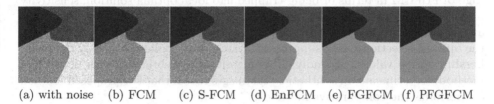

(a) with noise   (b) FCM   (c) S-FCM   (d) EnFCM   (e) FGFCM   (f) PFGFCM

Fig. 1. Synthetic image, FCM(SA=0.6286), S-FCM(SA=0.6963), EnFCM(SA=0.9824), FGFCM(SA=0.9812), PFGFCM(SA=0.9924)

4.2 Experiments on Real Images

In this experiments, we try to segment the real images into three classes: two original images, namely eight and MRI, and adding Gaussian noise are used and presented in Figs. 2 and 3. The images segmented by the comparative algorithms are shown in Figs. 2(b)–(f) and Figs. 3(b)–(f). From results of segmented images in Figs. 2(b)–(f), it can be seen that most noise in the images was removed by PFGFCM method, while most of image noise was not removed in the segmentation results of FCM, S-FCM, EnFCM, and FGFCM. From Figs. 3(b)–(f), the results show that FCM, S-FCM are very poor, while most noise was removed by EnFCM, FGFCM, and PFGFCM.

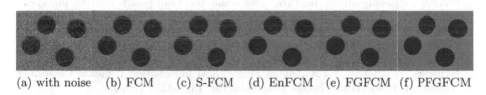

(a) with noise   (b) FCM   (c) S-FCM   (d) EnFCM   (e) FGFCM   (f) PFGFCM

Fig. 2. The results on the image eight with Gaussian noise

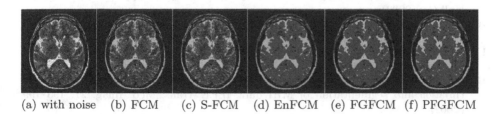

(a) with noise   (b) FCM   (c) S-FCM   (d) EnFCM   (e) FGFCM   (f) PFGFCM

Fig. 3. The results on the image MRI with Gaussian noise

5   Conclusions

In this paper, the proposed PFGFCM algorithm is proposed, in which PSO algorithm was introduced into FGFCM. The algorithm can overcome the disadvantage of FGFCM in terms of being trapped into local optima solution, sensitivity to initial points, especially in case of image heavily corrupted by noise. There are two contributions were introduced into the existing method FGFCM, firstly, the optimal-selection-based suppressed strategy was applied to modify the membership degree values. Secondly, PSO algorithm is hybridized with used FCM method. Experiments on synthetic, real images reveal that the proposed method outperforms the comparative methods.

Acknowledgments. This work was supported by the National Natural Science Foundation of China (No.: 61070008 and 61364025).

References

1. Pham, D., Xu, C., et al.: A survey of current methods in medical image segmentation. Annu. Rev. Biomed. Eng. 2, 315–337 (2000)
2. Bezdek, J.: Pattern Recognition with Fuzzy Objective Function Algorithms. Plenum, New York (1981)
3. Chen, S., Zhang, D.: Robust Image Segmentation using FCM with Spatial Constraints based on new Kernel-induced Distance Measure. IEEE Transactions on Systems, Man and Cybernetics 34(4), 1907–1916 (2004)
4. Fan, J.L., Zhen, W.Z., Xie, W.X.: Suppressed fuzzy c-means clustering algorithm. Pattern Recognition Letter 24, 1607–1612 (2003)
5. Szilagyi, L., Benyo, Z., et al.: MR brain image segmentation using an enhanced fuzzy C-means algorithm. In: Proc. 25th Annu. Int. Conf (IEEE EMBS), pp. 17–21 (2003)
6. Cai, W., Chen, S.: Fast and robust fuzzy c-means clustering algorithms incorporating local information for image segmentation. Pattern Recog. 40(3), 825–838 (2007)
7. Krinidis, S., Chatzis, V.: A Robust Fuzzy Local Information C-means Clustering Algorithm. IEEE Trans. Image Process. 5(19), 1328–1337 (2010)
8. Zhang, H., Jonathan Wu, Q.M., et al.: Image segmentation by a robust generalized fuzzy c-means algorithm. In: Proceedings of IEEE International Conference on Image Processing (ICIP 2013), pp. 4024–4028 (2013)
9. Klir, J.G., Yuan, B.: Fuzzy sets and fuzzy logic, theory and applications. Prentice-Hall Co. (2003)
10. Eberhart, R.C., Kennedy, J.: A new optimizer using particle swarm theory. In: Proceeding of Int. Symp. on Micro Machine and Human Science, pp. 39–43 (1998)
11. Kennedy, J., Eberhart, R.C.: Particle swarm optimization. In: Proceedings of International Conference on Neuron Networks, pp. 1942–1948 (1995)
12. Ahmed, M., Yamany, S.: A modified fuzzy C-means algorithm for bias field estimation and segmentation of MRI data. IEEE Trans. Med. Imag. 21, 193–199 (2002)
13. Zhao, F., Fan, J.: Optimal-selection-based suppressed fuzzy c-means clustering algorithm with self-tuning non local spatial information for image segmentation. Expert systems with applications 41, 4083–4093 (2014)

Evolutionary Learning and Stability
of Mixed-Rule Cellular Automata

Ryo Sawayama and Toshimichi Saito

Hosei University, Koganei, Tokyo, 184–8584 Japan
tsaito@hosei.ac.jp

Abstract. This paper studies the cellular automaton (CA) governed by combination of two rules. First, we analyze a class of CA that generates several isolated spatiotemporal patterns without transient phenomena. Second, we present an evolutionary algorithm that tries to optimize the combination of two rules to stabilize the desired isolated patterns. Performing basic numerical experiments, it is shown that the evolutionary algorithm can make transient phenomena and can stabilize the desired isolated patterns.

Keywords: cellular automaton, genetic algorithm, stability.

1 Introduction

The cellular automaton (CA) is a dynamical system in which time, space and state variables are all discrete [1]. The time evolution of the state variable is governed by a single rule. Depending on the rule and initial condition, the CA can generate a variety of spatiotemporal phenomena. The CAs have been studied not only in study of fundamental nonlinear dynamics but also in various engineering applications such as signal processing, information compressions, self-replications, and ciphers [2]-[5].

This paper studies the cellular automaton with mixed rules (MCA). Although the CAs are governed by a single rule, the MCA is governed by combination of two rules and can generate rich phenomena. Since general discussion of the MCA is hard, we focus on a class of MCA and consider stability of a class of steady state as the following. First, we analyze a class of the elementary CA (ECA, [1]) that generates several steady states without transient phenomena. Such a steady state is referred to as an isolated pattern. Using the mapping procedure [6], the number and kinds of the steady states are visualized. Second, we present an evolutionary algorithm that tries to optimize the combination of two rules to stabilize the desired isolated patterns given by the ECA. The algorithm is based on the genetic algorithm (GA, [7]). Performing numerical experiments for typical isolated patterns, it is shown that the GA-based algorithm can make transient phenomena and stabilize the desired isolated patterns. It means that the MCA can generate more robust patterns than ECA.

The results may be developed into analysis of wider class of MCAs and into engineering applications such as error correction and robust signal generation.

C.K. Loo et al. (Eds.): ICONIP 2014, Part II, LNCS 8835, pp. 271–278, 2014.
© Springer International Publishing Switzerland 2014

2   Mixed-Rule Cellular Automata

We define the ECA and MCA on the ring of N cells. Let $x_i^t \in \{0,1\} \equiv \boldsymbol{B}$ be the binary state of the i-th cell at discrete time t where $i = 1 \sim N$. The time evolution of x_i^t governed by the a Boolean function of x_i^t and its closest neighbors:

$$x_i^{t+1} = F_i(x_{i-1}^t, x_i^t, x_{i+1}^t) \tag{1}$$

where $x_{-1}^t = x_N^t$ and $x_{i+1}^t = x_1^t$ on the ring. F_i is referred to as a rule table. In the ECA, F_i does not depend on i ($F_i = F$) and the dynamics is defined by one rule table of 2^3 rules

$$
\begin{aligned}
f_0 &= F(0,0,0), \ f_1 = F(0,0,1), \ f_2 = F(0,1,0), \ f_3 = F(0,1,1) \\
f_4 &= F(1,0,0), \ f_5 = F(1,0,1), \ f_6 = F(1,1,0), \ f_7 = F(1,1,1)
\end{aligned}
\tag{2}
$$

where $f_j \in \boldsymbol{B}$ and $j = 0 \sim 7$. The rule table is equivalent to one of 8 bits binary number and the number is referred to as the rule number (RN). There exist 2^{2^3} rule tables of the ECA. Figure 1 (a) shows rule table of RN56 and a spatiotemporal pattern. Note that this pattern is a steady state with period 8 to which no transient phenomenon exists.

In the MCA, the rule table depends on the cell i, however, it is hard to consider all the combinations of N rules. For simplicity, this paper considers the case where the rule table is given by combination of two rules: F_i is either rule A

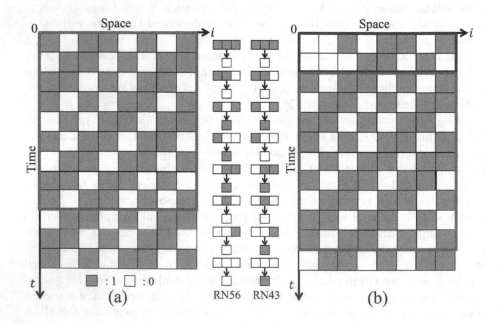

Fig. 1. Spatiotemporal patterns and rule tables (The red part is the steady state and the blue part is the transient state.) (a) ECA of RN56 (b) MCA of RN56 and 43

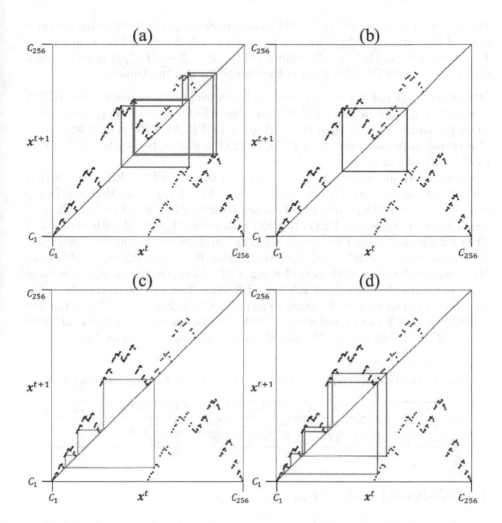

Fig. 2. Digital return map of RN56. (a) PEO with period 8 corresponding to Fig. 1 (a). (b) PEO with period 2 (c) PEO with period 4 (d) PEO with period 8.

or B. Figure 1 (b) shows a spatiotemporal pattern of MCA consisting of RN56 and RN43. Note that this MCA (RN56 and 43) has the same steady state as ECA (RN56) and that the steady state has transient phenomenon.

In order to visualize the dynamics of the MCA, we introduce the digital return map (Dmap, Fig. 2). The MCA of N cells is equivalent to the N-dimensional mapping from B^N to itself:

$$x^{t+1} = F_D(x^t), \ x^t \equiv (x_1^t, \cdots, x_N^t) \in B^N \tag{3}$$

Since B^N is equivalent to a set of 2^N lattice points $I_D \equiv \{C_1, \cdots, C_{2^N}\}$, F_D is equivalent to the mapping from I_D to itself. This is the Dmap and Fig. 2 shows examples. We give basic definition for the Dmap.

Definition 1 : A point $p \in I_D$ is said to be a periodic point (PEP) with period k if $p = F_D^k(p)$ and $p \neq F_D^l(p)$ for $0 < l < k$ where F_D^k is the k-fold composition of F_D. A sequence of the periodic points $\{F_D(p), F_D^2(p), \cdots, F_D^k(p)\}$ is said to be a periodic orbit (PEO). The PEO is the steady state of the Dmap.

Definition 2 : A point $q \in I_D$ is said to be an eventually periodic point (EPP) of a PEO if q is not a PEP and falls into the PEO. For an EPP q, there exists some positive integer m such that $F_D^m(q)$ is a PEP. An EPP is an initial point of a transient phenomenon to a PEO. A PEO is said to be isolated if it has no EPP (no transient state).

Figure 2 (a) shows Dmap of RN56 with an isolated PEO with period 8 that corresponds to the spatiotemporal pattern in Fig. 1 (a). Note that the Dmap has other eight PEOs, three of which are shown in Fig. 2 (b) to (d). That is, the RN56 can generate 9 PEOs (a PEO with period 1, a PEO with period 2, a PEO with period 4, six PEOs with period 8) and the ECA exhibits either PEO depending on the initial state. However, mixing RN56 and RN43, we obtain an MCA where the PEO with period 8 has EPPs as shown in Table 1(the same PEO is isolated in the CA of RN56 and RN43). The EPP corresponds to the transient phenomenon to the steady state of MCA in Fig. 1 (b). This MCA has 2 PEOs (period 2 and period 8) and 246 EPPs. The number of PEOs and EPPs of the ECA (RN56) and MCA (RN56 and 43) are summarized in Table 1.

Table 1. The typical example of ECA and MCA (TPEO = Teacher signal PEO)

cell index	1	2	3	4	5	6	7	8	#PEOs	#EPPs of TPEO
RN of ECA	56	56	56	56	56	56	56	56	9	0
RN of MCA	43	56	43	56	43	43	43	56	2	246

3   GA-Based Algorithm

In order to synthesize an MCA, we present an algorithm based on the GA. Since there exist a huge variety of MCA, this paper focuses on a fundamental case: MCA by mixture of two CAs with isolated PEOs. The teacher signal is one isolated PEO of the ECA. The purpose is generation of EPPs to the teacher signal PEO as many as possible (reinforcement of stability).

There exists various RNs that generate isolated PEOs, this paper focus on the isolated PEO in Table 2 (Fig. 1 (a)) that is generated by either element in the set of rule numbers (RNS).

$$RNS = \{ RN40, RN41, RN42, RN43, RN56, RN169, RN170\} \qquad (4)$$

These rules and their isolated PEOs are introduced in Ref. [1] as typical examples. In the GA-based algorithm, two rules are selected from the RNS and are referred to as rule A and rule B. The algorithm has K chromosomes with length 8 whose elements are either of rule A or B, e .g., "ABABAAAB" for the MCA in Table 2. The length of each chromosome is the number of cells in the ring space. The algorithm is defined by the following.

Step 1: Let generation $g = 0$. Let half of initial chromosomes be "AAAAAAAA" and let the other half be all "BBBBBBBB".

Step 2: Chromosomes are evaluated by

$$\text{Convergence rate: CR} = \frac{\#\text{EPPs of teacher signal PEO}}{2^N - \#\text{PEPs of teacher signal PEO}} \tag{5}$$

It can measure the domain of attraction to the teacher signal PEO. If CR=100 then all the initial states fall into the teacher signal PEO: the PEO is completely stable. If CR=0 then the PEO is isolated.

Step 3: Apply the one-point crossover with probability P_c and mutation with probability P_m. Chromosomes are selected by the elite strategy. Preserve some chromosome having the largest CR to the next generation.

Step 4: Let $g \leftarrow g+1$, go to Step 2 and repeat until the maximum generation G.

4   Numerical Experiments

In order to investigate the algorithm and the MCA, we have performed funda-mental numerical experiments. The teacher signal PEO is periodic point with period 8 in Table 2 (Fig. 1 (a)) and two rule number candidates are selected from the RNS (4).

We have fixed GA parameters: population size $K = 8$, crossover probability $P_c = 0.9$, mutation probability $P_m = 0.1$, and the maximum generation $G = 30$. Since all the rule number candidates can generate the teacher signal PEO, the storage of the PEO is guaranteed for all generations. Figure 3 shows evolution process for the MCA of RN56 and RN43. In this example, the CR increases rapidly in young generations and saturates as the algorithm evolves. As CR increases, #PEOs decreases as expected. After the saturation, the MCA has two

Table 2. Teacher signal; isolated PEO with period 8

z^1	$(1, 0, 1, 1, 0, 1, 0, 1)$
z^2	$(0, 1, 1, 0, 1, 0, 1, 1)$
z^3	$(1, 1, 0, 1, 0, 1, 1, 0)$
z^4	$(1, 0, 1, 0, 1, 1, 0, 1)$
z^5	$(0, 1, 0, 1, 1, 0, 1, 1)$
z^6	$(1, 0, 1, 1, 0, 1, 1, 0)$
z^7	$(0, 1, 1, 0, 1, 1, 0, 1)$
z^8	$(1, 1, 0, 1, 1, 0, 1, 0)$
$z^9 = z^1$	$(1, 0, 1, 1, 0, 1, 0, 1)$

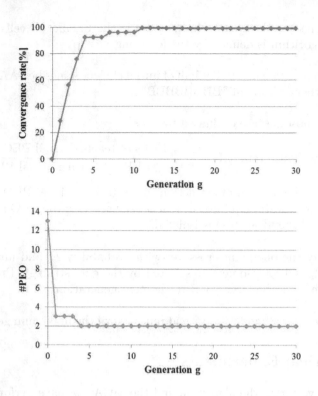

Fig. 3. CR and #PEOs in an evolution process for MCA of RN56 and RN43

PEOs as shown in the example of Dmap in Fig. 4. Using the Dmap, we can confirm that teacher signal PEO with period 8 is stabilized almost completely (#EPPs=246) and the other PEO with period 2 is isolated (#EPPs=0). In addition, we have confirmed that the algorithm exhibits such saturation characteristics for smaller/larger K as shown in Fig. 5: the increase becomes faster/slower for larger/smaller K.

We have applied the algorithm to all the combination of two rules in RNS and the results are summarized. In Table 3 (a), the diagonal components show #EPPs to the teacher signal PEO of ECA (rule A = rule B, #EPPs=0 for isolated PEOs). Non-diagonal components show #EPPs to the teacher signal PEO of MCA for all the combination of two RNs (rule A and rule B) at the maximum generation $g = G$. In Table 3 (b), the diagonal components show #PEOs of ECA and non-diagonal components show #PEOs of MCA. In almost all combinations, the MCA can have larger number of #EPPs and smaller number of #PEOs than the ECA. These results suggest that the MCA is convenient to generate stable spatiotemporal patterns.

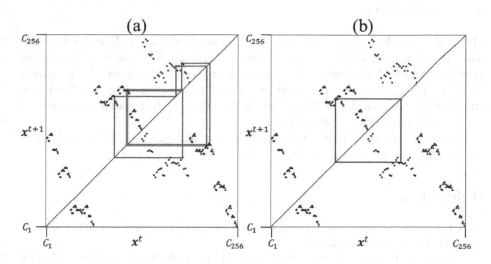

Fig. 4. Digital return map of MCA of RN56 and 43. (a) PEO with period 8, #EPPs to which is 246. (b) PEO with period 2, #EPPs to which is 0.

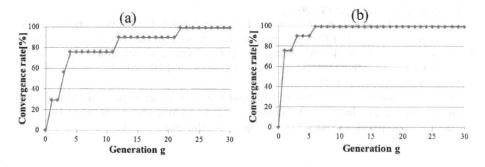

Fig. 5. GA results (a) $K = 6$ (b) $K = 10$

Table 3. Evolved MCA of two isolated rules in RNS (rule A and B)

(a) #EPPs of the teacher signal PEO

B \ A	40	41	42	43	56	169	170
40	0	2	0	2	0	21	20
41		0	2	0	147	21	20
42			0	2	136	21	20
43				0	246	21	20
56					0	169	168
169						0	1
170							0

(b) #PEOs

B \ A	40	41	42	43	56	169	170
40	3	4	3	4	3	6	3
41		9	4	6	5	8	3
42			19	8	3	4	19
43				13	2	6	7
56					9	4	3
169						19	9
170							36

5   Conclusions

MCA of two isolated rule tables are considered in this paper. In order to stabilize the isolated PEO of ECA, a GA-based algorithm is presented. In order to visualize the MCA dynamics, the Dmap is introduced. Performing basic numerical experiment, we have confirmed that the GA-based algorithm is effective to stabilize the teacher signal PEO. This is the first step to construct a variety of MCAs with interesting/desired dynamics.

Future problems include analysis of the dynamics of the MCA, analysis of evolution process, and engineering applications.

References

1. Chua, L.O.: A Nonlinear Dynamics Perspective of Wolfram's New Kind of Science, I, II. World Scientific (2005)
2. Wada, W., Kuroiwa, J., Nara, S.: Completely Reproducible Description of Digital Sound Data with Cellular Automata. Physics Letters A 306, 110–115 (2002)
3. Seredynski, M., Bouvry, P.: Block Cipher Based on Reversible Cellular Automata. In: Proc. of IEEE/CEC, pp. 2138–2143 (2004)
4. Rosin, P.L.: Training Cellular Automata for Image Processing. IEEE Trans. Image Process. 15(7), 2076–2087 (2006)
5. Lohn, J.D., Reggia, J.A.: Automatic Discovery of Self-Replicating Structures in Cellular Automata. IEEE Trans. Evolutionary Computation 1(3), 165–178 (1997)
6. Sawayama, R., Kouzuki, R., Saito, T.: Basic Dynamics of Elementary Cellular Automata with Mixed Rules: Periodic Patterns and Transient Phenomena. In: Proc. NOLTA, pp. 154–157 (2013)
7. Suzuki, S., Saito, T.: Synthesis of desired binary cellular automata through the genetic algorithm. In: King, I., Wang, J., Chan, L.-W., Wang, D. (eds.) ICONIP 2006. LNCS, vol. 4234, pp. 738–745. Springer, Heidelberg (2006)

Radical-Enhanced Chinese Character Embedding

Yaming Sun[1], Lei Lin[1,*], Nan Yang[2], Zhenzhou Ji[1], and Xiaolong Wang[1]

[1] Harbin Institute of Technology, Harbin, China
[2] Microsoft Research, Beijing, China
{ymsun,linl,wangxl}@insun.hit.edu.cn,
{nyang.ustc}@gmail.com, jzz@pact518.hit.edu.cn

Abstract. In this paper, we present a method to leverage radical for learning Chinese character embedding. Radical is a semantic and phonetic component of Chinese character. It plays an important role for modelling character semantics as characters with the same radical usually have similar semantic meaning and grammatical usage. However, most existing character (or word) embedding learning algorithms typically only model the syntactic contexts but ignore the radical information. As a result, they do not explicitly capture the inner semantic connections of characters via radical into the embedding space of characters. To solve this problem, we propose to incorporate the radical information for enhancing the Chinese character embedding. We present a dedicated neural architecture with a hybrid loss function, and integrate the radical information through softmax upon each character. To verify the effectiveness of the learned character embedding, we apply it on Chinese word segmentation. Experiment results on two benchmark datasets show that, our radical-enhanced method outperforms two widely-used context-based embedding learning algorithms.

Keywords: Chinese character embedding, radical, neural network.

1 Introduction

Chinese **radical** (部首) is a graphical component of Chinese character, which serves as an indexing component in Chinese dictionary[1]. In general, a Chinese character is phono-semantic, with a radical as its semantic and phonetic component suggesting part of its meaning. For example, " 氵 (water)" is the radical of "河 (river)", and "足 (foot)" is the radical of "跑 (run)".

Radical plays an important role for modelling the meaning of Chinese character. The reason lies in that characters with the same radical typically have similar semantic meanings and play similar grammatical roles. For example, the verbs " 打 (hit)" and "拍 (pat)" share the same radical " 扌 (hand)" and usually act as the subject-verb in sentences. Accordingly, the radical information can be used to enhance the meaning of characters by capturing the inner semantic connections of characters. In the vector space model, an effective character (or word) representation method is to encode each

---

* Corresponding author.
[1] http://en.wikipedia.org/wiki/Radical_(Chinese_character)

C.K. Loo et al. (Eds.): ICONIP 2014, Part II, LNCS 8835, pp. 279–286, 2014.
© Springer International Publishing Switzerland 2014

character into a dense, low-dimensional and real-valued vector, a.k.a character embedding [1,2]. The character embedding has been proven effective in many Chinese language processing tasks, such as word segmentation and POS-tagging [3,4]. However, to our best knowledge, existing studies for learning Chinese character representation typically only utilize the context information yet ignore the radical information of character. Unlike the previous studies that utilize the radical as discrete features in hypernym discovery [5] and similarity judgements of Chinese word pairs [6,7], we leverage the radical information to enhance the character embedding, which can be easily adopted off-the-shell in other Chinese language processing tasks like parsing [8].

In this paper, we extend an existing embedding learning algorithm [1] and propose a tailored neural architecture to leverage radical for enhancing the continuous representation of Chinese character. Our neural model integrates the radical information by predicting the radical of each character through a $softmax$ layer upon each character. The hybrid loss function is the linear combination of the loss of C&W model [1] and the cross-entropy error of $softmax$. To evaluate the effectiveness of the radical-enhanced character embedding, we apply it as the unique feature for Chinese word segmentation. Experiment results on two benchmark datasets show that, the radical-enhanced character embedding outperforms two widely-accepted embedding learning algorithms, C&W [1] and word2vec [2], which do not utilize the radical information. The major contributions of this paper are summarized as follows.

- To our best knowledge, this is the first work that leverages the radical information for learning Chinese character embedding.
- We report the results that our radical-enhanced method outperforms two existing context-based character embedding algorithms on Chinese word segmentation.

2   Related Work

The representation of word plays an important role in natural language processing [9,10]. In the early studies, a word is represented as a one-hot vector[2], whose main drawback is that it cannot reflect the semantic relations between words. With the revival of deep learning [11], many researchers focus on learning the continuous representation of words (word embedding). Bengio el al. [12] propose a feed-forward neural probabilistic language model to predict the next word based on its previous contextual words. Collobert et al. [1] propose a feed-forward neural network (C&W) which learns word embedding with a ranking-type cost. Mikolov et al. introduce the Recurrent neural network language models (RNNLMs) [13], Continuous Bag-of-Word (CBOW) and skip-gram model [2] to learn embedding for words and phrases. Huang et al. [14] propose a neural model to utilize the global context in addition to the local information. Tang et al. [15] propose a method to learn sentiment-specific word embedding.

We argue that the representation of words heavily relies on the characteristic of language. The linguistic feature of English has been studied and used in the word embedding learning procedure. Specifically, Luong et al. [16] utilize the morphological

---

[2] One-hot representation is a vector whose length is the size of vocabulary, and only one dimension is 1, others are 0.

property of English word and incorporate the morphology into word embedding. In this paper, we focus on learning Chinese character embedding by exploiting the radical information of Chinese character, which is tailored for Chinese language. Unlike Luong et al. [16] that initialize their model with pre-trained embedding, we learn Chinese character embedding from scratch.

3   Radical-Enhanced Model for Chinese Character Embedding Learning

In this section, we describe the details of leveraging the radical information for learning Chinese character embedding. Based on C&W model [17], we present a radical-enhanced model, which utilizes both radical and context information of characters. In the following subsections, we first briefly introduce the C&W model, and then present the details of our radical-enhanced model.

3.1   C&W Model

C&W model [17] is proposed to learn the continuous representation of a word from its context words. Its training objective is to assign a higher score to the reasonable ngram than the corrupted ngram. The loss function of C&W is a ranking-type cost:

$$loss_c(s, s^w) = max(0, 1 - score(s) + score(s^w))$$ (1)

where s is the reasonable ngram, s^w is the corrupted one with the middle word replaced by word w, and $score(.)$ represents the reasonability scalar of a given ngram, which can be calculated by its neural model.

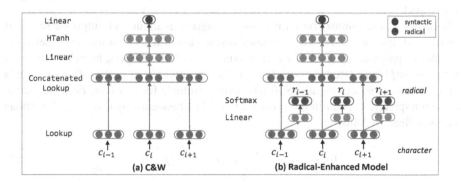

Fig. 1. C&W (a) and our radical-enhanced model (b) for learning character embedding

C&W is a feed-forward neural network consisted of four layers, as illustrated in Figure 1(a). The input of C&W is a ngram composed of n words, and the output is a score which evaluates the reasonability of the ngram. Each word is encoded as a column vector in the embedding matrix $\mathbf{W}_e \in \mathbb{R}^{d \times |V|}$, where d is the dimension of the vector, and V is the vocabulary. The *lookup* layer has a fixed window size n, and it maps each word of the input ngram into its embedding representation. The output $score(s)$ is computed as follows:

$$score(s) = \mathbf{W}_2 \mathbf{a} + b_2 \qquad (2)$$

$$\mathbf{a} = HTanh(\mathbf{W}_1[\mathbf{x}_1...\mathbf{x}_n] + \mathbf{b}_1) \qquad (3)$$

where $[\mathbf{x}_1...\mathbf{x}_n]$ is the concatenation of the embedding vectors of words $x_1, ...x_n$, $\mathbf{W}_1, \mathbf{W}_2, \mathbf{b}_1, b_2$ are the weights and biases of the two linear layers, and function $HTanh(.)$ is the $HardTanh$ function. The parameters of C&W can be learned by minimizing the loss through stochastic gradient descent algorithm.

3.2   Radical-Enhanced Model

In this part, we present the radical-enhanced model for learning Chinese character embedding. Our model captures the radical information as well as the context information of characters. The training objective of our radical-enhanced model contains two parts: 1) for a ngram, discriminate the correct middle character from the randomly replaced character; 2) for each character within a ngram, predict its radical. To this end, we develop a tailored neural architecture composed of two parts, context-based part and radical-based part, as given in Figure 1(b).

The context-based part captures the context information, and the radical-based part utilizes the radical information. The final loss function of our model is shown as follows:

$$Loss(s, s^w) = \alpha \cdot loss_c(s, s^w) + (1 - \alpha) \cdot \left(\sum_{c \in s} loss_r(c) + \sum_{c \in s^w} loss_r(c) \right) \qquad (4)$$

where s is the correct ngram, s^w is the corrupted ngram, $loss_c(s, s^w)$ is the loss of the context-based part, $loss_r(.)$ is the loss of the radical-based part, and α linearly weights the two parts.

Specifically, the context-based part takes a ngram as input and outputs a score, as described in Equation 1. The radical-based part is a list of feed-forward neural networks with shared parameters, each of which is composed of three layers, namely $lookup \rightarrow linear \rightarrow softmax$ (from bottom to top). The unit number of each $softmax$ layer is equal to the number of radicals. Softmax layer is suitable for this scenario as its output can be interpreted as conditional probabilities. The cross-entropy loss of each softmax layer is defined as follows:

$$loss_r(c) = -\sum_{i=0}^{N} \boldsymbol{p}_i^g(c) \times log(\boldsymbol{p}_i(c)) \qquad (5)$$

where N is the number of radicals; $\boldsymbol{p}^g(c)$ is the gold radical distribution of character c, with $\sum_i \boldsymbol{p}_i^g(c) = 1$; $\boldsymbol{p}(c)$ is the predicted radical distribution.

Our model is trained by minimizing the loss given in Equation 4 over the training set. The parameters are embedding matrix of Chinese characters, weights and biases of each linear layer. All the parameters are initialized with random values, and updated via stochastic gradient descent. Hyper-parameter α is tuned on the development set.

4 Neural CRF for Chinese Word Segmentation

In this section, we briefly introduce the neural CRF [4,10], which can be utilized for Chinese word segmentation with the character embedding as features. The illustration of neural CRF is shown in Figure 2.

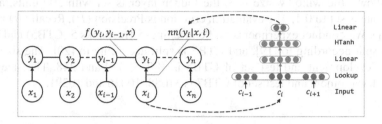

Fig. 2. Neural CRF for Chinese word segmentation. Each input character x_i is denoted with its embedding vector, and $window(x_i)$ is the input of the neural network

Given an observation sequence $\mathbf{x}$ and its gold tag sequence $\mathbf{y}$, neural CRF models their conditional probability as follows,

$$P(\mathbf{y}|\mathbf{x}) = \frac{exp\ \phi(\mathbf{y},\mathbf{x})}{\sum_{\mathbf{y}'} exp\ \phi(\mathbf{y}',\mathbf{x})} \tag{6}$$

$$\phi(\mathbf{y},\mathbf{x}) = \sum_i [f(y_i, y_{i-1}, \mathbf{x})\mathbf{w_1} + f(y_i, \mathbf{x})\mathbf{w_2}] \tag{7}$$

where $f(y_i, y_{i-1}, \mathbf{x})$ is a binary-valued indicator function reflecting the transitions between y_{i-1} and y_i, and $\mathbf{w_1}$ is its associated weight. $f(y_i, \mathbf{x})\mathbf{w_2}$ reflects the correlation of the input $\mathbf{x}$ and the i-th label y_i, which is calculated by a four-layer neural network as given in Figure 2. The neural network takes a ngram within a window of x_i as input, and outputs a distribution over all possible tags of x_i such as "B/I/E/S". The neural CRF is trained via maximizing the likelihood of $P(\mathbf{y}|\mathbf{x})$ over all the sentences in the training set. We use Viterbi algorithm [18] in the decoding procedure.

5 Experiments

In this section, we evaluate the radical-enhanced character embedding by applying it on Chinese word segmentation through neural CRF. We compare our model with C&W [17] and word2vec[3] [2], and learn Chinese character embedding with the same settings. To evaluate the embedding learned by different models, we only use the character embedding as features in neural CRF without feature engineering. We extract a radical mapping dictionary from an online Chinese dictionary[4], which contains 265

---

[3] Available at https://code.google.com/p/word2vec/. We utilize Skip-Gram as baseline.

[4] http://xh.5156edu.com/

radicals and 20,552 Chinese characters including both the simplified and traditional format. Each character listed in the radical dictionary is attached with its radical, such as ⟨吃(eat), 口(mouth)⟩. We use 1M randomly selected sentences from Sougou corpus[5] to train character embedding.

We empirically set the embedding size as 30, window size as 5, learning rate as 0.1, and the length of hidden layer as 30. The parameters of the neural CRF are empirically set as follows, the window size is 3, the hidden layer is set with 300 units, and the learning rate is set to 0.1. The evaluation criterion is Precision (P), Recall (R) and F1-score (F_1). We conduct experiments on Penn Chinese Treebanks 5 (CTB5) and CTB7. CTB5 is split according to [19], and CTB7 is split according to [20]. The size of the training, development and test set of CTB5 is 18085, 350 and 348. The size of the training, development and test set of CTB7 is 31088, 10036 and 10291.

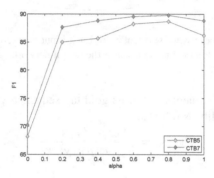

Fig. 3. F_1 on the development set of CTB5 and CTB7. $alpha$ is the weight of the context-based loss in our model.

We investigate how the alpha of the radical-enhanced model affects the performance of character embedding for word segmentation. The results on the development set is given in Figure 3. The trends of our model are consistent on CTB5 and CTB7. Both performances increase sharply at $alpha = 0.2$ because context information is crucial for this sequential labeling task yet not utilized in the purely radical-driven model ($alpha = 0$). The best performances are achieved with alpha in the range of [0.6,0.8], which is balanced between the radical and context information. We select alpha=0.8 in the following experiment.

We compare our radical-enhanced model with C&W model and Word2Vec in the framework of Neural CRF. Table 1 shows that, our model obtains better P, R and F_1 than C&W and word2vec on both CTB5 and CTB7. One reason is that the radial-enhanced model is capable to capture the semantic connections between characters with the same radical, which usually have similar semantic meaning and grammatical usage yet not explicitly modeled in C&W and word2vec. Another reason is that, the embeddings of lower-frequent characters are typically not well estimated by C&W and word2vec due to the lack of syntactic contexts. In the radical-enhanced model, their radicals bring important semantic information thus we obtain better embedding results.

---

[5] http://www.sogou.com/labs/dl/c.html

Table 1. Comparison of F_1 on the test set of CTB5 and CTB7

Method	CTB5			CTB7		
	P	R	F_1	P	R	F_1
NeuralCRF(C&W)	0.9215	0.9306	0.9260	0.8956	0.8974	0.8965
NeuralCRF(word2vec)	0.9132	0.9257	0.9194	0.8910	0.8896	0.8903
NeuralCRF(Our model)	**0.9308**	**0.9451**	**0.9379**	**0.9047**	**0.9022**	**0.9034**
CRF(character)	0.9099	0.9141	0.9120	0.8805	0.8769	0.8787
CRF(character + radical)	0.9117	0.9153	0.9135	0.8816	0.8778	0.8797

We also compare with two CRF-based baseline methods. $CRF(character)$ is the use of linear-chain CRF with character as its feature. In $CRF(character + radical)$, we utilize the radical information and the character as features with linear-chain CRF. Results of $CRF(character)$ and $CRF(character + radical)$ show that simply using radical as feature does not obtain significant improvement. Our radical-enhanced method outperforms two CRF-based baselines on both datasets, which further verifies the effectiveness of our radical-enhanced model that integrates the radical information for learning Chinese character embedding.

As a reference, we also report some existing studies on Chinese word segmentation. Wang et al. [20] use plenty of hand-crafted features together with features derived from large auto-analyzed data, and achieves 0.9811 in F_1 on CTB5, and 0.9565 in F_1 on CTB7 using linear-chain CRF. Mansur et al. [3] propose a feature-based neural language model and apply the learned feature embedding into a deep neural architecture. Their system achieves 0.94 in F_1 on Sighan 2005 PKU dataset. Zheng et al. [4] propose a neural network for Chinese word segmentation. They use pre-trained character embedding learned by C&W from 325MB Sina news, and achieve 0.9457 in F_1 on CTB3.

6   Conclusion

In this paper, we propose to leverage radical for enhancing the continuous representation of Chinese characters. To our best knowledge, this is the first work that utilizes the radical information for learning Chinese character embedding. A dedicated neural architecture with a hybrid loss function is introduced to effectively integrate radical information through the softmax layer. As a result, the radical-enhanced model is capable to capture the semantic connections between characters from both syntactic contexts and the radical information. The effectiveness of our method has been verified on Chinese word segmentation. Experiment results show that, our method outperforms two widely-accepted context-based embedding learning algorithms, which do not utilize the radical information.

Acknowledgements. We thank Duyu Tang, Furu Wei, Yajuan Duan and Meishan Zhang for their helpful discussions. This work is supported by National Natural Science Foundation of China (No. 61300114), Specialized Research Fund for the Doctoral Program of Higher Education (No. 20132302120047), China Postdoctoral Science special Foundation (No.2014T70340), China Postdoctoral Science Foundation (No. 2013M530156), the Key Basic Research Foundation of Shenzhen (JC201005260118A) and National Natural Science Foundation of China (61100094 & 61300114).

References

1. Collobert, R., Weston, J.: A unified architecture for natural language processing: Deep neural networks with multitask learning. In: Proceedings of the 25th International Conference on Machine Learning, pp. 160–167. ACM (2008)
2. Mikolov, T., Chen, K., Corrado, G., Dean, J.: Efficient estimation of word representations in vector space. In: ICLR (2013)
3. Mansur, M., Pei, W., Chang, B.: Feature-based neural language model and chinese word segmentation. In: IJCNLP (2013)
4. Zheng, X., Chen, H., Xu, T.: Deep learning for chinese word segmentation and pos tagging. In: EMNLP (2013)
5. Fu, R., Qin, B., Liu, T.: Exploiting multiple sources for open-domain hypernym discovery. In: EMNLP, pp. 1224–1234 (2013)
6. Jin, P., Carroll, J., Wu, Y., McCarthy, D.: Improved word similarity computation for chinese using sub-word information. In: 2011 Seventh International Conference on Computational Intelligence and Security (CIS), pp. 459–462. IEEE (2011)
7. Jin, P., Carroll, J., Wu, Y., McCarthy, D.: Distributional similarity for chinese: Exploiting characters and radicals. In: Mathematical Problems in Engineering (2012)
8. Zhang, M., Zhang, Y., Che, W., Liu, T.: Chinese parsing exploiting characters. In: Proc. ACL (Volume 1: Long Papers), Sofia, Bulgaria, pp. 125–134. Association for Computational Linguistics (August 2013)
9. Turney, P.D., Pantel, P., et al.: From frequency to meaning: Vector space models of semantics. Journal of Artificial Intelligence Research 37(1), 141–188 (2010)
10. Turian, J., Ratinov, L., Bengio, Y.: Word representations: a simple and general method for semi-supervised learning. ACL (2010)
11. Bengio, Y.: Deep learning of representations: Looking forward. arXiv preprint arXiv:1305.0445 (2013)
12. Bengio, Y., Ducharme, R., Vincent, P., Janvin, C.: A neural probabilistic language model. The Journal of Machine Learning Research 3, 1137–1155 (2003)
13. Mikolov, T., Karafiát, M., Burget, L., Cernocký, J., Khudanpur, S.: Recurrent neural network based language model. In: INTERSPEECH, pp. 1045–1048 (2010)
14. Huang, E.H., Socher, R., Manning, C.D., Ng, A.Y.: Improving word representations via global context and multiple word prototypes. In: Proc. ACL, pp. 873–882. Association for Computational Linguistics (2012)
15. Tang, D., Wei, F., Yang, N., Zhou, M., Liu, T., Qin, B.: Learning sentiment-specific word embedding for twitter sentiment classification. In: Proc. ACL (Volume 1: Long Papers), Baltimore, Maryland, pp. 1555–1565. Association for Computational Linguistics (June 2014)
16. Luong, M.-T., Socher, R., Manning, C.D.: Better word representations with recursive neural networks for morphology. In: CoNLL 2013, p. 104 (2013)
17. Collobert, R., Weston, J., Bottou, L., Karlen, M., Kavukcuoglu, K., Kuksa, P.: Natural language processing (almost) from scratch. The Journal of Machine Learning Research 12, 2493–2537 (2011)
18. Forney Jr., G.D.: The viterbi algorithm. Proceedings of the IEEE 61(3), 268–278 (1973)
19. Jiang, W., Huang, L., Liu, Q., Lü, Y.: A cascaded linear model for joint chinese word segmentation and part-of-speech tagging. In: Proc. ACL. Citeseer (2008)
20. Wang, Y., Jun'ichi Kazama, Y.T., Tsuruoka, Y., Chen, W., Zhang, Y., Torisawa, K.: Improving chinese word segmentation and pos tagging with semi-supervised methods using large auto-analyzed data. In: IJCNLP, pp. 309–317 (2011)

Conditional Multidimensional Parameter Identification with Asymmetric Correlated Losses of Estimation Errors

Piotr Kulczycki[1] and Malgorzata Charytanowicz[1,2]

[1] Polish Academy of Sciences, Systems Research Institute, Poland
[2] The John Paul II Catholic University of Lublin, Poland
{kulczycki,mchmat}@ibspan.waw.pl

Abstract. This paper is dedicated to the problem of the estimation of a vector of parameters, as losses resulting from their under- and overestimation are asymmetric and mutually correlated. The issue is considered from an additional conditional aspect, where particular coordinates of conditioning variables may be continuous, binary, discrete or categorized (ordered and unordered). The final result is an algorithm for calculating the value of an estimator, optimal in sense of expectation of losses using a multidimensional asymmetric quadratic function, for practically any distributions of describing and conditioning variables.

Keywords: parameters' vector identification, conditional factors, Bayes approach, asymmetric loss function, distribution free method, numerical algorithm.

1   Introduction

The proper identification (estimation) of parameters values, used in a model describing the reality under consideration, is always of fundamental significance in modern problems of science and practice. The need to consider implications of estimations errors different for under- and overestimations, leads directly to the concept of asymmetrical form of a loss function [Berger, 1980]. The significance of this problem has been investigated for simple cases of a single parameter [Zellner, 1985; McCullough, 2000]. It is also worth noting the results concerning the estimation of a single parameter with asymmetrical loss function, described in the paper [Kulczycki and Charytanowicz, 2013] in the conditional version, i.e. where the quantity under research is significantly dependent on conditional factors. If the actual value of factors of this type is available metrologically, their inclusion can make the model used considerably more precise. In this paper that research is generalized for the multidimensional case, where one identifies a few independent parameters, treated as a vector, and the losses resulting from the over- and underestimation may be asymmetrical and correlated. The concept presented here is based on the Bayes approach, which allows minimization of expected value of losses arising from estimation errors. For defining probability characteristics, the nonparametric methodology of statistical kernel

C.K. Loo et al. (Eds.): ICONIP 2014, Part II, LNCS 8835, pp. 287–294, 2014.
© Springer International Publishing Switzerland 2014

estimators was used, which freed the investigated procedure from forms of distributions characterizing both the identified parameters and conditioning quantities.

2   Statistical Kernel Estimators

Let the n-dimensional random variable X be given, with a distribution characterized by the density f. Its kernel estimator $\hat{f}: \mathbb{R}^n \to [0, \infty)$, calculated using experimentally obtained values for the m-element random sample

$$x_1, x_2, \ldots, x_m,$$ (1)

in its basic form is defined as

$$\hat{f}(x) = \frac{1}{mh^n} \sum_{i=1}^{m} K\left(\frac{x - x_i}{h}\right),$$ (2)

where $m \in \mathbb{N} \setminus \{0\}$, the coefficient $h > 0$ is called a smoothing parameter, while the measurable function $K: \mathbb{R}^n \to [0, \infty)$ of unit integral $\int_{\mathbb{R}^n} K(x)\,dx = 1$, symmetrical with respect to zero and having a weak global maximum in this place, takes the name of a kernel. The method presented here uses the one-dimensional Cauchy kernel, in the n-dimensional case generalized to the product kernel. For fixing the smoothing parameter value, the plug-in method is recommended, with the modification of this parameter. Details are found in the books [Kulczycki, 2005; Silverman, 1986; Wand and Jones, 1994].

The above concept will now be generalized for the conditional case. Here, besides the basic (termed the describing) n_Y-dimensional random variable Y, let also be given the n_W-dimensional random variable W, called hereinafter the conditioning random variable. Their composition $X = \begin{bmatrix} Y \\ W \end{bmatrix}$ is a random variable of the dimension $n_Y + n_W$. Assume that distributions of the variables X and, in consequence, W have densities, denoted below as $f_X : \mathbb{R}^{n_Y + n_W} \to [0, \infty)$ and $f_W : \mathbb{R}^{n_W} \to [0, \infty)$, respectively. Let also be given the so-called conditioning value, that is the fixed value of conditioning random variable $w^* \in \mathbb{R}^{n_W}$ such that $f_W(w^*) > 0$, and also the random sample

$$\begin{bmatrix} y_1 \\ w_1 \end{bmatrix}, \quad \begin{bmatrix} y_2 \\ w_2 \end{bmatrix}, \ldots, \begin{bmatrix} y_m \\ w_m \end{bmatrix},$$ (3)

obtained from the variable X. The particular elements of this sample are interpreted as the values y_i taken in measurements from the random variable Y, when the conditioning variable W assumes the respective values w_i. The kernel estimator of conditional density of the random variable Y distribution for the conditioning value w^*, i.e. $\hat{f}_{Y|W=w^*} : \mathsf{R}^{n_Y} \to [0, \infty)$, can be given by the following form helpful in practice:

$$\hat{f}_{Y|W=w^*}(y) = \hat{f}_{Y|W=w^*}\left(\begin{bmatrix} y_1 \\ y_2 \\ \vdots \\ y_{n_Y} \end{bmatrix}\right) = \tag{4}$$

$$= \frac{1}{h_1 h_2 \dots h_{n_Y} \sum_{i=1}^{m} d_i} \sum_{i=1}^{m} d_i K_1\left(\frac{y_1 - y_{i,1}}{h_1}\right) K_2\left(\frac{y_2 - y_{i,2}}{h_2}\right) \dots K_{n_Y}\left(\frac{y_{n_Y} - y_{i,n_Y}}{h_{n_Y}}\right),$$

where h_1, $h_2, \dots, h_{n_Y + n_W}$ represent smoothing parameters for particular coordinates of the random variable X, while the coordinates of the vectors w^*, y_i and w_i are denoted as

$$w^* = \begin{bmatrix} w_1^* \\ w_2^* \\ \vdots \\ w_{n_W}^* \end{bmatrix} \quad \text{and} \quad y_i = \begin{bmatrix} y_{i,1} \\ y_{i,2} \\ \vdots \\ y_{i,n_Y} \end{bmatrix}, \quad w_i = \begin{bmatrix} w_{i,1} \\ w_{i,2} \\ \vdots \\ w_{i,n_W} \end{bmatrix} \quad \text{for} \quad i = 1, 2, \dots, m, \tag{5}$$

whereas the so-called conditioning parameters d_i for $i = 1, 2, \dots, m$ can be defined by

$$d_i = K_{n_Y+1}\left(\frac{w_1^* - w_{i,1}}{h_{n_Y+1}}\right) K_{n_Y+2}\left(\frac{w_2^* - w_{i,2}}{h_{n_Y+2}}\right) \dots K_{n_Y+n_W}\left(\frac{w_{n_W}^* - w_{i,n_W}}{h_{n_Y+n_W}}\right). \tag{6}$$

The value of the parameter d_i characterizes the "distance" of the given conditioning value w^* from w_i – that of the conditioning variable for which the i-th element of the random sample was obtained. Then estimator (4) can be interpreted as the linear combination of kernels mapped to particular elements of a random sample obtained for the variable Y, when the coefficients of this combination characterize how representative these elements are for the given value w^*.

3   An Algorithm

Consider the parameters, whose values are to be estimated, denoted in the form of the vector $y \in \mathsf{R}^{n_Y}$. It will be treated as the value of the n_Y-dimensional random variable Y. Let also the n_W-dimensional conditional random variable W be given. The availability is assumed of the metrologically achieved measurements of the parameters' vector y, i.e. y_1, $y_2, \dots, y_m \in \mathsf{R}^{n_Y}$, obtained for the values w_1, $w_2, \dots, w_m \in \mathsf{R}^{n_W}$ of the conditional variable, respectively. Finally, let $w^* \in \mathsf{R}^{n_W}$ denote any fixed conditioning value. The goal is to calculate the estimator of this parameter's vector, optimal in the sense of minimum expected value of losses arising from errors of estimation, for conditioning value w^*. In order to solve such a task, the Bayes decision rule will be used. Because of clarity of presentation, a two-dimensional case ($n_Y = 2$) will be considered here. The idea itself may be transposed for larger dimensions, although at a natural – in such a situation – cost of increasing complexity.

Let therefore the estimated parameters be treated as the two-dimensional vector $\begin{bmatrix} y_1 \\ y_2 \end{bmatrix}$, as their estimators $\begin{bmatrix} \hat{y}_1 \\ \hat{y}_2 \end{bmatrix}$. The loss function $l : R^2 \times R^2 \rightarrow R$, which in accordance with the decision theory principles [Berger, 1980] defines losses occurring when the value $\begin{bmatrix} \hat{y}_1 \\ \hat{y}_2 \end{bmatrix}$ has been taken, while in reality the hypothetical state was $\begin{bmatrix} y_1 \\ y_2 \end{bmatrix}$, is assumed in a quadratic and asymmetrical form:

$$l\left(\begin{bmatrix} \hat{y}_1 \\ \hat{y}_2 \end{bmatrix}, \begin{bmatrix} y_1 \\ y_2 \end{bmatrix}\right) = \begin{cases} a_l(\hat{y}_1 - y_1)^2 + a_{ld}(\hat{y}_1 - y_1)(\hat{y}_2 - y_2) + a_d(\hat{y}_2 - y_2)^2 \\ \qquad \text{if} \quad \hat{y}_1 - y_1 \leq 0 \text{ and } \hat{y}_2 - y_2 \leq 0 \\ a_r(\hat{y}_1 - y_1)^2 + a_{rd}(\hat{y}_1 - y_1)(\hat{y}_2 - y_2) + a_d(\hat{y}_2 - y_2)^2 \\ \qquad \text{if} \quad \hat{y}_1 - y_1 \geq 0 \text{ and } \hat{y}_2 - y_2 \leq 0 \\ a_l(\hat{y}_1 - y_1)^2 + a_{lu}(\hat{y}_1 - y_1)(\hat{y}_2 - y_2) + a_u(\hat{y}_2 - y_2)^2 \\ \qquad \text{if} \quad \hat{y}_1 - y_1 \leq 0 \text{ and } \hat{y}_2 - y_2 \geq 0 \\ a_r(\hat{y}_1 - y_1)^2 + a_{ru}(\hat{y}_1 - y_1)(\hat{y}_2 - y_2) + a_u(\hat{y}_2 - y_2)^2 \\ \qquad \text{if} \quad \hat{y}_1 - y_1 \geq 0 \text{ and } \hat{y}_2 - y_2 \geq 0 \end{cases} \tag{7}$$

where $a_l, a_r, a_u, a_d > 0$, $a_{ld}, a_{ru} \geq 0$ and $a_{lu}, a_{rd} \leq 0$. The coefficients a_{ld}, a_{ru}, a_{lu}, a_{rd} represent the complementary correlation of estimation errors for both parameters.

Assume conditional independence [Dawid, 1979] of the estimated parameters. Then the density $f_{Y|W=w^*}$ representing their uncertainty may be shown as the product of the one-dimensional densities $f_{Y_1|W=w^*} : R \rightarrow [0, \infty)$ and $f_{Y_2|W=w^*} : R \rightarrow [0, \infty)$ corresponding to particular composites, i.e.

$$f_{Y|W=w^*}(y_1, y_2) = f_{Y_1|W=w^*}(y_1) f_{Y_2|W=w^*}(y_2) . \tag{8}$$

Let also the functions $f_{Y_1|W=w^*}$ and $f_{Y_2|W=w^*}$ be continuous and such that $\int_{-\infty}^{\infty} y_1 f_{Y_1|W=w^*}(y_1) \, dy_1 < \infty$ as well as $\int_{-\infty}^{\infty} y_2 f_{Y_2|W=w^*}(y_2) \, dy_2 < \infty$.

Detailed analysis [Kulczycki and Charytanowicz, 2014] shows that the criterion minimizing the expectation value of losses takes on the form of equations unsolvable in practice. Although if estimation of the densities presented above is reached using the kernel estimators, then one can design an effective numerical algorithm to this end. Thus, with any fixed $i = 1, 2, \ldots, m$, one can define the functions $U_{1,i} : R \to R$, $U_{2,i} : R \to R$, $V_{1,i} : R \to R$ and $V_{2,i} : R \to R$, given as

$$U_{1,i}(\hat{y}_1) = \frac{1}{h_1} \int\limits_{-\infty}^{\hat{y}_1} K\left(\frac{y_1 - y_{i,1}}{h_1}\right) dy_1 \tag{9}$$

$$U_{2,i}(\hat{y}_2) = \frac{1}{h_2} \int\limits_{-\infty}^{\hat{y}_2} K\left(\frac{y_2 - y_{i,2}}{h_2}\right) dy_2 \tag{10}$$

$$V_{1,i}(\hat{y}_1) = \frac{1}{h_1} \int\limits_{-\infty}^{\hat{y}_1} y_1 K\left(\frac{y_1 - y_{i,1}}{h_1}\right) dy_1 \tag{11}$$

$$V_{2,i}(\hat{y}_2) = \frac{1}{h_2} \int\limits_{-\infty}^{\hat{y}_2} y_2 K\left(\frac{y_2 - y_{i,2}}{h_2}\right) dy_2 . \tag{12}$$

Norm also the conditioning parameters d_i by introducing the positive values

$$d_i^* = \frac{d_i}{\sum_{i=1}^{m} d_i} \qquad \text{for} \quad i = 1, 2, \ldots, m ; \tag{13}$$

note that $\sum_{i=1}^{m} d_i^* = 1$. Then criterion for the vector $\begin{bmatrix} \hat{y}_1 \\ \hat{y}_2 \end{bmatrix}$ ensuring the minimum of

expectation value of losses for loss function (7) takes the form of the equations

$$\sum_{i=1}^{m} d_i^* U_{1,i}(\hat{y}_1)\left[(a_{ru} - a_{rd} - a_{lu} + a_{ld})\sum_{i=1}^{m} d_i^*\left(\hat{y}_2 U_{2,i}(\hat{y}_2) - V_{2,i}(\hat{y}_2)\right) + (a_{rd} - a_{ld})(\hat{y}_2 - \sum_{i=1}^{m} d_i^* y_{i,2})\right] +$$

$$+ 2a_1(\hat{y}_1 - \sum_{i=1}^{m} d_i^* y_{i,1}) + 2(a_r - a_1)\sum_{i=1}^{m} d_i^*\left(\hat{y}_1 U_{1,i}(\hat{y}_1) - V_{1,i}(\hat{y}_1)\right) + a_{ld}(\hat{y}_2 - \sum_{i=1}^{m} d_i^* y_{i,2}) +$$

$$+ (a_{lu} - a_{ld})\sum_{i=1}^{m} d_i^*\left(\hat{y}_2 U_{2,i}(\hat{y}_2) - V_{2,i}(\hat{y}_2)\right) = 0 \tag{14}$$

$$\sum_{i=1}^{m} d_i^* U_{2,i}(\hat{y}_2)\left[(a_{ru} - a_{rd} - a_{lu} + a_{ld})\sum_{i=1}^{m} d_i^*\left(\hat{y}_1 U_{1,i}(\hat{y}_1) - V_{1,i}(\hat{y}_1)\right) + (a_{lu} - a_{ld})(\hat{y}_1 - \sum_{i=1}^{m} d_i^* y_{i,1})\right] +$$

$$+ 2a_d(\hat{y}_2 - \sum_{i=1}^{m} d_i^* y_{i,2}) + 2(a_u - a_d)\sum_{i=1}^{m} d_i^*\left(\hat{y}_2 U_{2,i}(\hat{y}_2) - V_{2,i}(\hat{y}_2)\right) + a_{ld}(\hat{y}_1 - \sum_{i=1}^{m} d_i^* y_{i,1}) +$$

$$+ (a_{rd} - a_{ld})\sum_{i=1}^{m} d_i^*\left(\hat{y}_1 U_{1,i}(\hat{y}_1) - V_{1,i}(\hat{y}_1)\right) = 0 . \tag{15}$$

If one denotes the left sides of the above equations as $L_1(\hat{y}_1, \hat{y}_2)$ and $L_2(\hat{y}_1, \hat{y}_2)$, their partial derivatives are given by

$$\frac{\partial L_1(\hat{y}_1, \hat{y}_2)}{\partial \hat{y}_1} = \sum_{i=1}^{m} d_i^* \frac{1}{h_1 s_{i,1}} K\left(\frac{\hat{y}_1 - y_{i,1}}{h_1 s_{i,1}}\right) \left[(a_{ru} - a_{rd} - a_{lu} + a_{ld}) \sum_{i=1}^{m} d_i^* \left(\hat{y}_2 U_{2,i}(\hat{y}_2) - V_{2,i}(\hat{y}_2)\right) + \right.$$

$$\left. + (a_{rd} - a_{ld})(\hat{y}_2 - \sum_{i=1}^{m} d_i^* y_{i,2}) \right] + 2(a_r - a_l) \sum_{i=1}^{m} d_i^* U_{1,i}(\hat{y}_1) + 2a_l \tag{16}$$

$$\frac{\partial L_1(\hat{y}_1, \hat{y}_2)}{\partial \hat{y}_2} = \sum_{i=1}^{m} d_i^* U_{1,i}(\hat{y}_1) \left[(a_{ru} - a_{rd} - a_{lu} + a_{ld}) \sum_{i=1}^{m} d_i^* U_{2,i}(\hat{y}_2) + (a_{rd} - a_{ld}) \right] + $$

$$+ (a_{lu} - a_{ld}) \sum_{i=1}^{m} d_i^* U_{2,i}(\hat{y}_2) + a_{ld} \tag{17}$$

$$\frac{\partial L_2(\hat{y}_1, \hat{y}_2)}{\partial \hat{y}_1} = \sum_{i=1}^{m} d_i^* U_{2,i}(\hat{y}_2) \left[(a_{ru} - a_{rd} - a_{lu} + a_{ld}) \sum_{i=1}^{m} d_i^* U_{1,i}(\hat{y}_1) + (a_{lu} - a_{ld}) \right] + $$

$$+ (a_{rd} - a_{ld}) \sum_{i=1}^{m} d_i^* U_{1,i}(\hat{y}_1) + a_{ld} \tag{18}$$

$$\frac{\partial L_2(\hat{y}_1, \hat{y}_2)}{\partial \hat{y}_2} = \sum_{i=1}^{m} d_i^* \frac{1}{h_2 s_{i,2}} K\left(\frac{\hat{y}_2 - y_{i,2}}{h_2 s_{i,2}}\right) \left[(a_{ru} - a_{rd} - a_{lu} + a_{ld}) \sum_{i=1}^{m} d_i^* \left(\hat{y}_1 U_{1,i}(\hat{y}_1) - V_{1,i}(\hat{y}_1)\right) + \right.$$

$$\left. + (a_{lu} - a_{ld})(\hat{y}_1 - \sum_{i=1}^{m} d_i^* y_{i,1}) \right] + 2(a_u - a_d) \sum_{i=1}^{m} d_i^* U_{2,i}(\hat{y}_2) + 2a_d. \tag{19}$$

Then the solution of equations (14)-(15) can be calculated through Newton's multi-dimensional algorithm [Stoer and Bulirsch, 2002] as the limit of the two-dimensional sequence $\left\{ \begin{matrix} \hat{y}_{j,1} \\ \hat{y}_{j,2} \end{matrix} \right\}_{j=0}^{\infty}$ defined by formulas

$$\hat{y}_{0,1} = \frac{\sum_{i=1}^{m} d_i y_{i,1}}{\sum_{i=1}^{m} d_i} \quad \text{and} \quad \hat{y}_{0,2} = \frac{\sum_{i=1}^{m} d_i y_{i,2}}{\sum_{i=1}^{m} d_i} \tag{20}$$

$$\begin{bmatrix} \hat{y}_{j+1,1} \\ \hat{y}_{j+1,2} \end{bmatrix} = \begin{bmatrix} \hat{y}_{j,1} \\ \hat{y}_{j,2} \end{bmatrix} - \begin{bmatrix} \dfrac{\partial L_1(\hat{y}_{j,1}, \hat{y}_{j,2})}{\partial \hat{y}_1} & \dfrac{\partial L_1(\hat{y}_{j,1}, \hat{y}_{j,2})}{\partial \hat{y}_2} \\ \dfrac{\partial L_2(\hat{y}_{j,1}, \hat{y}_{j,2})}{\partial \hat{y}_1} & \dfrac{\partial L_2(\hat{y}_{j,1}, \hat{y}_{j,2})}{\partial \hat{y}_2} \end{bmatrix}^{-1} \begin{bmatrix} L_1(\hat{y}_{j,1}, \hat{y}_{j,2}) \\ L_2(\hat{y}_{j,1}, \hat{y}_{j,2}) \end{bmatrix}$$

$$\text{for} \quad j = 0, 1, \ldots, \tag{21}$$

while the quantities in the above dependencies are given by equations (14)-(19), whereas a stop condition takes the form of the conjunction of the inequalities

$$|\hat{y}_{j,1} - \hat{y}_{j-1,1}| \le 0{,}01\, \hat{\sigma}_1 \quad \text{and} \quad |\hat{y}_{j,2} - \hat{y}_{j-1,2}| \le 0{,}01\, \hat{\sigma}_2, \tag{22}$$

where $\hat{\sigma}_1$ and $\hat{\sigma}_2$ denote the estimators of standard deviations for particular coordinates of the vector Y.

4   Final Comments and Summary

This paper presents the algorithm for calculating the conditional estimator of the vector of independent parameters, where losses resulting from under- and overestimation are asymmetrical and mutually correlated. The conditional approach allows in practice for refinement of the model by including the current value of the conditioning factors. Use of the Bayes approach ensures a minimum expected value of losses, a statistical kernel estimators methodology frees the investigated procedure from forms of distributions of the describing and conditioning factors.

The correct performance of the algorithm has been proved in many numerical tests with illustrative generated data, and also by simulations, as well as by applying them to practical problems from control engineering, biomedicine and marketing. Above all, the general rule was translations of the examined estimator values in directions associated with smaller losses resulting from estimation errors, defined by loss function (7). Specifically, an increase in the value of the parameter a_l with respect to a_r, and so a growth in the value of this function for positive estimation errors of the first parameter (its overestimation), implied an increase in the value of the obtained estimator for this parameter. In consequence it reduces the probability of an overestimation. The opposite occurs when the parameter a_r value is increased with respect to a_l: the value of the obtained estimator decreases, which lowers the probability of underestimation. The more the ratio a_l/a_r differed from 1, the more intensive are the above effects. Analogous dependences appeared for the parameters a_d and a_u when estimating the second parameter. Subsequently the increase in the parameter a_{ld} value resulted in the simultaneous growth in the two parameters' values, reducing in both cases the probability of overestimation. Converse effects implied changes in the parameter a_{ru}. And finally, an increase in the absolute value of the parameter a_{rd} reduces the probability of underestimation of the first parameter as well as overestimation of the second, through a decrease in the value of the obtained estimator for the first, and an increase for the second. The opposite applies to the parameter a_{lu}.

The conditional approach implied the appropriate correction to the estimator value according to the nature of the correlation between describing variables and conditioning factors. If a parameter was positively correlated to such factor, an increase/decrease in the condition value resulted in an increase/decrease in the estimator value for that parameter. The opposite occurred for a negative correlation. Such relation may be more complex, according to any potential form of the dependence of the conditional densities $f_{Y_1|W=w^*}$ and $f_{Y_2|W=w^*}$ on conditioning values w^*.

An acceptable quality of results was obtained from sample sizes of just 50-100 when the conditioning value was positioned close to the main modal value of a conditional variable, and 100-200 at distance of standard deviation. Taking into account the complex multidimensional character of the task, it does not seem to be an excessive requirement in practice. Thanks to the averaging properties of kernel estimators, the algorithm proved to be robust to a small number or even lack of data from the neighborhood of a conditioning value.

The procedure presented in this paper has been given in its basic form. A clear interpretation means it is possible to make individual modifications. Above all this allows the inclusion of conditional factors other than continuous (real). Similarly to the kernel estimation definition formulated above for continuous random variables, one can construct kernel estimators for binary, discrete and categorized (including ordered) variables, as well as any of their compositions, especially with continuous variables – for details see broad and varied literature in this subject. The above can be particularly useful for the modern data analysis tasks, which more and more often take advantage of the many different configurations for particular types of attributes.

A broad description of the methodology introduced in this paper is presented in the article [Kulczycki and Charytanowicz, 2014] together with results of detailed verifications. The algorithm itself is given there in its ready-to-use form and can be applied directly without deep subject knowledge or laborious research.

Acknowledgments. Our heartfelt thanks go to our colleague Dr. Aleksander Mazgaj, with whom we commenced the research presented here. With his consent, this tex t also contains results of joint research.

References

1. Berger, J.O.: Statistical Decision Theory. Springer, New York (1980)
2. Dawid, A.P.: Conditional Independence in Statistical Theory. Journal of the Royal Statistical Society, Series B 41, 1–31 (1979)
3. Kulczycki, P.: Estymatory jadrowe w analizie systemowej. WNT, Warsaw (2005)
4. Kulczycki, P., Charytanowicz, M.: Conditional Parameter Identification with Different Losses of Under- and Overestimation. Applied Mathematical Modelling 37, 2166–2177 (2013)
5. Kulczycki, P., Charytanowicz, M.: An Algorithm for Conditional Multidimensional Parameter Identification with Asymmetric and Correlated Looses of Under- and Overestimations (in press, 2014)
6. Lehmann, E.L.: Theory of Point Estimation. Wiley, New York (1983)
7. McCullough, B.D.: Optimal Prediction with a General Loss Function. Journal of Combinatorics, Information & System Sciences 25, ss.207–ss.221 (2000)
8. Silverman, B.W.: Density Estimation for Statistics and Data Analysis. Chapman and Hall, London (1986)
9. Stoer, J., Bulirsch, R.: Introduction to Numerical Analysis. Springer, New York (2002)
10. Wand, M.P., Jones, M.C.: Kernel Smoothing. Chapman and Hall, London (1995)
11. Zellner, A.: Bayesian Estimation and Prediction Using Asymmetric Loss Function. Journal of the American Statistical Association 81, 446–451 (1985, 1986)

Short Text Hashing Improved by Integrating Topic Features and Tags

Jiaming Xu, Bo Xu, Jun Zhao, Guanhua Tian, Heng Zhang, and Hongwei Hao

Institute of Automation, Chinese Academy of Sciences, 100190 Beijing, P.R. China
{jiaming.xu,boxu,guanhua.tian,hongwei.hao}@ia.ac.cn,
{jzhao,hzhang07}@nlpr.ia.ac.cn

Abstract. Hashing, as an efficient approach, has been widely used for large-scale similarity search. Unfortunately, many existing hashing methods based on observed keyword features are not effective for short texts due to the sparseness and shortness. Recently, some researchers try to construct semantic relationship using certain granularity topics. However, the topics of certain granularity are insufficient to preserve the optimal semantic similarity for different types of datasets. On the other hand, tag information should be fully exploited to enhance the similarity of related texts. We, therefore, propose a novel unified hashing approach that the optimal topic features can be selected automatically to be integrated with original features for preserving similarity, and tags are fully utilized to improve hash code learning. We carried out extensive experiments on one short text dataset and even one normal text dataset. The results demonstrate that our approach is effective and significantly outperforms baseline methods on several evaluation metrics.

Keywords: Similarity Search, Hashing, Topic Features, Short Text.

1 Introduction

With the explosion of social media, numerous short texts become available in a variety of genres, e.g. tweets, instant messages, and questions in Q&A (Question and Answer) websites. In order to conduct fast similarity search in those massive datasets, hashing, which tries to learn similarity-preserving binary codes for document representation, has been widely used to accelerate similarity search.

Unfortunately, many existing hashing methods based on keyword space usually fail to fully preserve the semantic similarity for short texts due to the sparseness of text representation. In recent years, some researchers attempt to address the challenge by latent semantic approach. For example, Wang et al. [10] preserve the similarity of documents by fitting certain level topic distributions.

However, topics of a certain level are not sufficient to represent the optimal semantic information for different types of datasets. For instance, there are two short texts: *"Rafael Nadal missed the Australian Open"* and *"Roger Federer won Grand Slam title"*. If the number of topics is not large enough, the topic-based methods usually preserve similarity in coarse semantic features, such as *"Sport"*

C.K. Loo et al. (Eds.): ICONIP 2014, Part II, LNCS 8835, pp. 295–302, 2014.

and *"Tennis"*, missing some more fine grained topics, such as, *"Tennis Open Progress"* or *"Tennis Star News"*. Thus, we should select the optimal topics for certain dataset, and as a reasonable assumption that the new feature space integrated with latent topic features and observed keyword features are more sufficient to set up an effective feature space for hashing learning.

On the other hand, tags are not fully utilized in many hashing methods. Actually, in many real-world applications, documents are often associated with multiple tags [10]. For instance, in Q&A websites, each question has tags assigned by its questioner. Another example is microblog, some tweets are labeled with hashtags. Thus, we should make use of these tags for hashing learning.

Based on the above observations, this paper proposes a unified *short text Hashing with Topics and Tags*, dubbed HTT for simplicity.

The main contributions of our approach are listed as follows:

a). A unified short text hashing is proposed. This is the first time of incorporating keyword features, topic features and tags all into a unified hashing approach to solve hashing problem as far as we know, and the experimental results indicate that our approach outperforms the existing hashing techniques;

b). The optimal topic features can be selected automatically for certain type of dataset, and the experimental results show that the topic feature selection can achieve a better performance for the proposed hashing learning, compared with no the optimal topics selection;

c). Flexible to extend. We may even consider incorporating multi-granularity topic features to improve binary code learning in our proposed approach.

2   Related Work

Hash-based methods can be mainly divided into two categories. One category is data-oblivious hashing. As the most popular hashing technique, Locality-Sensitive Hashing (LSH) [1] based on random projection has been widely used for similarity search. However, for they do not aware of data distribution, those methods may lead to generating quite inefficient hash codes in practice [12]. Recently, more researchers focus attention on the other category, data-aware hashing, by using machine learning algorithms. For example, the Spectral Hashing (SpH) [11] generates compact binary codes by forcing the balanced and uncorrelated constraints into the learned codes. Self-Taught Hashing (STH) [14] is a method which decomposes the learning procedure into two steps: generating binary code and learning hash function. A supervised version of STH is proposed in [12] denoted as STHs. However, the previous hashing methods, directly working in keyword feature space, usually fail to fully preserve semantic similarity for short texts due to shortness and sparseness of text representations.

More recently, the most similar to our study is probably Semantic Hashing using Tags and Topic Modeling (SHTTM) [10], which attempts to learn binary codes by incorporating both the tags and topics. However, the problems of SHTTM are that: Although the topic distributions are used to preserve the content similarity in hash codes, they do not utilize the topics to improve hashing

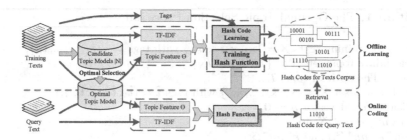

Fig. 1. The proposed approach HTT for short text hashing

function learning; Even the number of topics must keep consistent with dimensions of hash code, that this assumption is too strict to capture the optimal semantic features for different types of datasets.

3 Algorithm Description

The proposed HTT is depicted in Fig. 1. Given a dataset of n training documents denoted as: $\mathbf{X} = \{\mathbf{x}_1, \mathbf{x}_2..., \mathbf{x}_n\} \in \mathbf{R}^{m \times n}$. Denote their tags as: $\mathbf{t} = \{\mathbf{t}_1, \mathbf{t}_2..., \mathbf{t}_n\} \in \{0,1\}^{q \times n}$, where q is the total number of possible tags associated with each document. The goal of HTT is to obtain the optimal binary hash codes $\mathbf{Y} = \{\mathbf{y}_1, \mathbf{y}_2..., \mathbf{y}_n\}^T \in \{-1,1\}^{n \times l}$ for $\mathbf{X}$, and the hashing functions f: $\mathbf{R}^m \to \{-1,1\}^l$, which embed each document to binary code with l bits.

3.1 Estimate and Select the Optimal Topic Model

In order to select the optimal topic model, tags are utilized to evaluate the quality of topics. We estimate N different sets of topics, $\mathbf{T} = \{T_1, T_2, ..., T_N\}$. For each entry T_i, the probability topic distributions over documents are denoted as $\boldsymbol{\theta} = p(\mathbf{z}|\mathbf{x})$. $\boldsymbol{\mu} = \{\mu(T_1), \mu(T_2), ..., \mu(T_N)\}$ is the weight vector, where $\mu(T_i)$ indicate the importance of topic set T_i. The purpose is to select the optimal topic model T_O. Inspired by [5,4], we propose the optimal selection method as following. Firstly, a sub-set $\hat{\mathbf{X}} = \{\hat{\mathbf{x}}_1, \hat{\mathbf{x}}_2, ..., \hat{\mathbf{x}}_m\}$ with tags $\hat{\mathbf{t}} = \{\hat{\mathbf{t}}_1, \hat{\mathbf{t}}_2, ..., \hat{\mathbf{t}}_m\}$ is sampled from training dataset, and we find two groups of k nearest neighbors for each text $\hat{x}_i$: one group $(k^+ nn^+(\hat{\mathbf{x}}))$ is from the texts sharing any common tags, and the other $(k^- nn^-(\hat{\mathbf{x}}))$ is from the texts not sharing any common tags. Then the weight value is updated as follows:

$$\mu(T_i) = \mu(T_i) + \sum_{j=1}^{k^-} \frac{D_{KL}(T_i(\hat{\mathbf{x}}), T_i(nn_j^-(\hat{\mathbf{x}})))}{k^-} - \sum_{p=1}^{k^+} \frac{D_{KL}(T_i(\hat{\mathbf{x}}), T_i(nn_p^+(\hat{\mathbf{x}})))}{k^+} \quad (1)$$

Where, $D_{KL}(T_i(\hat{\mathbf{x}}), T_i(nn_j^-(\hat{\mathbf{x}}))) = \frac{1}{2} \sum_{z_k \in T_i} (p(z_k|\hat{\mathbf{x}}) \cdot log(\frac{p(z_k|\hat{\mathbf{x}})}{p(z_k|nn_j^-(\hat{\mathbf{x}}))}) +$

$p(z_k|nn_j^-(\hat{\mathbf{x}})) \cdot log(\frac{p(z_k|nn_j^-(\hat{\mathbf{x}}))}{p(z_k|\hat{\mathbf{x}})}))$, so is the value of $D_{KL}(T_i(\hat{\mathbf{x}}), T_i(nn_p^+(\hat{\mathbf{x}})))$.

Algorithm 1. The Optimal Topic Selection Procedure

Input:
   A set of n training texts $\mathbf{X} = \{\mathbf{x}_1, \mathbf{x}_2, ..., \mathbf{x}_n\}$ with tags $\mathbf{t} = \{\mathbf{t}_1, \mathbf{t}_2, ..., \mathbf{t}_n\}$,
   N candidate topic sets $\mathbf{T} = \{T_1, T_2, ..., T_N\}$.
Output: The optimal topic T_O.
1: Sample a sub-set $\hat{\mathbf{X}} = \{\hat{\mathbf{x}}_1, \hat{\mathbf{x}}_2, ..., \hat{\mathbf{x}}_m\}$ with tags $\hat{\mathbf{t}} = \{\hat{\mathbf{t}}_1, \hat{\mathbf{t}}_2, ..., \hat{\mathbf{t}}_m\}$;
2: Initialize weight vector $\mu \leftarrow \mathbf{0}$;
3: **for** each text $\hat{\mathbf{x}} \in \hat{\mathbf{X}}$ **do**
4:    Find $k^+ nn^+(\hat{\mathbf{x}})$ and $k^- nn^-(\hat{\mathbf{x}})$
5:    **for** $i \leftarrow 1$ to N **do**
6:       Update $\mu(T_i)$ by Eq. 1;
7:    **end for**
8: **end for**
9: **return** $T_O = \arg\max_{T_i \in \mathbf{T}} \mu(T_i)$;

After updating $\mu(T_i)$, we can obtain the optimal topic T_O with a highest weight value. In summary, the selection procedure is described in Algorithm 1.

3.2   Hash Codes Learning with Topic Features and Tags

Here, we adopt a simple but powerful way to combine original features and topics, similar as [8,4], and create new features $\tilde{\mathbf{x}} = [\mathbf{W}, \lambda\boldsymbol{\theta}]$, where $\mathbf{W}$ is TF-IDF (Term Frequency-Inverse Document Frequency) vector, $\boldsymbol{\theta}$ is topic distribution, and λ is a non-negative combination coefficient for balancing those two features. In order to utilize tags, we construct $n \times n$ local similarity matrix $\mathbf{S}$ as follows:

$$S_{ij} = \begin{cases} c_{ij} \cdot \frac{\tilde{\mathbf{x}}_i^T \tilde{\mathbf{x}}_j}{\|\tilde{\mathbf{x}}_i\| \cdot \|\tilde{\mathbf{x}}_j\|}, & if \ \mathbf{x}_i \in \mathbf{N}_k(\mathbf{x}_j) \ or \ vice \ versa \\ 0, & otherwise \end{cases} \tag{2}$$

where $\mathbf{N}_k(\mathbf{x})$ represents the set of k-nearest-neighbors of the document $\mathbf{x}$, and c_{ij} is a confidence coefficient. If two documents $\mathbf{x}_i$ and $\mathbf{x}_j$ share any common tag denoted as $T_{ij} = 1$, we set c_{ij} a higher value as follows:

$$c_{ij} = \begin{cases} a, \ if \ T_{ij} = 1 \\ b, \ if \ T_{ij} = 0 \end{cases} \tag{3}$$

where a and b are parameters satisfying $1 \geq a \geq b > 0$. For a particular dataset, the more trustworthy the tags are, the greater difference between a and b we set. As mentioned in [11], a natural way to measure the similarity quantity can be formulated as follows:

$$\sum_{i,j=1}^{n} S_{ij} \|\mathbf{y}_i - \mathbf{y}_j\|_F^2 \tag{4}$$

By introducing a diagonal $n \times n$ matrix $\mathbf{D}$ whose entries are given by $D_{ii} = \sum_{j=1}^{n} S_{ij}$, Eq. 4 can be rewritten as $tr(\mathbf{Y}^T(\mathbf{D} - \mathbf{S})\mathbf{Y}) = tr(\mathbf{Y}^T \mathbf{L} \mathbf{Y})$, where $\mathbf{L}$ is the graph Laplacian and $tr(\cdot)$ is the trace function. If we relax the discreteness condition $\mathbf{Y} \in \{-1,1\}^{n \times l}$, the optimal l-dimensional real-valued vector $\tilde{\mathbf{Y}}$ can be obtained by solving the following LapEig problem [2]:

$$\arg\min_{\tilde{\mathbf{Y}}} \quad tr(\tilde{\mathbf{Y}}^T \mathbf{L} \tilde{\mathbf{Y}})$$
$$\text{subject to} \quad \tilde{\mathbf{Y}}^T \mathbf{D} \tilde{\mathbf{Y}} = \mathbf{I} \tag{5}$$
$$\tilde{\mathbf{Y}}^T \mathbf{D} \mathbf{1} = \mathbf{0}$$

where $\tilde{\mathbf{Y}}^T \mathbf{D} \tilde{\mathbf{Y}} = \mathbf{I}$ requires the different dimensions to be uncorrelated, and $\tilde{\mathbf{Y}}^T \mathbf{D} \mathbf{1} = \mathbf{0}$ requires each dimension to achieve equal probability as positive or negative. Then, we can convert the l-dimensional real-valued vectors $\tilde{\mathbf{Y}}$ into binary codes $\mathbf{Y}$ via the median vector $\mathbf{m} = median(\tilde{\mathbf{Y}})$.

3.3   Training Hash Function

In order to generate hash code for query text, we are inclined to introduce linear SVM $f(\mathbf{x}) = sgn(\mathbf{w}^T \mathbf{x})$ which is consistent with previous hashing methods [14,7] and easy to learn the kernel SVM $f(\mathbf{x}) = sgn(\sum_{i=1}^{n} \alpha_i y_i^{(p)} K(\mathbf{x}, \mathbf{x}_i))$. Since we have obtained the their binary labels for the p-th bit $y_1^{(p)}, ... y_n^{(p)}$ of training texts $\mathbf{X}$, the corresponding linear SVM can be trained as solving quadratic optimization problem.

In accordance with the binary code learning, the combination features $\tilde{\mathbf{x}}$ are equally used to instead of the original features $\mathbf{x}$ for liner SVM classifiers training and predicting. The whole training and predicting procedure of the proposed approach HTT is described in Table 1.

4   Experiment and Analysis

Five alternative hashing methods compared with our proposed approach are STHs [12], STH [14], LSI [9], LCH [13] and SpH [11]. The parameter λ is tuned by 5-fold cross validation on the training set through the grid $\{0.125, 0.5, 1, 2, 4, 8, 16, 32\}$, and we fix the parameters $a = 1$ and $b = 0.01$ in Eq.3. The number of nearest neighbors is tuned to 25 when constructing the graph Laplacian for our approach, STHs and STH in all experiments. For LDA [3], the hyper-parameters are tuned as $\alpha = 50/K, \beta = 0.01$, and the candidate topics $\mathbf{T} = \{ 10,50,100,200\}$. We random sample 10% training texts which contain tags to select the optimal topic. For the original keyword feature space cannot well reflect the semantic similarity, even worse for short text, we simply test if the two documents share any common tag to decide whether a semantic similar text, and this methodology is used in [9,14,10]. The results reported are the average over 5 runs.

4.1   Data Set

We carried out extensive experiments on two publicly text datasets: one short text dataset, Search Snippets[1], and one normal text dataset, 20Newsgroups[2].

---

[1] http://jwebpro.sourceforge.net/data-web-snippets.tar.gz
[2] http://people.csail.mit.edu/jrennie/20Newsgroups/

Table 1. The training and predicting procedures of HTT approach

Training procedure:
Input:
1. A set of n training texts $\mathbf{X} = \{\mathbf{x}_1, \mathbf{x}_2, ..., \mathbf{x}_n\}$ with tags $\mathbf{t} = \{\mathbf{t}_1, \mathbf{t}_2, ..., \mathbf{t}_m\}$.
2. N candidate topic sets $\mathbf{T} = \{T_1, T_2, ..., T_N\}$, combination coefficient λ.
Output:
1. Hash codes $\mathbf{Y}$ for the training texts, l SVM classifiers for l-bit hash code.
2. The optimal topic T_O.
1. Compute TF-IDF weights $\mathbf{W}$ for $\mathbf{X}$ and normalize it.
2. Select the optimal topic T_O by Algorithm 1.
3. Obtain topic features $\boldsymbol{\theta} = LDA(\mathbf{X})$ by T_O and form new feature $\tilde{\mathbf{x}} = [\mathbf{W}, \lambda\boldsymbol{\theta}]$.
4. Construct confidence matrix $\mathbf{S}$ with tag information $\mathbf{t}$ by Eq. 2
5. Obtain the l-dimensional vectors $\tilde{\mathbf{Y}}$ with median vector $\mathbf{m} = median(\tilde{\mathbf{Y}})$.
6. Generate the hash codes $\mathbf{Y}$ by thresholding $\tilde{\mathbf{Y}}$ to the median vector $\mathbf{m}$.
7. Train l SVM classifiers by the generated hash codes $\mathbf{Y}$.
Predicting procedure:
Input:
1. A probe query example $\mathbf{q}$, combination coefficient λ.
2. l SVM classifiers and the optimal topic T_O.
Output:
The binary hash code $\mathbf{y}_q$ for $\mathbf{q}$.
1. Infer topic features by T_O and construct the combined features $\tilde{\mathbf{x}} = [\mathbf{W}, \lambda\boldsymbol{\theta}]$.
2. Obtain the hash code $\mathbf{y}_q$ by l SVM classifiers.

The **Search Snippets** dataset collected by Phan [8] was selected from the results of web search transaction using predefined phrases of 8 different domains. We further filter the stop words and stem the texts. 20139 distinct words, 10059 training documents and 2279 testing documents are left.

The **20Newsgroups** corpus was collected by Lang [6]. We use the popular 'bydate' version which contains 20 categories, 26214 distinct words, 11314 training documents and 7532 testing documents.

In those datasets, the original features are used to estimate the candidate topic models, and all category labels are treated as tags in our experiments.

4.2   Results and Analysis

We draw the precision-recall curves of retrieved texts in Fig. 2. For methods LCH, LSI and STH, although these methods try to preserve the similarity between documents in their learned hash codes, they do not utilize tag information. Although STHs achieves a better performance, the semantic similarity between documents that goes beyond keyword matching may not be fully reflected. Thus, the HTT method substantially outperforms these methods.

In order to verify the our unified hashing clearly, keyword features, topic features and tags are evaluated separately in second set of experiments. The results

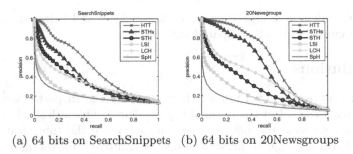

(a) 64 bits on SearchSnippets   (b) 64 bits on 20Newsgroups

Fig. 2. Precision-Recall curves on two datasets, with 64 hashing bits

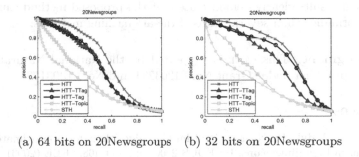

(a) 64 bits on 20Newsgroups   (b) 32 bits on 20Newsgroups

Fig. 3. Precision-Recall curves of various formal verification methods. (HTT utilizes TF-IDF, topic and tags; HTT-TTag contains topic and tags; HTT-Tag contains TF-IDF and Tags; HTT-Topic contains topic, and STH only contains TF-IDF)

are shown in Fig. 3. We can see that STH method just utilizing TF-IDF leads to inefficient hash code. On the other hand, HTT-Topic does not outperform STH-Tag that means that topic features do not outperform TF-IDF in some situations. In summary, the results indicate that the integration of as many as useful features is the key for preserving semantic information and improving the performance of hashing learning.

From the results in Tabel 2, we can see that the optimal topic set ($T50$ on both two datasets) outperforms other candidate topics. On 20Newsgroups

Table 2. Results of the retrieved examples within Hamming radius 3 on 32 bits

Dataset	SearchSnippets			20Newsgroups		
Metrics	Precision	Recall	F1	Precision	Recall	F1
HTT(T50)*	**0.9618**	**0.0709**	**0.1321**	0.9518	**0.2476**	**0.3930**
HTT(T10)	0.8926	0.0529	0.0999	0.8934	0.1947	0.3197
HTT(T100)	0.9312	0.0670	0.1250	0.9384	0.2248	0.3627
HTT(T200)	0.9013	0.0634	0.1185	0.9037	0.2017	0.3298
STHs	0.9337	0.0418	0.0801	**0.9608**	0.0732	0.1360
STH	0.8631	0.0167	0.0328	0.7541	0.0543	0.1013

dataset, we achieve a competitive result in precision and bring an more than 17% improvement to recall compared with STHs.

5   Conclusions

This paper proposes a novel unified short text hashing. In particular, the proposed approach can select the optimal topic features automatically from N candidate topic sets with the help of tags, and integrate the topic features with original features together to preserve semantic similarity. Furthermore, tags are fully utilized to adjust the pairwise similarity by a simple but effective method, which can deal with the situations that tags are multiple or incomplete. The experimental results clearly demonstrate that the proposed method can achieve better performance than several state-of-the-art hashing methods.

Acknowledgment. This work is supported by the National Natural Science Foundation of China under Grant No. 61203281 and No. 61303172.

References

1. Andoni, A., Indyk, P.: Near-optimal hashing algorithms for approximate nearest neighbor in high dimensions. In: FOCS 2006, pp. 459–468. IEEE (2006)
2. Belkin, M., Niyogi, P.: Laplacian eigenmaps for dimensionality reduction and data representation. Neural Computation 15(6), 1373–1396 (2003)
3. Blei, D.M., Ng, A.Y., Jordan, M.I.: Latent dirichlet allocation. The Journal of Machine Learning Research 3, 993–1022 (2003)
4. Chen, M., Jin, X., Shen, D.: Short text classification improved by learning multi-granularity topics. In: IJCAI, pp. 1776–1781. AAAI Press (2011)
5. Kononenko, I.: Estimating attributes: analysis and extensions of relief. In: Bergadano, F., De Raedt, L. (eds.) ECML 1994. LNCS, vol. 784, pp. 171–182. Springer, Heidelberg (1994)
6. Lang, K.: Newsweeder: Learning to filter netnews. In: ICML. Citeseer (1995)
7. Lin, G., Shen, C., Suter, D., van den Hengel, A.: A general two-step approach to learning-based hashing. In: ICCV, pp. 2552–2559 (2013)
8. Phan, X.H., Nguyen, L.M., Horiguchi, S.: Learning to classify short and sparse text & web with hidden topics from large-scale data collections. In: WWW, pp. 91–100. ACM (2008)
9. Salakhutdinov, R., Hinton, G.: Semantic hashing. IJAR 50(7), 969–978 (2009)
10. Wang, Q., Zhang, D., Si, L.: Semantic hashing using tags and topic modeling. In: SIGIR, pp. 213–222. ACM (2013)
11. Weiss, Y., Torralba, A.: Spectral hashing. In: NIPS, vol. 9, pp. 1753–1760 (2008)
12. Zhang, D., Wang, J., Cai, D., Lu, J.: Extensions to self-taught hashing: Kernelisation and supervision. Practice 29, 38 (2010)
13. Zhang, D., Wang, J., Cai, D., Lu, J.: Laplacian co-hashing of terms and documents. In: Gurrin, C., He, Y., Kazai, G., Kruschwitz, U., Little, S., Roelleke, T., Rüger, S., van Rijsbergen, K. (eds.) ECIR 2010. LNCS, vol. 5993, pp. 577–580. Springer, Heidelberg (2010)
14. Zhang, D., Wang, J., Cai, D., Lu, J.: Self-taught hashing for fast similarity search. In: SIGIR, pp. 18–25. ACM (2010)

Synthetic Test Data Generation for Hierarchical Graph Clustering Methods*

László Szilágyi[1,2], Levente Kovács[3], and Sándor Miklós Szilágyi[4]

[1] Dept. of Control Engineering and Information Technology,
Budapest University of Technology and Economics, Hungary
[2] Sapientia - Hungarian Science University of Transylvania, Romania
lazacika@yahoo.com
[3] Óbuda University of Budapest, Hungary
[4] Dept. of Informatics, Petru Maior University of Tîrgu Mureş, Romania

Abstract. Recent achievements in graph-based clustering algorithms revealed the need for large-scale test data sets. This paper introduces a procedure that can provide synthetic but realistic test data to the hierarchical Markov clustering algorithm. Being created according to the structure and properties of the SCOP95 protein sequence data set, the synthetic data act as a collection of proteins organized in a four-level hierarchy and a similarity matrix containing pairwise similarity values of the proteins. An ultimate high-speed TRIBE-MCL algorithm was employed to validate the synthetic data. Generated data sets have a healthy amount of variability due to the randomness in the processing, and are suitable for testing graph-based clustering algorithms on large-scale data.

Keywords: bioinformatics, fast Markov clustering, synthetic test data.

1 Introduction

Bioinformatics is one of the fields where there is an excessive need for clustering algorithms that are capable to handle large-scale protein sequence or interaction networks [2]. Graph-based algorithms usually need to store the matrix representation of the graph [5], which becomes prohibitively costly in memory storage above 10^4 nodes. Sparse matrix models made it possible to extend this limit towards one million nodes [12]. However, there are few publicly available large data sets, and even those existing ones are not suitable for a wide variety of algorithms.

Using synthetic test data is a frequently employed method, even if real data is also available (e.g. [4,13]). Our main goal is to provide synthetic but realistic test data for clustering algorithm designed to process large-scale protein sequence data sets. Our principal target is the Markov clustering algorithm, and

* Research supported by the Hungarian National Research Funds (OTKA), Project no. PD103921, the János Bolyai Fellowship Program of the Hungarian Academy of Sciences. Decision on support by Sapientia Institute for Research Programs is pending.

C.K. Loo et al. (Eds.): ICONIP 2014, Part II, LNCS 8835, pp. 303–310, 2014.
© Springer International Publishing Switzerland 2014

more exactly the TRIBE-MCL [5], which groups protein sequence data based on pairwise similarity measures stored in a similarity matrix. Synthetic data is created in two steps: first the main properties and attributes of the 11944-node similarity graph and corresponding similarity matrix of the SCOP95 data set [8] are identified, and then a random data set is generated, which has similar properties and the desired size. Properties considered by the proposed method include: four-level hierarchy of SCOP95, distribution of protein family sizes, density and distribution of nonzero values in various locations of the similarity matrix.

The rest of this paper is structured as follows. Section 2 presents background information on the structure and properties of the SCOP95 data set, and the TRIBE-MCL algorithm that will be used for test purposes. Section 3 presents the details of the proposed property identification and synthetic test data generation process. Section 4 produces a numerical analysis to support the validity of the produced synthetic data. Conclusions are given in the last section.

2 Background

2.1 The SCOP95 Database

The Structural Classification of Proteins (SCOP) database [10] contains protein sequences in order of tens of thousands, which are organized in a four-level hierarchy composed by classes, folds, superfamilies and families [2]. This hierarchy can be employed as ground truth for protein sequence clustering algorithms. The SCOP95 database that we use as input is a subset of SCOP (version 1.69), which contains 11944 proteins, exhibiting a maximum similarity of 95% among each other. Pairwise similarity matrices (e.g. BLAST [1], Smith-Waterman [9], Needleman-Wunsch [7]) are also available at the Protein Classification Benchmark Collection [8]. Our purpose is best served by the BLAST matrix due to its sparse nature: in its symmetrized version it has a density of 0.00387 indicating that an average node in the similarity graph is connected to 45 other nodes.

2.2 TRIBE-MCL Markov Clustering

TRIBE-MCL is an efficient clustering method based on Markov chain theory introduced by Enright et al [5]. TRIBE-MCL assigns a graph structure to the protein set such a way that each protein has a corresponding node. Edge weights are stored in the so-called similarity matrix S, which acts as a stochastic matrix. At any moment, edge weight s_{ij} reflects the posterior probability that protein i and protein j have a common evolutionary ancestor. TRIBE-MCL is an iterative algorithm, performing in each loop two main operations on the similarity matrix: inflation and expansion. Inflation raises each element of the similarity matrix to power r, which is a previously established fixed inflation rate. Due to the constraint $r > 1$, inflation favors higher similarity values in the detriment of lower ones. Expansion, performed by raising matrix S to the second power, is aimed to favor longer walks along the graph. Further operations like column or row normalization, and matrix symmetrization are included to serve the stability and

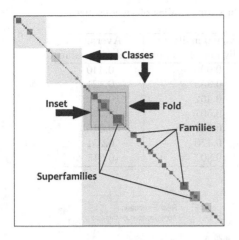 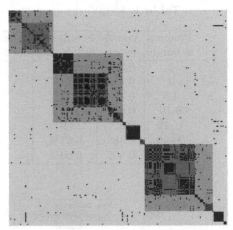

Fig. 1. A selected part (classes e-g) of the BLAST similarity matrix of the SCOP95 data set indicating the hierarchy of classes, folds, superfamilies and families (left), and the magnified view of superfamilies g.3.6-g.3.14 within the inset (right). Black pixels on the right image represent the nonzero values in the matrix.

robustness of the algorithm, and to enforce the probabilistic constraint. Similarity values that fall below a previously defined threshold value ε are rounded to zero. Clusters are obtained as connected subgraphs in the graph. Further details on TRIBE-MCL operations are available in [5,11].

3   Methods

3.1   Identification of SCOP95's Properties

The identification of the SCOP95 matrix properties is performed in several steps. Proteins are grouped into families of various sizes (1-557 members), each represented by a square block situated on the matrix diagonal. These diagonal blocks are not very sparse as they contain over two thirds of all nonzero values in the matrix. Superfamilies are small groups of families represented by larger diagonal blocks that include the blocks of contained families. Similarity values within the superfamily blocks but outside the family blocks are significantly less dense, and they become sparser as the distance from the diagonal grows. The structure of the similarity matrix is depicted in Fig. 1.

According to the recently developed theory of natural networks [3], the number of connections the graph nodes have follows a negative power distribution. This is also valid in case of the SCOP95 graph: both the distribution of connections and the distribution of family sizes share this attribute. Figure 2 exhibits the approximation of this distribution in the range of families with up to 80 proteins. A few larger families are also present in SCOP95, their distribution also follows the rule of the power distribution.

Table 1. Identified parameter values for families of various sizes

Proteins in family	Average density	Families of full density	Average density in not fully dense families	Average similarity value
2	0.827	82.7 %	0.000	0.440
3	0.809	70.4 %	0.353	0.421
4-5	0.733	51.5 %	0.452	0.382
6-10	0.659	31.3 %	0.507	0.358
11-19	0.617	14.9 %	0.547	0.318
20-99	0.485	6.67 %	0.470	0.251
100+	0.807	0.00 %	0.807	0.315

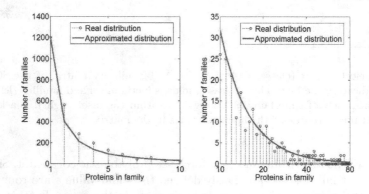

Fig. 2. Family sizes follows a negative power distribution: the amount of families of size n is proportional with n^{-k}

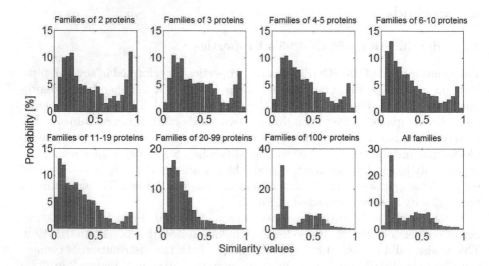

Fig. 3. Distribution of nonzero similarity values in the SCOP95 matrix, in case of protein couples situated in the same family: various distributions for all family sizes

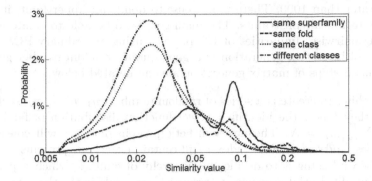

Fig. 4. Distribution of nonzero similarity values in the SCOP95 matrix, in case of protein couples situated in different families, but in the same superfamily, fold, or class, and also for proteins from different classes

Table 2. Identified densities in various parts of the SCOP95 BLAST similarity matrix

Same family	Different families same superfamily	Different superfamilies same fold	Different folds same class	Different classes
0.7211	0.0376	0.00274	0.00161	0.00103

Table 1 exhibits some identified parameters concerning matrix density and protein families. Average density gives us the probability that the similarity value s_{ij} with $i \neq j$ but proteins i and j chosen from the same family is a nonzero. Some part of the families are represented by fully dense blocks in the similarity matrix. Larger families of such property are usually rare. The average nonzero s_{ij} value ($i \neq j$) present in the families are also indicated in Table 1. The unit values situated on the diagonal are not counted into these averages. The distribution of the nonzero similarities within families of various sizes is exhibited in Fig. 3. These similarity values cover the whole range between 0 and 1. On the other hand, Fig. 4 shows the distribution of nonzero similarities between proteins of different families of the same superfamily, proteins of different superfamilies situated in the same fold, proteins of different folds situated in the same class, and proteins of different classes, respectively. These parts of the similarity matrix have a decreasing density in the enumerated order. Such similarity values rarely exceed 1/3.

The properties enumerated above are all taken in consideration when new matrices are generated.

3.2 Generating New Large Matrices

When a new matrix is generated, there is a single input parameter to set, namely the number of proteins (N) in the synthetic data set. This number N is supposed to be greater than 1000. There is no sense to define an upper limit, it will be forced by technical constraints. The main goal is to be able to create matrices describing pairwise similarities of 10^6 proteins using an ordinary PC. One important tool in matrix generation is a good-quality random number generator [6]. The main steps of matrix generation are enumerated below:

1. First thing to create is a series of random numbers $n_1, n_2, \ldots n_F$ such a way that they follow the identified power function distribution of family sizes, and $\sum_{i=1}^{F} n_i = N$. The new synthetic protein data set will consist of F families, and family with index i will contain exactly n_i proteins.
2. The second thing is to decide the hierarchy of families, which is performed sequentially. Initially we need to create a class, a fold in the class, a superfamily in the fold, and assign the first family to the newly created superfamily. For each further family there is a p_c probability to add a new class; if no new class is created then there is a p_f probability to add a new fold in an existing class; if no new fold is created then there is a p_s probability of add a new superfamily into an existing fold and place the new family there. Otherwise the new family is included into the existing superfamilies, following the popularity rule introduced in network theory [3]. The new family will be assigned to popular classes, folds and superfamilies with higher probability. The current version uses $p_c = 0.99$, $p_f = 0.8$, $p_s = 0.75$, but there is also an upper limit for the number of classes, which logarithmically grows with N.
3. Nonzero similarity values are generated as random numbers in the $(0, 1]$ interval. Diagonal values are all 1 by default. Further nonzeros within the blocks representing families follow the identified distribution functions shown in Fig. 3. Nonzero values in other regions of the matrix follow the corresponding distribution function indicated in Fig. 4. The density of nonzeros in various regions of the matrix corresponds to the values given in Table 2. The generated matrix is perfectly symmetrical.
4. Finally the randomly generated nonzeros are sorted according to row and column and are transferred into the output file. The header of the output file contains information on the number of proteins and the hierarchical structure of the generated data set so that it can serve as ground truth at testing.

4   Results and Discussion

The first-order validation of the proposed method consisted of inspection of properties and attributes of output matrices, and functional testing using the TRIBE-MCL algorithm.

For the sake of property inspection, we have created synthetic test data of sizes varying from 10,000 to 250,000 proteins, 25 instances of each. Table 3 indicates the average and standard deviation of the hierarchy attributes, namely the number of classes, folds, superfamilies, and families, and finally the density if the matrix as well. The table reflects that the randomness within the generation process produces a considerable amount of variance among matrices of the same

Table 3. Properties of the generated synthetic protein graphs, for various sizes of the data set

Proteins	Classes	Folds	Superfam.	Families	Density ($\times 10^{-3}$)
10k	7	280 ± 25	557 ± 45	1389 ± 129	5.06 ± 1.80
15k	7	392 ± 23	770 ± 42	1926 ± 105	4.16 ± 1.17
20k	8	515 ± 45	1024 ± 58	2526 ± 137	3.56 ± 0.59
30k	9	726 ± 28	1441 ± 57	3593 ± 103	3.03 ± 0.37
50k	11	1136 ± 34	2254 ± 68	5677 ± 184	2.44 ± 0.36
70k	12	1550 ± 65	3087 ± 124	7657 ± 295	2.15 ± 0.16
100k	13	2138 ± 53	4283 ± 91	10671 ± 227	1.72 ± 0.13
150k	14	3066 ± 95	6111 ± 195	15270 ± 404	1.63 ± 0.15
200k	15	3992 ± 122	7964 ± 233	19915 ± 624	1.54 ± 0.15
250k	16	4900 ± 88	9762 ± 154	24360 ± 420	1.47 ± 0.17

size. Each generated matrix is different of all others and can be used to test the TRIBE-MCL algorithm.

In order to run a set of simple functional tests, we have created test matrices of sizes between 10,000 and 50,000 varying in small steps, 15 instances of each size. All these matrices were fed to our ultimate high-speed and memory saving version [11] of the TRIBE-MCL algorithm, using inflation rate $r = 2.0$ and similarity threshold $\varepsilon = 10^{-3}$. Simple statistical parameters were extracted from the total runtime values, and are exhibited in Fig. 5. Generated matrices contain a considerable amount of variability, which seemingly reduces as the size of the matrix grows. Considerably wider test suites and detailed results using data generated by the proposed method are exhibited in [11].

The main limitation of the proposed method is the fact that it builds on information extracted from a single protein data set designed to test clustering

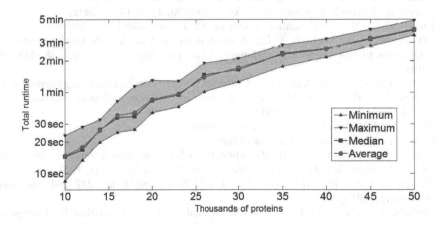

Fig. 5. Total runtime of TRIBE-MCL clustering process plotted against the protein count in the synthetic protein graph: minimum, maximum, median and average values

algorithms. Further efforts will be made to provide the user the opportunity to manually tune the attributes of the output data set and similarity matrix.

Creating synthetic protein data sets and corresponding pairwise similarity matrices of up to 250 thousand items can be performed on an ordinary Pentium4 PC with 2GB RAM in less than one minute. Creating larger data sets of up to 10^6 items is also possible, but it requires more memory and time.

5   Conclusions

In this paper we proposed a novel method to create test data to hierarchical clustering methods based on pairwise similarity measures. The proposed method was applied to generate synthetical protein data sets, their four-level hierarchical structure and sparse similarity matrix that contains BLAST-like pairwise alignment scores. Test matrices were fed to the TRIBE-MCL algorithm, which proved the validity of the synthetical data. The proposed method can efficiently support the validation process of hierarchical clustering algorithms on large-scale data.

References

1. Altschul, S.F., Madden, T.L., Schaffen, A.A., Zhang, J., Zhang, Z., Miller, W., Lipman, D.J.: Gapped BLAST and PSI-BLAST: a new generation of protein database search program. Nucl. Acids Res. 25, 3389–3402 (1997)
2. Andreeva, A., Howorth, D., Chadonia, J. M., Brenner, S. E., Hubbard, T. J. P., Chothia, C., Murzin, A. G.: Data growth and its impact on the SCOP database: new developments. Nucl. Acids Res. 36, D419–D425 (2008)
3. Barabási, A.L.: Linked: The New Science of Networks. Perseus Book Group, New York (2002)
4. BrainWeb: Simulated Brain Database,
   http://brainweb.bic.mni.mcgill.ca/brainweb/
5. Enright, A.J., van Dongen, S., Ouzounis, C.A.: An efficient algorithm for large-scale detection of protein families. Nucl. Acids Res. 30, 1575–1584 (2002)
6. Knuth, D.A.: Random number generator. US Patent 3548174A (1970)
7. Needleman, S.B., Wunsch, C.D.: A general method applicable to the search for similarities in the amino acid sequence of two proteins. J. Mol. Biol. 48, 443–453 (1970)
8. Protein Classification Benchmark Collection, http://net.icgeb.org/benchmark
9. Smith, T.F., Waterman, M.S.: Identification of common molecular subsequences. J. Mol. Biol. 147, 195–197 (1981)
10. Structural Classification of Proteins database,
    http://scop.mrc-lmb.cam.ac.uk/scop
11. Szilágyi, L., Szilágyi, S.M., Hirsbrunner, B.: A fast and memory-efficient hierarchical graph clustering algorithm. In: Loo, C.K., Yap, K.S., Wong, K.W., Teoh, A., Huang, K. (eds.) ICONIP 2014, Part I. LNCS, vol. 8834, pp. 247–254. Springer, Heidelberg (2014)
12. Szilágyi, S.M., Szilágyi, L.: A fast hierarchical clustering algorithm for large-scale protein sequence data sets. Comput. Biol. Med. 48, 94–101 (2014)
13. Várady, P., Benyó, Z., Benyó, B.: An open architecture patient monitoring system using standard technologies. IEEE Trans. Inform. Technol. Biomed. 6, 95–98 (2002)

Optimal Landmark Selection
for Nyström Approximation

Zhouyu Fu

School of Computing, Engineering & Mathematics
University of Western Sydney, Penrith, NSW 2751, Australia
z.fu@uws.edu.au

Abstract. The Nyström method is an efficient technique for large-scale
kernel learning. It provides a low-rank matrix approximation to the full
kernel matrix. The quality of Nyström approximation largely depends on
the choice of landmark points. While standard method uniformly samples columns of the kernel matrix, improved sampling techniques have
been proposed based on ensemble learning [1] and clustering [2]. These
methods are focused on minimizing the approximation error for the original kernel. In this paper, we take a different perspective by minimizing
the approximation error for the input vectors instead. We show under
some restrictive condition that the new formulation is equivalent to the
standard Nyström solution. This leads to a novel approach for optimizing
landmark points for the Nyström approximation. Experimental results
demonstrate the superior performance of the proposed landmark optimization method compared to existing Nyström methods in terms of
lower approximation errors obtained.

1 Introduction

Kernel methods have been very popular in machine learning for the past two
decades [3]. By exploiting the "kernel" trick, input vectors are implicitly mapped
to a higher dimensional reproducing kernel Hilbert space (RKHS). Although the
exact feature maps usually can not be determined in closed form, inner products
in RKHS can be expressed as kernel function values over input vectors. This is
useful in dealing with nonlinear data as the form of feature transformation can
be readily controlled by using different types of kernel functions [3].

A main issue with kernel methods is the computational cost. As a standard
approach, one has to evaluate and save the kernel matrix consisting of kernel
function values between all pairs of input examples. Many problems in kerel
learning also involve calculating the inverse of the kernel matrix [3]. This leads
to an $O(n^3)$ time and $O(n^2)$ space complexity, making kernel methods infeasible
for large-scale applications.

The Nyström method is an important technique for approximating the full
kernel matrix with a low-rank representation to achieve reduced complexity [4].
Rather than accessing every entry of the full matrix, the Nyström approximation
only needs to select a subset of landmark points and use the kernel evaluations

C.K. Loo et al. (Eds.): ICONIP 2014, Part II, LNCS 8835, pp. 311–318, 2014.
© Springer International Publishing Switzerland 2014

between training examples and the landmarks to interpolate entries of the full kernel matrix. This leads to a low-rank approximation of the original kernel matrix as the outer product of a feature matrix with reduced number of columns and its transpose. Instead of applying kernel learning algorithms on the full kernel matrix directly, one can simply treat rows of the feature matrix as feature vectors and apply any linear algorithms to them. This leads to significant speedups for large-scale applications.

Despite effective, the quality of the Nyström method largely depends on the landmarks selected for computing the approximation. Standard Nyström implementation simply selects the landmarks from the input data randomly without replacement [4]. This is equivalent to uniformly sampling the columns of the kernel matrix and imputing the remaining entries based on the columns sampled. Different sampling strategies have been proposed in the literature to improve uniform sampling of columns. A non-uniform sampling technique was proposed in [5] for selecting columns with probabilities proportional to the column norms. An ensemble Nyström method was presented in [1] by combining individual Nyström approximators each with different uniform sampling on columns of the kernel matrix. The ensemble method was shown to outperform the single Nyström approximator with either uniform or non-uniform sampling. Apart from column sampling technique that selects landmarks directly from training data, a clustering based sampling technique was proposed in [2]. It achieves improved approximation quality by employing the cluster centroids as landmarks from the clustering results.

The Nyström method is essentially a low-rank kernel approximation technique. Thus, the majority of existing methods and established theoretical bounds were focused on minimizing the approximation error at the kernel matrix level. We take a very different perspective in this paper. Instead of aiming at error minimization at kernel level, we present a new approach for landmark selection that minimizes the approximation error at feature vector level. The optimal landmarks should provide a set of linear basis vectors in RKHS capable of reconstructing the training examples with small reconstruction errors in RKHS. We show that this new formulation is equivalent to the standard Nyström formulation when the rank of the approximation matrix equals the number of landmarks. This unifies the two seemingly different perspectives to achieve the same goal. The formulated problem involves two sets of variables, the landmarks and the reconstruction weights for input vectors. It can be reformulated as an optimization problem over optimal value function and solved effectively using a perturbed optimization technique [6].

2   The Nyström Method

Let $\mathcal{X} = \{\mathbf{x}_i | i = 1, \ldots, n\}$ denote the input data set of n vectors with $\mathbf{x}_i \in \mathcal{R}^d$. $\mathbf{K} \in \mathcal{R}^{n \times n}$ is the corresponding kernel matrix consisting of kernel function evaluations for input data. The (i, j)th element of kernel matrix $\mathbf{K}$ is denoted by $\kappa(\mathbf{x}_i, \mathbf{x}_j)$, where κ represents the kernel function being used. With large sample size n, manipulating $\mathbf{K}$ would be computationally prohibitive.

The Nyström method uses a small subset of landmark points to obtain a low-rank approximation of matrix $\mathbf{K}$. Without loss of generality, assume that the first m points in the input data set were chosen as the landmarks with $m \ll n$, we can rewrite $\mathbf{K}$ into a block matrix in the following

$$\mathbf{K} = \begin{bmatrix} \mathbf{W} & \mathbf{G}^T \\ \mathbf{G} & \mathbf{B} \end{bmatrix} \qquad \mathbf{C} = \begin{bmatrix} \mathbf{W} \\ \mathbf{G} \end{bmatrix} \tag{1}$$

where $\mathbf{C}$ represents the first m columns taken from $\mathbf{K}$ storing kernel values calculated between the full data set and the landmarks, $\mathbf{W}$ represents the first m rows of $\mathbf{C}$ with kernel evaluations between landmarks, $\mathbf{G}$ and $\mathbf{B}$ denote the remaining blocks of $\mathbf{C}$ and $\mathbf{K}$.

We denote the eigen-value decomposition (EVD) of matrix $\mathbf{W}$ by $\mathbf{W} = \mathbf{U}\mathbf{\Sigma}\mathbf{U}^T$, where $\mathbf{\Sigma} = diag\,(\sigma_1, \ldots, \sigma_m)$ is a diagonal matrix whose diagonal elements are given by eigen-values sorted in descending order, $\mathbf{U} = [\mathbf{u}_1, \ldots, \mathbf{u}_m]$ is a matrix that contains the corresponding eigen-vectors in columns. The rank-r $(r < m)$ Nyström approximation of the kernel matrix $\mathbf{K}$ is then given by

$$\tilde{\mathbf{K}}_{nys}^{(r)} = \mathbf{C}\mathbf{W}_r^+\mathbf{C}^T \tag{2}$$

where $\mathbf{W}_r^+ = \mathbf{U}_r\mathbf{\Sigma}_r^{-1}\mathbf{U}_r^T$ is the rank-r pseudo-inverse of matrix $\mathbf{W}$, with $\mathbf{U}_r = [\mathbf{u}_1, \ldots, \mathbf{u}_r]$ and $\mathbf{\Sigma}_r^{-1} = diag\,(1/\sigma_1, \ldots, 1/\sigma_r)$.

The approximate kernel matrix can be decomposed by $\tilde{\mathbf{K}}_{nys}^{(r)} = \mathbf{Q}\mathbf{Q}^T$, where $\mathbf{Q} = \mathbf{C}\mathbf{U}_r\mathbf{\Sigma}_r^{-1/2}$. One can treat the rows of matrix $\mathbf{Q}$ as feature vectors and apply linear algorithms readily to the feature vectors. This avoids dealing with the full kernel $\mathbf{K}$ directly, which can be computationally demanding for large n.

With special case $r = m$, the rank-m Nyström approximation is given by

$$\tilde{\mathbf{K}}_{nys}^{(m)} = \begin{bmatrix} \mathbf{W} & \mathbf{G}^T \\ \mathbf{G} & \mathbf{G}\mathbf{W}^{-1}\mathbf{G}^T \end{bmatrix} \tag{3}$$

where $\mathbf{W}^{-1}$ denotes the inverse of matrix $\mathbf{W}$. The only difference between the original kernel matrix $\mathbf{K}$ and the approximation $\tilde{\mathbf{K}}_{nys}^{(m)}$ is in the bottom right corner, where $\tilde{\mathbf{K}}_{nys}^{(m)}$ approximates $\mathbf{B}$ with $\mathbf{C}\mathbf{W}^{-1}\mathbf{C}^T$.

3 Optimal Landmark Selection

3.1 Alternative View of Nyström Approximation

In this section, we present an alternative view of the Nyström approximation from the perspective of minimizing reconstruction errors of input data in RKHS. Existing Nyström methods and theoretical analyses have focused on the minimization of the following approximation error for the kernel matrix

$$\epsilon_1 = \|\mathbf{K} - \tilde{\mathbf{K}}_{nys}^{(r)}\|_F^2 \tag{4}$$
$$= \sum_{i,j} \left(\varphi(\mathbf{x}_i)^T\varphi(\mathbf{x}_j) - \mathbf{q_i^T}\mathbf{q_j}\right)^2$$

where $\varphi(.)$ denotes the explicit feature map from the input space to RKHS induced by the kernel function. $\mathbf{q_i}$ is the feature vector for point i. It is taken from the ith row of matrix $\mathbf{Q}$ based on the rank r approximation $\tilde{\mathbf{K}}_{nys}^{(r)} = \mathbf{QQ}^T$.

From a feature point of view, Eq. 4 basically measures the discrepancy between feature similarities derived from the two feature representations induced by kernels $\mathbf{K}$ and $\tilde{\mathbf{K}}_{nys}^{(r)}$. Alternatively, one can target at the differences between the feature representations. The key issue here is that the features are not directly comparable, as $\varphi(\mathbf{x}_i)$ and $\mathbf{q_i}$ have different feature dimensions. By noting that matrix $\mathbf{Q}$ is indirectly controlled by the landmarks used to generate the approximation, we can employ an alternative representation of the approximate features as linear expansions of the landmarks in RKHS. This leads to the following error function for feature approximation.

$$\epsilon_2 = \sum_{\mathbf{x} \in \mathcal{X}} \|\varphi(\mathbf{x}) - \tilde{\varphi}(\mathbf{x})\|^2 \tag{5}$$

$$\tilde{\varphi}(\mathbf{x}) = \sum_{j=1}^{m} \alpha_j \varphi(\mathbf{z}_j) \tag{6}$$

where $\mathbf{z}_j$ denotes the jth landmark, α_j denotes the jth coefficient of the linear expansion. The purpose here is to minimize the reconstruction error for each $\varphi(\mathbf{x})$ using RKHS mapping of landmarks $\varphi(\mathbf{z}_j)$'s as basis.

Given fixed landmarks, Eq. (5) simply reduces to the following least squares problem for each $\mathbf{x}$

$$\min_{\alpha} \|\varphi(\mathbf{x}) - \sum_{j=1}^{m} \alpha_j \varphi(\mathbf{z}_j)\|^2 \tag{7}$$

The solution of the above problem is given by

$$\alpha = \mathbf{W}^{-1}\mathbf{g} \tag{8}$$

where $\alpha = [\alpha_1, \ldots, \alpha_m]^T$ is the coefficient vector, $\mathbf{W} \in R^{m \times m}$ is the kernel matrix evaluated on landmarks $\mathbf{z}$'s with $W_{i,j} = \kappa(\mathbf{z}_i, \mathbf{z}_j)$, and $\mathbf{g} \in R^m$ is a vector of kernel evaluations between example $\mathbf{x}$ and landmarks whose jth element is given by $\kappa(\mathbf{x}, \mathbf{z}_j)$.

Based on the above result, we can compute the kernel value for any two examples $\mathbf{x}_i$ and $\mathbf{x}_j$ in the training data set by

$$\hat{K}(\mathbf{x}_i, \mathbf{x}_j) = \tilde{\varphi}(\mathbf{x}_i)^T \tilde{\varphi}(\mathbf{x}_j) = \mathbf{g}_i^T \mathbf{W}^{-1} \mathbf{g}_j \tag{9}$$

where $\mathbf{g}_i$ and $\mathbf{g}_j$ are defined similarly with $\mathbf{g}$ but restricted to $\mathbf{x}_i$ and $\mathbf{x}_j$ respectively.

Suppose the first m points from the input data were taken as landmarks. It can be easily verified that the kernel matrix of approximate features for the remaining $n - m$ points is equal to $\mathbf{GW}^{-1}\mathbf{G}^T$, where $\mathbf{G} = [\kappa(\mathbf{x}_i, \mathbf{x}_j)]_{i,j=m+1}^{n}$. Note that $\mathbf{W}$ and $\mathbf{G}$ have the same definitions here as in Section 2, the resulting approximate kernel is identical to the bottom-right block of rank-m Nyström approximation with m landmarks in Eq. (3). Hence, we provide a novel interpretation of the Nyström approximation here by viewing it from the perspective of approximating each input example with a linear combination of landmarks in RKHS with minimum reconstruction error in the least squares sense.

3.2 Model Formulation and Optimization

The alternative view of Nyström approximation presented above motivates a different approach for landmark selection. Instead of selecting landmarks from training examples directly, we can treat them as target variables to be optimized in an optimization framework. This not only allows more flexibility for landmark selection, but more importantly enables further minimization of approximation errors at feature and kernel level.

We formulate landmark selection as the following optimization problem

$$\min_{\mathbf{Z}} f(\mathbf{Z}) = \sum_{i=1}^{n} \min_{\alpha_i} r(\mathbf{Z}, \alpha_i) \qquad (10)$$

$$r(\mathbf{Z}, \alpha_i) = \|\varphi(\mathbf{x}_i) - \sum_{j=1}^{m} \alpha_{i,j}\varphi(\mathbf{z}_j)\|^2 + \lambda\|\alpha_i\|^2 \qquad (11)$$

where $\alpha_i = [\alpha_{i,j}]_{j=1}^{m}$ is the weight vector for example i, $\mathbf{Z} = [\mathbf{z_j}]_{j=1}^{m}$ is the target variable for optimization that contains all landmarks in columns. λ is a regularization parameter to control overfitting and ensure numerical stability.

The optimization problem defined above is special since it contains n subproblems in the objective function. The problem is similar to the reduced set problem [3] which involves searching for a subset of landmarks that best approximate the input data. However, since the landmarks need to be jointly optimized for all input examples, we can not apply the fixed-point iteration procedure in [3] to solve the problem.

Rather, we take a direct approach to solving $f(\mathbf{Z})$. Note that to evaluate the value of $f(\mathbf{Z})$ for fixed value of $\mathbf{Z}$, one needs to first solve the minimization of $g(\mathbf{z}, \alpha_i)$ w.r.t. α_i for each i. Functions like this are known as optimal value functions in the optimization literature [6]. According to Theorem 4.1 of [6], an optimal value function is differentiable if each sub-problem has a unique solution and is differentiable w.r.t. the target variable. The uniqueness of optimal solution here is ensured by the fact that each $r(\mathbf{Z}, \alpha_i)$ is strongly convex in α_i given that $\lambda > 0$ and thus a globally minimum solution can be attained. Differentiability w.r.t. to $\mathbf{Z}$ can be established given that the kernel function is differentiable w.r.t. each $\mathbf{z_j}$. This is true for the Gaussian kernel function employed in this work with $\kappa(\mathbf{x}_i, \mathbf{x}_j) = \exp\left(-\gamma\|\mathbf{x}_i - \mathbf{x}_j\|^2\right)$, where γ is the band-width parameter that controls the decay of the kernel value with increasing distances.

For a differentiable optimal value function $f(\mathbf{Z})$, we can calculate its derivative w.r.t. $\mathbf{Z}$ by substituting the optimal values from each sub-problem into f as if the function value does not depend on the α_i variables. Hence we can simply employ

Table 1. Benchmark data sets used in the experiments

	mushrooms	pendigits	satimage	segment	splice	usps
data	8124	7494	4435	2310	2175	7291
feat	112	16	36	19	60	256

a gradient descent algorithm to minimize the optimal value function. Note that similar techniques on optimal value functions have been used in solving different problems in machine learning before, such as multiple kernel learning [7] and multiple instance learning [8].

Specifically, let $\overline{\alpha}_i$ be the solution of sub-problem $r(\mathbf{Z}; \alpha_i)$ given by

$$\overline{\alpha}_i = (\mathbf{W} + \lambda \mathbf{I})^{-1} \mathbf{g_i} \tag{12}$$

where $\mathbf{W}$ and $\mathbf{g}_i$ are same as defined in the previous section.

We can calculate the gradient of f w.r.t. each $\mathbf{z}_j$ by

$$
\begin{aligned}
\frac{\partial g}{\partial \mathbf{z}_j} =& 4\gamma \sum_{i=1}^{n} \sum_{l=1}^{m} \overline{\alpha}_{i,l} \overline{\alpha}_{i,j} \kappa(\mathbf{z}_l, \mathbf{z}_j)(\mathbf{z}_l - \mathbf{z}_j) \\
& - 4\gamma \sum_{i=1}^{n} \overline{\alpha}_{i,j} \kappa(\mathbf{x}_i, \mathbf{z}_j)(\mathbf{x}_i - \mathbf{z}_j)
\end{aligned} \tag{13}
$$

Note that for each iteration of gradient descent, the value of $\overline{\alpha}$ needs to be updated before recomputing the gradient. We also adopt a simple line search strategy similar to the one presented in [8] to adaptively choose the appropriate step size in each descent step.

4   Experimental Results

In this section, we provide experimental comparisons of the proposed Nyström method against existing Nyström implementations on benchmark data sets. The following Nyström implementations are compared in our experiments.

- OptNys - Nyström with optimized landmark selection proposed in this paper
- RandNys - standard Nyström implementation with landmarks randomly selected from data points without replacement [4]
- ClustNys - Nyström with landmarks selected from the cluster centroids by applying K-means clustering to the training data points [2]
- EnsNys - Ensemble of 10 RandNys approximations with different random landmarks selected for each single approximator [1]

Six benchmark data sets were selected from the UCI Machine Learning Repository [1] for performance comparison. The statistics of each data set used are shown

---

[1] http://archive.ics.uci.edu/ml

in Table 1. Gaussian kernel has been used throughout our experiments, with the bandwidth parameter chosen to be the average distance between all pairs of input examples. We have fixed parameter λ in Eq. (11) to 10^{-6} for all experiments. For each data set, we tested different Nyström implementations with different numbers of landmarks m, ranging from 100 to 500 with step of 100. We have fixed the value of r, the rank of the final approximator, to the same value as m for each individual test. Figure 1 shows the performances of different methods being tested on the six data sets as measured by the approximation errors defined in Eq. (4) over different landmark sizes. Curves with different markers correspond to the approximation errors obtained by four different methods being compared in the experiment.

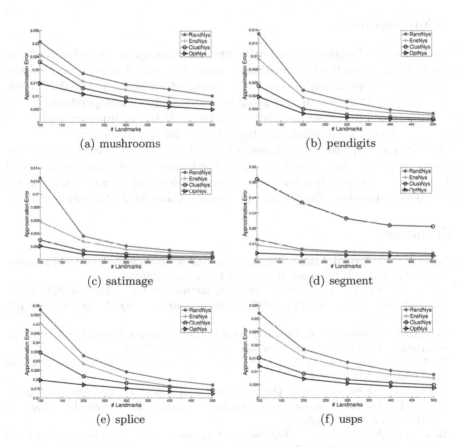

(a) mushrooms

(b) pendigits

(c) satimage

(d) segment

(e) splice

(f) usps

Fig. 1. Performance comparison of different Nyström methods

From the figure, it can be clearly seen that OptNys achieves better performances in terms of lower approximation errors than other Nyström implementations across all benchmark data sets being tested. For the other methods, RandNys produces the highest approximation errors and has poor performances overall. ClustNys obtains better results than EnsNys on five out of the six data sets, but produces the worst result on the segment data. OptNys is quite consistent in performance and gains a margin of advantages over ClustNys and EnsNys, the two state-of-the-art Nyström methods. The performance gap is more pronounced with smaller landmark sizes, where the approximation quality is sacrificed due to insufficient number of landmarks used for the approximator. OptNys can better deal with such scenarios as it aims to directly minimize the approximation errors at feature level. The results clearly demonstrate the importance of appropriate landmark selection for the Nyström approximation and effectiveness of the proposed optimized landmark selection algorithm for the improvement of approximation quality.

5   Conclusions

We present a novel algorithm for optimizing landmarks to improve the Nyström approximation. In contrast to existing Nyström methods that focus on minimizing the matrix approximation error, the proposed approach is based on minimizing the error of reconstructing input vectors from landmark points in RKHS. This leads to a reduced set problem formulated on the training examples and can be effectively solved using a perturbed optimization technique.

The Nyström approximation belongs to the broad category of techniques for low-rank kernel matrix approximation. In the future, we will explore the use of similar techniques for other low-rank approximation techniques such as sparse greedy kernel approximation [3] and incomplete Cholesky factorization [9].

References

1. Kumar, S., Mohri, M., Talwalkar, A.: Sampling methods for the nyström method. Journal of Machine Learning Research 13, 981–1006 (2012)
2. Zhang, K., Kwok, J.T.: Clustered nyström method for large scale manifold learning and dimension reduction. IEEE Transactions on Neural Networks 21(10), 1576–1587 (2010)
3. Scholkopf, B., Smola, A.J.: Learning with Kernels: Support Vector Machines, Regularization, Optimization and Beyond. MIT Express (2002)
4. Williams, C.K.I., Seeger, M.: Using the nyström method to speed up kernel machines. In: NIPS (2001)
5. Drineas, P., Mahoney, M.W.: On the nyström method for approximating a gram matrix for improved kernel-based learning. Journal of Machine Learning Research 6, 2153–2175 (2005)
6. Bonnans, J.F., Shapiro, A.: Optimization problems with pertubation: A guided tour. SIAM Review 40(2), 202–227 (1998)
7. Rakotomamonjy, A., Bach, F.R., Canu, S., Grandvalet, Y.: Simplemkl. Journal of Machine Learning Research 9(11), 2491–2521 (2008)
8. Fu, Z., Lu, G., Ting, K.M., Zhang, D.: Learning sparse kernel classifiers for multi-instance classification. IEEE Trans. Neural Networks 24(9), 1377–1389 (2013)
9. Bach, F.R., Jordan, M.I.: Predictive low-rank decomposition for kernel methods. In: ICML, pp. 33–40 (2005)

Privacy Preserving Clustering:
A k-Means Type Extension

Wenye Li

Macao Polytechnic Institute,
Rua de Luís Gonzaga, Macao SAR, China
wyli@ipm.edu.mo
http://staff.ipm.edu.mo/~wyli

Abstract. We study the problem of r-anonymized clustering and give a k-means type extension. The problem is partition a set of objects into k different groups by minimizing the total cost between objects and cluster centers subject to a constraint that each cluster contains at least r objects. Previous work has reported an approach when the cluster centers are constrained to be a real member of the objects. In this paper, we release the constraint and allow a center to be the mean of the objects in its group, similar to the settings of the classical k-means clustering model. To address the inherent computational difficulty, we exploit linear program relaxation to find high quality solutions in an efficient manner. We conduct a series of experiments and confirm the effectiveness of the method as expected.

Keywords: r-Anonymity, k-Means Clustering, Linear Programming.

1 Introduction

Clustering is to divide a set of objects into groups such that objects in the same group, or cluster, are more similar in some sense to each other than to those in different groups. It is a main task of exploratory data mining, and a common technique for statistical data analysis used in many fields, including machine learning, pattern recognition, image analysis, information retrieval, bioinformatics, and so on [1, 2].

A number of clustering models and algorithms have been developed, among which k-means clustering provides a generic method and has been applied in different domains. The model assumes all observations are from a vector space, in which the *mean* of a number of vectors makes sense. It partitions m observations into k clusters in which each observation belongs to the cluster with the nearest mean. The objective is to minimize the total squared distances between each object and its centers, and results in a partitioning of the data space into k cells [3].

A related model, in this paper we call k-exemplars clustering, operates in a similar way as k-means clustering except that it does not require the observations are from a vector space, but requires each center (or exemplar) be a real object

C.K. Loo et al. (Eds.): ICONIP 2014, Part II, LNCS 8835, pp. 319–326, 2014.

in the data set, instead of the mean of some objects which often does not exist in reality. Although the two clustering models seem to be similar, they usually require quite different computational approaches to find the solutions in practice.

Among the applications of clustering techniques, we are particularly interested in the application of data publishing. With publicly available data, people are able to use analytical methods for intelligent data processing and hopefully yield important new discoveries. In such applications, clustering provides a principled way in organizing the data to be published, as well as other benefits. Unfortunately, when publishing certain data (such as health diagnosis records) publicly, non-trivial personal information is inevitably involved. The disclosure of such personal data obviously raises serious concerns about privacy [4–6].

Given these risks, significant research effort has been devoted to developing methods for privacy preserving data analysis. A central idea in the area of privacy preserving data publishing has been to remove personally-identifying information that could be used for backtracking, based on the principle of r-anonymity [1] [7]. That is, reduce the granularity of data representation by mapping each record to an equivalence class of at least $r - 1$ other records in the data collection. In this way, the probability of determining the identity of an individual (re-identification probability) is decreased.

For a given data set, producing the optimal r-anonymized version is NP-hard [8]. Researchers have resorted to approximating this by heuristic methods. Common implementations of r-anonymity use transformation techniques such as generalization and suppression [7, 9, 5].

In this paper, we investigate the problem of achieving r-anonymity in given data. The approach we consider is based on clustering the data: mapping data points to cluster centers while ensuring that each cluster has at least r members to ensure r-anonymity in the resulting data representation. This is an alternative to the common techniques of generalization and suppression that have been employed in the data mining literature.

2   Related Work

Our work is related to several recent lines of investigation. Aggarwal et al. [10] originally linked the concept of r-anonymity to clustering. In their work they developed a model, called the "r-gathering" problem, where the goal is to cluster a set of objects from some metric space, such that each cluster has at least r objects. The specific objective they considered was to minimize maximal cluster radius subject to the r-anonymity requirement. This problem is proved to be NP-hard, but they show the existence of a polynomial time algorithm that produces a 2-approximation.

Wieland et al. [11] designed an efficient data anonymization algorithm. Their work considers in particular how to release information regarding the spatial distribution of a disease, while preserving privacy. The algorithm operates on

---

[1] The original name of the principle is k-anonymity. To avoid the confusion with the use of k in k-means, here we use r-anonymity instead of k-anonymity.

groups of data objects by computing a transition probability matrix between each pair of groups, with the objective of achieving a low group-to-group transition cost while maintaining a low risk of revealing a single object's profile. Once the transition probability is obtained via linear programming, the method uses a multi-nomial distribution to determine which group each object should be moved to. For a user-specified threshold $0 < \xi \leq 1$, the algorithm guarantees an expected re-identification probability no larger than ξ.

More recently, Li [12] formulated an r-anonymized clustering model and developed an iterative rounding based solution. The model assigns a set of objects into different clusters, with the objective of minimizing the object-to-center cost while ensuring that each cluster has no fewer than r objects, and the re-identification probability is strictly no larger than $\xi = \frac{1}{r}$. The work is an exemplar-based clustering model, which requires each center to be a real member of the data set.

3 A k-Means Type Model

We develop a model for solving the r-anonymized clustering problem in this paper. It is based on modelling a k-means approach.

3.1 Problem Formulation

Given a set of points $P = \{p_1, \cdots, p_m\} \subseteq R^d$, the problem is to partition the points into different clusters. Let c_i denote the center of the cluster (the mean of all points in the cluster) that data point p_i belongs to. The objective is to

$$\min \sum_{i=1}^{m} \|p_i - c_i\|^2 , \tag{1}$$

such that each cluster has at least r points.

Note that, unlike the classical k-means model, this problem does not specify the number of clusters. Instead, it implicitly confines the maximal cluster number by constraining the minimal number of points in each cluster. To overcome the difficulty of not knowing the exact cluster number, a natural way is to proceed is to solve the problem iteratively. For example, one could begin with the upper bound $\lfloor \frac{m}{r} \rfloor$, and decrease the cluster number gradually until the objective value cannot be improved. Observe that the optimization objective tends to favor creating more clusters to reduce the cost. Thus, an unconstrained iteration will only require a few iterations, but it will tend to produce a final cluster number equal to the upper bound. So throughout the paper, we also assume the k ($\leq \lfloor \frac{m}{r} \rfloor$) is given. With the addition of this constraint, the problem becomes equivalent to the minimization of the objective (1) such that we have k clusters where each cluster has at least r points.

3.2 Solution

To tackle the problem, we introduce decision variables $X = [x_{ij}] \in R^{m \times m}$ where x_{ij} is one if p_i and p_j are assigned to the same cluster, and zero otherwise. Using the fact that $\sum_{i=1}^{m} p_i = \sum_{i=1}^{m} c_i$ and $c_i = \frac{\sum_{j=1}^{m} x_{ij} p_j}{\sum_{j=1}^{m} x_{ij}}$, the objective (1) becomes

$$\min \sum_{i=1}^{m} \|p_i\|^2 - \sum_{i=1}^{m} \frac{\sum_{j=1}^{m} x_{ij} p_i^T p_j}{\sum_{j=1}^{m} x_{ij}}.$$

Then the problem can be equivalently expressed as

$$\min \quad -\sum_{i=1}^{m} \frac{\sum_{j=1}^{m} x_{ij} p_i^T p_j}{\sum_{j=1}^{m} x_{ij}}. \tag{2}$$

Here the matrix X can be further decomposed by $X = YY^T$, where $Y = [y_{i\ell}] \in R^{m \times k}$, $y_{i\ell} \in \{0, 1\}$. Note that $y_{i\ell} = y_{j\ell} = 1$ indicates that points p_i and p_j are assigned to the same cluster ℓ. Thus we also have

$$Y^T Y = diag \left(\sum_{i=1}^{m} y_{i1}, \cdots, \sum_{i=1}^{m} y_{ik} \right).$$

Let $Z = [z_{ij}] = Y \left(Y^T Y \right)^{-1} Y^T$, we can write the objective in (2) as

$$-\sum_{i=1}^{m} \sum_{j=1}^{m} z_{ij} p_i^T p_j.$$

Obviously, Z is symmetric and satisfies the transition relation: $Z^2 = Z$; moreover each element $z_{ij} \in [0, \frac{1}{r}]$. Other constraints on Z include $Ze^m = ZY e^k = e^m$ (e^m and e^k are vectors of ones in R^m and R^k respectively), which guarantees each point is assigned to exactly one cluster. Finally, the trace of Z should be equal to the number of clusters $tr(Z) = k$.

3.3 LP Relaxation

Following the work of [13], for example, the problem could be relaxed and solved by a semi-definite program. However, given our concerns of computational efficiency, we are more interested in pursuing a linear programming (LP) relaxation:

$$\min \quad -\sum_{i=1}^{m} \sum_{j=1}^{m} z_{ij} p_i^T p_j \tag{3}$$

subject to

$$\sum_{j=1}^{m} z_{ij} = 1, \ for \ all \ i \tag{4}$$

$$\sum_{i=1}^{m} z_{ii} = k \tag{5}$$

$$0 \leq z_{ij} \leq \frac{1}{r}, \; for \; all \; i,j \tag{6}$$

$$z_{ij} = z_{ji}, \; for \; all \; i,j \tag{7}$$

$$z_{ij} + z_{i\ell} \leq z_{ii} + z_{j\ell}, \; for \; all \; i,j,\ell \tag{8}$$

Note that in (8), the transition relation of Z is relaxed by the MET condition [14], which partially characterizes the relationship that if p_i and p_j, p_i and p_ℓ are in the same cluster, then p_j and p_ℓ should also be in the same cluster.

This LP problem (3) is still not scalable due to the large number ($O\left(m^3\right)$) of constraints in (8). To address this difficulty, we propose an efficient but exact incremental algorithm, similar to the idea of [15]: First we divide the constraints in (8) into base and extended constraints, in a manner described below. Then, we solve the problem with base constraints only. Given the relaxed solution, we then check extended constraints. If no violations are detected, the solution is reached and we terminate. Otherwise, all violations are moved to base constraints, and the problem is re-solved. The procedure repeats until no violations exist.

The point is that we can exploit structure in the problem to guide how to choose base constraints. Our separation is based on a simple yet key lemma.

Lemma 1. *For any data set P, there exists an optimal partition such that each cluster has at most $2r - 1$ points.*

The proof is direct. Suppose for an optimal partition, there exists a cluster with $2r$ or more points. Then we can always divide the cluster into two clusters with one of them having r points, such that the new objective value is no larger than the original one.

Since the maximal number of points of each cluster is bounded, one point is unlikely to be in the same cluster with the other if they are not close neighbors. Therefore, it is natural to define the base constraints only over decision variables x_{ij} where p_i is within p_j's $2r$ nearest neighbors, or vice versa. Using this technique, the number of constraints is typically reduced to $O\left(mr^2\right)$, which allows the algorithm to run very efficiently in practice.

We now summarize our LP relaxation algorithm: Let C and C' denote the set of base and extended constraints in (8) respectively. Then

1. Solve the LP problem (3) with constraints (4-7) and C;
2. Check the solution by C'. Move all violated constraints from C' to C.
3. Repeat Steps 1 and 2 until no violations found.

After each iteration, the linear program provides a lower bound on the objective. Upon completion, we can recover a hard cluster assignment. The classical k-means algorithm may be applied, or after a single value decomposition of Z [16]. However, the solution might not satisfy the r-anonymity criterion. In our work, we use the iterative rounding procedure developed in [17, 12, 18] that maintains the lower bound requirement.

Table 1. Comparison of different algorithms on UCI data sets, showing (objective value/maximal re-identification probability) among the clusters

Dataset(m)	r	k-means	k-exemplars	RAKM	RAKE
Pyrim(74)	10	119./.333	124./.250	123./.100	129./.100
	20	149./.0588	154./.0500	152./.0500	154./.0500
Iris(150)	10	30.0/.333	30.6/.333	32.2/.100	33.2/.100
	20	41.4/.0833	44.1/.100	47.3/.0500	53.1/.0500
Wine(178)	10	129./.500	141./.500	134./.100	144./.100
	20	153./.250	167./.0770	154./.0500	167./.0500
Triazines(186)	10	245./100	248./.333	279./.100	292./.100
	20	326./.200	347./.200	360./.0500	385./.0500
Sonar(208)	10	299./.500	331./1.00	315./.100	364./.100
	20	342./.167	386./.167	357./.0500	425./.0500
Glass(214)	10	63.1/1.00	62.8/1.00	76.6/.100	78.0/.100
	20	88.1/1.00	86.7/.500	95.0/.0500	96.2/.0500
Liver(345)	10	92.3/1.00	94.2/1.00	95.0/.100	100./.100
	20	111./.200	114./.333	118./.0500	119./.0500
Ionosphere(351)	10	474./1.00	469./1.00	514./.100	527./.100
	20	550./.333	555./1.00	572./.0500	586./.0500
Housing(506)	10	214./1.00	215./.333	232./.100	240./.100
	50	362./.125	383./.0430	417./.0200	438./.0200
Breast(683)	10	310./1.00	305./1.00	344./.100	348./.100
	50	469./.125	479./.083	490./.0200	505./.0200
Diabetes(768)	10	273./1.00	283./1.00	291./.100	297./.100
	50	393./.143	412./.0370	407./.0200	420./.0200

4   Evaluation

To evaluate the performance of the proposed algorithm, we conducted a series of evaluations. The experiments compared the re-identification probabilities between the conventional clustering algorithms (including k-means clustering and k-exemplars clustering) and the privacy preserving clustering algorithms (including r-anonymized k-exemplars clustering [12], or RAKE, and our proposed r-anonymized k-means clustering, or RAKM) on UCI data sets [19]. For k-means, we used the standard subroutine in MATLAB with 100 random starts, and the best cost values were reported. For k-exemplars, we used affinity propagation algorithm [20–22]. For two privacy preserving clustering algorithms, both need to solve a series of linear programs and we used off-the-shelf CPLEX software as the underlying solver.

As a preprocessing step, we replaced any missing values in the data by 0 and then the data were linearly normalized within $[-1, 1]$. The cost between a pair of objects was defined by their squared Euclidean distance. In these experiments we chose different r values, ranging from 10 to 50, depending on problem size. For each value of r, we set $k = \lfloor \frac{m}{r} \rfloor$.

The results are given in Table (1). Each result has two terms: the first is the objective value, (ref. equation (1)); the second is the maximal re-identification probability among the clusters. From these results one can see that, although the k-means and k-exemplars achieve lower costs, the two algorithms do not guarantee the r-anonymity bound on the privacy re-identification risk. Given r, the risk is often beyond the preferred threshold $\frac{1}{r}$. Thus the two algorithms cannot be applied directly where privacy issues are a concern.

For RAKM and RAKE, the performance is just as expected. Both of them have achieved a strong guarantee on the maximal re-identification probability. Comparing these two, RAKM achieves smaller objective values, also as expected.

5 Conclusion

Publishing data without releasing sensitive information is an important problem. Motivated by the idea of k-anonymity, we proposed a clustering-based method for privacy-preserving data publishing. Our model provide a general framework for a variety of applications. We demonstrated algorithmic approaches that provide strong guarantees that the re-identification risk is not larger than a user-specified threshold, if no more information is available to help identify individuals.

Although the optimal solution to the models is computationally hard, we developed an efficient technique based on linear program relaxation and iterative rounding. Although theoretical analysis of our work remains a challenge, empirical evaluations showed successful results as expected.

Acknowledgement. The work is partially supported by The Science and Technology Development Fund, Macao SAR, China.

References

1. Jain, A., Murty, M., Flynn, P.: Data clustering: A review. ACM Computing Surveys 31(3), 264–323 (1999)
2. Basu, S., Davidson, I., Wagstaff, K.: Constrained Clustering: Advances in Algorithms, Theory, and Applications. Chapman & Hall (2008)
3. MacQueen, J.: Some methods for classification and analysis of multivariate observations. In: Proceedings of the Fifth Berkeley Symposium on Mathematical Statistics and Probability, vol. 1. University of California Press (1967)
4. Sweeney, L.: Uniqueness of simple demographics in the U.S. population (2000), http://privacy.cs.cmu.edu/
5. Aggarwal, C., Yu, P.: Privacy-Preserving Data Mining: Models and Algorithms. Springer (2008)
6. Froomkin, A.: The death of privacy? Stanford Law Review 52 (2000)
7. Sweeney, L.: k-anonymity: A model for protecting privacy. Int. J. Uncertainty Fuzziness Knowledge Based Syst. 10(5) (2002)
8. Meyerson, A., Williams, R.: On the complexity of optimal K-anonymity. In: Proceedings of PODS 2004. ACM (2004)

9. Yu, T., Jajodia, S. (eds.): Secure Data Management in Decentralized Systems, vol. 33. Springer (2007)
10. Aggarwal, G., Feder, T., Kenthapadi, K., Khuller, S., Panigrahy, R., Thomas, D., Zhu, A.: Achieving anonymity via clustering. In: Proceedings of PODS 2006. ACM (2006)
11. Wieland, S., Cassa, C., Mandl, K., Berger, B.: Revealing the spatial distribution of a disease while preserving privacy. Proc. Natl. Acad. Sci. USA 105(46) (2008)
12. Li, W.: r-anonymized clustering. In: Huang, T., Zeng, Z., Li, C., Leung, C.S. (eds.) ICONIP 2012, Part I. LNCS, vol. 7663, pp. 455–464. Springer, Heidelberg (2012)
13. Peng, J., Wei, Y.: Approximating k-means-type clustering via semidefinite programming. SIAM Journal on Optimization 18(1) (2007)
14. Lisser, A., Rendl, F.: Graph partitioning using linear and semidefinite programming. Mathematical Programming 95(1) (2003)
15. Riedel, S., Clarke, J.: Incremental integer linear programming for non-projective dependency parsing. In: Proceedings of EMNLP 2006 (2006)
16. Xing, E., Jordan, M.: On semidefinite relaxation for normalized k-cut and connections to spectral clustering. Technical report, University of California, Berkeley (2003)
17. Li, W., Schuurmans, D.: Modular community detection in networks. In: Proceedings of the 22nd International Joint Conference on Artificial Intelligence, pp. 1366–1371. AAAI (2011)
18. Li, W.: Modularity segmentation. In: Lee, M., Hirose, A., Hou, Z.-G., Kil, R.M. (eds.) ICONIP 2013, Part II. LNCS, vol. 8227, pp. 100–107. Springer, Heidelberg (2013)
19. Asuncion, A., Newman, D.: UCI machine learning repository (2007)
20. Frey, B.J., Dueck, D.: Clustering by passing messages between data points. Science 315 (2007)
21. Li, W., Xu, L., Schuurmans, D.: Facility locations revisited: An efficient belief propagation approach. In: 2010 IEEE International Conference on Automation and Logistics, pp. 408–413. IEEE (2010)
22. Li, W.: Clustering with uncertainties: An affinity propagation-based approach. In: Huang, T., Zeng, Z., Li, C., Leung, C.S. (eds.) ICONIP 2012, Part V. LNCS, vol. 7667, pp. 437–446. Springer, Heidelberg (2012)

Stream Quantiles via Maximal Entropy Histograms

Ognjen Arandjelović, Ducson Pham, and Svetha Venkatesh

Centre for Pattern Recognition and Data Analytics, Deakin University, Geelong, Australia

Abstract. We address the problem of estimating the running quantile of a data stream when the memory for storing observations is limited. We (i) highlight the limitations of approaches previously described in the literature which make them unsuitable for non-stationary streams, (ii) describe a novel principle for the utilization of the available storage space, and (iii) introduce two novel algorithms which exploit the proposed principle. Experiments on three large real-world data sets demonstrate that the proposed methods vastly outperform the existing alternatives.

1 Introduction

The problem of quantile estimation is of pervasive importance across a variety of signal processing applications. It is used in data mining, simulation modelling [1], database maintenance, risk management in finance [2], and understanding computer network latencies [3], amongst others. A particularly challenging form of the quantile estimation problem arises when the desired quantile is high-valued (i.e. close to 1, corresponding to the tail of the underlaying distribution) and when data needs to be processed as a stream, with limited memory capacity. An illustrative practical example of when this is the case is encountered in CCTV-based surveillance systems. In summary, as various types of low-level observations related to events in the scene of interest arrive in real-time, quantiles of the corresponding statistics for time windows of different durations are needed in order to distinguish 'normal' (common) events from those which are in some sense unusual and thus require human attention. The amount of incoming data is extraordinarily large and the capabilities of the available hardware highly limited both in terms of storage capacity and processing power.

1.1 Previous Work

Unsurprisingly, the problem of estimating a quantile of a set has received considerable research attention, much of it in the realm of theoretical research. In particular, a substantial amount of work has focused on the study of asymptotic limits of computational complexity of quantile estimation algorithms [4, 5]. An important result emerging from this corpus of work is the proof by Munro and Paterson [5] which in summary states that the working memory requirement of any algorithm that determines the median of a set by making at most p sequential passes through the input is $\Omega(n^{1/p})$ (i.e. asymptotically growing at least as fast as $n^{1/p}$). This implies that the exact computation of a quantile requires $\Omega(n)$ working memory. Therefore a single-pass algorithm, required to process

C.K. Loo et al. (Eds.): ICONIP 2014, Part II, LNCS 8835, pp. 327–334, 2014.

streaming data, will necessarily produce an estimate and not be able to guarantee the exactness of its result.

Most of the quantile estimation algorithms developed for the use in practice are not single-pass algorithms i.e. cannot be applied to streaming data [6]. On the other hand, many single-pass approaches focus on the exact computation of the quantile and thus demand $O(n)$ storage space which is clearly an unfeasible proposition in the context we consider in the present paper. Amongst the few methods described in the literature and which satisfy our constraints are the histogram-based method of Schmeiser and Deutsch [7] (also by McDermott *et al.* [8]), and the P^2 algorithm of Jain and Chlamtac [1]. Schmeiser and Deutsch maintain a preset number of bins, scaling their boundaries to cover the entire data range as needed and keeping them equidistant. Jain and Chlamtac attempt to maintain a small set of *ad hoc* selected key points of the data distribution, updating their values using quadratic interpolation as new data arrives. Lastly, random sample methods, such as that described by Vitter [9], and Cormode and Muthukrishnan [10], use different sampling strategies to fill the available buffer with random data points from the stream, and estimate the quantile using the distribution of values in the buffer.

In addition to the *ad hoc* elements of the previous algorithms for quantile estimation on streaming data, which itself is a sufficient cause for concern when the algorithms need to be deployed in applications which demand high robustness and well understood failure modes, it is also important to recognize that an implicit assumption underlying these approaches is that the data is governed by a stationary stochastic process. The assumption is often invalidated in real-world applications. As we will demonstrate in Sec. 3, a consequence of this discrepancy between the model underlying existing algorithms and the nature of data in practice is a major deterioration in the quality of quantile estimates. Our principal aim is thus to formulate a method which can cope with non-stationary streaming data in a more robust manner.

2   Proposed Algorithms

We start this section by formalizing the notion of a quantile. This is then followed by the introduction of the key premise of our contribution and finally a description of two algorithms which exploit the underlying idea in different ways. The algorithms are evaluated on real-world data in the next section.

2.1   Quantiles

Let p be the probability density function of a real-valued random variable X. Then the q-quantile x_q of p is defined as:

$$\int_{-\infty}^{x_q} p(x)\, dx = q. \tag{1}$$

Similarly, the q-quantile of a finite set D can be defined as:

$$|\{x \ : \ x \in D \text{ and } x \leq x_p\}| \leq q \times |D|. \tag{2}$$

In other words, the q-quantile is the smallest value below which q fraction of the total values in a set lie.

2.2   Maximal Entropy Histograms

A consequence of the non-stationarity of data streams that we are dealing with is that at no point in time can it be assumed that the historical distribution of data values is representative of its future distribution, regardless of how much data has been seen. Thus, the value of a particular quantile can change greatly and rapidly, in either direction (i.e. increase or decrease). To be able to adapt to such unpredictable variability in input it is therefore not possible to focus on only a part of the historical data distribution but rather it is necessary to store a 'snapshot' of the entire distribution. We achieve this using a histogram of a fixed length, determined by the available working memory. In contrast to the previous work which either distributes the bin boundaries equidistantly or uses *ad-hoc* adjustments, our idea is to maintain bins in a manner which maximizes the entropy of the corresponding estimate of the historical data distribution.

2.3   Method 1: Interpolated Bins

The first method we describe readjusts the boundaries of a fixed number of bins after the arrival of each new data point d_{i+1}. Without loss of generality let us assume that each each datum is positive i.e. that $d_i > 0$. Furthermore, let the upper bin boundaries before the arrival of d_i be $b_1^i, b_2^i, \ldots, b_n^i$, where n is the number of available bins. Thus, the j-th bin's catchment range is $(b_{j-1}^i, b_j^i]$ where we will take that $b_0^i = 0$ for all i. We wish to maintain the condition that the piece-wise uniform probability density function approximation of the historical data distribution described by this histogram has the maximal entropy of all those possible with the histogram of the same length. This is achieved by having equiprobable bins. Thus, before the arrival of d_{i+1}, the number of historical data points in each bin is the same and equal to i/n. The corresponding cumulative density is given by:

$$p^i(d) = \frac{1}{n} \times \left[j + \frac{d - b_{j-1}^i}{b_j^i - b_{j-1}^i} \right] \quad \text{and} \quad b_{j-1}^i < d \le b_j^i. \tag{3}$$

After the arrival of d_i but before the readjustment of bin boundaries, the cumulative density becomes:

$$\tilde{p}^i(d) = \begin{cases} \frac{i}{i+1} \times \frac{1}{n} \times \left[j + \frac{d - b_{j-1}^i}{b_j^i - b_{j-1}^i} \right] & \text{for } d < d_i \\[2ex] \frac{i}{i+1} \times \frac{1}{n} \times \left[j + \frac{d - b_{j-1}^i}{b_j^i - b_{j-1}^i} \right] + \frac{1}{i+1} & \text{for } d \ge d_i \end{cases} \tag{4}$$

Lastly, to maintain the invariant of equiprobable bins, the bin boundaries are readjusted by linear interpolation of the corresponding inverse distribution function.

2.4   Method 2: Data-Aligned Bins

The algorithm described in the proceeding section appears optimal in that it always attempts to realign bins so as to maintain maximal entropy of the corresponding approximation for the given size of the histogram. However, a potential source of errors can emerge cumulatively as a consequence of repeated interpolation, done after every new datum. Indeed, we will show this to be the case empirically in Sec. 3. We now introduce an alternative approach which aims to strike a balance between some unavoidable loss of information, inherently a consequence of the need to readjust an approximation of the distribution of a continually growing data set, and the desire to maximize the entropy of this approximation.

Much like in the previous section, bin boundaries are potentially altered each time a new datum arrives. There are two main differences in how this is performed. Firstly, unlike in the previous case, bin boundaries are not allowed to assume arbitrary values; rather, they are constrained to the values of the seen data points. Secondly, only at most a single boundary is adjusted for each new datum. We now explain this process in detail.

As before, let the upper bin boundaries before the arrival of a new data point be $b_1^i, b_2^i, \ldots, b_n^i$. Since unlike in the case of the previous algorithm in general the bins will not be equiprobable we also have to maintain a corresponding list $c_1^i, c_2^i, \ldots, c_n^i$ which specifies the corresponding data counts. Each time a new data point arrives a new, an $(n+1)$-st bin is created temporarily. If the value of the new datum is greater than b_n^i (and thus greater than any of the historical data), a new bin is created after the current n-th bin, with the upper boundary set at $d(i)$. The corresponding datum count c of the bin is set to 1. Alternatively, if the value of the new data point is lower than b_n^i then there exists j such that $b_{j-1}^i < d \leq b_j^i$ and the new bin is inserted between the $(j-1)$-st and j-th bin. Its datum count is estimated as follows:

$$c = c_j \times \frac{d - b_{j-1}^i}{b_j^i - b_{j-1}^i} + 1. \tag{5}$$

Thus, regardless of the value of the new data point, temporarily the number of bins is increased by 1. The original number of bins is then restored by merging exactly a single pair of neighbouring bins. For example, if the k-th and $(k+1)$-st bin are merged, the new bin has the upper boundary value set to the upper boundary value of the former $(k+1)$-st bin, i.e. b_{k+1}^i, and its datum count becomes the sum of counts for the k-th and $(k+1)$-st bins, i.e. $c_k^i + c_{k+1}^i$. The choice of which neighbouring pair to merge, out of n possible options, is made according to the principle stated in Sec. 2.2, i.e. the merge actually performed should maximize the entropy of the new n-bin histogram. This is illustrated conceptually in Fig. 1(a).

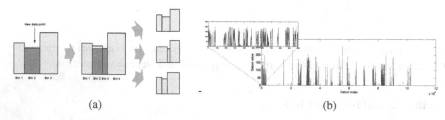

(a)   (b)

Fig. 1. (a) The update step in our data-aligned adaptive histogram algorithm. (b) One of the three large data streams used in our evaluation.

3 Evaluation and Results

To assess the effectiveness of the proposed algorithms, we evaluated their performance on three large 'real-world' data streams. The streams correspond to motion statistics used by an existing CCTV surveillance system for the detection of abnormalities in video footage. It is important to emphasize that the data we used was not acquired for the purpose of the present work nor were the cameras installed with the same intention. Rather, we used data which was acquired using existing, operational surveillance systems. In particular, our data comes from three CCTV cameras, two of which are located in Mexico and one in Australia. We next explain the source of these streams and the nature of the phenomena they represent.

3.1 Real-World Surveillance Data

Computer-assisted video surveillance data analysis is of major commercial and law enforcement interest. On a broad scale, systems currently available on the market can be grouped into two categories in terms of their approach. The first group focuses on a relatively small, predefined and well understood subset of events or behaviours of interest such as the detection of unattended baggage, violent behaviour, etc [11, 12]. The narrow focus of these systems prohibits their applicability in less constrained environments in which a more general capability is required. In addition, these approaches tend to be computationally expensive and error prone, often requiring fine tuning by skilled technicians. This is not practical in many circumstances, for example when hundreds of cameras need to be deployed as often the case with CCTV systems operated by municipal authorities. The second group of systems approaches the problem of de tecting suspicious events at a semantically lower level [13–15]. Their central paradigm is that an unusual behaviour at a high semantic level will be associated with statistically unusual patterns (also 'behaviour' in a sense) at a low semantic level – the level of elementary image/video features. Thus methods of this group detect events of interest by learning the scope of normal variability of low-level patterns and alerting to anything that does not conform to this model of what is expected in a scene, without 'understanding' or interpreting the nature of the event itself. These methods uniformly start with the same procedure for feature extraction. As video data is acquired, firstly a dense optical flow field is computed. Then, to reduce the amount of data that needs to be processed, stored, or transmitted, a thresholding operation is performed. This results in a sparse optical flow field whereby only those flow vectors whose magnitude exceeds a certain value are retained; non-maximum suppression is applied here as well. Normal variability within a scene and subsequent novelty detection are achieved using various statistics computed over this data. A typical data stream, shown partially in Fig. 1(b), corresponds to the values of such a statistic. Observe the non-stationary nature of the data streams which is evident both on the long and short time scales.

3.2 Results

We started by comparing the performance of our algorithms with the three alternatives from the literature described in Sec. 1.1: (i) the P^2 algorithm of Jain and Chlamtac [1],

(ii) the random sample based algorithm of Vitter [9], and (iii) the uniform adjustable histogram of Schmeiser and Deutsch [7]. Representative results, obtained using the same number of bins $n = 500$, for 0.95-quantile on stream 1 are shown Fig. 2 – the running quantile estimate of the algorithm (purple) is superimposed to the ground truth (cyan). Firstly, compare the performances of the two proposed algorithms. We found that in all cases and across time, the data-aligned bins algorithm produced a more reliable estimate. Thus, the argument put forward in Sec. 2.4 turned out to be correct – despite the attempt of the interpolated bins algorithm to maintain exactly a maximal entropy approximation to the historical data distribution, the advantages of this approach are outweighed by the accumulation of errors caused by repeated interpolations. The data-aligned algorithm consistently exhibited outstanding performance on all three data sets, its estimate being virtually indistinguishable from the ground truth. This is witnessed and more easily appreciated by examining the plots showing its running relative error. In most cases the error was approximately 0.2%; the only instance when the error would exceed this substantially is transiently at times of sudden large change in the quantile value (as in the case of stream 1), quickly recovering thereafter.

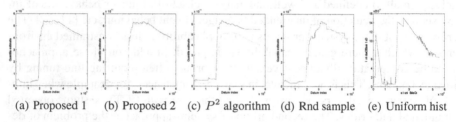

(a) Proposed 1   (b) Proposed 2   (c) P^2 algorithm   (d) Rnd sample   (e) Uniform hist

Fig. 2. Running estimate (purple) of the 0.95-quantile on stream 1 (ground truth is shown in cyan)

All of the algorithms from the literature performed significantly worse than both of the proposed methods. The limitations of the assumption of stationary data statistics implicitly made in the P^2 algorithm and discussed in Sec. 1.1 is readily evident by its observed performance. Following the initially good estimates when the true quantile value is relatively large, the algorithm is unable to adjust sufficiently to the changed data distribution and the decreasing quantile value. Across the three data sets, the random sample algorithm of Vitter [9] overall performed best of the existing methods, never producing a grossly inaccurate estimate. Nonetheless its accuracy is far lower than that of the proposed algorithms, as easily seen by the naked eye and further witnessed by the corresponding plots of the relative error, with some tendency towards jittery and erratic behaviour. The adaptive histogram based algorithm of Schmeiser and Deutsch [7] performed comparatively well on streams 2 and 3. On this account it may be surprising to observe its complete failure at producing a meaningful estimate in the case of stream 1. In fact the behaviour the algorithm exhibited on this data set is most useful in understanding the algorithm's failures modes. Notice at what points in time the estimate would shoot widely. After inspecting the input data it is readily observed that in each case this behaviour coincides with the arrival of a datum which is much larger than any of the historical data (and thus the range of the histogram). What happens then is that in re-scaling the histogram by such a large factor, many of the existing bins get 'squeezed' into only a single bin of the new histogram, resulting in a major loss of information. When this behaviour is contrasted with the performance of the algorithms we proposed in this paper, the importance of the maximal entropy principle as the foundational idea is easily appreciated; although our algorithms too readjust their bins upon the arrival of each new datum, the design of our histograms ensures that no major loss of information occurs regardless of the value of new data.

Considering the outstanding performance of our algorithms, and in particular the data-aligned histogram-based approach, we next sought to examine how this performance is affected by a gradual reduction of the working memory size. To make the task more challenging we sought to estimate the 0.99-quantile on the largest of our three data sets (stream 2). Our results are summarized in Table 1. This table shows the variation in the mean relative error as well as the largest absolute error of the quantile estimate for the proposed data-aligned histogram-based algorithm as the number of available bins is gradually decreased from 500 to 12. For all other methods, the reported result is for $n = 500$ bins. It is remarkable to observe that the mean relative error of our algorithm does not decrease at all. The largest absolute error does increase, only a small amount as the number of bins is reduced from 500 to 50, and more substantially thereafter. This shows that our algorithm overall still produces excellent estimates with occasional and transient difficulties when there is a rapid change in the quantile value. Plots in Fig. 3 corroborate this observation.

Table 1. Summary of experimental results for the estimation of 0.99-quantile on stream 2

Method		Mean relative error	Absolute L_∞ error
Proposed data-aligned bins w/ bin no.	500	0.5%	2.43
	100	0.5%	2.45
	50	0.5%	3.01
	25	0.4%	14.48
	12	0.5%	28.83
P^2 algorithm [1]		45.6%	112.61
Random sample [9]		17.5%	64.00
Equispaced bins [7]		0.9%	76.88

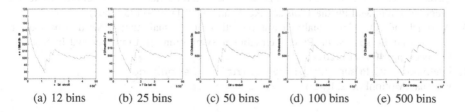

(a) 12 bins   (b) 25 bins   (c) 50 bins   (d) 100 bins   (e) 500 bins

Fig. 3. Running estimate (purple) of the 0.99-quantile on stream 2 produced using our data-aligned adaptive histogram algorithm (ground truth is shown in cyan)

4   Summary and Conclusions

We introduced two novel algorithms for the estimation of a quantile of a data stream when the available working memory is limited. The proposed algorithms were evaluated and compared against the existing alternatives described in the literature using three large data streams. The highly non-stationary nature of our data was shown to cause major problems to the existing algorithms, often leading to grossly inaccurate quantile estimates; in contrast, our methods were virtually unaffected by it. Our experiments demonstrate that the superior performance of our algorithms can be maintained effectively while drastically reducing the working memory size in comparison with the methods from the literature.

References

1. Jain, R., Chlamtac, I.: The P^2 algorithm for dynamic calculation of quantiles and histograms without storing observations. Communications of the ACM 28(10), 1076–1085 (1985)
2. Sgouropoulos, N., Yao, Q., Yastremiz, C.: Matching quantiles estimation. Technical report, London School of Economics (2013)
3. Buragohain, C., Suri, S.: Quantiles on Streams. In: Encyclopedia of Database Systems, pp. 2235–2240 (2009)
4. Guha, S., McGregor, A.: Stream order and order statistics: Quantile estimation in random-order streams. SIAM Journal on Computing 38(5), 2044–2059 (2009)
5. Munro, J.I., Paterson, M.: Selection and sorting with limited storage. Theoretical Computer Science 12, 315–323 (1980)
6. Gurajada, A.P., Srivastava, J.: Equidepth partitioning of a data set based on finding its medians. TR 90-24, University of Minnesota (1990)
7. Schmeiser, B.W., Deutsch, S.J.: Quantile estimation from grouped data: The cell midpoint. Communications in Statistics: Simulation and Computation 6(3), 221–234 (1977)
8. McDermott, J.P., Babu, G.J., Liechty, J.C., Lin, D.K.J.: Data skeletons: simultaneous estimation of multiple quantiles for massive streaming datasets with applications to density estimation. Bayesian Analysis 17, 311–321 (2007)
9. Vitter, J.S.: Random sampling with a reservoir. ACM Transactions on Mathematical Software 11(1), 37–57 (1985)
10. Cormode, G., Muthukrishnany, S.: An improved data stream summary: the count-min sketch and its applications. Journal of Algorithms 55(1), 58–75 (2005)
11. Philips Electronics N.V.: A surveillance system with suspicious behaviour detection. Patent EP1459272A1 (2004)
12. Lavee, G., Khan, L., Thuraisingham, B.: A framework for a video analysis tool for suspicious event detection. Multimedia Tools and Applications 35(1), 109–123 (2007)
13. Arandjelović, O.: Contextually learnt detection of unusual motion-based behaviour in crowded public spaces. In: Proc. International Symposium on Computer and Information Sciences, pp. 403–410 (2011)
14. intellvisions. iQ-Prisons, http://www.intellvisions.com/
15. iCetana. iMotionFocus, http://www.icetana.com/

A Unified Framework for Thermal Face Recognition

Reza Shoja Ghiass[1], Ognjen Arandjelović[2], Hakim Bendada[1],
and Xavier Maldague[1]

[1] Laval University, Canada
[2] Deakin University, Australia

Abstract. The reduction of the cost of infrared (IR) cameras in recent years has made IR imaging a highly viable modality for face recognition in practice. A particularly attractive advantage of IR-based over conventional, visible spectrum-based face recognition stems from its invariance to visible illumination. In this paper we argue that the main limitation of previous work on face recognition using IR lies in its *ad hoc* approach to treating different nuisance factors which affect appearance, prohibiting a unified approach that is capable of handling concurrent changes in multiple (or indeed all) major extrinsic sources of variability, which is needed in practice. We describe the first approach that attempts to achieve this – the framework we propose achieves outstanding recognition performance in the presence of variable (i) pose, (ii) facial expression, (iii) physiological state, (iv) partial occlusion due to eye-wear, and (v) quasi-occlusion due to facial hair growth.

1 Introduction

Following the overly optimistic expectations of early research, after several decades of intense research on the one hand and relatively disappointing practical results on the other, face recognition technology has been finally started to enjoy some success in the consumer market. An example can be found within the online photo sharing platform in Google Plus which automatically recognizes individuals in photographs based on previous labelling by the user. It is revealing to observe that the recent success of face recognition has been in the realm of data and image retrieval [1], rather than security, contrasting the most often stated source of practical motivation driving past research [2]. This partial paradigm shift is only a recent phenomenon. In hindsight, that success would first be achieved in the domain of retrieval should not come as a surprise considering the relatively low importance of type-II errors in this context: the user is typically interested in only a few of the retrieved matches and the consequences of the presence of false positives is benign, more often being mere inconvenience.

Nevertheless, the appeal of face recognition as a biometric means that the interest in its security applications is not waning. Face recognition can be performed from a distance and without the knowledge of the person being recognized, and is more readily accepted by the general public in comparison with other biometrics which are regarded as more intrusive. In addition, the acquisition of data for face recognition can be performed readily and cheaply, using widely available devices. However, although the interest in security uses of face recognition continues, it has become increasingly the case

C.K. Loo et al. (Eds.): ICONIP 2014, Part II, LNCS 8835, pp. 335–343, 2014.

that research has shifted from the use of face recognition as a sole biometric. Instead, the operational paradigm is that of face recognition as an element of a multi-biometric system [3].

Of particular interest to us is the use of infrared imaging as a modality complementary to the 'conventional', visible spectrum-based face recognition. Our focus is motivated by several observations. Firstly, in principle an IR image of a face can be acquired whenever a conventional image of a face can. This may not be the case with other biometrics (gait, height etc). Secondly, while certainly neither as cheap nor as widely available as conventional cameras, in recent years IR imagers have become far more viable for pervasive use. It is interesting to note the self-enforcing nature of this phenomenon: the initial technology advancement-driven drop in price has increased the use of IR cameras thereby making their production more profitable and in turn lowering their cost even further.

In a distal sense, the key challenges in IR-based face recognition remain to be pose invariance, robustness to physiological conditions affecting facial IR emissions, and occlusions due to facial hair and accessories (most notably eyeglasses). In a more proximal sense, as argued in a recent review [4], the main challenge is to formulate a framework which is capable of dealing with all of the aforementioned factors affecting IR 'appearance' in a unified manner. While a large number of IR-based face recognition algorithms have been described in the literature, without exception they all constrain their attention to a few, usually only a single, extrinsic factor (e.g. facial expression or pose). None of them can cope well with a concurrent variability in several extrinsic factors. Yet, this is the challenge encountered in practice.

In this paper our aim is to describe what we believe to be the first IR-based face recognition method which is able to deal with all of the major practical challenges. Specifically, our method explicitly addresses (i) the variability in the user's physiological state, (ii) pose changes, (iii) facial expressions, (iv) partial occlusion due to prescription glasses, and (v) quasi-occlusion due to facial hair.

2   Unified Treatment of Extrinsic Variability

In this section we detail different elements of our system. We start by describing the Dual Dimension Active Appearance Model Ensemble framework and then demonstrate how it can be extended to perform model selection which allows for the handling of partial occlusion due to prescription glasses, and quasi-occlusion due to facial hair.

2.1   Dual Dimension AAM Ensemble (DDAE)

At the coarsest level, the Dual Dimension Active Appearance Model Ensemble algorithm comprises three distinct components. These are: (i) a method for fitting an active appearance model (AAM) [5] particularly designed for fast convergence and reliable fitting in the IR spectrum, (ii) a method for selecting the most appropriate AAM from a trained ensemble, and (iii) the underlying extraction of person-specific discriminative information. The ultimate functions of these elements within the system as a whole are respectively pose normalization within a limited yaw range, invariance across the full range of yaw, and invariance to physiological changes of the user.

AAM Fitting. There are two crucial design aspects of the design and deployment of AAMs that need to be considered in order to ensure their robustness. These are: (i) the model initialization procedure, and (ii) the subsequent iterative refinement of model parameters. In the context of the problem considered in this paper, the former is relatively simple. Given that we are using thermal imaging background clutter is virtually non-existant applying simple thresholding to the input image allows the face to be localized and its spatial extent estimated. Reliable initialization of the AAM is then readily achieved by appropriately positioning its centroid and scale. A much greater challenge in the use of the AAM model for the normalization of pose and facial expression concerns the subsequent convergence – the model is notoriously prone to convergence towards a local mininum, possibly far from the correct solution. This problem is even more pronounced when fitting is performed on face images acquired in the IR spectrum; unlike in the visible spectrum, in thermal IR human faces lack face-specific detail that is crucial in directing iterative optimization. This is the likely explanation for the absence of published work on the use of AAMs for faces in IR until the work of Ghiass *et al.* [6]. Their key idea was to perform fitting by learning and applying an AAM not on raw IR images themselves but rather on images automatically processed in a manner which emphasizes high-frequency detail. Specifically the detail-enhanced image correction I_e is computed by anisotropically diffusing the input image I:

$$\frac{\partial I}{\partial t} = \nabla . (c(\|\nabla I\|) \, \nabla I) = \nabla c . \nabla I + c(\|\nabla I\|) \, \Delta I, \tag{1}$$

using a spatially varying and image gradient magnitude-dependent parameter $c(\|\nabla I\|) = \exp\{-\|\nabla I\|/400\}$, subtracting the result from the original image and applying histogram equalization $I_e = \mathrm{histeq}(I - I_d)$. Warped examples are shown in Fig. 1.

Fig. 1. Examples of automatically pre-processed thermal images warped to the canonical geometric frame using the described AAM fitting method

Person and Pose-Specific Model Selection. Applied to faces, the active appearance model comprises a triangular mesh with each triangle describing a small surface patch. To use this model for the normalization of pose, it is implicitly assumed that the surface patches are approximately planar. While this approximation typically does not cause fitting problems when pose variation is small this is not the case when fitting needs to be performed across a large range of poses (e.g. ranging from fully frontal to fully profile orientation relative to the camera's viewing direction). This is particularly pronounced when the applied AAM is generic (rather than person-specific) and needs to have sufficient power to describe the full scope of shape and appearance variability across faces. The DDAE method overcomes these problems by employing an ensemble of AAMs.

Each AAM in an ensemble 'specializes' to a particular region of IR face space. The space is effectively partitioned both by pose and the amount of appearance variability, making each AAM specific to a particular range of poses and individuals of relatively similar appearance (in the IR spectrum). In training, this is achieved by first dividing the training data corpus into pose-specific groups and then applying appearance clustering on IR faces within each group. A single AAM in an ensemble is trained using a single cluster. On the other hand, when the system is queried with a novel image containing a face in an arbitrary and unknown pose, the appropriate AAM from the ensemble needs to be selected automatically. This can be readily done by fitting each AAM from the ensemble and then selecting the best AAM as the one with the greatest maximal likelihood, that is, the likelihood achieved after convergence.

Discriminative Representation. A major challenge to IR-based face recognition in practice emerges as a consequence of thermal IR appearance dependence on the physiological state of the user. Factors which affect the sympathetic or parasympathetic nervous system output (e.g. exercise or excitement) or peripheral blood flow (e.g. ambient temperature or drugs) can have a profound effect. This is illustrated in Fig. 2. With the notable exception of the work by Buddharaju *et al.* [7] there has been little research done on developing an IR-based person-specific representation invariant to these changes.

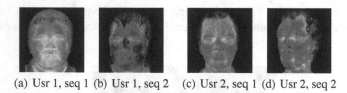

(a) Usr 1, seq 1   (b) Usr 1, seq 2   (c) Usr 2, seq 1   (d) Usr 2, seq 2

Fig. 2. Thermal IR appearance before ('seq 1') and after ('seq 2') mild exercise

We adopt the use of a representation which is dependent on the distribution of superficial blood vessels. Unlike the vessel network based representation [7] our representation is not binary i.e. a particular image pixel is not merely classified as a blood vessel or not. Rather, the extracted pseudo-probability map allows for the level of confidence regarding the saliency of a particular pixel to be encoded. Additionally, unlike the representation extracted by various blood perfusion methods [8], our representation is based on temperature gradients, rather than the absolute temperature. As such, it is less affected by physiological changes which influence the amount of blood flow, and is rather a function of the invariant distribution of underlying blood vessels.

Our representation is extracted using the so-called vesselness filter [9], originally developed for use on 3D MRI data for the extraction of tubular structures from images. Just as in 3D, in 2D this is achieved by considering the of the Hessian matrix $H(I, x, y, s)$ computed at the image locus (x, y) and at the scale s:

$$H(I, x, y, s) = \begin{bmatrix} L_x^2(x, y, \sigma) & L_x L_y(x, y, \sigma) \\ L_x L_y(x, y, \sigma) & L_y^2(x, y, \sigma) \end{bmatrix} \tag{2}$$

where $L_x^2(x,y,s)$, $L_y^2(x,y,s)$, and $L_x L_y(x,y,\sigma)$ are the second derivatives of $L(x,y,s)$, resulting from the smoothing the original image $I(x,y)$ with a Gaussian kernel $L(x,y,s) = I(x,y)*G(s)$. If the two eigenvalues of $H(I,x,y,s)$ are λ_1 and λ_2 and if without loss of generality we take $|\lambda_1| \leq |\lambda_2|$, two statistics which characterize local appearance and which can be used to quantify the local vesselness of appearance are $\mathcal{R}_A = |\lambda_1|/|\lambda_2|$ and $\mathcal{S} = \sqrt{\lambda_1^2 + \lambda_2^2}$. The former of these measures the degree of local 'blobiness' which should be low for tubular structures, while $\mathcal{S}$ rejects nearly uniform, uninformative image regions which are characterized by small eigenvalues of the Hessian. For a particular scale of image analysis s, the two measures, $\mathcal{R}_A$ and $\mathcal{S}$, are unified into a single measure $\mathcal{V}(s) = (1 - e^{-\frac{\mathcal{R}_A}{2\beta^2}}) \times (1 - e^{-\frac{\mathcal{S}}{2c^2}})$ for $\lambda_2 > 0$ and $\mathcal{V}(s) = 0$ otherwise, where β and c are the parameters that control the sensitivity of the filter to $\mathcal{R}_A$ and $\mathcal{S}$. The overall vesselness of a particular image locus can be computed as the maximal vesselness across scale $\mathcal{V}_0 = \max_{s_{min} \leq s \leq s_{max}} \mathcal{V}(s)$.

2.2 Robustness to Eye-Wear and Facial Hair Changes

A significant challenge posed to face recognition algorithms, conventional and IR-based ones alike, is posed by occlusions [5,10]. Of particular interest to us are specific commonly encountered occlusions – these are occlusions due to prescription glasses [11] (for practical purposes nearly entirely opaque to the IR frequencies in the short, medium and long wave sub-bands) and facial hair. To prevent them having a dramatic effect on intra-personal matching scores, it is of paramount importance to detect and automatically exclude from comparison the corresponding affected regions of the face. The DDAE framework can be extended to achieve precisely this.

In particular, we extend the existing AAM ensemble with additional models which are now occlusion-specific too. In the training stage this can be achieved by including AAMs which correspond to the existing ones but which are geometrically truncated. Note that this means that the new AAMs do not need to be re-trained – it is sufficient to adopt the already learnt appearance and shape modes and truncate them directly. In particular, we created two truncated models – one to account for the growth of facial hair (beard and moustache) and one to account for the presence of eye-wear. These two can also be combined to the produce the third truncated model for the handling of differential facial hair and eye-wear between two images hypothesized to belong to the same person. We created the two baseline truncated models manually; this is straightforward to do using high-level domain knowledge, as the nature of the specific occlusions in question constrains them to very specific parts of the face. An example of a fitted geometrically truncated AAM is shown in Fig. 3.

The application of the new ensemble in the classification of a novel image of a face, and in particular the process of model selection needed to achieve this, is somewhat more involved. In particular, the strategy whereby the highest likelihood model is selected is unsatisfactory as it favours smaller models (in our case the models which describe occluded faces). This is a well-known problem in the application of models which do not seek to describe jointly the entire image, i.e. both the foreground and the background, but the foreground only. Thus, we overcome this by not selecting the highest likelihood model but rather the model with the highest log-likelihood normalized by

Fig. 3. A geometrically truncated AAM. The truncated portion (red) is not used for fitting – here it is shown using the average shape after the fitting of the remainder of the mesh.

the model size [12]. Recall that the inverse compositional AAM fitting error is given by $e_{icaam} = \sum_{\text{All pixels } \mathbf{x}} [I_e(\mathbf{W}(\mathbf{x}; \mathbf{p})) - A_0(\mathbf{W}(\mathbf{x}; \mathbf{p}))]^2$, where $\mathbf{x}$ are pixel loci, A_i the retained appearance principal components and $\mathbf{W}(\mathbf{x}; \mathbf{p})$ the location of the pixel warped using the shape parameters $\mathbf{p}$. Noting that by design of the model the error contributions of different pixels are de-correlated, in our case the highest normalized log-likelihood model is the one with the lowest mean pixel error $\bar{e}_{icaam}$.

3   Evaluation and Results

In this section we describe the experiments we conducted with the aim of assessing the effectiveness of the proposed framework, and analyze the empirical results thus obtained. We start by summarizing the key features of our data set, follow that with a description of the evaluation methodology, and finally present and discuss our findings.

Evaluation Data. In Sec. 1 we argued that a major limitation of past research lies in the *ad hoc* approach of addressing different extrinsic sources of IR appearance variability. Hand in hand with this goes the similar observation that at present there are few large, publicly available data sets that allow a systematic evaluation of IR-based face recognition algorithms and which contain a gradation of variability of all major extrinsic variables. We used the recently collected Laval University Thermal IR Face Motion Database [13]. The database includes 200 people, aged 20 to 40, most of whom attended two data collection sessions separated by approximately 6 months. In each session a single thermal IR video sequence of the person was acquired using FLIR's Phoenix Indigo IR camera in the 2.5–5μm wavelength range. The duration of all video sequences in the database is 10 s and they were captured at 30 fps, thus resulting in 300 frames of 320×240 pixels per sequence. The imaged subjects were instructed to perform head motion that covers the yaw range from frontal ($0°$) to approximately full profile ($\pm 90°$) face orientation relative to the camera, without any special attention to the tempo of the motion or the time spent in each pose. The pose variability in the data is thus extreme. The subjects were also asked to display an arbitrary range of facial expressions. Lastly, a significant number of individuals were imaged in different physiological conditions in the two sessions. In the first session all individuals were imaged in a relatively inactive state, during the course of a sedentary working day and after a prolonged stay indoors, for the second session some users were asked to come straight from the exposure to cold ($< 0°C$) outdoors temperatures, alcohol intake, and/or exercise; see Fig. 2. In addition, individuals who wore prescription glasses in the first session were now asked to take them off. Several participants were also asked to allow for the growth of facial hair (beard, moustache) between the two sessions.

Evaluation Methodology. We evaluated the proposed algorithm in a setting in which the algorithm is trained using only a single image in an arbitrary pose and facial expression. The querying of the algorithm using a novel face is also performed using a single image, possibly in a different pose and/or facial expression. Pose, facial expression changes and partial occlusion due to eye-wear or hair all present a major challenge to the current state of the art, and the consideration of all of the aforementioned in concurrence make our evaluation protocol extremely challenging (indeed, more so than any attempted by previous work), and, importantly, representative of the conditions which are of interest in a wide variety of practical applications. In an attempt to perform a comprehensive comparative evaluation we contacted a number of authors of previously proposed approaches. However none of them was able or willing to provide us with the source code or a working executable of their methods. Thus herein we constrain ourselves to the comparison of the proposed method with the DDAE algorithm which was compared with a thermal minutia points [14] and vascular networks [15] methods in [13].

3.1 Results

The key results evaluation are summarized in Table 1 and Fig. 4. These show respectively the recognition rates achieved by our system in different experiments we conducted, and the receiver-operator characteristic curves corresponding to the recognition experiments in the presence of occlusion (in all subjects) due to facial hair or prescription glasses.

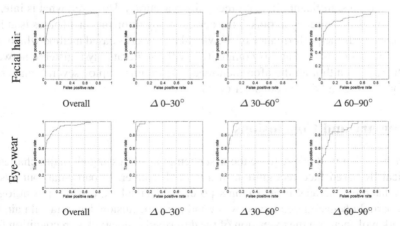

Fig. 4. Performance in the presence of occlusion across different extents of pose changes

Table 1. Average recognition rate. In experiments with partial occlusion the occlusion was differential (e.g. if a training image was acquired with eye-wear on, the test image was acquired with it off, and *vice versa.*

	Unoccluded	Facial hair	Eye-wear
Rank 1	100%	87%	74%
Rank 2	100%	94%	84%
Rank 3	100%	95%	92%

To start with, consider the results in Table 1. It is interesting to notice that performance deterioration is greater when occlusion is caused by eye-wear, rather than facial hair growth. This may be particularly surprising considering that in our data the area occluded by facial hair was larger in extent. One possible explanation of this finding may be rooted in different discriminative abilities of different facial components. Further work is needed to ascertain this; although the eye region appears to be highly informative in the visible spectrum [16] this may not be the case in the IR spectrum as suggested by evidence from some previous work [17]. However, there are alternative possibilities which may explain or contribute to the explanation of the observed performance differential. For example, the choice to grow a beard (say) is not arbitrary but rather a conscious decision made with aesthetic considerations in mind. It is possible that individuals who choose to grow facial hair have more characteristic faces. It is also possible that the explanation is of a more practical nature – perhaps the accuracy of AAM fitting is more affected by the absence of the information around the eyes, rather than those areas of the face typically covered by facial hair. We could not examine this quantitatively as it was prohibitively laborious to obtain the ground truth AAM parameters for the entire database. More research is certainly needed to establish the contribution of each of the aforementioned factors.

As Table 1 shows, both types of occlusions, those due to eye-wear and those due to facial hair, have a significant effect on recognition accuracy. However, what is interesting to observe is that already at rank-3 the correct recognition rate in all cases is at least 92%. This exceeds the performance of the vascular networks based method which used thermal minutia points [14] and is competitive with the iteratively registered networks approach [15], even though the aforementioned algorithms employ several images per person for training and do not consider occlusions.

4   Summary and Conclusions

We described what we believe to be the first attempt at addressing all major challenges in practical IR-based face recognition in a unified manner. In particular, our system explicitly handles changes in a person's pose, mild facial expression, physiological changes, partial occlusion due to eye-wear, and quasi-occlusion due to facial hair. Our future work will focus on the extension of the described framework to recognition from video and the utilization of partial information available in the regions of the face covered by facial hair.

References

1. Shan, C.: Face Recognition and Retrieval in Video. In: Schonfeld, D., Shan, C., Tao, D., Wang, L. (eds.) Video Search and Mining. SCI, vol. 287, pp. 235–260. Springer, Heidelberg (2010)
2. Chellappa, R., Wilson, C.L., Sirohey, S.: Human and machine recognition of faces: A survey. Proceedings of the IEEE (1995)
3. Muramatsu, D., Iwama, H., Makihara, Y., Yagi, Y.: Multi-view multi-modal person authentication from a single walking image sequence. In: ICB (2013)
4. Ghiass, R.S., Arandjelović, O., Bendada, A., Maldague, X.: Infrared face recognition: a literature review. In: IJCNN (2013)
5. Gross, R., Matthews, I., Baker, S.: Active appearance models with occlusion. In: IVC (2006)
6. Ghiass, R.S., Arandjelović, O., Bendada, A., Maldague, X.: Vesselness features and the inverse compositional AAM for robust face recognition using thermal IR. In: AAAI (2013)
7. Buddharaju, P., Pavlidis, I.T., Tsiamyrtzis, P., Bazakos, M.: Physiology-based face recognition in the thermal infrared spectrum. In: PAMI (2007)
8. Seal, A., Nasipuri, M., Bhattacharjee, D., Basu, D.K.: Minutiae based thermal face recognition using blood perfusion data. In: ICIIP (2011)
9. Frangi, A.F., Niessen, W.J., Vincken, K.L., Viergever, M.A.: Multiscale vessel enhancement filtering. In: Wells, W.M., Colchester, A.C.F., Delp, S.L. (eds.) MICCAI 1998. LNCS, vol. 1496, pp. 130–137. Springer, Heidelberg (1998)
10. Martinez, A.M.: Recognizing imprecisely localized, partially occluded and expression variant faces from a single sample per class. In: PAMI (2002)
11. Heo, J., Kong, S.G., Abidi, B.R., Abidi, M.A.: Fusion of visual and thermal signatures with eyeglass removal for robust face recognition. In: CVPRW (2004)
12. Arandjelović, O., Cipolla, R.: Automatic cast listing in feature-length films with anisotropic manifold space. In: CVPR (2006)
13. Ghiass, R.S., Arandjelović, O., Bendada, A., Maldague, X.: Illumination-invariant face recognition from a single image across extreme pose using a dual dimension AAM ensemble in the thermal infrared spectrum. In: IJCNN (2013)
14. Buddharaju, P., Pavlidis, I., Tsiamyrtzis, P.: Pose-invariant physiological face recognition in the thermal infrared spectrum. In: CVPRW (2006)
15. Buddharaju, P., Pavlidis, I.: Physiological face recognition is coming of age. In: CVPR (2009)
16. de Campos, T.E., Feris, R.S., Cesar Junior, R.M.: Eigenfaces versus eigeneyes: First steps toward performance assessment of representations for face recognition. In: Cairó, O., Cantú, F.J. (eds.) MICAI 2000. LNCS, vol. 1793, pp. 193–201. Springer, Heidelberg (2000)
17. Arandjelović, O., Hammoud, R.I., Cipolla, R.: Thermal and reflectance based personal identification methodology in challenging variable illuminations. In: PR (2010)

Geometric Feature-Based Facial Emotion Recognition Using Two-Stage Fuzzy Reasoning Model

Md. Nazrul Islam and Chu Kiong Loo[*]

Department of Artificial Intelligence
Faculty of Computer Science & Information Technology
University of Malaya
Lembah Pantai, 50603 Kuala Lumpur, Malaysia
nazrul_cse@siswa.um.edu.my, ckloo.um@um.edu.my

Abstract. Facial Emotion recognition is a significant requirement in machine vision society. In this sense, this paper utilizes geometric facial features and calculates displacement of feature points between expressive and neutral frames and finally applies a two-stage fuzzy reasoning model for facial emotion recognition and classification. The prototypical emotion sequence according to the Facial Action Coding System (FACS) is formed analyzing small, medium and large displacement. Furthermore geometric displacements are fuzzified and mapped onto an Action Units (AUs) by employing first-stage fuzzy reasoning model and later AUs are fuzzified and mapped onto an Emotion space by employing second-stage fuzzy relational model. The overall performance of the proposed system is evaluated on the extended Cohn-Kanade (CK+) database for classifying basic emotions like surprise, sadness, fear, anger, and happiness. The experimental results on the task of facial emotion analysis and emotion recognition are shown to outperform other existing methods available in the literature.

Keywords: Facial Emotion Recognition, Geometric Feature Extraction, Action Unit Detection, Fuzzy Reasoning Model, Facial Action Measurement.

1 Introduction

The task of social communication is carried out by the implicit and non- verbal form of signal which in terms expressed through hand gesture, head posture, and facial actions for defining the spoken message in a non-ambiguous approach [1]. Till-to-date psychological researcher has revealed that throughout communication, the verbal portion of the message is partly responsible for only 7% of the impact of the message as a total, and the vocal portion 38%, while the largest portion through facial expression is 55% of the impact of the speaker's message [2].

As of now, the term affective computing [3] is achieving more popularity in the research arena of Human Computer Interfaces (HCI) which enables computers to

---

[*] Corresponding author.

C.K. Loo et al. (Eds.): ICONIP 2014, Part II, LNCS 8835, pp. 344–351, 2014.
© Springer International Publishing Switzerland 2014

observe, understand and synthesize emotions, and to behave vividly. Apart from HCI, emotion recognition by machine vision has many interesting and real-life applications like virtual reality, video-conferencing, lie detection, and anti-social intentions, and so on. Regarding to these applications arena, facial emotion recognition through machine vision has attracted much attention just now.

The very first proposal on automatic measurements of facial muscle movement (Action Unit) was developed by Paul Ekman & Friesen (1978) called Facial Action Coding System (FACS) [4]. In that research, six basic emotional facial emotions amongst forty six are defined as the persistent across cultures: happiness, sadness, anger, fear, surprise and disgust, while other taxonomies are categorized as the extension of those basic emotions.

In general two major parts are distinguished for facial emotion recognition: feature extraction and emotion classification. Two different research trends are considered in the previous research of facial feature extraction: Appearance-based and Model-based. Appearance-based models are linear-nonlinear and fully person dependent where model-based provides 2D-3D person-independent face fitting model. Some of the well-known approaches are based on Gabor Wavelets [5, 6], Principal Components Analysis [7], Active Appearance and Geometric Models [8], Active Shape Model [9], Optical Flow and Deformable Models [10, 11], Discrete Cosine Transform in combining with Neural Networks [12], Online Clustering [13], Bayesian Networks [14], Reducing Dimensions [15] and others. The most challenging issues related to these techniques are occlusion, precise region selection, robustness, manual initialization of feature points, tracking loss, and finally the illumination effect which can degrade the recognition system performance.

On the other hand classification techniques are well practiced by different authors for techniques such as: General Rule-base [9], Fuzzy Reasoning Model [16], Multiple Adaptive Neuro-fuzzy Inferencing Model [17], and Support Vector Machine [18], etc.

The main contribution of this paper is to employ CLM-based face tracking [11] for feature extraction and propose a two-stage fuzzy reasoning model to classify facial emotions using standard Ekman´s FACS [4].

The paper is organized as follows. The tracking system is described in Section 2. Geometric-feature based measurement vector is presented in Section 3. The two-stage fuzzy reasoning model is presented in Section 4. Section 5 evaluates the performance. Finally, Section 6 concludes the proposed facial emotion recognition model.

2   CLM Based Face Tracking

The vision system employs CLM-based tracking technique [11] because of its feature set which can address and solve most of the typical tracking problems as follows:

- Automatic initialization of feature points.
- Person independent, i.e., No per-user calibration is required.
- Accurate feature point positioning during frontal face tracking.
- Real-time, robust face tracking.

One of the contributions of this paper is to increase the robustness during tracking with proper modification of the existing Tracker system to map only 17 feature points (as shown in Fig. 1) instead of the 66 originally mapped. We organize the 2D positions of the feature points in a way so that it can address all the frontal face regions where most of the emotions are observable as: 4 feature points mapping eyebrows, 4 feature points mapping eyes, 3 feature points mapping nose, and 6 feature points mapping mouth. Fig. 1 shows the emotion specific feature point localization on different facial regions. Considering 17 feature points only on frontal face parts, robustness can be achieved with continuous tracking and without shape distortion even if face turns a little bit.

3   Geometric Measurements

Fig. 1 visualizes the proposed feature point (red dots) sets, illustrated with an image. The distance between two features points is indicated by a double arrow with a dashed type line. The label $d_{1,2}$ represents the distance between feature points 1 and 2 for example. Table 1 represents the measurement vector of the proposed model.

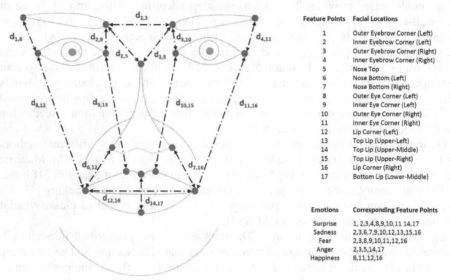

Feature Points	Facial Locations
1	Outer Eyebrow Corner (Left)
2	Inner Eyebrow Corner (Left)
3	Outer Eyebrow Corner (Right)
4	Inner Eyebrow Corner (Right)
5	Nose Top
6	Nose Bottom (Left)
7	Nose Bottom (Right)
8	Outer Eye Corner (Left)
9	Inner Eye Corner (Left)
10	Outer Eye Corner (Right)
11	Inner Eye Corner (Right)
12	Lip Corner (Left)
13	Top Lip (Upper-Left)
14	Top Lip (Upper-Middle)
15	Top Lip (Upper-Right)
16	Lip Corner (Right)
17	Bottom Lip (Lower-Middle)

Emotions	Corresponding Feature Points
Surprise	1, 2,3,4,8,9,10,11 14,17
Sadness	2,3,6,7,9,10,12,13,15,16
Fear	2,3,8,9,10,11,12,16
Anger	2,3,5,14,17
Happiness	8,11,12,16

Fig. 1. Visualization of feature points with specific emotions and distance vector

4   Two-Stage Fuzzy Reasoning Model

For each input, a measurement variable with three linguistic phrases, small, medium and large is defined, which represent to the degree of output. In order to give a shape of the membership functions, triangular form is selected to represent small and medium range while trapezoid form is selected to represent large range.

4.1 Fuzzification of Inputs

In the first stage of proposed system Measurements and AUs are considered as inputs and outputs respectively. Several rules are applied on Measurements in this stage to measure the corresponding degree of AU-intensity distinguished by Ekman. In the second stage, for each Emotion a separate rule set is defined based on FACS and our experimental investigation. Each rule is employed with the fuzzified AUs as input and returns a fuzzy Emotion value comprising with small, medium and large as output. Table 2 visualizes an example of fuzzification process of Measurement M_5.

Table 1. Measurement vector of distance displacement between expressive (E) face frame and neutral (N) face frame

Code	Measurement Name	Measurements
M_1	Left inner eyebrow displacement	$E(d_{2,9}) - N(d_{2,9})$
M_2	Right inner eyebrow displacement	$E(d_{3,10}) - N(d_{3,10})$
M_3	Left outer eyebrow displacement	$E(d_{1,8}) - N(d_{1,8})$
M_4	Right outer eyebrow displacement	$E(d_{4,11}) - N(d_{4,11})$
M_5	Mouth width vertical displacement	$E(d_{14,17}) - N(d_{14,17})$
M_6	Mouth corners horizontal displacement	$E(d_{12,16}) - N(d_{12,16})$
M_7	Left mouth corner-eyelid displacement	$E(d_{8,12}) - N(d_{8,12})$
M_8	Right mouth corner-eyelid displacement	$E(d_{11,16}) - N(d_{11,16})$
M_9	Inner eyebrow corners displacement	$E(d_{2,3}) - N(d_{2,3})$
M_{10}	Left upper lip vertical displacement	$E(d_{9,13}) - N(d_{9,13})$
M_{11}	Right upper lip vertical displacement	$E(d_{10,15}) - N(d_{10,15})$
M_{12}	Left inner eyebrow-nose root displacement	$E(d_{2,5}) - N(d_{2,5})$
M_{13}	Right inner eyebrow-nose root displacement	$E(d_{3,5}) - N(d_{3,5})$
M_{14}	Left mouth to nose displacement	$E(d_{6,12}) - N(d_{6,12})$
M_{15}	Right mouth to nose displacement	$E(d_{7,16}) - N(d_{7,16})$

Table 2. Fuzzification of Measurement M_5 (Mouth width vertical displacement)

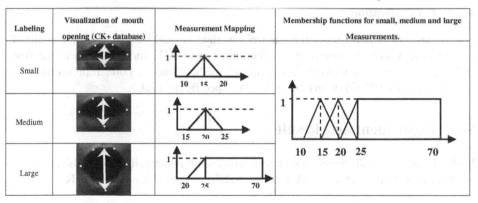

Labeling	Visualization of mouth opening (CK+ database)	Measurement Mapping	Membership functions for small, medium and large Measurements.
Small		10 15 20	
Medium		15 20 25	10 15 20 25 70
Large		20 25 70	

4.2 Fuzzy Inference System

As we consider two-stage fuzzy reasoning model, the fuzzy inferencing system works for each of the stage as pipelining. In the first stage, for each AU a separate rule set is defined based on FACS and our experimental investigation. Each rule is employed with the fuzzified Measurements M_j, j = 1, ..., 15 as input and returns a fuzzy AU value comprising with small, medium and large as the output. In this stage AUs intensity can be calculated based on the defuzzified value of each AUs. In the second stage, for each Emotion a separate rule set is defined based on FACS and our experimental investigation. Each rule is employed with the fuzzified AUs as input and returns a fuzzy Emotion value comprising with small, medium and large as output. An example of rules for deriving surprise emotion is schematically shown in Table 3.

Table 3. Two-stage fuzzy rules for AUs and Emotions (e.g., Surprise Case) Recognition

First-Stage Rules			Second-Stage Rules			
Measurements		AU	Action Units			Emotion
M₁	M₂	AU1	AU1	AU2	AU27	Surprise
Large	Large		Large	Large	Large	
Large	Medium	Large	Large	Large	Medium	
Medium	Large		Large	Medium	Large	Large
Medium	Medium		Medium	Large	Large	
Large	Small	Medium	Medium	Medium	Medium	
Small	Large		Large	Medium	Medium	
Small	Small		Medium	Large	Medium	
Small	Medium	Small	Medium	Medium	Large	
Medium	Small		Large	Large	Small	
			Large	Small	Large	
M₃	M₄	AU2	Small	Large	Large	Medium
Large	Large		Large	Medium	Small	
Large	Medium	Large	Medium	Large	Small	
Medium	Large		Large	Small	Medium	
Medium	Medium		Medium	Small	Large	
Large	Small	Medium	Small	Large	Medium	
Small	Large		Small	Medium	Large	
Small	Small		Small	Small	Small	
Small	Medium	Small	Small	Small	Large	
Medium	Small		Small	Small	Medium	
M₅		AU27	Small	Large	Small	Small
Large		Large	Small	Medium	Small	
Medium		Medium	Large	Small	Small	
Small		Small	Medium	Small	Small	

4.3   Defuzzification

For the process of defuzzification, centroid method is used by projecting the centroid value to the X-axis for intensity calculations of each AUs and Emotions in the first and second stage respectively. The centroid method finds a point representing the center of gravity (COG) of any fuzzy set on a specific interval.

5   Experiment and Result

In this section we will present the experimental setup and results used in the proposed work along with the analysis and comparison of the state-of-the-art methods.

5.1   Experiment

The experiments were performed using Visual C++ language. Both face tracking and emotion classification using two-stage fuzzy were written in the same language. The data analyzed is based on Extended Cohn-Kanade (CK+) [19] database which is publicly available and able to display facial emotions and action coded according to the Facial Action Coding System (FACS). All images in the database are in frontal view, and expressing one of 6 basic emotions. These categorized versions of image sequences are employed by tracking algorithm.

5.2   Result

In conducting our experiment, we firstly select a total of 150 image sequences from the database, where each category contains at least 30 image sequences for basic facial emotion recognition. Each image sequence contains at least 5 basic facial emotions beginning with a neutral face and ends with a specific emotion. Fig. 2 shows an overall of 90% of the 96% were correctly classified (right category), 6% were incorrectly classified (false category), while 4% were not classified at all (Unclassified) due to mouth corner and lower lip tracking error.

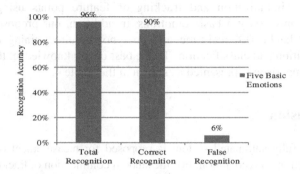

Fig. 2. Overall emotion recognition accuracy for five basic emotion categories

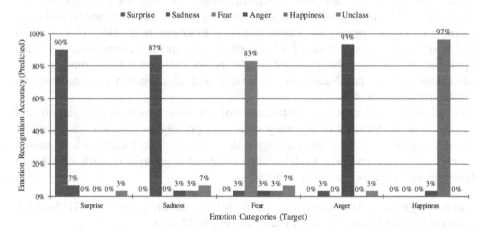

Fig. 3. Emotion recognition accuracy, Target vs. Predicted

Table 4. Performance comparison of the proposed and similar state-of-the-art prototype

Name	Input of the system	Feature Tracking Approach	Emotion Classification Approach	Land marks	Required Training	Emotion Recognition	Recognition Accuracy
D. Ghimire [20]	Facial Image sequence	EBGM method	SVM Classifier	52	Yes	6 basic Emotions	97.00%
R. Contreras [16]	Action Units or FDP	Not focused	Fuzzy Classifier	21	No	5 basic Emotions	81.40%
F. Tsalakanidou [9]	Facial Image sequence	ASM	General rule-based Classifier	81	No	4 basic Emotions	89.50%
Proposed System	Facial Image sequence	CLM	Fuzzy Classifier	17	No	5 basic Emotions	90.00%

The target emotion versus predicted emotion categories has shown in Fig. 3. Surprise image sequences have been recognized with an accuracy of 90%, while the sadness, fear, anger and happiness images with an accuracy of 87%, 83%, 93% and 97%, respectively. The tracking robustness boost up recognition accuracy to a high degree.

5.3 State-of-the-Art Comparisons

Using Face Tracker [11] and two-stage fuzzy classifier, our model is able to achieve 90% accuracy of facial emotion recognition which takes facial image sequence as input, automatic initialization and tracking of feature points using CLM based technique and can classify 5 basic emotions. In addition, our proposed model offer less number of landmarks, real-time and does not require training on the task of emotion recognition and classification. To the best of our knowledge, the state-of-the-art accuracy comparison is presented as shown in the Table 4.

6   Conclusion

In this paper, a fully automated system is proposed for measurement of facial muscle movement and emotion recognition. For the smooth continuation of tracking task a novel real-time CLM-based face tracker has been employed and a two-stage fuzzy reasoning model is developed for the task of classification. The first stage is designed for classifying of AUs from geometric measurement vector. On the other hand second stage is designed for recognition and classification of Emotions from AUs. The proposed system supports the recognition of 5 basic emotions surprise, sadness, fear, anger and happiness. In addition, the proposed model is an evidence for real-time, person-independent, robust facial emotion recognition and classification demonstrating an increased accuracy of 90%.

Future work will exploit investigations of all basic emotions recognition and classification. As localization accuracy is the first priority for emotion classification, we will improve accuracy of tracking system for accurate localization of mouth corner. Finally, the proposed model will be extended to incorporate with a clustering algorithm for better localization performance.

Acknowledgments. This work was supported by research project RG115-12ICT under the UMRG research grant of the University of Malaya.

References

1. Lewis, M., Haviland-Jones, J.M., Barrett, L.F.: Handbook of emotions. Guilford Press (2010)
2. Mehrabian, A.: Communication without words. Psychological Today 2, 53–55 (1968)
3. Picard, R.W.: Affective computing: challenges. International Journal of Human-Computer Studies 59(1), 55–64 (2003)

4. Ekman, P., Friesen, W.V.: Facial action coding system: A technique for the measurement of facial movement. Consulting Psychologists Press, Palo Alto (1978)
5. Bashyal, S., Venayagamoorthy, G.K.: Recognition of facial expressions using Gabor wavelets and learning vector quantization. Engineering Applications of Artificial Intelligence 21(7), 1056–1064 (2008)
6. He, X., et al.: Face recognition using Laplacianfaces. IEEE Transactions on Pattern Analysis and Machine Intelligence 27(3), 328–340 (2005)
7. Cho, K.-S., Kim, Y.-G., Lee, Y.-B.: Real-Time Expression Recognition System Using Active Appearance Model and EFM. In: Wang, Y., Cheung, Y.-m., Liu, H. (eds.) CIS 2006. LNCS (LNAI), vol. 4456, pp. 1078–1084. Springer, Heidelberg (2007)
8. Senechal, T., et al.: Facial Action Recognition Combining Heterogeneous Features via Multikernel Learning. IEEE Transactions on Systems, Man, and Cybernetics, Part B: Cybernetics 42(4), 993–1005 (2012)
9. Tsalakanidou, F., Malassiotis, S.: Real-time 2D+ 3D facial action and expression recognition. Pattern Recognition 43(5), 1763–1775 (2010)
10. Lin, D.-T.: Facial expression classification using PCA and hierarchical radial basis function network. Journal of Information Science and Engineering 22(5), 1033–1046 (2006)
11. Saragih, J.M., Lucey, S., Cohn, J.F.: Face alignment through subspace constrained mean-shifts. In: 2009 IEEE 12th International Conference on Computer Vision. IEEE (2009)
12. Kim, S.-P., et al.: Neural control of computer cursor velocity by decoding motor cortical spiking activity in humans with tetraplegia. Journal of Neural Engineering 5(4), 455 (2008)
13. Nuevo, J., Bergasa, L.M., Jiménez, P.: RSMAT: Robust simultaneous modeling and tracking. Pattern Recognition Letters 31(16), 2455–2463 (2010)
14. Wood, F., Black, M.J.: A nonparametric Bayesian alternative to spike sorting. Journal of Neuroscience Methods 173(1), 1–12 (2008)
15. Libralon, G.L., Romero, R.A. F.: Investigating Facial Features for Identification of Emotions. In: Lee, M., Hirose, A., Hou, Z.-G., Kil, R.M. (eds.) ICONIP 2013, Part II. LNCS, vol. 8227, pp. 409–416. Springer, Heidelberg (2013)
16. Contreras, R., Starostenko, O., Alarcon-Aquino, V., Flores-Pulido, L.: Facial feature model for emotion recognition using fuzzy reasoning. In: Martínez-Trinidad, J.F., Carrasco-Ochoa, J.A., Kittler, J., et al. (eds.) MCPR 2010. LNCS, vol. 6256, pp. 11–21. Springer, Heidelberg (2010)
17. Gomathi, V., Ramar, K., Jeevakumar, A.S.: Human Facial Expression Recognition using MANFIS Model. Proceedings of World Academy of Science: Engineering & Technology 50 (2009)
18. Kharat, G., Dudul, S.: Human emotion recognition system using optimally designed SVM with different facial feature extraction techniques. WSEAS Transactions on Computers 7(6), 650–659 (2008)
19. Lucey, P., et al.: The Extended Cohn-Kanade Dataset (CK+): A complete dataset for action unit and emotion-specified expression. In: 2010 IEEE Computer Society Conference on Computer Vision and Pattern Recognition Workshops, CVPRW (2010)
20. Ghimire, D., Lee, J.: Geometric feature-based facial expression recognition in image sequences using multi-class adaboost and support vector machines. Sensors 13(6), 7714–7734 (2013)

Human Activity Recognition by Matching Curve Shapes

Poorna Talkad Sukumar and K. Gopinath

Department of Computer Science and Automation,
Indian Institute of Science, Bangalore, India
poorna.t.s@gmail.com, gopi@csa.iisc.ernet.in

Abstract. In this paper, we present a new method for Human Activity
Recognition (HAR) from body-worn accelerometers or inertial sensors
using comparison of curve shapes.

Simple motion activities have characteristic patterns that are visible
in the time series representations of the sensor data. These time series
representations, such as the 3D accelerations or the Euler angles (roll,
pitch and yaw), can be treated as curves and activities can be recognized
by matching patterns (shapes) in the curves using curve comparison and
alignment techniques.

We transform the sensor signals into cubic B-splines and parametrize
the curves with respect to arc length for comparison. We tested our algorithm on the accelerometer data collected at Cleveland State University
[1]. The 3D acceleration signals were segmented at high-level and subject-
dependent 'representative' curves for the activities were constructed with
which test curves were compared and labeled with an overall accuracy
rate of 88.46% by our algorithm.

Keywords: Activity Recognition, Cubic B-splines, Arc-Length Parametrization, Curve Comparison.

1 Introduction

Human Activity Recognition (HAR) using body-worn accelerometers or inertial
sensors is an area of active research and has been found to benefit several real-
world applications. Most approaches to HAR consist of feature extraction and
classification using machine learning techniques [8]. We present a new method
to recognize simple motion activities that are usually periodic or performed
repeatedly for purposes such as rehabilitation [1] or sports training.

Simple motion activities, such as those involving a particular type of limb
motion, exhibit characteristic patterns in their sensor data plots. These activities
can be uniquely identified by their 3-dimensional time series signatures, such as
3D accelerations or Euler angles (roll, pitch and yaw). We transform these time
series data into curves and use shape matching techniques to identify the activity
whose curve shapes are closest to those of the given curve set.

We use curve comparison and alignment methods that are generally used
in handwriting analysis, topographic maps and many applications of computer

C.K. Loo et al. (Eds.): ICONIP 2014, Part II, LNCS 8835, pp. 352–360, 2014.

graphics, for HAR. We transform the sensor signals into cubic B-splines and parametrize the curves with respect to arc length which generates 'geometrically' uniform subdivision of the curves and hence used for curve matching. Our curve matching method is comparable to Dynamic Time Warping (DTW) [2] which is also used for finding shape similarities in temporal sequences (which may vary in time and speed).

The $(x, y$ and $z)$ curves corresponding to the segment of the signal containing a relatively long instance of an activity are chosen as the 'representative' curves for that activity. Test curves corresponding to the remaining segments are compared with the representative curves of the different activities. We use the least total sum of the squares of the differences in tangent angles at corresponding points on the three pairs of curves to determine the activity for a given test curve set. Test curves may not align exactly with the representative curves of the activities to which they belong. But our algorithm aligns the test curve with that region of the representative curve whose shape is closest to the test curve and this ensures that the differences in the tangent angles will be the least for the comparison with the representative curves to which the test curve set belongs. Since the activity signatures are different, the shape of a test curve belonging to one activity will not match well with that of a representative curve belonging to another activity. This is illustrated in figures 1 and 2.

Any 2-dimensional $(x(t), y(t))$ or 3-dimensional $(x(t), y(t), z(t))$ time series sensor data whose signatures can be used to differentiate among the activities can be used for HAR using this algorithm.

The rest of the paper is organized as follows. Transforming the time series sensor data into functions, arc length parametrization of the curves and the curve comparison method used are described in Section 2. Section 3 describes the testing of our method on the dataset collected at Cleveland State University [1] and the results obtained. Conclusion and Future work is discussed in Section 4.

2   Implementation of the Curve Matching Algorithm

2.1   Transforming the Sensor Data into Functions and Arc-Length Parametrization of the Curves

We transform the 3-dimensional time series representations into cubic B-splines using the Functional Data Analysis (FDA) package developed by Ramsay et al [3,4]. All implementations have been developed using Matlab.

B-spline curves can be approximately parameterized by arc length and we use the parametrization procedure described by Wang et al [5] in our algorithm. Let $Q(t)$ be a 3D representation of a cubic B-spline curve parameterized by time t given by

$$Q(t) = (x(t), y(t), z(t)), \tag{1}$$

where t ranges from t_0 to t_n with n being the number of spline segments and $\{t_0, t_1, ..., t_n\}$ the break points.

The arc length of each spline segment in the input spline curve, $Q(t)$, is computed as follows

$$l_i = \int_{t_i}^{t_{i+1}} \sqrt{(x'(t))^2 + (y'(t))^2 + (z'(t))^2} dt, \qquad (2)$$

where i varies from 0 to $n-1$. The above integral cannot be computed analytically and has to be approximated using numerical integration methods. We use Simpson's rule [7] to compute equation 2. Let

$$f(t) = \sqrt{(x'(t))^2 + (y'(t))^2 + (z'(t))^2}. \qquad (3)$$

Then equation 2 is approximated by

$$l_i \approx \frac{(t_{i+1} - t_i)}{6}[f(t_i) + 4f\left(\frac{t_i + t_{i+1}}{2}\right) + f(t_{i+1})]. \qquad (4)$$

$x'(t)$, $y'(t)$ and $z'(t)$ for any t between t_0 and t_n can be computed using the FDA package [3,4]. The total arc length of the curve is $L = \sum_{i=0}^{n-1} l_i$.

We now divide the curve into m segments spaced at equal arc length distances, $0, \tilde{l}, 2.\tilde{l}, ..., m.\tilde{l}$ from the start of the curve where $\tilde{l} = L/m$ is the arc length of each segment in the output curve. We now have to find the t values $\tilde{t}_0, \tilde{t}_1, ..., \tilde{t}_m$ corresponding to the $m+1$ points that divide the curve into equal arc length segments.

The value of $\tilde{t}_i$ can be computed using the bisection method described by Wang et al [5]. We first find the spline segment such that $\sum_{k=0}^{j-1} l_k \leq i.\tilde{l} < \sum_{k=0}^{j} l_k$, so that we have $t_j \leq \tilde{t}_i < t_{j+1}$. We set $t_{left} = t_j$ and $t_{right} = t_{j+1}$ and calculate $t_{mid} = (t_{left} + t_{right})/2$. $\tilde{t}_i$ lies in the segment $[t_{left}, t_{right}]$. We then calculate the arc length of the segment $[t_{left}, t_{mid}]$ given by $\int_{t_{left}}^{t_{mid}} f(t)dt$ using equation 4. Let $\Delta s = \sum_{k=0}^{j-1} l_k + \int_{t_j}^{t_{mid}} f(t)dt$. If $|\Delta s - i.\tilde{l}| < e$, where e is the error threshold, then $\tilde{t}_i = t_{mid}$. Else, if $\Delta s < i.\tilde{l}$, then $\tilde{t}_i$ lies in the segment $[t_{mid}, t_{right}]$ and we set $t_{left} = t_{mid}$. If $\Delta s > i.\tilde{l}$, then $\tilde{t}_i$ lies in the segment $[t_{left}, t_{mid}]$ and we set $t_{right} = t_{mid}$. We repeat the above steps till $\tilde{t}_i$ is found.

Using $\tilde{t}_0, \tilde{t}_1, ..., \tilde{t}_m$, we then compute the 3-dimensional function values, $(\tilde{x}_0, \tilde{y}_0, \tilde{z}_0)$, $(\tilde{x}_1, \tilde{y}_1, \tilde{z}_1)$, ..., $(\tilde{x}_m, \tilde{y}_m, \tilde{z}_m)$, at the equal arc length distances $s_0 = 0, s_1 = \tilde{l}, s_2 = 2.\tilde{l}, ..., s_m = m.\tilde{l}$. Using $[(s_0, \tilde{x}_0), (s_1, \tilde{x}_1), ..., (s_m, \tilde{x}_m)]$, $[(s_0, \tilde{y}_0), (s_1, \tilde{y}_1), ..., (s_m, \tilde{y}_m)]$ and $[(s_0, \tilde{z}_0), (s_1, \tilde{z}_1), ..., (s_m, \tilde{z}_m)]$, we build arc length parameterized cubic B-spline curves using the FDA package.

2.2 Partial Curve Matching

We use the method described by Femiani et al [6] for partial curve matching. Partial matching is comparing a curve (test curve) with a region of a longer curve (representative curve). We use tangent angles for comparing two curve segments because they are an invariant property that capture the shape of a curve (curvature, which is the derivative of the tangent angle with respect to arc length, can also be used for comparison).

We choose a value δ of arc length small enough to capture the curve details. Let $\hat{c}$ and c be the total arc lengths of the (longer) representative curve and the test curve, respectively. We divide the longer curve at $p = \lfloor \frac{\hat{c}}{\delta} \rfloor$ points and the test curve at $q = \lfloor \frac{c}{\delta} \rfloor$ points at equal arc length (δ) intervals. We label the $p+1$ and $q+1$ points (including the starting points) on the two curves as $\{\hat{c}_1, \hat{c}_2, ..., \hat{c}_{p+1}\}$ and $\{c_1, c_2, ..., c_{q+1}\}$, respectively and precompute the unit tangent vectors for the two curves at these sets of points. The x, y and z tangent vectors at arc length s is given by,

$$\left\langle T_x(s), T_y(s), T_z(s) \right\rangle = \left\langle \frac{x'(s)}{|r'(s)|}, \frac{y'(s)}{|r'(s)|}, \frac{z'(s)}{|r'(s)|} \right\rangle \tag{5}$$

where $|r'(s)|$ is given by,

$$|r'(s)| = \sqrt{(x'(s))^2 + (y'(s))^2 + (z'(s))^2} . \tag{6}$$

$x'(s)$, $y'(s)$ and $z'(s)$ used in the above equations can be computed using the FDA package [3,4].

Since $c < \hat{c}$, $p \geq q$ and there are $p - q + 1$ spans of $q + 1$ points in the longer curve. The test curve is compared with every span of the $p - q + 1$ spans of $q + 1$ points of the representative curve and the total sum of squares of the differences in the tangent vectors for every comparison is calculated.

$$\arg\min_{j} \sum_{i=1}^{q+1}((T_x(c_i) - T_x(\hat{c}_{j+i-1}))^2 + (T_y(c_i) - T_y(\hat{c}_{j+i-1}))^2 + (T_z(c_i) - T_z(\hat{c}_{j+i-1}))^2),$$

where j ranges from 1 to $p - q + 1$, gives the index of the region of the representative curves whose shapes match closest with those of the test curves. The time complexity for the curve matching algorithm is $O(pq)$.

3   Experimental Evaluation of Our Approach

3.1   Dataset Used

We evaluated our method on the accelerometer data collected at Cleveland State University [1]. The data was collected on five normal subjects and two subjects with post-stroke hemiparesis performing five activities. One of these activities (Reaching) was a traditional rehabilitation activity and the other four were video games (BubblePop, Mr.Chef, Batting, Kung 2) played on either the Sony Playstation2 or the Nintendo Wii platform. Two of the activities (Batting and Reaching) involved particular kinds of limb motion while the other activities were combinations of different movements but each activity was composed of movements different from those of the others. Reaching activity involved reaching to objects on a table and Batting activity consisted of a baseball swing. BubblePop, Mr.Chef and Kung 2 activities were not as consistent and required the player to respond to game play actions on the screen. In BubblePop, the player had to pop all blue colored bubbles on the screen and in Kung 2, figures appearing in various areas of the screen had to be 'punched' out of the way by the player. Mr.Chef involved the player performing many tasks such as grating cheese, salting fries, putting together orders and so on. The reader is referred to the paper cited above for details of the activities. These activities were performed repeatedly and intended for post-stroke rehabilitation. A 3-axis accelerometer was attached to the dominant wrist for all the subjects.

3.2   Activity Recognition Using Our Approach

The activities performed were clearly separated by pauses as seen in the acceleration plots and hence it was easy to segment out the activity instances in the signal streams. The acceleration norm signal was used for segmentation. We used a sliding window approach (with a small window size) and disregarded sequences whose linear regressions had slopes close to zero (i.e. sequences corresponding to the pauses). In this way, we obtained the start and end points of the all the activity instances.

A relatively long instance of an activity was chosen to build the representative curves to ensure that any test curve belonging to an activity was always shorter than the representative curve for that activity. The representative curves are required to contain segments corresponding to all the movements belonging to an activity. Separate segments containing different movements belonging to an activity can simply be concatenated to form the representative curve for an activity so that when compared with a test curve containing a particular movement of the activity, the region of the representative curve containing that movement is matched with it. Segmentation plays a very important role when curve matching techniques are used. Segments to be treated as test curves should neither be too short nor too long. Ideally, a segment should contain a few repetitions of any of the movements contained in an activity. If a segment is too short, it may not contain enough detail to recognize the activity. On the other hand, if the segment is too long, it will be matched with nearly the entire length of the representative curve and the patterns in the curve will not be aligned with their counterparts in the representative curve. Some useful time series segmentation approaches are discussed in the paper by Keogh et al [9].

Total arc-lengths of curves are indicative of the activities they contain. For example, a curve built using a few seconds of 'running' activity will have greater arc-length compared with a curve built using the same duration of 'standing' or 'walking' activity.

We chose subject-dependent learning since there was variability in the movements performed by the different subjects in terms of acceleration magnitude and movement plane.

Post-segmentation, the segments (of the 3-axis accelerations) containing the instances of activities for building both representative and test curves are transformed into arc-length parameterized cubic B-splines using the method described in Section 2.1. Test curves belonging to a subject are then matched with the representative curves of all the activities of that subject using the method described in Section 2.2. A test curve is classified as belonging to that activity corresponding to the comparison for which the total sum of squares of the differences in the tangent vectors is minimum.

3.3   Results

The test curves were labeled with activities with an overall accuracy of 88.46%. The individual accuracies were 83.33%, 100%, 88.24%, 76.47% and 100% for the activities BubblePop, Batting, Mr.Chef, Kung 2 and Reaching, respectively. The individual accuracies for the seven subjects were 91.67%, 90%, 88.89%, 90.91%,

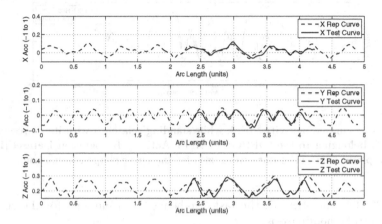

Fig. 1. A test curve compared with the representative (rep) curve for 'reaching' activity of subject 5. It is aligned with that region of the representative curve that has the minimum total sum of squares of differences in the tangent vectors (134.25).

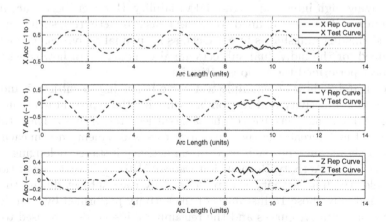

Fig. 2. The same test curve (as in fig 1) compared with the representative (rep) curve for 'batting' activity of subject 5. It is aligned with that region of the representative curve that has the minimum total sum of squares of differences in the tangent vectors (324.23).

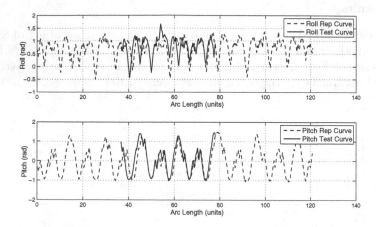

Fig. 3. A Test curve of activity 'drink while standing' matched with its representative curve, both belonging to the 'OPPORTUNITY Activity Recognition Dataset'[10,11]. 'Roll' and 'pitch' time series data were used for the curve shape comparison.

92.31%, 72.72% and 91.67%, the last two being the accuracies for the subjects with post-stroke hemiparesis.

Batting and Reaching activities were classified with 100% accuracy. This is probably because these two activities were each composed of a single kind of motion and were consistent across trials whereas the other activities (BubblePop, Kung 2 and Mr.Chef) were each composed of different kinds of movements. It is non-trivial to get better recognition rates for activities like BubblePop and Kung 2 having high inter-and intra-trial variability. However, better accuracies can be obtained if we treat defined component tasks of activities, such as grating cheese and shaking milkshakes in Mr.Chef, as individual activities. The accuracies obtained for the individual subjects are also indicative of how consistently each subject performed the various activities.

Figures 1 and 2 show an example test curve belonging to subject 5 and containing 'reaching' activity and its comparison with the representative curves of 'reaching' and 'batting' activities of the same subject, respectively. The total sum of squares of the differences in the tangent vectors is also mentioned in each case.

Figure 3 shows our approach used on an activity instance belonging to the 'OPPORTUNITY Activity Recognition Dataset' [10,11]. Twenty repetitions of 'drink while standing' activity was performed by each subject during the drill run. We randomly chose 15 instances of the activity performed by a subject to form the representative curves and the remaining 5 instances were used to form the test curves. Since the data was collected using inertial sensors, we chose to use the 'roll' and 'pitch' time series signals for matching curve shapes.

4   Conclusion and Future Work

We have implemented a new method for Human Activity Recognition with accelerometers or inertial sensors using matching of curve shapes. The method exploits the fact that simple motion activities exhibit characteristic patterns in their sensor data plots.

In order to use our method on periodic activities with higher acceleration magnitudes, such as walking or running, preprocessing of the time series data may be required such as applying signal processing algorithms to reduce noise or remove spikes in the accelerometer signals. The method described in this paper can be used, in general, to perform partial curve matching on any time series signals, including Electrocardiogram (ECG) and handwriting data.

We are currently looking at FDA techniques to derive qualitative information from the time series data such as the correctness of an activity performed so that we can provide the subjects with feedback on their performance.

We are also interested in combining outputs from multiple body-worn sensors.

Acknowledgments. We thank Ann Reinthal and Nigamanth Sridhar for overseeing the data collection process at Cleveland State University and for making the collected data available for our use. We also thank the Robert Bosch Centre for Cyber Physical Systems (RBCCPS) at the Indian Institute of Science (IISc), Bangalore, for funding our research.

References

1. Sanka, S., Reddy, P.G., Alt, A., Reinthal, A., Sridhar, N.: Utilization of a wrist-mounted accelerometer to count movement repetitions. In: Fourth International Conference on Communication Systems and Networks, pp. 1–6. IEEE (2012)
2. Muscillo, R., Schmid, M., Conforto, S., D'Alessio, T.: Early recognition of upper limb motor tasks through accelerometers: real-time implementation of a DTW-based algorithm. J. Comput. Biol. Med. 41, 164–172 (2011)
3. Ramsay, J., Silverman, B., Hooker, G., Graves, S.: Functional Data Analysis Home Page, http://www.psych.mcgill.ca/misc/fda/
4. Ramsay, J., Hooker, G., Graves, S.: Functional Data Analysis with R and MATLAB. Springer, New York (2009)
5. Wang, H., Kearney, J., Atkinson, K.: Arc-length parameterized spline curves for real-time simulation. In: Fifth International Conference on Curves and Surfaces, pp. 387–396 (2002)
6. Femiani, J.C., Razdan, A., Farin, G.: Curve Shapes: Comparison and Alignment
7. Faires, J., Burden, R.: Numerical Methods, 3rd edn. Brooks Cole (2002)
8. Bulling, A., Blanke, U., Schiele, B.: A Tutorial on Human Activity Recognition Using Body-worn Inertial Sensors. ACM Computing Surveys 46(3), 33:1–33:33 (2014)
9. Keogh, E., Chu, S., Hart, D., Pazzani, M.: An online algorithm for segmenting time series. In: International Conference on Data Mining, pp. 289–296. IEEE (2001)
10. Roggen, D., Calatroni, A., Rossi, M., Holleczek, T., Forster, K., Troster, G., Lukowicz, P., Bannach, D., Pirkl, G., Ferscha, A., Doppler, J., Holzmann, C., Kurz, M., Holl, G., Chavarriaga, R., Sagha, H., Bayati, H., Creatura, M., del, R., Millan, J.:

Collecting complex activity data sets in highly rich networked sensor environments. In: Seventh International Conference on Networked Sensing Systems, pp. 233–240. IEEE (2010)

11. OPPORTUNITY Activity Recognition Dataset Home Page, https://archive.ics.uci.edu/ml/datasets/OPPORTUNITY+Activity+Recognition

Sentiment Analysis of Chinese Microblogs Based on Layered Features

Dongfang Wang and Fang Li

Key Laboratory of Shanghai Education Commission for Intelligent Interaction
and Cognitive Engineering,
Department of Computer Science and Engineering,
Shanghai Jiao Tong University, China
{mickey,fli}@sjtu.edu.cn

Abstract. Microblogging currently becomes a popular communication way and detecting sentiments of microblogs has received more and more attention in recent years. In this paper, we propose a new approach to detect the sentiments of Chinese microblogs using layered features. Three layered structures in representing synonyms and highly-related words are employed as extracted features of microblogs. In the first layer, "extremely close" synonyms and highly-related words are aggregated into one set while in the second and the third layer, "very close" and "close" synonyms and highly-related words are aggregated respectively. Then in every layer, we construct a binary vector as a feature. Every dimension of a feature indicates whether there are some words in the microblog falling into that aggregated set. These three features provide perspectives from micro to macro. Three classifiers are respectively built from these three features for final prediction. Experiments demonstrate the effectiveness of our approach.

Keywords: Layered Features, Sentiment Analysis, Microblogs.

1 Introduction

As a social communication tool, microblogging is very popular among Internet users. More and more people post their opinions towards products or political views. Therefore sentiment analysis of microblogs can help social studies or marketing, and becomes a quite hot topic. The features used in microblogs sentiment analysis are of great impact to the classifying performance.

Mohammad et al. [1] implemented a number of features including Ngrams, characters in upper case, lexicons, POS (Part-of-Speech) tags, hashtags, punctuations, emoticons, negations and so on. They concluded that sentiment lexicon features along with Ngrams features led to the most gain in performance. Pak and Paroubek [2] analyzed the distribution of words frequencies in positive, negative and neutral sets. They concluded that POS tags are strong indicators therefore they use Ngrams and POS tags as features. In the work of Huang et al. [3], using sentiment words, POS tags, punctuations, adversative words, and emoticons as features achieved their best results. However, extracting these features needs a lot of computational and linguistic work.

C.K. Loo et al. (Eds.): ICONIP 2014, Part II, LNCS 8835, pp. 361–368, 2014.
© Springer International Publishing Switzerland 2014

Unigrams are widely used for simplicity and excellent performance. In the experiments of Pang et al. [4], using an SVM trained on Unigrams achieved the best result. Read [5], Bora [6], Purver and Battersby [7] also adopted Unigrams as features. Go et al. [8] concluded that Unigrams are simple but useful while only using Bigrams as features will cause sparseness of the feature space. However, even for Unigrams, sparseness is very severe. Due to the large quantity of Unigrams in the corpus, one word will easily drown in the sea of Unigrams and its characteristic cannot be well detected. Saif et al. [9] used semantic feature set and sentiment-topic feature set to alleviate the sparseness.

By analyzing thousands of Chinese microblogs, we find that many synonyms and highly-related words exist among microblogs. The meanings of these synonyms are quite close and the highly-related words usually appear in the same topics. Therefore synonyms and highly-related words can be regarded as the same when detecting sentiments.

In this paper, we propose a novel approach based on the observations above. We construct three layered features by aggregating synonyms and highly-related words. These three layered features provide perspectives in different levels, from micro to macro. By combining them we get a more complete perspective and achieve our best results. For comparison, we also conduct experimental comparison with other methods. Unlike [9], our method can be applied on any corpus even though they do not contain series of product entities and topics.

The rest of this paper is organized as follows. In section 2, we introduce feature extraction. The voting method is described in section 3. Section 4 shows the experimental results and discussion. Finally we conclude this paper in section 5.

2   Feature Extraction

2.1   Preprocessing

Before extracting features, we firstly split words using the tool ICTCLAS[1] because there is no space between Chinese words. URLs, hashtags, emoticons, users' names (with a character "@" before names), stop-words and words that appear less than 3 times in the training set are removed.

2.2   Synonyms and Highly-Related Words

Synonyms. Synonyms are words expressing the same or similar meanings, like 高兴 (happy) and 开心(joyful). According to how "close" the meanings of synonymous words are, synonyms are grouped into three levels respectively. The structure is from fine to coarse.

Highly-Related Words. Highly-related words are words that their meanings are different but the meanings are highly-related. For example, 棉农(cotton grower) and 茶

---

[1] http://www.ictclas.org/

农(tea grower) means different but they are both plant growers. According to how "close" the relationships of highly-related words are, they are grouped into three levels respectively. The structure is from fine to coarse.

We use the HIT IR-Lab Tongyici Cilin[2] to find highly-related words and synonyms in Chinese. Other dictionary like SentiWordNet[3], or formulas like PMI can also be used to find synonyms or highly-related words in different languages. Examples of synonyms and highly-related words are shown in Table 1.

Table 1. Examples of Synonyms and Highly-related Words

Notation (How close they are)	Examples of Synonyms	Examples of Highly-related Words
α (Extremely close)	1. 忧愁(worry), 犯愁 (worry);	1. 侨胞(countryman residing abroad), 港胞(countryman residing Hong Kong);
	2. 忧闷(depressed), 忧郁(gloomy);	2. 爱国同胞(patriotic fellow-countryman), 爱国者(patriot);
	3. 烦闷(anguish), 烦乱(upset);	3. 非洲人(African), 亚洲人(Asian);
	4. 委屈(grievance), 憋屈(grievance);	4. 意大利人(Italian), 美国人(American);
β (very close)	1.忧愁(worry), 犯愁(worry), 忧闷(depressed), 忧郁(gloomy);	1. 侨胞(countryman residing abroad), 港胞(countryman residing Hong Kong), 爱国同胞(patriotic fellow-countryman), 爱国者(patriot);
	2.烦闷(anguish), 烦乱(upset), 委屈(grievance), 憋屈(grievance);	2. 非洲人(African), 亚洲人(Asian), 意大利人(Italian), 美国人 (American);
γ (close)	忧愁(worry), 犯愁(worry), 忧闷(depressed), 忧郁(gloomy), 烦闷(anguish), 烦乱(upset), 委屈(grievance), 憋屈(grievance);	侨胞(countryman residing abroad), 港胞(countryman residing Hong Kong), 爱国同胞(patriotic fellow-countryman), 爱国者(patriot), 非洲人(African), 亚洲人(Asian), 意大利人(Italian), 美国人(American);

2.3   Reduction of Feature Dimension

Words in microblogs are of large quantity which leading to the sparseness of vector space of Unigrams. Compressing feature dimensions using synonyms and highly-related words will relieve this dilemma. In our method, a word and all of its synonyms and its highly related words are all expressed in only one feature dimension. For example, the word 火星(Mars), 荧惑(Mars), and 土星(Saturn) are all represented by one feature dimension, 荧惑(Mars) is synonymous word to 火星(Mars) while 土星(Saturn) is highly-related word to 火星(Mars). Also, the synonymous words and highly-related words of 荧惑(Mars) and 土星(Saturn) are included. That is, the words in one feature dimension consist of a closure set. We define the calculation for finding the synonyms and highly-related words of a word w_i and of a word set W in equation (1-3). And the algorithm for calculating a closure word set $W_{closure}$ for a word w_i, and for finding the set of all the word sets $\mathcal{A}$ is shown in Algorithm 1.

After all the words are split into disjoint closure sets, we use the results to construct features. Each dimension of our feature vector represents whether there are some

---

[2] http://www.datatang.com/data/42306/
[3] http://sentiwordnet.isti.cnr.it/download.php

words contained in the microblog falling into that set. As shown in Fig. 1, the feature vector is a binary vector and the dimension of the vector equals to the number of sets.

$$S(w_i) = \{Synonyms\ of\ w_i\}, \quad H(w_i) = \{Highly\text{-}realted\ words\ of\ w_i\} \qquad (1)$$

$$f(w_i) = S(w_i) \bigcup H(w_i) \qquad (2)$$

$$f(W) = \bigcup_{i=1}^{n} f(w_i),\ W = \{w_1, w_2, w_3, ..., w_n\} \qquad (3)$$

Algorithm 1. Calculating Closure Word Sets

Calculate_Closure (word w_i){	*Split_Words_into_Disjoint_Sets* (corpus m){
$W_{closure} = \{w_i\}$; $W_{new} = \{w_i\}$;	$M = \{m_1, m_2, ..., m_k\} = \{words\ in\ m\}$;
	$\mathcal{A} = \varnothing$;
While ($W_{new} \neq \varnothing$){	While ($M \neq \varnothing$){
$W_{new} = f(W_{closure}) - W_{closure}$;	$C = Calculate\_Closure(m_0)$, $m_0 \in M$;
$W_{closure} = W_{closure} \bigcup W_{new}$;	$M = M - C$;
}	$\mathcal{A} = \mathcal{A} \bigcup \{C\}$;
Return $W_{closure}$;	}
}	Return $\mathcal{A}$;
	}

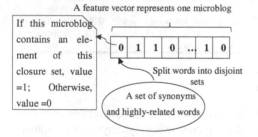

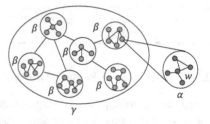

Fig. 1. Description of One Feature Vector   **Fig. 2.** Structure of the Layered Word Sets

2.4   Layered Feature

There are three layers of compressed features in our approach. The difference between these three layers is how "close" the meanings of synonymous words or how "close" the relationships of related words in one feature dimension. We group words that are "extremely close", "very close", "close" respectively and denote one aggregated set as α, β, γ respectively.

Note that one γ consists of many β and one β consists of many α. The aggregation is the most precise in the first layer while the aggregation is the least precise in the third layer, satisfying the inequality (4). The structure of the layered word sets is shown in Fig. 2. In every layer, words are split into disjoint sets. We construct a feature use the method shown in section 2.3 for every layer. These features provide perspectives from micro to macro. Therefore combing all of them, we get a more complete view of the Microblogs and it can provide more information than one single layer.

$$max\{distance_\alpha(w_{\alpha i},w_{\alpha j})\} \prec max\{distance_\beta(w_{\beta i},w_{\beta j})\} \prec max\{distance_\gamma(w_{\gamma i},w_{\gamma j})\} \quad (4)$$

3   Voting SVM Classifiers

Previous research [4] has shown that SVM performs excellently for sentiment analysis. Therefore we employ SVM classifier in our approach. Once we construct the three layered features, three SVM classifiers are built from each layer of features respectively. To combine the three classifiers, voting strategy is used. We consider that the features of the three layers are equally important, so we set the weights of these three classifiers to be equal when voting. The final classified result will be the majority predicted label [10]. The strategy of the classification is shown in Algorithm 2.

Algorithm 2. The Strategy of the Classification

```
Voting(microblog m){
    Classifier  C_i  is trained from features in layer  i
(1 ≤ i ≤ 3) ;
    votingResult = 0 ;
    for  (i = 1; i ≤ 3; i++)
    {if (Ci predicts the m is positive) votingResult ++ ;}
    if  (votingResult ≥ 2) return positive;
    else return negative;
}
```

Table 2. Component of the Cilin

	Number of word sets
In the 1st layer	13440
In the 2nd layer	3880
In the 3rd layer	1423
In the 4th layer	95
In the 5th layer	12

4   Experiments and Result Analysis

To reveal the effectiveness of our approach, in this section we carry out a series of experiments on two datasets and compare our results with other well-known approaches. All the data used in this paper comes from Xinlang Weibo[4], a large Twitter-like Chinese microblogging website.

4.1   Dictionary and Datasets

HIT IR-Lab Tongyici Cilin. We use the cilin to extract layered synonyms and highly-related words. In the cilin, there are five layers. We only make use of the first three layers because the last two layers are too coarse which will cause over-aggregation of words, shown in Table 2.

MoodLens Dataset [11]. There are four categories of sentiments: angry, disgusted, joyful and sad in this dataset. In our experiments, we take joyful as positive sentiment

---

[4] http://www.weibo.com

and the other three emotions as negative sentiments. We randomly select 5000 positive microblogs and 5000 negative microblogs from this dataset.

NLP&CC2013 Dataset[5]. Eight categories of sentiments serve as the labels. We take angry, disgusted, afraid and sad as negative sentiments and take happiness and like as positive sentiments. We use 965 positive microblogs and 965 negative microblogs.

4.2   Experimental Results

In the experiments, we compare our proposed approach with methods using Unigrams, Bigrams, Unigrams + POS tags, Bigrams + POS tags as features respectively. We also compare with lexicon method and the method proposed in [2]. In lexicon method, a microblogs will be labeled according to the number (the first deciding factor) and the strength (the second deciding factor) of the positive and negative words, using Dalian Ligong lexicon[6]. In [2], Bigrams and POS tags are used as the features with salience-feature-selection, and Naïve Bayes classifier is trained for classification. The performances of three classifiers that built from features of different layers are also examined respectively. In all experiments, 5-fold cross validation is conducted. And we use the criterion (5-7) to evaluate our system. Experimental results (%) on both datasets are shown in Table 3 & 4, respectively.

$$Precision_{emotion=i} = \frac{\#system\_correct(emotion=i)}{\#system\_proposed(emotion=i)} \quad i \in \{Pos, Neg\} \tag{5}$$

$$Recall_{emotion=i} = \frac{\#system\_correct(emotion=i)}{\#manually\_label(emotion=i)} \quad i \in \{Pos, Neg\} \tag{6}$$

$$F-measure = \frac{2*Precision*Recall}{Precision+Recall} \tag{7}$$

Table 3. Results on MoodLens Dataset

Method	Positive			Negative		
	Precision	Recall	F-measure	Precision	Recall	F-measure
Unigrams	60.44	62.18	61.30	61.06	59.30	60.17
Bigrams	59.24	51.08	54.86	57.00	64.85	60.68
Unigrams + POS tags	60.32	61.87	61.09	60.86	59.30	60.07
Bigrams + POS tags	60.24	53.23	56.52	58.11	**64.87**	61.30
Lexicon Method	54.66	44.96	49.34	53.26	62.71	57.60
Method of [2]	59.62	51.90	55.49	57.41	64.85	60.91
The 1st layered classifier	61.73	61.67	61.70	61.71	61.77	61.74
The 2nd layered classifier	61.62	62.38	62.00	61.91	61.15	61.53
The 3rd layered classifier	59.72	61.87	60.78	60.45	58.27	59.34
Combination of three layered classifier	**62.94**	**62.49**	**62.71**	**62.76**	63.21	**62.98**

---

[5] http://tcci.ccf.org.cn/conference/2013/
[6] http://ir.dlut.edu.cn/EmotionOntologyDownload.aspx

Table 4. Results on NLP&CC2013 Dataset

Method	Positive			Negative		
	Precision	Recall	F-measure	Precision	Recall	F-measure
Unigrams	74.16	72.93	73.54	73.37	74.59	73.98
Bigrams	65.54	53.59	58.97	60.75	71.82	65.82
Unigrams + POS tags	71.88	71.38	71.63	71.58	72.08	71.83
Bigrams + POS tags	62.84	61.05	61.93	62.13	63.90	63.00
Lexicon Method	59.69	57.64	58.65	59.05	61.07	60.04
Method of [2]	60.17	73.55	66.19	65.99	51.31	57.73
The 1st layered classifier	74.59	74.59	74.59	74.59	74.59	74.59
The 2nd layered classifier	73.86	71.82	72.83	72.58	74.58	73.57
The 3rd layered classifier	69.01	65.19	67.05	67.02	70.73	68.82
Combination of three layered classifier	**75.56**	**75.14**	**75.35**	**75.28**	**75.70**	**75.49**

From Table 3 & 4, we can find that on both datasets, our proposed method combining three layered classifiers achieves the best F-measure on both positive and negative sentiments. The classifier of the first layer also largely outperforms other comparison methods. It indicates that aggregating synonyms and highly-related words in the microblogs is a feasible approach for sentiment analysis. The results also reveal that the final combined classifier outperforms all the single classifiers. This may due to features in different layers provide different perspectives, from "micro" to "macro". Through voting, all the information gets together and improves the performance.

Observations of the classifying results indicate that layered features can handle sparseness more effectively. Examples like the microblog 他真令我们伤悲 (he makes us so sad) containing the word 伤悲 (sad) is predicted right in our approach but it is predicted wrong when using other features. And in the case that replacing the word 伤悲 (sad) with 悲伤 (sad), all the approaches predict right. The reason is, generally people use the word 悲伤 (sad) to express sad mood and the word 伤悲 (sad) is rarely used. However, these two words are grouped into one feature dimension when using layered feature. Therefore, layered features will make better use of rare words.

4.3   Further Discussion

Why Keep the Classifier of the Third Layer as One of the Voters? As shown in Table 3&4, the classifier of the third layer doesn't perform as well as the classifiers of the first and the second layers. However, we still keep this classifier. Because words not strongly related can have similarities, and the third layer provides a "macro" perspective to express similarities of the words. Meanwhile, the third layer provides the least sparse feature space.

Why not Combine the Three Layered Features as a Single Feature Vector? We have tested and the performance is not better than the voting method. Using a single feature vector will increase the dimensions of the feature space. And the second and the third layer will have less effects because their feature dimensions are less than the first layer.

5   Conclusions

In this paper, we propose a novel approach based on layered features. In this method, we assume that synonyms and highly-related words play equal roles in sentiment analysis and can be regarded as the same. Based on this assumption, three layered features are constructed by grouping synonyms and highly-related words. The difference among features of the three layers is how close the meanings or the relationships of the words are. And three classifiers are trained from features of different layers respectively. The final results will be a major vote comes from the three classifiers. In the comparison experiments, our approach achieves the best results.

Acknowledgement. This work was supported by the National Natural Science Foundation of China (Grant No. 61375053 and No. 60873134). We thank professor Liqing Zhang greatly for his valuable comments and suggestions.

References

1. Mohammad, S.M., Kiritchenko, S., Zhu, X.D.: NRC-Canada: Building the State-of-the-Art in Sentiment Analysis of Tweets. In: 7th International Workshop on Semantic Evaluation, pp. 321–327 (2013)
2. Pak, A., Paroubek, P.: Twitter as a Corpus for Sentiment Analysis and Opinion Mining. In: Proceedings of the International Conference on Language Resources and Evaluation, pp. 1320–1326 (2010)
3. Huang, S., You, J.P., Zhang, H.X., Zhou, W.: Sentiment Analysis of Chinese Micro-blog Using Semantic Sentiment Space Model. In: 2nd IEEE International Conference on Computer Science and Network Technology, pp. 1443–1447 (2012)
4. Pang, B., Lee, L., Vaithyanathan, S.: Thumbs up? Sentiment Classification using Machine Learning Techniques. In: Proceedings of the ACL 2002 Conference on Empirical Methods in Natural Language Processing, vol. 10, pp. 79–86 (2002)
5. Read, J.: Using Emoticons to Reduce Dependency in Machine Learning Techniques for Sentiment Classification. In: Proceedings of the ACL Student Research Workshop, pp. 43–48 (2005)
6. Bora, N.N.: Summarizing Public Opinions in Tweets. International Journal of Computational Intelligence and Applications, 41–55 (2012)
7. Purver, M., Battersby, S.: Experimenting with Distant Supervision for Emotion Classification. In: Proceedings of the 13th Conference of the European Chapter of the ACL, pp. 482–491 (2012)
8. Go, A., Bhayani, R., Huang, L.: Twitter Sentiment Classification Using Distant Supervision. CS224N Project Report, pp. 1-12. Stanford University (2009)
9. Saif, H., He, Y., Alani, H.: Alleviating Data Sparsity for Twitter Sentiment Analysis. In: 2nd Workshop of CEUR, pp. 2–9 (2012)
10. Tsutsumi, K., Shimada, K., Endo, T.: Movie Review Classification Based on a Multiple Classifier. In: The 21th PACLIC, pp. 481–488 (2007)
11. Zhao, J., Dong, L., Wu, J., Xu, K.: MoodLens: an Emoticon-Based Sentiment Analysis System for Chinese Tweets in Weibo. In: Proceedings of the 18th ACM SIGKDD International Conference on Knowledge Discovery and Data Mining, pp. 1528–1521 (2012)

Feature Group Weighting and Topological Biclustering

Tugdual Sarazin[1,2], Mustapha Lebbah[1], Hanane Azzag[1], and Amine Chaibi[1]

[1] Paris 13 University, Sorbonne Paris City - CNRS
LIPN-UMR 7030
99, av. J-B Clement - 93430, Villetaneuse
{firstname.secondname}@lipn.univ-paris13.fr
[2] ALTIC
25 rue Bergre, 75009 Paris
tugdual.sarazin@altic.org

Abstract. This paper proposes a new method to weight feature groups for biclustering. In this method, the observations and features are divided into biclusters, based on their characteristics. The weights are introduced to the biclustering process to simultaneously identify the relevance of feature groups in each bicluster. A new biclustering algorithm wBiTM (Weighted Biclustering Topological Map) is proposed. The new method is an extension to self-organizing map algorithm by adding the weight parameter and a new prototype for bicluster. Experimental results on synthetic data show the properties of the weights in wBiTM.

1 Introduction

Biclustering has become popular not only in the field of biological data analysis but also in other applications such as partitioning model [1], bayesian biclustering [2], exhaustive enumeration [3], self-organizing [4]. Similarly to clustering, biclustering is an unsupervised method for detecting meaningful structure in data. Recently, the dimensionality of the data involved in machine learning and data mining tasks has increased explosively. Data with extremely high dimensionality has presented serious challenges to existing learning methods [5]. In this paper, we propose a new approach of feature group weighting based on biclustering and topological maps model (wBiTM : Weighted Biclustering using Topological Map). Our model assigns for each feature group a new weighting parameter. The main difference between our approach and existing methods is that the weight is not associated to a single feature, but to feature groups.

The remaining of this paper is organized as follows, we briefly review the biclustering and feature weighting methods in section 2. We present the model and the associated algorithm in section 3. Section 4 is devoted to the methodology and experimental results. Finally, we draw some conclusions and outline future works in section 5.

2 Related Works

Biclustering methods have become an apparent need in many applications. These methods seeks to find simultaneously sub-matrices or blocks, that represent clusters of rows and clusters of columns. The term biclustering was first used by Cheng and Church

C.K. Loo et al. (Eds.): ICONIP 2014, Part II, LNCS 8835, pp. 369–376, 2014.

[6] in gene expression data analysis. Terms such as co-clustering, bidimensional clustering and subspace clustering, among others, are often used in the literature to refer to the same problem formulation. In the direct clustering approach (Block Clustering) [1], the data matrix is divided into several sub-matrices corresponding to blocks. The division of a block depends on the variance of its values. Indeed, more the variance is low, more the block is constant. Furthermore, author proposes two other algorithms: the first one "One-Way Splitting" is based on clustering the observations with features that have an intra-class variance greater than a given threshold in order to split the associated class. The second named "Two-Way Splitting" divides successively rows and columns, and compute for each iteration a large number of variance. Croeuc algorithm [7] seeks simultaneously a row and a column partition. The obtained blocks centers constitute a small matrix with a summary of the data. Coupled two-way clustering [8] defines a generic scheme for transforming a one-dimensional clustering algorithm into a biclustering algorithm. The algorithm relies on having a one-dimensional (standard) clustering algorithm that can discover significant clusters. Given such an algorithm, the coupled two-way clustering procedure will recursively apply the one-dimensional algorithm to sub-matrices, aiming to find subsets of observations giving rise to significant clusters of features and subsets of features giving rise to significant observations clusters. There are some other biclustering methods based on self-organizing maps (SOM) such as DCC (Double Conjugated Clustering) [9] and KDISJ (Kohonen for Disjonctive Table) [10]. The drawback of DCC method is the use of two maps (map of observations and map of features). These maps are built independently with the same size. Concerning KDISJ, it is only dedicated to categorical data.

Feature weighting/selection methods have been used somewhat successfully to improve cluster quality [11]. These algorithms find a subset of dimensions on which to perform clustering by removing irrelevant and redundant dimensions [12]. Feature weighting, on the other hand, is thought of as a generalization of feature selection [5]. In feature selection, a feature is assigned a binary weight, where 1 means the feature is selected and 0 otherwise. However, feature weighting, assigns weights to different features to indicate if a feature is most important than another. Feature weighting outperform traditional feature selection approaches because it adds a relevance feature score [13] instead of simply select feature. Feature weighting could be, also, reduced to feature selection if a threshold is set to select features based on their weights. Many subspace clustering algorithms have been proposed to handle high-dimensional data, aiming at finding clusters from subspaces of data, instead of the entire data space. Authors of [14] propose a feature group weighting method for subspace clustering of high-dimensional data. In this method, the features of high-dimensional data are divided into feature groups using two types of weights during a clustering process to simultaneously identify the importance of feature groups and observation features in each cluster. A new optimization model is given to define the optimization process and a new clustering algorithm based on k-means [15] is proposed to optimize the optimization model. This approach is an extension to k-means by adding two additional steps to automatically calculate the two types of subspace weights. In this work, we introduce a new approach that allows to organize a data matrix into homogeneous biclusters by considering simultaneously the set of rows and the set of columns and learns during the topological biclustering process a new parameter to identify the importance of feature groups.

3 Biclustering and Feature Group Weighting

Throughout the paper, we denote a matrix by bold capital letters such as $\mathbf{G}$. Vectors are denoted by small boldface letters such as $\mathbf{g}$ and matrix and vector elements are represented respectively by small letters such as g_i^j. As traditional self-organizing maps, which is increasingly used as tools for clustering and visualization, wBiTM consists of a discrete set of cells C called map with K cells. This map has a discrete topology defined as an undirected graph, it is usually a regular grid in 2 dimensions. For each pair of cells (c, r) on the map, the distance $\delta(c, r)$ is defined as the length of the shortest chain linking cells r and c on the grid. For each cell c, this distance defines a neighbor cell. Let $\Re^d$ be the euclidean data space and $\mathbf{D}$ the matrix of data, where each observation $\mathbf{x}_i = (x_i^1, x_i^2, ..., x_i^j, .., x_i^d)$ is a vector in $\mathbf{D} \subset \Re^d$. The set of rows (observations) is denoted by $I = \{1, ..., N\}$. Similarly, the set of columns (features) is denoted by $J = \{1, ..., d\}$. We are interested in simultaneously clustering observation I into K clusters $\{P_1, P_2, ..., P_k, .., P_K\}$, where $P_k = \{\mathbf{x}_i, \phi_z(\mathbf{x}_i) = k\}$ and features J into L clusters $\{Q_1, Q_2, ..., Q_l, .., Q_L\}$ where $Q_l = \{\mathbf{x}^j, \phi_w(\mathbf{x}^j) = l\}$. We denote by ϕ_z the assignment function of row (observation) and ϕ_w the assignment function of column (feature). The main purpose of wBiTM is to transform a data matrix $\mathbf{D}$ into a block structure organized in a topological map. In wBiTM, each cell $r \in C$ is associated with a prototype $\mathbf{g}_k = (g_k^1, g_k^2..., g_k^l, ..., g_k^L)$, and weight vector $\pi_k = (\pi_k^1, \pi_k^2..., \pi_k^l, ..., \pi_k^L)$ where $L < d$ and $g_k^l \in \Re$. $\mathbf{G} = \{\mathbf{g}_1, ..., \mathbf{g}_k\}$ and $\mathbf{\Pi} = \{\pi_1, ..., \pi_k\}$ denotes respectively the set of prototype and the weight vector. To facilitate formulation, we define two binary matrices $\mathbf{Z} = [z_i^k]$ and $\mathbf{W} = [w_j^l]$ to save the assignment associated respectively to observations and features:

$$z_i^k = \begin{cases} 1 \text{ if } \mathbf{x}_i \in P_k, \\ 0 \text{ else} \end{cases} \quad w_j^l = \begin{cases} 1 \text{ if } \mathbf{x}^j \in Q_l \\ 0 \text{ else} \end{cases}$$

Without using the weight parameter, biclustering topological maps, proposes to minimize the following cost function:

$$\mathcal{R}_{BiTM}(\mathbf{W}, \mathbf{Z}, \mathbf{G}) = \sum_{k=1}^{K} \sum_{l=1}^{L} \sum_{i=1}^{N} \sum_{j=1}^{d} \sum_{r=1}^{K} \mathcal{K}^T(\delta(r, k)) z_i^k w_j^l (x_i^j - g_r^l)^2$$

wBiTM extends the cost function with additional parameter π to control the feature group weights at each iteration of the biclustering process. To cluster $\mathbf{D}$ into K and L clusters in both observations and features, we propose the new following objective function to optimize in the biclustering process of wBiTM:

$$\mathcal{R}_{wBiTM}(\mathbf{W}, \mathbf{Z}, \mathbf{G}, \mathbf{\Pi}) = \sum_{k=1}^{K} \sum_{l=1}^{L} \sum_{i=1}^{N} \sum_{j=1}^{d} \sum_{r=1}^{K} \mathcal{K}^T(\delta(r, k)) z_i^k w_j^l (\pi_r^l x_i^j - g_r^l)^2 \quad (1)$$

This cost function can be written as follows:

$$\mathcal{R}_{wBiTM}(\mathbf{W}, \mathbf{Z}, \mathbf{G}, \mathbf{\Pi}) = \sum_{k=1}^{K} \sum_{l=1}^{L} \sum_{x_i^j \in B_k^l} \sum_{r=1}^{K} \mathcal{K}^T(\delta(r, k))(\pi_r^l x_i^j - g_r^l)^2$$

We can detect the bicluster of data denoted by $B_k^l = \{x_i^j | z_i^k w_j^l = 1\}$. Typically the neighborhood function $\mathcal{K}^T(\delta) = \mathcal{K}(\delta/T)$ is a positive function, which decreases as the distance between two cells in the latent space $\mathcal{C}$ increases and where T controls the width of the neighborhood function. Thus T is decreased between two values T_{max} and T_{min}. In practice, we use the neighborhood function defined as $\mathcal{K}^T(\delta(c, r)) = \exp\left(\frac{-\delta(c,r)}{T}\right)$ and $T = T_{max}(\frac{T_{min}}{T_{max}})^{\frac{t}{t_f-1}}$, where t is the current epoch and t_f the number of epoch.

The objective function (Eq. 1) can be locally minimized by iteratively solving the following three minimization problems:

– Problem 1: Fix $\mathbf{G} = \hat{\mathbf{G}}$, $\mathbf{W} = \hat{\mathbf{W}}$ and $\mathbf{\Pi} = \hat{\mathbf{\Pi}}$, solve the reduced problem $\mathcal{R}_{wBiTM}(\hat{\mathbf{W}}, \mathbf{Z}, \hat{\mathbf{G}}, \hat{\mathbf{\Pi}})$;
– Problem 2: Fix $\mathbf{G} = \hat{\mathbf{G}}$, $\mathbf{Z} = \hat{\mathbf{Z}}$ and $\mathbf{\Pi} = \hat{\mathbf{\Pi}}$, solve the reduced problem $\mathcal{R}_{wBiTM}$ $(\mathbf{W}, \hat{\mathbf{Z}}, \hat{\mathbf{G}}, \hat{\mathbf{\Pi}})$;
– Problem 3: Fix $\mathbf{W} = \hat{\mathbf{W}}$, $\mathbf{Z} = \hat{\mathbf{Z}}$ and $\mathbf{\Pi} = \hat{\mathbf{\Pi}}$, solve the reduced problem $\mathcal{R}_{wBiTM}(\hat{\mathbf{W}}, \hat{\mathbf{Z}}, \mathbf{G}, \hat{\mathbf{\Pi}})$;
– Problem 4: Fix $\mathbf{W} = \hat{\mathbf{W}}$, $\mathbf{Z} = \hat{\mathbf{Z}}$ and $\mathbf{G} = \hat{\mathbf{G}}$, solve the reduced problem $\mathcal{R}_{wBiTM}(\hat{\mathbf{W}}, \hat{\mathbf{Z}}, \hat{\mathbf{G}}, \mathbf{\Pi})$;

In order to reduce the computational time, we assign each observation and feature without using neighborhood cell as the traditional topological map.
Problem 1 is solved by defining z_i^k as:

$$z_i^k = \begin{cases} 1 \text{ if } \mathbf{x}_i \in P_k, k = \phi_z(\mathbf{x}_i) \\ 0 \text{ else} \end{cases}$$

Where each observation $\mathbf{x}_i$ is assigned to the closest prototype $\mathbf{g}_k$ using the assignment function, defined as follows:

$$\phi_z(\mathbf{x}_i) = \arg\min_c \sum_{j=1}^{d} \sum_{l=1}^{L} w_j^l (\pi_c^l x_i^j - g_c^l)^2 \tag{2}$$

Problem 2 is solved by defining w_j^l as

$$w_j^l = \begin{cases} 1 \text{ if } \mathbf{x}^j \in Q_l, l = \phi_w(\mathbf{x}^j) \\ 0 \text{ else} \end{cases}$$

Where each feature $\mathbf{x}^j$ is assigned to the closest prototype $\mathbf{g}^l$ using the assignment function, defined as follows:

$$\phi_w(\mathbf{x}^j) = \arg\min_l \sum_{i=1}^{N} \sum_{k=1}^{K} z_i^k (\pi_r^l x_i^j - g_r^l)^2 \tag{3}$$

Problem 3 is resolved for the numerical features by :

$$g_r^l = \frac{\sum_{k=1}^{K} \sum_{i=1}^{N} \sum_{j=1}^{d} \mathcal{K}^T (\delta(k,r) z_i^k w_j^l \pi_r^l x_i^j}{\sum_{k=1}^{K} \sum_{i=1}^{N} \sum_{j=1}^{d} \mathcal{K}^T (\delta(k,r)) z_i^k w_j^l}$$

This value is obtained by resolving the gradients $\frac{\partial \mathcal{R}_{wBiTM}}{\partial g_r^l} = 0$

$$g_r^l = \frac{\sum_{k=1}^{K} \mathcal{K}^T (\delta(k,r)) \sum_{i=1}^{N} \sum_{j=1}^{d} z_i^k w_j^l \pi_r^l x_i^j}{\sum_{k=1}^{K} \mathcal{K}^T (\delta(k,r)) \sum_{i=1}^{N} \sum_{j=1}^{d} z_i^k w_j^l} \tag{4}$$

For the problem 4, the component π_r^l of $\Pi = (\pi_r^1, \pi_r^2, ..., \pi_r^l, ..., \pi_r^d)$ is computed as follows:

$$\pi_r^l = \frac{\sum_{i=1}^{N} \sum_{j=1}^{d} \mathcal{K}^T (\delta(r, \phi_z(\mathbf{x}_i))) w_j^l x_i^j g_r^l}{\sum_{i=1}^{N} \sum_{j=1}^{d} \mathcal{K}^T (\delta(r, \phi_z(\mathbf{x}_i))) w_j^l (x_i^j)^2} \tag{5}$$

This value is obtained by resolving the gradients $\frac{\partial \mathcal{R}_{wBiTM}}{\partial \pi_r^l} = 0$. The main phases of wBiTM algorithm are presented in Algorithm 1.

Algorithm 1. wBiTM Algorithm
___
1: **Inputs:**

  – The data $\mathbf{D}$, prototypes $\mathbf{G}$ (Initialization).
  – t_f : the maximum number of iterations.

2: **Outputs:** Assignment matrix $\mathbf{Z}$, $\mathbf{W}$. Prototypes $\mathbf{G}$ and Π
3: **while** $t \le t_f$ **do**
4:   **for all** $\mathbf{x}_i \in \mathbf{D}$ **do**
5:     **Observation assignment phase**: Each observation $\mathbf{x}_i$ is assigned to the closest proto-
      type $\mathbf{g}_k$ using the assignment function, defined in equation 2
6:     **Features assignment phase**: Each feature $\mathbf{x}^j$ is assigned to the closest prototype $\mathbf{g}^l$
      using the assignment function, defined in equation 3
7:     **Quantization phase**: The prototype vectors are updated using following expression
      defined in equation 4
8:     **Weighting phase**: The weight vectors are updated using following expression defined
      in equation 5
9:   **end for**
10:   Update T
      $\{ T$ varies from T_{max} until $T_{min} \}$
11:   t++
12: **end while**
___

4   Experiments

To investigate the performance of the wBiTM, we used datasets extracted from UCI repository [16]. We also used in these experiments synthetic binary datasets [1]. Table 1 lists public and synthetic dataset parameters (number of observations, number of features, real class number and map size used in the learning phase). In order to compare wBiTM with biclustering methods, we selected the following approaches: BiTM [17], CTWC ([8]) and Croeuc [7]. Tables 2 and 3 present the experimental results. The size of wBiTM maps and BiTM maps is chosen according to Kohonen heuristic. The number of feature groups (columns of the matrix D) is exactly the same for all approaches wBiTM, BiTM, CTWC and Croeuc. However, in order to have the same number of clusters (rows of the matrix D), we chose the same map size of wBiTM and BiTM. But for CTWC and Croeuc approaches, we choose a proportional size of no-empty wBiTM and BiTM cells. For example, in the case of the simulated 1 dataset, the size of wBiTM map is identical to BiTM $4 \times 4 = 16$ map. The number of empty cells for wBiTM and BiTM maps is equal to 4. Thus, the size of rows partition for CTWC and Croeuc is equal to $16 - 4 = 12$. Although we can choose better initialization, we selected a random initialization with the same values for all approaches wBiTM, BiTM, CTWC and Croeuc. We computed two criteria, purity and rand index [18]. We selected the best performance obtained after 10 experiments.

Table 1. UCI datasets description

Datasets	# Observations	# Features	Map size	# Class
Isolet5	1559	617	12×12	26
Movement Libras	45	90	5× 5	15
Breast	699	10	7×7	2
Sonar Mines	208	60	6×6	2
Lung Cancer	32	56	4×4	2
Spectf 1	349	44	4×4	2
Cancer Wpbc Ret	198	33	6×6	2
Horse Colic	300	27	5×5	2
Heart	270	13	5×5	2
Glass	214	9	5×5	7
Simulated 1	2000	500	8×8	3
Simulated 2	2000	500	10×10	3
Simulated 3	2000	500	12×12	3
Simulated 4	2000	500	14×14	3

4.1   Performance Comparison between wBiTM and Biclustering Approaches

For purposes of comparisons with the other biclustering approaches, we have selected three methods: Croeuc and BiTM, the third one is CTWC (Coupled Two-Way Clustering) [8]. The license software is provided by Pr. Assif Yitzhaky and Pr. Eytan Domany. Detailed results are shown in table 2 and 3. Table 2 lists the purity index obtained with wBiTM, BiTM, CTWC and Croeuc approaches. We observe that for all datasets,

---

[1] These datasets are provided by Dr. Lazhar Labiod
https://sites.google.com/site/lazharlabiod/

wBiTM, BiTM, CTWC and Croeuc provide equivalent results. Also, from table 2, we can see in some datasets a large difference between wBiTM, BiTM, CTWC and Croeuc. For example, the performances of isolet 5 dataset in table 2 for wBiTM, BiTM, CTWC and Croeuc are respectively: 0.495, 0.316, 0.103 and 0.384. Observing purity index, we conclude that the use of weight parameter in the learning process, does not affect the biclustering. Table 3 lists the rand index obtained with wBiTM, BiTM, CTWC and Croeuc approaches. We note a slight decrease of rand performance index. To our knowledge, this is due to the limits of our approach, where we fix the map size and the feature group size.

Table 2. Purity criteria obtained with wBiTM, BiTM, CTWC and Croeuc

Datasets	wBiTM	BiTM	CTWC	Croeuc
isolet5	**0.495**	0.316	0.103	0.384
Breast	0.971	**0.978**	0.655	0.968
Sonar Mines	**0.727**	0.769	0.548	0.680
Lung Cancer	**0.821**	0.734	0.718	0.775
Spectf 1	**0.767**	0.759	0.727	0.749
Cancer Wpbc Ret	0.772	**0.787**	0.762	0.762
Horse Colic	**0.723**	0.719	0.67	0.671
Heart	0.814	**0.883**	0.555	0.798
glass	0.575	0.618	0.523	**0.590**
Simulated 1	0.983	**0.991**	0.901	0.828
Simulated 2	**0.901**	0.881	**0.902**	0.891
Simulated 3	0.983	**0.999**	0.910	0.989
Simulated 4	0.901	0.881	0.891	**0.913**

Table 3. Rand criteria obtained with wBiTM, BiTM, CTWC and Croeuc

Datasets	wBiTM	BiTM	CTWC	Croeuc
isolet5	0.858	0.926	0.91	**0.932**
Breast	**0.790**	0.687	0.505	0.643
Sonar Mines	0.494	**0.508**	0.502	0.508
Lung Cancer	**0.687**	0.459	0.556	0.446
Spectf 1	0.416	0.418	**0.513**	0.446
Cancer Wpbc Ret	0.435	0.435	**0.524**	0.430
Horse Colic	0.47	**0.472**	0.463	0.459
Heart	**0.615**	0.56	0.498	0.534
glass	0.607	0.653	**0.69**	0.506
Simulated 1	0.928	**0.930**	0.910	0.492
Simulated 2	**0.887**	0.889	0.718	0.831
Simulated 3	**0.882**	0.875	0.682	0.721
Simulated 4	0.853	**0.889**	0.812	0.691

5   Conclusion et Perpectives

In this paper, we have proposed a new feature group weighting using biclustering topological maps approach. The main novelty of our model is the use of topological model to organize the data matrix into homogeneous biclusters by considering simultaneously rows and columns, and learning new parameter of feature group weighting. A series of experiments are conducted to validate the proposed method. Experimental results demonstrate that our algorithm is promising and identify meaningful biclusters. Our algorithm inherits all the classical visualization of topological maps and provides a new visualizations to better data understanding. In future work, we will test and improve our method on further real applications. Further investigation is necessary to understand the relationship between weight feature group and feature group selection.

References

1. Hartigan, J.A.: Direct Clustering of a Data Matrix. Journal of the American Statistical Association 67(337), 123–129 (1972)
2. Shan, H., Banerjee, A.: Residual bayesian co-clustering for matrix approximation. In: SDM, pp. 223–234 (2010)
3. Tanay, A., Sharan, R., Shamir, R.: Discovering statistically significant biclusters in gene expression data. In: Proceedings of ISMB 2002, pp. 136–144 (2002)
4. Benabdeslem, K., Allab, K.: Bi-clustering continuous data with self-organizing map. Neural Computing and Applications (7-8), 1551–1562 (2012)
5. Alelyani, S., Tang, J., Liu, H.: Feature selection for clustering: A review. In: Data Clustering: Algorithms and Applications, pp. 29–60 (2013)
6. Cheng, Y., Church, G.M.: Biclustering of expression data (2000)
7. Govaert, G.: Simultaneous clustering of rows and columns. Control and Cybernetics 24(4), 437–458 (1995)
8. Getz, G., Levine, E., Domany, E.: Coupled two-way clustering analysis of gene microarray data. Proceedings of the National Academy of Sciences of the United States of America 97(22), 12079–12084 (2000)
9. Busygin, S., Jacobsen, G., Kremer, E., Ag, C.: Double conjugated clustering applied to leukemia microarray data. In: 2nd SIAM ICDM, Workshop on Clustering High Dimensional Data (2002)
10. Cottrell, M., Ibbou, S., Letrémy, P.: Som-based algorithms for qualitative variables. Neural Netw. 17(8-9), 1149–1167 (2004)
11. Benabdeslem, K., Hindawi, M.: Constrained laplacian score for semi-supervised feature selection. In: Gunopulos, D., Hofmann, T., Malerba, D., Vazirgiannis, M. (eds.) ECML PKDD 2011, Part I. LNCS, vol. 6911, pp. 204–218. Springer, Heidelberg (2011)
12. Parsons, L., Haque, E., Liu, H.: Subspace clustering for high dimensional data: a review. SIGKDD Explor. Newsl. 6(1), 90–105 (2004)
13. Wettschereck, D., Aha, D.W., Mohri, T.: A review and empirical evaluation of feature weighting methods for a class of lazy learning algorithms. Artif. Intell. Rev. 11(1-5), 273–314 (1997)
14. Chen, X., Ye, Y., Xu, X., Huang, J.Z.: A feature group weighting method for subspace clustering of high-dimensional data. Pattern Recogn. 45(1), 434–446 (2012)
15. Hartigan, J.A., Wong, M.A.: A k-means clustering algorithm. JSTOR: Applied Statistics 28(1), 100–108 (1979)
16. Bache, K., Lichman, M.: UCI machine learning repository (2013)
17. Chaibi, A., Lebbah, M., Azzag, H.: A new bi-clustering approach using topological maps. In: IJCNN, pp. 1–7 (2013)
18. Strehl, A., Ghosh, J., Cardie, C.: Cluster ensembles - a knowledge reuse framework for combining multiple partitions. Journal of Machine Learning Research 3, 583–617 (2002)

A Label Completion Approach to Crowd Approximation

Toshihiro Watanabe[1] and Hisashi Kashima[2]

[1] The University of Tokyo, Tokyo, Japan
toshihiro_watanabe@mist.i.u-tokyo.ac.jp
[2] Kyoto University, Kyoto, Japan
kashima@i.kyoto-u.ac.jp

Abstract. Majority vote is one of the most common methods for crowd-sourced label aggregation to get higher-quality labels. In this paper, we extend the work of Donmez et al. that estimates majority labels with a small subset of crowdsourcing workers in order to reduce financial and time costs. Our proposed method estimates the majority labels more accurately by completing missing labels to approximate the whole crowds even if some workers do not answer labels. Experimental results show that the proposed method approximates crowds more accurately than the method without label completion.

1 Introduction

Crowdsourcing is an idea to outsource human-intelligence tasks to a large group of unspecified people via Internet. Since crowdsourcing enables us to process a lot of tasks in a short time at low cost, it is used as a new tool for various fields in computer science. There are several emergence of online crowd-labor marketplaces such as Amazon's Mechanical Turk.

Labeling is one of the most popular and important use of crowdsourcing in computer science. An example is a task to request workers to judge whether a person is smiling or not in a given image. Since crowdsourcing tasks are executed by an unspecified number of workers, the quality of the obtained labels is uneven depending on workers' abilities or motivations. Majority vote is one of the commonly used methods to aggregate workers' labels to improve the label quality; however, a large number of workers are required to achieve a certain quality level, which results in large financial and time costs. Therefore, there are several attempts to reduce the number of workers required to get higher-quality labels by approximating whole crowds with a small number of workers. For example, IEThresh [2] estimates the confidence interval of each worker's majority score (i.e. how often the worker belongs to the majority), and CrowdSense [3] evaluates each worker's majority score using the number of agreements to the approximated label. Both algorithms actively decide which workers to assign tasks based on the estimated majority scores. One of the major drawbacks of these methods is that they cannot estimate the majority score of the workers to

C.K. Loo et al. (Eds.): ICONIP 2014, Part II, LNCS 8835, pp. 377–385, 2014.

whom the requester did not assign tasks. Moreover, these method assume that all of the requested workers return labels, which does not hold in practice.

In this paper, we propose a novel method for crowd approximation which addresses the missing label problem. Our proposed method completes such unobserved labels by using GroupLens method [6] which is a popular approach to recommendation systems, and estimates majority scores of workers by using the completed labels. Experiments show that the proposed method approximates crowds more accurately than the method without label completion.

2   Crowd Approximation Problem

The problem setting we address in this paper follows [3]. Let m and n be the number of workers and the number of items, respectively. In this paper, we consider the binary labeling tasks (the label for each item is $+1$ or -1). We now assume that items arrive sequentially, that is, an item arrives at each time $t = 1, \ldots, n$. At each time $t = 1, \ldots, n$, the requester chooses which worker to assign labeling task, and we denote the set of selected workers to be S_t. The size of S_t should be much smaller than the total number of workers m because we consider about approximating the majority label with small number of crowds. The selected worker $i \in S_t$ assigns a label $y_{it} \in \{+1, -1\}$ to item t (we will use the notation t for time as well as the indices of items). The requester aggregates the labels from selected workers. The goal of this problem is that the aggregated label should be correspond to the majority label. We summarize the notations and the problem setting for the following discussions.

Notation	Description	Notation	Description
m	the number of workers	t	time, the index for items
i	the index for workers	S_t	the set of selected workers
n	the number of items	y_{it}	worker i's label for item t

Problem setting

Input: m workers, n items
For each item $t = 1, \ldots, n$,

1. The requester selects the set of workers $S_t \subset \{1, \ldots, m\}$ whose labels will be same to the majority label.
2. The selected worker $i \in S_t$ assigns a label $y_{it} \in \{+1, -1\}$ to item t.
3. The requester aggregates the labels from selected workers.
4. Replace the majority label with the aggregated label gained in Step 3.

The crowd approximation problem contains some kind of dilemma structure. Since we consider about approximating the majority label by using small number of workers, it is not efficient to select all workers equally, and the concentration of selection on some workers is necessary. It means, however, that the requester cannot gain much information about other workers, which leads to the rise of the risk that the requester does not select the workers whose labels is closer to the majority labels. Therefore the requester should balance the two following actions in a trade-off relationship: (1) exploration: Selecting the workers whose information is not enough for the requesters, and (2) exploitation: Selecting the workers whose label will be same to the majority label based on the information so far The two existing methods (IEThresh, CrowdSense) balance the two actions by different approaches.

3   Existing Methods

IEThresh [2] is originally proposed for active learning from multiple workers. It first selects an item is from the pool of unlabeled items by using uncertainty sampling [5], and then choose the workers whose labels are expected to be the same as the majority label. This algorithm can be applied to our problem setting by omitting the selection of items in the original IEThresh.

IEThresh selects the workers who have large upper bound of the confidence interval on the majority score, and uses the majority vote of the selected workers for crowd approximation. The upper bound of worker i is given as

$$U_i = m_i + t_{\alpha/2}^{(n_i-1)} \frac{s_i}{\sqrt{n_i}}, \tag{1}$$

where m_i is the sample mean of worker i's reward (worker i's majority score), s_i is the unbiased variance of worker i's reward, and n_i is the sum of the number of times where worker i is selected by the algorithm and the algorithm parameter. $t_{\alpha/2}^{(n_i-1)}$ is the critical value of Student's t-distribution with $n_i - 1$ degrees of freedom at $\alpha/2$ confidence level. There are two reasons for selecting the workers with large U_is. One reason is that the workers with large sample mean of rewards are more likely to assign the label correspond to the majority label, and the other reason is that the requester wants to gain the knowledge about the workers with little information on the majority scores. The "fluctuation" of the worker's reward should be considered because the requester does not know which workers have higher majority scores in the earlier stage of the algorithm. The second term of Equation (1) is equivalent to the fluctuation. The requester can avoid to overlook the workers with higher majority scores by considering the fluctuation. As the algorithm processes, the fluctuation needs to decrease because the requester gradually gain the information about which workers are more likely to assign the labels same to the majority labels. The second term of the right hand of equation (1) satisfies this condition.

One of the other methods for crowd approximation is CrowdSense [3]. The requester considers the worker i's majority score as $Q_i = (a_{it} + K)/(c_{it} + 2K)$, where c_{it} is the number of times worker i is selected, a_{it} is the number of times the label y_{it} matches the approximating label, and $K > 0$ is the algorithm parameter. In CrowdSense, the workers with larger Q_is are selected; however, a worker is selected randomly in order to balance the two actions: exploitation and exploration.

In IEThresh and CrowdSense, the majority scores of the selected workers are computed, while those of the unselected workers are not updated. Moreover, these methods assume that all the selected workers return labels. Since this assumption does not necessarily holds in more practical settings, just applying these methods is not sufficient for crowd approximation in real world.

4   Crowd Approximation with Label Completion

The existing methods do not update the majority score of the workers who is not selected by the requester or is selected but does not return labels. However, it is expected to improve the approximation of the majority labels if we can estimate the majority levels of these workers and use them to decide which workers to select. Jung et al. [4] used the probabilistic matrix factorization [7] to fill in the missing labels, whose idea is also expected to be useful in our setting. At each time step t, our proposed algorithm completes the missing labels by using observed labels, and then computes each worker's upper confidence bound of the majority score based on the estimated labels, and decides which workers to select.

4.1   Label Completion Method

As the label completion method, we employ GroupLens method [6], a widely-used algorithm used for recommendation systems. We estimate missing labels by assuming that the workers whose labeling tendency are similar to each other are likely to assign the same labels. Suppose that m workers label N items. Let I_i and W_t be the set of items labeled by worker i and the set of workers who labeled item t, respectively. Denote by $y_{it} \in \{+1, -1\}$ the label that worker i assigned to item t. Notice that y_{it} is unknown if $t \notin I_i$. The mean label of worker i is given as $\bar{y}_i = \frac{1}{|I_i|} \sum_{t \in I_i} y_{it}$. We set $\bar{y}_i = 0$ when $|I_i| = 0$. We define the similarity between workers i and j as

$$r_{ij} = \frac{\sum_{t \in I_i \cap I_j} (y_{it} - \bar{y}_i)(y_{jt} - \bar{y}_j)}{\sqrt{\sum_{t \in I_i \cap I_j} (y_{it} - \bar{y}_i)^2} \sqrt{\sum_{t \in I_i \cap I_j} (y_{jt} - \bar{y}_j)^2}}.$$

Since each item is labeled by a limited number of workers, we estimate the missing labels $\hat{y}_{it}(t \notin I_i)$ from the known labels $y_{it}(t \in I_i)$ using the worker similarities by

$$\hat{y}_{it} = \bar{y}_i + \frac{\sum_{j \in W_t} r_{ij}(y_{jt} - \bar{y}_j)}{\sum_{j \in W_t} |r_{ij}|}.$$

4.2   Upper Confidence Bound

The crowd approximation problem is a variant of Multi-Armed Bandit Problem (MAB), which is the problem of identifying the reward-optimal candidate (arm) from a lot of arms of unknown properties. At each time, the player of MAB chooses one arm and receive a reward. The goal of MAB is to maximize the expected reward; however, since concentrating on arms whose current estimation of expected rewards are large might cause a problem of missing potentially better arms, we need to balance two actions, that are, exploration and exploitation. By replacing the arms and the reward with workers and the consensus between the majority label and the approximate label, respectively, the crowd approximation problem is regarded as MAB.

   IEThresh uses the confidence interval of the true reward to balance the two actions. However, we need to compute the critical value of Student's t-distribution, which cannot be computed common software and their computation is not so fast. Our proposed method uses Upper Confidence Bound (UCB) [1] instead of the confidence interval with t-distribution. The UCB of the expected reward in selecting arm a at time t is computed like

$$UCB(a) = m(a) + \sqrt{\frac{2\log t}{n(a)}}, \tag{2}$$

where $n(a)$ is the number of selecting the arm a, and $m(a)$ is the sample mean of rewards in selecting the arm a. In the earlier stage of the algorithm, we cannot identify which arm has the largest expected reward, so we need to consider the "fluctuation" of the reward, that is, how much the difference between the sample mean reward and the true reward is. The second term of (2) coincide with the fluctuation, and we can avoid the problem of ignoring the arms with larger expected rewards by considering this. Moreover, we can identify which arms have larger expected rewards as the algorithm proceeds, so the fluctuation term needs to decrease as t becomes larger. The second term of (2) satisfies this condition.

4.3   Proposed Algorithm

Our proposed algorithm computes the majority labels by processing the four actions below at each time $t = 1, \ldots, n$.

1. Completing missing labels
2. Compute UCBs on majority scores
3. Selecting the workers with larger UCBs
4. Aggregating labels within the selected workers

   In the first step, we complete the missing labels from the labels observed so far by GroupLens method. We can compute more accurately the majority scores of those workers who were not assigned tasks or were assigned tasks but return labels. Notice that we do not complete missing labels at $t = 1$ because there are no label information so far.

In the second step, we compute the UCB on each worker's majority score. For the computation of UCBs, we first compute the aggregated labels $\tilde{y}_s$ at $s = 1, \ldots, t - 1$ as follows:

$$
\tilde{y}_s = \begin{cases} +1 & \text{if } \sum_{i \in W_s} y_{is} > 0 \\ -1 & \text{if } \sum_{i \in W_s} y_{is} < 0 \\ x \sim \text{Rademacher} & \text{otherwise,} \end{cases}
$$

where W_s is the set of workers who we identify that give labels to item s (which we call observed worker set below). In short, we just take the majority vote of the observed workers at $t = s$. Rademacher is the probability distribution over $\{+1, -1\}$ with $P(X = +1) = P(X = -1) = 1/2$. After computing the aggregated labels, we compute the reward that worker i gained at $t = s$ like

$$
r_i(s) = \begin{cases} 1 - \frac{1}{2}|y_{is} - \tilde{y}_s| & \text{if } i \in W_s \\ 1 - \frac{1}{2}|\hat{y}_{is} - \tilde{y}_s| & \text{if } i \notin W_s. \end{cases}
$$

We give the observed workers reward 1 if they give the same label to the aggregated label $\tilde{y}_s$, and 0 otherwise. The rewards for the non-observed workers are computed by regarding the estimated labels $\hat{y}_s$ as observed labels. The larger the reward is, the more likely to approximate the majority labels. We then define worker i's reward list as follows:

$$
\mathbf{R}_i = [\overbrace{0, \ldots, 0}^{K}, \overbrace{1, \ldots, 1}^{K}, r_i(1), \ldots, r_i(t - 1)].
$$

The reward list represents the worker's reward history, and the worker is more likely to belong the majority if the list has more numbers close to 1. The K zeros and K ones at the head of the reward list play the role of avoiding zero-division in computing UCBs. Finally, we compute worker i's UCB on majority score like

$$
UCB_i = m_i + \sqrt{\frac{2 \log t}{2K + |I_{i,t-1}|}}, \tag{3}
$$

where m_i is the mean of the elements of $\mathbf{R}_i$, and $I_{i,t-1}$ is the set of items which we identify that worker i give labels to until time $t - 1$ (which we call observed item set).

In the third step, we define the set of workers who are likely to give majority labels as

$$
S_t = \{j \mid UCB_j \geq \varepsilon \cdot \max_i UCB_i\}.
$$

Note that ε controls the number of the selected workers. There are two reasons for selecting workers with larger UCBs. One reason is that the workers who we empirically know that belong to the majority are more likely to give the majority labels. Since the m_i in the first term of the equation (3) is nearly equal to the sample mean of the reward gained so far, the workers with larger empirical reward are more likely to be selected. The other reason is gaining the information about the majority levels for the workers about whom we have little knowledge. If the number of worker i is selected is small, that is, $|I_{i,t-1}|$ appearing in the second term of the equation (3) is small, the whole second term becomes large and worker i becomes more likely to be selected.

In the fourth step, we want to approximate the majority label with the aggregated label from the workers selected in the previous step, while some workers in S_t do not return labels. Hence we define the set of workers who return labels at time t as W_t, and approximate the majority label with the majority label $\tilde{y}_t$ in W_t as follows:

$$\tilde{y}_t = \begin{cases} +1 & \text{if } \sum_{i \in W_t} y_{it} > 0 \\ -1 & \text{if } \sum_{i \in W_t} y_{it} < 0 \\ x \sim \text{Rademacher} & \text{otherwise.} \end{cases}$$

We update the set of observed item set until time t as

$$I_{i,t} = I_{i,t-1} \cup \{t\}.$$

Notice that we do not give label to item t if $W_t = \varnothing$.

5 Experiments

We show that our proposed method approximates the majority more accurately than the method without label completion through experiments.

We used two datasets; one is the Recognizing Textual Entailment (RTE) dataset [8], and the other is the MovieLens dataset. Since the original RTE dataset is very sparse, we first chose the 40 most active workers who labeled many items, and then chose 400 most popularly labeled items by them. MovieLens is the data that each user (worker) gives 1–5 rating to each movie (item). In our experiments, rating 1–2 are converted to label -1, and rating 3–5 are converted to label $+1$. This data is also very sparse, so we chose 60 workers 300 items in a similar way to the RTE dataset.

We compared our proposed method and the method without label completion. We used "Accuracy" as an evaluation criteria. The accuracy is defined as the ratio of the number of items whose approximating label and the majority label matched to the total number of labeled items (i.e. $W_t \neq \varnothing$) las Accuracy $= \frac{\#\{t | W_t \neq \varnothing, \tilde{y}_t = y_{L_t}\}}{\#\{t | W_t \neq \varnothing\}}$. L_t is the set of workers who labeled item t and y_{L_t} is the majority label in L_t. L_t has both the workers selected by the algorithm and the workers not selected, so the observed worker set W_t is the subset of L_t.

By varying ε, we plotted use rates (x-axis, the ratio of the number of labels used by the algorithm to the number of total available labels) and accuracies (y-axis) in Fig. 1. Note that, in order to remove the effect of the order of item, we

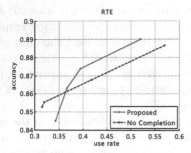
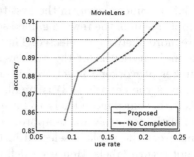

Fig. 1. Use rates and accuracies of each data set(Left: RTE, Right: MovieLens). The red solid line is our proposed method and the blue dashed line is the method without label completion. The dots on each line represent the use rate and accuracy at each ε.

changed the order of items 50 times in each ε and take the mean accuracy. The result of RTE suggested that our proposed method approximated the majority more accurately when we use the same number of labels, especially at higher use rate. This suggestion also applied to the result of MovieLens. Moreover, it can be said that we need less labels in approximating crowds with same accuracies.

6   Conclusions

We proposed the method for approximating the majority with small number of crowds even in the presence of missing labels. Our proposed method enabled estimating the majority scores of those workers who are not assigned a task to or are assigned a task but did not return a label. Experimental evaluations showed that our proposed method approximated the majority labels more accurately than the method without label completion.

References

1. Auer, P., Cesa-Bianchi, N., Fischer, P.: Finite-time analysis of the multiarmed bandit problem. Machine Learning 47(2-3), 235–256 (2002)
2. Donmez, P., Carbonell, J.G., Schneider, J.: Efficiently learning the accuracy of labeling sources for selective sampling. In: Proceedings of the 15th ACM SIGKDD International Conference on Knowledge Discovery and Data Mining, pp. 259–268 (2009)
3. Ertekin, S., Hirsh, H., Rudin, C.: Learning to predict the wisdom of crowds. In: Proceedings of Collective Intelligence (2012)
4. Jung, H.J., Lease, M.: Improving quality of crowdsourced labels via probabilistic matrix factorization. In: Proceedings of the 4th Human Computation Workshop at AAAI, pp. 101–106 (2012)
5. Lewis, D.D., Gale, W.A.: A sequential algorithm for training text classifiers. In: Proceedings of the 17th Annual International ACM SIGIR Conference on Reserch and Development in Information Retrieval, pp. 3–12 (1994)

6. Resnick, P., Iacovou, N., Suchak, M., Bergstrom, P., Riedl, J.: Grouplens: an open architecture for collaborative filering of netnews. In: Proceedings of the 1994 ACM Conference on Computer Supported Cooperative Work, pp. 175–186 (1994)
7. Salakhutdinov, R., Mnih, A.: Probabilistic matrix factorization. In: Advances in Neural Information Processing Systems (NIPS), vol. 20, pp. 1257–1264 (2007)
8. Snow, R., O'Connor, B., Jurafsky, D., Ng, A.Y.: Cheap and fast - but is it good? evaluating non-expert annotations for natural language tasks. In: Proceedings of the Conference on Empirical Methods in Natural Language Processing (EMNLP), pp. 254–263 (2008)

Multi-label Linear Discriminant Analysis with Locality Consistency

Yuzhang Yuan, Kang Zhao, and Hongtao Lu

Key Laboratory of Shanghai Education Commission for Intelligent Interaction and
Cognitive Engineering
Department of Computer Science and Engineering
Shanghai Jiao Tong University, Shanghai, 200240, China
{yuanyuzhang,sjtuzk,htlu}@sjtu.edu.cn

Abstract. Multi-label classification is common in many domains such as text categorization, automatic multimedia annotation and bioinformatics, etc. Multi-label linear discriminant analysis (MLDA) is an available algorithm for solving multi-label problems, which captures the global structure by employing the forceful classification ability of the classical linear discriminant analysis. However, some latest studies prove that local geometric structure is crucial for classification. In this paper, we present a new method called *Multi-label Linear Discriminant Analysis with Locality Consistency* (MLDA-LC) which incorporates local structure into the framework of MLDA. Specifically, we employ a graph regularized term to preserve the local structure for multi-label data. In addition, an efficient computing method is also presented to reduce the time and space cost of computation. The experimental results on three benchmark multi-label data sets demonstrate that our algorithm is feasible and effective.

Keywords: Multi-label Classification, Dimensionality Reduction, Locality Consistency.

1 Introduction

In traditional classification, each object belongs to only one category. However, an object may have several labels in many practical cases of classification. For example, a picture can be assigned to multiple labels such as *face*, *person*, and *entertainment*, and a web page can be annotated with *sports*, *basketball*, and *Rockets*. Such problems that multiple labels are related to a special sample are known as multi-label classification problems. The labels in multi-label classification are usually interrelated, while those in the traditional classification are mutually exclusive. Multi-label classification is common in real-world, and has attracted more and more attention recently [2,6].

In recent year, a new method called Multi-label Linear Discriminant Analysis (MLDA) has been proposed by extending the classical linear discriminant analysis (LDA) [10]. LDA, a notable dimensionality reduction algorithm, works well for traditional single-label classification problems, but can not handle these

C.K. Loo et al. (Eds.): ICONIP 2014, Part II, LNCS 8835, pp. 386–394, 2014.

situations with multiple labels, since each data point in the multi-label data sets may belong to several categories simultaneously [5]. MLDA makes use of the forceful classification ability of the classical LDA and incorporates the label correlations simultaneously. However, MLDA only captures the global structure of data by simultaneously maximizing the class-wise between-class distance and minimizing the class-wise within-class distance, while ignoring the local structure that has lately been demonstrated to be crucial for classification [3,9]. In this paper, we present a new Multi-label Linear Discriminant Analysis with Locality Consistency (MLDA-LC) method to capture the global structure and the local structure simultaneously. In addition, we present an efficient computing method to reduce the cost of computation for high-dimensional data. The experimental results on three multi-label data sets clarify that the new method is effective.

The following is the arrangement of this paper: Section 2 reviews the MLDA method; Section 3 introduces the new method we proposed; the experiment results on three multi-label data sets are shown and discussed in Section 4; the last section summarizes this paper.

2   A Brief Review of MLDA

Let $\{X_i, Y_i\}_{i=1}^{n}$ be a data set with multiple labels, where $X_i \in \mathbb{R}^d$ represents a data point, and $Y_i \in \{0,1\}^K$ denotes the class vector of X_i. $Y_i\{k\} = 1$ if X_i is a member of the k-th class, and 0 else. MLDA finds a linear transformation G that projects X into a lower dimensional space by $G^T X$.

The class-wise between-class and class-wise within-class scatter matrices in MLDA are formulated as:

$$S_b = \sum_{k=1}^{K} S_b^{(k)}, S_b^{(k)} = \sum_{i=1}^{n} Y_{ik}(\mu_k - \mu)(\mu_k - \mu)^T, \tag{1}$$

$$S_w = \sum_{k=1}^{K} S_w^{(k)}, S_w^{(k)} = \sum_{i=1}^{n} Y_{ik}(X_i - \mu_k)(X_i - \mu_k)^T, \tag{2}$$

where μ_k is the centroid of k-th class and μ is the multi-label global centroid, which are formulated as follows:

$$\mu_k = \frac{\sum_{i=1}^{n} Y_{ik}X_i}{\sum_{i=1}^{n} Y_{ik}}, \mu = \frac{\sum_{k=1}^{K}\sum_{i=1}^{n} Y_{ik}X_i}{\sum_{k=1}^{K}\sum_{i=1}^{n} Y_{ik}}. \tag{3}$$

MLDA also effectively utilize the label correlations as follows:

$$C_{kl} = \cos(Y_{(k)}, Y_{(l)}) = \frac{\langle Y_{(k)}, Y_{(l)} \rangle}{\|Y_{(k)}\|\|Y_{(l)}\|}. \tag{4}$$

Similar to classical LDA, the optimized objective function of MLDA is specified as follows:

$$\max_{G} \frac{tr(G^T S_b G)}{tr(G^T S_w G)}, \tag{5}$$

It is well-known that high-dimensional data usually lies in low-dimensional manifold of the ambient space, hence, the manifold of data is crucial for dimension reduction. However, MLDA only preserves the global structure while ignoring the local structure. In the next section, we proposed MLDA with Locality Consistency that incorporates local structure into the framework of MLDA.

3   MLDA-LC

In this section, we introduce our Multi-label Linear Discriminant Analysis with Locality Consistency (MLDA-LC) algorithm which captures global and local geometric structures of data. We employ a graph regularized term to preserve the local structure for multi-label data.

3.1   The New Objective Function

In order to capture the local geometric structure, nearby data points in the original space should have a small distance in the projection space. We can implement this through graph Laplacian. Connecting nodes X_i and X_j if X_i is within the κ closest neighbors of X_j or X_j is within the κ closest neighbors of X_i, where $\kappa > 0$ is a specified parameter [1]. Then an adjacency graph can be constructed with n data points. The weight of the edge connecting X_i and X_j is represented by E_{ij}, which is 0 if there is no edge between X_i and X_j, otherwise,

$$E_{ij} = exp(-\frac{\|X_i - X_j\|^2}{\epsilon}),$$ (6)

where $\epsilon > 0$ is the bandwidth parameter [9]. Let p_i be a projection of X_i, local structure for multi-label data can be preserved through minimizing the following objective function:

$$\sum_{ij} \|p_i - p_j\|^2 E_{ij},$$ (7)

The Laplacian matrix is defined as $L = D - E$, where D is a diagonal matrix whose entries are column sum of E, that is, $D_{ii} = \sum_j E_{ji}$. Let $P = [p_1, ..., p_n]$, it is simple to prove that

$$\frac{1}{2} \sum_{i=1}^{n} \sum_{j=1}^{n} \|p_i - p_j\|^2 E_{ij} = tr(P^T L P)^T.$$ (8)

Because the local structure information can be maintained through minimizing the function defined in Eq. (8), thus we proposed our Multi-label Linear Discriminant Analysis with Locality Consistency (MLDA-LC) by incorporating the graph Laplacian matrix into the MLDA framework. Our MLDA-LC finds an optimal transformation G via maximizing the following objective function:

$$\max_{G} \frac{tr(G^T S_b G)}{tr(G^T S_w G + \alpha G^T X L X^T G)},$$ (9)

where α is a balancing coefficient to adjust the weigh between global and local structures. It is costly to solve this objective function directly for multi-label data in high-dimensional space. Next, we propose an efficient method to handle this problem.

3.2 Efficient Computation Method

Uncorrelated Linear Discriminant Analysis (ULDA) is an improvement of classical LDA for single-label data, and it has been proved to have better performance and efficiency than LDA by removing the null space [11]. However, the null space can not be obtained directly from the formulas in MLDA-LC, so the class-wise scatter matrices in MLDA-LC should be rewritten firstly. Define

$$\tilde{X} = X - \mu e^T, \tag{10}$$

$$F = diag(\sqrt{f_1}, ..., \sqrt{f_n}), \tag{11}$$

$$B = diag(\sqrt{b_1}, ..., \sqrt{b_K}), \tag{12}$$

where $e = [1, ..., 1]^T$, $f_i = \sum_{k=1}^{K} Y_{ik}$, and $b_k = \sum_{i=1}^{n} Y_{ik}$. Then

$$
\begin{aligned}
S_b &= \sum_{k=1}^{K} \sum_{i=1}^{n} Y_{ik}(\mu_k - \mu)(\mu_k - \mu)^T \\
&= \sum_{k=1}^{K} b_k(\mu_k - \mu)(\mu_k - \mu)^T \\
&= \sum_{k=1}^{K} b_k\left(\frac{\sum_{i=1}^{n} Y_{ik}X_i}{b_k} - \mu\right)\left(\frac{\sum_{i=1}^{n} Y_{ik}X_i}{b_k} - \mu\right)^T \\
&= \sum_{k=1}^{K} \frac{1}{b_k} \sum_{i=1}^{n} Y_{ik}\tilde{X}_i \sum_{i=1}^{n} Y_{ik}\tilde{X}_i^T.
\end{aligned}
\tag{13}
$$

B is a diagonal matrix, so $B = B^T$, and S_b can be rewritten as

$$S_b = \tilde{X}YB^{-1}(B^{-1})^T Y^T \tilde{X}^T = H_b H_b^T, \tag{14}$$

where $H_b = \tilde{X}YB^{-1}$. Similarly, S_t can be expressed as

$$S_t = \tilde{X}FF^T \tilde{X}^T = H_t H_t^T, \tag{15}$$

where $H_t = \tilde{X}F$.

H_t is decomposed through SVD, thus $H_t = U\Sigma V^T$, and $U = (U_1, U_2)$, where $U_1 \in \mathbb{R}^{d \times t}$, $U_2 \in \mathbb{R}^{d \times (d-t)}$, $t = rank(H_t)$. It can be proved that $U_2^T S_b U_2 = 0$

and $U_2^T S_w U_2 = 0$. Thus let $G = U_1 Q$, and the problem Eq. (9) is equivalent to the following optimization problem:

$$\max_{G} \frac{tr(Q^T U_1^T S_b U_1 Q)}{tr(Q^T U_1^T S_w U_1 Q + \alpha Q^T U_1^T X L X^T U_1 Q)}. \tag{16}$$

According to this new formula, we can remove the null space U_2 firstly before we solve the original eigenproblem. Let

$$\tilde{H}_t = U_1^T H_t, \tilde{H}_b = U_1^T H_b. \tag{17}$$

Then

$$\tilde{S}_t = \tilde{H}_t \tilde{H}_t^T, \tilde{S}_b = \tilde{H}_b \tilde{H}_b^T, \tilde{S}_w = \tilde{S}_t - \tilde{S}_b. \tag{18}$$

We get Q by solving the eigenproblem, $(\tilde{S}_w + \alpha U_1^T X L X^T U_1)^{-1} \tilde{S}_b q = \lambda q$. In the Eq. (9), S_w and S_b are $d \times d$ matrices, while $\tilde{S}_w$ and $\tilde{S}_b$ in Eq. (18) are $t \times t$ matrices, where d is the dimensionality of data, $t = rank(H_t)$, and $t \ll d$ when the dimensionality of data is much larger than the amount of data points. As a result, we avoid computing eigen problem on the $d \times d$ matrices, and the cost of computation is thus reduced.

4   Experiment Results

In this section, we evaluate the proposed method in terms of classification performance and computational performance on three multi-label data sets.

4.1   Experimental Setup

In this subsection, we will introduce the setup of the experiment from three aspects, that is, Data Sets, Compared Algorithms and Performance Measures.

Data Sets. Three multi-label data sets are used in our experiment, including gene expression data (Yeast), images data (Scene), and web pages data (Yahoo).

The Yeast data set is a genetic data set. It contains 2417 genes, and the dimensionality of each data point is 103 [12]. The functional labels of a gene are probably more than 190, while only the 14 top hierarchical labels are selected in the experiment for simplicity.

The Scene data set includes 2407 natural scene pictures with six available labels. Each picture in this data set is described as a 294-dimensional vector [2].

The Yahoo data set is a high-dimensional multi-label data set with 11 top-level categories [7]. Each top-level category can be viewed as a dataset, thus it has 11 datasets.

Compared Algorithms. We compare our algorithm with other four multi-label classification algorithms.

MLDA-LC. Our proposed multi-label classification method based on MLDA. In this experiment, the parameter κ and the parameter ϵ for building the graph

Laplacian are specified to 5, and the balancing parameter α is tuned from the candidate set $[10^{-6}, 10^{-5}, 10^{-4}, 10^{-3}, 10^{-2}, 10^{-1}, 1, 10]$.

MLDA. It is a multi-label classification method by extending the classical LDA [10], which has been specifically reviewed in section 2.

ML-LS. Multi-Label Least Square (ML-LS) is a learning framework that extracts a shared subspace within multi-labels, so that the label correlations is acquired [6].

CCA-SVM. Canonical correlation analysis (CCA) is a conventional method for seeking the correlations among sets of samples [6]. CCA-SVM can be divided into two steps. First, CCA is used to reduce the dimensionality of data, then linear SVM is used for classification.

CCA-ridge. CCA-ridge [6] can also be divided into two steps, the first step is same as CCA-SVM. Then ridge regression is adopted after dimensionality reduction by CCA.

Performance Measures. AUC, macro F1 and micro F1 scores are used as the performance measures in the experiments.

AUC. Receiver operating characteristic (ROC) curve reflects classifier performance. Calculating the area under the ROC curve (AUC) can reduce ROC performance to a value representing expected performance [4]. A algorithm has greater AUC, and therefore better performance.

F1. F1 measure is the harmonic mean of the precision and recall. The macro F1 measure is the traditional arithmetic mean of the F1 measure computed for each problem. The micro F1 measure is an average weighted by the class distribution [8].

4.2 Results

Classification Performance. The classification performance for the three multi-label data sets are shown in Table 1-2. Because of the limited space, only seven datasets of Yahoo are shown in the Table 2. From the experiment results, it can be found that the classification performance of the four compared algorithms (MLDA, ML-LS, CCA-SVM, CCA-ridge) is related to specific data set, for example, CCA-ridge is better in the Yeast data set, and MLDA is better in the Sence data set. However, in all multi-label data sets, our MLDA-LC almost outperforms the other four algorithms. Therefore, capturing both global and local structures in MLDA-LC is helpful for multi-label classification performance.

Besides, we evaluate the sensitivity of MLDA-LC to the balancing parameter α. When the parameter α is 0, MLDA-LC is equivalent to MLDA. In a certain range, with the increasing of parameter α, the performance is improved. Keep increasing the parameter α, the performance may be decreasing.

Computational Performance. We evaluate the computational performance via comparing our method with the most related method MLDA. We show the results on the *Business* data set and the *Health* data set. The dimensionality of both data sets are high, the former is 16621, and the latter is 18430. The same

Table 1. Results on the Yeast Data set (left) and the Scene Data set (right)

Algorithms	AUC	macro F1	micro F1	AUC	macro F1	micro F1
MLDA-LC	**0.6369**	**0.4410**	**0.6136**	**0.9065**	**0.6872**	**0.6802**
MLDA	0.6248	0.4348	0.5876	0.8730	0.6276	0.6243
ML-LS	0.6126	0.4322	0.5638	0.8552	0.6087	0.6020
CCA-SVM	0.6349	0.4355	0.5946	0.8503	0.6095	0.5825
CCA-ridge	0.6355	0.4363	0.5582	0.8674	0.6215	0.6050

Table 2. Results on the yahoo datasets, the top section is the results of AUC, the middle is macro F1, and the bottom is micro F1

Algorithms	Arts	Business	Education	Health	Science	Computers	Society
MLDA-LC	**0.7522**	**0.8287**	**0.7555**	**0.8557**	**0.8083**	**0.7828**	**0.7061**
MLDA	0.7416	0.8210	0.7404	0.8518	0.7984	0.7798	0.6193
ML-LS	0.6939	0.8206	0.7171	0.8413	0.7464	0.7366	0.6493
CCA-SVM	0.7393	0.7775	0.7476	0.8517	0.8010	0.7779	0.6933
CCA-ridge	0.7396	0.8017	0.7479	0.8544	0.8036	0.7797	0.7017
MLDA-LC	**0.3441**	**0.3920**	**0.4222**	**0.6103**	**0.4265**	**0.2788**	**0.3226**
MLDA	0.2646	0.3876	0.3573	0.5325	0.3601	0.2377	0.1656
ML-LS	0.2826	0.2902	0.3362	0.5219	0.3022	0.2253	0.2475
CCA-SVM	0.3031	0.3486	0.3366	0.5406	0.3709	0.2628	0.2785
CCA-ridge	0.3124	0.3486	0.3379	0.5015	0.3490	0.2413	0.2766
MLDA-LC	**0.4441**	**0.7473**	**0.4673**	0.6360	**0.5105**	**0.5280**	**0.4691**
MLDA	0.3400	0.7004	0.3983	0.5418	0.4124	0.3888	0.3035
ML-LS	0.4059	0.6683	0.4368	**0.6448**	0.4116	0.4647	0.3852
CCA-SVM	0.4305	0.7084	0.4141	0.6229	0.4716	0.4663	0.4517
CCA-ridge	0.4409	0.7233	0.4144	0.6110	0.4632	0.5229	0.4630

Table 3. Computation Performance, the second column is the time of computation (in sec), the third column is the memory of computation (in MB)

Algorithm	Business	Health	Business	Health
MLDA-LC	**7.24**	**12.41**	**3863**	**3884**
MLDA	5259.48	5692.54	14982	15013

hardware was used to conduct the experiments, the CPU is Intel(R) Core(TM) i5-4430 CPU @ 3.00GHz, and the RAM is 16.0GB. In Table 3, the second column is the results on computation time, and the third column is the results on memory. It is obvious that MLDA takes much longer time and more memory than MLDA-LC, especially for computation time, MLDA-LC is faster by hundreds times than MLDA. This result show that our computing method reduce the cost of computation efficiently for high-dimensional data.

5 Conclusion

In this paper, we present a new method called Multi-label Linear Discriminant Analysis with Locality Consistency (MLDA-LC) for multi-label classification. MLDA-LC captures the global and local geometric structures to enhance the classification performance. In addition, an efficient computing method is also presented to reduce the time and space cost of computation. We compare the new method with other four multi-label classification methods on three data sets, the experiment results demonstrate that our algorithm is feasible and effective.

Acknowledgment. This work is supported by NSFC (No.61272247 and 60873133), the Science and Technology Commission of Shanghai Municipality (Grant No. 13511500200), 863 (No.2008AA02Z310) in China and the European Union Seventh Frame work Programme (Grant no. 247619).

References

1. Belkin, M., Niyogi, P.: Laplacian eigenmaps and spectral techniques for embedding and clustering. In: NIPS, vol. 14, pp. 585–591 (2001)
2. Boutell, M.R., Luo, J., Shen, X., Brown, C.M.: Learning multi-label scene classification. Pattern Recognition 37(9), 1757–1771 (2004)
3. Cai, D., He, X., Zhou, K., Han, J., Bao, H.: Locality sensitive discriminant analysis. In: IJCAI, pp. 708–713 (2007)
4. Fawcett, T.: An introduction to roc analysis. Pattern Recognition Letters 27(8), 861–874 (2006)
5. Fisher, R.A.: The use of multiple measurements in taxonomic problems. Annals of Eugenics 7(2), 179–188 (1936)
6. Ji, S., Tang, L., Yu, S., Ye, J.: A shared-subspace learning framework for multi-label classification. ACM Transactions on Knowledge Discovery from Data 4(2), 8 (2010)
7. Kazawa, H., Izumitani, T., Taira, H., Maeda, E.: Maximal margin labeling for multi-topic text categorization. In: Advances in Neural Information Processing Systems, pp. 649–656 (2004)
8. Lewis, D.D., Yang, Y., Rose, T.G., Li, F.: Rcv1: A new benchmark collection for text categorization research. The Journal of Machine Learning Research 5, 361–397 (2004)
9. Niyogi, X.: Locality preserving projections. In: Neural Information Processing Systems, vol. 16, p. 153 (2004)
10. Wang, H., Ding, C., Huang, H.: Multi-label linear discriminant analysis. In: Daniilidis, K., Maragos, P., Paragios, N. (eds.) ECCV 2010, Part VI. LNCS, vol. 6316, pp. 126–139. Springer, Heidelberg (2010)

11. Ye, J.: Characterization of a family of algorithms for generalized discriminant analysis on undersampled problems. Journal of Machine Learning Research, 483–502 (2005)
12. Zhang, M., Zhou, Z.: Ml-knn: A lazy learning approach to multi-label learning. Pattern Recognition 40(7), 2038–2048 (2007)

Hashing for Financial Credit Risk Analysis

Bernardete Ribeiro[1] and Ning Chen[2]

[1] CISUC - Department of Informatics Engineering, University of Coimbra, Portugal
[2] GECAD, Instituto Superior de Engenharia do Porto, Portugal
bribeiro@dei.uc.pt, ningchen74@gmail.com

Abstract. Hashing techniques have recently become the trend for accessing complex content over large data sets. With the overwhelming financial data produced today, binary embeddings are efficient tools of indexing big datasets for financial credit risk analysis. The rationale is to find a good hash function such that similar data points in Euclidean space preserve their similarities in the Hamming space for fast data retrieval. In this paper, first we use a semi-supervised hashing method to take into account the pairwise supervised information for constructing the weight adjacency graph matrix needed to learn the binarised Laplacian EigenMap. Second, we train a generalised regression neural network (GRNN) to learn the k-bits hash code. Third, the k-bits code for the test data is efficiently found in the recall phase. The results of hashing financial data show the applicability and advantages of the approach to credit risk assessment.

Keywords: hashing method, financial credit risk, generalised regression neural network, k-bits hash code.

1 Introduction

Hashing methods have become very popular to handle the overwhelmed big data that we face today. In particular they have been largely used for accessing millions of images or documents in the Internet. The idea is that they map high-dimensional representation of objects (images, documents, $\cdots$, etc.) into a binary representation, i.e., a few code bits. This is very cost efficient both from time and space complexity point of view since only a simple bitwise XOR operation is needed to compute the Hamming distance between two binary codes. When it comes to binary embeddings the learning process takes in general two stages: one for codeworks generation using training data and another for searching using test data. The crucial goal is then to learn the hash function in such a way that similar data in the sense of geometrical topological data points in high-dimensional space can be hashed such that they endow similarities in a binary topological space. There has been recent work on finding the hash function to generate the hash code either using random projections [10] [1] [12] or learning the hash function [16] [17] [9]. In this work, in the settings of a financial risk analysis problem, we follow the approach presented by Zhang et al. [17] Self-Taught Hashing (STH) for similarity search. First, we solve a more

C.K. Loo et al. (Eds.): ICONIP 2014, Part II, LNCS 8835, pp. 395–403, 2014.

general binarised Laplacian EigenMap [3] after the relaxation by removing the binary constraints. In this step the data are first embedded in an intermediate real-valued space. Second, we perform the thresholding in this space to obtain binary outputs. Then, instead of Zhang's 1-bit code linear SVM k classifiers for obtaining the codes for test data, we propose to learn k-bits code with one generalised regression neural network (GRNN) with k outputs. Finally, the k-bits code for the testing data is quickly generated through the learned neural network weights. We compare our approach using k-bits GRNN with 1-bit linear SVM classifier [17]. Furthermore we extend to 1-bit code non-linear SVM classifier and also to 1-bit code GRNN classifier. The paper presents three contributions: (i) extends to non-linear hashing (Gaussian kernel SVM); (ii) incorporates class label information of training data; (iii) uses parsimony with one GRNN classifier for learning k bits code at once instead of k SVM classifiers to learn 1 bit at a time (Zhang's approach in STH). The experimental evaluation on real world financial data of French companies is performed using several state-of-the-art methods.

The paper is organized as follows. In Section 2 we give the notation. In Section 3 we review the recent work on hashing for information retrieval and data mining. In Section 4 we present the fundamentals on hashing spectral methods in the financial background. In this context we illustrate the block diagram hashing procedure proposed in this paper. We describe the experimental design including dataset, evaluation metrics and results in Section 5. Finally, in Section 6 we give a brief summary of the main findings and address future lines of work.

2   Notation

Let $\mathcal{X} = \{x_1, \cdots, x_n\} \in \mathbb{R}^m$ be the set of n financial feature vectors extracted from training data companies and represented in a m-dimensional space. The goal is to learn a binary embedding function of k bits, $f : \mathbb{R}^m \mapsto \{-1, +1\}^k$, where the binary symbols have been defined as -1 and $+1$. The training set $\mathcal{X}$ allows us to produce the set of binary codes $\mathcal{Y} = \{y_1, \cdots, y_n\} \in \{-1, +1\}^k$. Given an input test sample, the learned mapping function allows to find the corresponding binary code, and then the Hamming distance can be used to determine the nearest neighbors from the training set. Additionally, in the supervised learning mode we assume that each training sample x_i has an associated label $t \in \{-1, +1\}$ whose value indicates whether a given company is healthy (-1) or bankrupt ($+1$). The hashing function f should be topologically consistent by assigning similar binary codes to similar data points and dissimilar otherwise.

3   Related Work

The problem of finding the nearest neighbor in high-dimensional space is itself of importance in many applications from science to industry. The complexity of most of the algorithms such as K-tree or R-tree grows exponentially with the data dimensionality m making them impracticable for dimension higher than, say, 15 [11]. In the majority of the applications the closest point is of interest if it lies within a specified ball distance ϵ. By exploring hashing techniques the search in big data is efficient both in terms of computational cost and memory. Although the previous work on hashing techniques (e.g. [10] [1] [12], [16] [17] [9]) has had impact, the recent rise of the prevalence of Big Data has leveraged the work presented in [16]. The spectral hashing technique (SpH) assigns binary hash keys to data points via thresholding the eigenvectors of the graph Laplacian [16]. Another popular unsupervised hashing method for fast document retrieval is Semantic Hashing [14] where the real-valued low-dimensional vectors obtained from Latent Semantic Indexing (LSI) [8] are binarised via thresholding. Recently, in [4] a method of linear spectral hashing solves an optimization problem based on spectral clustering which improves the out-of-sample extension analytical procedure described in [16]. In [9] an expectation-based asymmetric distance and a lower-bound-based asymmetric distance are proposed for binary embeddings and their applicability was demonstrated to several hashing methods.

Self-taught hashing [17] uses a two-step approach for codeword generation. The first step consists of the unsupervised learning of binary codes using a similar objective function to spectral hashing. To generalize the obtained mappings, in the second step it considers the set of training examples together with their labels, obtained in the previous step, and trains a support vector machine for each bit of the codeword. In our work, instead, we train a Generalised Regression Neural Network (GRNN) for learning k bits of codework which endows parsimony in the models (used to generate the test code in the next step) with improved performance and comparative computational cost in out-of-sample extension. We apply the technique to a data set of French financial data containing financial descriptors of companies with binary labels, namely, bankrupt and healthy. In the next section we give the context and describe the steps in this setting.

4   Hashing with Financial Data

We are given a set of n companies which are represented as m-dimensional vectors - the financial descriptors- by $\{\mathbf{x}_i\}_{i=1}^n \in \mathbb{R}^m$. Suppose that the length of our binary embedding is k bits. The company $\mathbf{x}_i$ has the corresponding binary code that we represent by $\mathbf{y}_i \in \{-1, +1\}^k$. The k-element of $\mathbf{y}_i^{(k)}$ is $+1$ if the bit is 'on' and -1 otherwise. Let Y denote the $n \times k$ matrix whose i-the row is the code for the i-th company, i.e. $[\mathbf{y}_1, \mathbf{y}_2, \cdots, \mathbf{y}_n]^T$.

4.1   Binary Embedding

Given an undirected graph G with n nodes, each node representing a data point, let W be a symmetric $n \times n$ matrix where W_{ij} is the connection weight between node i and j. We aim at representing each node of the graph as a low-dimensional vector where the similarities between pairs of data (in the original high-dimensional space) are preserved. The Laplacian matrix [7] is defined as:

$$L = D - W, \quad D_{ii} = \sum_{j \neq i} W_{ij} \quad \forall i \tag{1}$$

where D is a diagonal matrix whose entries are sums of the row elements of the matrix W. Let the low-dimensional embedding of the nodes be $Y = [\mathbf{y}_1, \mathbf{y}_2 \cdots \mathbf{y}_n]^T$, where the column $\mathbf{y}_i$ vector is the embedding for the vertex $\mathbf{x}_i$. We need to solve the following optimization problem:

$$\tilde{Y} = \arg\min \sum_{i,j}^n W_{ij} \|\mathbf{y}_i - \mathbf{y}_j\|^2 \tag{2}$$

$$s.t.\ \mathbf{y}_i \in \{-1, +1\}^k \quad \sum_{i=1}^n \mathbf{y}_i = \mathbf{0}, \quad \frac{1}{n}\sum_{i=1}^n \mathbf{y}_i \mathbf{y}_i^T = \mathbf{I} \tag{3}$$

The first constraint allows to maintain the balance between the bits $\{-1, +1\}$ and the second one ensures the bits are uncorrelated. After some mathematical transformation and using the L matrix the objective function can be written as:

$$\tilde{Y} = \arg\min Tr(Y^T LY) \tag{4}$$

$$s.t.\ Y^T D\mathbf{1} = 0 \quad Y^T LY = \mathbf{I} \tag{5}$$

where $Tr(.)$ means the matrix trace. By relaxing the constraint $\mathbf{y}_i \in \{-1, +1\}^k$ the $\tilde{Y} = [\mathbf{v}_1, \mathbf{v}_2, \cdots, \mathbf{v}_k]$ gives the real solution for the above generalised optimization problem. The k eigenvectors corresponding to the k smallest eigenvalues are computed by solving:

$$L\mathbf{v} = \lambda D\mathbf{v} \tag{6}$$

The binary vectors are obtained by thresholding the real-valued k dimensional vectors $\tilde{\mathbf{y}}_1, \cdots, \tilde{\mathbf{y}}_k$. A common procedure to perform the binary embedding is to use the median since it corresponds to maximizing an entropy criterion in information theory [2] [17]. The median works as a threshold thereby rendering an even distribution of $+1$'s and -1's. If the p element of the vector $\tilde{\mathbf{y}}_i^{(p)}$ is larger than the threshold the p-th bit of the i-th code is $+1$, otherwise is -1.

4.2 Building the Affinity Graph Matrix

The affinity graph matrix W is built by assuming that each i-th node corresponds to a given firm $\mathbf{x}_i$. In the K-nearest neighbor graph an edge between nodes i and j exists if $\mathbf{x}_i$ and $\mathbf{x}_j$ are nearby points, i.e., if $\mathbf{x}_i$ is among the K-nearest neighbors of $\mathbf{x}_j$ and $\mathbf{x}_j$ is among the K-nearest neighbors of $\mathbf{x}_i$. We also considered the supervised mode where the class information is available. As soon as the affinity graph is constructed, the weight matrix W can be specified by means of weighting schemes such as binary, heat kernel [7] and dot-product [5].

4.3 Learning the Hash Function

An interesting idea has been used in [17] where each bit of the binary codes (codeworks) are learned with a linear SVM. Herein instead of Zhang's 1-bit code linear SVM classifier for retrieving test data, we propose to learn k-bits code with one generalised regression neural network (GRNN) with k outputs. Finally the k-bits code for the testing data is quickly generated through the learned neural network weights. The learning stage may also be improved if a non-linear kernel is used, in particular, in the case of a real application. Therefore we compared

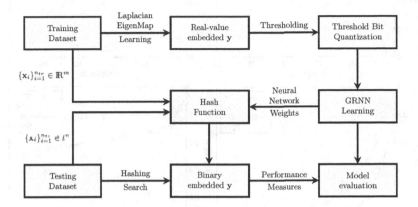

Fig. 1. Block diagram of Hashing Procedure (n_{tr} is the nr. of training examples; and n_{te} is the nr. of test queries for the financial data).

our approach using k-bits GRNN with 1-bit linear SVM classifier [17] and also with 1-bit non-linear SVM classifier. In the latter we used the Gaussian kernel. We also compared with 1-bit code GRNN classifier. In Figure 1 we illustrate the main blocks of the hashing procedure for financial credit risk analysis.

5   Experimental Design

In the experiments, we used the Diane French financial data set whose 30 attributes are characterized in [13]. French data set is a balanced subset of Diane financial database (600 bankruptcy or distressed companies and 600 healthy), which originally contains the financial statements of $110,723$ companies in small or middle size during the year 2003 to 2006. The class indicator gives the state of companies in the year 2007. For the above data set, the numeric attributes are normalized to unity range $[0,1]$ in the preprocessing phase. In order to evaluate a

Table 1. Contingency table for binary classification

	Class Positive	Class Negative
Assigned Positive	a (True Positives)	b (False Positives)
Assigned Negative	c (False Negatives)	d (True Negatives)

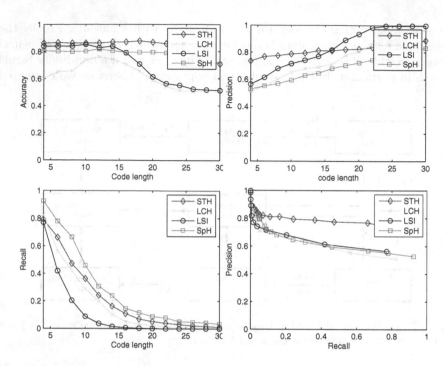

Fig. 2. Performance measures in Diane Financial Data Set for hashing with STH, LCH, LSI and SpH within Hamming distance $d \leq 2$

binary decision task, the contingency matrix representing the possible outcomes of the classification is defined as in Table 1. Several measures have been defined, such as, error rate ($\frac{b+c}{a+b+c+d}$), recall ($R = \frac{a}{a+c}$), and precision ($P = \frac{a}{a+b}$), as well as combined measures, such as, F_β measure, which combines recall and precision in a single score. When $\beta = 1$, $F_1 = \frac{2 \times P \times R}{P+R}$ is an harmonic average between precision and recall. In the experiments the parameter K-Nearest Neigbhors varied from 0 to 50 for constructing the graph for the Laplacian EigenMap. We also have used the heat kernel with $\sigma = 1$ in the supervised mode. According to Chung's [7] all the information about a graph is contained in its heat kernel. The Hamming ball radius ϵ was varied from 0 to 3 ($d \leq 3$). All the experimental results are averaged over 10 random training/test partitions. In Figure 2 we compared the performance measures in Financial Dataset for hashing with STH [17][1], LCH [10], LSI [8] and SpH [16][2] within Hamming distance $d \leq 2$. The two plots in the top and the one in the bottom left illustrate the accuracy, precision and recall in terms of the code length. In the plots, the code length

---

[1] We used the Self-Taught Hashing code available at http://www.dcs.bbk.ac.uk/ ~dell/publications/dellzhang_sigir2010_suppl.html slightly adapted.

[2] We used the Spectral Hashing Code code available at http://www.cs.huji.ac.il/~yweiss/SpectralHashing/

changed from 4 to 30 bits although we also performed experiments with 32, 64, 128 and 256 bits. In the fourth graph the precision-recall curve is plotted. As shown through the performance measures the STH method performed the best. Therefore, it was selected as the hashing method for the financial data. As there was space for improvement we refined the step 2. using instead of the SVM linear classifier for learning the 1-bit sample code, the Generalised Regression Neural Network (GRNN). For the GRNN we used the Matlab NN Toolbox, and for the SVM we used the LibSVM [6]. The spread for the radial basis function in the GRNN was varied from 0.1 to 1. The parameter γ for the non-linear SVM with Gaussian kernel was set in the range 0.01 to 0.1 and the parameter $C = 1$. In Figure 3 the accuracy and F1 measure for retrieving same class data as a function of the number of K Nearest Neighbors within Hamming Distance $d \leq 1$ and $d \leq 2$ are presented. We compare the results with GRNN 1-bit and with non-linear SVM and conclude that the hashing technique GRNN k-bits performs better. The GRNN [15] is a memory-based network with one-pass learning algorithm, with a highly parallel structure, which provides smooth transitions due to the non-parametric estimators of probability density functions. The neural network shows superior ability in learning with multiple outputs. The out-of-extension

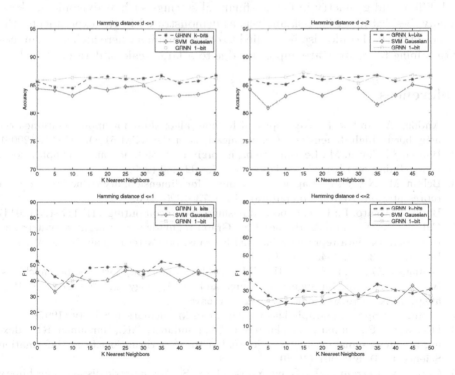

Fig. 3. Accuracy and F1 measures (max values) for retrieving same class data as a function of the number of K Nearest Neighbors within Hamming Distance $d \leq 1$ and $d \leq 2$. Notice the max values were obtaining by running k (code bits) from 4 to 32.

test code is simply recalled from the weights of the GRNN. This ability is not shown by any other classifier. The retrieved F1 measure for Hamming distance $d = 2$ is higher than for the other two methods. The average computational cost of training (in sec) is as follows: k linear SVM classifiers (0.21 ± 0.092), k GRNN classifiers (0.13 ± 0.03), k non-linear SVM classifiers (0.21 ± 0.087) and finally for one GRNN k-bits (0.03 ± 0.03). The latter cost improved by 85% w.r.t. the linear SVM classifier which shows that the proposed approach is also rather fast.

6   Conclusions

The paper shows the usefulness of the hashing technique in thousands of samples in a financial setting of data descriptors featuring the status of (bankrupt or healthy) companies. The hashing approach comprises two steps, the first one finds the binarised Laplacian EigenMap that preserves the similarity local structure of data; the second uses a generalised regression neural network with the ability to learn the k-bits code for the test samples. The retrieval of bankrupt (or healthy) companies is fast, computationally efficient and has shown to yield good performance measures. It would be interesting to use other manifold structures and differential geometry in further financial settings such as dynamic markets. Finally, besides financial risk analysis as demonstrated in the experiments, the proposed method could also be applied to other domains where high performance data mining techniques are important due to a large scale of data produced.

References

1. Andoni, A., Indyk, P.: Near-optimal hashing algorithms for approximate nearest neighbor in high dimensions. Communications of the ACM 51(1), 117–121 (2008)
2. Baluja, S., Covell, M.: Learning to hash: forgiving hash functions and applications. Data Mining and Knowledge Discovery 17, 402–430 (2008)
3. Belkin, M., Niyogi, P.: Laplacian eigenmaps for dimensionality reduction and data representation. Neural Computation 15, 1373–1396 (2002)
4. Bodo, Z., Csato, L.: Linear spectral hashing. Neurocomputing 141, 117–123 (2014)
5. Cai, D., He, X., Han, J., Huang, T.S.: Graph regularized non-negative matrix factorization for data representation. IEEE Trans. on Pattern Analysis and Machine Intelligence 33(8), 1548–1560 (2011)
6. Chang, C.C., Lin, C.J.: LIBSVM: A library for support vector machines. ACM Trans. on Intelligent Systems and Technology 2, 27:1–27:27 (2011), http://www.csie.ntu.edu.tw/~cjlin/libsvm
7. Chung, F.: Spectral Graph Theory. American Mathematical Society (1997)
8. Deerwester, S., Dumais, S.T., Furnas, G.W., Landauer, T.K., Harshman, R.: Indexing by latent semantic analysis. Journal of the American Society for Information Science 41(6), 391–407 (1990)
9. Gordo, A., Perronnin, F., Gong, Y., Lazebnik, S.: Asymmetric distances for binary embeddings. IEEE Trans. on Pattern Analysis and Machine Intelligence 36(1), 33–47 (2014)
10. Indyk, P., Motwani, R.: Approximate nearest neighbors: towards removing the curse of dimensionality. In: 30th STOC, pp. 604–613. ACM Press (1998)

11. Nene, S.A., Nayar, S.K.: A simple algorithm for nearest neighbor search in high dimensions. Tech. Rep. CUCS-030-95, CS Dep, University of Columbia, USA (1995)
12. Raginsky, M., Lazebnik, S.: Locality sensitive binary codes from shift-invariant kernels. In: Adv. in Neural Information Proc. Sys. (NIPS), pp. 1509–1517 (2009)
13. Ribeiro, B., Chen, N.: Graph weighted subspace learning models in bankruptcy. In: Proc. of Int. J. Conf. on Neural Networks (IJCNN), pp. 2055–2061. IEEE (2011)
14. Salakhutdinov, R., Hinton, G.: Semantic hashing. Int. J. Approx. Reasoning 50(7), 969–978 (2009)
15. Specht, D.F.: A general regression neural network. IEEE Transactions on Neural Networks 2(6), 568–576 (1991)
16. Weiss, Y., Torralba, A., Fergus, R.: Spectral hashing. In: Adv. in Neural Information Proc. Sys. 21 (NIPS), pp. 1753–1760 (2009)
17. Zhang, D., Wang, J., Cai, D., Lu, J.: Self-taught hashing for fast similarity search. In: Proc. of the 33rd Int. ACM SIGIR Conf. on Research and Development in Information Retrieval, pp. 18–25. ACM (2010)

MAP Inference with MRF by Graduated Non-Convexity and Concavity Procedure*

Zhi-Yong Liu, Hong Qiao, and Jian-Hua Su

State Key Laboratory of Intelligent Control and Management of Complex Systems,
Institute of Automation, Chinese Academy of Sciences, Beijing, 100190, China
zhiyong.liu@ia.ac.cn

Abstract. In this paper we generalize the recently proposed graduated non-convexity and concavity procedure (GNCCP) to approximately solve the maximum a posteriori (MAP) inference problem with the Markov random field (MRF). Unlike the commonly used graph cuts or loopy brief propagation, the GNCCP based MAP algorithm is widely applicable to any types of graphical models with any types of potentials, and is very easy to use in practice. Our preliminary experimental comparisons witness its state-of-the-art performance.

Keywords: Markov random field, Energy maximization, GNCCP.

1 Introduction

As a convenient and consistent way of modeling the context-dependent structures in images, the Markov random field (MRF) or undirected graphical model has been receiving extensive attention in computer vision and machine learning. Many problems such as image segmentation, edge detection, optical flow, active contours, and object matching can be formulated by MRF. In the context of MRF, the problem is firstly cast as a labeling problem which involves assigning each pixel/feature/region a label (we consider only discrete labeling problems in this paper), and is then accomplished by solving the inference problem of the maximum a priori (MAP) configuration, or equivalently an energy maximization problem. Although the MAP problem can be solved in polynomial time for some restricted cases, such as tree-structured MRF's and binary MRF's with submodular potentials, it is in general a non-polynomial (NP) hard problem.

In literature, many approximate MAP algorithms for MRF have been proposed since the 1970's, including typically loopy belief propagation and its variants, graph cuts, and relaxation techniques. The belief propagation and its variants [4] are optimal if the MRF's are tree-structured, or otherwise their performances decline greatly or even fail to converge [10]. The graph cuts [1] work quite successfully on low-level tasks (early vision) defined by the MRF's with regular sites and submodular potentials, especially on binary MRF's for which graph cuts are provably optimal. However, the graph cuts by definition

---

* This work was supported by the National Science Foundation of China(61375005).

C.K. Loo et al. (Eds.): ICONIP 2014, Part II, LNCS 8835, pp. 404–412, 2014.

can hardly cope with other types of graphical models with irregular sites or with non-submodular potentials.

On the other hand, the relaxation techniques relax the discrete MAP problem to be a continuous one, and then optimize it over the continuous set. One of the key advantages of relaxation techniques over the discrete methods mentioned above is that relaxations typically have no restriction on the potential type or graph structure. In this paper we are to generalize the recently proposed Graduated Non-Convexity and Concavity Procedure (GNCCP) [6], which was originally proposed to solve the assignment problem under one-to-one constraint, to approximately solve the MAP problem.

The MAP problem and some related work are briefly reviewed in the next section, and section 3 presents the proposed GNCCP MAP algorithm in detail. We discuss our experimental results in Section 4, and finally conclude this paper in Section 5.

2   Problem Formulation and Related Work

Given an MRF $G = (V, E)$ comprising a set V of vertices together with a set E of edges, and also a discrete label set $\mathcal{L} = \{1, ..., k\}$, its MAP problem is frequently studied from the perspective of energy maximization as an integer quadratic programming (IQP) as follows [12],

$$\text{max. } F(\mathbf{x}) = \mathbf{x}^\top \mathbf{W} \mathbf{x} + \mathbf{v}^\top \mathbf{x},$$
$$\text{s.t. } \mathbf{x} \in \Pi \tag{1}$$

where

$$\Pi := \{\mathbf{x}_{ia} = \{0, 1\}, \sum_a \mathbf{x}_{ia} = 1, i = 1, \ldots, n, a = 1, \ldots, k\}, \tag{2}$$

$\mathbf{W} \in R^{nk \times nk}$ with $\mathbf{W}_{ia,jb}$ describing how well the vertex i with label a agrees with the vertex j with label b, $\mathbf{v} \in R^{nk \times 1}$ with $\mathbf{v}_{ia}$ describing how well label a agrees with the data at site i. The problem is in general a NP-hard problem involving a $O(k^n)$ complexity, making consequently some approximations necessary in practice.

Below we describe briefly the approximate methods based on relaxation techniques especially the quadratic relaxation that is related to our works closely. The relaxation techniques firstly relax the discrete problem to be a continuous one to formulate a quadratic program (QP) by relaxing the set of constraints from Π to Ω as

$$\text{max.} F(\mathbf{x}), \text{s.t.} \mathbf{x} \in \Omega \tag{3}$$

where

$$\Omega := \{\mathbf{x}_{ia} \geq 0, \sum_a \mathbf{x}_{ia} = 1, i = 1, \ldots, n, a = 1, \ldots, k\} \tag{4}$$

is the convex hull of Π.

It is interesting to notice that the original integer quadratic program is equivalent to the relaxed quadratic program in the sense that both problems possess the same global maximum [11,2]. However, because the relaxed quadratic program is in general non-convex, it remains still a NP-hard problem to globally find its solution. The quadratic relaxation in [11] proposed a concave relaxation of (1) by replacing $\mathbf{W}$ and $\mathbf{v}$ respectively by $\mathbf{W} - \text{diag}(\mathbf{d})$ and $\mathbf{v} + \mathbf{d}$ where $\mathbf{d}_{ia} = \sum_{j,b} |\mathbf{W}_{ia,jb}|$ making $\mathbf{W} - \text{diag}(\mathbf{d})$ diagonally dominant and thus negative definite. In stead of constructing a concave relaxation, [5] relaxed the constraint $\sum_a \mathbf{x}_{ia} = 1$ to be a L2 one $\sum_a \mathbf{x}_{ia}^2 = 1$, which can be then solved globally for nonnegative $\mathbf{W}$ and $\mathbf{v}$, though it is still not concave. In [2] a spectral relaxation was proposed and the problem is then solved by finding the leading eigenpair. Though the global maximum of the relaxation is the same as the original discrete problem, it is not the case for the concave relaxation in [11], L2 relaxation in [5], or spectral relaxation [2], whose maximum are generally in Ω but not Π. Thus, one more step is further needed to project the solution from Ω back to Π. The MLA criterion was used by [11] and the softassign method was adopted by [5,2].

Below we generalize the GNCCP for the MAP problem, which exhibited a much better performance than softassign on graph matching problem [6].

3   GNCCP Based MRF MAP Estimation

In this section we will first generalize the GNCCP to the MAP problem, and then give some discussions on the algorithm.

3.1   GNCCP Based Algorithm

The GNCCP [6] was originally proposed to approximately solve the optimization problems defined on the set of partial permutation matrix $\mathcal{P}$, implying a one-to-one constraints different from the one-to-more one defined by Π (2). Given an IQP over $\mathcal{P}$, the GNCCP is proposed to approximately solve it by maximizing a series of objective functions as follows [1],

$$F_\zeta(\mathbf{x}) = (1 - |\zeta|)F(\mathbf{x}) + \zeta \mathbf{x}^\top \mathbf{x}, -1 \leq \zeta \leq 1, \ \mathbf{x} \in \mathcal{D}, \tag{5}$$

where $\mathcal{D}$ denotes the set of doubly stochastic matrix, i.e., the convex hull of $\mathcal{P}$, and ζ is increased from -1 gradually to 1.

Actually, (5) was shown to realize exactly a Convex-Concave Relaxation Procedure (CCRP), but without involving constructing the convex or concave relaxation, which are typically very difficult to construct [14,7,9,8]. To generalize the GNCCP to solve the MAP problem for MRF, we need just replace the domain $\mathcal{P}$ by Π, and consequently the relaxed domain $\mathcal{D}$ by Ω.

---

[1] It is noted that in its original formulation the GNCCP was proposed to handle *minimization* problem.

For each fixed ζ, the objective function $F_\zeta(\mathbf{x})$ is maximized by the Frank-Wolfe algorithm [3], by taking the previous result gotten by maximizing $F_{\zeta-d\zeta}(\mathbf{x})$ as the starting point. Specifically, the algorithm is summarized as follows,

Algorithm 3.1. GNCCP MAP ALGORITHM()

$\zeta \leftarrow -1, \mathbf{x} \leftarrow 1/m$
repeat
repeat
  $\mathbf{y} = \arg\max_\mathbf{y} \nabla F_\zeta(\mathbf{x})^\top \mathbf{y}, \text{s.t. } \mathbf{y} \in \Omega$
  $\alpha = \arg\max_\alpha F_\zeta(\mathbf{x} + \alpha(\mathbf{y} - \mathbf{x})), \text{s.t. } 0 \le \alpha \le 1$
  $\mathbf{x} \leftarrow \mathbf{x} + \alpha(\mathbf{y} - \mathbf{x})$
until converged
$\zeta \leftarrow \zeta + d\zeta$
until $\zeta > 1 \vee \mathbf{x} \in \Pi$
return $(\mathbf{x})$

In the algorithm, $\mathbf{y}$ can be efficiently found by a sequential assignment as follows. For each $i = 1..n$, find $c = \arg\max_a \nabla F_\zeta(\mathbf{x})_{ia}$, and then set $\mathbf{y}_{ia} = 1$ for $a = c$ and $\mathbf{y}_{ia} = 0$ otherwise, where $\nabla F_\zeta(\mathbf{x})$ takes the following form,

$$\nabla F_\zeta(\mathbf{x}) = (1 - |\zeta|)\nabla F(\mathbf{x}) + 2\zeta\mathbf{x}, -1 \le \zeta \le 1, \tag{6}$$

and $\nabla F(\mathbf{x}) = (\mathbf{W} + \mathbf{W}^\top)\mathbf{x} + \mathbf{v}$. α can be found in a closed form. It is also noted that in case $\mathbf{x}$ becomes discrete, which implies that $F_\zeta(\mathbf{x})$ becomes already convex, the algorithm terminates, even if ζ has not reached 1.

3.2 Discussions

Complexity Analysis. In each iteration, the computational complexity results mainly from the calculation of derivation, i.e., $\nabla F_\zeta(\mathbf{x})$, and also the line search, which typically involves the calculation of the objective function $F_\zeta(\mathbf{x})$ itself. For a dense $\mathbf{W}$, the calculation of $\nabla F_\zeta(\mathbf{x})$ or $F_\zeta(\mathbf{x})$ involves generally a $\mathcal{O}(n^2 k^2)$ complexity. However, in practice $\mathbf{W}$ is usually (quite) sparse, depending on the structure of MRF, the complexity can be further decreased. For instance, in case the smoothness energy takes a general form where different pairings of adjacent labels lead to different costs, $\mathbf{W}$ consists of at most $|E|k^2$ nonzero elements, resulting thus in a $\mathcal{O}(|E|k^2)$ complexity per iteration. For some more restricted smoothness energies such as the Potts model, the complexity is further decreased to be $\mathcal{O}(|E|k)$.

It is also worth noting that in case $\mathbf{W}$ can be specified by the adjacency matrix $\mathbf{A}$ together with one label cost matrix[2] $\mathbf{C} \in \mathbb{R}^{k \times k}$, which implies that $\mathbf{W}$ cannot be set in an arbitrary way and is the most frequently setting used by graph cuts and LBP [12], the complexity of each iteration of GNCCP becomes $\mathcal{O}(nk^2)$ and is further decreased to be $\mathcal{O}(nk)$ for Potts model.

---

[2] For instance, when $\mathbf{A}$ contains only 0 and 1, then $\mathbf{W}$ can be constructed by expanding each 0 by $\mathbf{0}$ and 1 by $\mathbf{C}$.

Error Bound Analysis. It can be shown (see [6] for details) that the GNCCP implicitly realizes the following concave relaxation

$$F_c(\mathbf{x}) := F(\mathbf{x}) - \lambda_{max}\mathbf{x}^\top\mathbf{x}, \tag{7}$$

where λ_{max} denotes the maximal eigenvalues of the Hessian matrix of $F(\mathbf{x})$. Thus, we can get the error bound of GNCCP by finding that of $F_c(\mathbf{x})$, as given by Theorem 1.

Theorem 1. *Denoting by e^* the optimal value of the IQP given by (1), and by $\mathbf{x}^* \in \Omega$ the optimal solution of the concave relaxation (7), there always exists a discrete configuration $\mathbf{y}^* \in \Pi$ resulting from $\mathbf{x}^*$ that satisfies*

$$F(\mathbf{y}^*) \geq e^* - \lambda_{max}n(1 - 1/k). \tag{8}$$

where λ_{max} denotes the maximal eigenvalue of $\mathbf{W} + \mathbf{W}^\top$.

Proof: Based on the equivalence between the original IQP (1) and relaxed QP (2), given any $\mathbf{x} \in \Omega$, we can always construct a $\mathbf{y} \in \Pi$ such that $F(\mathbf{y}) \geq F(\mathbf{x})$ by an efficient assignment strategy (see, e.g., the Proposition 2.1 in [2]). Therefore, we can get a discrete solution $\mathbf{y}^*$ from $\mathbf{x}^*$ such that $F(\mathbf{y}^*) \geq F(\mathbf{x}^*)$.

On the other hand, the optimal value of $F_c(\mathbf{x}^*)$ in (7) satisfies

$$F_c(\mathbf{x}^*) = F(\mathbf{x}^*) - \lambda_{max}\mathbf{x}^{*\top}\mathbf{x}^* + \lambda_{max}n \geq e^*$$
$$\Rightarrow F(\mathbf{x}^*) \geq e^* - \lambda_{max}n + \lambda_{max}\mathbf{x}^{*\top}\mathbf{x}^*$$
$$\geq e^* - \lambda_{max}n(1 - 1/k)$$

Thus, we have

$$\Rightarrow F(\mathbf{y}^*) \geq e^* - \lambda_{max}n(1 - 1/k). \square$$

A simple example is given below to get a feel of the bound $B_\lambda = \lambda_{max}n(1 - 1/k)$, by a comparison to that of the concave relaxation in [11], i.e., $B_d = \frac{1}{4}\sum_{ia,jb}|\mathbf{W}_{ia,jb}|$. A MRF with $n = 32 \times 32 = 1024$ nodes and two types of neighborhood systems, i.e., the first-order system with 4 neighbors and second-order with 8 neighbors are used in the example, and $\mathbf{W}$ is set as follows: $\mathbf{W}_{ia,jb} = 1$ if $\{i,j\} \in E, a = b$, or $\mathbf{W}_{ia,jb} = 0$ otherwise (Potts model). On each of the two neighbor systems the size of label set (k) increases from 2 to 10 to evaluate the two error bounds. The changes of B_λ and B_d with respect to k are plotted in Fig. 1, where $B_\lambda(4), B_d(4)$ and $B_\lambda(8), B_d(8)$ denote the error bounds on the first and second order neighbor systems respectively. It is observed that as k increases, B_d becomes larger linearly; by contrast, B_λ increases by a much smaller rate, which is specifically $(1 - 1/k)$. Therefore, on such a frequently encountered MAP problem, the error bound of GNCCP is much tighter than the one in [11].

It is also noted that the convergence of GNCCP is always guided by both the convex and concave relaxations, implying that the error bound above provides only initial estimation of the final result. However, since the objective function will become no longer convex, it is hard to analyze its error bound during the process. Here we give a typical convergence process of GNCCP shown by Figure 2, which shows how the GNCCP enhances the energy as ζ increases.

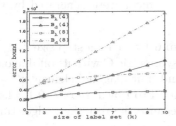

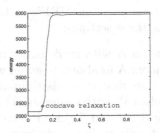

Fig. 1. Illustration on the error bounds B_λ and B_d versus the label set size, using the 4 (first) and 8 (second) order neighborhood systems. It is observed that B_λ is tighter than B_d except only for the case as $m = 2$.

Fig. 2. A typical convergence process of GNCCP based MAP. The convex relaxation is realized as $\zeta \approx -0.74$ where the energy is bound by (8). As the algorithm proceeds, the energy is further enhanced greatly.

4   Experimental Results

The proposed GNCCP MAP algorithm was evaluated by comparing with a variety of competing techniques, including ICM, max-product loopy belief propagation (LBP) [13], expansion-move graph cuts (GC_EXP) [1], and integer projected fixed point algorithm (IPFP) [5]. In particular, we evaluate their performance on the following three sets of experiments:

1. on regular MRF's with 4-neighbor and 8-neighbor systems;
2. on random MRF's with random neighbor systems, and random potentials;
3. on image segmentation.

In the first experiment we set $n = 256(16 \times 16)$, $k = 3$, the label cost $\mathbf{C} = \mathbf{I}_3$, and data cost $\mathbf{v}_{ia} = k - |d(i) - a|$. The neighbor system was set as 4-neighbor and 8-neighbor respectively, and on each of them we generated 20 images whose intensity was randomly fetched from $\{1, .., k\}$. The experimental results are listed in Table 1, where 'SEM' denotes the standard error of the mean. It is observed that on both regular RMF's GNCCP_MAP exhibited a comparable performance with GC_EXP, and both of them outperformed the other three competitors. Meanwhile, IPFP got better results than both ICM and LBP.

Table 1. The experimental results on regular MRF's

graph types	energy	ICM	LBP	GC_EXP	IPFP	GNCCP_MAP
4-neighbor	mean	611.00	637.00	790.00	753.20	789.60
	SEM	6.9314	31.2396	2.2757	7.7629	2.1937
8-neighbor	mean	1511.2	1537.9	1691.8	1680.5	1691.8
	SEM	29.053	22.653	3.0251	7.1899	3.0251

The three random MRF's in the second experiment were generated by the following three settings:

(a) random **A** with its densities increasing from 0.1 to 1 by a step size 0.1;
(b) random **A** as above, together with random symmetric **C** with its off-diagonal elements uniformly distributed with $[0, 1]$ and unit diagonal elements;
(c) random **W** with its densities increasing from 0.1 to 1 by a step size 0.1, where only IPFP and GC_EXP were evaluated because the rest three algorithms can hardly tackle a totally random **W**.

The remainder settings of the experiments were the same as those in the first experiment. The experimental results on the three MRF's are shown in Figure 3, where for a better visibility the five energies were subtracted by the minimal one (thus the minimal one becomes 0). Four observations could be summarized from the results. First, on MRF's with random **A** but regular **C**, all of the algorithms exhibited comparable performance, except for LBP on dense graphs. Second, the performance of GC_EXP declined quickly in the case of random **C**, which is in general non-submodular. Third, GNCCP_MAP showed a better performance than another relaxation technique, i.e., IPFP, on the MRF's with low dense **W**. Last, GNCCP_MAP got the best results on almost all of the random MRF's.

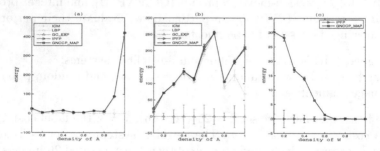

Fig. 3. Experimental results on the three random MRF's.(a): with random **A**; (b): with both random **A** and random **C**; (c) with random **W**

The image segmentation experiment was conducted on the input image (with the size 64 × 64) shown in Figure 4 , where we adopted the same setting as the first experiment by using 4-neighbor MRF. The segmented images as well as the obtained energies are shown in Figure 4. It is observed that GNCCP_MAP got the highest energy. The time-costs are plotted in Figure 4, which reveals that ICM was the fastest one, and meanwhile GNCCP_MAP was slightly faster than LBP in this experiment.

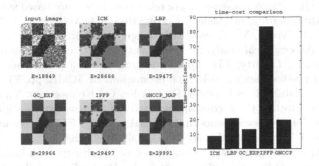

Fig. 4. Experimental results and comparative time-cost on image segmentation

5 Conclusions

In this paper we proposed the graduated non-convexity and graduated concavity procedure (GNCCP) based MAP algorithm for MRF. Compared with the discrete techniques such as loopy belief propagation and graph cuts, it can be generally used on any graphs with any types of potential. Involving only the gradient of the objective function, the GNCCP MAP algorithm is very easy to use in practice. Some experiments witnessed its simplicity and state-of-the-art performance.

References

1. Boykov, Y., Veksler, O., Zabih, R.: Fast approximate energy minimization via graph cuts. IEEE Transactions on Pattern Analysis and Machine Intelligence 23(11), 1222–1239 (2001)
2. Cour, T., Shi, J.: Solving markov random fields with spectral relaxation. Journal of Machine Learning Research - Proceedings Track, 75–82 (2007)
3. Frank, M., Wolfe, P.: An algorithm for quadratic programming. Naval Research Logistics Quarterly 3(1-2), 95–110 (1956)
4. Freeman, W., Pasztor, E., Carmichael, O.: Learning low-level vision. International Journal of Computer Vision 40(1), 25–47 (2000)
5. Leordeanu, M., Herbert, M., Sukthankar, R.: An integer projected fixed point method for graph matching and map inference. In: NIPS (2009)
6. Liu, Z.Y., Qiao, H.: Gnccp - graduated nonconvexity and concavity procedure. IEEE Transactions on Pattern Analysis and Machine Intelligence (2014), doi:10.1109/TPAMI.2013.223
7. Liu, Z.Y., Qiao, H., Xu, L.: An extended path following algorithm for graph matching problem. IEEE Transactions on Pattern Analysis and Machine Intelligence 34(7), 1451–1456 (2012)
8. Liu, Z.Y., Qiao, H., Yang, X., Hoi, C.S.: Graph matching by simplified convex-concave relaxation procedure. International Journal of Computer Vision (2014), doi:10.1007/s11263-014-0707-7

9. Liu, Z.Y., Qiao, H.: A convex-concave relaxation procedure based subgraph matching algorithm. Journal of Machine Learing Research: W&CP 25, 237–252 (2012)
10. Murphy, K.P., Weiss, Y., Jordan, M.I.: Loopy belief propagation for approximate inference: An empirical study. In: Uncertainty in Artificial Intelligence (1999)
11. Ravikumar, P., Lafferty, J.D.: Quadratic programming relaxations for metric labeling and markov random field map estimation. In: ICML, pp. 737–744 (2006)
12. Szeliski, R., Zabih, R., Scharstein, D., Veksler, O., Kolmogorov, V., Agarwala, A., Tappen, M., Rother, C.: A comparative study of energy minimization methods for markov random fields with smoothness-based priors. IEEE Transactions on Pattern Analysis and Machine Intelligence 30(6), 1068–1080 (2008)
13. Tappen, M.F., Freeman, W.T.: Comparison of graph cuts with belief propagation for stereo, using identical mrf parameters. In: Proceedings of the Ninth IEEE International Conference on Computer Vision, pp. 900–906. IEEE (2003)
14. Zaslavskiy, M., Bach, F., Vert, J.P.: A path following algorithm for the graph matching problem. IEEE Transactions on Pattern Analysis and Machine Intelligence 31(12), 2227–2242 (2009)

Two-Phase Approach to Link Prediction

Srinivas Virinchi * and Pabitra Mitra

Dept of Computer Science and Engineering,
Indian Institute of Technology
Kharagpur-721302, India
{virinchimnm,pabitra}@cse.iitkgp.ernet.in

Abstract. Link prediction deals with predicting edges which are likely
to occur in the future. The clustering coefficient of sparse networks is
typically small. Link prediction performs poorly on networks having low
clustering coefficient and it improves with increase in clustering coefficient. Motivated by this, we propose an approach, wherein, we add
relevant non-existent edges to the sparse network to form an auxiliary
network. In contrast to the classical link prediction algorithm, we use
the auxiliary network for link prediction. This auxiliary network has
higher clustering coefficient compared to the original network. We formally justify our approach in terms of Kullback-Leibler (KL) Divergence
and Clustering Coefficient of the social network. Experiments on several
benchmark datasets show an improvement of upto 15% by our approach
compared to the standard approach.

Keywords: Graph Mining, Local Similarity, KL Divergence, Clustering
Coefficient, Power-law degree distribution.

1 Introduction

A social network may be viewed as a graph where the nodes of the graph represent users of the social network and the edges correspond to the relationship
between the users. The link prediction problem, can be formally stated as, given
a network $G_t = (V, E_t)$ at the current time instance t, we need to predict the
new links added to G_t to form $G_{t'} = (V, E_{t'})$ for $t' > t$. Here, V is the set of
nodes in the network. E_t and $E_{t'}$ are the set of edges of the network at time t
and t' respectively.

Computing similarity between two unconnected nodes is essential as similar
nodes tend to form new links in the future. [1], [2] and [3] present a survey of
various similarity measures used in link prediction. They conclude that often
network topology is adequate to extract the potential future interactions. [4], [5]
and [6] present some of the popular local similarity measures used in link prediction. Common Neighbors (CN) computes the similarity between two nodes as the
number of common neighbors they share. In [4] and [6] the Adamic-Adar (AA)

---

* Corresponding author.

C.K. Loo et al. (Eds.): ICONIP 2014, Part II, LNCS 8835, pp. 413–420, 2014.

and Resource-Allocation (RA) similarity measures have been proposed respectively; similarity between two nodes is equal to the sum of inverse of a measure of the degree of each common neighbor. The difference is that RA penalizes the high-degree common nodes more than AA.

We formally describe how the score between two nodes x and y is computed using these similarity measures where $x, y \in V$ and $(x, y) \notin E_t$. Let $N(x)$ be the set of nodes adjacent to x.

Common Neighbors (CN) [5] : $CN(x, y) = | N(x) \cap N(y) |$

Adamic-Adar Index (AA) [4] : $AA(x, y) = \sum_{z \in N(x) \cap N(y)} \frac{1}{log(degree(z))}$

Resource-Allocation Index (RA) [6]: $RA(x, y) = \sum_{z \in N(x) \cap N(y)} \frac{1}{degree(z)}$

In [7] they show that embedding community information enhances the performance of the link prediction algorithms. In [8] a modification to the existing similarity measures has been proposed which exploits the power-law degree distribution which gives a higher weightage to the low-degree common neighbors and lower weightage to the higher degree common neighbors compared to the existing similarity.

Social networks are typically sparse in nature. The Clustering Coefficient (CC) [9] is a good indicator of density of the network. Hence, using the existing similarity measures between nodes in a sparse network works as a major challenge; the connectivity structure of a sparse graph does not provide sufficient local neighborhood information.

According to [10], in principle new links are created which are likely to form cliques or near-cliques in a given network. Increase in the number of cliques increases the Clustering Coefficient [9] of the network. Thus, new links are added in such a way that the CC of the network increases. In this paper, we propose an approach for predicting links by paying attention to CC. Motivated by this idea, in our work, we add some relevant non-existent edges to the sparse graph G_t which gives us an auxiliary graph $\hat{G}_t^*$. We use the connectivity structure of $\hat{G}_t^*$ in computing the similarity between the node pairs for predicting new links. We observed an improvement of upto 18% in classification accuracy on standard benchmark datasets using the proposed approach. Our specific contributions in this paper are as follows:

1. We show how link prediction can be performed on sparse networks by exploiting Clustering Coefficient.
2. We formulate our approach as an optimization problem involving KL Divergence and Clustering Coefficient by exploiting the power-law degree distribution of the networks.
3. The generality of our approach makes it feasible to use it along with any similarity measure for link prediction.

The rest of the paper is organized as follows: in section 2 we formally introduce the required background. In Section 3, we present our approach. We explain the experimental methodology in 4 and discuss the results in section 5. Finally we conclude in section 6.

2   Background

According to the *power-law degree distribution* [9], the probability of finding a k degree node in the network, denoted by p_k, is directly proportional to $k^{-\alpha}$ where, α is some positive constant.

Kullback-Leibler (KL) Divergence is used in the context of link prediction to measure the distance between degree distributions of the networks. If the KL Divergence is low, it means the distributions are very similar and vice-versa. The KL divergence between graphs G_1 and G_2 having probability (degree) distributions q and p respectively denoted by D_{KL} can be calculated as follows:

$$D_{KL}(p||q) = \sum_x p(x)log\frac{p(x)}{q(x)}$$

Note that KL Divergence measure takes two arguments as input. However, in the rest of the paper we assume that one of the arguments is fixed; so, we explicitly indicate one argument as input. The fixed argument is the degree distribution of $G_{t'}$. Hence, in the rest of the paper, KL(G) represents the KL Divergence between the degree distributions of graph G and $G_{t'}$.

Clustering Coefficient (CC) [9] is the average of the local CC of the nodes. The local CC of a node x is the fraction of the number of links between the neighbors of node x to the maximum possible number of links between them.

3   Motivation and Proposed Approach

In the link prediction problem, we need to predict the edges that are present in $G_{t'}$ but not in G_t. In general, the link prediction algorithm uses the structure of graph G_t to predict new links. Instead, it would be better to add relevant edges to G_t to form an auxiliary graph $\hat{G}_t^*$; $\hat{G}_t^*$ will have smaller KL value compared to G_t and larger than that of $G_{t'}$. Similarly, $\hat{G}_t^*$ will have higher CC value compared to G_t and smaller than that of $G_{t'}$. Thus, we can use $\hat{G}_t^*$ to predict the edges of the graph $G_{t'}$ more effectively as $\hat{G}_t^*$ has more relevant edges compared to G_t. Theoretically, we show the existence of such a G_t^* next. Note that G_t^* is theoretical; in practice we end up with some $\hat{G}_t^*$ using the proposed approach.

3.1   Existence of G_t^*

As discussed above, we need to find a G_t^* which satisfies the following constraints:

$$CC(G_t) < CC(G_t^*) < CC(G_{t'})$$
$$KL(G_t) > KL(G_t^*) > KL(G_{t'})$$

Consider the following notation:

* $KL1$ - KL Divergence of G_t (KL(G_t))
* α, β - Power-law coefficients of graph $G_{t'}$ and G_t^* respectively

* $p(x), q(x)$ - Degree Distribution of $G_{t'}$ and G respectively; $p(x) \propto x^{-\alpha}$, $q(x) \propto x^{-\beta}$

In this section, we show the theoretical existence of an optimal G_t^* which corresponds to the solution of the **optimization problem**:

$$G_t^* = \operatorname*{argmax}_{G} \quad CC(G)$$

$$\text{subject to} \quad KL(G) + \epsilon \le KL1$$

Solution: The Lagrangian for the above problem can be written as,

$$L(\beta, \lambda) = CC(G) + \lambda(KL1 - \epsilon - KL(G)) \tag{1}$$

From equation 1, we can observe that λ is the balancing factor which controls the rate of increase of CC value and rate of decrease of KL Divergence. So, we need to find the optimal λ.

For sparse graphs, the clustering coefficient correlates negatively with the degree [11] i.e., in specific it varies as k^{-1} for a k degree node and the network follows a power-law degree distribution. Thus, we can write the CC value as follows :

$$CC(G) = \int_{k=2}^{k=\infty} \frac{q(k)}{k} dk = \int_{k=2}^{k=\infty} \frac{k^{-\beta}}{k} dk = \frac{k^{-\beta}}{-\beta}\bigg|_2^{\infty} = \frac{1}{\beta 2^{\beta}} \tag{2}$$

Similarly, we can solve for KL Divergence of G with respect to $G_{t'}$ as follows:

$$KL(G) = \int_{k=2}^{k=\infty} p(k) log \frac{p(k)}{q(k)} dk = (\beta - \alpha)[-\frac{k^{1-\alpha}((\alpha-1)logk + 1)}{(\alpha-1)^2}\bigg|_2^{\infty}]$$

We can simplify the above equation as follows:

$$KL(G) = \frac{(\beta - \alpha)\alpha}{(\alpha-1)^2 \, 2^{\alpha-1}} = (\beta - \alpha)C, \tag{3}$$

where $C = \frac{\alpha}{(\alpha-1)^2 \, 2^{\alpha-1}}$. From equations 1, 2 and 3 we get,

$$L(\beta, \lambda) = \frac{1}{\beta \, 2^{\beta}} + \lambda(KL1 - \epsilon - (\beta - \alpha)C) \tag{4}$$

Considering $\frac{\partial L}{\partial \lambda} = 0$ and simplifying for optimal β (β^*) we get,

$$\beta^* = \alpha + \frac{(KL1 - \epsilon)}{C} \tag{5}$$

Considering $\frac{\partial L}{\partial \beta} = 0$ and equation 5 and simplifying for optimal λ (λ^*) we get,

$$\lambda^* = \frac{(1 + \beta^*)}{C\beta^{*2} \, 2^{\beta^*}} \tag{6}$$

From equations 5 and 6 we can write λ^* in terms of α and KL1 only and thus we can find λ^* so that we can maximize CC(G) and minimize KL(G) and thus end up with optimal G_t^* having degree distribution coefficient β^*. This shows that there exists a G_t^* which satisfies the constraints.

3.2   Two-Phase Link Prediction Algorithm

We propose a link prediction algorithm for sparse networks. Let $lp\text{-}factor$ ($0 \leq lp\text{-}factor \leq 1$) represent the fraction of edges we want to add to graph G_t to form $\hat{G}_t^*$. Let $score$ (any of CN, AA or RA) be a measure which takes two nodes x and y as input and returns the similarity between x and y. Let lp represent the number of links the user wants to predict. The steps of our approach are as follows:

1. $\forall (x,y) \notin E_t$ compute $score(x,y)$ to form the edge set E.
2. Sort E based on the descending order of score.
3. Add the top $lp\text{-}factor \times lp$ number of edges from E to G_t to form $\hat{G}_t^*$.
4. $\forall (x,y) \notin E_t$ compute $score(x,y)$ using $\hat{G}_t^*$ instead of G_t. ((x,y) may exist in E but we still compute $score(x,y)$).
5. Output the top lp edges.

In contrast to the standard link prediction algorithm, steps 3 and 4 are the extra steps in our proposed algorithm. Steps 3 and 4 together form the *second phase* of the *Two-Phase* algorithm. Let CN^2, AA^2 and RA^2 refer to the modifications of CN, AA and RA respectively using the *Two-Phase algorithm*.

4   Dataset and Experimental Methodology

For our experiment, we consider collaboration graphs from [12], where, each node is an author and an edge represents collaboration between them. We present relevant statistics on the datasets in table 1.

Table 1. Graph Datasets

Dataset	$|V|$	$|E_{t'}|$	$|E_t|$
GrQc	5241	14484	2896
HepTh	9875	25973	5194
AstroPh	18771	198050	39610
CondMat	23133	93439	18687

We randomly partition the graph into five parts by removing the edges randomly where each part has 20% of the edges. Now, we use one part as training data (G_t) and the remaining part for testing ($G_{t'}$) to examine the efficacy of our algorithm on sparse networks. We repeat this five times each time taking a different part to be the test data; we report the average values. We are interested in predicting only the top 10% and 20% of the missing links similar to the experimental setup used in [1] and [7]. In this context, accuracy is defined as the fraction of correctly predicted edges against the total number of edges predicted.

$$\text{Accuracy} = \frac{\text{No. of correctly predicted edges}}{\text{No. of edges predicted}}$$

5   Results and Discussion

We observe an improvement in results on all the datasets. For the results shown in table 2 we have set $lp\text{-}factor$ to 0.2 for CN^2 and to 0.7 for AA^2 and RA^2. CN^2 shows good improvement compared to CN on all the datasets. It shows a maximum improvement of 15% when predicting the top 10% edges on hep-th dataset. For predicting the top 20% edges using CN^2, we observe that we have a maximum improvement of upto 11% on the cond-mat dataset. Similarly, AA^2 shows good improvement compared to AA on all the datasets. It shows a maximum improvement of 5% when predicting the top 10% edges on both hep-th and cond-mat dataset. For predicting the top 20% edges using AA^2, we observe that we have a maximum improvement of upto 8% on the hep-th dataset. Similarly, RA^2 shows good improvement compared to RA on all the datasets. It shows a maximum improvement of 8% when predicting the top 10% edges on gr-qc, hep-th and astro-ph dataset. For predicting the top 20% edges using RA^2, we observe that we have a maximum improvement of upto 11% on the astro-ph dataset.

Table 2. Accuracy on predicting links

Dataset	Predicting 10%						Predicting 20%					
	CN	CN²	AA	AA²	RA	RA²	CN	CN²	AA	AA²	RA	RA²
gr-qc	60.44	**61.33**	68.48	**70**	61.29	**69.67**	35.3	**36.27**	42.08	**43.18**	41.6	**44.18**
hep-th	47	**62.5**	47.02	**52.2**	45.76	**51.1**	33.03	**42.59**	44.19	**52.05**	40.47	**45.61**
cond-mat	59.15	**61.36**	67.77	**72.8**	67.76	**71.44**	38.09	**49.56**	60.22	**66.17**	58.16	**63.45**
astro-ph	83.51	**83.8**	88.42	**93.93**	78.64	**84.7**	68.13	**72.51**	78.97	**84.94**	69.05	**81.12**

Table 3. CC and KL values

		Predicting 10% links				Predicting 20% links			
		gr-qc	hep-th	astro-ph	cond-mat	gr-qc	hep-th	astro-ph	cond-mat
CN^2	CC(G)	0.056	0.4735	0.106	0.514	0.06245	0.0585	0.125	0.1205
	KL(G)	0.2945	0.414	0.4095	0.585	0.3155	0.4245	0.3799	0.5105
	$L(\beta,\lambda)$	**0.043**	**0.044**	**0.123**	**0.533**	**0.0476**	**0.053**	**0.1574**	**0.1538**
AA^2	CC(G)	0.358	0.3695	0.43	0.506	0.418	0.489	0.504	0.5785
	KL(G)	0.171	0.242	0.213	0.2385	0.1365	0.1265	0.1005	0.075
	$L(\beta,\lambda)$	**0.359**	**0.394**	**0.447**	**0.623**	**0.425**	**0.538**	**0.655**	**0.72**
RA^2	CC(G)	0.381	0.379	0.488	0.514	0.491	0.491	0.535	0.581
	KL(G)	0.1325	0.2245	0.198	0.243	0.066	0.1285	0.105	0.073
	$L(\beta,\lambda)$	**0.388**	**0.4071**	**0.611**	**0.63**	**0.51**	**0.5407**	**0.653**	**0.727**
G_t	CC(G)	0.047	0.032	0.08	0.057	0.047	0.032	0.08	0.057
	KL(G)	0.18	0.3895	0.212	0.674	0.18	0.3895	0.441	0.674
	$L(\beta,\lambda)$	0.047	0.032	0.08	0.057	0.047	0.032	0.08	0.057
$G_{t'}$	CC(G)	0.5297	0.471	0.63	0.633	0.5297	0.471	0.63	0.633

We framed our approach as an optimization problem where we maximize the CC value of the graph and minimize the KL value of the graph. We showed earlier that there exists an optimal graph $\hat{G}_t^*$ which maximizes the required value $L(\beta, \lambda)$ which are shown in table 3. For the results in table 3, we use the value of $lp\text{-}factor$ to 0.2 for CN and 0.7 for AA and RA respectively. In the table, we show CC(G) and KL(G) which represent the CC and KL value of the graph $\hat{G}_t^*$ which we attain using CN^2, AA^2 and RA^2. From the table, we observe that the value of $L(\beta, \lambda)$ increases for $\hat{G}_t^*$ which we attain by using CN^2, AA^2 and RA^2. We observe that CC value increases most for AA^2 and RA^2 and the KL value is the least for them when compared to CN^2. Hence, on most of the datasets, we observe AA^2 and RA^2 having higher $L(\beta, \lambda)$ value compared to CN^2. Further, it shows significant improvement compared to base link prediction methods. We also observe that by using our approach (CN^2, AA^2 or RA^2) we increase the $L(\beta, \lambda)$ value when compared to the graph G_t.

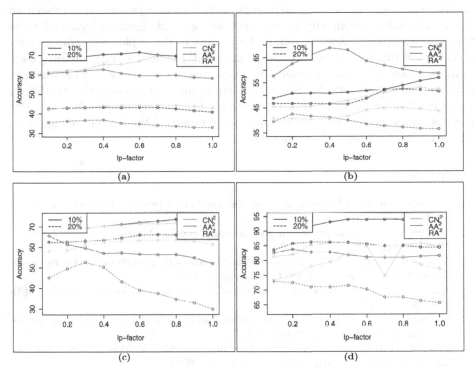

Fig. 1. Accuracy vs lp-factor plots. (a) gr-qc dataset. (b) hep-th dataset. (c) cond-mat dataset. (d) astro-ph dataset.

We observe from our results that $lp\text{-}factor$ is an important parameter in the proposed approach. We show the results by plotting accuracy versus $lp\text{-}factor$ on various datasets on predicting top 10% and 20% links in figure 1. In general, we observe that the accuracy varies by varying $lp\text{-}factor$ uniformly across all the datasets. We observe that for CN^2, AA^2 and RA^2 the accuracy reduces as the value of $lp\text{-}factor$ increases. In particular, on most of the datasets, we observe that the accuracy of CN^2 drops at $lp\text{-}factor = 0.2$. In a similar manner, AA^2 and RA^2 show a fall in accuracy for $lp\text{-}factor = 0.7$. The reason for the early fall of accuracy for CN^2 is that CN similarity measure is not as effective as AA and RA in terms of accuracy. So, we set the $lp\text{-}factor$ for CN^2 to a smaller value when compared to AA^2 and RA^2. *This indicates that CN is more effective on sparse graphs while AA and RA are effective on denser graphs.* From figure 1, we observe that as the $lp\text{-}factor$ increases the accuracy increases till a particular point and then starts decreasing which shows a direct connection between $lp\text{-}factor$ and accuracy similar to overfitting.

6   Conclusion

In this paper, we presented a two-phase approach to link prediction which shows significant improvement compared to the standard link prediction approach on sparse networks. Specifically, in our approach we add similar relevant edges to the existing sparse network and make it denser. We exploit the connectivity structure of this dense network for effective link prediction. The exhaustive experimentation using our approach demonstrates the superiority and robustness of our approach. Specifically, we show a significant improvement of upto 15% on benchmark datasets. In the future, we would like to test for the efficacy of this method in terms of increase in the number of phases of the two-phase algorithm to form a multi-phase link prediction.

References

1. Liben-Nowell, D., Kleinberg, J.: The link-prediction problem for social networks. In: Proc. of CIKM (2003)
2. Lü, L., Zhou, T.: Link prediction in complex networks: A survey. Physica A 390(6), 1150–1170 (2011)
3. Al Hasan, M., Zaki, M.J.: A survey of link prediction in social networks. In: Social Network Data Analytics, pp. 243–275. Springer (2011)
4. Adamic, L.A., Adar, E.: Friends and neighbors on the web. Social Networks 25(3), 211–230 (2003)
5. Newman, M.E.J.: Clustering and preferential attachment in growing networks. Physical Review E 64(2), 025102 (2001)
6. Zhou, T., Lü, L., Zhang, Y.: Predicting missing links via local information. The European Physical Journal B 71(4), 623–630 (2009)
7. Soundarajan, S., Hopcroft, J.: Using community information to improve the precision of link prediction methods. In: Proc. of WWW (2012)
8. Virinchi, S., Mitra, P.: Similarity measures for link prediction using power law degree distribution. In: Lee, M., Hirose, A., Hou, Z.-G., Kil, R.M. (eds.) ICONIP 2013, Part II. LNCS, vol. 8227, pp. 257–264. Springer, Heidelberg (2013)
9. Newman, M.E.J.: Networks: an introduction. Oxford University Press (2009)
10. Liu, Z., He, J., Srivastava, J.: Cliques in complex networks reveal link formation and community evolution. arXiv preprint arXiv:1301.0803 (2013)
11. Bloznelis, M.: Degree and clustering coefficient in sparse random intersection graphs. The Annals of Applied Probability 23(3), 1254–1289 (2013)
12. Leskovec, J., Kleinberg, J., Faloutsos, C.: Graph evolution: Densification and shrinking diameters. ACM TKDD 1(1), 2 (2007)

Properties of Direct Multi-Step Ahead Prediction of Chaotic Time Series and Out-of-Bag Estimate for Model Selection

Shuichi Kurogi, Ryosuke Shigematsu, and Kohei Ono

Kyushu Institute of technology, Tobata, Kitakyushu, Fukuoka 804-8550, Japan
kuro@cntl.kyutech.ac.jp, shigematsu@kurolab.cntl.kyutech.ac.jp
http://kurolab.cntl.kyutech.ac.jp/

Abstract. This paper examines properties of direct multi-step ahead (DMS) prediction of chaotic time series and out-of-bag (OOB) estimate of the prediction performance for model selection. Although previous studies of DMS estimation suggest that the DMS technique allows us accuracy improvements from iterated one-step ahead (IOS) prediction. However, it has not considered chaotic time series which has long-term unpredictability as well as short-term predictability, where the boundary of the horizon of long-term and short-term is not known previously. As a result of the model selection, the CAN2 with a large number of units are selected, which is supposed to be useful for avoiding unpredictable data of chaotic time series. We examine the relationship between the OOB prediction and the prediction for the test data, and we suggest that there is a mixed distribution of very small and very big magnitude of prediction errros owing to chaotic time series. We show the effectiveness and the properties of the present method by means of numerical experiments.

Keywords: Direct multi-step ahead prediction, chaotic time series, competitive associative net, out-of-bag estimate, model selection.

1 Introduction

This paper examines the property of direct multi-step ahead (DMS) prediction of chaotic time series (see [1]) and out-of-bag (OOB) estimate of the prediction performance for model selection. A previous study of DMS estimation [2] suggests that the DMS technique allows us accuracy improvements from iterated one-step ahead (IOS) prediction. However, the study has not considered chaotic time series which has long-term unpredictability as well as short-term predictability and the boundary of the horizon of long-term and short-term is not known previously. Thus, it is one of the problems of DMS prediction to avoid unpredictable data of chaotic time series for learning and prediction.

On the other hand, we have studied moments of predictive deviations as ensemble diversity for model selection in time series prediction [3], where we execute IOS prediction of chaotic time series by using bagging CAN2 (bagging competitive associative net). Here, the CAN2 is an artificial neural net intended for learning piecewise linear approximation of nonlinear function, and the bagging CAN2 is a bagging version [4,5] of the CAN2. We have shown that the bagging CAN2 has achieved good predictions

C.K. Loo et al. (Eds.): ICONIP 2014, Part II, LNCS 8835, pp. 421–428, 2014.

and the model selection method shows better performance than the conventional hold-out model selection method. However, a problem remains. Namely, it is hard to decide whether a prediction obtained by the selected model is good (without diverged error) or not, although the predictions are good on average for different initial states and different horizons.

As an another approach to time series prediction and model selection, we examine the DMS prediction and OOB estimate of the prediction performance in this paper. Here, the OOB estimate is a method to evaluate the generalization performance of the prediction and known to have smaller bias than the K-fold cross validation [6]. In the next section, we formulate the prediction of chaotic time series. In **3**, we describe the CAN2 and OOB estimate. In **4**, we show numerical experiments and analysis.

2   Prediction of Chaotic Time Series

Let y_t ($\in \mathbb{R}$) denote a chaotic time series for a discrete time $t = 0, 1, 2, \cdots$ satisfying

$$y_t = r(\boldsymbol{x}_t) + e(\boldsymbol{x}_t) \tag{1}$$

where $r(\boldsymbol{x}_t)$ is a nonlinear target function of a vector $\boldsymbol{x}_t = (y_{t-1}, y_{t-2}, \cdots, y_{t-k})^T$. Here, we suppose that y_t can be obtained not analytically but numerically, and then y_t involves an error $e(\boldsymbol{x}_t)$. The embedding dimension k should be selected properly (see the theory of chaotic time series [1] for details). By means of applying the above equation recursively, we have the value of y_{t+K} for $K = 0, 1, 2, \cdots$ as a function of $\boldsymbol{x}_t$. Namely, we have a function $r_K(\cdot)$ given by

$$y_t = r_K(\boldsymbol{x}_{t-K}) + e_K(\boldsymbol{x}_{t-K}), \tag{2}$$

where $e_K(\boldsymbol{x}_{t-K})$ represents propagation error through K times of recursive numerical computation of (2). Here, from short-term predictability and long-term unpredictability of chaotic time series, $e_K(\boldsymbol{x}_{t-K})$ remains small for short-term with small K but increases exponentially after the short-term. Now, let $y_{t:h} = y_t y_{t+1} \cdots y_{t+h-1}$ denote a time series with the initial time t and the horizon h. For a given and training time series $y_{t_g:n_g}$, we are supposed to predict succeeding time series $y_{t_p:h_p}$ for $t_p \geq t_g + h_g$. Then, we can make the training dataset $D^n = \{(\boldsymbol{x}_t, y_t) | t \in I^n\}$ for $n = h_g - k$ and $I^n = \{t | t_g + k \leq t < t_g + h_g\}$, and train a learning machine. Then, with the prediction function $f_K(\cdot)$ of the machine after the learning, we can execute iterated prediction as

$$\hat{y}_t = f_K(\hat{\boldsymbol{x}}_{t-K}) \tag{3}$$

for $t = t_p, t_p + 1, \cdots$, recursively, where the elements of $\hat{\boldsymbol{x}}_t = (x_{t1}, x_{t2}, \cdots, x_{tk})$ is given by

$$x_{tj} = \begin{cases} y_{t-j} \ (t - j < t_p) \\ \hat{y}_{t-j} \ (t - j \geq t_p). \end{cases} \tag{4}$$

Here, we suppose that y_t for $t < t_p$ is known for the prediction. In the following, the prediction by (3) is called iterated direct multi-step ahead (IDMS) prediction which involves IOS prediction with $K = 0$ for $t = t_p, t_p + 1, \cdots$, and DMS prediction with $K > 0$ for $t = t_p, t_p + 1, \cdots, t_p + K$.

3   Single CAN2, Bagging CAN2 and OOB Estimate

A single CAN2 has N units. The jth unit has a weight vector $\boldsymbol{w}_j \triangleq (w_{j1}, \cdots, w_{jk})^T \in \mathbb{R}^{k\times 1}$ and an associative matrix (or a row vector) $\boldsymbol{M}_j \triangleq (M_{j0}, M_{j1}, \cdots, M_{jk}) \in \mathbb{R}^{1\times(k+1)}$ for $j \in I^N \triangleq \{1, 2, \cdots, N\}$. The CAN2 after learning a training dataset $D^n = \{(\boldsymbol{x}_t, y_t)| t \in I^n\}$ approximates the target function $r(\boldsymbol{x}_t)$ by

$$\widehat{y}_t = \widetilde{y}_{c(t)} = \boldsymbol{M}_{c(t)}\widetilde{\boldsymbol{x}}_t, \tag{5}$$

where $\widetilde{\boldsymbol{x}}_t \triangleq (1, \boldsymbol{x}_t^T)^T \in \mathbb{R}^{(k+1)\times 1}$ denotes the (extended) input vector to the CAN2, and $\widetilde{y}_{c(t)} = \boldsymbol{M}_{c(t)}\widetilde{\boldsymbol{x}}_t$ is the output value of the $c(t)$th unit of the CAN2. The index $c(t)$ indicates the unit who has the weight vector $\boldsymbol{w}_{c(t)}$ closest to the input vector $\boldsymbol{x}_t$, or $c(t) \triangleq \underset{j\in I^N}{\text{argmin}} \, \|\boldsymbol{x}_t - \boldsymbol{w}_j\|$. The above function approximation partitions the input space $V \in \mathbb{R}^k$ into the Voronoi (or Dirichlet) regions

$$V_j \triangleq \{\boldsymbol{x} \mid j = \underset{i\in I^N}{\text{argmin}} \, \|\boldsymbol{x} - \boldsymbol{w}_i\|\} \tag{6}$$

for $j \in I^N$, and performs piecewise linear prediction for the function $r(\boldsymbol{x})$. Note that we have developed an efficient batch learning method shown in [7], which we have used in this application.

The the bagging CAN2 is obtained as follows (see [4,6] for details); let $D^{n\alpha^\#,j} = \{(\boldsymbol{x}_t, y_t)| \ t \in I^{n\alpha^\#,j}\}$ be the jth bag (multiset, or bootstrap sample set) involving $n\alpha$ elements, where the elements in $D^{n\alpha^\#,j}$ are resampled randomly with replacement from the training dataset D^n. Here, $\alpha \ (> 0)$ indicates the bag size ratio to the given dataset, and $j \in J^{\text{bag}} \triangleq \{1, 2, \cdots, b\}$. Here, note that $\alpha = 1$ is used in many applications (see [5],[6]), but we use variable α for improving generalization performance (see [6] for the effectiveness and the validity), and we use $\alpha = 2$ in the experiments shown below. Statistically, the data in D^n is out of the bag $D^{n\alpha^\#,j}$ with the probability $(1 - 1/n)^\alpha \simeq e^{-\alpha}$. Inversely, the data in D^n is in $D^{n\alpha^\#,j}$ with the probability $1 - e^{-\alpha} \simeq 0.632$ for usual $\alpha = 1$ and 0.865 for $\alpha = 2$, and they are used for training the learning machine. With multiple learning machines $\theta^j \ (\in \Theta^{\text{bag}} \triangleq \{\theta^j | j \in J^{\text{bag}}\})$ which have learned $D^{n\alpha^\#,j}$, the bagging for estimating the target value $r_t = r(\boldsymbol{x}_t)$ is done by

$$\widehat{y}_t^{\text{bag}} \triangleq \widehat{y}^{\text{bag}}(\boldsymbol{x}_t) \triangleq \frac{1}{b}\sum_{j\in J^{\text{bag}}} \widehat{y}_t^j \equiv \left\langle \widehat{y}_t^j \right\rangle_{j\in J^{\text{bag}}} \tag{7}$$

where $\widehat{y}_t^j \triangleq \widehat{y}^j(\boldsymbol{x}_t)$ denotes the prediction by the jth machine θ^j. The angle brackets $\langle\cdot\rangle$ indicate the mean, and the subscript $j \in J^{\text{bag}}$ indicates the range of the mean. For simple expression, we sometimes use $\langle\cdot\rangle_j$ instead of $\langle\cdot\rangle_{j\in J^{\text{bag}}}$ in the following.

The OOB estimate is used to estimate the generalization error of bagging ensemble as follows (see [6] for details): For the given training dataset $D^n = \{(\boldsymbol{x}_t, y_t)|t \in I^n\}$, we define the out-of-bag prediction error by the RMSE (root mean square prediction error)

$$L^{\text{ob}} \triangleq \left(\langle (\widehat{y}_t^{\text{ob}} - y_t)^2 \rangle_{t\in I^n} \right)^{1/2}, \tag{8}$$

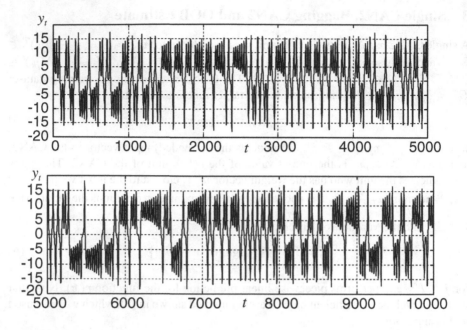

Fig. 1. Lorenz time series $y(t)$ for $t = 0, 1, 2, \cdots, 10000$

where $\hat{y}_t^{\text{ob}} \triangleq \langle \hat{y}_t^j \rangle_{j \in J^{\text{ob}}}$ is the mean of the out-of-bag predictions $\hat{y}_t^j = \hat{y}^j(\boldsymbol{x}_t)$ by the ensemble members which have learned the bags not involving the tth data $(\boldsymbol{x}_t, y_t)$.

4   Numerical Experiments and Analysis

4.1   Experimental Settings

As a chaotic time series, we employ Lorenz time series given by

$$\frac{dx}{dt_c} = -\sigma x + \sigma y, \quad \frac{dy}{dt_c} = -xz + rx - y, \quad \frac{dz}{dt_c} = xy - bz, \quad (9)$$

for $\sigma = 10$, $b = 8/3$, $r = 28$ (see [1]). Here, we use t_c for continuous time and $t \ (= 0, 1, 2, \cdots)$ for discrete time related by $t_c = tT$ with the sampling period T. We have generated 10,000 data points from the initial state $(x(0), y(0), z(0)) = (-8, 8, 27)$ with the sampling period $T = 25$ms via Runge-Kutta method with 128 bit precision of GMP (GNU multi-precision library). We use $y(t)$ for the time series to be processed (see Fig. 1). Here, note that we have observed three time series generated with $T = 250$ms, 25ms and 2.5ms, respectively, and they are all the same until 20s and the latter two time series with $T = 25$ms and 2.5ms are the same until 30s, while the difference increases exponentially after then. Furthermore, with the precision less than 128 bit, the difference of the above time series increases after shorter duration of time. This result is considered to be related to the property of chaotic time series with short-term predictability and long-term unpredictability owing to finite computational precision.

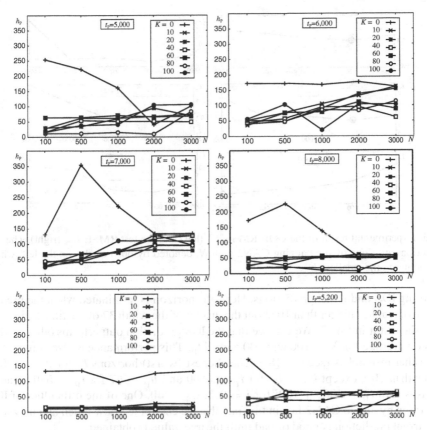

Fig. 2. Experimental result of predictable horizon h_{p} vs. the number of units N of the CAN2 by IDMS prediction for different prediction delay K and different prediction start time t_{p}. The line segments for $K = 0$ indicate the IOS prediction, while the data points for $K > 0$ satisfying $h_{\mathrm{p}} \leq K + 1$ indicate DMS prediction and $h_{\mathrm{p}} > K + 1$ indicate iterated DMS prediction.

From another point of view, the result indicates that the computational error by Runge-Kutta method decreases by reducing the sampling period, and $y(t)$ for each duration of time less than 1200 steps (=30s/25ms) in Fig. 1, or $y_{t_0:1200}$ for each initial time $t_0 = 0, 1, 2, \cdots$ with initial state $(x(t_0), y(t_0), z(t_0))$, is supposed to be almost correct, while cumulative computational error may increase exponentially after the duration.

We show the results for the embedding dimension being $k = 10$, the given and training time series being $y_{t_{\mathrm{g}}:h_{\mathrm{g}}} = y_{0:5000}$, and the multi-step ahead prediction of $y_{t_{\mathrm{p}}:n_{\mathrm{p}}}$ with the prediction start time $t_{\mathrm{p}} = 5000, 6000, 7000, 8000, 9000$ and 5200, and prediction horizon $h_{\mathrm{p}} = 10, 20, 40, \cdots, 100$.

4.2   Results and Analysis

Predictable Horizon By IDMS Prediction Using Single CAN2. Using single CAN2, we have examined the predictable horizon by means of IDMS predictions using (3)

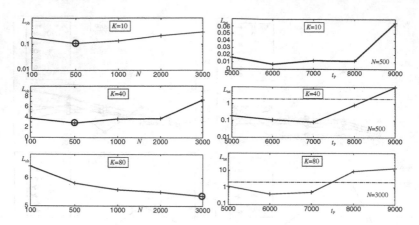

Fig. 3. Experimental result of the OOB RMSE L^{ob} (left) and the test RMSE L_{test} (right). The test predictions are done by the bagging CAN2 with N, denoted by open circles on the left, which have achieved the smallest L^{ob}.

involving IOS and DMS predictions. Here, the horizon is terminated when the prediction error reaches bigger than 10 about the quarter of the width 37 of the time series. We show the result in Fig. 2. We can see that the IOS prediction outperforms other predictions on average for $N = 100$ and 500 and all t_p. This performance is also competitive with other researches (see e.g. [8]). The shortest (worst) horizons t_p for $N = 500$ is bigger than 100, except $h_p = 66$ for $t_p = 5200$ and $h_p = 67$ for $t_p = 5400$ among 41 cases with $t_p = 5000 + 100i$ for $i = 0, 1, 2, \cdots, 40$. One of the difficulties of IOS prediction by single CAN2 lies in this fact that we cannot tell how much and how long the current prediction is good or bad until the true value is obtained.

From another point of view, we empirically have good function approximation results with the number of units N less than $n/k (\simeq 500$ in this case) for the number of training data $n (= h_g - k \simeq h_g = 5,000)$ with the embedding dimension $k (= 10)$ because a unit of the CAN2 requires k linear independent data to training the associative matrix or k-dimensional linear coefficient vector. When a Voronoi region V_j given by (6) has smaller number of training data than k, the present learning method [7] compensates the training data in V_j from the given training data nearest to the center vector w_j of V_j until the number of the data being k. Thus, in order to approximate a complicated function as of the chaotic time series, a large N bigger than n/k may be useful. Precisely, the present learning method tries to equalize the MSE (mean square prediction error) for all Voronoi regions because it is asymptotically optimal [7]. However, when there is an inevitable large error in a small input region, such as the error for long-term unpredictable data, it may be useful to divide the input region into a lot of Voronoi regions as much as possible more than $n/k = 500$ but smaller than $n = 5,000$, in order to exclude the unpredictable data from the learning of piecewise linear approximation in each Voronoi region.

OOB Estimate of DMS Prediction Error and Model Selection. Using bagging CAN2, we have OOB estimate of the performance of DMS prediction for model se-

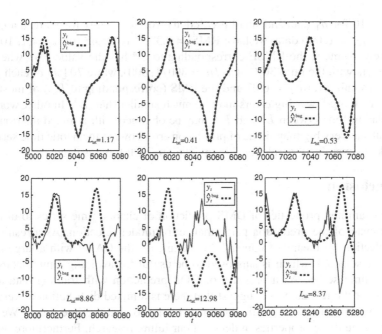

Fig. 4. Example of the target y_t and DMS prediction for $K = 80$ by the selected bagging CAN2, $\hat{y}_t^{\mathrm{bag}}$, which corresponds to the bottom right in Fig. 3

lection. Here, for the bagging CAN2, we use the number of bags being $b = 20$ for reducing computational cost and the bag size ratio being $\alpha = 2$ for obtaining smaller prediction error. Note that $b = 20$ is smaller than the values used for usual bagging methods [5], but we have had almost the same result of the model selection using the OOB estimate with $b = 100$. Furthermore, in our previous study [6], bigger α is useful when the number of training data is small (for learning analytical functions). Therefore, we use 5,000 training data, which we suppose sufficient data, in this experiment, while 2,000 data is used in [3]. However, the big $\alpha = 2$ works well in the experiments. The reason of this result is supposed to be owing to chaotic time series, and we are going to examine the reason in detail in our future research.

We have obtained OOB RSE L^{ob} given by (8) for $N = 100, 500, 1000, 2000$ and 3000 and obtained the best N with minimum L^{ob}. We show three examples of the result in Fig. 3 (left). As explained above, we have $K + 1$ DMS predictions for the prediction delay K. We use them as test predictions and we obtain the RMSE, denoted by L^{test}, to evaluate the prediction performance as shown in Fig. 3 (right). Furthermore, we show some examples of the target y_t and the prediction $\hat{y}_{t_c}^{\mathrm{bag}}$ for $K = 80$ corresponding to the bottom right in Fig. 3 in Fig. 4. From the figure, the predictions for $t_{\mathrm{p}} = $ 8000 and 9000 are not good, but other predictions with L^{test} less than 2 look not bad. In Fig. 3, the circled values of L^{ob} for large $K (\geq 40)$ seems very bigger than the average of L^{test} on the right, although there are good predictions with L^{test} less than 2. This is considered that there are a larger number of big prediction errors for larger K, and they are overestimated by the RMSE criterion. For example, when we have big prediction error $e_1 = 10$ for n_1 data and small prediction error $e_2 = 0.01$ for n_2 data among all $n = n_1 + n_2 = 1,000$ data, we have RMSE= 0.32, 3.16 , 5.48 for $n_1 = 1, 100, 300$, respectively. Namely, the RMSE corresponding to L^{ob} has the value 5.48 when there are big errors with the ratio 30[%] $(n_1/n = 300/1,000)$, while 70 [%] of each prediction shows small error $e_2 = 0.01$ and the RMS (corresponding to L_{test}) of most of the samples involving fewer big errors may be much smaller than L^{ob}. In other words, the above relationship between L^{ob} and L^{test} may be obtained with a mixed distribution of very small and very big magnitude of prediction errors owing to chaotic time series. We would like to examine this relationship in detail in our future research.

5   Conclusion

We have examined properties of DMS prediction of chaotic time series to utilize the OOB estimate of the prediction performance for model selection. As a result of the model selection by means of numerical experiments, the CAN2 with a large number of units are selected, which is supposed to be useful for avoiding unpredictable data of chaotic time series. As a property of OOB prediction with training data and the prediction for the test data, we suggest that there is a mixed distribution of very small and very big magnitude of prediction errors owing to chaotic time series. We would like to examine these properties in detail in our future research. Furthermore, we have shown that IOS prediction using single CAN2 has achieved longer predictable horizon than DMS prediction in many cases. We would like to examine the reason and develop a prediction method achieving longer predictable horizon as well as whose horizon or the uncertainty of the prediction can be estimated.

Acknowledgement. This work was supported by JSPS KAKENHI Grant Number 24500276.

References

1. Aihara, K.: Theories and applications of chaotic time series analysis. Sangyo Tosho, Tokyo (2000)
2. Chevillon, G.: Direct multi-step estimation and forecasting. Journal of Economic Surveys 21(4), 746–785 (2007)
3. Kurogi, S., Ono, K., Nishida, T.: Experimental analysis of moments of predictive deviations as ensemble diversity measures for model selection in time series prediction. In: Lee, M., Hirose, A., Hou, Z.-G., Kil, R.M. (eds.) ICONIP 2013, Part III. LNCS, vol. 8228, pp. 557–565. Springer, Heidelberg (2013)
4. Breiman, L.: Bagging predictors. Machine Learning 26, 123–140 (1996)
5. Efron, B., Tbshirani, R.: Improvements on cross-validation: the .632+ bootstrap method. J. American Statistical Association 92, 548–560 (1997)
6. Kurogi, S.: Improving generalization performance via out-of-bag estimate using variable size of bags. J. Japanese Neural Network Society 16(2), 81–92 (2009)
7. Kurogi, S., Ueno, T., Sawa, M.: A batch learning method for competitive associative net and its application to function approximation. In: Proc. of SCI 2004, vol. V, pp. 24–28 (2004)
8. Espinoza, M., Suykens, J.A.K., Moor, B.D.: Short term chaotic time series prediction using symmetric LS-SVM regression. In: NOLTA 2005, Bruges, Belgium, pp. 18–21 (2005)

Multi-document Summarization Based on Sentence Clustering

Hai-Tao Zheng*, Shu-Qin Gong, Hao Chen, Yong Jiang, and Shu-Tao Xia

Tsinghua-Southampton Web Science Laboratory,
Graduate School at Shenzhen, Tsinghua University, Shenzhen, China
{zheng.haitao,jiangy,xiast}@sz.tsinghua.edu.cn,
{gongshuqin90,jerrychen1990}@gmail.com

Abstract. A main task of multi-document summarization is sentence selection. However, many of the existing approaches only select top ranked sentences without redundancy detection. In addition, some summarization approaches generate summaries with low redundancy but they are supervised. To address these issues, we propose a novel method named Redundancy Detection-based Multi-document Summarizer (RDMS). The proposed method first generates an informative sentence set, then applies sentence clustering to detect redundancy. After sentence clustering, we conduct cluster ranking, candidate selection, and representative selection to eliminate redundancy. RDMS is an unsupervised multi-document summarization system and the experimental results on DUC 2004 and DUC 2005 datasets indicate that the performance of RDMS is better than unsupervised systems and supervised systems in terms of ROUGE-1, ROUGE-L and ROUGE-SU.

Keywords: Multi-document summarization, sentence clustering, representative selection, redundancy detection.

1 Introduction

With the number of information grows exponentially, how can we obtain useful information from the mass of resources has become increasingly important. Multi-Document Summarization (MDS), which products a summary for a set of topic-related documents, is an effective tool to solve the problem. Sentence selection is a widely-accepted approach to generate a summary for multiple documents. Goldstein et al. developed Maximal Marginal Relevance (MMR) method to detect redundant sentences, minimize redundancy, and maximize relevance of a summary [1]. Seno et al. proposed a sentence clustering method to cluster the sentences using an incremental clustering algorithm [2]. In the algorithm, the first sentence from the first document is set as the first cluster, the following sentences are clustered based on the judgement of whether the sentence should belong to an existing cluster or a new cluster. Wang et al. suggested a system based on Latent Dirichlet Allocation (LDA) [3]. In the system, a new

---

* Corresponding author.

C.K. Loo et al. (Eds.): ICONIP 2014, Part II, LNCS 8835, pp. 429–436, 2014.
© Springer International Publishing Switzerland 2014

Bayesian sentence-based topic model is generated by making use of both term-document and term-sentence associations. There are some approaches generating summaries with relative low redundancy, such as [4–6]. However, most of these methods are supervised and labor-intensive.

Generally speaking, most existing approaches either could not effectively recognize redundancy, or used deep natural language analysis leading to supervised fashion. To address these issues, we present a novel method named Redundancy Detection-based Multi-document Summarizer (RDMS). The aim of RDMS is to automatically produce a summary for a set of topic-related documents with low redundancy. We first generate an informative sentence set, then we uses an improved $k-means$ algorithm to cluster sentences. After sentence clusters generated, we rank the clusters, and select candidates to get a candidate summary set. Finally, we select the representative with redundancy detection. To sum things up, the main contributions of the paper are:

1. We propose a novel method RDMS for multi-document summarization based on sentence clustering, which detects redundancy when selecting representative sentences. As a result, our method enhances the summarization process by eliminating redundancy.

2. For the sentence clustering process, we develop an improved $k-means$ clustering algorithm.

3. Based on two datasets, we evaluate RDMS comprehensively by comparing eight existing summarization methods in terms of ROUGE-1, ROUGE-L and ROUGE-SU.

2 Redundancy Detection-Based Multi-document Summarizer

2.1 Overview

Let us take a glance at the procedure of our multi-document summarization system: There is a set of documents $D = \{d_1, d_2, \ldots, d_m\}$, in which d_1, d_2,$\cdots$, d_m are topic-related documents, m is the number of documents. By extracting informative sentences from D, a set of sentences $S = \{s_1, s_2, \ldots, s_n\}$ is generated. We cluster S into S_1, S_2,$\cdots$, S_k, in which k is the number of sentence clusters. After getting S_1, S_2,$\cdots$, S_k, we rank the clusters to get S_1', S_2',$\cdots$, S_k' and select a representative sentence from each cluster to form the candidate sentence set $R_c = \{r_1, r_2, \ldots, r_k\}$. Finally, we choose the representative sentences from R_c in a controlled fashion with redundancy detection until the summary R_r reaches the required length. The procedure of our multi-document summarization system is described in Fig.1.

2.2 RDMS Method

Informative Sentence Set Generating. In this step, our framework uses the sentence boundary detector RASP [7] to split a set of topic-related documents

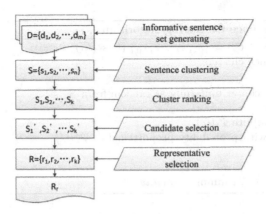

Fig. 1. The procedure of RDMS system

into sentences. Considering short sentences contain less information, we remove the sentences with length less than five words to generate informative sentence set $S = \{s_1, s_2, \ldots, s_n\}$.

Sentence Clustering. First, we use $S = \{s_1, s_2, \ldots, s_n\}$ as the input of sentence clustering algorithm and $\sqrt{n}$ as the number of clusters. Shanlin et al. proved that the maximum number of actual clusters is smaller than $\sqrt{n}$ [8]. The algorithm chooses $\sqrt{n}$ initial centers one by one in a controlled fashion for clustering, in which the current set of chosen centers will have an impact on the choice of the next center. Specifically, for each sentence s and the set of cluster centers $C = \{c_1, c_2, \ldots, c_p\}$, we compute the minimum similarity between s and C as follows:

$$\Psi_s(C) = min_{i=1}^{p} \frac{s \cdot c_i}{\| s \| \| c_i \|} \tag{1}$$

Where p is the number of centers existing in C, and $\frac{s \cdot c_i}{\|s\|\|c_i\|}$ means the cosine similarity between sentence s and sentence c_i. Then we choose s with probability proportional to $\Psi_s^2(C)$. After getting the whole initial center set C, we adopt the $k\text{-}means$ clustering algorithm. The whole clustering algorithm is shown in Algorithm 1.

Cluster Ranking. Different clusters represent different topics of the whole document set. Since some clusters could be noisy, we need to rank clusters to filter the clusters. We compute the sum of similarities that between sentences in one cluster and S. Then, we calculate the weight of cluster S_i to rank the clusters as follows:

$$W(S_i) = \sum_{i=1}^{m} \sum_{j=1}^{n} \frac{s_i \cdot s_j}{\| s_i \| \| s_j \|} \tag{2}$$

Where m is the number of sentences in cluster S_i and $s_i \in S_i$, n is the number of sentences in the whole sentence set S and $s_j \in S$. We are aiming to generate

---

Algorithm 1. Sentence Clustering Algorithm

Input: The sentence set $S = \{s_1, s_2, \ldots, s_n\}$ (k is $\sqrt{n}$)
Output: The sentence clusters.
Method:

  1: $C \leftarrow$ choose one center uniformly at random from S
  2: while $|C| < k$ do
  3:     for each s, compute $\Psi_s(C)$
  4:     choose $s \in S$ with probability proportional to $\Psi_s^2(C)$
  5:     $C \leftarrow C \cup \{s\}$
  6: end while
  7: initial centers $\leftarrow C$, running $k\text{-}means$

---

the summary based on the whole sentence set S, to some extent we take the global importance of sentence into consideration.

Candidate Selection. After clusters are ranked, we choose a representative sentence for each cluster based on the sentence weight. The score of s in S_i is defined as follows:

$$\Upsilon(s, S_i) = \sum_{j=1}^{m} \frac{s \cdot s_{ij}}{\| s \| \| s_{ij} \|} \tag{3}$$

Where s_{ij} is the jth sentence in S_i and m is the number of sentences in cluster S_i. Then we select the representative sentence with the highest score for each cluster to construct the candidate set $R_c = \{r_1, r_2, \ldots, r_k\}$, where k is the number of clusters. In the process, we consider the importance of sentence within the cluster, that is to say we take the local importance of sentence into consideration.

Representative Selection. First, we select the representative sentence of the cluster with the highest weight in R_c, and add it to summary. Then we choose the sentence s with the next highest weight and compute the redundancy score $\Phi_s(R_r)$ as follows:

$$\Phi_s(R_r) = max_{i=1}^{p} \frac{s \cdot r_i}{\| s \| \| r_i \|} \tag{4}$$

Where p is the number of sentences existing in R_r, and $r_i \in R_r$. $\Phi_s(R_r)$ measures the degree of redundancy by obtaining the maximum of the similarity between s and the set of chosen representative sentences. If $\Phi_s(R_r)$ is lower than the threshold λ (a parameter used to measure redundancy), then we add it to summary, otherwise not. Repeat the step until the length of summary meets the requirement. Note that λ is an important parameter to leverage the redundancy in R_r and we will set the value of λ based on empirical experiments.

We interpret our RDMS method as follows: First, we extract informative sentences from a set of topic-related documents. Then we run our sentence clustering algorithm which is based on the improved $k-means$ clustering algorithm. As we know, the selection of initial centers is crucial to the clustering quality of $k-means$. Hence, we select initial centers in a careful manner rather than start with a random set of k centers, which makes better clustering results than the original $k-means$. Due to different clusters make a difference contribution to the documents, we rank the sentence clusters after sentence clustering. Then we select a representative sentence for each cluster to form the candidate set of summary. During cluster ranking and candidate selection, we take both local importance and global importance of sentence into consideration, in order to get representative sentences with high quality. We also conduct representative selection to strengthen the redundancy eliminating. Representative sentences are chosen one by one in a controlled way, where the current set of chosen sentences will have an impact on the choice of the next sentence. We introduce parameter λ as a threshold, if the maximum similarity between the current sentence and the sentences existing in summary is lower than λ, then we add the current sentence to the summary, otherwise not. We repeat the procedure of selecting representative sentences until the length of summary is satisfied.

3   Evaluation

The Document Understanding Conference (DUC)[1] is an annual evaluation for summarization. In the experiments, we use DUC 2004 and DUC 2005 corpus to evaluate our RDMS method. Table 1 gives a brief description of the datasets.

Table 1. Details of DUC 2004, DUC 2005 dataset

	DUC 2004	DUC 2005
number of document collections	50	50
number of articles in each document	10	25-50
data source	TDT	TREC
summary length	665 bytes	250 words

For evaluation, we use the ROUGE[2] (Recall-Oriented Understudy for Gisting Evaluation) system. To properly evaluate the summary we use ROUGE-1, ROUGE-L and ROUGE-SU based measures. Other details of the measures are available in [9].

We compare our RDMS system with several state-of-the-art summarization approaches described briefly as follows: (1) **DUC Best:** It means the system performing best on the official datasets, which is based on Hidden Markov Model (HMM). (2) **Random:** The approach selects sentences randomly from the document set. (3) **Centroid:** The method uses MEAD algorithm to extract sentences according to the following three parameters: centroid value, positional value and

---

[1] http://duc.nist.gov/
[2] http://www.berouge.com/Pages/default.aspx

first-sentence overlap [10]. (4) **LexPageRank**: The method first constructs a sentence connectivity graph based on cosine similarity and then selects important sentences based on the concept of eigenvector centrality [11]. (5) **LSA**: The method applies the singular value decomposition (SVD) on the terms by sentences matrix to select highest ranked sentences [12]. (6) **NMF**: The method performs non-negative matrix factorization (NMF) to cluster sentences into groups and selects sentences from each group for summarization [13]. (7) **KM**: The method performs $k-means$ algorithm on terms by sentences matrix to cluster the sentences and then chooses the centroids for each sentence cluster to construct a summary. (8) **FGB**: The FGB approach utilizes the mutual influence of the document for clustering and summarization [14].

Since our system selects the initial centers with probability, we conduct experiments ten times and obtain the average score. We also stem summaries when we compare them to the reference summaries. First, we determine the parameter λ by set its values $0.1, 0.2, \cdots, 1.0$ respectively as shown in Fig.2. We found that $\lambda = 0.3$ can get the best performance. That is to say, $\lambda > 0.3$ makes summary contain more redundancy. If $\lambda < 0.3$ the summary covers noisy words leading to low ROUGE scores. Therefore we set $\lambda = 0.3$ in our later experiments.

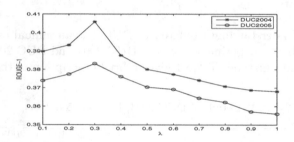

Fig. 2. The effect of λ in DUC 2004 and DUC 2005

For fair comparison, a summary of 665 bytes is generated for each document set on DUC 2004 dataset, and a summary of 250 words is generated for each document set on DUC 2005 dataset. The comparison results between RDMS and other eight systems are presented in Table 2.

From experimental results, it is clear that RDMS performs better than other compared systems on different evaluation measures. We also have the following observations: (1) It is obvious that Random has the worst performance among all the methods. (2) LSA and NMF and KM are the systems based on sentence clustering, all of them first generate sentence clusters and then select representative sentences from each cluster. (3) Different with LSA, NMF and KM, Centroid takes positional value and first-sentence overlap into consideration which are not used in clustering-based summarization, which leads to better result. (4) LexPageRank performs better than Centroid. It is due to the fact that LexPageRank ranks the sentence with eigenvector centrality, which implicitly accounts for relation among all sentences. (5) FGB outperforms LexPageRank. FGB makes use of term-document and term-sentence matrices, which enhancing the generated

Table 2. Evaluation results on DUC 2004 dataset and DUC 2005 dataset

Dataset	DUC 2004			DUC 2005		
Systems	ROUGE-1	ROUGE-L	ROUGE-SU	ROUGE-1	ROUGE-L	ROUGE-SU
DUC Best	0.38224	0.38687	0.13233	0.38036	0.34764	0.10012
Random	0.31865	0.34521	0.11779	0.29012	0.26395	0.09066
Centroid	0.36728	0.36182	0.12511	0.3535	0.32562	0.11007
LexPageRank	0.37842	0.37531	0.13097	0.3760	0.33179	0.12021
LSA	0.34145	0.34973	0.11946	0.30461	0.26476	0.10806
NMF	0.36747	0.36749	0.12918	0.32026	0.28716	0.11278
KM	0.34872	0.35882	0.12115	0.31762	0.29107	0.10806
FGB	0.38724	0.38423	0.12957	0.38175	0.35018	0.12006
RDMS	**0.40589**	**0.40072**	**0.14219**	**0.38319**	**0.35376**	**0.12297**

summary. (6) DUC Best uses HMM, it requires a large amount of training data and is a supervised system. (7) Our RDMS outperforms all other systems. For example, on DUC 2004 dataset, RDMS gets **0.40589** ROUGE-1 score, **0.40072** ROUGE-L score and **0.14219** ROUGE-SU score, while DUC Best (supervised system) gets 0.38224 ROUGE-1 score, 0.38687 ROUGE-L score and 0.13233 ROUGE-SU score. We attribute this to the fact that we take both local importance and global importance of sentence into consideration to get representative sentences with high quality. The summary covers more comprehensive keywords through sentence clustering and representative selection. Additionally, it benefits from our system detecting redundancy twice. The first time is during the procedure of sentence clustering, and the second time is during the procedure of representative selection. By setting a parameter λ, we leverage the whole degree of redundancy to get a summary with high quality. Different with supervised methods, our parameter setting is disposable and time efficiency is high, while supervised methods need iterative optimization and time efficiency is low. In conclusion, RDMS system has three advantages: avoiding redundancy, covering more significant words and being an automatic system.

4   Conclusion and Future Work

In this paper, we propose RDMS system for multi-document summarization, it is composed by five steps: informative sentence set generating, sentence clustering, cluster ranking, candidate selection and representative selection. Meanwhile, the RDMS system is unsupervised. Compared with the state-of-the-art systems, the experimental results indicate that our proposed system performs better than unsupervised systems and comparable with supervised systems of this area. We believe that our method will play an important role for multi-document summarization in an automatic fashion.

In the future, we will improve our system by exploiting more similarity measurement between sentences. In addition, we will explore more effective clustering methods to improve the efficiency of the proposed method.

Acknowledgments. This research is supported by the 863 project of China (2013AA013300), National Natural Science Foundation of China (Grant No. 61375054) and Tsinghua University Initiative Scientific Research Program (Grant No. 20131089256).

References

1. Goldstein, J., Mittal, V., Carbonell, J., Kantrowitz, M.: Multi-document summarization by sentence extraction. In: Proceedings of the 2000 NAACL-ANLP Workshop on Automatic Summarization, vol. 4, pp. 40–48. Association for Computational Linguistics (2000)
2. Seno, E.R.M., das Graças Volpe Nunes, M.: Some experiments on clustering similar sentences of texts in portuguese. In: Teixeira, A., de Lima, V.L.S., de Oliveira, L.C., Quaresma, P. (eds.) PROPOR 2008. LNCS (LNAI), vol. 5190, pp. 133–142. Springer, Heidelberg (2008)
3. Wang, D., Zhu, S., Li, T., Gong, Y.: Multi-document summarization using sentence-based topic models. In: Proceedings of the ACL-IJCNLP 2009 Conference Short Papers, pp. 297–300. Association for Computational Linguistics (2009)
4. Conroy, J.M., Schlesinger, J.D., Goldstein, J., O'leary, D.P.: Left-brain/right-brain multi-document summarization. In: Proceedings of the Document Understanding Conference (2004)
5. Kumar, N., Srinathan, K., Varma, V.: Using wikipedia anchor text and weighted clustering coefficient to enhance the traditional multi-document summarization. In: Gelbukh, A. (ed.) CICLing 2012, Part II. LNCS, vol. 7182, pp. 390–401. Springer, Heidelberg (2012)
6. He, Z., Chen, C., Bu, J., Wang, C., Zhang, L., Cai, D., He, X.: Document Summarization Based on Data Reconstruction. In: AAAI (2012)
7. Briscoe, T., Carroll, J., Watson, R.: The second release of the RASP system. In: Proceedings of the COLING/ACL on Interactive Presentation Sessions, pp. 77–80. Association for Computational Linguistics (2006)
8. Yang, S.L., Li, Y.S., Hu, X.X., Pan, R.Y.: Optimization Study on k Value of K-means Algorithm. Systems Engineering-Theory and Practice 2, 012 (2006)
9. Lin, C.Y.: Rouge: A package for automatic evaluation of summaries. In: Text Summarization Branches Out: Proceedings of the ACL 2004 Workshop, pp. 74–81 (2004)
10. Radev, D.R., Jing, H., Stys, M., Tam, D.: Centroid-based summarization of multiple documents. Information Processing Management 40(6), 919–938 (2004)
11. Erkan, G., Radev, D.R.: LexPageRank: Prestige in Multi-Document Text Summarization. In: EMNLP, vol. 4, pp. 365–371 (2004)
12. Gong, Y., Liu, X.: Generic text summarization using relevance measure and latent semantic analysis. In: Proceedings of the 24th Annual International ACM SIGIR Conference on Research and Development in Information Retrieval, pp. 19–25. ACM (2001)
13. Lee, D.D., Seung, H.S.: Algorithms for non-negative matrix factorization. In: Advances in Neural Information Processing Systems, pp. 556–562 (2001)
14. Wang, D., Zhu, S., Li, T., Chi, Y., Gong, Y.: Integrating clustering and multi-document summarization to improve document understanding. In: Proceedings of the 17th ACM Conference on Information and Knowledge Management, pp. 1435–1436. ACM (2008)

An Ontology-Based Approach to Query Suggestion Diversification

Hai-Tao Zheng*, Jie Zhao, Yi-Chi Zhang, Yong Jiang, and Shu-Tao Xia

Tsinghua-Southampton Web Science Laboratory,
Graduate School at Shenzhen, Tsinghua University, Shenzhen, China
{zheng.haitao,jiangy,xiast}@sz.tsinghua.edu.cn,
angelahai@126.com, zhangyichi0604@gmail.com

Abstract. Query suggestion is proposed to generate alternative queries and help users explore and express their information needs. Most existing query suggestion methods generate query suggestions based on document information or search logs without considering the semantic relationships between the original query and the suggestions. In addition, existing query suggestion diversifying methods generally use greedy algorithm, which has high complexity. To address these issues, we propose a novel query suggestion method to generate semantically relevant queries and diversify query suggestion results based on the WordNet ontology. First, we generate the query suggestion candidates based on Markov random walk. Second, we diversify the candidates according the different senses of original query in the WordNet. We evaluate our method on a large-scale search log dataset of a commercial search engine. The outstanding feature of our method is that our query suggestion results are semantically relevant belonging to different topics. The experimental results show that our method outperforms the two well-known query suggestion methods in terms of precision and diversity with lower time consumption.

Keywords: Query suggestion, search logs, semantic relationships, diversify.

1 Introduction

Since it is getting difficult for search engines to satisfy users' information needs directly, query suggestion is proposed to generate alternative queries and help users explore and express their information needs. Query suggestion plays an important role in improving the usability of search engines and it has been well studied in academia as well as industry. However, much efforts focus on how to suggest queries relevant to the original query without considering the diversity of query suggestion results. A study shows that queries submitted to search engines are usually short consisting of two or three words on average [15] . When a user knows little about the search topic, it is difficult to construct a good query.

---

* Corresponding author.

C.K. Loo et al. (Eds.): ICONIP 2014, Part II, LNCS 8835, pp. 437–444, 2014.

The two situations make search intend ambiguous, and query suggestion results diversity is necessary. Recently, some methods has been proposed to diversify query suggestion results. For example, Ma et al. [10] diversify the results by employing the Markov random walk and hitting time.

However, existing query suggestion diversification methods generally use greedy algorithm, which has high complexity, resulting relative low efficiency. In addition, many existing query suggestion methods only use search logs to obtain query suggestion without considering semantic relationships between queries, resulting in some query suggestion results unrelated to the original query semantically.

To address these issues, we propose a novel method, called Ontology-based Diversifying Query Suggestion (ODQS), to diversify query suggestion results using an ontology - WordNet [16]. First, we generate query suggestion candidates from search logs based on the forward random walk analysis. Second, we diversify the candidates based on the different senses of the original query in the WordNet. The paper has three main contributions as follows:

1) We exploit search logs and semantic relationships between queries based on WordNet to diversify query suggestion results and increase the relevance of query suggestion results.

2) Our method improves the efficiency of the query suggestion diversification by employing the WordNet ontology.

3) We evaluate the effectiveness and efficiency of the proposed method by comparing two existing query suggestions methods.

The rest of this is organized as follows. Section 2 is devoted to a detailed description of our method to diversify query suggestion results. We conduct experiment on AOL search logs to evaluate our proposal section 3. We discuss related work of query suggestion in Section 4. Finally, we conclude the paper with future work in Section 5.

2   Ontology-Based Diversifying Query Suggestion

In this section, we elaborate how to leverage search logs and semantic relationships to improve the quality and diversification of query suggestions. Fig.1 shows the ontology-based diversifying query suggestion model. We first generate suggestion candidates from search logs based on the forward random walk. A query-URL bipartite graph is constructed as $G = (V, E)$, in which vertexes are composed by two parts $V = V_1 \bigcup V_2$. V_1 represents the set of all unique queries and V_2 represents the set of all unique URLs. There exits an edge from node $x \in V_1$ to node $y \in V_2$ if y is a clicked URL of query x in the search logs. The weight $\omega(x, y)$ is the total number of times that y is clicked when query x is issued. Note that since the edge is an undirected edge, the weight $\omega(y, x)$ is the same as $\omega(x, y)$. The transition probabilities from node i to node j in this paper is defined as follows:

$$P_{t+1|t}(j \mid i) = \frac{\omega(i,j)}{\sum_k \omega(i,k)} \tag{1}$$

where k ranges over all nodes. $P_{t+1|t}(j \mid i)$ denotes the transition probability from node i at step t to node j at time step $t+1$.

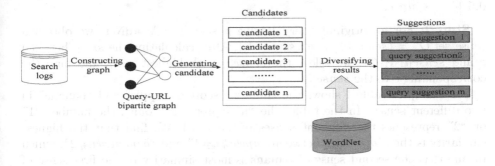

Fig. 1. Ontology-based diversifying query suggestion model

We represent the transitions as a sparse matrix A and perform the random walk using A. We calculate the probability of transition from node i to node j in t steps as $P_{t|0}(j \mid i) = [A^t]_{ij}$, right here A^t means t-step random walk transition matrix. It gives a measure of the volume of paths between these two nodes. If there are many paths the transition probability will be high. The larger the transition probability $P_{t|0}(j \mid i)$ is, the more the node j is similar to the node i. We select the top n largest $P_{t|0}(j \mid i)$ as candidates. In this study, we set the $n = 100$ empirically. In fact, $n = 100$ is large enough because we generally select 20 query suggestions at most.

To improve the quality of query suggestions and diversify query suggestions, we exploit semantic relationships between queries based on the WordNet ontology. In the study, we map the suggestion candidates based on the different senses of original queries. There are three steps for diversifying query suggestions based on WordNet as follows:

Step 1: For each query suggestion candidate c, we compute similarity scores between the different senses of an original query q and candidate c. In this way, we exploit the semantic relationships between query suggestion c and the orignal query q. We use the Lin similarity method [5], because it is a probabilistic model and is a universal similarity function, not tied to a particular application or a form of knowledge representation. For concept c_1 and c_2, their Lin similarity is defined as follows:

$$sim_L(c_1, c_2) = \frac{2 \times \lg p(lcs(c_1, c_2))}{\lg p(c_1) + \lg p(c_2)} \tag{2}$$

where $lcs(c_1, c_2)$ is the least common parent node of c_1 and c_2; $p(c)$ is probabilities of concept c and $p(c) = \frac{\sum_{w \in words(c)} count(w)}{N}$, where $words(c)$ is a set of words which concept c contains, $count(w)$ is the number of w in Brown corpus and N is the number of words in the Brown Corpus.

For two phrases consisted multiple terms, we calculate their similarity sim_P between the last noun of the two phrases as the final score. If the last noun of the two phrases are same, we calculate the similarity sim_L between the penultimate word of candidate and the last noun of the original query. The final score $sim_P = 0.5 + 0.5 \times sim_L$.

Step 2: For an original query having n senses in WordNet, we obtain a sense set $Q_s = \{sense[0], \dots sense[n-1]\}$. After calculating the sim_L between c and $sense[i] \in Q_s$, we select the $sense[i]$ with highest sim_L and map the corresponding c to the $sense[i]$.

For example, Table 1 shows the similarity score of "apple" and "banana" in two different senses. In the table, the "n" represents "noun", the number "1" or "2" represents the different senses of the word. We find that the highest similarity is the sim_L score between "$apple\#n\#1$" and "$banana\#n\#2$", which means that the second sense of banana is most similarity to the first sense of apple. For an original query "apple", "banana" is a suggestion candidate, then we map "banana" to the first sense of "apple".

Table 1. The similarity results of "apple" and "banana" using Lin similarity

combinations of different senses	similarity score
$apple\#n\#1, banana\#n\#1$	0.11802556069890623
$apple\#n\#1, banana\#n\#2$	0.6867056880240358
$apple\#n\#2, banana\#n\#1$	0.0
$apple\#n\#2, banana\#n\#2$	0.0

Step 3: We rank suggestion candidates according to the similarity score for each $sense[i] \in Q_s$. Based on the ranking results, we select the queries which have the top k highest similarity score as suggestions. The value of k is determined on how many query suggestions we need. Since the candidates are selected from different groups based on $sense[i] \in Q_s$, our method ensures that the final query suggestions are semantically diversified and related to the Q_s.

3   Evaluation

3.1   Experimental Setup

In this section, we conduct empirical experiments to show the effectiveness of our proposed algorithm. We select AOL Search Data as our data set, which is a collection of real search log data based on real users. The data set consists of 20M web queries collected from 650k users over three months. We clean the data by keeping those frequent, well-formatted, English queries (queries which only contain characters 'a', 'b',...,'z' and space, and appear more than 5 times). After cleaning, we obtain a total of 9,752,848 records, 604,982 unique queries and 785,012 unique URLs.

We construct a query-URL bipartite graph on our data set, and randomly sample a set of 50 queries from our data set as the testing queries. For each testing query, we obtain six query suggestions. In order to evaluate the quality of the results, three experts are requested to rate the query suggestion results with "0" or "1", where "0" means "irrelevant" and "1" means "relevant". We use the precision measurement in our experiment, i.e., precision at position n is defined as:

$$p@n = \frac{rn}{n} \tag{3}$$

where rn is the number of relevant queries in the first n results.

In order to evaluate the diversity of query suggestion results, we propose a diversity measurement method. Intuitively, if similarity score between two queries was high, then the diversity of the two queries are low. We select the first sense for each noun of q_i in WordNet, and then we use Bag of Words model to indicate q_i, denoting $\vec{qt_i} = (qt_{i1}, \ldots, qt_{in})$. We calculate the similarity (q_i, q_j) by Cosine similarity. Hence, we measure the diversity of two queries as follows:

$$D(q_i, q_j) = 1 - cos(\vec{qt_i}, \vec{qt_j}) = 1 - \frac{\sum_{k=1}^{n} qt_{ik} \times qt_{jk}}{\sqrt{\sum_{k=1}^{n} qt_{ik}^2} \times \sqrt{\sum_{k=1}^{n} qt_{jk}^2}} \tag{4}$$

We evaluate the diversity over a query set S_q as follows:

$$SD(S_q) = \frac{\sum_{i=1}^{K} \sum_{j=1, i \neq j}^{K} D(q_i, q_j)}{K \times (K-1)} \tag{5}$$

where K is the size of the query set S_q. We compare our method with Diversifying Query Suggestion (DQS) method [10] and Forward Random Walk (FRW) method [3]. We realize all the methods on a Core i5 computer, with CPU clock rate of 2.8 GHz, 4GB RAM, and running Windows 7.

3.2   Experimental Results

The comparison results are shown in Fig.2 and Fig.3. We found that our method is superior to FRW and DQS in precision as well as diversity. For example, when six query suggestions are returned, average precisions for the three methods are ODQS 0.59, DQS 0.54 and FRW 0.50. ODQS is 5% higher than DQS and 9% higher

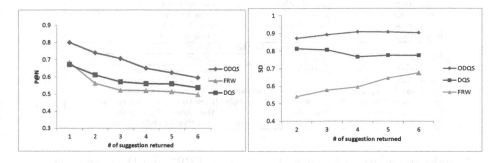

Fig. 2. The average precision   **Fig. 3.** The average of diversity

than FRW. The average diversities for the three methods are ODQS 0.9, DQS 0.78 and FRW 0.68. ODQS is 12% higher than DQS and 22% higher than FRW.

We list three cases for three queries in Table 2. Given a query, we discovered that our suggestion results cover much more latent topics intuitively. For example, given a query "birthday cakes", our method suggested "kids birthday cakes" and "princess birthday cakes" which is the most relevant to the given query. It also suggests "birthday invitations" and "kids craft", which is be closely related to birthday. "wedding cakes" as a kind of cake is also suggested, which means that the diversified suggestion are discovered by our method. DQS returns diversified but not so relevant query suggestions, such as "electric power washers". The results of FRW are relevant, i.e., most suggestions are related to the "cakes". However, most suggestions are redundant and represent the same topic "cakes". The three cases in Table 2 show that suggestion results of our method are relevant as well as diversified.

Different from FRW and DQS, which only leverage search logs to generate suggestions, our method exploits semantic relationships between original query and suggestions based on WordNet. We re-ranking the suggestion candidates based on their relevance between different senses of the original queries. Therefore, the suggestions are more semantically related to the original queries. We also use WordNet to find latent topics - different senses of the original query, and map suggestion candidates to these latent topics. Our method is able to discover the query suggestions in different topics intrinsically. As a result, our method diversifies the query suggestions better than the FRW and DQS method.

Table 2. Query Suggestion Comparisons between ODQS and otherMethods

Query = travel		
ODQS	FRW	DQS
air fares	travelocity	travelocity
air travel	expedia	lonely planet
aarp passport	orbitz	travel texas
hotels	airline tickets	frommers california travel guide
Yahoo travel	airfare	frommers maui
travel channel	hotels	haunted travels
Query = world		
ODQS	FRW	DQS
world map	world map	world map
child labor	atlas	the world fact book
time zones	world atlas	cia united states
world bank	cnn	time in london
costa rica	child labor	world clock
education	time zones	time in spain
Query = birthday cakes		
ODQS	FRW	DQS
kids birthday cakes	birthday cakes	baby shower cakes
baby shower cakes	cakes	electric power washers
birthday invitations	baby shower cakes	wedding invitations
princess birthday cakes	princess birthday cakes	how to make icing roses
wedding cakes	wedding cakes	the cheese cake factory
kids crafts	birthday invitations	creative birthday gifts

In order to compare efficiency of the ODQS, FRW, and DQS, we implement three methods on the same computer and found the average times of three methods are DQS 721,469ms, FRW 220,277ms and ODQS 269,832ms. It is obvious that DQS consumes much more time than FRW and ODQS. It is chiefly because DQS algorithm has an dense matrix operations and iterative operation, which is time-consuming. We also found that our method has similar running time with FRW. It is because our method generate query suggestion candidates using the sparse transition matrix, and diversify query suggestions based on WordNet without using iterative operation. However, our method is able to generate diversified suggestions while FRW is not.

We analyze the time complexity of the three method here. To calculate a forward random walk, we encode the start distribution as a row vector ν_i with a unit entry at query node i, and obtain $P_{t|0}(j \mid i) = \nu_j[A]^t$. Then the time complexity of FRW is $O(n^2)$. The time complexity of DQS is $O(n^2) + (k - 1)O(n \times m)$, where k is the number of returned query suggestions and m is the number of neighbors of node i ($m << n$). The time complexity of our method is $O(n^2)+O(m)$ ($m << n$). To conclude, the experimental results show our method can suggest the diversified queries and improve the precision of suggestions in lower time consumption.

4 Related Work

Query suggestion techniques are used to improve search experience and help users to express these information needs. Most early query suggestion methods leverage document information or corpus [1, 2]. Recently, query suggestion using search logs has been widely studied.

Craswell et al. [3] proposed a backward random walk on query-URL bipartite. Mei et al. [4] introduced the concept of hitting time to iterative compute the hitting time. Beeferman et al. [6] used a hierarchical agglomerative on a query-URL bipartite to discover similar queries. Cao et al. [7] clustered similar queries into a concept and build the concept sequence suffix tree to realize query suggestion. Boldi et al. [8] proposed a concept of query-flow graph to obtain query suggestion results. There are some methods to improve the query-flow graph to suggest queries [9, 12–14]. In recent years, researchers presented the concept of diversifying query suggestions [10, 11]. However, most existing methods only use the search logs without considering the semantic relationships between queries, which may reduce the quality and the diversification of query suggestions.

5 Conclusion and Future Work

In this paper, we propose an ontology-based diversify query suggestion method. Query suggestion candidates are generated using Markov random walk. We rank the suggestion candidates and diversify the results based on an ontology - WordNet. Unlike existing query suggestions methods, we diversify query suggestions using not only search logs but also semantic relationships based on WordNet. The experimental results show that our method is able to suggest the diversified queries and outperform the FRW and DQS methods in terms of precision with lower time consumption.

In the future, we will focus on the latent topics mining and the optimization of the similarity algorithm. Since WordNet does not contain all words, especially spoken words and some names, we will try a comprehensive ontology to discover the latent topics related to the original query.

Acknowledgments. This research is supported by the 863 project of China (2013AA013300), National Natural Science Foundation of China (Grant No. 61375054) and Tsinghua University Initiative Scientific Research Program Grant No. 20131089256.

References

1. Meij, E., Bron, M., Hollink, L., Huurnink, B., de Rijke, M.: Learning Semantic Query Suggestions. In: Bernstein, A., Karger, D.R., Heath, T., Feigenbaum, L., Maynard, D., Motta, E., Thirunarayan, K. (eds.) ISWC 2009. LNCS, vol. 5823, pp. 424–440. Springer, Heidelberg (2009)
2. Gong, Z., Cheang, C.W., Hou U, L.: Web Query Expansion by WordNet. In: Andersen, K.V., Debenham, J., Wagner, R. (eds.) DEXA 2005. LNCS, vol. 3588, pp. 166–175. Springer, Heidelberg (2005)
3. Craswell, N., Szummer, M.: Random walks on the click graph. In: SIGIR 2007, pp. 239–246 (2007)
4. Mei, Q.Z., Zhou, D.Y., Church, K.: Query Suggestion Using Hitting Time. In: CIKM 2008, pp. 469–478 (2008)
5. Lin, D.: An Information-Theoretic Definition of Similarity. In: ICML 1998, pp. 296–304 (1988)
6. Beeferman, D., Berger, A.: Agglomerative clustering of a search engine query log. In: KDD 2000, pp. 407–416 (2000)
7. Cao, H.H., Jiang, D.X., Pei, J., Liao, Z., Chen, E., Li, H.: Context-Aware Query Suggestion by Mining Click-Through and Session Data. In: KDD 2008, pp. 875–884 (2008)
8. Boldi, P., Bonchi, F., Castillo, C., Donato, D., Gionis, A.: The Query-flow Graph: Model and Applications. In: CIKM 2008, pp. 56–63 (2008)
9. Anagnostopoulos, A., Becchetti, L., Castillo, C., Gionis, A.: An optimization framework for query recommendation. In: WSDM 2010, pp. 609–618 (2010)
10. Ma, H., Lyu, M.R., King, I.: Diversifying Query Suggestion Results. In: AAAI 2010, pp. 1399–1404 (2010)
11. Song, Y., Zhou, D., He, L.: Post-Ranking Query Suggestion by Diversifying Search Results. In: SIGIR 2011, pp. 815–824 (2011)
12. Baraglia, R., Nardini, F., Castillo, C., Perego, R., Donato, D., Silvestri, F.: The effects of time on query flow graph-based models for query suggestion. In: RIAO 2010, pp. 182–189 (2010)
13. Zhang, Z., Nasraoui, O.: Mining search engine query logs for query recommendation. In: WWW 2006, pp. 1039–1040 (2006)
14. Sadikov, E., Madhavan, J., Wang, L., Halevy, A.: Clustering Query Refinements by User Intent. In: WWW 2010, pp. 841–850 (2010)
15. Spink, A., Jansen, B.J.: A Study of Web Search Trends. Webology 1(2) (2004)
16. Miller, G.A., Beckwith, R., Fellbaum, C.D., Gross, D., Miller, K.: WordNet: An online lexical database. Int. J. Lexicograph. 3(4), 235–244 (1990)

Sensor Drift Compensation Using Fuzzy Interference System and Sparse-Grid Quadrature Filter in Blood Glucose Control*

Péter Szalay[1], László Szilágyi[1], Zoltán Benyó[1], and Levente Kovács[2]

[1] Department of Control Engineering and Information Technology,
Budapest University of Technology and Economics, Budapest, Hungary
{szalaip,lszilagyi,benyo}@iit.bme.hu
[2] Applied Informatics Institute, John von Neumann Faculty of Informatics,
Obuda University, Budapest, Hungary
kovacs.levente@nik.uni-obuda.hu

Abstract. Diabetes mellitus is a serious chronic condition of the human metabolism. The development of an automated treatment has reached clinical phase in the last few years. The goal is to keep the blood glucose concentration within a certain region with minimal interaction required by the patient or medical personnel. However, there are still several practical problems to solve. One of these would be that the available sensors have significant noise and drift. The latter is rather difficult to manage, because the deviating signal can cause the controller to drive the glucose concentration out of the safe region even in the case of frequent calibration. In this study a linear-quadratic-Gaussian (LQG) controller is employed on a widely used diabetes model and enhanced with an advanced Sparse-grid quadratic filter and a fuzzy interference system-based calibration supervisor.

Keywords: Diabetes, LQG control, Sparse-grid quadratic filter, fuzzy inference system.

1   Introduction

The blood glucose concentration is regulated through a complex endocrine system of the human body, where insulin plays a key role. If the glucose-insulin interaction is impaired, diabetes is diagnosed. Artificial Pancreas (AP) is a mean to provide an automated treatment of the insulin dependent type-1 diabetes (T1DM) by keeping the blood glucose levels of the patient in normoglycemic range (3.9 - 7.8 mmol/L). It consists of three main parts: a continuous glucose monitor (CGM), an insulin pump for subcutaneous delivery and a control algorithm for closed-loop control.

* Research supported by the Hungarian National Research Funds (OTKA), Project no. PD103921 and by the European Union TÁMOP-4.2.2.A-11/1/KONV-2012-0073 project. Levente Kovács is Bolyai Fellow of the Hungarian Academy of Sciences.

C.K. Loo et al. (Eds.): ICONIP 2014, Part II, LNCS 8835, pp. 445–453, 2014.

The aim of the AP is to automatically regulate the blood glucose concentration of T1DM patients with minimal interaction required, ensuring reasonable safety in all times. There are various methods to choose from: classical PID [1]; run-to-run control [2]; exact linearization based nonlinear control [3]; $\mathcal{H}_\infty$ control [4,5]; Model Predictive Control (MPC) [6,7]; Linear Parameter Varying (LPV) control [8,9], among others. Soft computing based methods, such as fuzzy logic control [10] and model-free soft computing-based control [11] are gaining popularity as well. Most of these techniques require signals beyond what is physically measurable; hence, accurate estimation of the state variables is needed.

However, most commercially available CGMs do not measure the glucose concentration directly, but the time derivative instead, therefore the readings will slowly drift from the actual values. Hence, calibration is needed. In this work it will be investigated how nonlinear stochastic filtering and fuzzy interference-based timing of the calibration can ensure satisfactory operation of the AP.

2   Diabetes Model

The employed model is described by the following differential equations [12]:

$$\dot{C}(t) = -k_{a,int}C(t) + \frac{k_{a,int}}{V_G}Q_1(t)$$

$$\dot{Q}_1(t) = -\left(\frac{F_{01}}{Q_1(t) + V_G} + x_1(t)\right)Q_1(t) + k_{12}Q_2(t)-$$

$$-R_{cl}\max\{0, Q_1(t) - R_{thr}V_G\} - Phy(t)+$$

$$+EGP_0\max\left\{0, 1 - \frac{S_{IE}k_{b1}k_a}{V_Ik_e}S_2(t)\right\} + \min\left\{U_{G,ceil}, \frac{G_2(t)}{t_{max}}\right\}$$

$$\dot{Q}_2(t) = x_1(t)Q_1(t) - \left(k_{12} + \frac{S_{ID}k_{b1}k_a}{V_Ik_e}S_2(t)\right)Q_2(t) \tag{1}$$

$$\dot{x}_1(t) = -k_{b1}x_1(t) + \frac{S_{IT}k_{b1}k_a}{V_Ik_e}S_2(t)$$

$$\dot{S}_2(t) = -k_aS_2(t) + k_aS_1(t)$$

$$\dot{S}_1(t) = -k_aS_1(t) + u(t)$$

$$\dot{G}_2(t) = (G_1(t) - G_2(t))/\max\{t_{max}, G_2(t)/U_{G,ceil}\}$$

$$\dot{G}_1(t) = -G_1(t)/\max\{t_{max}, G_2(t)/U_{G,ceil}\} + D(t)$$

where the state variables are: $C(t)$ glucose concentration in the subcutaneous tissue [mmol/L]; $Q_1(t)$, $Q_2(t)$ the masses of glucose in accessible and non-accessible compartments [mmol]; $x_1(t)$ remote effect of insulin on glucose distribution [1/min]; $S_1(t)$, $S_2(t)$ insulin masses in the accessible and non-accessible compartments [mU]; $G_1(t)$, $G_2(t)$ glucose masses in the accessible and non-accessible compartments [mmol]. $u(t)$ injected insulin flow of rapid-acting insulin [mU/min] is the input of the system, while $D(t)$ amount of ingested carbohydrates [mmol/min], and $Phy(t)$ effect of physical activity [mmol/min] represent the disturbances [12]. All parameters are assumed to be time-invariant.

Based on [12] and [13] the model is reduced to a 9th order one (three neglected states) as the linear transfers of the reduced states have time constants comparable to the sampling time (T_s) of the used sensor. As CGM measurements are available every 5 minutes, the following discrete-time sensors model is used:

$$x_d[k+1] = x_d[k] + w_d[k] \qquad y_1[k] = x_d[k] + C[k] + z_1[k] \qquad (2)$$

where $C[k] = C(k \cdot T_s)$, $x_d[k]$ is the state variable associated with sensor drift, $y_1[k]$ is the output of the sensor, finally $w_d[k]$ and $z_1[k]$ are white noises with Gaussian distribution. The former represents the disturbance that drives the sensor drift with 0.1667 $mmol^2/L^2$ variance, while the latter is additive measurement noise with 0.25 $mmol^2/L^2$ variance. Because of the drift, the sensor must be re-calibrated using a manual glucose measurement $y_2[k] = C[k] + z_2[k]$, which is assumed to have only additive measurement noise $z_2[k]$ with very small variance 0.0025 $mmol^2/L^2$ compared to sensor noise. However, these manual measurements cannot be done very often.

3   Closed-Loop Control

Here a relatively simple control method is proposed as the main focus of this paper is not on the control algorithm: Linear-quadratic-Gaussian (LQG) control [14]. It is basically a moving horizon model predictive control, which minimizes a quadratic constraint for the state variables and control signal, which is as follows:

$$J(x,u) = \tfrac{1}{2} \sum_{k=0}^{N-1} (<\mathbf{Q}_k x[k], x[k]> + <\mathbf{R}_k u[k], u[k]>) + \tfrac{1}{2} <\mathbf{Q}_N x[N], x[N]> \qquad (3)$$

where $J(x,u)$ is defined for an N sample-long time horizon, $x[k]$ represents the vector of state variables and $u[k]$ the controlled input. $< . >$ denotes the scalar product, while $\mathbf{Q}_k$ and $\mathbf{R}_k$ are appropriately chosen symmetric and positive definite matrices. In the case of a linear time-invariant system given in state-space form with matrices $\mathbf{A}$, $\mathbf{B}$ and $\mathbf{C}$ and time-invariant weighting matrices $\mathbf{Q}$ and $\mathbf{R}$ as well as infinite time horizon $(N = \infty)$, $J(x,u)$ can be minimized using state-feedback control $u[k] = \mathbf{K}x[k]$. The gain for the feedback can be acquired by solving Discrete-time algebraic Riccati equation:

$$\begin{aligned} \mathbf{X} &= \mathbf{Q} + \mathbf{A}^T \mathbf{X} \mathbf{A} - \mathbf{A}^T \mathbf{X} \mathbf{B} (\mathbf{B}^T \mathbf{X} \mathbf{B} \mathbf{R})^{-1} \mathbf{B}^T \mathbf{X} \mathbf{A} \\ \mathbf{K} &= (\mathbf{B}^T \mathbf{X} \mathbf{B} \mathbf{R})^{-1} \mathbf{B}^T \mathbf{X} \mathbf{A} \end{aligned} \qquad (4)$$

Since model (1) is nonlinear, it is approximated with a linear model in every step k, then the gain for the state feedback $K[k]$ is computed. For better disturbance rejection integral control has been included [14] and the linear model extended accordingly. The control law is as follows:

$$\begin{aligned} x_i[k+1] &= x_i[k] + y[k] - 4.9 \\ u[k] &= K[k] \left(\left(x_b \; 0 \right)^T - \left(x[k] \; x_i[k] \right)^T \right) + u_b \end{aligned} \qquad (5)$$

where x_b is the steady state value of the state variables, $x_i[k]$ is the state variable corresponding to integral control, and u_b is the steady state control input for zero meal intake and 4.9 $mmol/L$ normoglycemic blood glucose concentration denoted with $y[k] = Q_1[k]/V_G$. For $n \times n$ matrix $\mathbf{Q}$ and matrix $\mathbf{R}$ (that is a scalar) used in the cost function (3) the following values were selected:

$$\mathbf{Q} = e_1 e_1^T + e_n e_n^T \qquad \mathbf{R} = 10 \qquad (6)$$

where e_i denotes the i-th unit vector in $\mathbb{R}^n$. The controller was tuned for extreme meal intake and physical activity scenario detailed later in Section 6.

4   Observer Design

Since both significant disturbances, measurement noise and sensor drift are present state estimation is required using Kalman filtering technique. Sigma-point filters are an increasingly popular option in the case of nonlinear models [15]. These use a special set of points - called sigma points - to estimate the mean and covariance of distributions needed in the Kalman filter algorithm. A good example would be the Gauss-Hermite Quadrature Filter (GHQF) [16]. It offers high accuracy when all disturbances and measurement noises have Gaussian distribution, but requires a large number of sigma-points. Hence, it needs relatively large computational power undesirable in certain practical applications. Sparse-grid quadrature nonlinear filtering (SGQF) can overcome this dimensionality problem [15], and consequently was chosen for our application as well.

4.1   Sigma-Point Selection

Let us introduce the notation χ for a set of sigma-points. This set contains N sigma points denoted as ξ_i, $i = 1, \ldots, N$. The sigma-points represent the stochastic variable μ with mean $\hat{\mu}$ and covariance matrix Σ, and can be written in the form: $\xi_i = \Sigma^{\frac{1}{2}} \varphi_i + \hat{\mu}$. $\Sigma^{\frac{1}{2}}$ is the factor of Σ so that $\Sigma = \Sigma^{\frac{1}{2}} \Sigma^{\frac{T}{2}}$, and since Σ is positive definite the Cholesky decomposition is used. φ is an additional vector used for sigma point determination [15]. μ is not limited to state variables only, it can contain the disturbances and measurement noises as well. The dimension of μ will be denoted with L. GHQF requires m^L sigma points, where m is usually set to 3. Level-3 SGQF provides a good approximation of the results of GHQF, but requires only $2L^2 + 4L + 1$ or less sigma-points. The exact number depends on how the three parameters of sigma-point filter (p_1, p_2, p_3) are set [15]. They reflect the case of univariate estimation, where the points $\mu + \{-p_1, 0, p_1\}$ and $\mu + \{-p_3, -p_2, 0, p_2, p_3\}$ are used to estimate certain moments of a univariate Gaussian distribution transformed by a nonlinear function. If all parameters are different the sigma-points used in the level-3 SGQF are shown in equations (7) and (8), where $C = L(L-1)/2$, while $\hat{\omega}_1, \ldots, \hat{\omega}_5$ are defined from the parameters $p_1, \ldots, p_3$ using moment matching method.

$$\omega_i = \begin{cases} \frac{(L-1)(L-2+L\hat{\omega}_i^2)}{2} - L(L-1)\hat{\omega}_1 + L\hat{\omega}_3 & i = 1 \\ (L-1)\hat{\omega}_2(\hat{\omega}_1 - 1) & i = 2, \ldots, 2L+1 \\ \hat{\omega}_4 & i = 2L+2, \ldots, 4L+1 \\ \hat{\omega}_5 & i = 4L+2, \ldots, 6L+1 \\ \hat{\omega}_2^2 & i = 6L+2, \ldots, 6L+1+4C \end{cases} \quad (7)$$

$$\varphi_i = \begin{cases} 0 & i = 1 \\ e_i p_1 & i = 2, \ldots, L+1 \\ -e_i p_1 & i = L+2, \ldots, 2L+1 \\ e_i p_2 & i = 2L+2, \ldots, 3L+1 \\ -e_i p_2 & i = 3L+2, \ldots, 4L+1 \\ e_i p_3 & i = 4L+2, \ldots, 5L+1 \\ -e_i p_3 & i = 5L+2, \ldots, 6L+1 \\ e_i p_1 + e_j p_1, & i = 6L+2, \ldots, 6L+1+C \ j \neq i \\ -e_i p_1 + e_j p_1, & i = 6L+2+C, \ldots, 6L+1+2C \ j \neq i \\ e_i p_1 - e_j p_1, & i = 6L+2+2C, \ldots, 6L+1+3C \ j \neq i \\ -e_i p_1 - e_j p_1, & i = 6L+2+3C, \ldots, 6L+1+4C \ j \neq i \end{cases} \quad (8)$$

The complete discrete-time nonlinear system created from the T1DM model (1) using Euler method, and the sensor model (2) is of the following form:

$$x[k+1] = f(x[k]) + \mathbf{B}w[k] \qquad y[k] = \mathbf{C}x[k] + z[k] \qquad (9)$$

where $x[k] = (C[k], Q_1[k], Q_2[k], x_1[k], S_2[k], S_1[k], G_2[k], G_1[k], x_d[k])^T$ and $y[k] = (y_1[k], y_2[k])^T$ or only $y_1[k]$ depending on whether manual measurement is available or not.

5 Fuzzy Calibration Supervisor

Since the CGM has drift, it needs to be repeatedly re-calibrated. The used filter can compensate to some degree; however, the state estimation will eventually deviate from the true value, and the controller relying on those estimations will malfunction as well. The goal is to keep the glucose concentration levels within normoglycemic range, and most importantly to avoid hypoglycemia. However, the manual measurements needed for calibration cannot be made too frequently. Furthermore, the patient must sleep, and hence calibration is only allowed at night if it is to avoid a potentially life-threatening situation. Mamdani Fuzzy Interference System was used to determine the timing of the calibration. There are six inputs:

1. Time passed since the last manual measurement. The fuzzy sets associated with this input represent that the last calibration was: not long ago (< 60-90 *min*), really long ago (> 12 *hours*), or neither;

2. Actual time. The fuzzy sets associated are that the patient is probably sleeping (from around 11 pm to around 6 am), or that the patient is about to go to sleep;

3. $7.8 - \hat{C}[k] - \sqrt{[\Sigma_{xx}]_{1,1}[k]}$, where $\hat{C}[k]$ is the estimated glucose concentration in the subcutaneous tissue, $[\Sigma_{xx}]_{1,1}[k]$ is the variance of the estimation error $C[k] - \hat{C}[k]$. $7.8\ mmol/L$ is the lower bound of severe hyperglycemia. There are three fuzzy sets: one associated with severe hyperglycemia, one representing when the signal might be close to the border, finally one capturing the case when the signal is far even considering the estimation error;

4. The time derivative of the above signal; The decreasing and increasing trend of this signal is captured by two fuzzy sets.

5. $\hat{C}[k] - \sqrt{[\Sigma_{xx}]_{1,1}[k]} - 3.9$, since the glucose concentration decreasing below $3.9\ mmol/L$ is considered hypoglycemia. The fuzzy sets are the same as for the third input;

6. The time derivative of the above signal. Same fuzzy sets as in the case of the 4th input;

The rules used are as follows:

- If a lot of time has passed since the last calibration, a manual measurement is needed, however
- If the last measurement was only a short time ago, there must not be a measurement.
- If the patient is probably asleep, and the estimation does not strongly indicate hypoglycemia, there should be no measurement.
- If the patient is about to go to sleep, there should be a measurement, unless both hypo-, and hyperglycemia is highly unlikely.
- If there is a hyperglycemic episode, and the blood glucose levels are assumed to be increasing, calibration is needed.
- If there is a hypoglycemic episode, or the patient is close to a hypoglycemic episode, calibration is needed.
- If the patient is in or close to a hyperglycemic episode, and the blood glucose levels are assumed to be increasing, calibration is needed.
- If the patient is close to a hyperglycemic episode, but the blood glucose levels are assumed to be decreasing, measurement is not needed.
- If the patient is close to a hypoglycemic episode, but the blood glucose levels are assumed to be increasing, measurement is not needed.

It is needless to see that the actual parameters of this module can depend on various factors, such as lifestyle or sensor properties. Furthermore, the parameters of the fuzzy sets should reflect the uneven significance of hypo- and hyperglycemia.

6   Results

Simulations were conducted to show the capabilities of the proposed system. The numerical values of the model parameters were randomly chosen from the

6 patient parameter sets presented in [12] and 50 virtual patient parameter sets generated using the parameter bounds also presented in [12]. The meal intake and physical activity was randomized as well. Each simulation covered 48 hours and assumed unusually high carbohydrate (CHO) intake (180g CHO to 310g CHO). The details are summarized in Table 1. Furthermore there is 50 % chance of physical activity starting between 9:00-12:00 for 1-4 hours. Uniform distribution was used in all cases. The SGQF is assumed to have 10 % initial estimation error, and that the patient does not miss any of the manual measurements requested by the so-created Calibration Supervisory unit. Fig. 1 shows an example for one of the simulations. The acquired results are summarized in Table 2.

Table 1. High carbohydrate intake simulation parameters

	Breakfast	Snack 1	Lunch	Snack 2	Dinner	Snack 3
Chance of occurrence	100 %	50 %	100 %	50 %	100 %	50 %
Amount [g]	40-60	5-25	70-110	5-25	55-75	5-15
Time [hour]	6-10am	8-11am	11am-3pm	3-6pm	6-10pm	10pm-12am

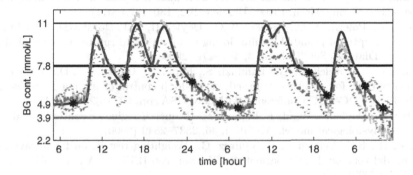

Fig. 1. Example of simulation results. The real (solid blue line), measured (dashed green line) and estimated (dash-dotted magenta line) blood glucose concentration are displayed, as well as the estimated error bounds (dotted red line) and the instances of the manual measurement (black asterisk). The three solid horizontal lines in the background mark the upper (7.8 mmol/L) and lower (3.9 mmol/L) bounds of the normoglycemic region, as well as the threshold for severe hyperglycemia (11 mmol/L).

Table 2. Simulation results for high CHO intake scenario. The percentage represents the time spent in the designated regions through all 500 Monte Carlo simulations.

Hypoglycemia ($< 3.9\ mmol/L$)	1.10 %
Normoglycemia ($3.9\text{-}6\ mmol/L$)	33.56 %
Mild hyperglycemia ($6\text{-}7.8\ mmol/L$)	26.01 %
Hyperglycemia ($6\text{-}11\ mmol/L$)	63.26 %
Severe hyperglycemia ($>11.1\ mmol/L$)	2.08 %
Average number of calibrations	10.91

7 Conclusion

The proposed controller, aided by the sparse-grid quadrature filter and the calibration supervisor could reduce both hypoglycemic and severe hyperglycemic episodes for all virtual patients in the case of extreme meal intake and sensor drift. A more sophisticated control algorithm and the optimization of filter parameters and the calibration supervisor could improve these results further. In future works the robustness of the solution must be examined, since it is safe to assume that the used T1DM model is inaccurate, let it be parameter inaccuracy or additive/multiplicative uncertainty defined in frequency domain. Furthermore, additional safety measures must be taken if the patient is less cooperative and reliable with the sensor calibration. It is also possible to use the filter for prediction, hence providing more information for the supervisor unit.

References

1. Palerm, C.: Physiologic insulin delivery with insulin feedback: A control systems perspective. Comp. Meth. Progr. Biomed. 102, 130–137 (2011)
2. Zisser, H., Palerm, C.C., Bevier, W.C., Doyle III, F.J., Jovanovic, L.: Clinical update on optimal prandial insulin dosing using a refined run-to-run control algorithm. J. Diabetes Sci. Technol. 3, 487–491 (2009)
3. Palumbo, P., Pizzichelli, G., Panunzi, S., Pepe, P., Gaetano, A.D.: Tests on a virtual patient for an observer-based, closed-loop control of plasma glycemia. In: 50th IEEE CDC-ECC Conference, Orlando, USA, pp. 6936–6941 (2011)
4. Parker, R., Doyle, F., Ward, J., Peppas, N.: Robust $\mathcal{H}_\infty$ glucose control in diabetes using a physiological model. AIChE J. 46, 2537–2549 (2000)
5. Femat, R., Ruiz-Velazquez, E., Quiroz, G.: Weighting restriction for intravenous insulin delivery on T1DM patient via $\mathcal{H}_\infty$ control. IEEE T. Autom. Sci. Eng. 6, 239–247 (2009)
6. Hovorka, R., Canonico, V., Chassin, L., Haueter, U., Massi-Benedetti, M., Federici, M.O., Pieber, T., Schaller, H., Schaupp, L., Vering, T., Wilinska, M.: Nonlinear model predictive control of glucose concentration in subjects with type 1 diabetes. Physiol. Meas. 25, 905–920 (2004)
7. Kovatchev, B., Cobelli, C., Renard, E.: Multi-national study of subcutaneous model-predictive closed-loop control in type 1 diabetes: summary of the results. J. Diabetes Sci. Technol. 4, 1374–1381 (2010)
8. Kovács, L., Benyó, B., Bokor, J., Benyó, Z.: Induced $\mathcal{L}_2$-norm minimization of glucose-insulin system for type I diabetic patients. Comp. Meth. Progr. Biomed. 102, 105–118 (2011)
9. Pena, R.S., Ghersin, A., Bianchi, F.: Time-varying procedures for insulin-dependent diabetes mellitus control. J. Electr. Comp. Eng. 2011, 1–10 (2011)
10. Phillip, M., Battelino, T., Atlas, E., Kordonouri, O., Bratina, N., Miller, S., Biester, T., Stefanija, M., Muller, I., Nimri, R., Danne, T.: Nocturnal glucose control with an artificial pancreas at a diabetes camp. N. Engl. J. Med. 368, 824–833 (2013)
11. Zarkogianni, K., Vazeou, A., Mougiakakou, S., Prountzou, A., Nikita, K.: An insulin infusion advisory system based on autotuning nonlinear model-predictive control. IEEE J. Biomed. Eng. 58, 2467–2477 (2011)

12. Wilinska, M., Chassin, L., Acerini, C., Allen, J., Dunger, D., Hovorka, R.: Simulation environment to evaluate closed-loop insulin delivery systems in type 1 diabetes. J. Diabetes Sci. Technol. 4, 132–144 (2010)
13. Szalay, P., Eigner, G., Kozlovszky, M., Rudas, I., Kovács, L.: The significance of LPV modeling of a widely used T1DM model. In: EMBC 35th Annual International Conference of the IEEE, Osaka, Japan, pp. 3531–3534 (2013)
14. Lantos, B.: Theory and design of control systems II. Akademia, Budapest (2003)
15. Jia, B., Xin, M., Cheng, Y.: Sparse-grid quadrature nonlinear filtering. Automatica 48, 327–341 (2012)
16. Arasaratnam, I., Haykin, S., Elliot, R.: Discrete-time nonlinear filtering algorithms using Gauss–Hermite quadrature. Proc. IEEE 95, 953–977 (2007)

Webpage Segmentation Using Ontology and Word Matching

Huey Jing Toh and Jer Lang Hong

School of Computing and IT, Taylor's University, Subang Jaya, Malaysia
{hueyjing.toh,jerlang.hong}@taylors.edu.my

Abstract. Webpage segmentation is a non trivial task and is a promising research area in the field of computing study. Webpage segments demarcate informative and non-informative content in a webpage through the extraction of text and image. Not only that, segments can also distinguish different types of information between segments. Webpage segmentation is certainly useful in web ranking, classification, and other web mining applications. Segments identification is also useful in page display for constraint limited screen devices such as smart phones, PDAs etc. Recent research focused on using ontology tool to segment a webpage. However, this tool only supports English language. In this paper, we propose a multilingual ontology tool for segmenting webpage. Our tool has shown higher accuracy than existing methods for webpage segmentation.

Keywords: Webpage Segmentation, Ontology, Word Matching.

1 Introduction

In this modern era, society has been increasingly reliant on the constantly advancing technology for the smooth and reliable on-goings of most of their daily activities. With the introduction of hardware and software from the field of IT and technology to the homes and offices of the global community, countless data are created, processed, stored, and transferred on a regular basis, regardless from a local machine or from one device to another. With numerous types of information to be displayed and manipulated by technology users, an abundant variety of data formats have been created, whereby whichever format proven to be most suitable to the intended purpose and circumstances may be opted to be used by the individual or organization in order to portray the data as accurately as possible. For storing and displaying static data, documents such as text files, Word documents, and Excel spreadsheets are put to use. On the other hand, web-based formats such as HTML pages and XML files are also utilized to display a different set of information to the public.

With the abundance of data formats for selection, digital device users are provided with the freedom to create and display as much information in whichever format as they see fit. However, this in turn causes the existence of a troubling issue when a third party wishes to extract and make use of particular components within the data

C.K. Loo et al. (Eds.): ICONIP 2014, Part II, LNCS 8835, pp. 454–461, 2014.

files as used and portrayed by its author(s). First of all, this data appears homogenous to the users and it is difficult to differentiate them apart using existing tools. Secondly, this data is in huge quantities, therefore processing them manually is labor intensive and time consuming. Finally, there is no standard convention is presenting this data formatted in its own language. For example, HTML language is ambiguous and lacks uniformity in its design. English language on the other hand, has a rather loose grammar and is highly ambiguous in its representation. To resolve this issue, researcher tend to segment webpages into demarcate segments. These segments, which contain valuable information, will be very helpful to separate the contextual information apart. This in turn, will help to speed up the processing power and accuracy in further research such as webpage classification, indexing, and extraction.

In this paper, we propose a novel ontology segmentation technique to segment a webpage into segments. Unlike previous work, our tool provides multilingual support and could effectively segments a webpage with high accuracy. Experiment results show that our tool is highly efficient in segmenting webpages compared to existing state of the art tools.

2   Related Work

Generally, the typical webpages that are regularly viewed online can be segregated into multiple sections or specific domains. These particular areas can encompass, but are not limited to, advertisement displays, navigational links, headers, footers, as well as the content area itself. In the case of some of these sections, further break down and partitioning may also occur. Therefore, the term *webpage segmentation* refers to the process of conducting the meaningful partitioning of a single webpage into the previously mentioned areas or section, as deemed understandable to a human viewer of the page itself. Currently, there exist a variety of approaches which are catered to conducting the segmentation of online webpages, most of which are based on the webpage's own DOM Tree and is used to accommodate applications for information retrieval, such as de-duplication detection [4], automatic page adaption [6] [12], information or content extraction [18], as well as web searches which are keyword-based, amongst other similar applications. Classified as a form of image segmentation problem [17], webpage segmentation approaches are typically considered to be computationally expensive.

In terms of the DOM trees, they are usually casted as weighted graphs. By doing so, with the utilization of the meta-heuristic Graph-Theoretic approach [4], the weights may be used to determine the placement of certain sets of nodes, such as if they should be located separately or within the same segment. As for the sub-trees of the DOM trees, individual blocks are identified and manipulated, by taking the entropies of the terms within the blocks [13], then conducting a thorough comparison between them. Furthermore, according to the work published by Cai et al. [3], the VIPS algorithm which was used in their paper takes on a computer vision perspective when addressing the issue of webpage segmentation. This means that DOM tree based heuristics as well as additional visual cues are key components for the optimal performance of the segmentation algorithm, not to mention strategies involving quantitative linguistics strategies are also involved in certain cases. First seen as being utilized amongst image retrieval systems, VIPS algorithm works on segmenting webpages by extracting contextual-based information from the particular webpage [3] [19]. With the addition of studies done by Liu at al. [16] and He et al. [9], it was elaborated on editable pDOC values and its contribution in resolving problems relating to extracting image segments on webpages.

According to the papers by Hua et al. [10] and Feng et al. [8], webpages which undergo the segmentation process are essentially performed based on the structural and separator tags found within the DOM tree. Unfortunately, a restriction of the procedure is that it is known to be successful on only a limited amount of online webpages. On the other hand, Li et al.'s research focuses more on visual cues, such as the size and the position of the different domains found on a webpage in order to perform the segmentation, while a linguistic based approach to extract contextual information and images is preferred by Joshi and Liu [11] in their paper. In the case of the later of the two papers, several assumptions were made, including the belief that valid images only consist of images which are accompanied by captions. Lastly, the latest survey done on the topic of webpage segmentation is by Alcic and Conrad [1], which revolves around another different approach, namely clustering-based. Nevertheless, performing any segmentation of webpages on a large scale, as long as an algorithm which is visual-based is being utilized, it has been proven to relatively computationally expensive.

3   Problem Formulation

Segmenting a webpage is a non trivial task. Various approaches have been used to segment a webpage efficiently. Some of the well known state of the art methods are utilizing DOM Tree and visual cue, and recently, a few attempts have been made on using ontologies. All of these methods are efficient, but they are not without problems. The biggest concern and flaws in all these methods is that they failed to translate machine understandable form to human understandable form. Machines recognized HTML language as well as visual information such as width and height. Recent ontology tools attempt to address semantic gap, they are proven to be helpful to certain extent. However, they may not addressed the problem fully. This is because human perception on webpages covers a larger amount of information than what do the machine is capable of, from visual view, to language, and most importantly, human intelligence. Due to the huge complexity of semantic gap, our research attempts to leverage the problem mentioned previously, by addressing the semantic gap using common ontology WordNet.

4   Motivation

We use common ontology WordNet to segment webpage. Inspired by the extensive libraries provided by WordNet and the multilingual as well as cross platform supports, we believe that using WordNet is the best choice for segmenting webpage efficiently. Though research on using ontology tools for segmenting webpage is still in its infancy, ontologies research has shown a dramatic increase in its application unlike other approaches. This is because ontology tools are able to achieve higher precision than other conventional methods. Secondly, recent research on ontologies provide support for higher intelligence, that is they provide more semantic capabilities and larger knowledge domains for the researchers. Suggested Upper Merged Ontology (SUMO) for example, gave upper ontology support for the common ontology WordNet by providing a formal ontology containing 25,000 terms and 80,000 axioms. On the other hand, research has been widely carried out to provide multilingual support for WordNet. This is a significant advantage for us, as different languages have different syntaxes and structures in their representation, making them complicated to be included in the WordNet library. An alternative implementation is therefore required which involves complete reimplementation from the syntaxes, synsets, structure, and word relationships. Due to the fact that WordNet contains multilingual support for 90% of the languages in this world, it is certainly possible for us to use ontology to segment a webpage without much difficulty. We are able to segment most webpages if not all, written in any languages having different structure and layout.

5   Proposed Solution

Overview
To properly segment a webpage, we have to carefully define the boundary of each segment in the webpage. We conduct a user study to identify segments in a webpage. A sample of 5 users are chosen to identify segments in each webpage. If more than 3 users identify the same piece of segments for that particular webpage, that segments are considered as correct segment. Otherwise, further validation is carried out by having additional users to correctly identify segments in a webpage. This step is repeated until the majority of users identify a segment correctly. Once the user study is completed, we use the same set of webpages to test our method. We develop a novel ontology based method to segment a webpage. To achieve this, we use DOM Parser (ICE Browser) to obtain the DOM Tree and common ontology WordNet to analyze the semantic properties of data in the webpage.

Constructing Domain Ontology
First, our method parses through the webpage and constructs DOM Tree accordingly. Once the DOM Tree is obtained, we traverse through the DOM Tree in a depth first search manner and locate each of the text nodes in that tree. We compare the text in each of these nodes. To compare them, we tokenize the text into individual words. Once the text is tokenized, we match each of the words with each other using WordNet similarity check (Jiang and Conrath algorithm). A pair of words is considered matched if the similarity is more than 70%. Otherwise they are considered not matched. We also match a bag of words (2 to 4 words) with other bag of words using

WordNet similarity check. When bags of words are matched, we stored them in a list. This list will then be used to construct the ontology domain of our methods.

Cluster Groups

Once we obtained the list of similar keywords from the webpage, we identify the location of these keywords. If more than one keywords are located within a text node, we consider that text node as highly similar. If a node which contains 5 text nodes, of which at least 3 of the text nodes contain similar keywords, we label that node as potential segment. Otherwise, we traverse the tree upwards until a higher match is obtained for the similar keywords in that particular subtree. Once a potential segment is found, we stop traversing the tree further, and we continue with finding other potential segments. When a list of segments are found, we named this list as cluster groups.

Initial Segmentation

We use ontology tool to segment a webpage. This tool checks for the semantic properties of data (words which contain similar semantics) in the segment to identify segments which are semantically related. We categorized segments into two types, they are segments which are semantically related and segments without semantic properties. Segments with semantic properties are those segments with high level of similar keywords, whereas segments without semantic properties are those with low level of similar keywords. To differentiate these two types of segments, we check and locate the subtree which contains text nodes. Once a subtree with at least two text nodes are located, we identify keywords with similar semantic properties in these text nodes. If these text nodes contain a high degree of similar keywords, we treat these subtree as a segment. Otherwise we traverse the tree upwards to locate text nodes with similar keywords. The traversal is repeated until 10 text nodes are found. At this stage, we consider the subtree as segment without semantic properties.

Iterative Segmentation

To enhance the segmentation process of our method, we repeat the previous procedure with an additional step of clustering the segments in an iterative manner so that a higher accuracy could be obtained. We adopt SOM based clustering technique to group segments which are semantically similar. Unlike other existing clustering techniques, SOM clustering technique is a multi objective based clustering, which make it suitable for detecting and differentiating groups of data with similar properties due to its varying degree of classifying groups. Two segments may have similarity score with differences of only 5% within the same segment group. For example, the segments containing information related to *Cat* and *Dog* can be grouped together as both of them are *canine, mammal,* and *animal*. However, the segment containing information related to *Cat* and *House* are not semantically similar as one is related to *animal* while the other is related to *building*. Once segments with similar semantic properties are identified, we can then extract out the segments using the conventional approach by extracting out the segment's DOM Tree.

Multilingual Support

We use common ontology WordNet for our method, which has multilingual support. Multilingual support is an added advantage as webpages are written in a number of languages, hence any ontology tool which can support up to 85% of these webpages are considered good. Though WordNet does have support for multilingual features, these languages are structured differently with different syntaxes and format. Therefore, it requires the users to handle these languages differently using WordNet. Fortunately, WordNet provides common interface and support for multi languages. To identify webpage segments written in other languages, we use multilingual Word-Net and repeat the previous steps so that segments containing important information could be extracted.

To check for semantic similarity between keywords written in other languages, we need to implement the similarity methods in WordNet to cater for other languages. Fortunately, it is not difficult to map the implementation of Word Matching in English to that of other languages of WordNet as the functionalities provided by WordNet across other languages are almost similar though the accuracy returned by all these different methods may not be exactly similar. For example, a match between Cat and Dog in English WordNet nay return 75% similar while that of Chinese WordNet may return 73% similar. Once we have implemented all the similarity check methods for WordNet written in other languages, we repeat the similarity check procedure used previously. Unlike other existing techniques, our approach is capable of segmenting a webpage regardless of its layout and structure. This is a significant advantage as our approach is able to solve the complexity and ambiguity inherent in the HTML language.

6   Experimental Tests

We conduct our experimental tests on a wide range of datasets. We collect a random sample of 200 pages from the deep web repositories. To measure the effectiveness of our algorithm, we use precision and recall which are formulated as follows:

Recall=Correct/Actual*100
Precision=Correct/Extracted*100

We compare our method against the state of the art segmentation algorithm by [5]. Before we use our method on the webpage, we conduct a user study to correctly identify an image segment and their surrounding contextual information. A segment is considered correctly extracted if the surrounding contextual information for an image is identical with that of the user study. Once our system extracts an image segment from the webpage, we consider that segment as correctly extracted. Images with no surrounding contextual information are ignored from our study and evaluation. The actual value signifies the actual amount of image segments in our study.

Table 1. Experimental Tests

	OntoSegment [5]	Our method
Actual	983	983
Extracted	785	866
Correct	604	824
Recall	61.44	83.83
Precision	76.94	95.15

Table 1 shows our experimental tests conducted on the sample pages. Our method outperforms the methods of [5] both in terms of recall and precision rates. This could be attributed to the fact that our method is able to detect the semantic properties of image segments rather than their visual boundaries. Detecting the semantic properties of image segments help to improve the accuracy of the system as this technique reflects more on the human perception of image segments. In addition to that, we use multilingual Word-Net to detect segments from webpages written in various languages, which could cover larger domain than other ontology tools which could only support English language.

7 Conclusions

We have shown that it is possible to segment a webpage using common ontology WordNet. Unlike other existing work, we have segmented webpages written in multi languages. This is a significant improvement as study has shown that webpage written in English language covers 80% of the total existing webpages in the world, with the remaining webpages written in other languages. Any method which could effectively segment webpages regardless of their written languages will be very helpful. Our experiment shows that our method could outperform existing methods in segmenting webpages. Our experiment also show that our method is able to cover larger language domain than existing ontology based segmentation tool.

References

1. Alcic, S., Conrad, S.: A Clustering-based Approach to Web Image Cotext Extraction. In: Proceedings of the Third International Conferences on Advances in Multimedia, pp. 74–79 (2011)
2. Cai, D., Yu, S., Wen, J.-R.: VIPS: a Vision-based Page Segmentation Algorithm VIPS: a Vision-based Page Segmentation Algorithm (2003)
3. Cai, D., He, X., Li, Z., Ma, W.Y., Wen, J.R.: Hierarchical clustering of WWW image search results using visual, textual and link information. In: Proceedings of the 12th Annual ACM International Conference on Multimedia, pp. 952–959. ACM (2004)
4. Chakrabarti, D., Kumar, R., Punera, K.: A graph-theoretic approach to webpage segmentation. In: Proceeding of the 17th International Conference on World Wide Web, pp. 377–386. ACM, New York (2008)

5. Chan, C.S., Adel, J., Hong, J.L., Goh, W.W.: Ontological based Webpage Segmentation. In: IEEE International Conference on Fuzzy Systems, and Knowledge Discovery, pp. 883–887 (2013)
6. Chen, Y., Ma, W.Y., Zhang, H.J.: Detecting web page structure for adaptive viewing on small form factor devices. In: Proceedings of the 12th International Conference on World Wide Web, pp. 20–24 (2003)
7. Fauzi, F., Hong, J.L.: Webpage Segmentation for Extracting Images and Their Surrounding Contextual Information. In: ACM International Conference on Multimedia, pp. 649–652 (2009)
8. Feng, H., Shi, R., Chua, T.-S.: A bootstrapping framework for annotat-ing and retrieving WWW images. In: Proceedings of the 12th Annual ACM International Conference on Multimedia, MULTIMEDIA 2004, p. 960. ACM Press, New York (2004)
9. He, X., Cai, D., Wen, J.R., Ma, W.Y., Zhang, H.J.: Clustering and searching WWW images using link and page layout analysis. ACM Transactions on Multimedia Computing, Communications, and Applications (TOMCCAP) 3(2), 10 (2007)
10. Hua, Z., Wang, X.J., Liu, Q., Lu, H.: Semantic knowledge extraction and annotation for web images. In: Proceedings of the 13th Annual ACM International Conference on Multimedia, pp. 467–470. ACM (2005)
11. Joshi, P.M., Liu, S.: Web document text and images extraction using DOM analysis and natural language processing. In: Proceedings of the 9th ACM Symposium on Document Engineering, DocEng 2009, p. 218. ACM Press, New York (2009)
12. Kang, J., Yang, J., Choi, J.: Repetition-based web page segmentation by detecting tag patterns for small-screen devices. IEEE Transactions on Consumer Electronics 56(2), 980–986 (2010)
13. Kao, II.-Y., Ho, J.-M., Chen, M.-S.: WISDOM: Web Intra page Informative Structure Mining based on Document Object Model. IEEE Transactions on Knowledge and Data Engineering 17(5), 614–627 (2005)
14. Kohlschutter, C., Nejdl, W.: A densitometric approach to web page segmentation. In: Proceeding of the 17th ACM Conference on Information and Knowledge Management (2008)
15. Li, J., Liu, T., Wang, W., Gao, W.: A broadcast model for Web image annotation. In: Proceedings of the 7th Pacific Rim Conference on Multimedia (2006)
16. Liu, J., Li, M., Liu, Q., Lu, H., Ma, S.: Image annotation via graph learning. Pattern Recognition 42(2), 218–228 (2009)
17. Pnueli, A., Bergman, R., Schein, S., Barkol, O.: Web Page Layout Via Visual Segmentation (2009)
18. Spengler, A., Gallinari, P.: Learning to extract content from news webpages. In: Proceedings of 2009 International Conference on Advanced Information Networking and Applications Workshops, pp. 709–714. IEEE (2009)
19. Wang, C., Zhang, L., Zhang, H.-J.: Learning to reduce the semantic gap in web image retrieval and annotation. In: Proceedings of the 31st Annual International ACM SIGIR Conference on Research and Development in Information Retrieval, SIGIR 2008, p. 355. ACM Press, New York (2008)

Continuity of Discrete-Time Fuzzy Systems

Takashi Mitsuishi[1], Takanori Terashima[2], Koji Saigusa[3], Nami Shimada[1],
Toshimichi Homma[4], Kiyoshi Sawada[1] and Yasunari Shidama[5]

[1] University of Marketing and Distribution Sciences, Kobe, Japan
[2] Hokkaido Hakodate Technical High School, Hakodate, Japan
[3] Kyushu Sangyo University, Fukuoka, Japan
[4] Osaka University of Economics, Osaka, Japan
[5] Shinshu University, Nagano, Japan
takashi_mitsuishi@red.umds.ac.jp

Abstract. The purpose of this study is to prove the existence of IF-THEN fuzzy rules which minimize the performance functional of the nonlinear discrete-time feedback control. In our previous study, the problem of fuzzy optimal control was considered as the problem of finding the minimum (maximum) value of the performance function with fuzzy approximate reasoning. This study analyzes a discrete-time system to make numerical simulation of a real model more simple and fast. A continuity of fuzzy approximate reasoning on the compact set of membership functions selected from continuous function space guarantees an optimal control.

Keywords: Approximate reasoning method, discrete-time fuzzy systems, functional analysis, calculus of variations.

1 Introduction

Recent years have seen a rapid development in information technology and computing systems. Accordingly, much attention has been given to the accurate and quick decision making systems. Various robots and application programs behave like a human being. Fuzziness is essential in those situations and becomes more important. Fuzzy approximate reasoning could apply not only to the field of engineering but also the field of business administration and social psychology [1, 2]. Although many researchers studied and proposed application examples of fuzzy logic control, it has not been analysed systematically and mathematically like usual control engineering. This study shows that the mathematical analysis of membership functions and approximate reasoning in the fuzzy logic control with the object of constructing automatically IF-THEN rules give optimal control.

The optimization of fuzzy control discussed in this paper is different from the conventional methods such as classical control and modern control. In our previous study, we considered fuzzy optimal control problems as problems of finding the minimum (maximum) value of the performance function with feedback law constructed by fuzzy rules through fuzzy approximate reasoning [3–5]. This study analyzed a discrete-time system to make numerical simulation of a

C.K. Loo et al. (Eds.): ICONIP 2014, Part II, LNCS 8835, pp. 462–469, 2014.

real model more simple and faster. The study also analyzed the existence of a unique solution of the state equation in the discrete-time system. To guarantee the convergence of the optimal solution, we proved the compactness of the set of membership functions. The fuzzy inference is continuous by assuming that it is a functional on the set. Then, we showed that the system has an optimal feedback control by essential use of compactness of the set of fuzzy membership functions. The pair of membership functions which is synonymous with IF-THEN rules and minimizes an integral performance function of fuzzy logic control exists.

2 Discrete-Time Fuzzy Systems, IF-THEN Rules and Approximate Reasoning

Throughout this paper, $\mathbb{R}^l$ denotes the l-dimensional Euclidean space with the usual Euclidean norm. Let $f(v_1, v_2) : \mathbb{R}^l \times \mathbb{R} \to \mathbb{R}^l$ be a (nonlinear) vector valued function which is Lipschitz continuous. In addition, assume that there exists a constant $M_f > 0$ such that $\|f(v_1, v_2)\| \leq M_f\,(\|v_1\| + |v_2| + 1)$ for all $(v_1, v_2) \in \mathbb{R}^l \times \mathbb{R}$.

Consider a system given by the following state equation:

$$x(n + 1) = f(x(n), u(n)), \tag{1}$$

where $x(n)$ is the state and the control input $u(n)$ of the system is given by the state feedback $u(n) = \rho(x(n))$. For a sufficiently large $r > 0$, $B_r = \{x \in \mathbb{R}^l : \|x\| \leq r\}$ denotes a bounded set containing all possible initial states x_0 of the system. Let N be a sufficiently large final time. Then, we have

Proposition 1. *Let $\rho : \mathbb{R}^l \to \mathbb{R}$ be a Lipschitz continuous function and $x_0 \in B_r$. Then, the state equation $x(n + 1) = f(x(n), \rho(x(n)))$ has a unique solution $x(n, x_0, \rho)$ $(n = 0, 1, 2, \cdots, N)$ with the initial condition $x(0) = x_0$ such that the mapping $(n, x_0) \in \{0, 1, 2, \cdots, N\} \times B_r \mapsto x(n, x_0, \rho)$ is continuous.*

For any $r_2 > 0$ and $\Delta > 0$, put

$$\Phi_\Delta = \{\rho : \mathbb{R}^l \to \mathbb{R};\ \sup_{u \in \mathbb{R}^l} |\rho(u)| \leq r_2,$$

$$\forall \in u, u' \in \mathbb{R}^l,\ |\rho(u) - \rho(u')| \leq \Delta\|u - u'\|\}.$$

Then, the following a) and b) hold.
a) For any $x_0 \in B_r$ and $\rho \in \Phi_\Delta$, for all $n = 0, 1, 2, \cdots, N$, there exists d_n, such that $\|x(n, x_0, \rho)\| \leq d_n \leq r_1$, where

$$r_1 = \max_{n=0,1,2,\cdots,N} \{d_n\}. \tag{2}$$

b) Let $\rho_1, \rho_2 \in \Phi_\Delta$. For all $n = 0, 1, 2, \cdots, N$, there exists $E(\Delta, L_f)$, such that

$$\|x(n, x_0, \rho_1) - x(n, x_0, \rho_2)\| \leq E(\Delta, L_f) \sup_{u \in [-r_1, r_1]^l} |\rho_1(u) - \rho_2(u)|, \tag{3}$$

where L_f is Lipschitz constant of f.

Proof. The existence and the uniqueness of the solution of the state equation (1) assured by the continuity and Lipschits condition of the function f respectively [6]. The continuity of the solution on $\{0, 1, 2, \cdots, N\} \times B_r$ can be proved obviously. We shall verify a) by induction. If $n = 0$, put $d_0 = r$, noting $x(0, x_0, \rho) = x_0 \in B_r$, then we have $\|x(0, x_0, \rho)\| \le r = d_0$. Thus it is valid for $n = 0$. Assuming it valid for $n = k$, that is, for $x_0 \in B_r$ and $\rho \in \Phi_\Delta$, there exist d_k such that $\|x(k, x_0, \rho)\| \le d_k$. We observe that

$$\|x(k+1, x_0, \rho)\| = \|f(x(k, x_0, \rho), \rho(x(k, x_0, \rho)))\|$$
$$\le M_f (\|x(k, x_0, \rho)\| + |\rho(x(k, x_0, \rho))| + 1) \le M_f(d_k + r_2 + 1) = d_{k+1}.$$

Put $d_{k+1} = M_f(r_2 + d_k + 1)$, then it valid for $n = k + 1$. Therefore it hold for $n = 0, 1, 2, \cdots, N$.

b) As the same way above, we shall proof b) by induction. Let $\rho_1, \rho_2 \in \Phi_\Delta$. If $n = 0$, put $E_0(\Delta, L_f) = 0$, noting that $x(0, x_0, \rho_1) = x_0$, $x(0, x_0, \rho_2) = x_0$, then we have

$$\|x(0, x_0, \rho_1) - x(0, x_0, \rho_2)\| = 0 \le E_0(\Delta, L_f) \sup_{u \in [-r_1, r_1]^l} |\rho_1(u) - \rho_2(u)|.$$

Assuming that for $x_0 \in B_r$, there exists $E_k(\Delta, L_f)$ such that

$$\|x(k, x_0, \rho_1) - x(k, x_0, \rho_2)\| \le E_k(\Delta, L_f) \sup_{u \in [-r_1, r_1]^l} |\rho_1(u) - \rho_2(u)|.$$

Then we have

$$\|x(k+1, x_0, \rho_1) - x(k+1, x_0, \rho_2)\|$$
$$\le L_f(E_k(\Delta, L_f) + \Delta E_k(\Delta, L_f) + 1) \sup_{u \in [-r_1, r_1]^l} |\rho_1(u) - \rho_2(u)|.$$

Put $E_{k+1}(\Delta, L_f) = L_f\{(1+\Delta)E_k(\Delta, L_f)+1\}$, then it valid for the case $n = k+1$. Thus the proof is completed with

$$E(\Delta, L_f) = \max_{k=0,1,2,\ldots,N} E_k(\Delta, L_f).$$

In this section we briefly explain the approximate reasoning using Mamdani method and product-sum-gravity method [2] which decides feedback output in the previous nonlinear system for the convenience of the reader.

Assume the feedback law ρ consists of the following m IF-THEN type fuzzy control rules.

RULE i: IF x_1 is $\underline{A}_{i1}$ and x_2 is $\underline{A}_{i2}$ ... and x_l is $\underline{A}_{il}$

THEN y is $\underline{B}_i$ $(i = 1, 2, \ldots, m)$ (4)

Here, m is the number of fuzzy production rules, and l is the number of premise variables $x_1, x_2, \ldots, x_l$. Let $A_{ij}(x_j)$ and $B_i(y)$ $(i = 1, 2, \ldots, m; j = 1, 2, \ldots, l)$ be fuzzy grade of each fuzzy set $\underline{A}_{ij}$ and $\underline{B}_i$ for input x_j and consequent output y

in the i-th rule respectively. For simplicity, we write "IF" and "THEN" parts in the rules by the following notation:

$$\mathcal{A}_i = (A_{i1}, A_{i2}, \ldots, A_{il}) \quad (i = 1, 2, \ldots, m),$$

$$\mathcal{A} = (\mathcal{A}_1, \mathcal{A}_2, \ldots, \mathcal{A}_m), \quad \mathcal{B} = (B_1, B_2, \ldots, B_m).$$

Then, the IF-THEN type fuzzy control rules (4) is called a fuzzy controller, and is denoted by $(\mathcal{A}, \mathcal{B})$. In the rules, the tuple of premise variable $x = (x_1, x_2, \ldots, x_l)$ is called an input information given to the fuzzy controller $(\mathcal{A}, \mathcal{B})$, and y is called an control variable.

In this study, when an input information $x = (x_1, x_2, \ldots, x_l) \in \mathbb{R}^l$ is given to the fuzzy controller $(\mathcal{A}, \mathcal{B})$ in other words, the IF-THEN rules (4), then one can obtain the amount of operation from the controller through the following procedures:

Procedure 1: The degree of each of the rules is calculated by

$$\alpha_{\mathcal{A}_i}(x) = \bigodot_{j=1}^{l} A_{ij}(x_j) \quad (i = 1, 2, \ldots, m).$$

Here, $\odot$ means scaling down calculation "product $\prod$" or clipping calculation "minimum $\wedge$" for the fuzzy intersection.

Procedure 2: The inference result of each i-th rule is calculated by

$$\beta_{\mathcal{A}_i B_i}(x, y) = \alpha_{\mathcal{A}_i}(x) \odot B_i(y) \quad (i = 1, 2, \ldots, m).$$

Procedure 3: The aggregated output of all rules is following.

$$\gamma_{\mathcal{A}\mathcal{B}}(x, y) = \bigoplus_{i=1}^{m} \beta_{\mathcal{A}_i B_i}(x, y).$$

Here, $\oplus$ means scaling down calculation "sum $\sum$" or "maximum $\bigvee$" for the fuzzy union.

Procedure 4: In the defuzzification stage the center of gravity method is adopted.

$$\rho_{\mathcal{A}\mathcal{B}}(x) = \frac{\int y \gamma_{\mathcal{A}\mathcal{B}}(x, y) dy}{\int \gamma_{\mathcal{A}\mathcal{B}}(x, y) dy}.$$

3 Compactness of a Family of Sets of Membership Functions

Fix a sufficiently large $r > 0$, $r_2 > 0$ and a final time N of the control (1) according to Section 2. Put r_1 be the positive constant determined by (2). We also fix $\Delta_{ij} > 0$ $(i = 1, 2, \ldots, m; \; j = 1, 2, \ldots, l)$. Let $C[-r_1, r_1]$ and $C[-r_2, r_2]$

be the Banach space of all continuous real functions on $[-r_1, r_1]$ and $[-r_2, r_2]$ respectively. We consider the following two sets of fuzzy membership functions.

$$F_{A_{ij}} = \{\mu \in C[-r_1, r_1]; \ 0 \le \mu(x) \le 1 \text{ for } \forall x \in [-r_1, r_1],$$
$$|\mu(x) - \mu(x')| \le A_{ij}|x - x'| \text{ for } \forall x, x' \in [-r_1, r_1]\},$$

$$G = \{\mu \in C[-r_2, r_2]; \ 0 \le \mu(y) \le 1 \text{ for } \forall y \in [-r_2, r_2]\}.$$

The set $F_{A_{ij}}$ above, which is more restrictive than G, contains triangular, trapezoidal and bell-shaped fuzzy membership functions with gradients less than positive value A_{ij}. Consequently, if $A_{ij} > 0$ is taken large enough, $F_{A_{ij}}$ contains almost all fuzzy membership functions which are used in practical applications. In this study, we shall assume that the fuzzy membership functions $\underline{A}_{ij}$ in premise parts of the IF-THEN rules (4) belong to the set $F_{A_{ij}}$ for all $i = 1, 2, \ldots, m$ and $j = 1, 2, \ldots, l$. On the other hand, we assume that the membership function $\underline{B}_i$ in consequent part belongs to G.

In the following, we endow the space $F_{A_{ij}}$ and G with norm topology on the space of continuous functions. Then, for all $i = 1, 2, \ldots, m; \ j = 1, 2, \ldots, l, F_{A_{ij}}$ and G are compact [3]. Put

$$\mathcal{L} = \prod_{i=1}^{m} \left(\prod_{j=1}^{l} F_{A_{ij}} \right) \times G^m.$$

Then, every element $(\mathcal{A}, \mathcal{B})$ of $\mathcal{L}$ is fuzzy controller given by the IF-THEN rules (4). By the Tychonoff theorem [8], we can have the following proposition.

Proposition 2. *$\mathcal{L}$ is compact and metrizable with respect to the product topology on $\left(C[-r_1, r_1]^l \times C[-r_2, r_2] \right)^m$.*

To avoid making the denominator of the fractional expressions in the defuzzification stage in the previous section equal to 0, for any $\delta > 0$, consider the set:

$$\mathcal{L}_\delta = \left\{ (\mathcal{A}, \mathcal{B}) \in \mathcal{L}; \ \forall x \in [-r_1, r_1]^l, \int_{-r_2}^{r_2} \gamma_{AB}(x, y) dy \ge \delta \right\}, \tag{5}$$

which is a slight modification of $\mathcal{L}$. If δ is taken small enough, it is possible to consider $\mathcal{L} = \mathcal{L}_\delta$ for practical applications. An element $(\mathcal{A}, \mathcal{B})$ of $\mathcal{L}_\delta$ is an admissible fuzzy controller. Since $\mathcal{L}_\delta$ is a closed subset of $\mathcal{L}$, we can have the following proposition trivially.

Proposition 3. *The set $\mathcal{L}_\delta$ of all admissible fuzzy controllers is compact and metrizable with respect to the product topology.*

4 Continuity of Approximate Reasoning for Existence of Solution and Optimization

For any $(\mathcal{A}, \mathcal{B})$ in $\mathcal{L}_\delta$ from (5), we define the feedback control low $u(x) = \rho_{AB}(x)$ of the state equation (1) at certain time $n \in \{0, 1, 2, \ldots, N\}$ on the basis of the IF-THEN rules (4).

$$\rho_{AB}(x) = \frac{\int_{-r_2}^{r_2} y\gamma_{AB}(x,y)dy}{\int_{-r_2}^{r_2} \gamma_{AB}(x,y)dy},$$

where

$$\gamma_{AB}(x,y) = \bigoplus_{i=1}^{m} \beta_{A_i B_i}(x,y), \quad \beta_{A_i B_i}(x,y) = \alpha_{A_i}(x) \odot B_i(y)$$

$$\text{and} \quad \alpha_{A_i}(x) = \bigodot_{j=1}^{l} A_{ij}(x_j) \quad (i = 1,2,\ldots,m).$$

Then, the following proposition about ρ_{AB} for Proposition 1 is obtained.

Proposition 4. *Let* $(A,B) \in \mathcal{L}_\delta$. *Then, the following* a) *and* b) *hold.*
a) ρ_{AB} *is Lipschitz continuous on* $[-r_1,r_1]^l$.
b) $|\rho_{AB}(x)| \le r_2$ *for all* $x \in [-r_1,r_1]^l$.

Proof. In this paper, this proposition is proved only about the case that ρ_{AB} is constructed by product-sum-gravity method [2]. In the case of Mamdani method, it is proved similarly.
a) Since ρ_{AB} is the composite mapping of α_{A_i}, $\beta_{A_i B_i}$ and γ_{AB}, Lipschitz continuities of α_{A_i}, $\beta_{A_i B_i}$ and γ_{AB} on $[-r_1,r_1]^l$ are already proved in previous studies [3, 4]. Fix $(A,B) \in \mathcal{L}_\delta$, for all $x = (x_1,x_2,\ldots,x_l), x' = (x_1',x_2',\ldots,x_l') \in [-r_1,r_1]^l$, we have

$$|\rho_{AB}(x') - \rho_{AB}(x)| \le \frac{4r_2{}^3}{\delta^2} \sum_{i=1}^{m} \sum_{j=1}^{l} \Delta_{ij} \|x' - x\|,$$

where Δ_{ij} is Lipschitz constant defined by previous section. This inequality shows Lipschitz continuity of ρ_{AB} on $[-r_1,r_1]^l$.
b) Omitted.

By Proposition 1 and 4, the state equation (1) for the feedback law ρ_{AB} has an unique solution $x(n,x_0,\rho_{AB})$ with the initial condition $x(0) = x_0$ [6].
    The performance index of this fuzzy feedback control system is evaluated with following integral performance function:

$$J = \int_{B_r} \sum_{n=0}^{N} w(x(n,\zeta,\rho_{AB}), \rho_{AB}(x(n,\zeta,\rho_{AB})))d\zeta \qquad (6)$$

where $w : \mathbb{R}^l \times \mathbb{R} \to \mathbb{R}$ is a positive continuous function. The performance function J depends on feedback ρ which consists of IF-THEN rules through the fuzzy approximate reasoning. Since admissible range of initial state B_r and final time N are known, the mapping J is a functional on the family of membership functions. The optimal control problem in this study is considered to be the calculus of variations by treating J as functional on function space. The following proposition guarantees the existence of admissible fuzzy controller which minimizes (maximizes) the previous performance functional (6).

Proposition 5. *The performance functional*

$$(\mathcal{A}, \mathcal{B}) \in \mathcal{L}_\delta \mapsto \int_{B_r} \sum_{n=0}^{N} w(x(n, \zeta, \rho_{AB}), \rho_{AB}(x(n, \zeta, \rho_{AB})))d\zeta$$

has a minimum (maximum) value on the compact space $\mathcal{L}_\delta$ defined by (5).

Proof. For this proposition, only a case of Mamdani method [1] is proved. The proof for the case of product-sum-gravity method can be given similarly and briefly. It suffices to prove the continuity of the mapping above on $\mathcal{L}_\delta$ as functional. We firstly have following inequality by induction. For all $i = 1, 2, \cdots, m$,

$$\left| \alpha_{\mathcal{A}_i{}^k}(x) - \alpha_{\mathcal{A}_i}(x) \right| = \left| \bigwedge_{j=1}^{l} A_{ij}^k(x_j) - \bigwedge_{j=1}^{l} A_{ij}(x_j) \right| \le \sum_{j=1}^{l} \left\| A_{ij}^k - A_{ij} \right\|_\infty,$$

here

$$\left\| A_{ij}^k - A_{ij} \right\|_\infty = \sup_{x_j \in [-r_1, r_1]} \left| A_{ij}^k(x_j) - A_{ij}(x_j) \right| \quad (i = 1, 2, \cdots, m; \; j = 1, 2, \cdots, l).$$

Next we can have easily

$$\left| \beta_{\mathcal{A}_i{}^k \mathcal{B}_i^k}(x, y) - \beta_{\mathcal{A}_i \mathcal{B}_i}(x, y) \right| \le \left\| A_i^k - A_i \right\|_\infty + \left\| B_i^k - B_i \right\|_\infty,$$

here $\left\| A_i^k - A_i \right\|_\infty = \sum_{j=1}^{l} \left\| A_{ij}^k - A_{ij} \right\|_\infty$. As the same way above, we ca have the following inequality by induction.

$$\left| \gamma_{\mathcal{A}^k \mathcal{B}^k}(x, y) - \gamma_{AB}(x, y) \right| = \left\| A^k - A \right\|_\infty + \left\| B^k - B \right\|_\infty,$$

here $\left\| A^k - A \right\|_\infty = \sum_{i=1}^{m} \left\| A_i^k - A_i \right\|_\infty, \left\| B^k - B \right\|_\infty = \sum_{i=1}^{m} \left\| B_i^k - B_i \right\|_\infty$. Noting that $\int_{-r_2}^{r_2} \gamma_{AB}(x, y)dy \ge \delta$, routine calculation gives the estimate

$$\left| \rho_{\mathcal{A}^k \mathcal{B}^k}(x) - \rho_{AB}(x) \right| \le \frac{4r_2{}^2}{\delta^2} \left(\left\| A^k - A \right\|_\infty + \left\| B^k - B \right\|_\infty \right).$$

Assume that $(\mathcal{A}^k, \mathcal{B}^k) \to (\mathcal{A}, \mathcal{B})$ $(k \to \infty)$ in $\mathcal{F}_\delta$. Then it follows from the estimate above that

$$\lim_{k \to \infty} \sup_{x \in [-r_1, r_1]^l} \left| \rho_{\mathcal{A}^k \mathcal{B}^k}(x) - \rho_{AB}(x) \right| = 0. \tag{7}$$

Fix $(n, \zeta) \in \{0, 1, 2, \cdots, N\} \times B_r$, by b) of Proposition 1, we have

$$\lim_{k \to \infty} \left\| x(n, \zeta, \rho_{\mathcal{A}^k \mathcal{B}^k}) - x(n, \zeta, \rho_{AB}) \right\| = 0. \tag{8}$$

Further, it follows from (7), (8) and a) of Proposition 1 that

$$\lim_{k \to \infty} \rho_{A^k B^k}(x(n, \zeta, \rho_{AB})) = \rho_{AB}(x(n, \zeta, \rho_{AB})). \tag{9}$$

Since w is continuous we have

$$\lim_{k \to \infty} \sum_{n=0}^{N} w(x(n, \zeta, \rho_{A^k B^k}), \rho_{A^k B^k}(x(n, \zeta, \rho_{A^k B^k})))$$

$$= \sum_{n=0}^{N} w(x(n, \zeta, \rho_{AB}), \rho_{AB}(x(n, \zeta, \rho_{AB}))). \tag{10}$$

It follows from (10) and the Lebesgue's dominated convergence theorem [7, 8] that the mapping is continuous on the compact metric space $\mathcal{F}_\delta$. As a result it has a minimum (maximum) value on $\mathcal{F}_\delta$, and the proof is complete.

5 Conclusion

A mathematical analysis of nonlinear discrete-time feedback control system that the feedback is constructed by fuzzy approximate reasoning has described. The existence of a unique solution of the state equation in the discrete-time system resulted from the continuity of inference method on the state variable. According to the mathematical considerations, we conclude that there exists minimum value of performance function on the family of sets of membership functions. Future research will focus on the method of solution by the calculus of variations using successive approximate with compactness of the set of membership functions and utilize the fuzziness in the context of social sciences.

References

1. Mamdani, E.H.: Application of fuzzy algorithms for control of simple dynamic plant. Proc. IEE 121(12), 1585–1588 (1974)
2. Mizumoto, M.: Improvement of fuzzy control (IV) - Case by product-sum-gravity method. In: Proc. 6th Fuzzy System Symposium, pp. 9–13 (1990)
3. Mitsuishi, T., Kawabe, J., Wasaki, K., Shidama, Y.: Optimization of Fuzzy Feedback Control Determined by Product-Sum-Gravity Method. Journal of Nonlinear and Convex Analysis 1(2), 201–211 (2000)
4. Mitsuishi, T., Endou, N., Shidama, Y.: Continuity of Nakamori Fuzzy Model and Its Application to Optimal Feedback Control. In: Proc. IEEE International Conference on Systems, Man and Cybernetics, pp. 577–581 (2005)
5. Mitsuishi, T., Terashima, T., Shidama, Y.: Optimization of SIRMs Fuzzy Model Using Łukasiewicz Logic. In: Huang, T., Zeng, Z., Li, C., Leung, C.S. (eds.) ICONIP 2012, Part II. LNCS, vol. 7664, pp. 108–116. Springer, Heidelberg (2012)
6. Miller, R.K., Michel, A.N.: Ordinary Differential Equations. Academic Press, New York (1982)
7. Riesz, F., Sz.-Nagy, B.: Functional Analysis. Dover Publications, New York (1990)
8. Dunford, N., Schwartz, J.T.: Linear Operators Part I: General Theory. John Wiley & Sons, New York (1988)

Sib-Based Survival Selection Technique for Protein Structure Prediction in 3D-FCC Lattice Model

Rumana Nazmul[1] and Madhu Chetty[2]

[1] Faculty of Information Technology, Monash University, Australia
rumana.nazmul@monash.edu
[2] Faculty of Science and Technology, Federation University, Australia
madhu.chetty@federation.edu.au

Abstract. Protein Structure Prediction (PSP) is a challenging optimization problem in computational biology. A large number of non-deterministic approaches such as Evolutionary Algorithms (EAs) have been have been effectively applied to a variety of fields though, in the rugged landscape of multimodal problem like PSP, it can perform unsatisfactorily, due to premature convergence. In EAs, selection plays a significant role to avoid getting trapped in local optima and also to guide the evolution towards an optimal solution. In this paper, we propose a new Sib-based survival selection strategy suitable for application in a genetic algorithm (GA) to deal with multimodal problems. The proposed strategy, inspired by the concept of crowding method, controls the flow of genetic material by pairing off the fittest offspring amongst all the sibs (offspring inheriting most of the genetic material from an ancestor) with its ancestor for survival. Furthermore, by selecting the survivors in a hybridized manner of deterministic and probabilistic selection, the method allows the exploitation of less fit solutions along with the fitter ones and thus facilitates escaping from local optima (minima in case of PSP). Experiments conducted on a set of widely used benchmark sequences for 3D-FCC HP lattice model, demonstrate the potential of the proposed method, both in terms of diversity and optimal energy in regard to various state-of-the-art selection methods.

Keywords: Protein Structure Prediction, Diversity, Survival Selection.

1 Introduction

In applying optimization to complex real-world problems e.g., Protein Structure Prediction (PSP), the main challenge is caused by the complex landscape of multimodal problem [1]. The search process must be able to escape the local minima and to obtain several good solutions simultaneously, to guide the search towards global minima. However, like any other population-based EAs, Genetic Algorithms (GAs) starting with an initial population of sufficiently distinct individuals, are also compelled to converge prematurely, due to the stochastic error induced by the genetic operators. Since the selection operators affect the

C.K. Loo et al. (Eds.): ICONIP 2014, Part II, LNCS 8835, pp. 470–478, 2014.
© Springer International Publishing Switzerland 2014

population of the next generation either by selecting parents for reproduction (parental selection) or by determining the individuals to survive (survival selection), they have an important role to play. Usually selection methods widely used for both reproduction and survival selection, control the direction of the search by following the law of "survival of the fittest". Even though, favoring the fitter individuals help to speed up the convergence process, it causes a decrease in the possibility of finding global optima [2]. This is because, it restricts the participation of less-fit candidates, instead of giving them the opportunity to flourish, which could have lead the search towards better direction. Further, prematurely removing individuals reduce the diversity in the parent population resulting in offspring containing similar (and hence redundant) genetic material [2]. The lack of diverse genetic material causes trapping in a state that prevent genetic operators (such as cross-over and mutation) to produce offspring which are superior to parents and results in premature stagnation of evolution. Thus, to deal with multimodal problems, the selection mechanism must be able to balance between two inversely related factors: *selection pressure* and *diversity*.

Here, in this paper, we propose a new survival strategy, namely "Sib-based Survival Selection", inspired by the concept of crowding techniques. In the proposed strategy, each offspring is tagged by a number that denotes the identity of the ancestor; the individual from which it has derived the larger part of its genetic material. All the offspring with the same tag (i.e., having same ancestor) compete with each other and the winner offspring (i.e., with best fitness) is selected for competing next with the ancestor. The winner between these two competitors (the parent and the selected offspring) is determined either deterministically or stochastically depending on the ancestor, that not only ensures the preservation of the best individual in the next generation, but it also implements the "downhill movement" concept to allow escaping from local optima.

2   Background

2.1   HP Lattice Model

Protein Structure Prediction (PSP) can be defined as the problem of finding the native structure of a protein having the lowest possible free energy from its amino acid sequence. As classification studies [3] have used predictor sets to overcome computational complexity, similarly researches have been carried out with simplified models [4] to deal with the NP-hard problem. In this paper, we use Hydrophobic-Polar (HP) energy model [4] and 3D-FCC lattice for the modelling purpose. HP model is the most widely used model for lattice simulation, which classifies the amino acids either as *hydrophobic (H) or polar (P)* based on their affinity for water. On the the other hand, Face-centred-cubic lattice (FCC) is one of the most compactly packed lattices, in which each lattice point has 12 neighbors with 12 directional vectors. A more detail about the HP and 3D-FCC can be found from [5].

2.2   Methods in-vogue for Survival Selection

In this section we present a brief overview of commonly used selection algorithms; an extensive analysis can be found in [6]. In literature, widely used selection methods such as fitness proportional selection, tournament selection, etc. are designed to select individual solutions through a fitness-based process that favors the solutions with high fitness over less fit ones. For example, tournament selection technique randomly selects k individuals from the selection pool using a uniform probability distribution with replacement and then designates the one with the best fitness as the winner of the tournament [7]. On the other hand, $(\mu+\lambda)$-strategy is based on deterministic selection of the best μ individuals from a set of λ offspring and μ parent individuals [2]. In fitness proportional selection mechanism [7], (also known as "roulette-wheel" selection), the probability of an individual's survival is directly proportional to its fitness value.

Furthermore, crowding methods [8] have been introduced for survival selection of GA to preserve diversity in population. These methods are based on the idea of pairing each offspring with a similar individual in the current population to compete for survival in the next generation. In deterministic crowding [9], individuals are paired for recombination in such a way that each individual participates exactly once in reproduction and the two offspring produced from each pair compete with their closest similar parent and replace, if it has higher fitness than the corresponding parent. Although, this method can maintain multiple solutions in multimodal landscape, it may fail to generate some more fit individuals that could be done by allowing fit individuals to recombine with others more than once. In probabilistic crowding [10] the surviving individual from the parent-offspring pair is selected by a probabilistic formula based on their fitness values.

3   The Method: Sib-Based Survival Selection Technique

The proposed survival selection technique called "Sib-based Survival Selection" or, in short S3 prevents premature convergence by restricting the rapid flow of genetic material from highly fit members of the population to others and thereby directs the search towards global optimization. The S3 mechanism works based on any parental selection that allows an individual to participate in reproduction more intensely instead of restricting it to partake only once. Initially, each individual involved in reproduction is tagged by a Parent IDentification (PID) number and those with fitness values residing in the range of best $x\%$, are considered as the elite members ($Elite\_ancestor$) of the population. The parents selected for a reproduction operation then produce two offspring that are then tagged by $ancestor\_tag$ indicating the identification (PID) of the ancestor individual (Λ_{PID}) with which it resembles closely. Although ancestor tagging can be done based on similarity measurement among the offspring and the parent individuals, here it has been pursued in a different manner that suits the PSP problem more perceptively. Since PSP involves optimization problem wherein satisfaction of Self Avoiding Walk (SAW) constraint is mandatory, a SAW validation technique is applied with every cross-over operation. If the new conformation is non-SAW, a repairing method needs to be applied on the concatenated segment from the cut-point, to repair the conformation as a SAW with the minimum possible changes of the parent individual. However, instead of applying the SAW repairing process on the succeeding segment blindly, here we propose to apply this on the shorter segment to achieve two objectives: i) to reduce the repairing cost since the success rates are higher on shorter length, ii) to satisfy the purpose of exchanging information by cross-over operation with minimum disruption of that provided by the parents.

After each cross-over operation, based on the position of the cut-point, each offspring (O_i) is identified by its *ancestor_tag* denoting the PID of the ancestor (Λ_{PID}) from which the offspring derives the longer segment of its genetic material. The offspring are then inserted into the descendant list (denoted as ∂_{PID}) of the corresponding ancestor. Subsequently, amongst all the descendants in a non-empty descendant list (∂_{PID}), one having the best fitness value is selected as the *Challenger* (denoted by ζ_{PID}) by *SelectChallenger*() according to Eqn. (1) and paired off with the ancestor (Λ_{PID}) for competition.

$$\forall_{PID=1...pop}\zeta_{PID} = Best(\partial_{PID,1}, \partial_{PID,2}, \ldots, \partial_{PID,\kappa}) \tag{1}$$

where, κ denotes the number of offspring in the ∂_{PID}. While selecting the survivors for the next generation, the winner in the competition between *Challenger* (ζ_{PID}) and the ancestor (Λ_{PID}) is determined as the *Survivor* ($\Im_{PID}$) either deterministically or stochastically, depending on whether the ancestor (Λ_{PID}) belongs to the group of *Elite_ancestor* or not. According to the fitness of the ancestor (Λ_{PID}), if it belongs to the group of elite members (*Elite_ancestor*) in the current population, then the one with better fitness between the *Challenger* (ζ_{PID}) and the ancestor (Λ_{PID}) is selected to be passed to the next generation by *FindSurvivorDet*() according to Eqn.(2).

$$\Im_{PID} = \begin{cases} \zeta_{PID} & \text{if } f(\Lambda_{PID}) > f(\zeta_{PID}) \\ \Lambda_{PID} & \text{else} \end{cases} \tag{2}$$

Otherwise, the *Survivor* ($\Im_{PID}$) is selected stochastically by *FindSurvivorSch*() according to Eqn.(3) as accomplished in probabilistic crowding [10].

$$\Im_{PID} = \begin{cases} \Im_{PID} & \text{if } Rnd \leq P(\zeta_{PID}) \\ \Lambda_{PID} & \text{else} \end{cases} \tag{3}$$

where, $Rnd \in [0,1]$ is a random number and $P(\zeta_{PID})$ is the probability of selecting the challenger which is calculated by Eqn.(4).

$$P(\zeta_{PID}) = \frac{f(\zeta_{PID})}{f(\zeta_{PID}) + f(\Lambda_{PID})} \tag{4}$$

Thus, the proposed approach helps to reduce the loss of diversity by putting constraint on rapid flow of genes from the fittest members of the population. Furthermore, the hybridization of the deterministic and the probabilistic selection (as described above) helps to preserve the best individuals in the current population along with implementing the concept of "downhill movement". This concept, by allowing the fitness to be worse temporarily before further improvement can occur, helps to exploit the less-fit individuals that are unexposed yet. The *AncestorTagging*, as a companion of crossover operation , is shown in *Algorithm 1*, while *Algorithm 2* describe the technique to select the survivors for the next generation.

Algorithm 1. AncestorTagging $(OldPop, \partial)$

Input: $OldPop$=Current Population
∂=The descendant list of each ancestor (initially empty)
Output: Updated $DescendantList$ (∂)

1: $\{P_1, P_2\} = SelectParent()$
2: $\{C_1, C_2\}=Crossover(P_1, P_2)$
3: $\{PID_1, PID_2\} = TagProcessing(CutPoint)$ ▷ PID_1 and PID_2 are the Parent Identification number of the ancestors for C_1 and C_2
4: $\partial_{PID_1}=\partial_{PID_1} \cup C_1$
5: $\partial_{PID_2}=\partial_{PID_2} \cup C_2$

Algorithm 2. SurvivalSelection $(OldPop, pop)$

Input: $OldPop$= Current Population, pop= population size
Output: $NewPop$=Population for the next generation

1: **For** $PID = 1$ to pop **do**
2:     **if** ∂_{PID} is *non empty* **then**
3. $\zeta_{PID}= SelectChallenger(PID)$ ▷ using Eqn. (1)
4:         **if** $\Lambda_{PID} \in Elite\_ancestor$ **then**
5:             $\Im_{PID}= FindSurvivorDet(PID)$ ▷ using Eqn. (2)
6.   **else**
7:             $\Im_{PID} = FindSurvivorSch(PID)$ ▷ using Eqn. (3)
8:         **End if**
9:     **else**
10:         $\Im_{PID}= OldPop_{PID}$
11:     **End if**
12:     $NewPop_{PID}= \Im_{PID}$
13: **End For**

---

4   Experimental Results and Discussion

To appraise the efficacy of the proposed "Sib-based Survival Selection" (S3) technique, we compare the method with three other survival selection techniques (i.e., binary tournament, roulette-wheel and $(\mu+\lambda)$-strategy) and with the deterministic crowding method [9]. To provide a uniform basis for comparison, four existing methods and the proposed one have been implemented with the same parametric settings in the same evolutionary framework (cross-over rate, parental selection), except the survival selection techniques. Method1, Method2 and Method3 are based on binary tournament, roulette-wheel and (μ+λ)-strategy, respectively to select the individuals to survive from the selection pool. Method4 uses the proposed S3 technique and Method5 works according to the deterministic parental selection strategy used for deterministic crowding [9]. The selection pool for Method1-Method3 consists of the best *pop* number of individuals with highest fitness chosen from the combined pool of parents and newly created offspring. Furthermore, to substantiate the performance of the proposed selection scheme on different parental selection techniques, we have compared Method1-Method4 using two different parental selection strategies: binary tournament and ASAM [11].

Experiments using 3D-FCC lattice have been conducted on a data set consisting of 11 benchmark sequences [5]. Since for all the short length sequences (B1-B4) in data set, without any exception proposed method reaches the known optimal energies for all the cases, in this paper we present the experimental results for the relatively long sequences (i.e., B5-B11). The results are the average outcome of 25 independent runs for each method. For all simulations, the population size is set to 100 and the cross-over rate is set as 1.0. To get an empirical observation on the influence of the proposed and existing survival selection techniques, we have executed a sample run of 500 generations without applying any mutation and diversification. In every 10th generation, we have separately calculated the diversity by pair-wise Hamming distance calculation over the population for the proposed and existing methods in comparison.

Figure 1 portrays the curves of the normalized diversity value over the evolution for the best runs of the methods stated earlier using ASAM parental selection technique along with Method5 for the two longest sequences (i.e., B10 and B11). Clearly, in all cases the diversity curves demonstrate the superiority of Method4 (proposed) and Method5, while Method3 preforms better than Method1 and Method2 for almost all the instances.

Further, we have investigated the effect of S3 in terms of optimal energy. The best and average energy values (along with standard deviations) obtained by Method1-Method4 using binary tournament and ASAM as the parental selection techniques are recorded in Table 1and in Table 2, respectively. The best energy value we report here is the lowest energy value obtained amongst all independent runs, while the reported mean energy value corresponds to the average of the best energies found in each independent run. The results depict the fact that, for all the sequences, Method4 outperforms the other three methods (Method1-Method3) both in terms of average and best energy. Moreover, all the methods using ASAM performs better than the corresponding one with binary tournament as the parental selection technique for all the instances except for $B8$ in Method1 and $B5$ and $B7$ in Method4.

The performance of the proposed method (Method4) has been compared against method5 and the comparative results have been shown in Table 3.

From the results, we observe that Method4 and Method5 take the advantage of maintaining diversity over the evolution and ends up with better energy for all the instances than the other three methods (shown in Table 1 and Table 2). However, it is note worthy that the energies obtained by the proposed method (Method4) are better than that of Method5, as shown in Table 3.

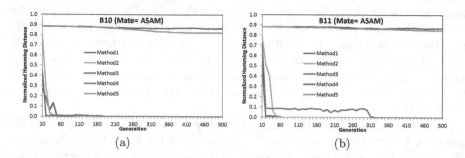

(a)   (b)

Fig. 1. Diversity Curves for Method1-Method5

Table 1. Comparison of the methods in terms of optimal energy using Binary Tournament (BT) as the parental selection technique

	\multicolumn{2}{c}{Method1}		\multicolumn{2}{c}{Method2}		\multicolumn{2}{c}{Method3}		\multicolumn{2}{c}{Method4}	
	Best	Avg±STD	Best	Avg±STD	Best	Avg±STD	Best	Avg±STD
B5	-52	-48.2± 2.59	-51	-49.6± 1.34	-54	-53.0 ± 0.71	-64	-61.4 ± 1.95
B6	-50	-48.8± 1.30	-51	-46.2± 2.77	-52	-51.8 ± 0.45	-61	-59.4 ± 1.14
B7	-104	-98.0± 7.52	-102	-98.8± 2.17	-105	-103.0 ± 2.12	-117	-113.8 ± 2.59
B8	-97	-91.6± 4.93	-95	-93.0± 1.41	-103	-96.6 ± 4.39	-110	-106.2 ± 2.17
B9	-137	-133.6± 3.36	-135	-133.6± 1.14	-141	-133.2 ± 4.49	-160	-155.2 ± 2.77
B10	-118	-114.8± 2.77	-124	-116.4± 5.41	-129	-124.8 ± 3.96	-142	-139.2 ± 2.59
B11	-116	-112.6± 2.88	-122	-114.6± 6.27	-128	-126.8 ± 1.10	-148	-142.8 ± 3.90

Table 2. Comparison of the methods in terms of optimal energy using ASAM as the parental selection technique

	\multicolumn{2}{c}{Method1}		\multicolumn{2}{c}{Method2}		\multicolumn{2}{c}{Method3}		\multicolumn{2}{c}{Method4}	
	Best	Avg±STD	Best	Avg±STD	Best	Avg±STD	Best	Avg±STD
B5	-58	-52.8 ± 3.27	-57	-51.0 ± 3.94	-56	-53.2 ± 2.59	-61	-60.2 ± 0.84
B6	-53	-50.6 ± 2.30	-51	-48.4 ± 1.95	-52	-51.4 ± 0.89	-62	-59.4 ± 1.52
B7	-108	-102.0 ± 4.06	-103	-100.8 ± 1.92	-106	-99.0 ± 4.30	-116	-113.8 ± 1.79
B8	-96	-90.8 ± 3.70	-101	-95.6 ± 4.34	-104	-98.6 ± 3.51	-111	-108.4 ± 2.19
B9	-140	-132.8 ± 6.10	-138	-132.6 ± 6.35	-142	-134.6 ± 4.56	-163	-157.8 ± 3.56
B10	-124	-121.8 ± 1.48	-128	-123.4 ± 5.08	-133	-129.6 ± 3.36	-144	-137.2 ± 4.28
B11	-120	-116.2 ± 2.68	-126	-121.0 ± 2.92	-130	-127.2 ± 2.68	-154	-147.4 ± 4.28

Table 3. Comparison of proposed method (Method4) using BT and ASAM as parental selection techniques with Method5 in terms of optimal energy

	Method4 (Mate=BT)		Method4 (Mate=ASAM)		Method5	
	Best	Avg±STD	Best	Avg±STD	Best	Avg±STD
B5	-64	-61.4 ± 1.95	-61	-60.2 ± 0.84	-59	-58.0± 0.71
B6	-61	-59.4 ± 1.14	-62	-59.4 ± 1.52	-58	-57.0± 1.22
B7	-117	-113.8 ± 2.59	-116	-113.8 ± 1.79	-112	-110.6± 1.14
B8	-110	-106.2 ± 2.17	-111	-108.4 ± 2.19	-107	-105.2± 2.68
B9	-160	-155.2 ± 2.77	-163	-157.8 ± 3.56	-158	-155.6 ± 1.92
B10	-142	-139.2 ± 2.59	-144	-137.2 ± 4.28	-134	-132.6± 1.67
B11	-148	-142.8 ± 3.90	-154	-147.4 ± 4.28	-143	-139.4± 2.51

5   Conclusion

The multimodal problems as compared with unimodal ones, is more susceptible to premature convergence. Here in this paper, we have proposed a new survival selection mechanism for genetic algorithms to deal with premature convergence, which is considered as one of the major difficulties in applying evolutionary algorithms on multimodal landscapes. The proposed selection mechanism attempts to get rid of this problem by restraining the proliferation of genetic material from highly fit individuals to others in the population. In addition, the proposed survival selection technique helps to exploit the disguised potential of less-fit individuals along with the preservation of the best individuals over the evolution. It also helps to discover unexplored region by allowing the fitness to be worse temporarily in order to move on to other more potential peaks. The experimental results demonstrate that the proposed method avoids premature convergence by maintaining a diverse population during the entire optimization process and outperforms in solution quality as compared with other survival selection methods.

References

1. Ursem, R.K.: Models for evolutionary algorithms and their applications in system identification and control optimization. In: BRICS (2003)
2. De Jong, K.A.: Evolutionary computation: a unified approach. MIT Press (2006)
3. Ooi, C., Chetty, M., Teng, S.: Differential prioritization in feature selection and classifier aggregation for multiclass microarray datasets. Data Mining and Knowledge Discovery 14(3), 329–366 (2007)
4. Lau, K., Dill, K.A.: A lattice statistical mechanics model of the conformational and sequence spaces of proteins. Macromolecules 22(10), 3986–3997 (1989)
5. Islam, M.K., Chetty, M.: Clustered memetic algorithm with local heuristics for ab initio protein structure prediction. IEEE Trans. Evolutionary Computation 17(4), 558–576 (2013)
6. Blickle, T., Thiele, L.: A comparison of selection schemes used in genetic algorithms (1995)
7. Goldberg, D.E., Deb, K.: A comparative analysis of selection schemes used in genetic algorithms. In: FOGA, pp. 69–93. Morgan Kaufmann (1991)

8. Mengshoel, O.J., Galan, S.F.: Generalized crowding for genetic algorithms. In: Genetic and Evolutionary Computation Conference 2010 (GECCO 2010), pp. 775–782 (2010)
9. Mahfoud, S.W.: Niching methods for genetic algorithms. Urbana, 51(95001) (1995)
10. Mengshoel, O.J., Goldberg, D.E.: Probabilistic crowding: Deterministic crowding with probabilistic replacement. In: Proc. of the Genetic and Evolutionary Computation Conference (GECCO 1999), pp. 409–416 (1999)
11. Nazmul, R., Chetty, M.: An adaptive strategy for assortative mating in genetic algorithm. In: IEEE Conference on Evolutionary Computation, June 20-23, pp. 2237–2244 (2013)

Tensor Completion Based on Structural Information

Zi-Fa Han, Ruibin Feng, Long-Ting Huang, Yi Xiao,
Chi-Sing Leung, and Hing Cheung So

Dept. of Electronic Engineering, City University of Hong Kong, Hong Kong
{zifahan,yeren7722}@gmail.com,
{rfeng4-c,longhuang4-c}@my.cityu.edu.hk,
eeleungc@cityu.edu.hk, hcso@ee.cityu.edu.hk

Abstract. In tensor completion, one of the challenges is the calculation of the tensor rank. Recently, a tensor nuclear norm, which is a weighted sum of matrix nuclear norms of all unfoldings, has been proposed to solve this difficulty. However, in the matrix nuclear norm based approach, all the singular values are minimized simultaneously. Hence the rank may not be well approximated. This paper presents a tensor completion algorithm based on the concept of matrix truncated nuclear norm, which is superior to the traditional matrix nuclear norm. Since most existing tensor completion algorithms do not consider of the tensor, we add an additional term in the objective function so that we can utilize the spatial regular feature in the tensor data. Simulation results show that our proposed algorithm outperforms some the state-of-the-art tensor/matrix completion algorithms.

Keywords: Tensor Completion, Truncated Nuclear Norm, Linearized Alternative Direction Method.

1 Introduction

The aim of matrix completion is to recover the missing entries of a low rank matrix via the known ones [1,2]. Since the objective function of low rank matrix completion is not convex, the original problem cannot be solved effectively. Fortunately, some theoretical studies [1,3,4] indicate that it is appropriate to replace the minimization of the matrix rank by the minimization of the matrix nuclear norm under some general constraints. However, as mentioned in [5], the major limitation of using the matrix nuclear norm is that the rank may not be well approximated, since all the singular values are minimized simultaneously in the optimization process. To solve this problem, the truncated matrix nuclear norm has been proposed in [5]. Liang et al. [4] have pointed out that when the matrix contains some regular patterns in the spatial domain, the transformed coefficients of the matrix are sparse in the transform domain. Hence an additional term of l_1-norm of the transformed coefficients is included in the objective function.

Low rank tensor completion has received considerable attention recently [6,7]. Unlike matrices, calculating the rank of a general tensor is a difficulty [7]. Liu et al. [7] have proposed the concept of tensor nuclear norm, called tensor trace norm. The tensor nuclear norm is the weighted sum of the matrix nuclear norms of all matrices unfolded along each mode. Since the concept of the tensor nuclear norm is based on the matrix nuclear norm, the tensor rank may not be well approximated too.

C.K. Loo et al. (Eds.): ICONIP 2014, Part II, LNCS 8835, pp. 479–486, 2014.
© Springer International Publishing Switzerland 2014

This paper introduces a tensor completion algorithm based on the concept of the truncated matrix nuclear norm. To further improve the performance, an additional cost term, which is equal to the sum of transformed coefficients of the tensor, is included in the objective function. Experimental results demonstrate that our proposed algorithm outperforms some state-of-the-art tensor completion algorithms.

In Section 2 we review the related works about matrix completion and tensor completion. Section 3 provides the details of our algorithm. Experimental results are shown in Section 4. Finally, conclusion is given in Section 5.

2 Matrix and Tensor Completions

Throughout this paper, matrices are denoted by upper case letters, e.g., X, Y, and lower case letters for the entries, e.g., x_{ij}. The Frobenius norm of the matrix X is defined as $\|X\|_F := (\sum_{ij} |x_{ij}|)^{\frac{1}{2}}$. The inner product is defined as $< X, Y >:= \sum_{i,j} x_{ij} y_{ij}$.

Given an incomplete low rank matrix $G \in \mathbb{R}^{m \times n}$, the matrix completion problem can be formulated as follows:

$$\min_X \text{rank}(X), \text{ s.t. } x_{ij} = g_{ij}, \ (i, j) \in \Omega, \tag{1}$$

where $X \in \mathbb{R}^{m \times n}$ and Ω is the set corresponding to the observed entries while the remaining elements are missing. Unfortunately, in general the rank minimization is an NP-hard problem. Candès et al. [1] have proved that under mild conditions, the low rank solution can be found by the following matrix nuclear norm minimization problem:

$$\min_X \|X\|_*, \text{ s.t. } \mathcal{P}_\Omega(X) = \mathcal{P}_\Omega(G), \tag{2}$$

where $\|X\|_* := \sum_{i=1}^{\min(m,n)} \sigma_i(X)$ is the matrix nuclear norm of X, σ_i is the i-th largest singular value of X, and $\mathcal{P}_\Omega$ is an operator indicates that the equality happens only on the entries that belong to Ω. For a matrix with a regular pattern, Liang et al. [4] have proposed the sparse low-rank texture inpainting (SLRTI) method, given by

$$\min_{X,W} \lambda\|W\|_1 + \|X\|_*, \text{ s.t. } \mathcal{P}_\Omega(X) = \mathcal{P}_\Omega(G), X = C_1 W C_2^T, \tag{3}$$

where W is the transformed coefficient matrix for some bases C_1 and C_2. In [4], C_1 and C_2 are DCT basis functions. In [4], for color images, Liang et al. have proposed to apply the matrix completion three times on the three color channels individually.

Since the values of the largest r nonzero singular values of a matrix do not affect the matrix rank, Hu et al. [5] have proposed the truncated matrix nuclear norm regularization method, given by

$$\min_X \|X\|_r, \text{ s.t. } \mathcal{P}_\Omega(X) = \mathcal{P}_\Omega(G), \tag{4}$$

where $\|X\|_r := \sum_{i=r+1}^{\min(m,n)} \sigma_i(X)$ is the truncated matrix nuclear norm. Problem (4) can be rewritten as

$$\min_X \|X\|_* - \max_{AA^T=I_{r \times r}, BB^T=I_{r \times r}} \text{Tr}(AXB^T), \text{ s.t. } \mathcal{P}_\Omega(X) = \mathcal{P}_\Omega(G), \tag{5}$$

where $A \in \mathbb{R}^{r \times m}$ and $B \in \mathbb{R}^{r \times n}$. In [5], the basic idea of the optimization procedure is as follows. At each iteration, we first fix X^k and compute A and B based on SVD of X^k. Afterwards, we use the updated A and B to update X.

2.1   Tensor Completion

Tensors are denoted by calligraphic uppercase letters, such as $\mathcal{X}$ and $\mathcal{T}$. An N-mode tensor (or N-order tensor) is denoted as $\mathcal{X} \in \mathbb{R}^{I_1 \times I_2 \times \cdots \times I_N}$, whose elements are $x_{i_1, \cdots, i_N}$. For instance, a vector is 1-mode tensor and a matrix is 2-mode tensor. A mode-n fiber of $\mathcal{X}$ is a vector obtained by fixing every index except the nth index. In some situations, it is convenient to unfold (matrcization) a tensor into a matrix. The mode-n matricization ("Unfold" operator) of a tensor is denoted by $\mathcal{X}_{(n)}$. In fact, the columns of the mode-n matricization are mode-n fibers of $\mathcal{X}$ in the lexicographical order. The reverse operator "Fold" is defined as $\text{Fold} \mathcal{X}_{(n)} := \mathcal{X}$.

Given a low rank incomplete tensor $\mathcal{T} \in \mathbb{R}^{I_1 \times \cdots \times I_N}$, finding its missing entries can be generalized by the matrix completion as

$$\min_{\mathcal{X}} \text{rank}(\mathcal{X}), \text{ s.t. } \mathcal{P}_\Omega(\mathcal{X}) = \mathcal{P}_\Omega(\mathcal{T}), \tag{6}$$

where $\mathcal{X}$ and $\mathcal{T}$ are n-mode tensors with the same size in each dimension. Replacing the tensor rank by tensor nuclear norm [7], we can rewrite (6) as

$$\min_{\mathcal{X}} \|\mathcal{X}\|_*, \text{ s.t. } \mathcal{P}_\Omega(\mathcal{X}) = \mathcal{P}_\Omega(\mathcal{T}), \tag{7}$$

where $\|\mathcal{X}\|_*$ is the tensor nuclear norm defined as $\|\mathcal{X}\|_* := \sum_{i=1}^{n} \alpha_i \|\mathcal{X}_{(i)}\|_*$ satisfying $\alpha_i \geq 0$ and $\sum_{i=1}^{n} \alpha_i = 1$. Based on this model, Ji et al. [7] have proposed the high accuracy low rank tensor completion (HaLRTC) method.

3   Proposed Method

To explore the spatial structure property of tensor, let us consider a natural color image, shown in Figure 1(a). It can be considered as a 3-mode tensor $\mathcal{T} \in \mathbb{R}^{I_1 \times I_2 \times 3}$. Its mode-1 and mode-2 matricizations are shown in Figure 1(b) and Figure 1(c), from which we can see that these two matricization still preserve the spatial structure to some extent. Therefore, we can utilize the spatial structure information of the low rank tensor through mode-1 or mode-2 matricization.

For the tensor data with some spatial regular patterns, we propose the following objective function:

$$\min_{\mathcal{X}, W} \lambda \|W\|_0 + \|\mathcal{X}\|_*, \text{ s.t. } \mathcal{P}_\Omega(\mathcal{X}) = \mathcal{P}_\Omega(\mathcal{T}), \mathcal{X}_{(j_0)} = C_1 W C_2^T, \tag{8}$$

where $\lambda > 0$, j_0 is the j_0-th dimension index of the tensor $\mathcal{X}$, $\mathcal{X}_{(j_0)}$ is mode-j_0 matricization of $\mathcal{X}$, and $\|W\|_0$ denotes the number of non-zero entries in the transformed coefficients W of $\mathcal{X}_{(j_0)}$.

(a)   (b)

(c)

Fig. 1. Mode-1 and mode-2 matricizations of a color image. (a) Original image. (b) mode-1 matricization. (c) mode-2 matricization.

In (8), the additional term $\|W\|_0$ is used to preserve the regular property. Since minimizing the l_0 norm cannot be calculated by an efficient way, we replace it by the l_1-norm [8]. Hence (8) is approximated as

$$\min_{\mathcal{X},W} \lambda\|W\|_1 + \|\mathcal{X}\|_*, \text{ s.t. } \mathcal{P}_\Omega(\mathcal{X}) = \mathcal{P}_\Omega(\mathcal{T}), \mathcal{X}_{(j_0)} = C_1 W C_2^T. \tag{9}$$

With the tensor nuclear norm, the optimization problem becomes

$$\min_{\mathcal{X},W} \lambda\|W\|_1 + \sum_{i=1}^{n} \alpha_i \|\mathcal{X}_{(i)}\|_*, \text{ s.t. } \mathcal{P}_\Omega(\mathcal{X}) = \mathcal{P}_\Omega(\mathcal{T}), \mathcal{X}_{(j_0)} = C_1 W C_2^T. \tag{10}$$

Instead of using the matrix nuclear norm $\|\mathcal{X}_{(i)}\|_*$, we replace it with the truncated matrix nuclear norm $\|\mathcal{X}_{(i)}\|_r$. Hence (10) becomes

$$\min_{\mathcal{X},W} \lambda\|W\|_1 + \sum_{i=1}^{n} \alpha_i \|\mathcal{X}_{(i)}\|_r, \text{ s.t. } \mathcal{P}_\Omega(\mathcal{X}) = \mathcal{P}_\Omega(\mathcal{T}), \mathcal{X}_{(j_0)} = C_1 W C_2^T. \tag{11}$$

Using (4), the optimization problem can be modified as

$$\min_{\mathcal{X},W} \lambda\|W\|_1 + \sum_{i=1}^{n} \alpha_i \|\mathcal{X}_{(i)}\|_* - \alpha_i \max_{A_i A_i^T = I, B_i B_i^T = I} \text{Tr}(A_i \mathcal{X}_{(i)} B_i^T),$$
$$s.t. \mathcal{P}_\Omega(\mathcal{X}) = \mathcal{P}_\Omega(\mathcal{T}), \mathcal{X}_{(j_0)} = C_1 W C_2^T. \tag{12}$$

Introducing dummy variables $M_i, i = 1, ..., n$, we can express (12) as

$$\min_{\mathcal{X},W,M_i} \lambda\|W\|_1 + \sum_{i=1}^{n} \alpha_i \|M_i\|_* - \alpha_i \text{Tr}(A_i M_i B_i^T)$$
$$s.t. \mathcal{P}_\Omega(\mathcal{X}) = \mathcal{P}_\Omega(\mathcal{T}), \mathcal{X}_{(j_0)} = C_1 W C_2^T, \mathcal{X}_{(i)} = M_i, i = 1, ..., n. \tag{13}$$

The augmented Lagrangian function of (13) is written as

$$L(\mathcal{X}, M_1, .., M_n, W, Y_1, .., Y_n, F)$$

$$= \lambda\|W\|_1 + \sum_{i=1}^{n}(\alpha_i\|M_i\|_* - \alpha_i \mathrm{Tr}(A_i M_i B_i^T) + <Y_i, \mathcal{X}_{(i)} - M_i>$$

$$+ \frac{\beta}{2}\|\mathcal{X}_{(i)} - M_i\|_F^2) + <F, \mathcal{X}_{(j_0)} - C_1 W C_2^T> + \frac{\beta}{2}\|\mathcal{X}_{(j_0)} - C_1 W C_2^T\|_F^2,$$

where $\{Y_1, .., Y_n, F\}$ are Lagrangian parameters.

With the objective function, we can iteratively update $\mathcal{X}$, M_i, W, Y_i and F. Based on the concept of linearized alternative direction method (LADM) [9], at each iteration, we first update $\mathcal{X}^{k+1}$.

Computing $\mathcal{X}^{k+1}$:

$$\mathcal{X}^{k+1} = \min_{\mathcal{X}} \frac{\beta^k}{2}\|\mathcal{X} - \mathrm{Fold}_{(j_0)}(C_1 W^k C_2^T - \frac{F^k}{\beta^k})\|_F^2$$

$$+ \sum_{i=1}^{n}\frac{\beta^k}{2}\|\mathcal{X} - \mathrm{Fold}_{(i)}(M_i^k - \frac{Y_i^k}{\beta^k})\|_F^2. \tag{14}$$

Therefore, taking the error support into consideration we obtain

$$\mathcal{X}^{k+1}$$

$$= \begin{cases} \frac{1}{n+1}[\mathrm{Fold}_{(j_0)}(C_1 W^k C_2^T - \frac{F^k}{\beta^k}) + \sum_{i=1}^{n}\mathrm{Fold}_{(i)}(M_i^k - \frac{Y_i^k}{\beta^k})], & (i_1, i_2, \cdots, i_n) \notin \Omega \\ (\mathcal{T}), & (i_1, i_2, \cdots, i_n) \in \Omega \end{cases} \tag{15}$$

Then we compte M_i^{k+1}.

Computing M_i^{k+1}:

$$M_i^{k+1} = \mathcal{D}_{\frac{\alpha_i}{\beta^k}}(\mathcal{X}_{(i)}^{k+1} + \frac{Y_i^k}{\beta^k} + \frac{\alpha_i}{\beta^k}A_i^T B_i), \tag{16}$$

where $\mathcal{D}_\tau(\cdot)$ is the singular value shrinkage operator [10], which is defined as $\mathcal{D}_\tau(X) = U\mathcal{D}_\tau(\Sigma)V^T$ with $X = U\Sigma V^T \in \mathbb{R}^{m\times n}$, $\Sigma = \mathrm{diag}(\{\sigma_i\}_{1\le i\le \min\{m,n\}})$ and $\mathcal{D}_\tau(\Sigma) = \mathrm{diag}(\max\{\sigma_i - \tau, 0\})$. Then we need to update W^{k+1}.

Computing W^{k+1}:

$$W^{k+1} = \min_{W} \lambda\|W\|_1 + \frac{\beta^k}{2}\|C_1 W C_2^T - (\mathcal{X}_{(j_0)}^{k+1} + \frac{F^k}{\beta^k})\|_F^2. \tag{17}$$

Note that, (17) is not easy to minimize with respect to W. Therefore, we apply the linearization technique [9] to overcome this difficulty. More specifically,

$$\frac{1}{2}\|C_1 W C_2^T - (\mathcal{X}_{(j_0)}^{k+1} + \frac{F^k}{\beta^k})\|_F^2$$

$$\approx \frac{1}{2}\|C_1 W^k C_2^T - (\mathcal{X}_{(j_0)}^{k+1} + \frac{F^k}{\beta^k})\|_F^2 + <g_k, W - W^k> + \frac{1}{2\eta}\|W - W^k\|_F^2,$$

(18)

where $g_k = C_1^T(C_1 W^k C_2^T - (\mathcal{X}_{(j_0)}^{k+1} + \frac{F^k}{\beta^k}))C_2$ is the gradient of $\frac{1}{2}\|C_1 W C_2^T - (\mathcal{X}_{(j_0)}^{k+1} + \frac{F^k}{\beta^k})\|_F^2$ at W^k, and $\eta > 0$ is a proximal parameter. With (18), (17) can be rewritten as

$$W^{k+1} = \mathcal{S}_{\frac{\lambda\eta}{\beta}}(W^k - \eta g_k),$$

(19)

where $\mathcal{S}_\tau(\cdot)$ represents the soft threshold operator [11] defined as $\mathcal{S}_\tau(x) = \text{sgn}(x) \cdot (|x| - \tau)$.

Computing Y_i^{k+1}:

$$Y_i^{k+1} = Y_i^k + \beta^k(\mathcal{X}_{(i)}^{k+1} - M_i^{k+1}).$$

(20)

Computing F^{k+1}:

$$F^{k+1} = F^k + \beta^k(\mathcal{X}_{(j_0)}^{k+1} - C_1 W^{k+1} C_2^T).$$

(21)

Computing β^{k+1}:

$$\beta^{k+1} = \rho\beta^k.$$

(22)

The algorithm is summarized in Algorithm 1.

Algorithm 1. Optimization framework via LADM

Input: Input tensor $\mathcal{T}$, error support Ω, transformation bases C_1, C_2.
Initialization: $\mathcal{X} = \mathcal{T}, Y_i = 0, W = 0, j_0, \lambda, \beta_0, \alpha, \eta, \rho$;
   while not convergence **do**
      Update $\mathcal{X}^{k+1}$ using (15);
      Update M_i^{k+1} using (16);
      Update W^{k+1} using (19);
      $Y_i^{k+1} = Y_i^k + \beta^k(\mathcal{X}_{(i)}^{k+1} - M_i^{k+1})$;
      $F^{k+1} = F^k + \beta^k(\mathcal{X}_{(j_0)}^{k+1} - C_1 W^{k+1} C_2^T)$;
      $\beta^{k+1} = \rho\beta^k$;
   end while
Output: $\mathcal{X}$

---

4   Experiment Results

In this section, several experiments are conducted on both synthetic data and real visual data to verify the effectiveness of the proposed algorithm for tensor completion. Two state-of-the-art algorithms, SLRTI [4] and HaLRTC [7], are also considered.

Table 1. RSE comparison on synthetic data.

Dataset	Missing %	90%	80%	70%	60%	50%	40%	30%	20%	10%
Date Set 1	HaLRTC	0.8288	0.4039	0.1046	0.0568	0.0409	0.0160	0.0114	0.0024	0.0005
	Our method	0.1772	0.0123	0.0026	0.0011	0.0011	0.0003	0.0002	0.0002	0.0002
Date Set 2	HaLRTC	0.6483	0.1101	0.0184	0.0104	0.0077	0.0059	0.0040	0.0031	0.0021
	Our method	0.0132	0.0076	0.0049	0.0041	0.0040	0.0035	0.0033	0.0028	0.0015

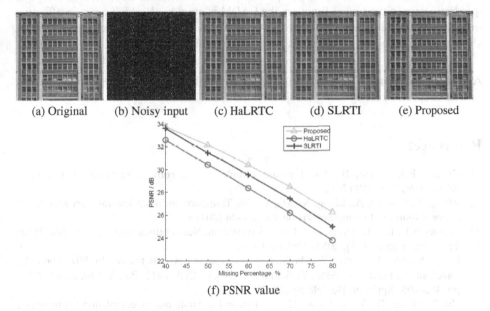

(a) Original   (b) Noisy input   (c) HaLRTC   (d) SLRTI   (e) Proposed

(f) PSNR value

Fig. 2. (a)-(e) Comparison of tensor completion with 80% of missing components. (f) PSNR values of the recovered building facade images versus different % of missing components.

We generate two synthetic tensor datasets to verify our algorithm. One 3-mode tensor $\mathcal{X} \in \mathbb{R}^{50 \times 50 \times 50}$ is generated by sampling 12 cubes in a 3D space (Dataset 1). The other is generated based on the method in [6] (Dataset 2). The Ω of the two tensors is generated uniformly at random. The missing ratio ranges from 10% to 90%. We take HaLRTC to compare with our tensor completion algorithm. The α is set to $[1/3, 1/3, 1/3]$ and the maximal iteration number is 500. For our algorithm, we select $\lambda \in [0.0005, 0.0019]$ and j_0 is set to 1. We measure the performance by $RSE = \|\mathcal{X} - \mathcal{T}\|/\mathcal{T}$. The experimental results are shown in Table 1. It is seen that our proposed algorithm outperforms the HaLRTC, and that it is more robust to missing components.

We also test our algorithm with real data. The parameters of our method, SLRTI and HaLRTC are tuned to achieve the best performance. A color image with resolution of 460×460 is used to verify our algorithm. That means, the tensor is with size of $460 \times 460 \times 3$. The image is then corrupted by randomly distributed missing entries ranging from 40 to 80 percentage. The results are provided in Figure 2. From these figures, we could see that the proposed algorithm has better visual effect and can achieve much higher PSNR values compared with the other two algorithms.

5 Conclusion

We have extended the concept of the matrix truncated nuclear norm to the tensor case. We present a tensor completion algorithm based on the matrix truncated nuclear norm. To explore the structure property (piece-wise smooth) of a tensor data, an l_1-norm term is added into a tensor completion optimization model. Then, in order to handle the quadratic term in the object function, the LADM is employed, which gives an effective numerical iteration scheme to solve this optimization problem. Experimental results show that the proposed algorithm is more robust than HaLRTC and SLRTI algorithms.

Acknowledgement. The work was supported by RGC Competitive Earmarked Research Grant from Hong Kong (Project No.: CityU 115612).

References

1. Candès, E.J., Recht, B.: Exact matrix completion via convex optimization. Commun. ACM 55(6), 111–119 (2012)
2. Zhang, Z., Ganesh, A., Liang, X., Ma, Y.: Tilt: Transform invariant low-rank textures. International Journal of Computer Vision 99(1), 1–24 (2012)
3. Candès, E.J., Tao, T.: The power of convex relaxation: Near-optimal matrix completion. IEEE Trans. Inf. Theor. 56(5), 2053–2080 (2010)
4. Liang, X., Ren, X., Zhang, Z., Ma, Y.: Repairing sparse low-rank texture. In: Fitzgibbon, A., Lazebnik, S., Perona, P., Sato, Y., Schmid, C. (eds.) ECCV 2012, Part V. LNCS, vol. 7576, pp. 482–495. Springer, Heidelberg (2012)
5. Hu, Y., Zhang, D., Ye, J., Li, X., He, X.: Fast and accurate matrix completion via truncated nuclear norm regularization. IEEE Trans. Pattern Anal. 35(9), 2117–2130 (2013)
6. Xu, Y., Hao, R., Yin, W., Su, Z.: Parallel matrix factorization for low-rank tensor completion (submitted)
7. Liu, J., Musialski, P., Wonka, P., Ye, J.: Tensor completion for estimating missing values in visual data. IEEE Trans. Pattern Anal. 35(1), 208–220 (2013)
8. Donoho, D.L., Elad, M.: Maximal sparsity representation via l_1 minimization. Proc. Nat. Aca. Sci. 100, 1297–2202 (2003)
9. Lin, Z., Liu, R., Su, Z.: Linearized alternating direction method with adaptive penalty for low-rank representation. In: Proc. NIPS 2011, pp. 612–620 (2012)
10. Cai, J.F., Candès, E.J., Shen, Z.: A singular value thresholding algorithm for matrix completion. SIAM J. Optim. 20(4), 1956–1982 (2010)
11. Donoho, C.L., Johnstone, I.M.: Adapting to unknown smoothness via wavelet shrinkage. J. Am. Stat. Assoc. 90(432), 1200–1224 (1995)

Document Versioning Using Feature Space Distances

Wei Lee Woon[1], Kuok-Shoong Daniel Wong[2], Zeyar Aung[1],
and Davor Svetinovic[1]

[1] EECS, Masdar Institute of Science and Technology,
P.O. Box 54224 Abu Dhabi, UAE
{wwoon,zaung,dsvetinovic}@masdar.ac.ae
[2] Daniel Wireless Software Pte Ltd, #03-04, 3 Science Park Drive,
Singapore Science Park, Singapore 118223
ksdwong@gmail.com

Abstract. The automated analysis of documents is an important task given the rapid increase in availability of digital texts. In an earlier publication, we had presented a framework where the edit distances between documents was used to reconstruct the version history of a set of documents. However, one problem which we encountered was the high computational costs of calculating these edit distances. In addition, the number of document comparisons which need to be done scales quadratically with the number of documents. In this paper we propose a simple approximation which retains many of the benefits of the method, but which greatly reduces the time required to calculate these edit distances. To test the utility of this method, the accuracy of the results obtained using this approximation is compared to the original results.

Keywords: String Matching, Text Processing, Data Mining, Versioning, Information Retrieval.

1 Introduction

The proliferation of network connectivity and web content have led to rapid increases in the quantity and availability of digital texts, motivating the development of new tools to cope with this sudden increase. In an earlier work [1], we demonstrated a procedure for inferring the version histories of unlabelled document collections using the edit distances between these documents. However, the calculation of the edit distances was computationally expensive. In this paper, we present a procedure for approximating inter-document distances without calculating the edit distances between all pairs of documents.

1.1 Relevant Background

Alternative text representations have been presented in the past. The idea of using string alignment to study documents is related to the *Levenshtein's distance* between documents [2]; however, this method uses alignments at the character

C.K. Loo et al. (Eds.): ICONIP 2014, Part II, LNCS 8835, pp. 487–494, 2014.

level, and is more appropriate for lower-level applications, for example in the study of text entry methods for mobile phones [3]. Our method operates at the level of words, which we believe represent the appropriate level of resolution for the majority of text documents.

Another similar approach in terms of document representation is the string kernel technique [4,5]. The original study examined the use of character level kernels but similar kernels calculated at the level of words have also been proposed [6]. Briefly, the string kernel method projects each document to a high dimensional feature space where each "dimension" is the weighted number of occurrences of a particular substring (which need not be contiguous) in a document. As the number of such substrings is potentially huge, the dimensionality of the resulting subspace is also large, motivating the use of the kernel procedure [7]. With respect to document alignments, these methods are similar to the related operation of *local* alignment, which have also been used for biological sequence analysis; in contrast, the method proposed here performs a global alignment of the documents concerned.

2   Experimental Methodology

2.1   Feature Extraction

Edit distances between pairs of documents and strings are found using dynamic programming. However, this is computationally expensive, and the addition of new documents requires comparisons with all other documents in the set. To address this, we propose a method by which the number of document comparison required may be greatly reduced. The proposed method exploits the kernel trick [8]; given a symmetric, positive semi-definite measure of similarity or distance $K(,)$, the theorem states that there is a high dimensional Hilbert space $\mathcal{H}$ (also known as the feature space) such that:

$$K(x_1, x_2) = \langle \Phi_K(x_1), \Phi_K(x_2) \rangle, \tag{1}$$

where $x_1, x_2 \in \mathcal{X}$, the input space, Φ_K is the mapping $\mathcal{X} \to \mathcal{H}$ and $<,>$ denotes the inner-product in the feature space.

In the present context, we note that the edit distance measure possesses the characteristics of a Mercer kernel [9]. A theoretical proof is beyond the scope of this paper, but this intuition is supported by a number of observations. As the minimum edit path is independent of the order of the documents in $K(,)$, the symmetry condition is satisfied. Positive semi-definiteness can be empirically tested by calculating the eigenvalues of the closely related similarity matrix Σ:

$$\Sigma = \begin{bmatrix} d_{max} - d_{1,1} & \dots & d_{max} - d_{1,n} \\ \vdots & \ddots & \vdots \\ d_{max} - d_{n,1} & \dots & d_{max} - d_{n,n} \end{bmatrix}; \tag{2}$$

where $d_{i,j}$ is the edit distance between documents i and j, $n = |\mathcal{D}|$ and $d_{max} = \max_{ij}\{d_{i,j} : i, j \in \mathcal{D}\}$. Our observations have been that the vast majority of the eigenvalues of Σ are positive, with a number of very large positive eigenvalues. Also, since the collection of documents being studied would occupy only a subspace $\mathcal{S}$ of this space, we select a subset $\mathcal{B} \subset \mathcal{D}$ for use as basis vectors in $\mathcal{H}$, and define $\tilde{\mathcal{S}}$ as the subspace spanned by these basis vectors. Finally, we obtain the coordinates of a given document in $\mathcal{S}$ by projecting it onto $\mathcal{B}$ via the edit distances:

$$\mu_i = \{d_{max} - d_{i,j} : j \in \mathcal{B}\}. \tag{3}$$

Distances between documents i and j can now be found quickly by calculating the Euclidean distances in this transformed space.

For the sake of comparison, two conventional vector-space representations, Term Frequency (*TF*) and Term Frequency Inverse Document Frequency (*TF-IDF*) [10] have been included in our experiments. In these cases, the cosine distance is used to compare different documents.

2.2 Scoring of Results

The version history of the documents can now be inferred by finding the *Hamiltonian Path* through all the documents then comparing them to the true version histories as follows [1]:

Algorithm $Score(\mathrm{w}, d_1, d_2, \dots, d_n)$
1. Accuracy $\leftarrow 0$
2. **for** $i \leftarrow 1$ **to** n
3.   TmpAccuracy $\leftarrow 0$
4.   **for** $j \leftarrow \max\{1, i - \mathrm{w}\}$ **to** $(i - 1)$
5.   **if** $Earlier(d_j, d_i)$ **then**
6.   TmpAccuracy $\leftarrow$ TmpAccuracy$+1$
7.   **for** $j \leftarrow (i + 1)$ **to** $\min\{n, i + \mathrm{w}\}$
8.   **if** $Earlier(d_i, d_j)$ **then**
9.   TmpAccuracy $\leftarrow$ TmpAccuracy$+1$
10.   Accuracy $\leftarrow$ Accuracy$+\dfrac{\mathrm{TmpAccuracy}}{\min\{n,i+\mathrm{w}\}-\max\{1,i-\mathrm{w}\}-1}$
11. **return** $100\times$Accuracy$/n\%$

where the function $Earlier(d_i, d_j)$ returns TRUE if d_1 occurs before d_j in the actual version history of the document. As can be seen, the procedure works by sliding a window of length w over the inferred document sequence. The score for the document in the center of this window is given by the proportion of the other documents in the window which are correctly sequenced with respect to the the document in question. The total score for the entire version history is the average over the scores of all the individual documents.

2.3 Data

Two sets of data are used in our current experiments [1]:

1. **Linux Kernel Source Code.** - In the first test, the proposed method is used to version C source files taken from the linux kernel. The source distribution consists of many files so testing was done with the following three randomly selected files: *loop.c,fork.c* and *eth.c*, for which 127, 216 and 58 unique versions were extracted respectively.

2. **"Wiki" Pages.** - versioned Wikipedia pages were collected using the "Way-back Machine"[1]. For our tests, we chose Wikipedia entries for *Fourier Transform* and *Perl*, as these are popular topics with regularly updated and comparatively lengthy entries.

   For our tests, we downloaded all available (at the time of writing) versions of these pages from the Internet Archive. For the Fourier Transform page, there were 24 versions available while for the Perl page, we were able to obtain 82 versions.

3   Results

3.1   Notation

For the results presented here, the following notation is used:

1. **Feature Extraction Schemes.** The method of feature extraction is written as $<feature>$, where *feature* is one of {DA (document alignment), FS (feature space), TFIDF, TF}. FS features are additionally written as FS-$< x\% >$, where $x\%$ refers to the percentage of documents in the collections which are used as "basis vectors" in the feature space. Further, for results corresponding to the Wikipedia pages, feature names will be written as $<feature>-<format>$, where *format* is one of {html,text}, and indicates if the procedure was applied to pure HTML, or if the HTML tags had been stripped in advance.

2. **Scoring Schemes.** As mentioned, the scoring scheme is tunable to different window sizes. These are denoted as **Global, Local-5,Local-1** for the case with all documents, window size of 5, and of 1.

3.2   Source Code Versioning Task

First, we study the performance of the algorithm with the Kernel source code. For the files *loop.c, fork.c* and *eth.c*, the results are presented in Tables 1, to 3. The following observations could be made:

1. The FS approach is able to produce very accurate results which approach those obtained using DA. Broadly speaking, accuracy was observed to improved as the number of basis vectors used was increased, as is expected, though the incremental benefit of increasing the number of basis vectors diminshed after a certain point (generally around $20 - 30\%$ of $|\mathcal{D}|$).

---

[1] From the "Internet Archive" - http://www.archive.org

This means that the proposed method allows comparable performance with less than a third of the computational effort of calculating the full distance matrix. However, as with the full distance matrices, local accuracy values are consistently less accurate compared to the global accuracy values.

2. The accuracy of the FS approach was significantly improved when the basis vectors were uniformly spaced along the version history. This makes sense as in this way there is a higher chance of obtaining a better coverage of the distribution of $\mathcal{D}$ in the feature space.

3. Interestingly, for the file *fork.c*, the accuracy obtained using FS *exceeded* that obtained with DA, even though the feature space projections uses information that is already contained in the full distance matrix. One possible explanation is that using the projections in this way helped to eliminate noise in the data by restricting the analysis to only the relevant subspace of the full feature space.

4. The larger the set of documents being analyzed, the better the accuracy of the method.

Table 1. Versioning scores (**loop.c**). Scores in bold denote highest scoring items in the respective columns.

Method/Class	Scores(%)			
	Global	Local-5	Local-1	Average
Original DA	**99.0**	**97.3**	**96.1**	**97.5**
TFIDF	97.1	82.7	70.5	83.4
TF	98.7	93.7	91.3	94.6
Randomized FS-10%	94.4	86.0	82.1	87.5
FS-30%	98.7	93.2	88.7	93.5
FS-50%	**99.0**	96.0	93.5	96.2
FS-75%	98.8	95.0	92.2	95.3
Uniform FS-10%	**99.0**	96.9	93.7	96.5
FS-30%	**99.0**	95.7	90.6	95.1
FS-50%	**99.0**	96.7	93.7	96.5
FS-75%	**99.0**	96.7	93.7	96.5

3.3 Wikipedia Entries

The same experiments were then repeated using the Wikipedia versions, and results have been tabulated in Table 4.

Table 2. Versioning scores (**fork.c**). Scores in bold denote highest scoring items in the respective columns.

	Method/Class	Scores(%)			
		Global	Local-5	Local-1	Average
Original	DA	99.5	92.1	90.7	94.1
	TFIDF	71.7	70.8	70.8	71.1
	TF	99.1	89.4	86.8	91.8
Randomized	FS-10%	98.8	88.6	86.8	91.4
	FS-30%	**99.6**	93.7	92.4	95.2
	FS-50%	**99.6**	**94.6**	**93.7**	**96.0**
	FS-75%	**99.6**	93.9	92.8	95.4
Uniform	FS-10%	99.4	92.4	89.4	93.7
	FS-30%	**99.6**	93.5	90.3	94.5
	FS-50%	**99.6**	93.6	91.2	94.8
	FS-75%	**99.6**	93.5	90.3	94.5

Table 3. Versioning scores (**eth.c**). Scores in bold denote highest scoring items in the respective columns.

	Method/Class	Scores(%)			
		Global	Local-5	Local-1	Average
Original	DA	**98.7**	93.9	**91.4**	**94.6**
	TFIDF	97.5	88.2	77.6	87.7
	TF	96.9	86.5	82.8	88.7
Randomized	FS-10%	96.4	86.5	77.2	86.7
	FS-30%	98.1	91.7	83.7	91.2
	FS-50%	98.2	92.2	85.3	91.9
	FS-75%	98.3	92.7	85.9	92.3
Uniform	FS-10%	97.2	86.0	78.4	87.2
	FS-30%	**98.7**	**94.2**	86.2	93.0
	FS-50%	95.3	91.8	85.3	90.8
	FS-75%	**98.7**	**94.2**	86.2	93.0

Table 4. Document versioning scores: Wikipedia entries. Scores in bold denote highest scoring items in the respective columns.

Method/	Fourier; Scores (%)				Perl; Scores (%)			
Class	Global	Local-5	Local-1	Average	Global	Local-5	Local-1	Average
DA	97.5	94.2	87.5	93.0	95.5	91.1	**87.2**	91.3
TFIDF	97.1	92.0	75.0	88.0	95.6	74.3	62.8	77.6
TF	97.1	92.4	79.2	89.6	94.1	78.9	68.3	80.4
FS-10%	73.3	63.6	62.3	66.4	93.1	82.3	75.6	83.6
FS-30%	96.1	89.9	77.5	87.8	96.1	89.2	82.1	89.1
FS-50%	96.2	90.2	79.2	88.5	96.8	90.9	81.7	89.8
FS-75%	97.6	94.2	86.7	92.8	**97.0**	**92.2**	82.7	90.6
FS-10%	**97.8**	**94.5**	87.5	**93.3**	96.6	90.9	81.1	89.5
FS-30%	97.1	93.3	83.3	91.2	96.9	91.5	86.0	**91.4**
FS-50%	89.1	86.6	75.0	83.6	96.9	91.5	84.8	91.0
FS-75%	90.2	89.8	**89.6**	89.9	96.5	89.8	78.7	88.3

Original rows: DA, TFIDF, TF. *Randomized* rows: FS-10%, FS-30%, FS-50%, FS-75%. *Uniform* rows: FS-10%, FS-30%, FS-50%, FS-75%.

1. The results exhibit the same broad trends as with the Kernel source code, though the performance of the FS-based methods showed some overall decline. Of the two Wikipedia files studied, better results were obtained when versioning the *Perl* wikipedia page. This is consistent with what was observed when versioning the Kernel source files where the scores obtained with largest collection of file - in that case *fork.c* - were the highest.
2. However, the size of the document collection was not the only factor - for example, the Wikipedia *Perl* dataset contained 82 versions compared to 58 for the *eth.c* collection, yet the versioning accuracy of the former was still markedly lower.

4   Conclusions and Future Plans

The results presented here demonstrate that the feature space approximation is a useful and efficient alternative to aligning all pairs of documents in the study set. Potential avenues for further development include:

1. A more systematic method could be found for the selection of the basis nodes. For example, a method based on clustering of the nodes, then selecting the cluster centers as the basis vectors might be more appropriate
2. Optimisation of the basic dynamic programming algorithm. For e.g. bioinformatics tools such as BLAST and FASTA have been devised which can

search massive biological databases very efficiently using a variety of simplifications (such as only focussing on promising regions of the alignment matrix).

References

1. Woon, W.L., Wong, K.-S.: String alignment for automated document versioning. Knowledge and Information Systems (2008)
2. Levenshtein, V.I.: Binary codes capable of correcting deletions, insertions, and reversals. Soviet Physics Doklady 10(8), 707–710 (1966)
3. Soukoreff, W.R., Mackenzie, S.I.: Measuring errors in text entry tasks: an application of the levenshtein string distance statistic. In: CHI 2001: CHI 2001 Extended Abstracts on Human Factors in Computing Systems, pp. 319–320. ACM Press, New York (2001)
4. Lodhi, H., Taylor, J.S., Cristianini, N., Watkins, C.J.C.H.: Text classification using string kernels. In: Advances in Neural Information Processing Systems (NIPS), pp. 563–569 (2000)
5. Lodhi, H., Saunders, C., Shawe-Taylor, J., Cristianini, N., WatkinsText, C.: classification using string kernels. J. Mach. Learn. Res. 2, 419–444 (2002)
6. Cancedda, N., Gaussier, E., Goutte, C., Renders, J.M.: Word sequence kernels. J. Mach. Learn. Res. 3, 1059–1082 (2003)
7. Cristianini, N., Taylor, S.: An introduction to support vector machines. Cambridge University Press, Cambridge (2000)
8. Mercer, J.: Functions of positive and negative type, and their connection with the theory of integral equations. Philosophical Transactions of the Royal Society of London. Series A, Containing Papers of a Mathematical or Physical Character 209, 415–446 (1909)
9. Aradhye, H., Dorai, C.: New kernels for analyzing multimodal data in multimedia using kernel machines. In: Proceedings of 2002 IEEE International Conference on Multimedia and Expo, ICME 2002, vol. 2, pp. 37–40 (2002)
10. Lan, M., Tan, C.-L., Low, H.-B., Sung, S.-Y.: A comprehensive comparative study on term weighting schemes for text categorization with support vector machines. In: WWW 2005: Special Interest Tracks and Posters of the 14th International Conference on World Wide Web, pp. 1032–1033. ACM Press, New York (2005)

Separation and Classification of Crackles and Bronchial Breath Sounds from Normal Breath Sounds Using Gaussian Mixture Model

Ali Haider, M. Daniyal Ashraf, M. Usama Azhar, Syed Osama Maruf,
Mehdi Naqvi, Sajid Gul Khawaja, and M. Usman Akram

College of Electrical and Mechanical Engineering,
National University of Sciences and Technology, Pakistan
haider_friendlyguy@hotmail.co.uk, daniyalashraf40@yahoo.com,
{usamaazhar32,sajid.gul.2009,usmakram}@gmail.com,
{maruf_jr1991,syedmehdihasan}@hotmail.com

Abstract. A computer aided diagnostic system capable of analyzing respiratory sounds can be very helpful in detection of pneumonia, asthma and tuberculosis as the Respiratory sound signal carries information about the underlying physiology of the lungs and is used to detect presence of adventitious lung sounds which are an indication of disease. Respiratory sound analysis helps in distinguishing normal respiratory sounds from abnormal respiratory sounds and this can be used to accurately diagnose respiratory diseases as is done by a medical specialist via auscultation. This process has subjective nature and that is why simple auscultation cannot be relied upon.In this paper we present a novel method for automated detection of crackles and bronchial breath sounds which when coupled together indicate presence and severity of Pneumonia. The proposed system consists of four modules i.e., pre-processing in which noise is filtered out, followed by feature extraction. The proposed system then performs classification to separate crackles and bronchial breath sounds from normal breath sounds.

1 Introduction

Respiratory sound analysis can be helpful in the diagnosis of respiratory diseases such as pneumonia as they are one of the major killers all over the world especially in the developing countries.Pneumonia is amongst the top 10 leading causes of death amongst asians living in the US [26] .According to the latest estimates it also accounts for 1.4 million deaths in children under-five years of age annually. This represents 18% of all annual under-five worldwide mortality [1] and about 98% of these deaths occur in developing countries. The economic cost of asthma, COPD, and pneumonia was $106 billion in 2009: $81 billion in direct health expenditures $25 billion in indirect cost of mortality [27]. Developed Countries can bear the such staggering costs but this is not the case for under developed countries and this lead to a high mortality rate of their citizens. Another reason for high mortality rate for people who acquire these diseases

C.K. Loo et al. (Eds.): ICONIP 2014, Part II, LNCS 8835, pp. 495–502, 2014.
© Springer International Publishing Switzerland 2014

is because of a lack of trained personnel as the ratio of doctors to patients is very low and another reason is late diagnosis of the disease. The medical specialist begins diagnosis of these diseases begins by recording history of the patient followed by a physical exam in which detection of adventitious lung sounds and abnormal lung sounds is undertaken with the help of a stethoscope; a process called auscultation. Adventitious lung sounds are additional respiratory sounds superimposed on breath sounds [3]. They are of two types 1) stationary and 2) non-stationary. The former category contains wheezes and rhonchi whereas the latter category contains crackles. Crackles are discontinuous explosive sounds which occur usually during inspiration [3]. Figure 1 shows the waveforms for normal breath , crackles and bronchial breaths.

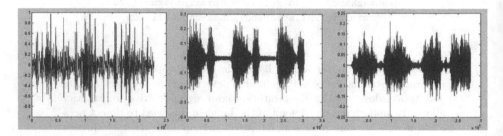

Fig. 1. Breath sound waveforms. From Left to right: Crackle sound waveform, Normal breath sound waveform; Bronchial sound waveform.

The occurrence of crackles is an indication of the severity of the pulmonary disease [4] and combined with the presence of bronchial breaths they confirm the presence of pneumonia [23].Bronchial Breath sounds are abnormal breath sounds detected at the posterior chest wall, containing higher frequency components and a higher intensity than that of normal breath sounds heard in the same region [3]. Therefore, simple auscultation cannot be relied upon as auscultation with a stethoscope is a subjective process since it depends on the individual's own hearing, experience and ability to distinguish different sounds patterns[2]. Thus there is a need for a system which accurately detects the presence of crackles and bronchial breaths in the respiratory sound of patient so that it can help in the accurate diagnosis of pneumonia.

This article consists of five sections. Section 2 highlights existing methods and related work for respiratory sound analysis. Section 3 describes a brief overview and all steps of the proposed system. The results are presented in Section 4 followed by conclusions in Section 5.

2   Related Work

Respiratory Sound Analysis is a relatively new area of interest for researchers and though few methods for detection and classification of crackles have been formulated, methods for the detection of bronchial breath sounds have not yet been given much attention. Presence of crackles and bronchial breaths is strong indicator of pneumonia and it is imperative that methods should be devised for their detection to help in its clinical diagnosis. Bronchial sounds are breath sounds with abnormally high frequencies and intensity, and with a loud and long expiratory phase. These sounds have frequency components of about 600-1000 Hz recorded over the posterior chest wall [24]. Gross V et al. [25] presented a method for the detection of bronchial breath sounds for pneumonia by using the ratio between the highest inspiratory and highest expiratory flow taken from spectral power for the 300-600 Hz frequency band. They found significant differences between the values of the healthy side and the side afflicted with pneumonia. The existing methods for detection of crackles are broadly categorized as the non-linear separation stationary non-stationary filter (ST-NST) [5] and its several modified version [6,7], the wavelet transform-based stationary-non-stationary filter (WTST-NST) [8] and the generalized fuzzy rule-based stationary non-stationary filter (GFST-NST)[9].

Mohammed Bahoura and Xiaoguang Lu [10,11] presented a model WPST-NST method based on double thresh-holding in wavelet domain by using time domain features which separates crackle's coefficients. Unlike simple wavelet transform, wavelet packet transform is obtained by applying wavelet transform at every level which is equivalent to multi-channel filtering. Using this model an accuracy of 93.9% is achieved [11].Fatma Ayari et al. [12] has discussed two methodologies to classify crackles and their extraction from the lung signal. The first one is statistics based methodology and the second is fuzzy nonlinear classifiers. Nine features have been selected to enhance behavior of crackle. These features relate to amplitude, time and waveform. Sensitivity of 98.34% and positive predictive value of 97.88% have been achieved. Martinez-Hernandez et al. [13] used lung sounds which are acquired by multi-channel microphone array on which feature extraction is done by multivariate AR model and after dimensionality reduction techniques like PCA and SVD are applied, the classification was done by SNN (Supervised Neural Network) using back-propagation method and Levenberg-Marquardt rule a.k.a. damped-least square.

3   Proposed Method

Over the past few years computerized methods of respiratory sound acquisition and analysis have overcome many shortcomings of simple auscultation [2].Computer aided diagnostic systems have brought new horizons in detection and treatment of many diseases. Thus, many different techniques have been used to acquire and then use respiratory sounds for diagnosis of pulmonary diseases. Figure 2 shows a complete flow diagram for all the phases of the proposed system starting from sound acquisition to classification of crackles and bronchial breath sounds. The proposed system is divided into four phases, i.e., acquisition of the lung sound and preprocessing, feature extraction and wavelet decomposition with removal of the silent phase and classification. A Gaussian Mixture Model based on a Bayes decision rule is used to differentiate crackles and bronchial breath sounds from normal respiratory sounds.

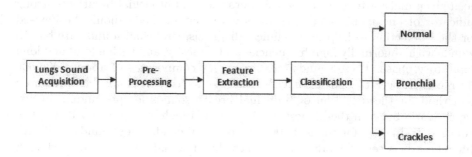

Fig. 2. Flow diagram of proposed system

3.1 Preprocessing

Pre-Processing of the lung sound is done with two objectives in mind, i.e., reduction of background noises and the other is to enhance the quality of the recorded sound. To cater for the former objective pre-filtering using a band-pass filter with cut-off frequencies at 100 Hz to reduce heart sounds and 2500 Hz to eliminate high- frequency noise is applied [14].

3.2 Feature Extraction

A feature set is formed to distinguish between normal respiratory sounds, crackles and bronchial breath sounds. The formation of the feature vector was done by the following method: The pre-processed sound files are used to calculate three spatial-temporal features namely, pitch, energy and spectrogram. These features are extracted from the entire spectrum after which the wavelet decomposition is done using Daubechies-8 wavelet for 5th level decomposition tree. The levels containing silent phase are then discarded and the level with the most information, i.e., the last level is then used to calculate other 9 features. The entire feature vector contains 12 features which are of the form $f_v = x_1, x_2, x_3, \ldots\ldots, x_{12}$. The description of the features in the feature vector is given below:

1. $Pitch(x_1)$ is associated with the frequency of sound wave and allows its ordering based on it [15], i.e., high pitch is attributed to sound having a high frequency and low pitch is associated with sound having a low frequency.
2. $Energy(x_2)$ is the energy of wavelet or wavelet packet decomposition.
3. $Range(x_3)$ is the difference between the maximum and minimum of the sound sample.
4. $IQR(x_4)$ is the inter-quartile range of the sound in time-domain.
5. $MAD(x_5)$ is the median absolute deviation of the sound. Moment(x9) central sample moment of sound specified by the positive integer order.
6. $Kurtosis(x_6)$ is a descriptor of the shape of the probability distribution of the real- valued random variable [17].

Feature extraction phase extracts different features out of which best features are selected by feature selection phase using rank tests.

The features showing the least $P - value$ are selected based on the results of rank sum tests. Top four common features with minimum $p - value$ are selected which are $Range(x_6)$, $IQR(x_7)$, $MAD(x_8)$ and $Kurtosis(x_{11})$and are then used to classify crackles from normal respiratory sounds. For bronchial breath sounds the Top three common features with minimum $p - value$ are selected which are $Pitch(x_1)$,$Energy(x_3)$ and $Kurtosis(x_11)$.

3.3   Classification

For the purpose of classification, we use a Bayesian classifier using Gaussian functions known as the Gaussian Mixture Model [20]. The Gaussian Mixture Model offers fast and accurate performance in the testing phase while being trained extensively in the training phase. In our context, each sound class is modeled by a Gaussian Mixture Model obtained by training [21]. The training is done by using a training data set with the respiratory sounds labeled as crackles and normal and side by side this procedure is done for bronchial breath sounds in which the training data set is labeled as bronchial breath sounds and normal. After this the Bayes decision rule is used in the estimation of a decision criteria from the training set. Two classes $X_1 = crackles$ and $X_2 = normal$ have been defined for classification of crackles and for the classification of Bronchial Breath sounds the classes are $X_3 = bronchialbreaths$ and $X_4 = normal$. The parameters for GMM are optimized using Expectation Maximization (EM) which is an iterative method and it chooses optimal parameters by finding the local maximum value of GMM distributions for training data [22].

4   Experimental Results

The evaluation of proposed system is done using respiratory sound files. We use a sound repository of 41 files out of which 14 have been categorized as crackles and 9 have been categorized as bronchial breath sounds by a human grader. The performance of the proposed system has been evaluated using sensitivity, specificity, positive predictive value (PPV) and accuracy as figures of merit. These parameters were calculated using equations 1-4 respectively.

$$sensitivity = \frac{TP}{(TP + FN)} \tag{1}$$

$$specificity = \frac{TN}{(TN + FP)} \tag{2}$$

$$PPV = \frac{TP}{(TP + FP)} \tag{3}$$

$$accuracy = \frac{(TP + TN)}{(TP + TN + FN + FP)} \tag{4}$$

where

- TP are true positives, meaning crackles are correctly classified.
- TN are true negatives, meaning normal respiratory sounds are correctly classified.
- FP are false positives, meaning normal respiratory sounds are wrongly classified as crackles regions.
- FN are false negatives, meaning crackles are wrongly classified as normal respiratory sounds.

Table-1 and 2 show the evaluation results of proposed system for crackle and bronchial breath detection respectively. They also highlight comparison of proposed system with other classifiers like Support Vector Machine (SVM) and Artificial Neural Network (ANN) consisting of 10 nodes in the input layer.

Table 1. Proposed system evaluation results for crackle detection

Classifier	TP	TN	FN	FP	Sensitivity	Specificity	PPV	Accuracy
GMM	13	27	1	0	92.85	100	100	97.56
SVM	13	25	1	2	92.8	92.5	86.6	92.6
ANN	10	25	4	2	85.3	71.2	92.5	85.3

Table 2. Proposed system evaluation results for bronchial breath detection

Classifier	TP	TN	FN	FP	Sensitivity	Specificity	PPV	Accuracy
GMM	9	27	0	5	100	84.37	64.28	87.85
SVM	8	27	1	5	88.8	84.37	61.54	85.37
ANN	5	30	2	4	71.43	88.24	55.56	85.37

Table 3. Performance comparison of proposed system with existing methods for crackle detection only

Method	Sensitivity	Specificity	PPV	Accuracy
Lu, Xiaoguang et al. [11]	92.9	-	94.4	93.9
Martinez-Hernandez et al. [13]	91.38	94.34	-	92.86
Fatima Ayari et al. [12]	98.34	-	97.88	-
PM	92.85	100	100	97.56

From Table-1 and 2 it can be seen that Gaussian Mixture Model out performs the other classifiers and is thus selected as the classifier of choice in our proposed method. The effectiveness of GMM can be ascertained from the high accuracy shown in the testing phase. For performance comparison with existing methods, we present the values of sensitivity, specificity, PPV and accuracy of the methods of Xiaoguang Lu et al. [11] , Martinez-Hernandez et al. [13] and Fatma Ayari et al. [12]. The results of these comparisons are shown in table-3. The table shows that the proposed system outperforms in terms of accuracy of classification.

5   Conclusion

Digital signal processing techniques can be applied to study lung sounds as an aid to clinical diagnosis [21]. New methods for identifying and measuring adventitious lung sounds are being developed in various research institutes [23]. We have introduced a new classification tool for Crackles and Bronchial Breath sound analysis as their detection is important for the evaluation of the severity of the respiratory disease and to diagnose pneumonia. The proposed system consisted of four phases i.e. preprocessing, feature extraction, feature selection and finally the classification. We have used sensitivity, specificity, PPV and accuracy to evaluate our proposed system. The system has achieved specificity of 100%, sensitivity of 92.85%, PPV of 100% and accuracy of 97.56% which are better than the recently published methods. It is evident from the comparison with previous systems that the proposed method has outperformed them and has classified crackles and Bronchial Breath sounds from other respiratory sounds with good accuracy.

References

1. UNICEF, Pneumonia and Diarrhea Tackling the Deadliest Diseases of the World's Poorest Children, UNICEF Division of Policy and Strategy, New York (June 2012)
2. Sovijarvi, A.R.A., Vanderschoot, J., Earis, J.E.: Standardization of computerized respiratory sound analysis. Eur. Respir. Rev. 10(77), 585 (2000)
3. Sovijrvi, A.R.A., Dalmasso, F., Vanderschoot, J., Malmberg, L.P., Righini, G., Stoneman, S.A.T.: Definition of terms for applications of respiratory sounds. Eur. Respir. Rev. 10(77), 597–610 (2000)
4. Epler, G.R., Carrington, C.B., Gaensler, E.A.: Crackles (rales) in the interstitial pulmonary diseases. Chest 73, 333–339 (1978)
5. Ono, M., Arakawa, K., Mori, M., Sugimoto, T., Harashima, H.: Separation of fine crackles from vesicular sounds by a nonlinear digital filter. IEEE Trans. Biomed. Eng. 36(2), 286–291 (1989)
6. Arakawa, K., Harashima, H., Ono, M., Mori, M.: Non-linear digital filters for extracting crackles from lung sounds. Front. Med. Biol. Eng. (3), 245–257 (1991)
7. Hadjileontiadis, L.J., Panas, S.M.: Nonlinear separation of crackles and squawks from vesicular sounds using third-order statistics. In: 18th International Conference of the IEEE Engineering in Medicine and Biology Society, vol. 5, pp. 2217–2219 (1996)
8. Hadjileontiadis, L.J., Panas, S.M.: Separation of discontinuous adventitious sounds from vesicular sounds using a wavelet-based filter. IEEE Trans. Biomed. Eng. 44(12), 1269–1281 (1997)
9. Tolias, Y.A., Hadjileontiadis, L.J., Panas, S.M.: Realtime separation of discontinuous adventitious sounds from vesicular sounds using a fuzzy rule-based filter. IEEE Trans. Inf. Technol. Biomed. 2(3), 204–215 (1998)
10. Bahoura, M., Lu, X.: Separation of Crackles from Vesicular Sounds Using Wavelet Packet Transform (2006)
11. Bahoura, M., Lu, X.: An Automatic System For Crackles Detection And Classification. In: IEEE CCECE/CCGEI (2006)

12. Ayari, F., Ksouri, M., Alouani, A.: A new scheme for automatic classification of pathologic lung sounds. IJCSI International Journal of Computer Science Issues 9(4(1)) (2012)
13. Martinez-Hernandez, H.G., Aljama-Corrales, C.T., Gonzalez-Camarena, R., Charleston-Villalobos, V.S., Chi-Lem, G.: Computerized Classification of Normal and Abnormal Lung Sounds by Multivariate Linear Autoregressive Model. In: Proceedings of the 2005 IEEE Engineering in Medicine and Biology 27th Annual Conference, Shanghai, China, September 1-4 (2005)
14. Earis, J.E., Cheetham, B.M.G.: Current methods used for computerized respiratory sound analysis. Eur. Respir. Rev. 10(77), 586–590 (2000)
15. Klapuri, A., Davy, M.: Signal processing methods for music transcription, p. 8. Springer (2006) ISBN 978-0-387-30667-4
16. Ihara, S.: Information theory for continuous systems, p. 2. World Scientific (1993) ISBN 978-981-02-0985-8
17. Dodge, Y.: The Oxford Dictionary of Statistical Terms, OUP (2003) ISBN 0-19-920613-9
18. Pappas, P.A., De Puy, V.: An Overview of Non-parametric Tests in SAS: When, Why, and How
19. Ansari, A.R., Bradley, R.A.: Rank-Sum Tests for Dispersions. Institute of Mathematical Statistics is collaborating with JSTOR to digitize, preserve, and extend access to The Annals of Mathematical Statistics
20. Usman Akram, M., Tariq, A., Almas Anjum, M., Younus Javed, M.: Automated detection of exudates in colored retinal images for diagnosis of diabetic retinopathy. Applied Optics 51(20(10)) (July 2012)
21. Bahoura, M., Pelletier, C.: Respiratory Sounds Classification using Cepstral Analysis and Gaussian Mixture Models. In: Proceedings of the 26th Annual International Conference of the IEEE EMBS San Francisco, CA, USA, September 1-5 (2004)
22. Duda, R.O., Hart, P.E., Stork, D.G.: Pattern Classification. Wiley (2001)
23. Earis, J.E., Cheetham, B.M.G.: Future perspectives for respiratory sound research. Eur. Respir. Rev. 10(77), 641–646 (2000)
24. Sovijarvi, A.R.A., Malmberg, L.P., Charbonneau, G., Vanderschoot, J., Dalmasso, F., Sacco, C., Rossi, M., Earis, J.E.: Characteristics of breath sounds and adventitious respiratory sounds. Eur. Respir. Rev. 10(77), 591–596 (2000)
25. Gross, V., Fachinger, P., Penzel, T., Koehler, U., von Wichert, P., Vogelmeier, C.: Detection of bronchial breathing caused by pneumonia. Biomed. Tech (Berl) 47(6), 146–150 (2002)
26. NHLBI Fact Book, Fiscal Year 2012, 38 p. (2012)
27. NHLBI Fact Book, Fiscal Year 2012, 51p. (2012)

Combined Features for Face Recognition
in Surveillance Conditions

Khaled Assaleh[1], Tamer Shanableh[2], and Kamal Abuqaaud[1]

[1] American University of Sharjah, Department of Electrical Engineering, Sharjah, UAE
kassaleh@aus.edu, kamal.abuqaaud@gmail.com
[2] American University of Sharjah, Department of Computer Science and Engineering,
Sharjah, UAE
tshanableh@aus.edu

Abstract. This paper addresses the challenging problem of face recognition in surveillance conditions based on the recently published database called SCface. This database emphasizes the challenges of face recognition in uncontrolled indoor conditions. In this database, 4160 face images were captured using five different commercial cameras of low resolution, at three different distances, both lighting conditions and face pose were uncontrolled. Moreover, some of the images were taken under night vision mode. This paper introduces a novel feature extraction scheme that combines parameters extracted from both spatial and frequency domains. These features will be referred to as Spatial and Frequency Domains Combined Features (SFDCF). The spatial domain features are extracted using Spatial Deferential Operators (SDO), while the frequency domain features are extracted using Discrete Cosine Transform (DCT). Principal Component Analysis (PCA) is used to reduce the dimensionality of the spatial domain features while zonal coding is used for reducing the dimensionality of the frequency domain features. The two feature sets were simply combined by concatenation to form a feature vector representing the face image. In this paper we provide a comparison, in terms of recognition results, between the proposed features and other typical features; namely, eigenfaces, discrete cosine coefficients, wavelet subband energies, and Gray Level Concurrence Matrix (GLCM) coefficients. The comparison shows that the proposed SFDCF feature set yields superior recognition rates, especially for images captured at far distances and images captured in the dark. The recognition rates using SFDCF reach 99.23% for images captured by different cameras at the same distance. While for images captured at different distances, SFDCF reaches a recognition rate of 93.8%.

Keywords: face recognition, feature extraction, SCface database, Surveillance cameras.

1   Introduction

Despite being one of the most convenient biometric systems, face recognition is not considered among the most reliable ones. This is especially true when images are captured by uncontrolled surveillance cameras [1]. Face recognition can be utilized in many applications including security, forensics and psychological assessments.

C.K. Loo et al. (Eds.): ICONIP 2014, Part II, LNCS 8835, pp. 503–514, 2014.
© Springer International Publishing Switzerland 2014

Unlike many other biometrics, face recognition offers the advantage of not requiring collaboration from the subject being identified.

Typically, an automatic face recognition system starts with a face detection stage [2], where the input image is scanned by sub-windows with different sizes to locate and crop the face to be recognized. Face detection is based on different features such as: facial appearance (eyes location), motion (in videos), skin color or combination of these methods [3]. Preprocessing of the detected face images is the second stage in a face recognition system. This stage tends to preserve and enhance the essential discriminant elements of visual appearance of each face image by emphasizing the interpersonal differences and deemphasizing the intrapersonal differences. One or more combinations of the following steps might be part of the preprocessing stage: color transformation and illumination processing [4], denoising [5], resizing and pose adjusting.

After preprocessing comes feature extraction, a feature extractor converts an image into a reduced and representative set of parameters. These parameters vary among the different known feature extraction algorithms; some algorithms follow appearance-based methods while others follow model-based approaches [15]. One of the earliest and most commonly used appearance-based approaches in face recognition is based on the concept of eigenfaces. Eigenfaces are extracted by performing Principle Component Analysis (PCA) on the preprocessed set of faces taken form the training part of a face recognition database. In PCA, each training face image is projected onto a subset of the eigenfaces to produce a sequence of coefficients that represent the feature vector of that face [6]. This face recognition method uses a simple minimum distance classifier to classify a face based on the distance between its feature vector and that of the reference face.

An alternative approach to eigenfaces-based feature extraction is the Discrete Cosine Transform (DCT). DCT transforms an image from the spatial domain into the frequency domain by projecting that image into the DCT basis images. Similar to the eigenface approach, this projection yields a set of coefficients called the DCT coefficients, of which a feature vector is derived. This transformation possesses excellent energy compaction properties, which makes it preserve desired features of the face in a relatively small number of DCT coefficients [7].

Gray level co-occurrence matrix (GLCM) and its statistical features have also been used in face recognition [8]. GLCM is used for extracting texture features by computing how frequently a pixel value appears in a certain direction in the spatial domain with respect to another pixel value. Differential operators that also reveal texture information were also used previous work [9].

The last stage in a face recognition system is classification which is comprised of two modes: training and testing. In the training mode, features are processed to generate face models, while in the testing mode, features are matched against the face models to find the best matching identity. Figure 1 shows a block diagram of a generic face recognition system.

In our proposed method, face detection was done in one of two approaches: color segmentation and variable-size cropping. The former was used for colored face images while the latter was used for greyscale images. After face detection, each face image was preprocessed by means down sampling, normalization and filtering.

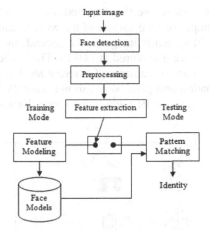

Fig. 1. A generic face recognition system

Features are then extracted in the spatial and frequency domains. More specifically, spatial domain features are based on spatial differential operators while frequency domain features were based on DCT. Each feature extraction scheme was followed by dimensionality reduction. PCA was used to reduce the dimensionality of the spatial domain features while zonal coding was used to reduce the dimensionality of the DCT features. Linear discriminant functions are used to classify the extracted features.

The remainder of this paper is organized as follows: Section 2 highlights the challenges in the SCface database with a brief description of the database. In section 3 we discuss the various steps of our proposed solution. Section 4 reviews the classification approach used. Results and discussions are presented in Section 5. Finally, the paper is concluded in Section 6.

2   SCface Database

For images taken by a relatively high resolution camera at a fixed distance with slight changes in pose and illumination, face recognition is relatively straightforward. Researchers have achieved more than 95% recognition rate for such conditions. For instance, high recognition rates were achieved on the Yale database [10] using wavelet transformation. Likewise, results reported on the Essex Grimace and the ORL databases reported recognition rates ranging between 94.5% to 98.5% [11]. On the other hand, SCface is a challenging face recognition database due to the fact that its images are of low resolution captured under different surveillance conditions [12]. SCface is a database of static images of 130 different people. Images were captured by five different video surveillance cameras. The cameras have varying quality and resolution levels. The capturing is done at three different distances. Two of these cameras were also used in the night vision mode. Therefore, a set of seven surveillance images per subject were collected at three different distances. In [12], the main objective of the authors was to introduce the SCface database for the face recognition community. However, to emphasize the challenging nature of the database they did provide some recognition results using standard eigenface algorithms where they reported a very low recognition rate of 10%. Figure 2 (a) shows a sample of face images of the SCface database, images in the first row were captured at 4.20 m, while images in the second row were captured at 2.60 m and images in the last row were captured at 1.00 m. The challenges in the SCface database are mainly due to the use of cameras of different resolutions capturing faces at different distances. Illumination and pose were also not controlled. Additionally, some images were taken in the dark using IR night vision mode. Examples of the IR images are shown in Figure 2 (b).

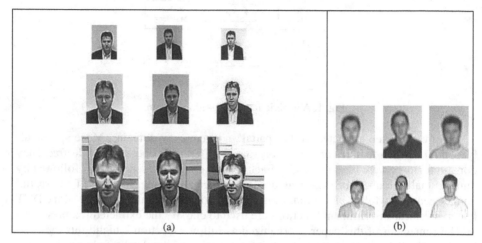

(a)   (b)

Fig. 2. Sample of face images taken from the SCface database

Table 1. Comparison between the SCface databse and other face recognitin databases

Database	Color	Image size	Subjects	Challenges
AR face database	Yes	576x768	126	1,3,8
Yale face database	No	320x243	15	1,3,8
Yale (B) face database	No	640x480	10	1,2,3,8
PIE face database	Yes	640x486	68	1,2,3,8
ORL face database	No	92 x 112	40	1,2,3,8
The Human scan database	No	384 x 286	23	1,2,3,8
FERET database	Yes	256 x 384	1199	1,2,3,7,8,9
SCface database	Mixed	Different sizes	130	1,2,3,4,5,6, 7,8

Challenge code	Challenge description
(1)	Facial expression
(2)	Pose to camera
(3)	Illumination
(4)	IR images (night mode vision)
(5)	Distance from camera
(6)	Camera quality (low resolution images)
(7)	Different cameras
(8)	w/no glasses, w/no scarf
(9)	Aging

To further emphasize the challenges introduced by the SCface database, Table 1 shows a comparison between the SCface database and other common face recognition databases.

3   Proposed Solution

The proposed solution includes several important steps prior to feature extraction. These steps include face detection, color transformation, down-sampling, normalization and filtering. These steps are needed to minimize the effect of the varying distances and camera types used in capturing the images as well as the effect of varying lighting conditions.

Two sets of features were extracted, one in the spatial domain and the other in the frequency domain. In the spatial domain, we extract features using Spatial Differential Operators (SDO) across the rows in a face image. More specifically, we compute the Euclidean distance between all the possible pairwise permutations of the image rows. For an image of m rows, $\sum_{i=1}^{m}(m - i) = \frac{1}{2}m(m - 1)$ distances are obtained.

Even for a small size image of 100 rows, it is noted that R is prohibitively large to be used as a feature vector. This warrants the need for dimensionality reduction. Hence, Principle Component Analysis (PCA) is used to obtain a manageable size feature vector. For the frequency domain feature extraction, each preprocessed image was transformed using DCT followed by zigzag scanning of coefficients known as zonal coding. Both spatial domain frequency domain feature vectors are concatenated to form one feature vector for each face image. Figure 3 shows the steps of transforming an input face image into a feature vector. The remainder of this section describes in more details each block in Figure 3. It also describes the classification techniques used in this work.

Face Detection
Face detection is required to extract a face from an image and to eliminate unnecessary background details. In our proposed solution, we used two different approaches for face detection; these are cropping with dynamic mask size and color segmentation. We used cropping with dynamic mask size approach with faces captured in the night vision mode. On the other hand, we used the color segmentation approach for the rest of the images. Dynamic mask size is needed to crop variable size faces captured at different distances from the cameras. In this face detection scheme, it is assumed that nose is almost located at the center of the face. The width of the dynamic mask is a function of the distance between the eyes; the height is function of the average distance between the mouth center and each eye. It should be noted that the nose and eyes locations are provided with the database. The other approach to face detection is based on color segmentation which does not require eye and nose coordinates. In this approach, an image is transformed from RGB into YCbCr color space for further thresholding. In [13], it was established that Cb and Cr values of skin color pixels fall within the following ranges: $77{\leq}Cb{\leq}127$ and $133{\leq}Cr {\leq}173$. Thresholding is followed by morphological operation to detect the face.

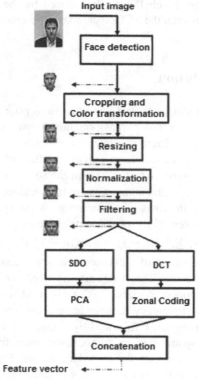

Fig. 3. Preprocessing and feature extraction steps

Preprocessing

After the face has been detected, several preprocessing steps were applied to eliminate or reduce the effects of resolution and lighting variability. These steps include converting each image into grey scale, resizing to one common size, normalization by replacing each greyscale with its corresponding z-score, and low pass filtering. Low pass filtering is done to reduce artifacts introduced by normalization.

Spatial Domain Feature Extraction

In the spatial domain, we apply differential operators across rows in a face image followed by PCA. A Differential operator refers to computing the distance between two rows. Several distance measures were investigated in this paper including Euclidean, standardized Euclidean, city block and cosine measures [14, 15]. It was found that standardized Euclidean distance yields the best feature vectors in terms of recognition rates. Standardized Euclidean distance between two n-dimensional row vectors $\mathbf{x}_i$ and $\mathbf{y}_j$ is given by:

$$d(x_i, y_j) = \left[(x_i - y_j) D^{-1} (x_i - y_j)^T \right]^{1/2} \tag{1}$$

Where D is an $n \times n$ diagonal matrix whose diagonal entries are the variances of the vector elements.

For an $m \times n$ image the SDO feature extraction scheme results in an R dimensional feature vector given by

$$d_k = d(x_i, y_j) \tag{2}$$

Where $= 1, 2, \ldots, \frac{1}{2}j(j-1)$, $i = 1, 2, \ldots m-1$, and $j = i+1, i+2, \ldots, m$, and $R = \sum_{i=1}^{m}(m-i) = \frac{1}{2}m(m-1)$

Figure 4 shows samples of two face images and their corresponding feature vectors using Equations 1 and 2 above.

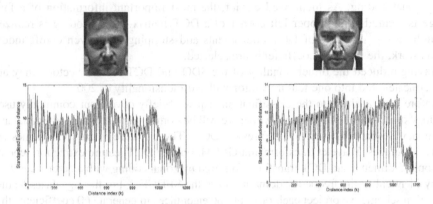

Fig. 4. Face images and their corresponding feature vectors

Frequency Domain Feature Extraction

As a feature extraction method, transformation into the frequency domain preserves the most important features of a face such as the hair line, the position of eyes, nose and mouth [7][12]. In the frequency domain, we extracted features by applying DCT followed by Zonal coding. The latter coding approach is used to convert a DCT matrix into a feature vector at a given frequency cutoff.

The DCT transform of an $M \times N$ image, $z(x,y)$, yields $M \times N$ DCT coefficients, $c(u,v)$, given by equations 3 and 4.

$$c(u, v) = \alpha(u)\alpha(v) \sum_{x=0}^{M-1} \sum_{y=0}^{N-1} Z(x,y) \cos\frac{\pi(2x+1)u}{2M} \cos\frac{\pi(2y+1)v}{2N} \tag{3}$$

$$\alpha(u) = \begin{cases} \sqrt{\frac{1}{M}} & u = 0 \\ \sqrt{\frac{2}{M}} & u \neq 0 \end{cases} \tag{4}$$

where $u,v = 0,1,2,3,4\ldots, N\text{-}1$.

Dimensionality Reduction

As mentioned previously, the feature extraction schemes used in this work as based on SDOs and 2D DCT. As noticed, the two schemes generate high dimensional feature vectors. Hence, we use PCA and zonal coding to reduce the dimensionality prior to generating the final feature vector of a given face image.

PCA uses an orthogonal linear transformation to convert, in our case, a set of feature vectors into another set of linearly uncorrelated variables. This set is known as principal components. Dimensionality reduction is possible because the number of principal components is less than or equal to the dimensionality of the input feature vectors. In this work, the dimensionality is reduced by retaining the largest 100 principal components.

On the other hand, the dimensionality of the DCT-based feature vectors is reduced using zonal coding. As mentioned earlier, the most important information of a face image is located in the upper left corner of a DCT matrix. Zonal coding is realized through zigzag scanning of DCT coefficients and stopping at a given cutoff index. In this work, the first 100 coefficients are selected.

Having reduced the dimensionality of the SDO and DCT feature vectors, they are then concatenated into one feature vector with a dimensionality of 200.

Before we transition to the classification step we briefly cover four commonly used feature extraction schemes against which we will be comparing our proposed Spatial and Frequency Domains Combined Features (SFDCF). These feature extraction schemes are DCT, eigenfaces, Wavelet transform and GLCM. For the DCT feature extraction scheme we apply 2-dimensioanl DCT transform followed by zonal coding with a certain cutoff in a way identical to the frequency domain part of the SFDCF.   For the eigenfaces feature extraction scheme we project each face into 60 eigenfaces to generate 60 coefficients that represent our 60-dimensional feature vector. For the Wavelet transform feature extraction scheme we decompose a face image into two levels to generate seven frequency bands, out of which we derive band-dependent statistical features to form our feature vector. Finally, for the GLCM feature extraction scheme we down quantize the image to 4bits per pixel and compute the GLCM which results in a relatively large number of parameter which we further reduce by applying PCA.

4   Classification

Since the main contribution of this this work is in the preprocessing and feature extraction part, we employ simple classification scheme based on linear discriminant functions (LDF). This classifier maps a set of N d-dimensional feature vectors, $\{\mathbf{x}_i; i = 1,2, ... N\}$, defined by a $N \times d$ matrix T_r into their corresponding c class labels (represented by the $N \times c$ matrix Y) via the following relationship:

$$T_r W = Y \tag{5}$$

Where,

$$T_r = [x_1, x_2, x_3, ... , x_N]^t \tag{6}$$

$$Y = [y_1, y_2, y_3, ... , y_c] \tag{7}$$

W is the weight matrix comprised of c weight vectors $\{w_j; j = 1, 2, \dots c\}$ that can be determined using the pseudo inverse solution such as

$$W = (T_r^t T_r)^{-1} T_r^t Y \qquad (8)$$

The matrix Y is comprised of c columns; each column has N entries that can be either 1 or 0 depending on the class assignment of the corresponding feature vector. The class of a given feature vector z can be determined by the following formula

$$class \ of \ z = arg \ max_j \ (zw_j) \qquad (9)$$

5   Experimental Results

In this section we provide an assessment, in terms of face recognition rates, of the proposed Spatial and Frequency Domains Combined Features (SFDCF). As such, we compare the recognition rates obtained by the SFDCF features to four well-known feature extraction schemes; these are: GLCM, Eigenfaces, DCT and Wavelet transform. To study the effect of the different cameras and the different distances at which images were captured, we establish three different sets of experiments. In the first set, we investigate the effect of training and testing on images captured with different cameras while keeping the distance from the camera fixed. In the second set, we study the effect of training and testing on images captured at different distances. In the third set, we use test images captured in the dark using the night vision mode. As mentioned previously, note that face images in the used dataset were captured at three different distances (4.20, 2.60 and 1.00 meters) with five different cameras. To compare the previously mentioned feature extraction schemes with SFDCF we established the following experiment setup for the first set of experiments:

For the first set of experiments, we have five images per distance captured by five different cameras for each one of the 130 subjects. In a leave-one-out manner, we train on four images and test on the fifth. This will result in five different combinations for each distance; each combination corresponds to training on 520 images and testing on 130. The recognition results of the five combinations are averaged and reported. This is done for each of the five different feature extraction schemes as shown in Figure 5.

Figure 5 shows that the proposed system based on SFDCF yields the highest recognition rates for all three distances. The obtained recognition results are also superior to previously published work based on eigenfaces [12] and our own DCT-based work in [16]. Also, it is concluded that the GLCM and Wavelet transform feature extraction schemes perform very poorly compared to the other three schemes. Therefore, we no longer use them in the subsequent experiments.

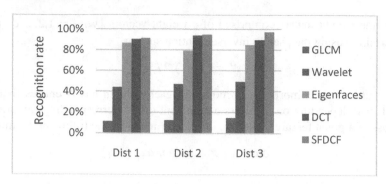

Fig. 5. Average recognition rates across 5 different cameras

To assess the effect of training and testing on images captured at different distances, we use the second set of experiments described above. In this set we train on all images captured at two of the three distances using all cameras (i.e. 10 images) and perform the test on the remaining images that were captured at the third distance (i.e. 5 images). As such, the three possible combinations are:

a) Train on all images captured at Dist1 and Dist2 by all five cameras (i.e. 10 images) and test on all images captured at Dist3 by all five cameras (i.e. 5 images). This is denoted by Ts3, Tr1, Tr2.

b) Train on all images captured at Dist1 and Dist3 by all five cameras (i.e. 10 images) and test on all images captured at Dist2 by all five cameras (i.e. 5 images). This is denoted by Ts2, Tr1,Tr3.

c) Train on all images captured at Dist2 and Dist3 by all five cameras (i.e. 10 images) and test on all images captured at Dist1 by all five cameras (i.e. 5 images). This is denoted by Ts1, Tr2, Tr3.

The average recognition rates are shown in Figure 6. Once again, the proposed SFDCF yields the best performance as compared to the DCT based and eigenfaces based methods.

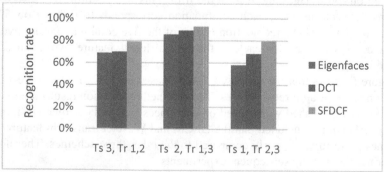

Fig. 6. Average recognition rates, testing at a distance and training at the other distances

It is worthwhile to mention that in [17] Jae Young et. al. have studied the effect of training and testing on different distances. They used a slightly different arrangement by training on 11 images (5 at Dist1 and 5 at Dist2 in addition to the mug shot image) and testing on the remaining 5 images at Dist3. In other words, they used one additional (high resolution) image in their training. Nevertheless, they reported a much lower recognition rate of 62.78% compared to our results shown in Figure 6.

The third set of experiments focus on testing on the images captured by the night-vision mode cameras (i.e. camera 6 and camera 7). In this set of experiments, training was done using all images captured by the five regular cameras at each distance while the testing is done on the images captured by the night-vision mode cameras at the same distance. The poor quality of the test images is expected to result in lower recognition rates. This is manifested by the results shown in Figure 7. Nonetheless, the proposed system based on SFDCF yields the highest recognition rates as compared to the other two schemes.

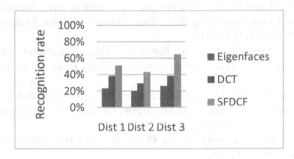

Fig. 7. Average recognition rates, testing on night vision mode images

6   Conclusion

In this paper we propose a face recognition system for uncontrolled indoor conditions evaluated on the SCface database. The proposed system is comprised of a series of steps of preprocessing and feature extraction followed by a simple classification scheme. These steps of preprocessing and feature extraction focus on maximizing the discriminability of the feature vectors. A novel set of Spatial and Frequency Domains Combined Features (SFDCF) is introduced and examined against other commonly used features in face recognition. The proposed system is evaluated using the SCface database in different scenarios across different cameras and different distances. The introduced SFDCF features are found to be superior to other existing features and the overall system is found to largely outperform previously published results on the same database. It should be noted that the face detection and preprocessing steps have greatly enhanced the discriminability of all the examined features and hence the recognition rates.

References

1. Chuu, T., Azmin, S.: A study on face recognition in video surveillance system using multi-class Support Vector Machines. In: TENCON 2011, IEEE Region 10 Conference, pp. 25–29 (2011)
2. Jain, A., Ross, A., Prabhakar, S.: An Introduction to Biometric Recognition. IEEE Transactions on Circuits and Systems for Video Technology 14(1), 4–20 (2004)
3. Li, S., Jain, K.: Hand book of face recognition. Springer, New York (2005)
4. Han, H., Shan, S., Chen, X., Gao, W.: A comparative Study on Illumination Preprocessing in Face Recognition. Pattern Recognition 46(6), 1691–1699 (2013)
5. Jiang, M., Feng, J.: Robust Low-rank Subspace Recovery and Face Image Denoising for Face Recognition. In: IEEE International Conference on Image Processing (ICIP), pp. 3033–3036 (September 2011)
6. Belhumeur, P., Hespanha, J., Kriegman, D.: Eigenfaces vs. Fisherfaces: Recognition Using Class Specific Linear Projection. IEEE Transactions on Pattern Analysis and Machine Intelligence, 711–720 (July 1997)
7. Hajiarbabi, M., Askari, J., Sadri, S., Saraee, M.: Face Recognition using Discrete Cosine Transform plus Linear Discriminant Analysis. World Congress on Engineering 1 (July 2007)
8. Eleyan, A., Demirel, H.: Co-occurrence matrix and its statistical features as a new approach for face recognition. Turkish Journal of Electrical Engineering & Computer Sciences 19(1), 97–107 (2011)
9. Perlibakas, V.: Distance measures for PCA-based face recognition. Pattern Recognition Letters 25(6), 711–724 (2004)
10. Ying, S., Yushi, Z., Cheng, Z., Xili, Z., Lihong, Z.: Face Recognition Based on Image Transformation. In: 2009 WRI Global Congress on Intelligent Systems, China, pp. 418–421 (May 2009)
11. Mandal, T., Wu, Q., Yuan, Y.: Curvelet based face recognition via dimension reduction. Signal Processing 89(12), 2345–2353 (2009)
12. Grgic, M., Delac, K., Grgic, S.: SCface - surveillance cameras face database. Multimedia Tools and Applications Journal 51(3), 863–879 (2011)
13. Douglas, C., King, N.: Face segmentation using skin-color map in videophone applications. IEEE Transactions on Circuits and Systems for Video Technology, 551–564 (1999)
14. Wang, L., Zhang, Y., Feng, J.: On the Euclidean distance of images. IEEE Transactions on Pattern Analyses and Machine Intelligence 27(8), 1334–1339 (2005)
15. Shi, J., Samal, A., Marx, D.: How effective are landmarks and their geometry for face recognition? Computer Vision and Image Understanding 102(2), 117–133 (2006)
16. Assaleh, K., Shanableh, T., Abuqaaud, K.: Face Recognition using different surveillance cameras. In: ICCSPA, Sharjah, UAE (February 2013)
17. YoungChoi, J., Ro, Y., Plataniotis, K.: A Comparative Study of Preprocessing Mismatch Effects in Color Image based Face Recognition. Pattern Recognition 44, 412–430 (2011)

Sparse Coding on Multiple Manifold Data

Hanchao Zhang and Jinhua Xu

Department of Computer Science and Technology
East China Normal University
Shanghai, China
jhxu@cs.ecnu.edu.cn

Abstract. Sparse coding has been widely used in computer vision. While capturing high-level semantics, the independent coding process neglects connections between data points. Some recent methods use Laplacian matrix to learn sparse representations with locality preserving on the manifold. Considering data points may lie in or close to multiple low dimensional manifolds embedded in the high dimensional descriptor space, we use sparse representations to code the local similarity between data points on each manifold and embed this topology to sparse coding algorithm. By keeping the locality of manifolds we can preserve the similarity and separability at the same time. Experimental results on several benchmark data sets show our algorithm is effective.

Keywords: Sparse coding, sparse representation, manifold learning, Laplacian matrix, locality preserving.

1 Introduction

Sparse coding has been attracting much attention since its emergence. Thousands of articles have been published forming a solid basis of its applications. Given a data set, sparse coding finds a dictionary and sparse coordinates of the data points in the set. It is believed that these representations can reflect some high-level semantic structures of the data set.

Recently, many researchers considered high-dimensional data points lie in or close to a manifold of intrinsic low-dimension. Variations of local manifold learning algorithms have been proposed, such as local linear embedding (LLE) [8], Hessian LLE [2], and Laplacian eigenmaps (LEM) [1], trying to preserve local relationships between points.

In the sparse coding field, it is also suggested that locality is more essential than sparsity [10], since locality can lead to sparsity but not necessary vice versa. Wang [9] used locality constraint instead of the sparsity constraint. Zheng *et al.* [11] and Gao *et al.* [4], proposed two similar sparse coding methods incorporating the Laplacian matrix to enhance the consistency of the representations. Lu [7] extended this algorithm by introducing the geometrical structure of the dictionary. These methods preserve local similarity of the data points in the ambient space but points lie in the same neighborhood defined in the feature space might not

C.K. Loo et al. (Eds.): ICONIP 2014, Part II, LNCS 8835, pp. 515–523, 2014.
© Springer International Publishing Switzerland 2014

be in the same manifold. In other words, a neighborhood of the high-dimensional space can not separate two data points lying in the different manifolds. Under the manifold assumption, we try to find a way to preserve the consistency and separability of the manifolds in the sparse coding framework.

Another motivation of our research is that we find two kinds of relationships might be essential for clustering and classification tasks. The first is the relationships between the data points and their own intrinsic characters, and the second is the relationships between data points. These two are related, but not necessary the same. Both of them can be reflected in representations. We aim at using connections between data points to improve sparse coding representations reflecting the first kind of relationships.

In this paper, we use sparse representation to encode relationships, local similarity more exactly, between data points, and then use a Laplacian matrix to incorporate this topology in sparse coding.

2   Related Work

2.1   Sparse Coding (SC)

Let X be a set of D-dimension feature vectors abstracted from images ie. $X = [x_1, x_2, \ldots, x_N] \in \mathbb{R}^{D \times N}$, $B = [b_1, b_2, \ldots, b_M] \in \mathbb{R}^{D \times M}$ be a dictionary and $S = [s_1, s_2, \ldots, s_N] \in \mathbb{R}^{M \times N}$ be sparse codes. The sparse coding problem can be formulated as

$$\min_{B,S} \|X - BS\|_F^2 + \lambda \|S\|_1. \tag{1}$$

This optimization problem can be solved by alternatively optimizing B and S while keeping the other one fixed. More specifically, it is solved by

$$\min_{s_i} f(s_i) = \|x_i - Bs_i\|^2 + \lambda \|s_i\|_1, \ i = 1, \ldots, N, \tag{2}$$

$$\min_{B} \|X - BS\|_F^2 \quad s.t. \ \|b_i\|^2 \leq 1, \ i = 1, \ldots, M. \tag{3}$$

By (2) and (3), we can say the learned dictionary B depends on the total data set X rather than any particular data points. Hence, the words in the dictionary B can be seen as intrinsic characters of the set and the coordinates S can be seen as representations that can capture these characters without considering of connections between data points due to independent encoding.

2.2   Graph Sparse Coding (GraphSC)

The Graph sparse coding proposed by Zheng [11] and the Laplacian sparse coding proposed by Gao [4] used Laplacian matrixes to smooth the manifold obtained by sparse coding, enhancing the local consistency. Their methods can be formulated as

$$\min_{B,S} \|X - BS\|_F^2 + \alpha \text{Tr}(SGS^T) + \beta \sum_{i=1}^{M} \|s_i\|_1. \tag{4}$$

G can be obtained by $G = W - U$ where U is the nearest neighbor's distance matrix, or called similarity matrix. Zheng [11] constructed U as follows,

$$U_{ij} = \begin{cases} 1 \text{ if } \boldsymbol{x}_j \in \mathrm{N_{knn}}\{\boldsymbol{x}_i\} \text{ or } \boldsymbol{x}_i \in \mathrm{N_{knn}}\{\boldsymbol{x}_j\} \\ 0 \text{ if } \boldsymbol{x}_j \notin \mathrm{N_{knn}}\{\boldsymbol{x}_i\} \text{ and } \boldsymbol{x}_i \notin \mathrm{N_{knn}}\{\boldsymbol{x}_j\} \end{cases}. \tag{5}$$

Gao [4] constructed U as follows

$$U_{ji} = U_{ij} = \begin{cases} \sum_{d=1}^{D} \min\left(X_{di}, X_{dj}\right) \text{ if } \boldsymbol{x}_j \in \mathrm{N_{knn}}\{\boldsymbol{x}_i\} \\ \qquad\qquad 0 \qquad\qquad \text{ if } \boldsymbol{x}_j \notin \mathrm{N_{knn}}\{\boldsymbol{x}_i\} \end{cases}. \tag{6}$$

Though these two U were different, both of them were calculated in the local feature space. W is a diagonal matrix, $W_{ii} = \sum_{j=1}^{N} u_{ij}$.

3 Proposed Sparse Coding

3.1 Sparse Representation Encoding Similarity

Here we use the Sparse Manifold Clustering and Embedding (SMCE) method proposed by Elhamifar [3] to learn sparse representations encoding local similarity between points. SMCE aims to find the neighborhood in the manifold and the weights by selecting the nearest low-dimensional affine subspace. It can be formulated as an optimization problem:

$$\min \|\boldsymbol{Q}_i \boldsymbol{c}_i\|_1; \ s.t. \ \|\boldsymbol{X}_i \boldsymbol{c}_i\|_2 \leq \varepsilon, \mathbf{1}^T \boldsymbol{c}_i = 1, \tag{7}$$

$$\boldsymbol{X}_i \triangleq \left[\frac{\boldsymbol{x}_1 - \boldsymbol{x}_i}{\|\boldsymbol{x}_1 - \boldsymbol{x}_i\|_2} \cdots \frac{\boldsymbol{x}_N - \boldsymbol{x}_i}{\|\boldsymbol{x}_N - \boldsymbol{x}_i\|_2} \right] \in \mathrm{I\!R}^{D \times N}. \tag{8}$$

Q is the proximity inducing matrix, which tends to select points close to $\boldsymbol{x}_i$. We simply choose Q to be $\frac{\|\boldsymbol{x}_j - \boldsymbol{x}_i\|_2}{\sum_{j \neq i} \|\boldsymbol{x}_j - \boldsymbol{x}_i\|_2} \in (0, 1]$. For each solution satisfying constraints, $\boldsymbol{x}_n$ corresponding to nonzero $\boldsymbol{c}_n$ can approximately span a low-dimensional affine subspace. And the solution to the objective function is the nearest affine subspace which can be seen as the nearest neighborhood of data point $\boldsymbol{x}_i$ in the manifold it belongs to.

Using the method of Lagrange multipliers, the above optimization problem can be changed to

$$\min \lambda \|\boldsymbol{Q}_i \boldsymbol{c}_i\|_1 + \frac{1}{2} \|\boldsymbol{X}_i \boldsymbol{c}_i\|_2^2, \ s.t. \ \mathbf{1}^T \boldsymbol{c}_i = 1. \tag{9}$$

This problem can be solved using the alternative direction method of multiplies algorithm(ADMM) and soft threshold algorithms [3]. Rewriting $\boldsymbol{x}_i \approx [\boldsymbol{x}_1 \boldsymbol{x}_2 ... \boldsymbol{x}_N] \boldsymbol{u}_i$, $\boldsymbol{u}_i$ can be seen as a representation of certain sparsity, where $\boldsymbol{u}_i$ is defined as

$$u_{ii} \triangleq 0, u_{ij} \triangleq \frac{c_{ij}/\|\boldsymbol{x}_j - \boldsymbol{x}_i\|_2}{\sum_{t \neq i} c_{it}/\|\boldsymbol{x}_t - \boldsymbol{x}_i\|_2}, j \neq i. \tag{10}$$

We use SMCE because firstly it can separate close manifolds, secondly it can choose the right size of the neighborhood without explicitly choosing K, thirdly it is robust to the parameter.

3.2 Sparse Coding Preserving Consistency and Separability

The similarity relationship is preserved during the sparse coding. The similarity matrix can be obtained by symmetrizing the similarity representation U using $U = \max\{|U|, |U^T|\}^1$. The geometry constraint is constructed as $G = W - U$ where $W_{ii} = \sum_{j=1}^{N} u_{ij}$ and the optimization problem can be formed as

$$\min_{B,S} \|X - BS\|_F^2 + \alpha \text{Tr}(SGS^T) + \beta \sum_{i=1}^{M} \|s_i\|_1. \tag{11}$$

We solve this problem using the feature sign algorithm [5].

It should be noticed that this formulation is similar to Zheng [11] and Gao [4]. But the size of the neighborhoods in their method was fixed as a parameter, which might be inappropriate considering the complexity of the real word. Besides, Zheng [11] used kNN neighborhoods but these neighborhoods could contain data points in the different manifold. Gao[4] used histogram intersection kernel plus kNN neighborhoods, which might be more separable than Zheng, but it neglected the multiple manifolds structure too. Instead, we use the neighbor points in an approximate affine subspace with x_i being its zero point, and the neighborhood of x_i and connection strength of the neighborhood are obtained automatically. Informally, this approximate affine subspace passes through the manifold and is close to the tangent space of the manifold at x_i. Hence our proposal method is more meaningful and flexible. The motivation of Liu [6] is similar to us. The major difference is that they aim at keeping the local decomposition property not the local similarity. Their method might be hard to extend to preserve more global property, while ours can be extended to hypergraph version as [4] naturally.

4 Experimental Results

In this section our proposed method is evaluated for both classification and clustering tasks on publicly available image data sets, including PIE[2], COIL20[3], USPS[4].

---

[1] $|U|$ denotes calculating absolute value of each element in U.
[2] http://vasc.ri.cmu.edu/idb/html/face
[3] http://www1.cs.columbia.edu/CAVE/software/softlib/coil-20.php
[4] http://www-i6.informatik.rwth-aachen.de/ keysers/usps.html

4.1   Clustering

For clustering we investigate our method on two data sets i.e. CMU-PIE and COIL20. And we compare five algorithms on these sets: K-means clustering algorithm, PCA plus K-means, sparse coding(SC) plus K-means, Graph sparse coding (GraphSC) plus K-means, and our proposal method plus K-means. In all experiments, we first apply PCA to reduce the data dimensionality by keeping 98% information. Next we use each of methods to learn representations and perform K-means in the learned space. The K-means is repeated 50 times, and the best result according to the objective function of K-means is recorded. Sparsity parameter β for all clustering experiment is set to 0.1.

Evaluation Metrics. As in [11], the accuracy (AC) and the normalized mutual information (NMI) are calculated to evaluate our proposal method.

Given a data point x_i, let c_i and r_i be its cluster label provided by ground truth and by the algorithm respectively. The AC can be defined as

$$AC = \frac{\sum_{i=1}^{N} \delta(c_i, map(r_i))}{N}, \tag{12}$$

where N is total number of samples, $\delta(x, y)$ is the delta function, and $map(r_i)$ maps the cluster r_i to its best label which can be found using the Kuhn-Munkres algorithm[11]. The mutual information is defined as

$$MI(C, C') = \sum_{c_i \in C, c'_j \in C'} p(c_i, c'_j) \cdot \log_2 \frac{p(c_i, c'_j)}{p(c_i) \cdot p(c'_j)}, \tag{13}$$

where C is the true cluster sets obtained from the ground truth and C' obtained from each algorithm, and $p(c_i)$ and $p(c'_j)$ are the probabilities that a sample arbitrarily selected from the data set belongs to the clusters c_i and c'_j respectively, and $p(c_i, c'_j)$ is the joint probability. The NMI can be defined as:

$$\overline{MI}(C, C') = \frac{MI(C, C')}{\max(H(C), H(C'))}. \tag{14}$$

CMU-PIE Face Database. For CME-PIE face database, there are 68 subjects with 41368 face images. The size of each image is 32×32, with 256 grey levels. Each image is represented by a 1024-dimensional vector. The data points are clustered to 68 categories. We choose a subset[5] of CMU-PIE with the fixed pose and expression as in [11] for comparison. Thus, there are 21 pictures for each subject under different conditions. After PCA projection, the dimensionality is reduced to 64. The dictionary dimensionality is set to 128 in all sparse coding algorithms. The results can be seen in Table 1. Our proposed algorithm achieves the highest scores.

---

[5] http://www.cad.zju.edu.cn/home/dengcai/Data/FaceData.html

Table 1. Performance on a subset of CME-PIE

(a) AC (%)					(b) NMI (%)				
K-means	PCA	SC	GraphSC	Ours	K-means	PCA	SC	GraphSC	Ours
31.7	32.9	60.4	84.7	86.5	63.3	66.6	78.8	94.9	96.1

COIL20 Database. For COIL20 database, there are 1440 32×32 gray scale images of 20 objects. Backgrounds have been discarded. Each image is represented by a 1024-dimensional vector. We cluster the data points into 20 categories. After PCA, the dimensionality is reduced to 175. So we use 256 dimensional dictionary in all three sparse methods, and the other experimental setup is the same as before. The results can be seen in Table 2. Our proposed algorithm achieves higher scores than the others in both AC and NMI.

Table 2. Performance on COIL20

(a) AC (%)					(b) NMI (%)				
K-means	PCA	SC	GraphSC	Ours	K-means	PCA	SC	GraphSC	Ours
62.0	63.7	74.5	78.6	80.9	73.4	74.7	84.5	86.8	89.1

Parameter Selection. In our algorithm there are two parameters, the geometry regularization parameter α in (11) and the sparsity regularization parameter λ in SMCE(9). We compare different values in clustering task. α is set to be $\{1, 5, 10, 15, 20, 25, 30\}$, λ is $\{10, 50, 70, 90, 100, 110, 130, 150\}$. We vary one parameter while keeping the other fixed. The results are illustrated in Fig. 1 and Fig. 2. These results show that our algorithm is robust in a large range. Besides α favors relatively large values. It might be because the similarity connections represented by SMCE is more reliable than GraphSC. It should be noted that when α is extremely small, for example, α is 1, the result is worse than sparse coding. This phenomenon might be due to the disparity between two kinds of relationships mentioned before.

4.2 Classification

We use USPS handwritten digits data set to evaluate our proposal method for classification. There are 7291 training images and 2007 test images of size 16×16 in the data set. Each picture is represented by a 256-dimensional vector. For a new data point x_t, we use (9) to calculate the new sparse representation coding the similarity u and the similarity matrix U can be expanded as

$$\begin{bmatrix} U & |u| \\ |u^T| & 0 \end{bmatrix}. \tag{15}$$

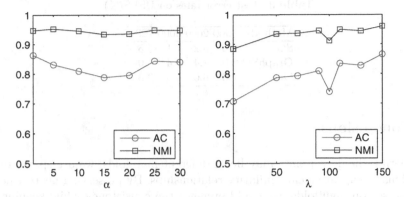

Fig. 1. Clustering performance on a subset of PIE

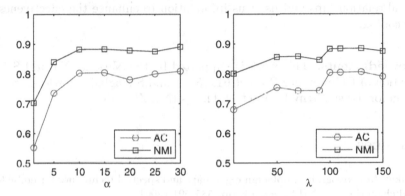

Fig. 2. Clustering performance on COIL20

The new geometry constraint matrix $\hat{G}$ can be constructed and the new representation s_t for x_t can be obtained by solving the optimization problem while keeping B and S fixed, that is,

$$\min_{s_t} f(s_t) = \|x_t - Bs_t\|^2 + \alpha \hat{G}_{tt} s_t^T s_t + 2\alpha s_t^T \left(\sum_{j \neq t} \hat{G}_{tj} s_j \right) + \beta \|s_t\|_1. \quad (16)$$

For classification task, we perform five-fold cross validation to find the best parameter pair (α, k) for GraphSC and parameter pair (α, λ) for our proposal method. The test values for α are {0.01, 0.1, 1, 10}, for k {2, 3, 4, 5, 6, 7, 8, 9, 10} and for λ {50, 100, 150, 200}. The test values for the size of dictionary are {32, 64, 128, 256} and for the sparsity β are {0.1, 0.2, 0.3, 0.4, 0.5}. The size of training sets are {1000, 2000, 5000, 7291}. We train linear SVM to obtain the classification results, which are illustrated in Table 3. Our proposal method performs best in all 4 conditions especially when training set is relatively small.

Table 3. Test error rates on USPS(%)

Methods	1000	2000	5000	7291
SC	9.6	8.2	7.1	5.5
GraphSC	8.0	7.4	6.4	5.0
Ours	7.0	6.0	4.8	3.9

5   Conclusions

Considering the data points may lie in different manifolds, we use sparse representations to capture local similarity relationships. By preserving nearest neighborhood set on manifolds, we could enhance the consistency while keeping the separability. Experimental results demonstrate the effectiveness of our proposal method. In the future, we will try to find representations which can capture other non-local relationships and use this information to enhance the effectiveness of sparse coding.

Acknowledgments. This work is supported by the National Natural Science Foundation of China under Project 61175116, and Shanghai Knowledge Service Platform for Trustworthy Internet of Things (No. ZF1213).

References

1. Belkin, M., Niyogi, P.: Laplacian eigenmaps and spectral techniques for embedding and clustering. In: NIPS, vol. 14, pp. 585–591 (2001)
2. Donoho, D.L., Grimes, C.: Hessian eigenmaps: Locally linear embedding techniques for high-dimensional data. Proceedings of the National Academy of Sciences 100(10), 5591–5596 (2003)
3. Elhamifar, E., Vidal, R.: Sparse manifold clustering and embedding. In: NIPS, pp. 55–63 (2011)
4. Gao, S., Tsang, I.H., Chia, L.T.: Laplacian sparse coding, hypergraph laplacian sparse coding, and applications. IEEE Transactions on Pattern Analysis and Machine Intelligence 35(1), 92–104 (2013)
5. Lee, H., Battle, A., Raina, R., Ng, A.Y.: Efficient sparse coding algorithms. Advances in Neural Information Processing Systems 19, 801 (2007)
6. Liu, B.D., Wang, Y.X., Zhang, Y.J., Shen, B.: Learning dictionary on manifolds for image classification. Pattern Recognition 46(7), 1879–1890 (2013)
7. Lu, X., Yuan, H., Yan, P., Yuan, Y., Li, X.: Geometry constrained sparse coding for single image super-resolution. In: 2012 IEEE Conference on Computer Vision and Pattern Recognition (CVPR), pp. 1648–1655. IEEE (2012)
8. Roweis, S.T., Saul, L.K.: Nonlinear dimensionality reduction by locally linear embedding. Science 290(5500), 2323–2326 (2000)
9. Wang, J., Yang, J., Yu, K., Lv, F., Huang, T., Gong, Y.: Locality-constrained linear coding for image classification. In: 2010 IEEE Conference on Computer Vision and Pattern Recognition (CVPR), pp. 3360–3367. IEEE (2010)

10. Yang, J., Yu, K., Gong, Y., Huang, T.: Linear spatial pyramid matching using sparse coding for image classification. In: IEEE Conference on Computer Vision and Pattern Recognition, CVPR 2009, pp. 1794–1801. IEEE (2009)
11. Zheng, M., Bu, J., Chen, C., Wang, C., Zhang, L., Qiu, G., Cai, D.: Graph regularized sparse coding for image representation. IEEE Transactions on Image Processing 20(5), 1327–1336 (2011)

Mutual Information Estimation
with Random Forests

Mike Koeman and Tom Heskes

Institute for Computing and Information Sciences,
Radboud University Nijmegen, The Netherlands

Abstract. We present a new method for estimating mutual information based on the random forests classifiers. This method uses random permutation of one of the two variables to create data where the two variables are independent. We show that mutual information can be estimated by the class probabilities of a probabilistic classifier trained on the independent against the dependent data. This method has the robustness and flexibility that random forests offers as well as the possibility to use mixtures of continuous and discrete data, unlike most other approaches for estimating mutual information. We tested our method on a variety of data and found it to be accurate with medium or large datasets yet inaccurate with smaller datasets. On the positive side, our method is capable to estimate the mutual information between sets of both continuous and discrete variables and appears to be relatively insensitive to the addition of noise variables.

Keywords: Mutual information, random forests, probabilistic classification trees.

1 Introduction

Mutual information is a concept from information theory that measures the mutual dependency between two random variables. It is used in a variety of applications with among others feature selection [1], as a cost function in decision tree learning [2] and to learn the structure of Bayesian networks [3]. Calculating the mutual information is non-trivial and many methods for estimating it have been proposed, among others [4], the Kraskov KNN method that uses nearest neighbors based estimation [5] and the kernel density estimator method [6] that first estimates the distribution function with the help of a kernel density estimator [7]. A problem of these methods is that they are not suited to handle combinations of discrete and continuous variables or cannot properly handle sets of variables for X and Y. We present a novel method that does not have these limitations.

Mutual information is defined by

$$I(X;Y) = \int_Y \int_X P(X,Y) \log \frac{P(X)P(Y)}{P(X,Y)} dY dX \qquad (1)$$

C.K. Loo et al. (Eds.): ICONIP 2014, Part II, LNCS 8835, pp. 524–531, 2014.
© Springer International Publishing Switzerland 2014

where $I(X;Y)$ is the mutual information between X and Y, $P(X,Y)$ is the joint probability and $P(X)$ and $P(Y)$ are the marginalized distributions. The natural logarithm is often used for the log.

Mutual information can be seen as the amount of information that is known about one variable if we measure the other. One of the main advantages of mutual information over simpler measures of dependency like covariance is that it captures nonlinear dependencies as well. For a more extensive treatment of mutual information and its properties we refer to an information theory textbook [8].

As long as the probability density functions X and Y are known, it is straightforward to calculate I with (1). If they, however, are unknown, it is impossible to directly use (1) and an alternative approach will have to be used. In Sect. 2 we will explain how estimating mutual information can be rephrased as a classification problem, which can then be solved by a non-linear classifier. Section 3 we present results on the comparison between our method, the KNN method and the KDE method. We end with with a conclusion and discussion in Sect. 4. For now we will consider the mutual information between two continuous variables. It is easy to see how our approach can be generalized to sets of discrete and continuous variables.

2 Estimating Mutual Information with Random Forests

2.1 Phrasing Mutual Information Estimation as a Classification Problem

Let **D** be a dataset consisting of N samples of the random variables X and Y. This dataset is described by an unknown $P(X,Y)$. We define **D'** which is equal to **D** except X has been randomly permuted. This permutation cancels all dependencies from **D** and is now described by $P(X)P(Y)$ instead of $P(X,Y)$. We now have two datasets in the same space. We combine these to create $\hat{\mathbf{D}}$ and assign a class label to each sample depending on the dataset it originated from. Each sample from **D** becomes label *foreground* and from **D'** label *background*. We now have a two class problem and can train a classifier on this problem to estimate $P(background \mid X,Y)$ and $P(foreground \mid X,Y)$. We can use these probabilities to rewrite (1) to an expression that uses these probabilities instead of $P(X,Y)$. We start by noting that $P(X,Y)$ approaches $P(X,Y \mid foreground)$ and $P(X)P(Y)$ approaches $P(X,Y \mid background)$ as $N \to \infty$ with a suitable classifier and substitute this into (1) to obtain

$$I(X;Y) = \int_X \int_Y P(X,Y) \log \frac{P(X,Y \mid foreground)}{P(X,Y \mid background)} dY \, dX. \qquad (2)$$

We now use Bayes law to rewrite (2) into

$$I(X;Y) = \int_X \int_Y P(X,Y) \log \frac{P(foreground \mid X,Y)P(background)}{P(background \mid X,Y)P(foreground)} dY \, dX. \quad (3)$$

Finally we replace the integral over $P(X, Y)$ by its sample estimate

$$I(X;Y) = \frac{1}{N} \sum_{i=1}^{N} \log \frac{P(foreground \mid x_i, y_i)P(background)}{P(background \mid x_i, y_i)P(foreground)}. \tag{4}$$

This equation can now be used with a suitable non-parametric probabilistic classifier. Note that if the dataset is governed by a certain parametric distribution it is often better to directly estimate this distribution from the data and apply (1). This classifier-based approach to mutual information has several advantages over other methods. Firstly the X and Y do not necessarily represent single variable, they can be a combination of several variables as well. This allows the method to estimate the mutual information between groups of variables. Secondly this method allows for both real-valued and discrete variables mixes of those, as long as the classifier used can handle such combinations.

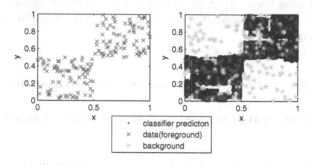

Fig. 1. An overview of the method. The plot on the left shows an example of a dataset. The plot on the right shows this same dataset but now with background samples and the classifier prediction trained on this data.

2.2 Random Forests

Random forests is an ensemble decision tree method proposed by Breiman [9]. It creates a subset of the variables for each tree in combination with bagging [10], which performs very well compared to other classifiers and received quite some attention lately [11]. Its non-parametric nature and robustness make random forests a potential canidate to use with (4). It is, however, non-trivial to obtain precise class probabilities from random forests, see [12], and work has been done to improve these probabilities [13]. For our probabilistic random forests we use Platt scaling [13] combined with a weighted average over all trees with the probabilities estimated coming from the out-of-bag samples. This approach is explained more thoroughly in the following scheme.

1. Train a random forests classifier.
2. For each leaf in each tree determine the amount of *foreground* and *background* samples from the out-of-bag samples for that tree.

3. For each test sample determine all leaves this sample ends up in, and consecutively sum up the amount of *background* and *foreground* from the out-of-bag calculated in step 2. The fraction of these will give an unscaled estimate of $P(\text{foreground} \mid X, Y)$.
4. The obtained probabilities are calibrated using Platt scaling to obtained corrected probabilities.

Platt scaling [13] is a method of calibrating probabilities and is defined by

$$\hat{p}(c|s) = \frac{1}{1 + e^{As+B}} \tag{5}$$

where $\hat{p}$ gives the probability that an example belongs to class c, given that it has obtained the score s, and where A and B are parameters of the function. We can use logistic regression on the data consisting of the combination of *foreground* and *background* samples to find the optimal values for A and B. For the score s we use the logit function on the probabilities obtained by step 3:

$$s_i = \log \frac{P(\text{foreground} \mid x_i, y_i)}{1 - P(\text{foreground} \mid x_i, y_i)} \tag{6}$$

where s_i is the score belonging to sample i. We calculate every $P(\text{foreground} \mid x_i, y_i)$ and use it together with (5) to obtain calibrated probabilities for each sample. These calibrated probabilities are then substituted into (4) to estimate the mutual information.

An alternative interpretation to this method of estimating mutual information is to view it as the average over an ensemble of adaptive binning models. Adaptive binning is a way of binning where the bin size is variable instead of constant. Each tree in the ensemble can be seen as a way to split the data into bins of varying size with each its own probability for *foreground*. Step 3 in our scheme then takes the weighted average over all these models.

2.3   Creating Foreground and Background Samples

To implement the above scheme successfully we need the *foreground* and *background* samples to be (close to) independent. In this section we describe a particular cross-validation approach that aims to achieve this. Random permutation of X produces **D'** where X and Y are independent of each other. However, **D'** is not independent of **D** because it uses the same numerical values for the variables X and Y. If $\hat{\mathbf{D}} = \{\mathbf{D}, \mathbf{D'}\}$ is separated in two subsets, these are no longer independent which creates a bias when trained on one subset and tested on the other. This leads to a severe underestimate of the mutual information. We have implemented and tested various solutions to remedy this situation. The following scheme was found to be the overall best performing.

1. Split the data into k folds.
2. Take $k - 1$ folds, the training set, and randomly split this in half. Next, permute X of one half and combine this with the other half to obtain $\hat{\mathbf{D}}$.

3. Train the classifier on $\hat{\mathbf{D}}$.
4. Take the left out fold, and obtain $P(\textit{foreground} \mid X, Y)$ for each sample by using the classifier trained on $\hat{\mathbf{D}}$.
5. Repeat step 2 to 4 k times so that each sample has been tested once.
6. Repeat step to 1 to 5 a number of times to average out the results and reduce the variance.

This scheme makes sure that all samples in the test set are indeed fully independent of the samples in the training set. Effectively, we build our models based on a little less than half of the available data. However, in practice this apparent loss of power is weakened by averaging over several random splits and is important to prevent the bias that would otherwise result from the dependencies between *foreground* and *background*.

3   Results

The method described was tested on three different artificial datasets.

1. **Non-linear Dependency:** A block shaped dataset where the probability is defined by

$$
P(x, y) = \begin{cases} 2 & \text{if } 0 \le x < \tfrac{1}{2} \text{ and } 0 > y < \tfrac{1}{2}, \\ 2 & \text{if } \tfrac{1}{2} \le x \le 1 \text{ and } \tfrac{1}{2} \le y \le 1, \\ 0 & \text{elsewhere.} \end{cases}
$$

2. **Real/Discrete Mixture:** A dataset where one variable is a discrete variable and the other is continuous. The possible values for the discrete variable are uniquely determined by the value of the real variable.
3. **Real/Discrete Mixture Noise Variables:** Same as 1 but now with a variable number of additional Gaussian noise variables added to X and Y to test the method with sets of variables for X and Y.

The true values for the mutual information have been obtained by taking the theoretical class probabilities for a perfect classifier and substituting these into (4). See Fig. 2 for an example of datasets 1 and 2.

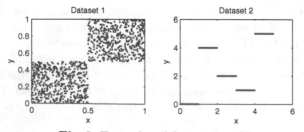

Fig. 2. Examples of dataset 1 and 2

The first two datasets have been simulated with a variable number of samples and have been tested with the presented method, the Kraskov KNN method [5] and the KDE version by Moon [6]. The third dataset has been simulated with 1000 objects and a variables number of noise variables. We used 100 trees per forest, 5-fold cross validation and repeated everything 6 times to reduce variance. The KNN method was performed with $k = 4$ and $k = 12$. For each setting, we considered 50 randomly generated data sets. These plots are shown in Fig. 3 to 5.

Figure 3 plots the estimated mutual information as a function of the number of samples. In this plot the deviation from the true value approximates the bias of the method and size of the error the variance. It can be seen that the random forests method does not perform very well with a low number of samples but approaches the accuracy of the Kraskov method with higher numbers of samples. At 1000 samples it has a little more bias than KNN with $k = 6$ but has reduced variance and has similar performance compared to KNN with $k = 12$. The KDE method performs reasonably well at a low number of samples with low variance, but cannot compare to the other methods at higher numbers of samples because of its bias. Figure 4 shows that the KNN method can handle the discrete variables surprisingly well, even though the method was developed primarily for

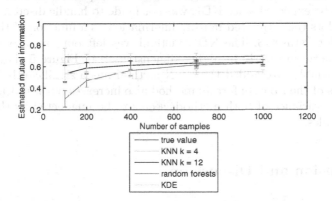

Fig. 3. Number of samples against estimated mutual information for dataset 1

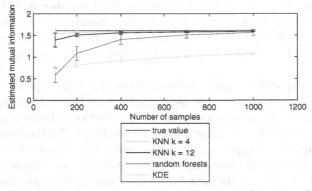

Fig. 4. Number of samples against estimated mutual information for dataset 2

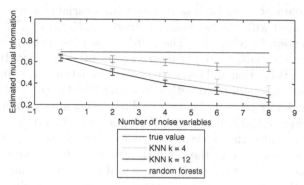

Fig. 5. Number of noise variables against estimated mutual information for dataset 3

real numbers. It has relatively low bias and variance at most numbers of samples. The random forests method is again able to give reasonable estimates at the higher numbers of samples, but has a relatively strong bias for lower number of samples. The KDE method has a large relative bias, especially for many samples, which was to be expected since KDE was not made to handle discrete variables.

Figure 5 plots the estimated mutual information as a function of the number of variables for dataset 3. The KDE method was left out on purpose since it is known to perform poorly in higher dimensions [14]. Figure 5 shows that as more noise variables are added the KNN method has a significantly increasing bias. The bias of the random forests method also increases but less dramatically. Curiously, the variance of both methods seems to be unaffected by the number of noise variables.

4   Conclusion and Discussion

We presented a novel method for estimating mutual information based on random forests. This method is able to estimate the mutual information for combinations of discrete and real-valued variables. We have shown that in datasets with a medium to high amount of samples it performs better than the kernel density estimator method and is roughly at par with the KNN method. Furthermore our method apears to be more robust to the addition of irrelevant variables.

Our method requires the accurate estimation of the class probabilities. Random forests are not necessarily the best classifiers to be used within our framework: they are primarily designed to obtain excellent classification performance, not so much to accurately estimate class probabilities. Furthermore, standard decision trees divide the input space in axis-parallel rectangles. For some distributions, e.g., multivariate Gaussians, oblique decision trees [15] appear to be a more natural choice. More generally, future work may therefore consider different nonlinear probabilistic classifiers within the same paradigm.

References

1. Fleuret, F.: Fast binary feature selection with conditional mutual information. Journal of Machine Learning Research 5, 1531–1555 (2004)
2. Sethi, I.K., Sarvarayudu, G.P.R.: Hierarchical classifier design using mutual information. IEEE Transactions on Pattern Analysis and Machine Intelligence 4(4), 441–445 (1982)
3. Cheng, J., Bell, D.A., Liu, W.: Learning belief networks from data: An information theory based approach. In: Proceedings of the Sixth International Conference on Information and Knowledge Management, pp. 325–331. ACM (1997)
4. Kwak, N., Choi, C.-H.: Input feature selection by mutual information based on parzen window. IEEE Transactions on Pattern Analysis and Machine Intelligence 24(12), 1667–1671 (2002)
5. Kraskov, A., Stögbauer, H., Grassberger, P.: Estimating mutual information. Physical Review E 69(6), 066138 (2004)
6. Moon, Y.-I., Rajagopalan, B., Lall, U.: Estimation of mutual information using kernel density estimators. Physical Review E 52(3), 2318 (1995)
7. Parzen, E., et al.: On estimation of a probability density function and mode. Annals of Mathematical Statistics 33(3), 1065–1076 (1962)
8. Bishop, C.M.: Pattern Recognition and Machine Learning, vol. 1. Springer, New York (2006)
9. Breiman, L.: Random forests. Machine Learning 45(1), 5–32 (2001)
10. Breiman, L.: Bagging predictors. Machine Learning 24(2), 123–140 (1996)
11. Criminisi, A., Shotton, J., Konukoglu, E.: Decision forests for classification, regression, density estimation, manifold learning and semi-supervised learning. Microsoft Research Cambridge, Tech. Rep. MSRTR-2011-114 5(6), 12 (2011)
12. Biau, G., Devroye, L., Lugosi, G.: Consistency of random forests and other averaging classifiers. Journal of Machine Learning Research 9, 2015–2033 (2008)
13. Bostrom, H.: Calibrating random forests. In: Seventh International Conference on Machine Learning and Applications, ICMLA 2008, pp. 121–126. IEEE (2008)
14. Hwang, J.N., Lay, S.R., Lippman, A.: Nonparametric multivariate density estimation: a comparative study. IEEE Transactions on Signal Processing 42(10), 2795–2810 (1994)
15. Murthy, S.K., Kasif, S., Salzberg, S.: A system for induction of oblique decision trees. Journal of Artificial Intelligence Research 2(1), 1–32 (1994)

Out-Of-Vocabulary Words Recognition Based on Conditional Random Field in Electronic Commerce

Yanfeng Yang[1], Yanqin Yang[1], Hu Guan[2], and Wenchao Xu[1]

[1]School of Information Science Technology
East China Normal University
500 Dongchuan Road, Shanghai 200241, China
[2]TravelSky Technology Limited
MinHang, Shanghai 200241, China

Abstract. Most previous researches on out-of-vocabulary words recognition are concentrating on traditional areas, while in electronic commerce area this recognition is rarely involved. In this paper, we focus on the out-of-vocabulary words recognition in the field of electronic commerce. Based on the unique characteristics of electronic commerce data, we introduce the Conditional Random Fields (CRFs) into the out-of-vocabulary words recognition, and use the electronic commerce text corpus for the experimental verification. The experimental results show that CRFs in the recognition of out-of-vocabulary words in the field of electronic commerce is effective.

Keywords: Electronic Commerce, Out-Of-Vocabulary Recognition, Conditional Random Fields.

1 Introduction

Word segmentation and labeling is one of the fundamental problems in the Natural Language Processing(NLP) research fields[3]. It has been actively studied by many researchers for several different natural languages especially for the Chinese words which are not delimited by white-space. In electronic commerce fields, the most common approach in word segmentation is lexicon-based one where the longest matching algorithm or the maximum matching algorithm is used[10]. For example, forward maximum matching algorithm, backward maximum matching algorithm, bi-direction matching algorithm, minimum size segmentation algorithm, etc. However, as dictionary cannot cover all words, this raise an issue of out-of-vocabulary(OOV) words[10].

In electronic commerce fields, the description of the product is also written without any delimiter, and this cause many issues in word segmentation, particularly when the user searching for the product. Identifying of OOV words such as brand, business, origin, specification, are also the main challenging tasks. Among the past researches, many scholars have attached great importance to

C.K. Loo et al. (Eds.): ICONIP 2014, Part II, LNCS 8835, pp. 532–539, 2014.
© Springer International Publishing Switzerland 2014

OOV words recognition and obtained some research results. But we have found most of these researches are aimed at common Chinese words, not for a particular field, such as electronic commerce. Therefore, this research aims to address this problem.

In this paper, we present a model based on CRFs to recognize out-of-vocabulary words in electronic commerce field. Using of the characteristics of CRFs, we not only use the context information of the words in part of speech tagging, but also take full advantage of the statistical information of the training set as the characteristics. Moreover the model provides more information for multi-category words tagging. The result shows that our model is valid to recognize the out-of-vocabulary words.

This paper is organized as follows. Section 2 reviews the background and summarizes related work. Section 3 presents CRFs and elaborates the design. Experiment results and discussion are given in section 4 and section 5 presents the conclusion and future research directions.

2   Background and Related Work

To segment and label the product description, the electronic commerce companies often adopt methods based on dictionary, such as forward maximum matching method, backward maximum matching method, bi-direction matching method, minimum size segmentation method. However,methods based on dictionary have significant limitations, the main one is not able to deal with out-of-vocabulary words[1]. In this paper, we will adopt CRF, a method based on the statistical frequency of the string in the corpus to determine whether they constitute a word. Just like HMM and MEMM, this method takes advantage of the laws of the Chinese group of words. Overall, there are still many exploration of space to apply CRF to segment and label sequence data in the electronic commerce field.

The basic methods used to identify OOV words are: rule-based approach and statistical-based approach.

Rule-based method uses the fields of knowledge to define the perception rules which is to test the semantic concepts in video[2]. It mainly uses the law and context characteristics of OOV words to observe the structure of the words and the relationship between the word and its tag position, and then sum up the suitable rules to identify OOV words. This method depends the completeness and rationality of the rules, but it lacks of adaptability.

Statistical-based method[7] uses the mathematical statistics to extract information from the corpus to recognize OOV words. Some of the more common algorithms are Hidden Markov Model(HMM) and Maximum Entropy Markov Model(MEMM). A brief introduction of HMM and MEMM is as follows.

Hidden Markov Model is a generative model, assigning a joint probability to paired observation and label sequences[4]. we can regard HMM as a directed graph, its node S_t represents the state at time t while O_t represents the observation at time t. The structure of HMM as shown in figure 1.

Compared to HMM, MEMM is a discriminative model that extends a standard maximum entropy classifier by assuming that the unknown values to be learnt are connected in a Markov chain rather than being conditionally independent of each other[8]. In MEMM, the HMM transition and observation function are replaced by a single function $P(s|s',o)$ which provides the probability of the current state s given the previous state s' and the current observation o[9]. In contrast to HMM, in which the current observation only depends on the current state, the current observation in MEMM may also depends on the previous states[8]. The structure of MEMM as shown in figure 1.

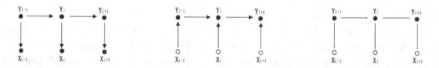

Fig. 1. Graphical structure of simple HMM(left), MEMM(center), and the chain-graph case of CRFs(right)

3   Conditional Random Field and Out-Of-Vocabulary Words Recognition

3.1   The Definition of CRFs

Definition. Let $G = (V, E)$ be a graph such that $Y = (Y_v)_{v \in V}$, so that Y is indexed by the vertices of G. Then (X, Y) is a conditional random field in case, when conditioned on X, the random variables Y_v obey the Markov property with respect to the graph: $p(Y_\nu|X, Y_\omega, \omega \neq \nu) = p(Y_\nu|X, Y_\omega, \omega \sim \nu)$, where $\omega \sim \nu$ means that ω and ν are neighbors in G. In the simplest and most important example for modeling sequences, G is a simple chain graph or line in which the nodes corresponding to elements of Y[6], as shown in figure 1.

The graphical structure of a conditional random field may be used to factorize the joint distribution over elements Y_ν of Y into a normalized product of strictly positive, real-valued potential functions, derived from the notion of conditional independence[5]. Each potential function operates on a subset of the random variables represented by vertices in G. The potential functions must therefore ensure that it is possible to factorize the joint probability such that conditionally independent random variables do not appear in the same potential function. The easiest way to fulfill this requirement is to require each potential function to operate on a set of random variables whose corresponding vertices form a maximal clique within G. In the case of a chain-structured CRF, such as that depicted in Figure 1, each potential function will operate on pairs of adjacent label variables Y_i and Y_{i+1}.

Lafferty et al.[6] define the the probability of a particular label sequence y given observation sequence x to be a normalized product of potential functions, each of the form

$$\exp(\sum_j \lambda_j t_j(y_{i-1}, y_i, x, i) + \sum_k \mu_k s_k(y_i, x, i)), \tag{1}$$

where $t_j(y_{i-1}, y_i, x, i)$ is a transition feature function of the entire observation sequence and the labels at positions i and $i-1$ in the label sequence; $s_k(y_i, x, i))$ is a state feature function of the label at position i and the observation sequence; and λ_j and μ_k are parameters to be estimated from training data. As the product of real-value functions does not satisfy the axioms of probability theory, CRFs use the observation-dependent normalization $Z(x)$ for conditional distribution.Using this notation, the conditional probability of a label sequence is written as

$$\frac{1}{Z(s, x)} \exp(\sum_j \lambda_j t_j(y_{i-1}, y_i, x, i) + \sum_k \mu_k s_k(y_i, x, i)) \tag{2}$$

3.2 The Model Training of CRFs

The model training is typically performed by maximum likelihood method to ensure the conditional probability $P_\lambda(y|x)$ largest. Given a fixed set $T = (x_k, y_k)_N^{k-1}$ to maximize the logarithmic likelihood L_λ by adjusting the weight vector λ

$$L_\lambda = \sum_k \log P_\lambda(y_k|x_k) = \sum_k [\lambda \cdot F(y_k, x_k) - \log Z_\lambda(x_k)] \tag{3}$$

In order to find out the maximum of L_λ, we should have a differential equation of L_λ

$$\nabla L_\lambda = \sum_k [F(y_k, x_k) - E_{P_\lambda(Y|x_k)} F(Y, x_k)] \tag{4}$$

That means, when the average value of the global feature vectors is equal to the mathematical expectation of the model, L_λ will obtain maximum. The mathematical expectation $E_{P_\lambda(Y|x_k)} F(Y, x_k)$ could be calculated by the forward-backward algorithm. Meanwhile, for each position i in the observation sequence x, we define the transfer matrix: $M_i[y, y'] = \exp \lambda \cdot f(y, y', x, i)$

Let f as local features, $f_i(y, y') = f(y, y', x, i), F(y, x) = \sum_k f(y_{i-1}, y_i, x, i)$, and let $\times$ represents the multiplication of a real number and matrix, thus:

$$E_{P_\lambda(Y|x)} F(Y, x) = \sum_y P_\lambda(y|x) F(y, x) = \sum_i \frac{\alpha_{i-1}(f_i \times M_i)\beta_i^T}{Z_\lambda(x)} \tag{5}$$

$$Z_\lambda(x) = a_n \cdot 1^T \tag{6}$$

where α_i, β_i are defined as follows: $\alpha_i = \begin{cases} \alpha_{i-1} M_i, & 0 < i \leqslant n \\ 1, & i = 0 \end{cases}$, $\beta_i^T = \begin{cases} M_{i+1}\beta_{i+1}^T, & 1 \leqslant i < n \\ 1, & i = n \end{cases}$, therefore, for each data series,the α_i and β_i can be calculated by a forward scanning and backward scanning, and then calculate the mathematical expectation.

3.3 Out-Of-Vocabulary Words Recognition

This paper mainly adopt the method based on CRFs to recognize out-of-vocabulary words. The experimental model as shown in figure 2,

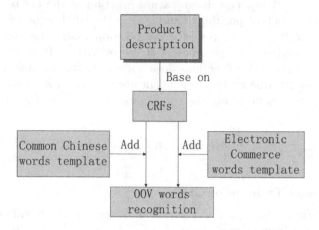

Fig. 2. The out-of-vocabulary in electronic commerce recognition model based on CRFs

According to the above experimental model, the experiment mainly includes the steps as follows:

(1) To obtain electronic commerce data, then use programs for statements coarse segmentation to get word and tag, and finally generate the documents which are suitable for CRFs.

(2) To mark the type of the text corpus. According to the experimental requirements, we focus on labeling products chunk(NPC), manufacturers chunk(NJ), brand chunk(N), size chunk(M), and the rule BX represents the start of the block X, IX represent the subsequent of the block X, O represents the other blocks, form the corpus A.

(3) On the basis of corpus A, we carry on the out-of-vocabulary annotations, tag B represents the start of OOV word, tag I represents the subsequent of the OOV word, tag O represents the others, form the corpus B.

(4) Select a part of the corpus randomly as the training data from corpus B, and build characteristic template T1. Then test the rest of the corpus, thus to calculate the accuracy of the result of OOV words recognition.

3.4 Experiment Corpus Description

This experiment uses the product description data in electronic commerce as the corpus data. According to the experiment process and mechanism above, this experiment corpus production is as follows:

(1) Preprocessing the corpus data and performing the words segmentation and tagging.

(2) Converting the treated data to the format suitable for CRFs, and then marking the block type especially focusing on labeling NPC, NJ, N, M.

(3) Labeling the boundary of the labeled corpus with B, I, O.

After the processing, the experimental corpus which is suitable for CRF has came into being. In this experiment, we select 2800 product description data, where 2000 as the training corpus and 800 as the test corpus. For the above corpus structure, the feature templates we use are as shown in Table 1:

Table 1. Different approaches

Feature	Description	Weight
$W = W_0$	The current character	1
$W = W_{-1}$	The previous character	1
$W = W_1$	The next character	1
$W = W_{-1}W_0$	The previous character and current character	1
$W = W_0W_1$	The current character and next character	1
$W = W_{-1}W_0W_1$	The previous,current and next character	1
$T - T_{-1}T_0$	The previous output tag and current output tag2	1

4   Experimental Results

In this paper, we adopt CRF++ tools which is simple for use and has been widely used in the field of natural language processing for the experimental operation. We use three evaluation indexes, P presents the precision, R presents the recall rate, and the F presents the F measure, defined as follows: $P = \frac{n}{k}$, $R = \frac{n}{d}$, $F = \frac{2 \cdot P \cdot R}{P+R}$, where n means the number of words in the correct identification, k means the number of words in identification, d means the number of correct words.

In order to verify the performance of our model which is based on CRFs and the characteristics of electronic commerce, we select the common template which is used for Chinese word segmentation and labeling as the contrast experiment, and finally obtain the experimental results as shown in Table 2 and figure 3.

Table 2. OOV words recognition results

Feature Template	Precision	Recall	F-measure
Common Template	68.57%	66.13%	67.33%
Electronic Commerce Template	70.85%	70.10%	70.47%

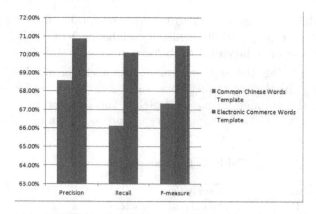

Fig. 3. OOV words recognition results

Based on the above experimental results, we get the conclusion as follows:

(1)Compared with the model based on the common template, the precision of the model based on CRFs and electronic commerce template reaches 70.85% and has 2.28% increase, the recall rate reaches 70.10% and has 3.97% increase, the F-measure reaches 70.47% and has 3.14% increase, overall, the performance of the model is effective.

(2)Under the condition of the same corpus, the result of the model based on the electronic commerce template achieves better than the model based on common template.

(3)Different templates produce different experimental results, and in accordance with the word segmentation and labeling in the field of electronic commerce, we could not use the common template.

5   Conclusions and Future Work

For out-of-vocabulary words recognition in the field of electronic commerce, is still in the exploratory stage. Firstly we need to develop a perfect training corpus, which is a difficult task, secondly we need to exert the characteristics of the electronic commerce field data.On the basis of previous studies, this paper presents the method based on CRFs to recognize the out-of-vocabulary in electronic commerce, which concludes brands, merchants,specifications, flavour in the product information, and analyses the experiment result.

Compared to HMM and MEMM, although CRFs can simultaneously use the n words before and the m words after the center word as the context information, the relationship between words is still so simple that results has not reached at the optimum. In the future work, we will try to consider joining the mutual information and co-occurrence frequency between the words on the basis of existing features.

Acknowledgment. This work was supported by the National Natural Science Foundation of China under Grant No.61300043.

References

1. Bazzi, I.: Modelling out-of-vocabulary words for robust speech recognition. PhD thesis, Massachusetts Institute of Technology (2002)
2. Brill, E.: A simple rule-based part of speech tagger. In: Proceedings of the Workshop on Speech and Natural Language, pp. 112–116. Association for Computational Linguistics (1992)
3. Chang, P.-C., Galley, M., Manning, C.D.: Optimizing chinese word segmentation for machine translation performance. In: Proceedings of the Third Workshop on Statistical Machine Translation, pp. 224–232. Association for Computational Linguistics (2008)
4. Freitag, D., McCallum, A.: Information extraction with hmm structures learned by stochastic optimization. In: AAAI/IAAI 2000, pp. 584–589 (2000)
5. Klinger, R., Tomanek, K.: Classical probabilistic models and conditional random fields. TU, Algorithm Engineering (2007)
6. Lafferty, J., McCallum, A., Pereira, F.C.N.: Conditional random fields: Probabilistic models for segmenting and labeling sequence data (2001)
7. Marcu, D., Wong, W.: A phrase-based, joint probability model for statistical machine translation. In: Proceedings of the ACL 2002 Conference on Empirical Methods in Natural Language Processing, vol. 10, pp. 133–139. Association for Computational Linguistics (2002)
8. McCallum, A., Freitag, D., Pereira, F.C.N.: Maximum entropy markov models for information extraction and segmentation. In: ICML, pp. 591–598 (2000)
9. Ratnaparkhi, A., et al.: A maximum entropy model for part-of-speech tagging. In: Proceedings of the Conference on Empirical Methods in Natural Language Processing, Philadelphia, PA, vol. 1, pp. 133–142 (1996)
10. Van, C., Kameyama, W.: Khmer word segmentation and out-of-vocabulary words detection using collocation measurement of repeated characters subsequences (2013)

Least Angle Regression in Orthogonal Case

Katsuyuki Hagiwara

Faculty of Education, Mie University
1577 Kurima-Machiya-cho, Tsu, 514-8507, Japan
hagi@edu.mie-u.ac.jp

Abstract. LARS(least angle regression) is one of the sparse modeling methods. This article considered LARS under orthogonal design matrix, which we refer to as LARSO. In this article, we showed that LARSO reduces to a simple non-iterative algorithm that is a greedy procedure with shrinkage estimation. Based on this result, we found that LARSO is exactly equivalent with a soft-thresholding method in which a threshold level at the kth step is the $(k + 1)$th largest value of the absolute values of the least squares estimators. For LARSO, C_p type model selection criterion can be derived. It is not only interpreted as a criterion for choosing the number of steps/coefficients in a regression problem but also regarded as a criterion for determining an optimal threshold level in LARSO-oriented soft-thresholding method which may be useful especially in non-parametric regression problems. Furthermore, in the context of orthogonal non-parametric regression, we clarified relationship between LARSO with C_p type criterion and several methods such as the universal thresholding and SUREshrink in wavelet denoising.

Keywords: LARS, orthogonal regression, soft-thresholding.

1 Introduction

In recent years, sparse modeling methods such as LARS(least angle regression) and LASSO(least absolute shrinkage and selection operator) are extensively studied[3,6,7]. As shown in [3], LASSO and classical forward stagewise can be implemented as modifications of LARS. An interesting point shown in [3] as an extreme case is that LARS algorithm for a special case of orthogonal regression that employs natural basis vectors reduces to soft-thresholding on output samples. In this article, we focus on LARS for a more general case of orthogonal design in details, which we refer to as LARSO(LARS in the Orthogonal case) for the sake of simplicity. Considering LARSO is not only meaningful for understanding the properties of LARS as an example but also useful especially in practical applications of non-parametric orthogonal regression or curve-fitting methods such as wavelet denoising[1,2]. In section 2, we present regression setting and original LARS algorithm. In section 3, we derive a simple LARS algorithm under orthogonality condition and consider the model selection problem. Section 3 also includes some discussions on LARSO in relation to the other methods especially in the context of non-parametric regression. Section 4 is devoted to conclusions and future works.

C.K. Loo et al. (Eds.): ICONIP 2014, Part II, LNCS 8835, pp. 540–547, 2014.

2   Review of LARS

2.1   Regression Problem

Let $x = (x_1, \ldots, x_m)$ and y be explanatory variables and response variable, for which we have n samples : $\{(x_{i,1}, \ldots, x_{i,m}, y_i) : i = 1, \ldots, n\}$. We define $x_j = (x_{1,j}, \ldots, x_{n,j})' \in \mathbb{R}^n$ for $j = 1, \ldots, m$, where $'$ stands for the transpose operator. We assume that $m \leq n$ holds and $x_1, \ldots, x_m$ are linearly independent. We also define $\mathbf{X} = (x_1, \ldots, x_m)$ and $y = (y_1, \ldots, y_n)'$. Let $e_1, \ldots, e_n$ be i.i.d. samples from $N(0, \sigma^2)$; i.e. normal distribution with mean 0 and variance σ^2. We define $e = (e_1, \ldots, e_n)'$ and assume $y = \eta + e$. We therefore have $\eta = \mathbb{E}y$, where $\mathbb{E}$ is the expectation with respect to the joint probability distribution of y. The case of $m = n$ corresponds to a non-parametric regression problem which we will mention later.

2.2   LARS Algorithm

For an index subset $A \subseteq \{1, \ldots, m\}$ and a vector of signs $s = (s_1, \ldots, s_m)$, $s_j \in \{-1, +1\}$, we define an $n \times |A|$ matrix $\mathbf{X}_{s,A}$ whose column vectors are given by $s_j x_j$, $j \in A$, where $|A|$ stands for the number of members of A. We define $\mathbf{G}_{s,A} = (\mathbf{X}'_{s,A}\mathbf{X}_{s,A})$ and $\alpha_{s,A} = (1'_A \mathbf{G}_{s,A}^{-1} 1_A)^{-1/2} \in \mathbb{R}$, where 1_A is a vertical vector of ones with a length of $|A|$. We define

$$u_{s,A} = \mathbf{X}_{s,A} w_{s,A} \tag{1}$$

$$w_{s,A} = \alpha_{s,A} \mathbf{G}_{s,A}^{-1} 1_A. \tag{2}$$

By the definition of $w_{s,A}$ and $\alpha_{s,A}$, we can see that $\|u_{s,A}\|^2 = 1$ and $\mathbf{X}'_{s,A} u_{s,A} = \alpha_{s,A} 1_A$. $u_{s,A}$ is thus a unit equiangular vector for all of x_j, $j \in A$.

We denote LARS estimate/output at the kth step by $\widehat{\mu}_k$. We define $\widehat{\mu}_0 = 0_n$, where 0_n is an n-dimensional zero vector. We define

$$\widehat{c}_k = (\widehat{c}_{k,1}, \ldots, \widehat{c}_{k,m})' = \mathbf{X}'(y - \widehat{\mu}_{k-1}) \tag{3}$$

$$\widehat{C}_k = \max_{1 \leq j \leq m} |\widehat{c}_{k,j}|. \tag{4}$$

We also define $s_k = (s_{k,1}, \ldots, s_{k,m})$, $s_{k,j} = \text{sign}(\widehat{c}_{k,j})$ and $A_k = \{j : |\widehat{c}_{k,j}| = \widehat{C}_k\}$, where sign is a sign function. A_k is called an active set. We denote a complement of A_k by $\overline{A}_k$. We define

$$a_k = (a_{k,1}, \ldots, a_{k,m})' = \mathbf{X}' u_{s_k, A_k}. \tag{5}$$

The update rule of LARS is given by

$$\widehat{\mu}_k = \widehat{\mu}_{k-1} + \gamma_k u_{s_k, A_k}, \tag{6}$$

where $\gamma_k > 0$ is a step size. By defining $\gamma_{k,j}^+ = (\widehat{C}_k - \widehat{c}_{k,j})/(\alpha_{s_k,A_k} - a_{k,j})$ and $\gamma_{k,j}^- = (\widehat{C}_k + \widehat{c}_{k,j})/(\alpha_{s_k,A_k} + a_{k,j})$, the step size of LARS algorithm is defined by $\gamma_k = \min_{j \in \overline{A}_k}\{\gamma^+_{k,j}, \gamma^-_{k,j}\}$. Let $\widehat{j} \in \overline{A}_k$ be an index at which this minimum is achieved. We assume that it exists. By this update, we can show that $A_{k+1} = A_k \bigcup\{\widehat{j}\}$ and

$$\widehat{C}_{k+1} = \widehat{C}_k - \gamma_k \alpha_{s_k, A_k} \tag{7}$$

hold at the $(k+1)$th step. This procedure is repeated for $k = 1, \ldots, m-1$ and $\widehat{\mu}_m$ is defined as the least squares estimate.

3   LARS in the Orthogonal Case

3.1   Derivation of Simple Algorithm

We consider the case of $\mathbf{X}'\mathbf{X} = \mathbf{I}_m$, where $\mathbf{I}_m$ is an $m \times m$ identity matrix; i.e. column vectors of $\mathbf{X}$ are orthonormal. We refer to this condition as orthonormality condition below. Note that $\mathbf{X}'_{s,A}\mathbf{X}_{s,A} = \mathbf{I}_{|A|}$ holds for any s and A under the orthonormality condition. Since orthogonal regression reduces to orthonormal regression under an appropriate normalization, normality is not essential in this paper. We refer to LARS under this orthonormality condition as LARSO (LARS in the Orthogonal case). As an extreme case, [3] has treated a special case of orthonormal regression in which x_j is a natural basis vector; i.e. $x_{i,j}$ is one if $i = j$ and zero otherwise. We consider a more general case in this paper.

We define

$$\widetilde{c} = (\widetilde{c}_1, \ldots, \widetilde{c}_m)' = \mathbf{X}'\mathbf{y}, \tag{8}$$

which is the least squares estimator of full model under the orthonormality condition. Let $p_1, \ldots, p_m$ be indices that satisfy $|\widetilde{c}_{p_1}| > |\widetilde{c}_{p_2}| > \cdots > |\widetilde{c}_{p_n}|$. We assume that there are no ties for the sake of simplicity while it can be handled as in [3]. This assumption is actually valid with probability one since $\widetilde{c}_j$'s have normal distributions as seen in later. We define $s_j = \text{sign}(\widetilde{c}_j)$, $j = 1, \ldots, m$.

Theorem 1. *LARSO estimate at the kth step is given by*

$$\widehat{\mu}_k = \sum_{j=1}^k \widetilde{b}_{j,k} x_{p_j} \tag{9}$$

for $k = 1, \ldots, m-1$, where

$$\widetilde{b}_{j,k} = (|\widetilde{c}_{p_j}| - |\widetilde{c}_{p_{k+1}}|)s_{p_j}, \ j = 1, \ldots, k. \tag{10}$$

For $k = m$, we define $\widehat{\mu}_m = \mathbf{X}\widetilde{c}$ by setting $|\widetilde{c}_{p_{m+1}}| = 0$.

A representation of $\widehat{\mu}_k$ in Theorem 1 implies that we do not need a repeated procedure for LARSO; i.e. we just need to calculate the order statistics among the absolute values of the least squares estimators. By this theorem, $\widetilde{b}_{j,k}$ is found to be a shrinkage estimator under a greedy procedure since $|\widetilde{b}_{j,k}| = ||\widetilde{c}_{p_j}| - |\widetilde{c}_{p_{k+1}}||$

and $\widetilde{c}_{p_j}$ is the least squares estimator. We need some facts for proving Theorem 1. At the kth step of LARSO, we simply have

$$\mathbf{G}_{\boldsymbol{s}_k, A_k} = \mathbf{I}_m \tag{11}$$

$$\alpha_{\boldsymbol{s}_k, A_k} = \frac{1}{\sqrt{k}} \tag{12}$$

$$\boldsymbol{u}_{\boldsymbol{s}_k, A_k} = \frac{1}{\sqrt{k}} \mathbf{X}_{\boldsymbol{s}_k, A_k} \mathbf{1}_{A_k} = \frac{1}{\sqrt{k}} \sum_{j \in A_k} s_{k,j} \boldsymbol{x}_j. \tag{13}$$

By (5), (13) and the orthonormality condition, we have

$$a_{k,j} = \begin{cases} \frac{s_{k,j}}{\sqrt{k}} & j \in A_k \\ 0 & j \in \overline{A}_k \end{cases}. \tag{14}$$

By the definition of $\gamma_{k,j}^+$ and $\gamma_{k,j}^-$ together with (12) and (14), we have $\gamma_{k,j}^+ = \sqrt{k}(\widehat{C}_k - \widehat{c}_{k,j})$ and $\gamma_{k,j}^- = \sqrt{k}(\widehat{C}_k + \widehat{c}_{k,j})$ for $j \in \overline{A}_k$. Thus, we have

$$\widehat{j} = \arg\max_{j \in \overline{A}_k} |\widehat{c}_{k,j}| \tag{15}$$

$$\gamma_k = \sqrt{k}(\widehat{C}_k - |\widehat{c}_{k,\widehat{j}}|). \tag{16}$$

By (3), (6) and (13), for $k \geq 2$, we have

$$\widehat{c}_k = \mathbf{X}'\boldsymbol{y} - \mathbf{X}'\widehat{\mu}_{k-1} = \mathbf{X}'\boldsymbol{y} - \mathbf{X}'\widehat{\mu}_{k-2} - \gamma_{k-1}\mathbf{X}'\boldsymbol{u}_{\boldsymbol{s}_{k-1}, A_{k-1}}$$

$$\cdots$$

$$= \mathbf{X}'\boldsymbol{y} - \sum_{l=1}^{k-1} \gamma_l \mathbf{X}'\boldsymbol{u}_{\boldsymbol{s}_l, A_l} = \widetilde{\boldsymbol{c}} - \sum_{l=1}^{k-1} \frac{\gamma_l}{\sqrt{l}}\mathbf{X}' \sum_{p \in A_l} s_{l,p}\boldsymbol{x}_p. \tag{17}$$

In component-wise fashion, we have

$$\widehat{c}_{k,j} = \widetilde{c}_j - \sum_{l=1}^{k-1} \frac{\gamma_l}{\sqrt{l}} \sum_{p \in A_l} s_{l,p}\boldsymbol{x}_j'\boldsymbol{x}_p, \quad j = 1, \ldots, m \tag{18}$$

Lemma 1. $A_k = \{p_1, \ldots, p_k\}$ *for any* k.

Proof. We consider the proof by induction. When $k = 1$, $\widehat{c}_1 = \mathbf{X}'\boldsymbol{y} = \widetilde{\boldsymbol{c}}$ by (3) and the definition of $\widehat{\mu}_0$. Therefore, $A_1 = \{p_1\}$ holds by the definition of A_k and p_1. We assume that $A_l = \{p_1, \ldots, p_l\}$ for any $l \in \{1, \ldots, k\}$. Under this assumption, if $j \in \overline{A}_k$ then $j \notin A_l$ for any $l \in \{1, \ldots, k-1\}$. Therefore, we have $\widehat{c}_{k,j} = \widetilde{c}_j$ by (18) and the orthonormality condition. We then have $\max_{j \in \overline{A}_k} |\widehat{c}_{k,j}| = \max_{j \in \overline{A}_k} |\widetilde{c}_j|$ in (15) and thus obtain $\widehat{j} = p_{k+1}$ by the definition of p_{k+1}. This implies that $A_{k+1} = A_k \bigcup \{p_{k+1}\}$ by the definition of LARS. □

Note that we have

$$\gamma_k = \sqrt{k}(\widehat{C}_k - |\widetilde{c}_{p+1}|), \tag{19}$$

by (16) and the fact that $\widehat{c}_{k,\widehat{j}} = \widetilde{c}_{p_{k+1}}$ holds in the above proof.

Lemma 2. *At each k, we have*

$$\gamma_k = \sqrt{k}(|\tilde{c}_{p_k}| - |\tilde{c}_{p_{k+1}}|). \tag{20}$$

Proof. We first show that $\widehat{C}_k = |\tilde{c}_{p_k}|$ holds for any $k \in \{1, \ldots, m-1\}$. When $k = 1$, $\widehat{c}_1 = \mathbf{X}'\mathbf{y} = \tilde{c}$ holds by (3). We thus have $\widehat{C}_1 = |\tilde{c}_{p_1}|$ by (4). We assume that $\widehat{C}_k = |\tilde{c}_{p_k}|$. Under this assumption, we have $\widehat{C}_{k+1} = |\tilde{c}_{p_{k+1}}|$ by (7), (12) and (19). Therefore, $\widehat{C}_k = |\tilde{c}_{p_k}|$ holds for any k by induction. Thus, (20) is obtained by (19). □

Lemma 3. $s_{k,j} = s_j$ *holds for* $j \in A_k$.

Proof. We first show that $s_{j,p_j} = s_{p_j}$ holds for any j. Since $\widehat{c}_1 = \mathbf{X}'\mathbf{y} = \tilde{c}$ holds by (3), we have $\widehat{c}_{1,p_1} = \tilde{c}_{p_1}$ for $j = 1$. Since $p_k \notin A_l$ for any $l \in \{1, \ldots, k-1\}$ by Lemma 1, $\widehat{c}_{k,p_k} = \tilde{c}_{p_k}$ holds for $k \geq 2$ by (18) and the orthonormality condition. This implies that $s_{k,p_k} = s_{p_k}$ for any $k \geq 2$. Therefore $s_{j,p_j} = s_{p_j}$ holds for any j. Fix a $j \leq k - 1$. By (18), Lemma 2 and orthonormality condition, we have

$$\widehat{c}_{k,p_j} = \tilde{c}_{p_j} - \sum_{l=j}^{k-1}(|\tilde{c}_{p_l}| - |\tilde{c}_{p_{l+1}}|)s_{l,p_j}, \tag{21}$$

where we used $p_j \notin A_l$ for any $l \in \{1, \ldots, j-1\}$ by Lemma 1. We assume that $s_{j,p_j} = \cdots = s_{k-1,p_j} = s_{p_j}$. Under this assumption, we have $\widehat{c}_{k,p_j} = |\tilde{c}_{p_k}|s_{p_j}$ by (21). This implies that $s_{k,p_j} = s_{p_j}$. By induction with this equation and $s_{j,p_j} = s_{p_j}$, we have $s_{k,p_j} = s_{p_j}$ for any $k \geq j$. Since j is arbitrarily fixed, $s_{k,p_j} = s_{p_j}$ always holds when $k \geq j$. If $p_j \in A_k$ then $j \leq k$ by Lemma 1. Thus $s_{k,p_j} = s_{p_j}$ holds. This completes the proof. □

Proof. (Proof of Theorem 1) By (1), (6), (13), Lemma 2 and Lemma 3, we have

$$\widehat{\boldsymbol{\mu}}_k = \widehat{\boldsymbol{\mu}}_{k-1} + (|\tilde{c}_{p_k}| - |\tilde{c}_{p_{k+1}}|)\sum_{j \in A_k} s_j \boldsymbol{x}_j. \tag{22}$$

By repeating this procedure with Lemma 1, we have

$$\begin{aligned}
\widehat{\boldsymbol{\mu}}_k &= \widehat{\boldsymbol{\mu}}_{k-1} + (|\tilde{c}_{p_k}| - |\tilde{c}_{p_{k+1}}|)\sum_{j \in A_k} s_j \boldsymbol{x}_j \\
&= \widehat{\boldsymbol{\mu}}_{k-2} + (|\tilde{c}_{p_{k-1}}| - |\tilde{c}_{p_k}|)\sum_{j \in A_{k-1}} s_j \boldsymbol{x}_j + (|\tilde{c}_{p_k}| - |\tilde{c}_{p_{k+1}}|)\sum_{j \in A_k} s_j \boldsymbol{x}_j \\
&= \widehat{\boldsymbol{\mu}}_{k-2} + (|\tilde{c}_{p_{k-1}}| - |\tilde{c}_{p_{k+1}}|)\sum_{j \in A_{k-1}} s_j \boldsymbol{x}_j + \tilde{b}_{k,k}\boldsymbol{x}_{p_k} \\
&\cdots \\
&= \sum_{j=1}^{k} \tilde{b}_{j,k}\boldsymbol{x}_{p_j}.
\end{aligned} \tag{23}$$

□

3.2   Relation to Soft-Thresholding Method and LASSO

We define a soft-thresholding operator on $\widetilde{c}_j$, $j = 1, \ldots, m$ by

$$\bar{b}_j(\theta) = \begin{cases} \widetilde{c}_j - \theta & \widetilde{c}_j > \theta \\ 0 & |\widetilde{c}_j| \leq \theta \\ \widetilde{c}_j + \theta & \widetilde{c}_j < -\theta \end{cases} \tag{24}$$

in which $\theta(> 0)$ is a threshold level. By Theorem 1, the following fact is straight-forward.

Corollary 1. $\widehat{\boldsymbol{\mu}}_k$ *in Theorem 1 can be rewritten by*

$$\widehat{\boldsymbol{\mu}}_k = \sum_{j=1}^{m} \bar{b}_j \left(|\widetilde{c}_{p_{k+1}}| \right) \boldsymbol{x}_j. \tag{25}$$

By Corollary 1, LARSO exactly reduces to soft-thresholding method in which a threshold level at the kth step is $|\widetilde{c}_{p_{k+1}}|$; i.e. the $(k+1)$th largest value among the absolute values of the least squares estimators. As is naturally expected, [3] has also obtained the same conclusion in the case of natural basis vector, which is obtained by setting $\widetilde{c}_j = y_j$ in our result. For denoising via wavelet, [1,2] have proposed soft-thresholding methods. Since discrete wavelet transform is one of the non-parametric orthogonal regression methods, our formulation applies with $m = n$ under an appropriate choice of a set of basis functions.

On the other hand, it has been known that LASSO[6] reduces also to a soft-thresholding on $\widetilde{c}$ under orthonormality condition, in which a threshold level is given by a regularization parameter. Therefore LARSO at the kth step is equivalent with LASSO in which a regularization parameter is determined by the $(k+1)$th largest value among the absolute values of the least squares estimators.

3.3 C_p Type Model Selection Criterion

For LARSO estimate, the risk normalized by σ^2 is shown to be given by

$$R(k) = \frac{1}{\sigma^2} \mathbb{E} \|\widehat{\boldsymbol{\mu}}_k - \boldsymbol{\eta}\|^2 = \frac{1}{\sigma^2} \mathbb{E} \|\widehat{\boldsymbol{\mu}}_k - \boldsymbol{y}\|^2 - n + 2\mathrm{DF}(k), \tag{26}$$

where $\mathrm{DF}(k) = \frac{1}{\sigma^2} \mathbb{E}(\widehat{\boldsymbol{\mu}}_k - \boldsymbol{\mu})'(\boldsymbol{y} - \boldsymbol{\eta})$ and $\boldsymbol{\mu} = \mathbb{E}\widehat{\boldsymbol{\mu}}_k$. $\mathrm{DF}(k)$ is known as the degree of freedom. The well known C_p[5] is the unbiased estimator of $R(k)$. $\mathrm{DF}(k)$ is equal to the number of adjustable coefficients for linear regression model; see [5,3]. In [3], it is shown that the degree of freedom of LARS is consistent with the number of steps under a relatively mild condition that is referred to as positive cone condition. Note that it may be consistent with the number of coefficients in almost cases. The orthonormal case here satisfies the positive cone condition as argued in [3]. We thus have the following fact.

Theorem 2. *Under the orthonormality condition, we have*

$$\mathrm{DF}(k) = k. \tag{27}$$

Note that a general proof in [3] under the positive cone condition is somewhat complicated but it is possible to give a simple proof under the orthonormality condition while we omit it; e.g. see [3] for an extreme case in which x_j's are natural basis vectors, thus, a special case of orthogonal regression. Based on this result, C_p type criterion for choosing an optimal k is given by $\widehat{R}(k) = \frac{1}{\widehat{\sigma}^2}\|\widehat{\boldsymbol{\mu}}_k - \boldsymbol{y}\|^2 - n + 2k$, where $\widehat{\sigma}^2$ is an appropriate estimate of noise variance. This C_p type criterion can be viewed as a criterion for determining the number of explanatory variables, or equivalently the number of coefficients in a regression problem. In LARS, it generally corresponds to the number of steps. From an another point of view, application of this C_p type criterion is a theoretically supported method for the selection of a threshold level in LARSO-oriented soft-thresholding.

3.4   Relation to Methods in Wavelet Denoising

Let $\{(\boldsymbol{u}_i, y_i) : \boldsymbol{u}_i \in \mathbb{R}^d,\ y_i \in \mathbb{R},\ i = 1,\ldots,n\}$ be input-output samples. Let $\phi_1,\ldots,\phi_n$ be functions on $\mathbb{R}^d$. We consider here a non-parametric regression problem by using $f(\boldsymbol{u}_i) = \sum_{k=1}^n c_k\phi_k(\boldsymbol{u}_i)$. Under a setting of $m = n$ and $x_{i,k} = \phi_k(\boldsymbol{u}_i)$, our results are directly applied when the orthonormality condition is satisfied. Wavelet denoising is a method based on discrete wavelet transform with model selection; see e.g. [1,2]. It is viewed as a method of non-parametric orthogonal regression and our result is applicable to wavelet denoising. Since $\boldsymbol{\eta} = \mathbb{E}y \in \mathbb{R}^n$, there exists a $\boldsymbol{c} = (c_1,\ldots,c_n)' \in \mathbb{R}^n$ such that $\boldsymbol{\eta} = \mathbf{X}\boldsymbol{c}$. We assume that there exists an index set A^* such that $c_j \neq 0$ for $j \in A^*$ and $c_j = 0$ for $j \notin A^*$; i.e. $j \in A^*$ is an index of component that relates to $\boldsymbol{\eta}$ and $j \notin A^*$ is an index of noise-related component. We define $k^* = |A^*|$. We assume that n is sufficiently larger that k^* and A^* is fixed for any assumed n. This corresponds to the case that invariant part of data is represented by a relatively small number of components. In other words, we have a sparse representation of a target function $\boldsymbol{\eta}$ in terms of $\mathbf{X}$. We note that the discussion below is based on the assumption of $\mathbf{X}'\mathbf{X} = n\mathbf{I}_n$ instead of orthonormality.

By the assumption of the normality and independency of additive noise, it is easy to show that $\widetilde{c}_k - c_k \sim N(0, \sigma^2/n)\ k =, 1\ldots, n$ and those are independent. On the other hand, a threshold level of LARSO-oriented soft-thresholding has previously shown to be $|\widetilde{c}_{p_{k+1}}|$. This is then the $(k+1)$th largest value among the absolute values of independent random variables according to the normal distribution. Based on this fact, roughly speaking, we can show that $|\widetilde{c}_{p_{k+1}}| \sim \sqrt{\sigma^2 \log n}$ for a large n in a probabilistic sense if $k+1 > k^*$; e.g. see [4] for details. This implies that LARSO is approximately equivalent with the universal soft-thresholding method in [1,2] if there is a sparse representation at a k and n is sufficiently large. In a soft-thresholding method, a threshold level works as a level for removing ineffective coefficients and simultaneously as a shrinkage parameter. Note that the universal threshold level that is given by $\sqrt{\sigma^2 \log n}$ is an uniform upper bound on $|\widetilde{c}_j|$ for any $j \notin A^*$. Therefore, it may surely removes delicate components and, also, it seems to be too large as a shrinkage parameter. LARSO-oriented soft-thresholding may have an advantage that the balance between thresholding and shrinkage can be selected by C_p type criterion depending on data.

On the other hand, [2] has proposed SURE (Stein's unbiased estimator for risk) for choosing an optimal threshold level in soft-thresholding. SURE and a risk derived for LARSO are apparently identical but the derivations are different. This is because a threshold level is fixed in SURE but it is a data-dependent statistic in LARSO. [2] suggests that candidates for a threshold level is chosen from $[0, \sqrt{2\sigma^2 \log n}]$. In application of SUREshrink, we choose a threshold level that minimizes SURE among these predetermined candidates. In LARSO, a set of candidates is $\{|\tilde{c}_{p_1}|, \ldots, |\tilde{c}_{p_n}|\}$ that may well cover the range suggested in SURE if we follow the previous discussion on the universal thresholding. Thus, we can say that LARSO may be free from a choice of candidates including the number of candidates.

4 Conclusions and Future Works

In this paper, we showed that LARSO(LARS in the Orthogonal case) reduces to a simple non-iterative algorithm that is a greedy procedure with shrinkage estimation. Based on this result, we found that LARSO is exactly equivalent with a soft-thresholding method, in which a threshold level at the kth step is given by the $(k + 1)$th largest value among the absolute values of the least squares estimators. For LARSO, we can derive C_p type criterion as in [3]. It can be applied to the choice of the number of steps/coefficients in orthogonal regression while, from an another viewpoint, it works as a criterion for choosing a threshold level in LARSO-oriented soft-thresholding. Furthermore, in the context of nonparametric orthogonal regression, we clarified the relationship between LARSO with C_p type criterion and the methods in wavelet denoising such as the universal thresholding and SUREshrink[1,2]. Although we confirmed a validity of LARSO with C_p type criterion in wavelet denoising application through simple numerical experiments, more detailed evaluation is left as a future work.

References

1. Donoho, D.L., Johnstone, I.M.: Ideal spatial adaptation via wavelet shrinkage. Biometrika 81, 425–455 (1994)
2. Donoho, D.L., Johnstone, I.M.: Adapting to unknown smoothness via wavelet shrinkage. Journal of the American Statistical Association 90, 1200–1224 (1995)
3. Efron, B., Hastie, T., Johnstone, I., Tibshirani, R.: Least angle regression. Ann. Statist. 32, 407–499 (2004)
4. Hagiwara, K.: Nonparametric regression method based on orthogonalization and thresholding. IEICE Trans. Inf. & Syst. E94-D, 1610–1619 (2011)
5. Mallows, C.L.: Some Comments on CP. Technometrics 15, 661–675 (1973)
6. Tibshirani, R.: Regression shrinkage and selection via the lasso. J. R. Statist. Soc. Ser. B. 58, 267–288 (1996)
7. Zou, H., Hastie, T., Tibshirani, R.: On the degree of freedom of the LASSO. Ann. Statist. 35, 2173–2192 (2007)

Evaluation Protocol of Early Classifiers over Multiple Data Sets

Asma Dachraoui[1,2], Alexis Bondu[1], and Antoine Cornuéjols[2]

[1]EDF R&D, 1 avenue du Général de Gaulle, 92140 Clamart, France
{asma.dachraoui,alexis.bondu}@edf.fr
[2]AgroParisTech - INRA, UMR-518-MIA, 16 rue Claude Bernard,
F-75005 Paris, France
antoine.cornuejols@agroparistech.fr

Abstract. Early classification approaches deal with the problem of reliably labeling incomplete time series as soon as possible given a level of
confidence. While developing new approaches for this problem has been
getting increasing attention recently, their evaluation are still not thoroughly considered. In this article, we propose a new evaluation protocol
for early classifiers. This protocol is generic and does not depend on
the criteria used to evaluate the classifiers. Our protocol is successfully
applied to 23 publicly available data sets.

Keywords: Early classification, Time series, Evaluation protocol.

1 Introduction

In recent years, the interest in adapting supervised and unsupervised machine
learning (ML) approaches to time series analysis has considerably increased.
A part of these studies focuses on applying these approaches on incomplete time
series, when they are progressively recorded. This article focuses on the particular
context of early classification of time series. Early classification techniques are
useful for many time-critical applications. For example, in medicine, earliest
diagnosis based on first signal outputs may help to remedy some diseases [1].
In air, road or marine traffic it is important to anticipate the risks of collision or
crash before receiving all signals [2]. These approaches allow one to make early
and reliable predictions based on incomplete time series.

The conventional time series classification problem consists in predicting the
labels of **complete** time series. In this case, the classifier is learned on a training
set composed by complete time series. Then the classifier can be applied on
new complete time series. By contrast, an early classifier can be applied on
incomplete time series after being learned on complete time series. In this
case, the objective is to predict the labels **as soon as possible** given a level of
confidence. Early classification can be considered as a multi-objective problem.
On the one hand, the objective of **quality** consists in reliably predicting the
labels in order to make an **appropriate** action. On the other hand, the objective

C.K. Loo et al. (Eds.): ICONIP 2014, Part II, LNCS 8835, pp. 548–555, 2014.
© Springer International Publishing Switzerland 2014

of **earliness** aims to make this action as soon as possible, before a deadline. At last, the early classification problem involves two conflicting objectives.

The evaluation of conventional classifiers has been well studied in the literature. The comparison of two classifiers over multiple data sets requires the use of a statistical test, in order to **objectively** and **reliably** decide if one classifier is better than the other. In [3] the authors recommend to use the Wilcoxon signed-rank test. The comparison of several early classifiers is a special case since it considers both quality and earliness objectives. Therefore, an approach could be better than another on specific objective and worse on the other. In the literature, the methods that optimize several objectives are mainly exploited for learning purposes. For instance, the scalarized multi-objective learning approach which aggregates two objectives into a single scalar objective function allows one to optimize the two objectives by setting a regularization parameter. Multi-Objective Evolutionary Algorithms are used to design classifiers in diverse problems with more than one objective [4]. The majority of these approaches involve one or more regularization parameters which are adjusted by a learning algorithm. These approaches are not suitable for the evaluation.

In this article, we propose a new evaluation protocol for early classifiers. This protocol is based on the Wilcoxon signed-rank test and the Pareto optimum. The proposed protocol is parameter-free and generic since it does not involve an evaluation criterion that "*mixes*" the two conflicting objectives. The remainder of this article is structured as follows: in Sect.2, we present a brief review of early classification of time series. In Sect.3 we propose a new evaluation protocol for early classifiers. Sect.4 presents our choices for evaluating the quality and the earliness of the classifiers. The experimental results are discussed in Sect. 5. Section 6 concludes this article and highlights limitations and future works.

2   Early Classification

We define a time series $x = \{(t_1, x_1), (t_2, x_2), ..., (t_L, x_L)\}$ of length L as a sequence of real values $\{x_{j \in [1,L]}\}$ associated with the timestamps $\{t_{j \in [1,L]}\}$. The input data set is a collection of N pairs $\{(x_i, y_i) | i \in [1, N]\}$ where x_i is the ith time series, $y_i \in \mathcal{Y}$ is its class value. $\mathcal{Y}$ is a finite set of class labels. By contrast with the conventional classification, the early classification consists in predicting the labels before the time series are completed. In the conventional case, the single objective is to maximize the quality of the prediction without considering the time constraint. In the case of early classification, there are two conflicting objectives to optimize: i) predict the labels of incomplete time series **as soon as possible**; ii) maximize the **quality** of the classifier. In practice, the predictions are triggered once a given level of **confidence** is reached.

Let $\tilde{x}_i$ be an incomplete time series, $H(.)$ is a prediction function which predicts the class label of $\tilde{x}_i$ (*see Eq.1*) and $C(.)$ is a function which measures the confidence of this prediction (*see Eq.2*).

$$H(\tilde{x}_i) \longrightarrow \hat{y} \ (1) \qquad C_H(\tilde{x}_i) \longrightarrow \sigma \ (2) \qquad \tilde{t}_i : C_H(\tilde{x}_i) \geq \tau \ (3)$$

The earliness of a classifier is evaluated based on the instants $\tilde{t}_i$ when the predictions are triggered. More formally, $\tilde{t}_i$ is the earliest timestamp such that the class label of $\tilde{x}_i$ is predicted with a confidence level exceeding a threshold τ (*see Eq.3*).

In the literature, the existing early classification approaches may be distinguished by the manner of setting the confidence. The confidence measures can be decomposed in two categories according to the type of the classifier:

1. The *generative classifiers* provide conditional probabilities of the class values on which are based the confidence measures. The simplest approach consists in triggering the predictions once the probability of $\hat{y}$ exceeds a fixed **threshold** [5][6]. This approach is improved in [7] by introducing the concept of the *reliability*. The key idea is to model the missing values of the time series as a random variable, conditionally to the observed data points. For a given $\tilde{x}_i$, the distribution of the possible ways to complete the time series is estimated. The confidence is then evaluated by the part of this distribution leading to the predicted class value.
2. The *discriminative classifiers* provide a class label based on a decision boundary. In this case, the confidence measures are defined using the distance from the decision boundaries [8].

Methods such as [9],[10], [11] implicitly introduce the notion of the confidence. In [11], the *Minimum Prediction Length* (MPL) is proposed in order to determine the earliest instant from which the prediction will be the same as if the full-length time series is used. An extended *1-Nearest Neighbor* (1NN) is proposed. In [9], the proposed approach uses a reject option to label as early as possible incomplete time series. It is based on the agreement of an ensemble of base classifiers to accept or reject an output.

3   Evaluation over Multiple Data Sets

The rest of this article focuses on the evaluation of early classifiers. In this section, we first discuss existing research work on comparing conventional classifiers. Thereafter, we propose a new protocol to compare early classifiers.

3.1   Comparison of Two Classifiers

The evaluation of conventional classifiers are thoroughly considered in the literature. Numerous evaluation criteria such as the accuracy, the *Balanced error Rate* (BER) or the *Area Under the ROC Curve* (AUC) [12] allow one to evaluate the classifiers on a specific data set. According to J. Demšar [3], a reliable evaluation of several classifiers should be done over multiple data sets, in order to reflect the general quality and not the quality on a specific data set. If the output value *(denoted by z)* is smaller than -1.96, the quality of both classifiers are significantly different.

3.2   A New Evaluation Protocol for Early Classification

Our objective is to **reliably** compare two early classifiers. In this section, we propose a generic evaluation protocol which does not depend on the criteria used to evaluate the quality and the earliness of the classifiers.

Step 1 - Prediction of the Labels Over Multiple Data Sets:
Let $\mathcal{D}$ be an ensemble of K data sets denoted by $\mathcal{D} = \{D_1, D_2, \ldots, D_K\}$. Each data set is divided into two disjoint training and test sets. Let $H_{(A)}$ and $H_{(B)}$ be the prediction functions associated with two early classifiers denoted by A and B. At learning phase, both classifiers are trained over the **same** training sets of complete time series. At test phase, for each test set, two pairs of scores $\{Q(.), T(.)\}$ are computed for the classifiers A and B. The two scores are given respectively by the criteria evaluating the quality and the earliness of the predictions *(see Sect.4.1 and Sect.4.2)*. $Q(.)$ and $T(.)$ are computed based on all time series of the test set. A confidence measure denoted by $C_H(.)$ is compared with a fixed threshold τ in order to trigger the predictions. If $C_H(.)$ does not exceed τ before the time series is complete, the prediction will be triggered at the last timestamp. As shown in Fig.1, the instants of the predictions vary for each time series. During this process, the following pieces of information are retained for each time series $\tilde{x}_i$: y_i the true label, $\hat{y}_i$ the predicted class value, $\tilde{t}_i$ the instant of the prediction and $\hat{P}_i(y_i, \tilde{x}_i)$ the predicted probabilities of the class values. These pieces of information, denoted by $\mathcal{I}$, are used to compute the values of $Q(.)$ and $T(.)$ over the entire test set.

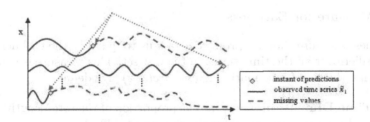

Fig. 1. Illustration of the labels predictions triggering

Step 2 - Independent Comparison Under Each Objective:
The scores $Q(.)$ and $T(.)$ are computed for each data set and for each classifier. Then, the Wilcoxon signed-rank test is independently performed for the both objectives: the quality and the earliness. Based on the statistic z, we can conclude if there is a statistically significant difference between the classifiers under the quality objective and/or under the earliness objective. The best classifier is designated by the most important sum of ranks.

Step 3 - Global Comparison by Using Pareto Optimum:
In order to compare two early classifiers, we drawn inspiration from the Pareto optimum [4]: a classifier is considered as better than an other if it improves at least one of the two objectives without degrading the other. Based on the values of the statistic z independently computed for the quality and the earliness objectives, there are three possible cases: *i)* A is better than B, *ii)* B is better than A and *iii)* A and B are indiscernible since there is no significant differences under the two objectives.

4　Evaluation Criteria for Early Classifiers

This section presents a possible implementation for the quality, the earliness and the confidence measures.

4.1　A Measure for Prediction Quality

The evaluation of the quality of the classifiers can be measured by diverse criteria. We choose to exploit the Multi-class Area Under the ROC Curve (AUC) criterion [13]. For a given class label y_j, the area under the ROC curve AUC_{y_j} relatively to y_j is computed based on the triggered predictions (*see Fig.1*):

$$Q(.) = \mathbb{E}_{j=1}^{J}[AUC_{y_j}] = \sum_{j=1}^{J} P_k(y_j)AUC_{y_j} \tag{4}$$

4.2　A Measure for Earliness

The earliness measure that we propose allows us to quantify the earliness of an early classifier over all the time series in the test set. This measure is computed by exploiting the set of information $\mathcal{I}$ (*see Sect.3.2*). We define:

1. **The Time Dimension:** $t = \frac{\tilde{t}_i}{L}$ as the proportion of the total length L (*with L is the length of a complete time series, and $t \in [0,1]$*).

2. **The Earliness:** $Pr(t) = \frac{n_t}{N}$ as the proportion of the triggered predictions at instant t (*with n_t is the number of the triggered predictions at the instant t, N is the size of the test set, and $Pr(t) \in [0,1]$*).

We denote $Pr(t)$ as the Earliness Curve of the classifier over the time dimension. The global earliness of the classifier (*denoted by $T(.)$*) is measured by the Area Under the Earliness Curve. Since t and $Pr(t)$ vary in $[0,1]$, $T(.)$ varies in $[0,1]$. When the area $T(.)$ is equal to 1, that means that all the predictions are triggered at the first instant. In this case the classifier performs a perfect earliness. Conversely, when $T(.)$ is equal to 0 that means that the classifier is too conservative and waits until the last instant to make the predictions.

4.3   A Measure for Confidence

In this section, we propose a possible way to fix the confidence level. The prediction of the label of the time series $\tilde{x}_i$ is triggered once the probability of the predicted class (*denoted by* $max_1(\hat{P}(y|\tilde{x}_i))$) exceeds k times the probability of the second most probable class (*denoted by* $max_2(\hat{P}(y|\tilde{x}_i))$):

$$\tilde{t}_i : max_1(\hat{P}(y|\tilde{x}_i)) \geq k * max_2(\hat{P}(y|\tilde{x}_i)) \tag{5}$$

5   Experiments

The objective of our experiments is to compare two early classifiers. In this section, we exploit the generic evaluation protocol proposed in Sect.3.2.

Implementation of the Early Classifiers:
In this article, early classification is implemented based on a collection of classifiers trained in parallel [14]. Each classifier corresponds to one instant of the time series. These classifiers do not exploit the same explicative variables. The progressive arrival of the time series is simulated by hiding the forthcoming data points: the input of the current classifier is only composed by the previous data points up to the current instant. In order to use the proposed evaluation protocol, we implement two different classifiers based on the above described implementation: i) a *Selective Naive Bayes* (SNB) using a the regularized approach MODL [15] which shows a capacity for a good discrimination between classes and ii) a *Naive Bayes* (NB) with 10 *EqualFrequency* as a baseline for comparison. The NB is expected to be less efficient than the SNB: the NB estimates $\hat{P}(y|\tilde{x}_i)$ based on the equal frequency discretization while the SNB gives an estimate based on a discretization optimized by the data.

Data Sets Description:
Our experiments were performed over 23 data sets selected from the UCR Time Series Classification and Clustering repository [16]. We randomly and disjointedly re-sampled each data set into a set of 70% of examples for the training set and the remainder for the test set.

Results:
In this paragraph, we evaluate our generic protocol by experimenting the classifiers under several data sets. We first run the SNB and the NB classifiers over all the data sets in order to compute the quality and the earliness scores. Then, we applied the Wilcoxon test for the two classifiers independently under the quality and the earliness objectives. The results are listed in Tab.1. The quality of the predictions $Q(.)$ and the earliness $T(.)$ results of each classifier over each data set are depicted with a varying value of k (*see Eq.5*). In our experiments, we varied k from 2 to 12 with a step equal to 1. But for simplicity, Tab.1 shows only the results for k ranging from 2 to 10 with a step equal to 4 (*the omitted results do not affect the drawn conclusions*). The last line in the table represents

Table 1. Empirical results of the SNB and NB early classifiers

k	Prediction quality						Earliness					
	2		6		10		2		6		10	
Data Set	Q_{SNB}	Q_{NB}	Q_{SNB}	Q_{NB}	Q_{SNB}	Q_{NB}	T_{SNB}	T_{NB}	T_{SNB}	T_{NB}	T_{SNB}	T_{NB}
50words	0.505	0.527	0.520	0.530	0.519	0.527	0.766	0.687	0.679	0.564	0.642	0.513
CBF	0.807	0.840	0.866	0.875	0.877	0.890	0.803	0.814	0.717	0.739	0.697	0.722
ChlorineC	0.5	0.549	0.764	0.648	0.799	0.655	0.993	0.988	0.6104	0.667	0.455	0.595
CinC	0.777	0.669	0.8007	0.677	0.854	0.699	0.758	0.988	0.706	0.977	0.622	0.971
CricketX	0.468	0.488	0.475	0.481	0.478	0.482	0.856	0.992	0.807	0.990	0.799	0.990
CricketY	0.523	0.478	0.515	0.470	0.496	0.470	0.985	0.927	0.956	0.920	0.933	0.920
CricketZ	0.483	0.514	0.491	0.511	0.489	0.510	0.95	0.376	0.922	0.339	0.903	0.339
ECG5Days	0.955	0.793	0.992	0.824	0.994	0.839	0.318	0.614	0.209	0.580	0.200	0.568
FaceAll	0.483	0.476	0.489	0.467	0.482	0.475	0.940	0.878	0.809	0.738	0.748	0.687
FaceUCR	0.503	0.536	0.496	0.524	0.497	0.533	0.969	0.989	0.959	0.969	0.938	0.941
Mallat	0.910	0.928	0.913	0.930	0.916	0.931	0.886	0.718	0.845	0.709	0.833	0.709
MedicalImg	0.603	0.529	0.587	0.520	0.589	0.519	0.733	0.560	0.324	0.475	0.180	0.424
MoteStain	0.994	0.962	0.994	0.962	0.994	0.962	0.006	0.085	0.006	0.084	0.006	0.084
StarLight	0.742	0.920	0.738	0.919	0.734	0.919	0.988	0.715	0.996	0.683	0.996	0.716
SwedishLeaf	0.539	0.484	0.527	0.482	0.530	0.480	0.942	0.742	0.919	0.721	0.903	0.720
Symbols	0.958	0.934	0.958	0.935	0.958	0.940	0.610	0.815	0.609	0.811	0.609	0.818
TwoLECG	0.983	0.921	0.986	0.936	0.987	0.945	0.3539	0.464	0.341	0.424	0.336	0.393
uWaveX	0.719	0.702	0.724	0.701	0.734	0.692	0.966	0.993	0.965	0.993	0.930	0.991
uWaveY	0.748	0.787	0.749	0.796	0.749	0.796	0.992	0.386	0.988	0.390	0.987	0.390
uWaveZ	0.775	0.727	0.780	0.726	0.781	0.723	0.751	0.799	0.728	0.797	0.726	0.795
wafer	0.958	0.925	0.981	0.924	0.98	0.940	0.923	0.909	0.889	0.840	0.883	0.824
WordsSyn	0.54	0.518	0.541	0.509	0.545	0.509	0.928	0.786	0.654	0.647	0.630	0.647
yoga	0.764	0.658	0.889	0.640	0.907	0.693	0.700	0.795	0.407	0.761	0.308	0.748
StatisticZ	-1.3686729		**-2.28112149**		**-3.10232523**		-0.36497944		-0.36497944		-0.69954392	

the values of the statistic z for overall predictions quality and earliness results. Significant values are reported in bold font.

Based on the values of the statistic z and the Pareto optimum, we can conclude that SNB is better than NB under the two objectives. In fact, by varying k we only observe two cases: *1)* SNB and NB are indiscernible (*for $k < 4$,*) and *2)* SNB is better than NB (*for $k \geq 4$*). In this second case, SNB is always better than NB under the quality objective and SNB and NB are always indiscernible under the earliness objective. Therefore, according to the Pareto optimum, we can conclude that the SNB classifier is better than the NB classifier. Finally, these results show that the proposed protocol performs as expected: the SNB classifier outperforms the NB baseline classifier.

6   Conclusion

In this article, we proposed a new evaluation protocol to compare two early classifiers under two conflicting objectives: the quality and the earliness. The protocol is generic and does not depend on the implementation choices made for computing the two objectives. Overall, the results suggest that this protocol could represent a useful evaluation technique to objectively and reliably compare early classifiers. Further, a limitation that we identified indicates that the Wilcoxon test does not ensure the concomitance of the results for all the data sets under the two objectives. It is possible that our protocol suggests that a classifier is better than another under both objectives without this may be true for the same data set. This is due to the fact that the Wilcoxon test is based on the differences over the two evaluation criteria. Thus, we only guarantee that the largest differences over the two criteria are observed for the same classifier. As future work, we intend to build a benchmark of early classification approaches in the literature based on our proposed protocol.

References

1. Nayak, S.G., Davide, O., Puttamadappa, C.: Classification of bio optical signals using k- means clustering for detection of skin pathology. International Journal of Computer Applications (IJCA) 1(2), 112–116 (2010)
2. Chiou, J.-M.: Dynamical functional prediction and classification, with application to traffic flow prediction.. The Annals of Applied Statistics 6(4), 1588–1614 (2012)
3. Demšar, J.: Statistical comparisons of classifiers over multiple data sets. J. Mach. Learn. Res. 7, 1–30 (2006)
4. Jin, Y., Sendhoff, B.: Pareto-based multiobjective machine learning: An overview and case studies. IEEE Transactions on Systems, Man, and Cybernetics, Part C 38(3), 397–415 (2008)
5. Trapeznikov, K., Saligrama, V.: Supervised Sequential Classification Under Budget Constraints. In: AISTATS, pp. 581–589 (2013)
6. Anderson, H.S., Parrish, N., Tsukida, K., Gupta, M.R.: Reliable early classification of time series. In: ICASSP, pp. 2073–2076 (2012)
7. Anderson, H.S., Parrish, N., Tsukida, K., Gupta, M.R.: Early Time-Series Classification with Reliability Guarantee. tech. rep., SANDIA Laboratories (2012)
8. Delany, S.J., Cunningham, P., Doyle, D., Zamolotskikh, A.: Generating estimates of classification confidence for a case-based spam filter. In: Muñoz-Ávila, H., Ricci, F. (eds.) ICCBR 2005. LNCS (LNAI), vol. 3620, pp. 177–190. Springer, Heidelberg (2005)
9. Hatami, N., Chira, C.: Classifiers With a Reject Option for Early Time-Series Classification. CoRR (2013)
10. Xing, Z., Pei, J., Dong, G., Yu, P.S.: Mining Sequence Classifiers for Early Prediction. In: SDM, pp. 644–655. SIAM (2008)
11. Xing, Z., Pei, J., Yu, P.S.: Early prediction on time series: A nearest neighbor approach. In: Boutilier, C. (ed.) IJCAI, pp. 1297–1302 (2009)
12. Fawcett, T.: ROC Graphs: Notes and Practical Considerations for Data Mining Researchers. tech. rep., HP Laboratories, Palo Alto (2003)
13. Bondu, A.: Active Learning using Local Models. PhD thesis, University of Angers (2008)
14. Dachraoui, A., Bondu, A., Cornuejols, A.: Early classification of individual electricity consumptions. In: RealStream2013 (ECML), pp. 18–21 (2013)
15. Boullé, M.: Data grid models for preparation and modeling in supervised learning. In: Hands-On Pattern Recognition: Challenges in Machine Learning, Microtome, vol. 1, pp. 99–130 (2011)
16. Keogh, E., Xi, X., Wei, L., Ratanamahatana, C.: The UCR Time Series Classification/Clustering Homepage (2006)

Exploiting Level-Wise Category Links
for Semantic Relatedness Computing

Hai-Tao Zheng*, Wenzhen Wu, Yong Jiang, and Shu-Tao Xia

Graduate School at Shenzhen, Tsinghua University, Guangdong, China
wuwz12@mails.tsinghua.edu.cn,
{zheng.haitao,jiangy,xiast}@sz.tsinghua.edu.cn

Abstract. Explicit Semantic Analysis(ESA) is an effective method that
adopts Wikipedia articles to represent text and compute semantic relat-
edness(SR). Most related studies do not take advantage of the semantics
carried by Wikipedia categories. We develop a SR computing framework exploiting Wikipedia category structure to generate abstract features for texts and considering the lexical overlap between a pair of text.
Experiments on three datasets show that our framework could gain better performance against ESA and most other methods. It indicates that
Wikipedia category graph is a promising resource to aid natural language
text analysis.

Keywords: Semantic Relatedness, Explicit Semantic Analysis, Wikipedia,
Hierarchy.

1 Introduction

Information accumulation is the most overt outcome of rapidly development
of the Internet, and text content accounted for the majority among the User-
generated Content [1] data in web. Automatically estimating the semantic relatedness of two text fragments is fundamental for many text mining and IR
applications. In this work, we develop a SR computing framework exploiting
Wikipedia category structure to generate abstract features for texts. Our Experiments show its superiority against ESA and and other methods.

Existing approaches measuring text semantic relatedness could be classified to
three categories. lexical overlap is a kind of straight-forward but weak method,
for mainly text relatedness is not based on common terms of given text pair.
The other kind of approaches are based on knowledge bases such as WordNet [2]
or Cyc [3]. Texts are initially mapped to taxonomy nodes in knowledge bases,
then the SR of text pairs could be computed by exploiting links and paths
between nodes [4][5][6][7]. Besides, LDA [8], LSA [9] and ESA [10] constitute the
$3rd$ category of schemes that generate feature vector representation for text by
exploring the words co-occurrence relationships or word occurrence statistics in
a corpus [11]. Normally cosine similarities are then utilized to measure the SR.

---

* Corresponding author.

C.K. Loo et al. (Eds.): ICONIP 2014, Part II, LNCS 8835, pp. 556–564, 2014.

It is known that when reading human can refer the content in text to things in real world and relations between them. For instance, the term *iPhone* should be linked to the popular smart phone, moreover, the information it conveys implicitly to human may also includes Apple Company which design the smart-phone, and *Samsung galaxy*, another cellphone brand dominating the market, etc. Analogously, ESA is served as an interpreter that refer given text to seman-tically related articles of Wikipedia. Hence, a semantic relatedness of a pair of texts sharing little common words could be captured by ESA. Prior work [12] shows that ESA achieve the state of the art performance on SR task.

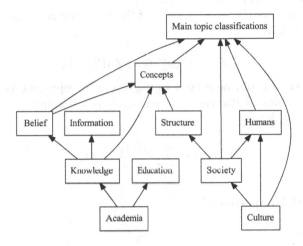

Fig. 1. Wikipedia category structure

In Wikipedia, except articles exploited in ESA, there is also a wealth of cat-egories which together form a gigantic tree-like graph(a miniature is shown in Figure 1). Actually Wikipedia categories offer a higher perspective to interpret text with human-created category nodes and their linking relationships. We be-lieved that category structure is a promising resource for measuring text semantic relatedness in higher abstraction levels.

In the paper,we will show that a level-wise propagation framework that gen-erate Wikipeida category features for text, combined with lexical overlap consid-eration, can be used in SR computing task and would gain better performance comparing ESA and most other methods.

The structure of this paper is organized as follow. The overview of ESA is stated in Section 2. Section 3 is devoted to a detailed description of our frame-work. We initially outline the procedure of Wikipedia category hierarchy con-struction, then present a level-wise propagation method to generate category features for the text pairs' SR computing. Preprocessing and experimental eval-uation will be elaborated in Section 4. Finally, we provide concluding remarks and suggestions for future work in Section 5.

2   Explicit Semantic Analysis

In this section we outline the ESA's principal fundamentals. ESA firstly maintain a database stores the mapping between terms and their relevant Wikipedia articles. Similar to ordinary IR task, all the Wikipedia articles with meaningful content are indexed by terms(except the stop words) they contain. Hence each non-stop term is mapped to a list of Wikipedia articles. Namely

$$Vec(t) = \{e_1, e_2, e_3, \ldots, e_n\} \, . \tag{1}$$

Where $e_i = \{e_i.title, e_i.score, e_i.fulltext\}$.

TF-IDF measure are chosen as a score of the mapping in original ESA implementation. So we have

$$e_i.score = tf\_idf(t_k, e_i.fulltext) \, . \tag{2}$$

So far, the $Vec(t)$ could be served as the semantic representation of a single term. For longer documents, the representation is the centroid of the vectors representing the each terms.

$$Vec(d) = \sum_{t \in d} \frac{tf(t)}{|d|} Vec(t) \tag{3}$$

3   Proposed Framework

3.1   Overview

As shown in Figure 2, the proposed framework is comprised of three components. 1)Categories hierarchy construction. We believe that the extent of abstraction

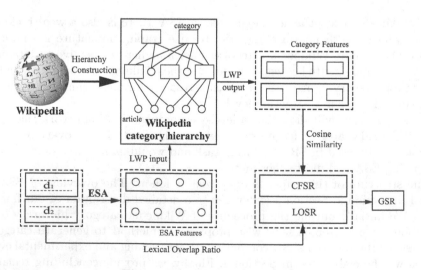

Fig. 2. The proposed framework

of category nodes are related to their hierarchical positions on the category graph. Hence we reconstruct the original categories graph to form a hierarchy on which; 2)Level-wise propagation(LWP), which propagating the feature scores from ESA feature nodes to category nodes on specific level in hierarchy; and 3)Lexical overlap semantic relatedness(LOSR), which complement the cosine-similarity of category vector(CFSR) to form the ultimate semantic relatedness score GSR for the given pair of texts.

3.2 Wikipedia Hierarchy Construction

Wikipedia's categories and its articles comprise a large structure that is not a strict hierarchical structure. It's hard to point out the abstraction extent of a given node in this structure. To meet this end we construct a hierarchical structure by automatically labeling each node with a level number by computer from the top down the original category graph. This process starts at the specific node "main_topics_classification", which serves as the first level of the hierarchy. Then iteratively visit the nodes in every level l, labeling their child nodes with the next level number $l + 1$ until all the nodes have been labeled. The number of nodes being labeled with each level of category is displayed on Table 1.

Table 1. The number of nodes in each level of category hierarchy

Level	1	2	3	4	5	6	7	8	9	10
# Node	1	26	834	12, 681	88, 152	257, 044	311, 540	18, 388	78, 235	16, 908

3.3 Level-Wise Propagation

In our system, ESA features are mapped to nodes of the category hierarchy described in Section 3.2. Inspired by the fundamentals of feed-forward neural network, we devise an scheme to propagate the scores of ESA features to specific level of category hierarchy.

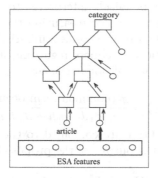

Fig. 3. The Level-wise propagation

The propagation process is illustrated in Figure 3. It begins at the leaf nodes that correspond to ESA features, followed by adding the scores to each node's parent nodes level by level, until the specific level. Similar to Equation (1), the feature representation of document d on hierarchical level l can be shown as follow:

$$Vec(d, l) = \{e_1, e_2, \ldots, e_n\}. \tag{4}$$

For each $e \in Vec(d, l)$

$$e.score = \sum_{link(e', e)} \frac{e'.score}{\log(1 + inlink(e))}. \tag{5}$$

where $link(e', e)$ signifies that e' is a child node of e, and $inlink(e)$ denotes the number of nodes linked to node e in Wikipedia category graph. For normalization, we update the scores as $e.score := \dfrac{e.score}{N}$ where

$$N = \sum_{e'' \in V(d, l)} e''.score. \tag{6}$$

The rationale behind Equation (5) is a simple common sense: a category in level k is more important when 1) there are more categories in level $k+1$ linking to it; 2) the categories linking to it are more important. However many categories contain a vast number of child nodes, hence they will gain a dominant scores that decrease the interpretation capability. To dampen this effect , the scores being propagated are divided by a logarithm to penalize the large categories. Thus, the semantic relatedness based on category features of text pair is defined as follows:

$$CFSR(d_1, d_2) = cosine\_similarity(Vec(d_1, l), Vec(d_2, l)). \tag{7}$$

In practice, we fix $l = 4$ for that in repeated experiments the system demonstrates the highest performance in the $4th$ level .

For category features are generated to interpret the text, comparing to ESA features, in a more abstract level, a pair of texts that are unrelated by ESA features could be interpreted to be related to some extent. On the other hand, texts de facto unrelated may be falsely interpreted to highly related. To address this problem, we introduction a factor measuring the ratio of common term in a given pair of texts to append to the cosine_similarity based on text's category features. We define semantic relatedness regarding lexical overlap(except the stopwords) between a given pair of document as:

$$LOSR(d_1, d_2) = \frac{\#(\text{common term occurrences})}{d_1.length + d_2.length}. \tag{8}$$

Finally, the CFSR and LOSR are combined to the General Semantic Relatedness score that served as the output of the proposed framework:

$$GSR(d_1, d_2) = \alpha * CFSR(d_1, d_2) + (1 - \alpha) * LOSR(d_1, d_2). \tag{9}$$

where $\alpha \in [0, 1]$.

4 Evaluation

4.1 Preprocessing

We acquire the snapshot of Wikipedia dump(static documents) of December 2nd, 2013 , then extract and transform data into relational database. Similar to the preprocessing in [12], the following phrases are involved.

- Extracting category and data, mainly including full text of article, from original Wikipedia dump XML files.
- Stemming and indexing the content field of pages.
- Filtering categories. Among the full collection of categories of Wikipedia, there are large amount of administration-related ones. Such like "Categories_for_merging" that doesn't offer any semantic association information. We remove this kind of categories and categories linking too much articles such as "living people".

4.2 Experimental Setup and Results

Initially, we utilize the Level-wise Propagation method for the representation of Barack Obama's second inauguration speech[1], which is illustrated briefly in Table 2. It outlines the top 10 concepts by ESA system, and by our propagation scheme applying to different levels in Wikipedia category hierarchy. It is noted that the category nodes are more abstract in higher levels(with greater level number) than those in lower levels. Moreover, intuitively those categories are fairly semantically related to the content of the speech. Those characteristics imply that Wikipedia Category feature representation may bridge the semantic gaps between texts, which is important for SR tasks.

Moreover,experiments on three datasets are employed to demonstrate the effectiveness of our framework. *Lee.50* [13] comprise 50 documents from the Australian Broadcasting Corporation's news mail service. Those documents, covering a variety of topics, are paired and then judged by volunteer students on their relatedness to each other. *MSRpar* and *MSRvid*[2] are stand for Microsoft Research Paraphrase Corpus and Microsoft Research Video Description respectively. They each consist of 750 text pairs with human-judged scores on their relatedness. Consistent with the prior works on this dataset, we employ Pearson correlation to measure the relatedness between human-judged scores and resulting score output by our system. Higher correlation indicates that the system output results are more likely in accord with gold standard scores.

Experimental result of *Lee.50* is presented in Figure 4 (a). It is seen that combining CFSR and LOSR can significantly enhance the performance by each single measure which is illustrated when $\alpha = 0$ and $\alpha = 1$ respectively. In addition,

---

[1] http://www.whitehouse.gov/the-press-office/2013/01/21/inaugural-address-president-barack-obama

[2] http://www.cs.york.ac.uk/semeval-2012/task6/data/uploads/datasets/train-readme.txt

Table 2. Category-based interpretation of Obama's inauguration speech

ESA Features	level 4	level 3	level 2
Second_inauguration_ of_Barack_Obama	Categories_by_nationality	Social_sciences	Humans
Great_Depression	American_studies	Sociology	Society
Liberalism	History_by_country	Social_philosophy	Culture
New_Deal	Subfields_of_political_science	Scientific_disciplines	Humanities
United_States_ federal_budget	Political_geography	Social_systems	Politics
Economics	United_States_federal_policy	Political_philosophy	Chronology
Oath_of_office	Society_by_nationality	Structure	History
Medicare_(United_ States)	20th_century	Social_institutions	Concepts
Health_insurance_in_ the_United_States	Political_ideologies	Fields_of_history	Law
Economic_democracy	People_by_occupation	Political_science	Nature

when $\alpha = 0.3$ the system gains the best performance that is also state-of-the-art performance on *Lee.50* dataset to the best of our knowledge. This parameter configuration($\alpha = 0.3$) is also adopted for experiments on other two datasets. In most cases, our system outperform ESA system by a significant margin. It is also noted that when $\alpha = 1$ our system still display better performance against the ESA system, which indicate that the category features generated by level-wise propagation could interpret texts better than ESA on SR tasks.

As to the other datasets, Figure 4 (b) shows that our system(GSR) gains a moderate and the best performance respectively in comparison with other methods. Especially for *MSRvid*, our method highly outperforms the LSA_{cos}. It is worth mention that out of a collection of methods [14], ESA_{align} and LSA_{cos} respectively achieve the best performances on the two datasets.

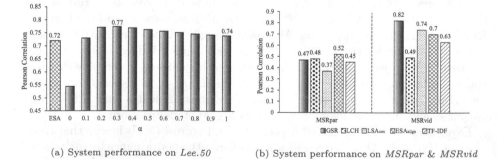

(a) System performance on *Lee.50*   (b) System performance on *MSRpar* & *MSRvid*

Fig. 4. The proposed system's performance comparing other methods on different datasets

5   Conclusion and Future Work

In the paper we propose a level-wise propagation method for feature generation in representation of text for SR tasks. It transform the ESA feature of text to higher level category nodes in the constructed hierarchical structure of Wikipedia category graph. To balance the relatively high abstraction property of category features generated by level-wise propagation, we append the lexical overlap relatedness measure to the framework. Experiments show its effectiveness against ESA and other methods.

However, the scoring scheme of category mapping adopted by us have not yet considered the different importance between a variety of link types of category. We will focus on this issue to improve our framework further in our future work.

Acknowledgments. This research is supported by the 863 project of China (2013AA013300), National Natural Science Foundation of China (Grant No. 61375054) and Tsinghua University Initiative Scientific Research Program Grant No.20131089256.

References

1. Krumm, J., Davies, N., Narayanaswami, C.: User-generated Content. IEEE Pervasive Computing 7(4), 10–11 (2008)
2. Miller, G.A.: WordNet: a lexical database for English. Communications of the ACM 38(11), 39–41 (1995)
3. Lenat, D.B.: CYC: A large-scale investment in knowledge infrastructure. Communications of the ACM 38, 33–38 (1995)
4. Wu, Z., Palmer, M.: Verbs semantics and lexical selection. In: Proceedings of the 32nd annual meeting on Association for Computational Linguistics, pp. 133–138. Association for Computational Linguistics (June 1994)
5. Leacock, C., Chodorow, M.: Combining local context and WordNet similarity for word sense identification. WordNet: An Electronic Lexical Database 49(2), 265–283 (1998)
6. Lin, D.: An information-theoretic definition of similarity. In: ICML, vol. 98, pp. 296–304 (July 1998)
7. Jiang, J.J., Conrath, D.W.: Semantic similarity based on corpus statistics and lexical taxonomy. arXiv preprint cmp-lg/9709008 (1997)
8. Blei, D.M., Ng, A.Y., Jordan, M.I.: Latent dirichlet allocation. The Journal of Machine Learning Research 3, 993–1022 (2003)
9. Dumais, S., Furnas, G., Landauer, T., Deerwester, S., Deerwester, S.: Latent semantic indexing. In: Proceedings of the Text Retrieval Conference (1995)
10. Gabrilovich, E., Markovitch, S.: Computing Semantic Relatedness Using Wikipedia-based Explicit Semantic Analysis. IJCAI 7 (2007)
11. Yazdani, M., Popescu-Belis, A.: Computing text semantic relatedness using the contents and links of a hypertext encyclopedia. Artificial Intelligence 194, 176–202 (2013)
12. Gabrilovich, E., Markovitch, S.: Wikipedia-based semantic interpretation for natural language processing. Journal of Artificial Intelligence Research 34(2), 443 (2009)

13. Lee, M.D., Pincombe, B., Welsh, M.: A comparison of machine measures of text document similarity with human judgments. In: 27th Annual Meeting of the Cognitive Science Society (CogSci 2005), pp. 1254–1259 (2005)
14. Banea, C., Hassan, S., Mohler, M., Mihalcea, R.: Unt: A supervised synergistic approach to semantic text similarity. In: Proceedings of SemEval 2012, pp. 635–642 (2012)

Characteristic Prediction of a Varistor in Over-Voltage Protection Application

Kohei Nagatomo[1,3], Muhammad Aziz Muslim[2], Hiroki Tamura[3],
Koichi Tanno[4], and Wijono[2]

[1]Graduate School of Electric and Electronic Engineering,
University of Miyazaki
1-1 Gakuenkibanadainishi, Miyazaki-City, Miyazaki, Japan
tc12025@student.miyazaki-u.ac.jp
[2]Department of Electrical Engineering, Faculty of Engineering,
University of Brawijaya
Jalan Veteran, Malang, Jawa Timur, Indonesia
{muh_aziz2,wijono}@ub.ac.id
Department of Environmental Robotics, Faculty of Engineering,
University of Miyazaki
[3]1-1 Gakuenkibanadainishi, Miyazaki-City, Miyazaki, Japan
htamura@cc.miyazaki-u.ac.jp
Department of Electrical and Systems, Faculty of Engineering,
University of Miyazaki
[4]1-1 Gakuenkibanadainishi, Miyazaki-City, Miyazaki, Japan
tanno@cc.miyazaki-u.ac.jp

Abstract. This paper presents a characteristic prediction of a varistor
for modeling a varistor in over-voltage protection applications. Variable
resistor (Varistor) is one of common devices to protect following electric devices from over-voltage. However, the principle of varistor is still
unclear due to its non-linear characteristics between amount of voltage
and current. To model the non-linearity, the prediction using adaptive
network-based-fuzzy inference system (ANFIS) will be used with several
datasets obtained by a high-voltage experiment concerning to the varistor. The result can be used to model the varistor as a voltage clipping
device.

Keywords: Overvoltage protection, metallic oxide varistor, non-
linearity, ANFIS.

1 Introduction

Surge protection devices are used widely to protect the electric devices from
destructions by an over-voltage such as the lightning and electrostatic. As typical
surge protection devices are arrestor which is put in the top of telephone pole and
metal oxide varistor which is used in the small scale circuits. These devices have
non-linearity correlation between voltage and current. If a huge surge voltage is
applied to the varistor, the resistance of varistor decrease substantially and large

C.K. Loo et al. (Eds.): ICONIP 2014, Part II, LNCS 8835, pp. 565–572, 2014.
© Springer International Publishing Switzerland 2014

amount of currents flow in the varistor [1]. The characteristic of surge protection devices is useful to control surge voltages.

However, these surge protection devices have time delay to control the surge voltages and the delay depends on the applied surge voltage value. This paper focuses on the varistor and its relation between the applied surge voltage and the time to control the surge voltage. Although the earlier papers attempt to make the non-linear characteristic between supplied surge voltage and current clear using individual processes [2], [3], the relation between supplied surge voltage and the time to control the surge voltage has not been referred. In this paper, the characteristic prediction of varistor will be conducted for modeling a varistor based on a prediction using adaptive neuro fuzzy inference system (ANFIS). In the Section 2, behavior of surge voltage will be confirmed through an high voltage experiment with metal oxide varistor. In the Section 3, 6 parameters will be defined from waveform obtained by the high voltage experiment. In addition, an adaptive neuro fuzzy inference system (ANFIS) will be trained and examined by defined 6 parameters. In the Section 4, Comparison between measurement data obtained by the high- voltage experiment and predicted data obtained by the trained ANFIS and Multi Regression Analysis (MRA), and the effectivity and repeatability of predicted data will be examined.

2   High-Voltage Experiments

2.1   Experimental Conditions

High-voltage experiments were conducted to obtain the dataset consisting of two parameters for an adaptive neuro fuzzy inference system. In this paper, a ZnO type ceramic varistor is used. The dataset of two parameters, i.e., input voltage and the controlled voltage caused by a varistor. The input voltage is a kind of impulse voltage generated by using specific high-voltage circuit, as shown in Fig.1. D_1 and D_2 are diode. C_s and C_b are capacitors. DSTM and DGM are facilities to measure high-voltage impulse and AC voltage. EZK is an electronic trigger spheres. ZAG is trigger unit. R_m works as DC divider. The impulse voltage is created from DC voltages, range of 3 to 10 kV with 0.5 kV step. The controlled voltage will be obtained by supplying each step of impulse voltage to the circuit containing the varistor connected in parallel with C_b in Fig.1.

2.2   Experimental Result

By using the impulse voltage generation circuit, dataset was obtained. Fig.2 shows examples of impulse voltage generated by the circuit without varistor and Fig.3 shows the voltage across varistor with the generator input voltage of 10 kV DC voltage.

As experimental results, V_{imp}, T_{half}, V_{cut}, T_{cut}, V_{min} and T_{min} were defined as parameters for the following analysis. V_{imp} indicates the maximum voltage in the circuit without varistor. T_{half} is a time from generation of impulse voltage

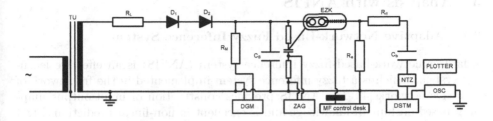

Fig. 1. The high-voltage circuit for generation of impulse voltage

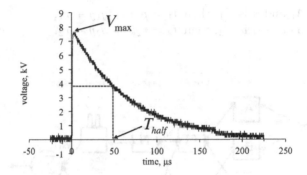

Fig. 2. Example of impulse voltage at 10 kV DC voltage

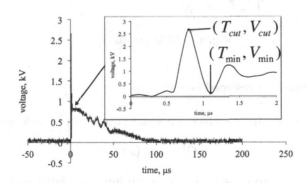

Fig. 3. Example of controlled voltage at 10 kV DC voltage

to decreasing until 50 percent of V_{imp}. T_{cut} and T_{cut} are voltage cut by the metal oxide varistor and its time in Fig.3, respectively. V_{min} and T_{min} are defined as a minimal voltage and its time, respectively. From Fig.2 and Fig.3, it can be seen that V_{imp} is controlled by the metal oxide varistor.

3   Analysis with ANFIS

3.1   Adaptive Network-Based Fuzzy Inference System

Adaptive network-based-fuzzy inference system (ANFIS) is an effective learning system, which is a fuzzy inference system implemented in the framework of adaptive networks[4]. The ANFIS provides construction of input-output mapping based on human knowledge and is excellent in non-linear prediction. Fig.4 shows ANFIS architecture proposed by Jang. ANFIS here provides 1 output from 2 input data. Supposed that if-then rules of Takagi and Sugeno's type[5] are used in this ANFIS architecture.

Rule 1: If x is A_1 and y is B_1, then $f_1 = p_1x + q_1y + r_1$,
Rule 2: If x is A_2 and y is B_2, then $f_2 = p_2x + q_2y + r_2$.

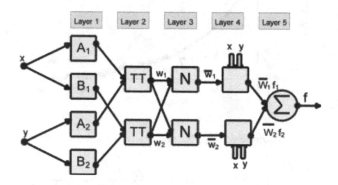

Fig. 4. ANFIS architecture

ANFIS has 5 layers represented below:

Layer 1. Each node i in this layer is a square node with a node function

$$O_i^1 = \mu_{A_i}(x) \tag{1}$$

where x is the input to node i, and A_i is the linguistic label such as *small* and *large* associated with this node function. O_i^1 is also called membership function and it specifies the degree to which the given x satisfies the quantifier A_i. The membership function is usually defined as a bell-shaped function

$$\mu_{A_i}(x) = \frac{1}{1 + \left[\left(\frac{x - c_i}{a_i}\right)^2\right]^{b_i}} \tag{2}$$

where $\{a_i, b_i, c_i\}$ is the parameter set.

Layer 2. Each node in this layer conducts the multiplication of the incoming signals and sends the product out as described below,

$$w_i = \mu_{A_i}(x) \times \mu_{B_i}(y), i = 1, 2. \tag{3}$$

Outputs of this layer indicate the firing strength of a rule.

Layer 3. Calculating the ratio of the ith rules firing strength to the sum of all rules firing strengths is conducted as described below,

$$\bar{w}_i = \frac{w_i}{w_1 + w_2}, i = 1, 2. \tag{4}$$

Layer 4. Parameters called consequent parameters are obtained from the equation

$$O_i^4 = \bar{w}_i f_i = \bar{w}_i (p_i x + q_i y + r_i) \tag{5}$$

where $\bar{w}_i$ is the output of layer 3, and $\{p_i, q_i, r_i\}$ is the parameter set.

Layer 5. This layer computes the overall output as the summation of all incoming signals, i.e.,

$$O_i^5 = \sum \bar{w}_i f_i = \frac{\sum_i w_i f_i}{\sum_i w_i} \tag{6}$$

In this paper, local search learning[6] for ANFIS parameter$\{a, b, c\}$ $\{p, q, r\}$ turning. By using local search, a higher probability system can be obtained.

3.2 Datasets

As represented in Sect 2.2, V_{imp}, T_{half}, V_{cut}, T_{cut}, V_{min} and T_{min} are used as dataset. T_{half}, T_{cut} and T_{min} obtained as microsecond order are normalized to be suitable data for the following analysis. For the ANFIS, 3 datasets, DC 4.0, 6.0 and 7.5 were used as testing data, and the other was used as training dataset for ANFIS.

In this paper, 2 pair of the experimental parameter, V_{cut} and T_{cut} or V_{min} and T_{min} will be supervised data and predicted by V_{imp} and T_{half}. All of 14 datasets will be divided into 2 groups, 11 dataset are used as training dataset and 3 datasets are test datasets.

4 Results

4.1 Characteristic Prediction of the Varistor

By using all of experimental datasets that consist of V_{imp} and T_{half} as input data, the predicted 2 pair of parameter V_{cut} and T_{cut} or V_{min} and T_{min} will be obtained. In this paper, the prediction of the 2 pair of parameter, V_{cut} and T_{cut} and V_{min} and T_{min} using multi-regression analysis (MRA) is also conducted to compare with ANFIS. Verifying the relations between actual and predicted V_{cut},

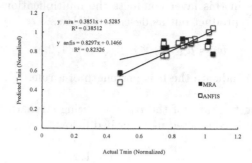

Fig. 5. Relationship between actual and predicted T_{min}

T_{cut}, V_{min}, T_{min}, V_{cut} and T_{min} had strong linearity and T_{cut} were almost same, 0.9 microsecond each. As the results, for some parameters, it was understood that V_{cut}, V_{min} and T_{cut} have linear-characteristics between the actual and predicted data. Therefore, V_{cut}, V_{cut} and T_{cut} can be predicted easily as linear parameter of varistor. However, for T_{cut}, only results of MRA show a weak linear characteristic as shown in Fig.5.

From Fig.5, the ANFIS using local search learning method provides higher accuracy than the MRA. In this paper, because of a few of the datasets, the fact that T_{cut} has non-linearity cannot be articulated.

4.2 Confirmation of the Generalization Ability

Consider to the results obtained from experiments, a relationship between the impulse voltage defined as V_{imp} and the predicted T_{min} by MRA and ANFIS is illustrated, as shown in Fig.6. Furthermore, Fig.7 shows same relationship with Fig.6 focused on 3 datasets defined as testing data of ANFIS. In Fig.6, the predicted T_{min} of ANFIS provides more accurate prediction than MRA.

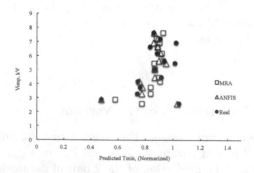

Fig. 6. Characteristic between impulse voltage V_{imp} and predicted T_{min}

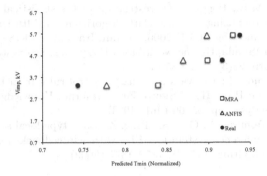

Fig. 7. Characteristic between impulse voltage V_{imp} and predicted T_{min} concerning ANFIS's testing data

However, for the result of testing data of ANFIS, as shown in Fig.7, the result of MRA is better than ANFIS. It is assumed that the increasing the number of training dataset make the reliability of prediction result of ANFIS become higher.

5   Conclusion

In this paper, a prediction of characteristic for modeling of a varistor in over-voltage protection application was presented. The varistor has a non-linearity resistance. The high-voltage input into the varistor and the voltage clipped by varistor is defined as important parameters. The high-voltage experiments was performed to obtain the dataset consisting of V_{imp}, T_{half}, V_{cut}, T_{cut}, V_{min} and T_{min} for the construction of ANFIS. The ANFIS trained by the experimental data was 2-input 1-output that 2 input are V_{imp} and T_{half} and 1 output is V_{cut}, T_{cut}, V_{min} or T_{min}. As the results, because three of predicted parameter, V_{cut}, T_{cut}, V_{min} had strong linearity, these parameters can be predicted easily. However, the characteristic between actual and predicted T_{min} had the non-linearity. From the relationship between actual impulse voltage V_{imp} and predicted T_{min}, it is possible to use one of the relation to predict characteristics of varistor. However, the number of dataset has to be increased more and more to improve ANFIS architecture and its learning results accuracy.

References

1. Harnden Jr., J.D., Martzloff, F.D., Morris, W.G., Golden, F.B.: 'Metal-oxide varistor: a new way to suppress transients. Reprint of reprint from Electronics (1972)
2. Meshkatoddini, M.R.: Metal Oxide ZnO-Based Varistor Ceramics. In: Sikalidis, C. (ed.) Advances in Ceramics - Electric and Magnetic Ceramics, Bioceramics, Ceramics and Environment. InTech (2011) ISBN: 978-953-307-350-7

3. Meshkatoddini, M.R., Boggs, S.: Investigation of the statistical behavior of thin ZnO- based varistors using a Monte Carlo Algorithm. In: 14th Iranian Conference on Electrical Engineering, ICEE 2006, Tehran, Iran, May 16-18 (2006)
4. Jang, J.R.: ANFIS: adaptive-network-based fuzzy inference system. IEEE Trans. Syst. Man Cybern. 23(3), 665–685 (1993)
5. Takagi, T., Sugeno, M.: Derivation of fuzzy control rules from human operator's control actions. In: Proc. IFAC Symp. Fuzzy Inform., Knowledge Representation and Decision Analysis, pp. 55–60 (July 1983)
6. Vairappan, C., Tamura, H., Gao, S., Tang, Z.: Bach type local search-based adaptive neuro-fuzzy inference system (ANFIS) with self-feedbacks for time-series prediction. Neurocomputing 72(7–9), 1870–1877 (2009)

Optimizing Complex Building Renovation Process with Fuzzy Signature State Machines

Gergely I. Molnárka[1] and László T. Kóczy[2]

[1] Department of Building Constructions and Architecture, Széchenyi István University, Egyetem tér 1, Győr, Hungary
[2] Department of Automation, Széchenyi István University, Egyetem tér 1, Győr, Hungary

Abstract. In contrary to recently built office and commercial buildings, the service life of the traditional European residential houses was not calculated. Some estimations exist about the life span of different types of building constructions, however, these estimations may not reassure the owners of urban-type residential houses that were built before the second world war. A thorough and professional renovation may extend the service life of buildings by decades, the question is how to prepare the most effective renovation procedure.

As a combination of the fuzzy signature structure and the principles of finite-state machine a new formal method is proposed for generating a tool for supporting the renovation planning, concerning the costs and importance of repair. With the support of information obtained from a given pre-war urban-type residential house, the available technical guides and the contractors' billing database an optimized renovation process of the roof structure is presented as a case study.

Keywords: urban-type residential house, building renovation, fuzzy signature, fuzzy state machine.

1 The Attributes of the Urban Residential Houses in Budapest

The distribution of almost one million apartments in Budapest, Hungary presents a fairly heterogeneous picture: among others, the apartments located in urban-type residential buildings (built before 1940) take 27% [1]. The criteria of the pre-war urban-type residential house are the decorative street façades, the existence of courtyard and air-shafts and the traditional masonry load bearing structures.

The average physical condition of these old residential buildings is below standard. The symptoms of overall physical obsolescence and deterioration are clearly observable:

- Missing wall and plinth damp proof courses;
- Uncertain load-bearing performance of the slab systems and side corridors;

C.K. Loo et al. (Eds.): ICONIP 2014, Part II, LNCS 8835, pp. 573–580, 2014.

- Out-dated mechanical and electrical systems;
- Broken and crumbling building envelope, floor covering and supplementaries;
- Cracked and perforated roof tiling and flashings;
- Low energetic performance of air-shaft partitioned walls and openings;
- Detached and missing plaster and finishing of the façade surfaces.

Due to the similarity it is supposable that a simple accumulated depreciation curve may illustrate well the state transition of any residential house that is applied for financial planning of office and commercial buildings at present [2]. The curve *a* in Fig.1 depicts an average amortisation of a building. The state transition of a building (as a complex structure of several building components) can be described with a *continuous function*, where the recovery works (or regularly performed maintenance interventions) and deteriorations influence its shape. In extreme situations, the physical condition of the examined building may change, therefore the state transition function may become fractionally continuous (as curve *b* and *c* in Fig.1). During the state transition evaluation it has to be mentioned that a state transition of a building component or group of components may have no effect to other building components (e.g. a roof renovation process does not improve the performance of the foundation). It means that the state of the building components determine the state of the given building together, while the state of components may change individually. The working stages of the group of components and constructions are characteristically overlapping, consecutive, or independent from each other. Nevertheless, these stages may also be sectioned (e.g. a stairway reconstruction process in more phases).

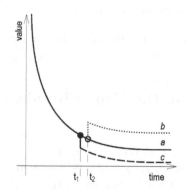

Fig. 1. Amortization curves as characteristic description of complex structures

The owners are responsible for the physical condition of their apartments and the jointly owned building elements. Their decision-making becomes more difficult if there is no suitable decision support tool at community's disposal.

As a consequence, the building maintenance and repair processes are prevailingly ineffective and therefore they are not long-term sustainable; only a consistent and well organized renovation plan might help determine the necessary decisions about the renovation steps, offering answers both financing and scheduling.

2 A Model Proposed for Supporting the Renovation

The proposed approach to be introduced here for handling the problem described above is a model and an attempt for solution based on the principles of fuzzy signatures and fuzzy state machines combined into what we call *Fuzzy Signature State Machines (FSSM)*.

2.1 Principles of Fuzzy Signatures and Signature State Machines

The most important reason for using fuzzy signatures here as the starting point is the fact that the sub-structures and components of each building are arranged in hierarchical tree-like structures, where the whole building might be presented by the root of the tree and each mayor sub-component is a first level branch, with further sub-branches describing sub-sub-components, etc.

Fuzzy sets of a universe X are defined by

$$A = \{X, \mu_A\}, \text{ where } \mu_A : X \to [0,1] \tag{1}$$

Vector Valued Fuzzy Sets (VVFS) [3] are a simple extension that may be considered as a special case of L-fuzzy sets [4]:

$$A_n = \{X, \mu_{A_n}\}, \text{ where } \mu_A n : X \to [0,1]^n \tag{2}$$

Thus a membership degree is a multi-component value here, e.g. $[\mu_1, \mu_2, \cdots, \mu_n]^T$. Fuzzy signatures represent a further extension of VVFS as here any component might be a further nested vector, and so on [7], [5]:

$$A_{fs} = \{X, \mu_{A_{fs}}\}, \text{ where } \mu_{A_{fs}} : X \to M_1 \times M_2 \times \cdots \times M_n,$$
$$\text{where } M_i = [0,1] \text{ or } [M_{i_1} \times M_{i_2} \times \cdots \times M_{i_n}]^T \tag{3}$$

The following simple example illustrates how a fuzzy signature operates:

$$\mu_{A_{fs}} = [\mu_1, \mu_2, [\mu_{3_1}, \mu_{3_2}, [\mu_{3_{3_1}}, \mu_{3_{3_2}}, \mu_{3_{3_3}}]], \mu_4, [\mu_{5_1}, \mu_{5_2}], \mu_6]^T \tag{4}$$

The first advantage of using fuzzy signatures rather than VVFS is that there any closer grouping and sub-grouping of fuzzy features may be given. Fuzzy signatures are associated with an aggregation system. Each sub-component set may be aggregated by its respective aggregation operation, thus reducing the sub-component to one higher level.

This kind of membership degree reduction is necessary when the data are partially of different structure, e.g. some of the sub-components are missing. The operations among fuzzy signatures with partially different structure may be carried out, by finding the *largest common sub-structure* and reducing all signatures up to that substructure. This might be necessary if the surveys referred to this paper are considered as often their depth and detail are different. As an example, maybe in "survey A" the roof structure is considered as a single component of the house and is evaluated by a single linguistic quality label, while in "survey B" this is done in detail and tiles, load-bearing structure, tinsmith work, chimney shafts and roof auxiliaries are described separately.

In our previous work we applied vector-valued fuzzy sets and fuzzy signatures [8] for describing sets of objects with uncertain features, especially when an internal theoretical structure of these features could be established. In [9] we presented an approach where the fuzzy signatures could be deployed for describing existing residential houses in order to support decisions of Local Authorities concerning when and how these buildings should be renovated involving non-measurable (and subjective) factors. In that research a series of theoretically arrangeable features were taken into consideration and eventually a single aggregated fuzzy membership value could be calculated on the basis of available detailed expert evaluation sheets. In that model, however, the available information does not support any decision strategy concerning actual sequence of the measures leading to complete renovation; and it is also insufficient to optimize the sequence from the aspect of local or global cost efficiency. In the following section, the mathematical model of the proposed maintenance protocol will be introduced.

2.2   Application of Fuzzy State Machines in the Modelling

Finite State Machines are determined by the sets of input states X, internal states Q, and the transition function f. The latter determines the transition that will occur when a certain input state change triggers state transition. There are several alternative (but mathematically equivalent) models known from the literature. For simplicity the following is assumed as the starting point of our new model:

$$A = \langle X, Q, f \rangle \tag{5}$$

$$f : X \times Q \to Q, \text{ where } X = \{x_i\} \text{ and } Q = \{q_i\} \tag{6}$$

Thus, a new internal state is determined by the transition function as follows:

$$q_{i+1} = f(x_i, q_i) \tag{7}$$

The transition function/matrix maybe interpreted with help of a relation R on $X \times Q \times Q$, where $R(x_i, q_j, q_k) = 1$, if $f(x_i, q_j) = q_k$ and $R(x_i, q_j, q_k) = 0$, if $f(x_i, q_j) \neq q_k$.

The states of the finite state machine are elements of Q. In the present application an extension to fuzzy states is considered in the following sense. Every aspect of the phenomenon to model is represented by a state universe of substates Q_i. The states themselves are (fuzzy) subsets of the universe of discourse state sets, so that within Q_i a frame of cognition is determined (its fineness depending on the application context and on the requirements toward the optimisation algorithm), so that typical states are considered. Any transition from one state to the other (improvement of the condition, refurbishment or renovation) involves a certain cost c. In the case of a transition from q_i to q_j it is expressed by a membership value $\mu_{ij} = c(q_i, q_j)$. In our model the added cost $\Sigma\mu_{i_j}$ along a path $q_{i_1} \rightarrow q_{i_2} \rightarrow \cdots \rightarrow q_{i_n}$ is not usually equivalent with the cost of the transition μ_{i_n} along the edge $q_{i_l} \rightarrow q_{i_n}$. This is in accordance with the non-additivity property of the fuzzy (possibility) measure and is very convenient in our application, as it is also not additive in the case of serial renovations. The Fig. 2 illustrates the initial (deteriorated), internal and the accept (renovated) state of a possible renovation process; the bottom diagram represents an internal (μ_{12})transition between q_1 and q_2 states.

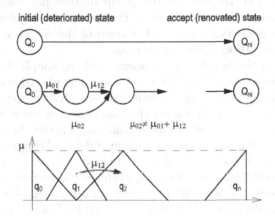

Fig. 2. Initial, internal and accept states of a renovation process

In the case of fuzzy signature machines each of the leaves contains a sub-automaton with the above property. The parent leave of a certain sub-graph is constructed from the child leaves, so that the sub-automaton $A^i = A^{i_1} \times A^{i_2} \times \cdots \times A^{i_m}$, and thus the states of A^i are $Q^i = Q^{i_1} \times Q^{i_2} \times \cdots \times Q^{i_n}$, so that the transition $Q^{j_1} \rightarrow Q^{j_2}$ in this case means the parallel (or subsequent) transitions $q_{j_{1_1}} \rightarrow q_{j_{1_2}} \times q_{j_{2_1}} \rightarrow q_{j_{2_2}} \times \cdots \times q_{j_{n_1}} \rightarrow q_{j_{n_2}}$. A special aggregation is associated with each leaf; similarly as it is in the fuzzy signatures, however, in this case the aggregation calculates the resulting cost $\mu_{j_{1_2}}$ of the transition $q_{j_1} \rightarrow q_{j_2}$, so that $\mu_{j_{1_2}} = c(q_{j_1}, q_{j_2}) = a_j(c(q_{j_{1_1}}, q_{j_{1_2}}), c(q_{j_{2_1}}, q_{j_{2_2}}), \cdots, c(q_{j_{n_1}}, q_{j_{n_2}}))$, where a stands for the respective aggregation.

The selection of aggregation operator is a key issue that may determine the final result of the model; however, the signature structure makes the application of different aggregation methods possible for each node .

3   Case Study: The Roof Structure of a Residential House

As an initial state an overall visual diagnostic survey of the given residential building is supposed that gives detailed determination and state description of each building component, disclosing the relation between causes and consequences. The obtained information helps determining the maintenance steps; observing their influencing attributes additional data can be given to each step.

With the knowledge of professional rules as they are clearly described in [10], [11], etc., the general maintenance procedure can be decomposed into eight distinct sequences that correspond to the previously mentioned supposition. Following the principles of signature structure, these automata represent the parent nodes of building component groups; while the total renovation process as a sum of these nodes represents the root of tree-like structure.

The state transition of roof structure is represented by A^1 automaton; its seven sub-automata correspond to the group of building components of the roof structure. Due to the complexity of the group of components (hierarchical relationship, dependencies, etc.) this section of the signature structure may demonstrate the operation of the entire model.

The attributes of the building components and the simplicity of the method verify alike the application of arithmetic average type of Ordered Weighted Averaging Aggregation (OWA) operator as it was presented by Yager in [12]. In this case the w weighting factor represents the importance of the renovation steps.

The state description of each group as child nodes is the starting point of the model. The available diagnostic survey contained detailed (linguistic) evaluations that determined the initial values of the child nodes. Some simplifications were realized for the evaluations that did not modified the results significantly, but the calculations were accelerated. The most important simplifications are: the universe was normalized to [0,1] that was covered by Ruspini partition-type [6] triangular and trapezoidal functions.

In the study, two simple state machines are discussed: the A^{14} (chimney shaft) and the A^{15} (auxiliaries) represent typical state transition processes with different structures. The observations indicate that an effective state transition in these two state machine may significant improvement in the physical condition of the total roof structure.

In A^{14} state machine the possible states were established ($Q_0^{14}; \cdots ; Q_4^{14}$; the Q_{-1}^{14} represents the demolished phase), and the possible transitions were examined, excluding the "restrained" events that may not occur in reality; the $\mu_{01}^{14}; \cdots ; \mu_{34}^{14}$ (cost) state change trigger the transitions. With a simple comparison the optimum (cost-effective) transitions were determined. The Fig. 3 illustrates frame of the complete roof structure fuzzy signature state machines and the A^{14}, A^{15} state transitions.

The official contractors' billing database were applied for determining each μ_i trigger; the cost of total reproduction of the roof structure was calculated as one unit.

The Table 1 summarizes the values of child nodes in the A^1 fuzzy signature structure both in initial state and in accept state. It is clearly visible that with

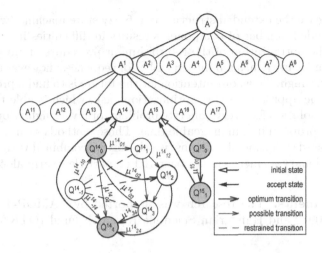

Fig. 3. Diagram representation of the roof structure FSSM

some simple and well-organized interventions the aggregated value of the roof structure may be improved effectively. The obtained results confirm the recommendations of the building diagnostic survey: the importance of the renovation of chimney shafts and the removal of unused aerials were emphasized.

Table 1. Attributes and aggregation of the roof fuzzy signature structure

Node (initial state)	Observation	Node (accept state)	Observation	Weight
A_0^{11}	$x_0^{11} = 0.70$			$w^{11} = 0.65$
A_0^{12}	$x_0^{12} = 0.35$			$w^{12} = 0.50$
A_0^{13}	$x_0^{13} = 0.55$			$w^{13} = 0.45$
A_0^{14}	$x_0^{14} = 0.75$	A_4^{14}	$x_4^{14} = 0.85$	$w^{14} = 0.30$
A_0^{15}	$x_0^{15} = 0.20$	A_1^{15}	$x_1^{15} = 0.90$	$w^{15} = 0.80$
A_0^{16}	$x_0^{16} = 0.25$			$w^{16} = 0.80$
A_0^{17}	$x_0^{17} = 0.60$			$w^{17} = 0.25$
A_0^1	$x_0^1 = 0.385$	A_1^1	$x_1^1 = 0.532$	

4   Conclusions and Future Work

As it is discussed in the Section 1, the attributes of the residential houses, their constructions and failure types make the development of an optimization method useful for supporting the decisions of owners: the characters of deteriorations, the hierarchically ordered structure of building components vindicate the use of fuzzy signature structure, while the application of a state machine system is indeed reasonable due to the continuous state transition of the entire building.

In general case the extended structure of a fuzzy state machine (with theoretically) unbounded number of components results in difficulties in optimization. In practice, the optimization of the refurbishment procedure of any sort of residential buildings always has a limited number of sequences; however this number might be rather high. Thus, our intention for the future is to find a proper heuristics at the basic approach to this optimization task, which is able to provide a quasi-optimal solution for every concrete problem, or a very lightly optimal solution for every problem in a manageable time. These methods seem to be various population-based evolution algorithms, which may be combined with local search cycles (evolutionary memetics), e.g. bacterial or particle swarm algorithms.

Acknowledgments. The research was supported by TÁMOP-4.2.2.A-11/1/ KONV-2012-0012 and Hungarian Scientific Research Fund (OTKA) K105529, K108405.

References

1. Csizmady, A., Hegedüs, J., Kravalik, Z., Teller, N.: Long-term housing conception and mid-term housing program of Budapest, Hungary (in Hungarian). Technical report, Local Government of Budapest, Hungary (2005)
2. Gibson, E.J.: Working with the Performance Approach in Building (CIB W060). Report, Rotterdam, The Netherlands (1982)
3. Kóczy, L.T.: Vector Valued Fuzzy Sets. J. Busefal 4, 41–57 (1980)
4. Goguen, J.: L-fuzzy sets. Journal of Mathematical Analysis and Applications 18, 145–174 (1967)
5. Pozna, C., Minculete, N., Precup, R.-E., Kóczy, L.T., Ballagi, Á.: Signatures: Definitions, operators and applications to fuzzy modelling. Fuzzy Sets and Systems 201, 86–104 (2012)
6. Ruspini, E.H.: A new approach to clustering. Information and Control 15, 22–32 (1969)
7. Kóczy, L.T., Vámos, T., Biró, G.: Fuzzy Signatures. In: EUROFUSE-SIC 1999, pp. 210–217 (1999)
8. Vámos, T., Kóczy, L.T., Biro, G.: Fuzzy signatures in data mining. In: Joint 9th IFSA World Congress and 20th NAFIPS International Conference, vol. 5, pp. 2842–2846 (2001)
9. Molnárka, G.I., Kóczy, L.T.: Decision Support System for Evaluating Existing Apartment Buildings Based on Fuzzy Signatures. Int. J. of Computers, Communications & Control 6, 442–457 (2011)
10. de Freitas, V.P. (ed.): A State-of-the-Art Report on Building Pathology (CIB W086). Technical report, Porto University-Faculty of Engineering, Porto, Portugal (2013)
11. Harris, S.Y.: Building pathology: deterioration, diagnostics, and intervention. Wiley, New York (2001)
12. Yager, R.R.: On ordered weighted averaging aggregation operators in multicriteria decision making. IEEE Transactions on Systems, Man and Cybernetics 18, 183–190 (1988)

News Title Classification with Support from Auxiliary Long Texts

Yuanxin Ouyang[1,2], Yao Huangfu[1], Hao Sheng[1,2], and Zhang Xiong[1,2]

[1] School of Computer Science and Engineering, Beihang University, Beijing, China
[2] Research Institute of Beihang University in Shenzhen, Shenzhen, China
{oyyx,shenghao,xiongz}@buaa.edu.cn, huangfuyao@cse.buaa.edu.cn

Abstract. The performance of short text classification is limited due to its intrinsic shortness of sentences which causes the sparseness of vector space model. Traditional classifiers like SVM are extremely sensitive to the features space, thereby making classification performance unsatisfying in short text related applications. It is believed that using external information to help better represent input data would possibly yield satisfying results. In this paper, we target on the problem of news title classification which is an essential and typical member in short text family and propose an approach which employs external information from long text to address the problem the sparseness. Afterwards Restricted Boltzman Machine are utilised to select features and then finally perform classification using Support Vector Machine. The experimental study on Reuters-21578 and Sogou Chinese news corpus has demonstrates the effectiveness of the proposed method.

1 Introduction

With explosion of information online, how to quickly get useful information from the overwhelming data has become a difficult task. One of typical applications is to obtain interesting points in large amount of news online. To meet this requirement, a lot of techniques have been proposed and one notable one is news title classification, which is a vital work as prior to many tasks such as theme extraction [7], social network feeding [6] and etc.

To classify news title, an intuitive way is to employ bag-of-words based methods, which have proven success in a lot of information retrieval tasks. However, due to the fact that news title are normally short, sometimes only with few words, bag-of-words based methods are not applicable in this kind of short text oriented classification. To overcome this problem, one major solution to use external information to enrich the information beneath just news title. This kind of methods are normally called feature extension and representative methods include feature extension based on N-Gram [12], using internet information like web engine [3], frequent items set like MMRFS [4], and etc. Besides these methods, there is an alternate approach which tried to employ long text to help short text classification as Long texts usually contain more information which would help short text classification to some extent [14]. Since helping learning tasks

C.K. Loo et al. (Eds.): ICONIP 2014, Part II, LNCS 8835, pp. 581–588, 2014.
© Springer International Publishing Switzerland 2014

with long texts in the same domain is a possible solution [9], it is believed in the news related application, the long text can be utilised from news content.

Despite the potential of using long text to help news title classification, simply combining the short texts with semantically unrelated long texts or topics could adversely affect the classification on the short texts [9]. To further address this problem, we propose a generative stochastic artificial neural network (s-ANN) to represent instances with a new set of features which are more denser on information level. This is where feature space transformation comes in and one typical and successful model is Restriced Boltzman Machine (DBM). In this paper, we use RBM as feature selection along with projecting tricks to avoid the inconsistency between short and long texts. Finally, we feed the selected features into the traditional classifier Support Vector Machine to conduct news title classification.

The rest of the paper is organised as follow. In section 2 the background of information enriching and RBM will be discussed. Section 3 we will elaborated the proposed methods and section 4 will presents the experimental study to discuss the method's potential. Section 5 will concludes the paper and point out possible future research direction.

2   Background

2.1   Information Enriching

Using external information to enrich the features representing short texts is also known as feature extension. Some well-known methods include natural language processing (NLP) based approach such as N-Gram [12], Word Senses [10], and Multi-Word [18]; background knowledge such as web information [3,2]; and frequent items set such as MMRFS [4], MFS [1]; and some other method like CFWS [13] and MC [19].

Using long texts to enrich the information is another external information based approach to help short text classification. Jin et al. successfully using the auxiliary long text to help short text clustering [9]. Phan et al. proposed to train topic models on a collection long text in the same domain and then make inference on short text to help the learning task on short texts [14,15]. Bollegala et al. came up with a measurement of word similarity using web information [3]. The core idea is to see the hit of individual Query A, Query B, and the overall hit of both Query A and B, where A and B are both key words. In [4], MMRFS is introduced as an algorithm of frequent feature selection with multi-parameters constrains. Coverage, interdependency and redundancy of each frequent feature are taken into consideration. MFS method further removes the duplication of words in multiple frequent sets, and is able to avoid the redundancy of information and preserve the context of the original text as the meantime [1]. Jin et al. proposed a class of topic model Dual Latent Dirichlet Al (DLDA), which jointly learns a set of target topics on the short texts and another set of auxiliary topics on the long texts while carefully modeling the attribution to each type of topics when generating documents [9]. Phan et al. collected a very large external data collection and built a classification model on both a small set of labeled training data and a rich set of hidden topics discovered from that data collection [15].

2.2 Restricted Boltzman Machine on Feature Selection

Restricted Boltzman Machine (RBM) is widely applied to tasks of speech recognition, text categorization, and computer vision, which are important set of application areas involve high-dimensional sparse input vectors [5]. RBM could be used as a feature selection model which can learn a set of feature and feed them to subsequent learning process such as classification and regression [16]. It can also be considered as a discriminative model [11], or used in other tasks such as ensemble classifier [17].

Generally RBM consists of two layers. One is the visible layer which has m nodes, which represents the dimension of the input data. The another one is hidden layer which has n nodes, which is the dimension of new features if this model is used in feature selecting. The object function to be maximised is the RBM energy function:

$$E(v, H) = -\sum_{i=1}^{n}\sum_{j=1}^{m} w_{ij}h_i v_j - \sum_{j=1}^{m} b_j v_j - \sum_{i=1}^{n} c_i h_i \qquad (1)$$

where h_i denotes the i-th hidden node, v_j denotes the j-th visible node and w_{ij} is the weight which connects those two nodes. b_j and c_i are the bias term of visible nodes and hidden nodes respectively. We can further turn Eq. (1) into a probability distribution [8] and we have:

$$p(v, h) = \frac{e^{-E(v,h)}}{\sum_{v,h} e^{-E(v,h)}} \qquad (2)$$

Here all of the weights and biases are updated iteratively just to maximise Eq. (2) which could be interpreted as to make the probability of each input data vector and its matching hidden nodes vector largest so that the new feature vector is the best representation of the original vector.

3 Long Text Based Information Enrichment Using RBM

The whole process of the proposed approach is depicted as in Fig. 1, which consists of four main steps, i.e., Preprocessing, Long text projection, Feature selection using RBM, and Classification.

1) Preprocessing

As elaborated in Fig. 1, the first step is to perform some basic preprocessing and build the original feature space. Afterwards we can remove the words which have low information gain from the original feature space and get the selected feature spaces.

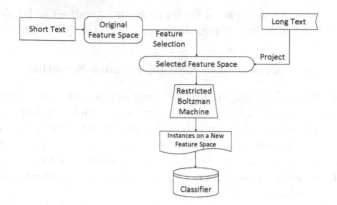

Fig. 1. Classifier with Features Generated by RBM with Long Texts

2) Long Text Projecting

In this step the long text is used to enrich information. The long text projecting process is as following: Suppose that $D_{L_i} = \{x_0, x_1, ..., x_m\}$ denotes a long text vector instance in corpus, $D_{S_i} = \{x_0, x_1, ..., x_n\}$ denotes a short text vector instance, the feature space $F = \{F_1, F_2, ..., F_z\}$ is generated using $D_{S_{1..I}}$, where I is the total number of short text instances. For every x_i in D_{L_i}, if x_i in the feature space F then x_i is put into the new instance $new\_D_{L_i}$, which has a dimension of the feature space F.

3) RBM Based Feature Selection

When we try to utilise the advantage of RBM in the mission of text classification since the traditional classifiers such as Support Vector Machine (SVM) and Bayesian classifier are extremely sensitive to the feature expression of an instance, there is another important challenge which needs to be attached much attention. The traditional representation of short texts would cause the sparseness problem, which makes the average non-zero feature number of an instance a bit small. This problem will probably gain poor outcome if we directly feed this feature space to SVM. It would also counteract the whole profiting performance of RBM by simple use the sparse instances as input data to that model. We would explain as follow:

In RBM, Weights updating algorithm by calculating derivatives on each weight is as follow:

$$\Delta w_{ij} \leftarrow \Delta w_{ij} + p(H_i = 1 \mid v) \cdot v_j - p(H_i = 1 \mid v') \cdot v'_j \tag{3}$$

$$\Delta b_j \leftarrow \Delta b_j + v_j - v'_j \tag{4}$$

$$\Delta c_j \leftarrow \Delta c_i + p(H_i = 1 \mid v) - p(H_i = 1 \mid v') \tag{5}$$

where

$$p(V_j = 1 \mid h) = \sigma(\sum_{i=1}^{n} w_{ij}h_i + b_j) \tag{6}$$

$$p(H_i = 1 \mid v) = \sigma(\sum_{j=1}^{m} w_{ij}v_j + c_i) \tag{7}$$

where v_j is the original vector in input dataset and v_j' is a sampled vector after applying 1-fold *Contrastive Divergence* in the model. In Eq. (7), it is seen that the left side of the equation would be mainly susceptible to the bias term c_i. If v_j is zero, it means that the hidden layer bias C would be dominant in probability distribution and directly influence the sampling progress. When it comes to the update of w_{ij} in Eq. (3), the according gradient would deviate towards the bias, which would ultimately fail to generate a good compressed representation of the original features.

From the discussion above, it comes clear that it is necessary to precede long text projecting. Afterwards, words which are abundant in long text would fit in original feature space which is the input data of RBM, hence the the number of v_j in an instance would decrease many and the model would be far less susceptible to the bias term.

4) Classification

Finally, instead of using the sampled binary represented by the hidden nodes. The un-squeezed probability (the input of sigmoid function) is used as the outcome because it would be better when the outcome is fed into a linear classifier. Intuitively, a continuously value assigned on each attribute would reflect more precisely on the how much this feature account for the instance.

4   Experimental Study

4.1   Datasets and Evaluation Metrics

In this research, two datasets are employed to conduct experimental study. The first one is the well-known corpus Reuters-21578[1], which is an English news corpus whose documents were assembled and indexed with categories by personnel from Reuters Ltd and Carnegie Group, Inc. The another one is Sogou Chinese news corpus[2], which is widely used Chinese NLP dataset.

To evaluate the performance of proposed method, in this research three commonly used evaluation metrics are employed, i.e., Precision, Recall, and F_1 to evaluate the performance of the proposed model. Precision is the fraction of the instances the model predicts are relevant, Recall is the fraction of the relevant instances the model predicts, while F_1 is considered as a critical evaluation

---

[1] http://www.daviddlewis.com/resources/testcollections/reuters21578/
[2] http://www.sogou.com/labs/dl/c.html

because it is the harmonic mean of recall and precision, which both of them are equivalently weighted. The three metrics definition are defined as below:

$$Precision = \frac{TruePositive}{TruePositive + FalsePostitive} \tag{8}$$

$$Recall = \frac{TruePositive}{TruePositive + FalseNegative} \tag{9}$$

$$F_1 = \frac{2 * Recall * Precision}{Recall + Precision} \tag{10}$$

where *TruePositive* means the number of the documents which are correctly classified to one class, *FalsePositive* reflects ones that are wrongly classified to that class by model, while *FalseNegative* indicates the number of documents the model fails to classify.

4.2 Experimental Result and Discussion

In this research, we conduct these experiments on two different datasets Reuters-21578 and Sogou Chinese new corpus. Then we compared the proposed method RBM trained with long texts against the traditional SVM method with original features and also RBM method only trained by short texts. To further reveal the advantage of proposed methods, we also conduct another contrastive experiments by respectively using the sampled binary output, probability output and unnormalized probability output of hidden layers in RBM. The performances and results are showed in Table 1, Table 2, Table 3 and Table 4.

Table 1. Performances of Three Different Models on Reuter-21578

Category	SVM			RBM (short texts)			RBM (long texts)		
	Recall	Precision	F_1	Recall	Precision	F_1	Recall	Precision	F_1
acq	0.936	0.771	0.845	0.920	0.819	0.867	0.943	0.862	**0.900**
coffee	0.889	0.851	0.869	0.8181	0.857	0.837	0.864	0.826	0.844
crude	0.635	0.814	0.713	0.651	0.814	0.724	0.72	0.773	**0.745**
earn	0.961	0.873	0.915	0.982	0.921	0.950	0.964	0.956	**0.970**
grain	0.803	0.854	0.828	0.796	0.813	0.804	0.873	0.880	**0.876**
sugar	0.821	0.881	0.850	0.864	0.905	0.884	0.818	0.878	0.847
trade	0.752	0.844	0.795	0.741	0.809	0.774	0.818	0.807	**0.812**
AVERAGE	0.828	0.841	0.834	0.861	0.858	0.854	0.892	0.891	**0.890**

Table 2. Results of three different output representations on Reuter-25178

Category	Binary	Probability	Unnormalized
acq	0.793	0.853	**0.900**
coffee	0.739	0.851	0.844
crude	0.606	0.760	0.745
earn	0.896	0.948	**0.970**
grain	0.731	0.845	**0.876**
sugar	0.727	0.863	0.847
trade	0.654	0.726	**0.812**
AVERAGE	0.778	0.867	**0.890**

Table 3. Performances of Three Different Models on Sougou Corpus

Category	SVM			RBM (short texts)			RBM (long texts)		
	Recall	Precision	F_1	Recall	Precision	F_1	Recall	Precision	F_1
Automobile	0.859	0.839	0.849	0.877	0.828	0.852	0.883	0.862	**0.872**
Business	0.720	0.618	0.665	0.779	0.632	0.698	0.761	0.646	**0.699**
IT	0.870	0.647	0.742	0.851	0.633	0.726	0.854	0.629	0.724
Education	0.870	0.922	0.895	0.892	0.914	0.903	0.915	0.922	**0.918**
Sport	0.872	0.924	0.897	0.899	0.933	0.916	0.904	0.943	**0.923**
Ent	0.779	0.875	0.824	0.812	0.884	0.846	0.810	0.880	0.843
AVERAGE	0.828	0.841	0.812	0.831	0.837	0.833	0.846	0.839	**0.842**

Table 4. Results of three different output representations on Sougou Corpus

Category	Binary	Probability	Unnormalized
Automobile	0.829	0.848	**0.872**
Business	0.631	0.655	**0.699**
IT	0.719	0.731	0.724
Education	0.865	0.909	**0.918**
Sport	0.871	0.908	**0.923**
Ent	0.827	0.861	0.843
AVERAGE	0.807	0.838	**0.842**

It is observed from Tables 1 and 2 that just using RBM trained by short texts as feature selection already improves the performance over traditional SVM. If training the RBM with corpus (long texts), the performance can be improved further, although the F_1 value in **coffee** category drops 2.5%, in **sugar** drops 0.3% in Reuter-25178, and in **IT** drops 1.8%, in **Ent** drops 0.3% in Sogou corpus comparing to the original SVM result. The results demonstrate the inner advantage of RBM as a feature selection method and could be further refined to achieve better performance.

In Tables 3 and 4, **Binary** means sampling from the normalized probability in hidden layers, **Probabiliy** denotes just outputting the probability of each node's value in hidden layers being equal to 1, and **unnormalized** is the proposed approach, which uses the unnormalized probability (the input of sigmoid function). From the tables it is seen that the performance of binary representation is unsatisfying, the probability representation is better and the unnormalized probability performs best in general though the F_1 value in categories such as **coffee**, **crude** and **sugar** drop 0.7%, 0.15% and 1.6% in Reuter-25178 and 0.7% percentage points in Sogou corpus compared against the probability output. But the unnormalized probability output still outperforms the binary output in each category. It can be concluded that using continuous value as output is better than the binary output because it reflects the weights more specifically.

5   Conclusion and Future Work

In this paper we have proposed a model to classify short texts with RBM using long texts. It utilised external information from long corpus to help address the sparseness of the traditional vector space model (VSM) representation of short text and RBM as feature selection, then we verify the effectiveness of the proposed approach by carrying out some experiments on benchmark Reuters-21578 and Sougou Corpus. In the future, we could refine our model to make it enable to take the weight of different features into consideration.

Acknowledgements. This work was supported by the National Natural Science Foundation of China (No. 61103095), the International S&T Cooperation Program of China (No. 2010DFB13350), and the Fundamental Research Funds for the Central Universities. We are grateful to Shenzhen Key Laboratory of Data Vitalization (Smart City) for supporting this research.

References

1. Ahonen-Myka, H.: Discovery of frequent word sequences in text. In: Hand, D.J., Adams, N.M., Bolton, R.J. (eds.) Pattern Detection and Discovery. LNCS (LNAI), vol. 2447, pp. 180–189. Springer, Heidelberg (2002)
2. Banerjee, S., Ramanathan, K., Gupta, A.: Clustering short texts using wikipedia. In: Proceedings of the 30th Annual International ACM SIGIR Conference on Research and Development in Information Retrieval, pp. 787–788. ACM (2007)

3. Bollegala, D., Matsuo, Y., Ishizuka, M.: Measuring semantic similarity between words using web search engines. In: Proceedings of the 16th International Conference on World Wide Web, pp. 757–766 (2007)
4. Cheng, H., Yan, X., Han, J., Hsu, C.W.: Discriminative frequent pattern analysis for effective classification. In: Proceedings of IEEE 23rd International Conference on Data Engineering, pp. 716–725. IEEE (2007)
5. Dauphin, Y., Bengio, Y.: Stochastic ratio matching of rbms for sparse high-dimensional inputs. In: Advances in Neural Information Processing Systems, pp. 1340–1348 (2013)
6. Dilrukshi, I., De Zoysa, K., Caldera, A.: Twitter news classification using svm. In: Proceedings of 8th International Conference on Computer Science Education, pp. 287–291 (April 2013)
7. Drury, B., Torgo, L., Almeida, J.: Classifying news stories to estimate the direction of a stock market index. In: Proceedings of 6th Iberian Conference on Information Systems and Technologies, pp. 1–4 (June 2011)
8. Hinton, G.: A practical guide to training restricted boltzmann machines. Momentum 9(1), 926 (2010)
9. Jin, O., Liu, N.N., Zhao, K., Yu, Y., Yang, Q.: Transferring topical knowledge from auxiliary long texts for short text clustering. In: Proceedings of the 20th ACM International Conference on Information and Knowledge Management, pp. 775–784. ACM (2011)
10. Kehagias, A., Petridis, V., Kaburlasos, V.G., Fragkou, P.: A comparison of word- and sense-based text categorization using several classification algorithms. Journal of Intelligent Information Systems 21(3), 227–247 (2003)
11. Larochelle, H., Bengio, Y.: Classification using discriminative restricted boltzmann machines. In: Proceedings of the 25th International Conference on Machine Learning, pp. 536–543. ACM (2008)
12. Li, R., Tao, X., Tang, L., Hu, Y.-F.: Using maximum entropy model for chinese text categorization. In: Yu, J.X., Lin, X., Lu, H., Zhang, Y. (eds.) APWeb 2004. LNCS, vol. 3007, pp. 578–587. Springer, Heidelberg (2004)
13. Li, Y., Chung, S.M., Holt, J.D.: Text document clustering based on frequent word meaning sequences. Data & Knowledge Engineering 64(1), 381–404 (2008)
14. Phan, X.H., Nguyen, C.T., Le, D.T., Nguyen, L.M., Horiguchi, S., Ha, Q.T.: A hidden topic-based framework toward building applications with short web documents. IEEE Transactions on Knowledge and Data Engineering 23(7), 961–976 (2011)
15. Phan, X.H., Nguyen, L.M., Horiguchi, S.: Learning to classify short and sparse text & web with hidden topics from large-scale data collections. In: Proceedings of the 17th International Conference on World Wide Web, pp. 91–100. ACM (2008)
16. Srivastava, N., Salakhutdinov, R.R., Hinton, G.E.: Modeling documents with deep boltzmann machines. arXiv preprint arXiv:1309.6865 (2013)
17. Zhang, C.-X., Zhang, J.-S., Ji, N.-N., Guo, G.: Learning ensemble classifiers via restricted boltzmann machines. Pattern Recognition Letters 36, 161–170 (2014)
18. Zhang, W., Yoshida, T., Tang, X.: Text classification based on multi-word with support vector machine. Knowledge-Based Systems 21(8), 879–886 (2008)
19. Zhang, W., Yoshida, T., Tang, X., Wang, Q.: Text clustering using frequent itemsets. Knowledge-Based Systems 23(5), 379–388 (2010)

Modelling Mediator Intervention in Joint Decision Making Processes Involving Mutual Empathic Understanding

Rob Duell

VU University Amsterdam, ASR Group
De Boelelaan 1081
1081 HV, Amsterdam
r.duell@vu.nl

Abstract. In this paper an agent model for mediation in joint decision-making processes is presented addressing a disputant-oriented intervention, specifically an education technique. By wielding an education intervention, a mediator can induce a learning process in a disputant. Through this learning process, the disputant may change orientation towards a specific action option. In this way the mediator agent assists two individual social agents in establishing and expressing empathic understanding, as a means to develop solidly grounded joint decisions.

Keywords: mediation, empathy, joint decision making.

1 Introduction

In many real world scenarios, human individuals jointly have to decide on action options. In these decision processes social aspects, such as communication and empathic understanding, play an important role. In conflict situations, when reaching a joint decision becomes difficult, the disputants may seek (often impartial) assistance from a human mediator (e.g., [12]). This paper is part of a series on mediator support of joint decision processes. In [5], an analysis is made on the possible outcomes of joint decision processes. Based on this, mediator assistance for establishing mutual empathic understanding has been analyzed, modeled and simulated in [6]. This paper extends the intervention repertoire of the mediator with a disputant-oriented intervention, for example the education technique as mentioned in [17]. The adopted aim of this intervention is to arrive at the most stable outcome of such joint decision processes, where the involved human individuals arrive at a common action option, with emotional grounding for each individual, and mutually acknowledged empathic understanding between all individuals (e.g., [5], [15]). In this research the models and simulations are constructed around two types of agents. Human individuals trying to reach a joint decision are modeled as instances of the social agent model in accordance with [15], with minor adaptations dependent on the context. The human in the role of mediator is modeled in this paper as a mediator agent that is able to educate one of the disputants in order to arrive at a stable joint decision.

C.K. Loo et al. (Eds.): ICONIP 2014, Part II, LNCS 8835, pp. 589–596, 2014.

2   Hebbian Learning

Mediation literature is in general vague on determinants for applying specific interventions (e.g., [17], [18]). Therefore, in this paper a scenario is adopted in which the two disputants differ in their response to an external trigger. The education intervention in this paper is modeled as a mediator agent providing so-called pseudo-experiences to one of the individual social agents in the joint decision process. Through these pseudo experiences the social agent goes through a Hebbian learning process [7]. The mediator's aim for providing these pseudo-experiences is to influence the social agent's orientation towards a specific action option in relation to a world stimulus. This orientation may concern the perceived relevance of an action option in relation to a stimulus, as well as the internal simulation of an action option as part of the emotional valuing process.

3   The Adapted Social Agent Model

In [15] a social agent model for joint decision making is presented addressing the role of mutually acknowledged emphatic understanding in decision making. This neurologically inspired cognitive agent model uses the following principles: mirroring (see also [11]), internal simulation (see also [8]) and emotion-related-valuing (see also [3]). Interacting social agents may develop mutual empathic understanding (see also [4]), which may be shown (nonverbal) and acknowledged (verbal). For further details, see [15].

This section describes the adaptations to the social agent model, to enable learning processes induced by receiving pseudo-experiences from a mediator agent. The mediator's aim for providing these pseudo-experiences is to influence the social agent's orientation towards a specific action option in relation to a world stimulus. This orientation may concern the perceived relevance of an action option in relation to a stimulus, as well as the internal simulation of an action option as part of the emotional valuing process. These two aspects of an agent's orientation relate to two connections in the social agent model: (1) the connection between the sensory representation state for the stimulus $SR(s)$ and the action preparation state $PS(a)$, and (2) the prediction link between the action preparation state $PS(a)$ and the effect state $SR(e)$. In order to support a learning process, these two connections are modified to support Hebbian learning (cf. [7], [14]), with an initial connection strength of zero. In addition to these modifications, the weight of the connection between $SR(e)$ and $PS(b)$ is lowered slightly in order to properly focus on the emotion-related valuing process. The adaptations are summarised in Table 1.

4   The Mediator Agent Model

For the mediator agent a dynamical systems perspective is adopted as advocated, for example, in [1], [13]. The quantitative aspects are modeled in a mathematical manner as in [1], [13], [16]. This fits in the domain of small continuous-time

Table 1. Social agent model modifications

From state	Connection		To state	Hebbian learning
SR(s)	ω_{31a}	0.0	PS(a)	X
PS(a)	ω_{21e}	0.0	SR(e)	X
SR(e)	ω_{31b}	0.7	PS(b)	

recurrent neural networks as advocated by [2] and inspired by e.g., [9], [10]; see also [16].

This section describes the dynamical systems model for the mediator agent. The mediator agent supports the expression and transfer of pseudo-experiences to a social agent, in order to influence the social agent's orientation towards a specific action option in relation to a world stimulus. The pseudo-experiences are expressed in terms of world states on the effects of actions, and aim to induce learning processes in the social agent. These learning processes can strengthen the connection between the sensory representation state for the stimulus and the action preparation state, as well as the prediction link between the action preparation state and the effect state (sensory representation of the effect of the action). The strength of these connections represent the relevance of an action option in relation to a stimulus, and the internal simulation of an action option as part of the valuing process respectively.

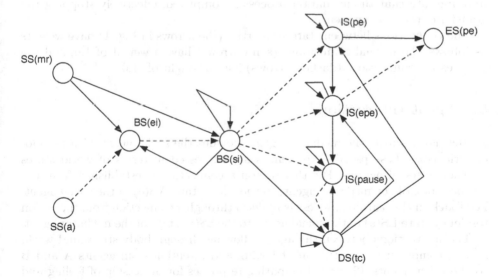

Fig. 1. Overview of the mediator agent model

In order to provide successive pseudo-experiences to other agents, the mediator agent model (Fig. 1) supports multiple cycles. The state properties used in the mediator agent model are summarised in Table 2. In each of these cycles a high activation level for state IS(pe) models the expression of a single pseudo-experience, followed by a pause between successive expressions. The state DS(tc) models the occurrence of a single transfer cycle by maintaining a high activation level. Within such a transfer cycle three other states sequentially obtain a high activation level, indicating the start of a single transfer (state IS(pe)), the end of a single transfer (state IS(epe)) and a pause in the transfer (state IS(pause)). In each cycle, a high activation level for state IS(pe) leads to a high activation level in state ES(pe), thereby expressing the pseudo-experience. Subsequently, a high activation level for state IS(epe) in turn suppresses state ES(pe), thereby ending the single transfer. Finally, a high activation level for state IS(pause) suppresses state DS(tc), indicating the end of a cycle. The states IS(pe), IS(epe) and IS(pause) can obtain a high activation level when:

- state DS(tc) has a high activation level, and
- the preceding state (in the sequence IS(pe), IS(epe), IS(pause)) has a high activation level, and
- suppression by state BS(si) is reduced.

State BS(si), in combination with state BS(ei), establishes a rhythm for both starting cycles and the timing interval within a cycle (transfer of pseudo-experience and pause). Both states BS(si) and BS(ei) depend on a high activation level of the mediation request sensor state SS(mr). Finally, state SS(a) may indicate that the mediation process is completed, effectively stopping the transfer of pseudo-experiences.

The connections between state properties (the arrows in Fig. 1) have weights as follows: the normal connections (solid arrows) have a weight of 1.0, and the suppressing connections (dashed arrows) have a weight of -1.0.

5   Agent Interaction

As discussed above, the mediator agent provides the social agent with pseudo-experiences. These pseudo-experiences are expressed in terms of world states on the effects of actions. For this reason a connection is established from the ES(pe) state of the mediator agent, to the effect state WS(e) of the social agent. Feedback on the learning process is realised through a connection from the action tendency state ES(a) of the social agent to the SS(a) state of the mediator agent.

The interactions between the agents, flowing through body states and world states, comprise: (1) expression of feeling and intention from agents A and B to the other agents, (2) verbal empathic responses for ownership of feeling and intention from agents A and B to the other agents, (3) communication of pseudo-experiences from the mediator agent to agent A.

Table 2. State properties used in the mediator agent model, and settings used for (initial activation) level, threshold τ and steepness σ parameters in the mediator agent model

Notation	Description	Level	τ	σ
SS(mr)	sensor state for detecting a mediation request	1.0	-	-
SS(a)	sensor state for detecting an intention expression for action option a by social agent	0.0	-	-
BS(si)	belief state on the start of a timing interval	1.0	1.0	10
BS(ei)	belief state on the end of a timing interval	0.0	1.5	10
DS(tc)	desire state for a single transfer cycle	0.0	0.5	10
IS(pe)	intention state for the start of a single pseudo-experience	0.0	0.6	10
IS(epe)	intention state for the end of a single pseudo-experience	0.0	1.6	32
IS(pause)	intention state for the pause after transferring a single pseudo-experience	0.0	1.6	20
ES(pe)	execution state for expressing pseudo experience	0.0	0.5	10

6   Simulation Results

In this section the simulation results are presented for one of the scenarios that have been explored. In this scenario the mediator agent provides pseudo-experiences to a social agent. The social agent incorporates Hebbian learning [e.g., 7, 14] for two connections: the connection between SR(s) and PS(a) and the internal simulation connection between the preparation state for action option a PS(a) and the predicted effect state SR(e). The simulation uses the following values for the Hebbian learning parameters: the learning rate is 0.03 and the extinction rate is 0.001. The settings used for the connection strength parameters in the mediator agent model are as described in Section 4. The settings used for the initial activation level, threshold and steepness parameters are provided in Table 2. The simulation results are shown in Fig. 2.

In the simulation presented in Fig. 2, the mediator agent provides pseudo-experiences to the social agent A. Initially, agent A does not express the tendency for action option a, nor does it express any emotion felt towards this action option nor does it develop its preparation states PS(a) and PS(b). Due to the provided pseudo-experiences the strength of the social agent connections between SR(s) and PS(a) and the connection between PS(a) and SR(e) increases. In the simulation results these connection strengths are shown as black lines: the lower black line shows the connection between PS(a) and SR(e), and the upper black line shows the connection between SR(s) and PS(a). As a result of these Hebbian learning processes, the social agent starts developing its preparation states PS(a) and PS(b), and after time point 215 starts expressing its tendency for action option a and its (positive) feeling towards action option a. Finally, the expression of the tendency for action option a indicates to the mediator that the process is completed, obviating the need for the transfer of additional pseudo-experiences.

Both agents may receive the world stimulus at different levels. In this scenario, agent A and agent B receive a world stimulus at level 1 and level 0.5 respectively. The latter means that agent B is dependent on agent A for its activation of preparation for action option a and the associated emotional response and feeling. Therefore the activation levels for agent B stay low until they increase due

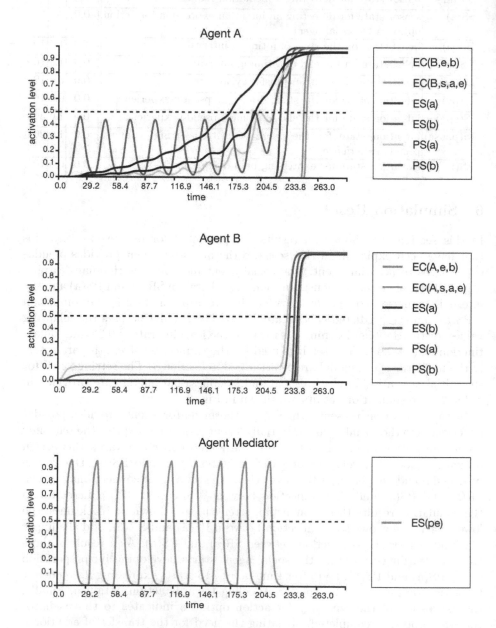

Fig. 2. Influencing social agent A's orientation towards action option a

to agent A, through mirroring, internal simulation and emotion-related valuing. Ultimately, both agents acknowledge their mutual empathic understanding.

7 Discussion

In this paper a mediator model was presented for wielding a disputant-oriented intervention in assisting social agents in reaching a joint decision involving mutually acknowledged empathic understanding.

The mediator model presented in this paper does what it is conceived to do, in that a realistic scenario from the domain of dispute resolution can be simulated. In the presented scenario the mediator is able to change one of the disputant's orientation towards an action option, and subsequently the disputants may be in the position to reach a joint decision with mutual empathic understanding.

Reaching a solidly grounded joint decision is an important factor in mediation as it may relate to the satisfaction of disputants with the mediation-result. Moreover, in situations where the disputants are assumed to have an ongoing relationship after mediation [18], the factor of empathic understanding may be important.

Mediation is a large and multifaceted domain. Literature on mediator interventions and techniques is mainly descriptive, and sparse on determinants for selecting an approach (e.g., [17], [18]). As mentioned in [18] it is difficult, time consuming and expensive to investigate ongoing mediations. For these reasons it is problematic to obtain data on effectiveness of mediation techniques, perhaps also because of confidentiality issues. Therefore validation with laboratory or field data for mediation processes falls outside the scope of this paper, and is left for future work.

The computational architecture as used in this paper does not mean that the results are dependent on the specific design and implementation aspects for the social agents. The mediator agent depends only on external aspects that are assumed to exist in human interaction, such as the expression of intention and emotion, and communication.

The computational model contributed in this paper may provide a basis to further explore the development of support for joint decision making processes, for example in the form of a mediation assistant agent. Such an assistant may provide analysis and give process advice in order to develop a joint decision, and take care that no escalating conflicts arise. This will be a direction of future research.

References

1. Ashby, W.R.: Design for a Brain. Chapman and Hall, London (1952)
2. Beer, R.D.: On the dynamics of small continuous-time recurrent neural networks. Adaptive Behavior 3, 469–509 (1995)
3. Damasio, A.R.: Descartes' Error: Emotion, Reason and the Human Brain. Papermac, London (1994)

4. De Vignemont, F., Singer, T.: The empathic brain: how, when and why? Trends in Cogn. Sciences 10, 437–443 (2006)
5. Duell, R., Treur, J.: A Computational Analysis of Joint Decision Making Processes. In: Aberer, K., Flache, A., Jager, W., Liu, L., Tang, J., Guéret, C. (eds.) SocInfo 2012. LNCS, vol. 7710, pp. 292–308. Springer, Heidelberg (2012)
6. Duell, R., Treur, J.: Modelling Mediator Assistance in Joint Decision Making Processes Involving Mutual Empathic Understanding. In: Hwang, D., Jung, J.J., Nguyen, N.-T. (eds.) ICCCI 2014. LNCS, vol. 8733, pp. 544–553. Springer, Heidelberg (2014)
7. Hebb, D.: The Organisation of Behavior. Wiley (1949)
8. Hesslow, G.: Conscious thought as simulation of behaviour and perception. Trends Cogn. Sci. 6, 242–247 (2002)
9. Hopfield, J.J.: Neural networks and physical systems with emergent collective computational properties. Proc. Nat. Acad. Sci (USA), 79, 2554–2558 (1982)
10. Hopfield, J.J.: Neurons with graded response have collective computational properties like those of two-state neurons. Proc. Nat. Acad. Sci (USA), 81, 3088–3092 (1984)
11. Iacoboni, M.: Mirroring People: the New Science of How We Connect with Others. Farrar, Straus & Giroux, New York (2008)
12. Moore, C.W.: The Mediation Process: Practical Strategies for Resolving Conflict. Wiley (2014)
13. Port, R.F., van Gelder, T.: Mind as motion: Explorations in the dynamics of cognition. MIT Press, Cambridge (1995)
14. Treur, J.: A Computational Agent Model for Hebbian Learning of Social Interaction. In: Lu, B.-L., Zhang, L., Kwok, J. (eds.) ICONIP 2011, Part I. LNCS, vol. 7062, pp. 9–19. Springer, Heidelberg (2011)
15. Treur, J.: Modelling Joint Decision Making Processes Involving Emotion-Related Valuing and Empathic Understanding. In: Kinny, D., Hsu, J.Y.-j., Governatori, G., Ghose, A.K. (eds.) PRIMA 2011. LNCS, vol. 7047, pp. 410–423. Springer, Heidelberg (2011)
16. Treur, J.: An Integrative Dynamical Systems Perspective on Emotions. Biologically Inspired Cognitive Architectures Journal 4, 27–40 (2013)
17. Wall, J. J.A., Stark, J.B., Standifer, R.L.: Mediation, A Current Review and Theory Development. The Journal of Conflict Resolution 45(3), 370–391 (2001), http://www.jstor.org/stable/3176150
18. Wall, J.A., Dunne, T.C.: Mediation Research: A Current Review. Negotiation Journal 28, 217–244 (2012), doi:10.1111/j.1571-9979.2012.00336.x.

Author Index

Abawajy, Jemal H.   I-559, II-237, II-245
Abbass, Hussein   III-571
Abbass, Hussein A.   I-570, III-68
Abdullah, Jafri Malin   II-186
Abdullah, Mohd Tajuddin   III-390
Abernethy, Mark   I-578
Abro, Altaf Hussain   I-59
Abuqaaud, Kamal   II-503
Adams, Samantha V.   III-563
Akaho, Shotaro   II-26
Akimoto, Yoshinobu   III-587
Akram, Muhammad Usman   III-226
Akram, M. Usman   II-495
Alahakoon, Damminda   I-519, I-551
Ali, Rozniza   III-103
Al-Jumaily, Adel   I-471, II-101
Allesiardo, Robin   I-374, I-405
Anam, Khairul   I-471, II-101
Araki, Shohei   III-325
Araki, Takamitsu   II-26
Arandjelović, Ognjen   II-327, II-335
Araujo Carmen Paz, Suárez   I-50
Asadi, Houshyar   III-474, III-483
Ashraf, M. Daniyal   II-495
Assaleh, Khaled   II-503
Aung, Zeyar   II-487, III-431
Awais, Mian.M.   I-199
Ayaz, Sadaf   III-226
Azhar, M. Usama   II-495
Aziz, Maslina Abdul   I-559, II-237,
    II-245
Azzag, Hanane   II-369
Azzag, Hanene   I-207

Bacciu, Davide   I-543
Báez Patricio, García   I-50
Bahrmann, Frank   III-543, III-553
Bai, Li   III-9
Baig, Mirza M.   I-199
Ban, Tao   III-365
Bapi, Raju S.   I-35
Bargiela, Andrzej   I-462
Ben Amar, Chokri   III-292
Bendada, Hakim   II-335

Bennani, Younès   II-52, II-60
Benyó, Zoltán   II-445
Bhatti, Asim   III-258, III-493, III-501
Böhme, Hans-Joachim   III-543, III-553
Bondu, Alexis   II-548
Botzheim, János   III-596
Bouguettaya, Athman   III-673
Bouneffouf, Djallel   I-374, I-405, III-373
Brohan, Kevin   III-563
Buf, J.M.H. du   I-511
Buruzs, Adrienn   III-447
Butt, Ayesha Javed   III-661
Butt, Naveed Anwer   III-661
Butt, Rabia Ghias   III-661

Cai, Ruichu   I-350
Cangelosi, Angelo   III-563
Carvalho, Desiree Maldonado   I-287
Chai, Kok Chin   III-407
Chaibi, Amine   II-369
Chandra, B.   I-535
Chang, Wui Lee   III-415
Charytanowicz, Malgorzata   II-287
Cheah, Wooi Ping   I-127, III-612
Chen, Badong   II-68
Chen, Gang   III-300, III-308
Chen, Hao   II-429
Chen, Mushangshu   III-202
Chen, Ning   II-395
Chen, Weiqi   I-350
Chen, Youxin   I-358
Cheng, Minmin   III-167
Cheng, Yun   II-126
Chetty, Madhu   I-446, II-470
Chiew, Fei Ha   III-407
Chin, Wei Hong   III-604
Chiu, Steve C.   I-103, I-215
Cho, Sung-Bae   I-42
Chong, Lee-Ying   III-653
Chong, Siew-Chin   III-653
Chotipant, Supannada   III-682
Chowdhury, Ahsan Raja   II-229
Chung, Yuk Ying   III-439
Cichocki, Andrzej   I-503, III-111

Cocks, Bernadine   I-1
Copeland, Leana   I-335, I-586
Cornuéjols, Antoine   I-159, II-60, II-548
Cui, Xiao-Ping   II-203

Dacey, Simon   III-333
Dachraoui, Asma   II-548
Dai, Yuli   III-365
Dammak, Mouna   III-292
Da San Martino, Giovanni   II-93
Ding, Tianben   I-175
Dong, Hai   III-673, III-682
Dong, Zheng   I-239
Douch, Colin   III-300
Doya, Kenji   I-35
Du, Jia Tina   I-1
Duchemin, Jean-Bernard   III-493
Duell, Rob   II-589

Egerton, Simon   III-596
El-Alfy, El-Sayed M.   I-199, I-397
Esfanjani, Reza Mahboobi   III-509

Faisal, Mustafa Amir   III-431
Fard, Saeed Panahian   I-135
Faye, Ibrahima   II-186
Feng, Ruibin   I-271, I-279, II-479
Féraud, Raphaël   I-374
Feraud, Raphael   I-405
Földesi, Péter   III-447
Fong, Pui Kwan   III-390
Fu, Zhouyu   II-311
Fujimoto, Yasunari   III-587
Fujisawa, Shota   I-619
Fujiwara, Kenzaburo   I-183
Fukushima, Kunihiko   I-78
Furber, Steve   III-563

Gaburro, Julie   III-493
Galluppi, Francesco   III-563
Garcez, Artur d'Avila   I-69
Gedeon, Tom   I-335, I-586
Geng, Bin   II-203
Ghazali, Rozaida   I-559
Ghesmoune, Mohammed   I-207
Ghiass, Reza Shoja   II-335
Goh, Khean Lee   III-276
Goh, Pey Yun   I-127
Goh, Sim Kuan   I-570
Gong, Shu-Qin   II-429

Gopinath, K.   II-352
Grama, Ion   III-9
Gu, Nong   III-535
Gu, Xiaodong   III-25
Gu, Yingjie   I-103, I-215
Gu, Yun   III-95, III-150
Guan, Hu   II-532
Gunawardana, Kasun   I-519
Guo, Lunhao   II-1
Guo, Ping   II-17, II-118

Haggag, Hussein   III-501
Haggag, Sherif   III-501
Hagiwara, Katsuyuki   II-540
Haider, Ali   II-495
Han, Jialin   III-398
Han, Pang Ying   III-628
Han, Zi-Fa   I-271, II-479
Handaga, Bana   I-421
Hao, Hongwei   II-295
Hao, Zhifeng   I-350
Harada, Hideaki   III-341
Harashima, Yayoi   II-9
Hatwágner, Miklós F.   III-447
Hayashi, Yoichi   I-619
He, Kun   I-335
He, Xiangjian   III-439
Hellbach, Sven   III-543, III-553
Herawan, Tutut   I-559, II-237, II-245
Heskes, Tom   II-524
Hettiarachchi, Imali   III-527
Hettiarachchi, Imali Thanuja   III-519
Higashi, Masatake   I-304
Hilmi, Ida   III-276
Himstedt, Marian   III-543
Hino, Hideitsu   II-26
Hirayama, Kotaro   II-160
Hirose, Akira   I-175, I-223
Hirsbrunner, Béat   I-247
Ho, Shiaw hooi   III-276
Homma, Toshimichi   II-462
Hong, Jer Lang   II-454
Horie, Teruki   III-85
Hou, Zeng-Guang   III-535
Hu, Jin   III-535
Hu, Yaoguang   III-398
Huang, Heyan   I-495
Huang, Kaizhu   II-84, III-317, III-349
Huang, Long-Ting   II-479
Huang, Tingwen   I-382

Huang, Weicheng   III-58
Huang, Xu   III-127
Huang, Yifeng   III-42
Huangfu, Yao   II-581
Hussain, Amir   III-103
Hussain, Farookh Khadeer   III-673,
    III-682, III-690
Hussain, Omar Khadeer   III-682, III-690
Hussain, Walayat   III-690
Hussin, Fawnizu Azmadi   III-276

Ibrahim, Zaidah   III-439
Ikram, Muhammad Touseef   III-661
Inoue, Hirotaka   I-143
Ishiguma, Yuki   III-58
Ishwara, Manjusri   I-551
Islam, Md. Nazrul   II-344
Islam, Sheikh Md. Rabiul   III-127
Ismaili, Oumaima Alaoui   I-159
Isokawa, Teijiro   I-527
Ito, Yoshifusa   I-602
Izawa, Tomoki   I-619
Izumi, Hiroyuki   I-602

Javed, Ehtasham   II-186
Ji, Hong   II-68
Ji, Zhenzhou   II-279
Jia, Wenjing   III-76
Jia, Zhenghong   I-86
Jia, Zhenhong   III-76, III-95, III-150
Jiang, Bo   III-103
Jiang, Shouxu   II-126
Jiang, Xiaofan   II-126
Jiang, Yong   II-429, II-437, II-556
Jin, Jie   III-119
Jin, Qiyu   III-9
Jin, Zhong   I-103, I-215
Jo, Hyunrae   II-110

Kabir Ahmad, Farzana   I-327
Kasabov, Nicola   I-86
Kasabov, Nikola   I-421, III-76
Kashima, Hisashi   II-377
Kaski, Samuel   II-135
Katayama, Yasunao   I-255
Keil, Andreas   II-68
Keil, Sabrina   III-553
Khan, Abdullah   II-237, II-245
Khawaja, Sajid Gul   II-495
Khosravi, Abbas   II-178

Khosravi, Rihanna   II-178
Kim, Jihun   I-454
Kim, Kyuil   III-357
Kinattukara, Tejy   III-183
King, Irwin   I-610, II-44
Kiong, Loo Chu   III-628
Klein, Michel C.A.   I-59
Ko, Sangjun   III-357
Kobayashi, Masakazu   I-304
Kóczy, László T.   II-573, III-447
Koeman, Mike   II-524
Koga, Kensuke   III-341
Kondo, Yusuke   I-295
Kõnnusaar, Tiit   I-19
Kovács, Levente   II-303, II-445
Kubota, Naoyuki   III-596
Kudo, Hiroaki   III-175
Kulczycki, Piotr   II-287
Kurematsu, Ken   III-135
Kurita, Takio   II-9
Kurogi, Shuichi   II-35, II-160, II-421,
    III-58

Laroche, Romain   I-405
Lau, See Hung   III-423
Le, Kim   III-127
Lebbah, Mustapha   I-207, II-369
Lee, Minho   I-11, I-454, II-110
Lee, Nung Kion   III-390
Lee, Younsu   III-357
Lemaire, Vincent   I-159
Leu, George   III-571
Leung, Chi-Sing   I-271, I-279, I-312,
    II-479
Li, Chao   I-413
Li, Fang   II-361
Li, Junhua   I-503
Li, Qiang   I-239
Li, Wenye   I-319, II-319
Li, Xiao   I-438
Li, Xiucheng   II-126
Li, Yanjun   II-118
Li, Zhe   I-438
Li, Zhihao   I-350
Li, Zhijun   II-126
Lian, Cheng   I-382
Lim, Chee Peng   I-151, III-381, III-415
Lin, Jie   III-33
Lin, Lei   II-279, III-17
Ling, Goh Fan   III-628

Liu, Anjin   I-263
Liu, Derong   I-389
Liu, Guisong   I-366
Liu, Lingshuang   I-366
Liu, Linshan   I-191
Liu, Peng   III-50
Liu, Quansheng   III-9
Liu, Wei   III-150
Liu, Wenqi   III-284
Liu, Xiabi   II-1
Liu, Ye   III-202
Liu, Yichen   III-159
Liu, Zhirun   I-495
Liu, Zhi-Yong   II-404
Lobato, David   I-511
Loo, Chu Kiong   I-151, II-344, III-381, III-604
López Pablo, Fernández   I-50
Lu, Bao-Liang   II-170, III-234
Lu, Deji   III-167
Lu, Hongtao   I-487, II-386
Lu, Jie   I-263, III-398
Lu, Zuhong   III-167
Luo, Bin   III-103
Luo, Zhenbo   I-358
Lyu, Michael R.   II-44

Ma, Shuyuan   III-398
Maldague, Xavier   II-335
Malik, Aamir Saeed   II-186, III-276
Mallipeddi, Rammohan   I-11
Mamun, Abdullah Al   I-570
Man, Mustafa   III-103
Man, Yuanyuan   I-610
Manzoor, Adnan R.   I-59
Maruf, Syed Osama   II-495
Masood, Ammara   II-101
Matsui, Nobuyuki   I-527
Matsumoto, Tetsuya   III-175
Matsuoka, Shouhei   III-135
Matsuyama, Yasuo   III-85
Mei, Jincheng   II-170
Mejdoub, Mahmoud   III-292
Mendis, Sumudu   I-586
Micheli, Alessio   I-543
Minemoto, Toshifumi   I-527
Mitra, Pabitra   II-194, II-413
Mitsuishi, Takashi   II-462
Miyapuram, Krishna Prasad   I-35
Mizobe, Yuta   II-35

Mohamed, Shady   III-258, III-474, III-483, III-501, III-509
Mohammadi, Arash   III-474, III-483
Mohammed, Shady   III-527
Molnárka, Gergely I.   II-573
Morie, Takashi   III-341
Moriwaki, Masafumi   III-85
Moriyasu, Jungo   I-231
Mountstephens, James   III-194
Murao, Hajime   I-119
Murli, Norhanifah   I-421
Muslim, Muhammad Aziz   II-565

Nadeem, Yasser   III-226
Nagatomo, Kohei   II-565
Nagayoshi, Masato   I-119
Nahavandi, Saeid   II-178, III-258, III-474, III-483, III-493, III-501, III-509, III-519, III-527
Nair, Ajay   I-446
Nakamura, Kiyohiko   I-183
Nakane, Ryosho   I-255
Nakano, Daiju   I-255
Nakazato, Junji   III-365
Nandagopal, D. (Nanda)   I-1
Naqvi, Mehdi   II-495
Nascimento, Mariá C.V.   I-287
Naseem, Rashid   I-559
Navarin, Nicolò   II-93
Nawi, Nazri Mohd   II-237, II-245
Nazmul, Rumana   II-470
Ng, Chee Khoon   III-407, III-423
Ng, Kai-Tat   I-279
Nguyen, Thanh Thi   III-519
Nie, Yifan   I-413
Ninomiya, Shota   III-85
Nishimura, Haruhiko   I-527
Niu, Xiamu   III-620, III-636
Nõmm, Sven   I-19
Nyhof, Luke   III-527

Ohnishi, Noboru   III-175
Okutsu, Takumi   I-111
Omori, Toshiaki   I-27
Ong, Thian-Song   III-644, III-653
Ono, Kohei   II-421
Ooi, Shih Yin   III-612
Osana, Yuko   I-111, I-342
Osendorfer, Christian   III-250

Ouyang, Yuanxin   II-581, III-284
Ozawa, Seiichi   III-365

Pamnani, Ujjval   I-35
Pang, Shaoning   III-300, III-308, III-333
Park, Ukeob   I-11
Parque, Victor   I-304
Patterson, Cameron   III-563
Peng, Liang   III-535
Peng, Long   III-535
Pérez-Carrasco, José-Antonio   III-563
Pham, Ducson   II-327
Phon-Amnuaisuk, Somnuk   III-579
Ping, Liew Yee   III-628
Príncipe, Jose C.   II-68

Qiao, Hong   I-191, II-404
Qu, Hong   I-366
Qu, Jiao   II-203

Rahim Zadeh, Delpak   III-483
Rai, Shri M.   I-578
Rajapakse, Jayantha   I-519, I-551
Rast, Alexander D.   III-563
Razzaq, Fuleah A.   III-258
Redko, Ievgen   II-52
Rehman, M.Z.   II-237, II-245
Resende, Hugo   I-287
Rezaei, Hossein   III-509
Ribeiro, Bernardete   II-395
Riedel, Martin   III-543
Rong, Wenge   I-413, III-284

Saha, Moumita   II-194
Sahar, Sadaf   III-226
Saigusa, Koji   II-462
Saito, Toshimichi   I-231, II-221, II-254,
    II-271
Sakai, Ko   III-135
Saleiro, Mário   I-511
Samiullah, Mohammad   II-229
Sarazin, Tugdual   II-369
Sarker, Tusher Kumer   II-212
Sato, Naoyuki   I-167
Sato, Takumi   II-221, II-254
Sato-Shimokawara, Eri   III-587
Sawada, Kiyoshi   II-462
Sawayama, Ryo   II-271
Seera, Manjeevan   I-151, III-381
Seldon, Henry Lee   I-95

Seng, Woo Chaw   III-218
Seth, Sohan   II-135
Shah, Habib   I-559
Shan, Lili   III-17
Shanableh, Tamer   II-503, III-266
Shanmuganathan, Subana   I-429
Shao, Di   III-17
Sharma, Rajesh Kumar   I-535
Shawe-Taylor, John   II-135
Shen, Shan-Chun   III-234
Shen, Yikang   I-413
Sheng, Hao   II-581
Shidama, Yasunari   II-462
Shigematsu, Ryosuke   II-421
Shikano, Akihiro   III-85
Shimada, Hajime   III-325
Shimada, Nami   II-462
Shimamura, Jumpei   III-365
Shovon, Md. Hedayetul Islam   I-1
Showkat, Dilruba   II-229
So, Hing Cheung   II-479
Song, Jungsuk   III-357
Song, Lei   III-333
Soyer, Hubert   III-250
Sperduti, Alessandro   I-543, II-93
Srinivasan, Cidambi   I-602
Su, Jian-Hua   II-404
Sublime, Jérémie   II-60
Sultana, Rezwana   II-229
Sum, John   I-279
Suman, Shipra   III-276
Sun, Yaming   II-279
Svetinovic, Davor   II-487, III-431
Syarif, Munalih Ahmad   III-644
Szalay, Péter   II-445
Szilágyi, László   I-247, II-303, II-445
Szilágyi, Sándor Miklós   I-247, II-303

Tabatabaei, Seyed Amin   I-59
Tada, Shunsuke   III-365
Takagi, Hiroki   III-175
Takakura, Hiroki   III-325
Takeguchi, Satoshi   II-35
Takeuchi, Yoshinori   III-175
Talkad Sukumar, Poorna   II-352
Tamaki, Hisashi   I-119
Tamukoh, Hakaru   III-341
Tamura, Hiroki   II-565
Tan, Kay Chen   I-570
Tan, Shing Chiang   I-127, III-612

Tanaka, Gouhei   I-255
Tanaka, Takuma   I-183
Tanaka, Yuki   I-619
Tang, Huiming   I-382
Tang, Jiangjun   III-571
Tang, Maolin   II-212
Tang, Tiong Yew   III-596
Tang, XiangLong   III-50
Tanno, Koichi   II-565
Tao, Ye   III-50
Tay, Kai Meng   III-407, III-415, III-423
Tee, Connie   III-644
Teo, Jing Xian   I-95
Teoh, Andrew Beng Jin   III-644, III-653
Terashima, Takanori   II-462
Terzić, Kasim   I-511
Tian, Guanhua   II-295
Tian, Jian   III-210
Tirumala, Sreenivas Sremath   III-308
Toh, Huey Jing   II-454
Tokunaga, Fumiaki   II-221
Toomela, Aaro   I-19
Tran, Dang Cong   II-143, II-151, II-263
Tran, Son N.   I-69
Tran, Van Hung   II-263
Treur, Jan   I-59
Tsang, Peter Wai Ming   I-312
Tu, Enmei   I-86
Turcsany, Diana   I-462

Ueki, Takuya   II-35
Ueyama, Yuki   II-76
Ungku Amirulddin, Ungku Anisa
    III-458
Urvoy, Tanguy   I-405

van der Smagt, Patrick   III-250
Venkatesh, Svetha   II-327
Verma, Brijesh   III-183
Vijayalakshmi, Ramasamy   I-1
Villmann, Thomas   III-543
Virinchi, Srinivas   II-413

Walker, Peter   III-493
Walter, Nicolas   III-276
Wan, Wai Yan   I-271
Wang, Dongfang   II-361
Wang, Haixian   III-167
Wang, Hui   II-143
Wang, Jianmin   I-438

Wang, Lijuan   I-350
Wang, Shen   III-620, III-636
Wang, Shuai   I-610
Wang, Xiaolong   II-279, III-17
Wang, Zhuo   I-487
Wangikar, Pramod P.   I-446
Watanabe, Toshihiro   II-377
Wei, Hui   I-239
Wei, Qinglai   I-389
Wen, Wen   I-350
Wennekers, Thomas   III-563
Wijono   II-565
Wong, Kok Wai   I-479
Wong, Kuok-Shoong Daniel   II-487
Wong, Man To   III-439
Wong, Shen Yuong   III-466
Wong, T.J.   I-610
Woon, Wei Lee   II-487, III-431
Wu, Minhua   I-358
Wu, Peng   II-17
Wu, Qiang   III-150
Wu, Wenzhen   II-556
Wu, Yangwei   III-142
Wu, Yi   II-1
Wu, Zhijian   II-143, II-151, II-263

Xia, Shu-Tao   II-429, II-437, II-556
Xiang, Wei   I-594
Xiao, Yi   II-479
Xie, Xiurui   I-366
Xin, Xin   I-495, II-118
Xiong, Zhang   I-413, II-581, III-284
Xu, Bo   II-295
Xu, Jiaming   II-295
Xu, Jinhua   II-515
Xu, Wenchao   II-532
Xu, Zenglin   II-44
Xue, Haoyang   III-95, III-150

Yamaguchi, Toru   III-587
Yamaguchi, Yukiko   III-325
Yamane, Toshiyuki   I-255
Yamashita, Yoichiro   II-160
Yamauchi, Koichiro   I-295
Yan, Xuehu   III-620, III-636
Yang, Cheng   III-25
Yang, Ching-Nung   III-636
Yang, Chun   III-317
Yang, Haiqin   II-44
Yang, Jian   III-33

Yang, Jie   I-86, III-9, III-76, III-95, III-150
Yang, Kyon-Mo   I-42
Yang, Nan   II-279
Yang, Xi   II-84
Yang, Xiong   I-389
Yang, Yanfeng   II-532
Yang, Yanqin   II-532
Yao, Wei   I-382
Yao, Xuan   III-242
Yap, Hwa Jen   III-458, III-466
Yap, Keem Siah   III-458, III-466
Yaw, Chong Tak   III-458
Yeh, Wei-Chang   III-439
Yin, Ooi Shih   III-628
Yin, Qian   II-17
Yin, Xu-Cheng   II-203, III-317
Yokota, Tatsuya   III-111
Yokote, Ryota   III-85
Yoshikawa, Hikaru   II-160
Yu, Ying   III-33
Yuan, Chang   I-350
Yuan, Yuzhang   II-386
Yuan, Zejian   II-68
Yue, Junwei   III-42
Yusoff, Nooraini   I-327

Zafar, Madeeha   III-226
Zainuddin, Zarita   I-135
Zare, Mohammad Reza   III-218
Zeng, Zhigang   I-382
Zhang, Bo-Wen   II-203

Zhang, Guangquan   I-263, III-398
Zhang, Hanchao   II-515
Zhang, Heng   II-295
Zhang, Liqing   III-1, III-42, III-119, III-142, III-202
Zhang, Mengjie   III-300
Zhang, Qing   I-358
Zhang, Rui   II-84, III-349
Zhang, Shufei   III-349
Zhang, Tianyu   I-610
Zhang, Wei   III-76
Zhang, Xinjian   III-1
Zhang, Yi-Chi   II-437
Zhang, Yuxi   III-42
Zhao, Haohua   III-142
Zhao, Jie   II-437
Zhao, Jun   II-295
Zhao, Kang   II-386
Zhao, Wei   III-50
Zhao, Youyi   I-438
Zheng, Hai-Tao   II-429, II-437, II-556
Zheng, Nanning   II-68
Zheng, Suiwu   I-191
Zheng, Wei-Long   III-234
Zhou, Fang   II-203
Zhou, Lei   III-76
Zhou, Yue   I-594, III-159, III-210, III-242
Zhu, Dengya   I-479
Zhu, Lei   III-333
Zhu, Xuanying   I-335